A Mathematics Companion for Science and Engineering Students

A Mathematics Companion for Science and Engineering Students

Jerome R. Breitenbach
CALIFORNIA POLYTECHNIC STATE UNIVERSITY, SAN LUIS OBISPO

New York Oxford
OXFORD UNIVERSITY PRESS
2008

Oxford University Press, Inc., publishes works that further Oxford University's
objective of excellence in research, scholarship, and education.

Oxford New York
Auckland Cape Town Dar es Salaam Hong Kong Karachi
Kuala Lumpur Madrid Melbourne Mexico City Nairobi
New Delhi Shanghai Taipei Toronto

With offices in
Argentina Austria Brazil Chile Czech Republic France Greece
Guatemala Hungary Italy Japan Poland Portugal Singapore
South Korea Switzerland Thailand Turkey Ukraine Vietnam

Published by Oxford University Press, Inc.
198 Madison Avenue, New York, New York 10016
http://www.oup.com

Library of Congress Cataloging-in-Publication Data
Breitenbach, Jerome R.
A mathematics companion for science and engineering students / Jerome R. Breitenbach.—1st ed.
p. cm.
Includes bibliographical references and index.
ISBN 978–0–19–532775–5 (alk. paper)
1. Algebra. 2. Functions. I. Title.
QA155.B74 2008
510—dc22 2007014502

9 8 7 6 5 4 3 2 1

Printed in the United States of America
on acid-free paper

In memory of my companion while writing this book,
Alex J. Beagle

Here is just a sampling of the questions answered in this book:

- ***Prologue:*** In most mathematical discourse, are the symbols "ε" and "δ" used to represent numbers that are *large* or *small*?
- ***Chapter 1:*** What do the logic symbols "$\forall$" and "$\exists$" mean?
- ***Chapter 2:*** For two sets A and B, what is the difference between having $A \in B$ and $A \subseteq B$?
- ***Chapter 3:*** If $-3 - i4 = 5e^{i\theta}$ for some real number θ, then what is the value of θ? (*Hint*: The answer is *not* $\tan^{-1} \frac{4}{3}$.)
- ***Chapter 4:*** What (precisely!) does it mean to say that an infinite sequence $\{x_i\}_{i=1}^{\infty}$ has a value x as its *limit*?
- ***Chapter 5:*** What is the *inverse image* of a function?
- ***Chapter 6:*** What is wrong with the following "proof" that $1 = -1$?

$$1 = \sqrt{1} = \sqrt{(-1)(-1)} = \sqrt{-1}\sqrt{-1} = i \cdot i = -1.$$

- ***Chapter 7:*** How can a logarithm be converted from one base to another?
- ***Chapter 8:*** What is the difference between averaging a collection of numbers with an *arithmetic mean* versus a *geometric mean*?
- ***Chapter 9:*** How can simultaneous linear equations be solved by hand *painlessly*? (*Hint*: Avoid division.)
- ***Chapter 10:*** In Euclidean geometry, what are *skew lines*?
- ***Chapter 11:*** How can one determine (without plotting!) that the equation $x^2 - 2xy + 3y^2 - 4x + 5y - 6 = 0$ describes an ellipse in the xy-plane?
- ***Chapter 12:*** In general, what is a *dimensionless physical unit* (e.g., radian)? And, when is such a unit equivalent to the number 1?
- ***Epilogue:*** How can it be that some infinities are *bigger* than others?

Brief Contents

Contents

Preface

Science and engineering are fields that require both a knowledge of, and a facility in, mathematics. For many careers in these fields, especially the most interesting and those that push forward the state of the art, an ability to use mathematics to solve analytical problems and to think creatively is a must.

At most colleges and universities, undergraduate programs in science and engineering begin immediately with rigorous coursework in calculus and differential equations, followed by further study in mathematics. Throughout, a student's adeptness at mathematical analysis can be the difference between success and failure.

The purpose of this book is to provide students in these programs with a convenient and reliable reference on a variety of topics in, and related to, basic mathematics. Professional scientists and engineers, as well as pre-college students who are interested in these professions, should also find it useful. Besides being a source for self-study, the book can be used as either a supplemental reader in mathematically intense courses or as a textbook itself (perhaps in conjunction with others). A particularly good place for its introduction is a freshman orientation course, to be used then and thereafter.

The bulk of the material herein is generally referred to as "precalculus". All major topics are thoroughly covered; but, care has been taken not to overwhelm the reader with mathematical terminology (e.g., "ring", "neighborhood") they are unlikely to encounter in practice. Some additional topics (e.g., Stirling's approximation, uniform convergence) have been included which, though unfamiliar to the typical high school graduate, might well become useful later on; thus, to some extent the book grows along with the reader. Also, the last chapter discusses practical matters regarding the application of mathematics to the measurement and manipulation of experimental data.

The level of the presentation is such that, upon careful reading, the intended reader should be able to assimilate all that is said. This book differs, though, from other treatments of basic math for nonmathematics students in two significant ways. First, being primarily a reference manual rather than a tutorial, many assertions are made without justification; the writing style is succinct; and, examples are used more to explain than to teach. Second, with few exceptions, all definitions and facts are stated completely, precisely, and in detail; the text does not shield the reader from the full story. (The alternative would be to make the material *seem* easier than it is, thereby leaving pitfalls for readers to stumble upon later and somehow resolve on their own.[1])

On the other hand, more is given here than just a compendium of facts and formulas. Much ancillary information—observations, explanations, warnings, techniques,

1. As just one case in point, see fn. 21 in Chapter 3.

tips—appears on nearly every page. And, although the book does not coddle the reader, it is not written in the formal theorem-proof style of a typical mathematics text. In a nutshell, rigor is not sacrificed to attain readability; rather, rigorous mathematics is made readable.

A large appendix provides 360 problems ranging from easy to challenging, many of which have arisen in science and engineering courses. Almost entirely, these are not mindless exercises (i.e., "drill" or "plug-in" problems), but opportunities for readers to test both their knowledge of the material and their skill in applying it. Additionally, some problems present facts and terminology beyond what is addressed in the chapters themselves, and thus might be read for that reason alone. Accompanying the problems are complete and fully detailed solutions, which also serve to demonstrate acceptable mathematical argumentation.

An up-to-date list of errata is maintained online at

www.oup.com/us/he/breitenbach

How to Use This Book

If desired, the book can be read from start to finish in the order presented. More likely, though, it will be referred to sporadically when information is needed on a particular topic, such as a definition (e.g., the precise meaning of "limit") or a formula (e.g., a trigonometric identity). For this purpose, three points of entry are provided:

- *Brief and Detailed Tables of Contents:* Use these two lists to hunt for a specific topic within a general area.[2]
- *List of Notation:* Use this list to spot a mathematical operation by its symbolic representation.
- *Index:* Use this list to locate a topic by its name.

Furthermore, after finding your topic in the text, be prepared to consult the list of notation or the index, should you encounter some unfamiliar symbolism or terminology .

While reading the text, you might want to simultaneously peruse the corresponding problems in the appendix. Besides being exercises, these problems can be viewed as additional examples showing how to apply the material in the chapters. (Of course, for maximum benefit, always attempt a problem—or, at least ponder it—before looking at the solution!)

Regardless of how the book is used, it is recommended that the prologue, which is helpful for all that follows, be read first. It contains various tidbits that are rarely taught formally anywhere, although everyone is expected to eventually "pick them up" somewhere. Finally, the epilogue provides a quick overview of some "higher" mathematics, from calculus on.

2. Perusing the detailed table of contents will not only show what topics are covered in the book, but may pique the reader's interest about areas to be investigated further.

Writing Conventions

- When a displayed item (e.g., a numbered equation, figure, or table) is referred to by its identifier, "E" stands for "epilogue", "P" for "problem", and "S" for "solution". Thus, just as (1.2) is expression number 2 in Chapter 1, so Figure E.3 is figure number 3 in the epilogue, Table P4.5.6 is table number 6 in Problem 4.5, and (S7.8.9j) is part j of expression number 9 in Solution 7.8.
- When a term is first defined, and when its definition is significantly enhanced at a later place, **bold** type is used.
- Parentheses are sometimes used in a definition to indicate words that are optional. Example:

  A **(straight) line** is a straight, unbounded, complete, one-dimensional geometric figure.

  Thus, although it is common practice to precede the word "line" with "straight" (perhaps for emphasis or clarity), every line *is* a straight line.
- Other conventions followed in this book are discussed in the first section of the prologue.

Acknowledgments

The author would like to thank Professor Kent E. Morrison of the Mathematics Department at the California Polytechnic State University, San Luis Obispo, for reviewing an early draft of this work. Also, a near-final version was read by Cal Poly students Kathryn Rowe, Hirofumi Takahashi, and Kalia Glassey, who provided many valuable suggestions. Additionally, further improvements were suggested and some corrections provided by several anonymous reviewers.

San Luis Obispo, California Jerome R. Breitenbach
June 2007

A Mathematics Companion for Science and Engineering Students

Prologue: Reading, Writing, and Thinking Mathematics

Having knowledge of mathematics is certainly helpful to scientists and engineers; however, also important are related *skills*. Specifically, it is often necessary to *assimilate* the mathematical statements of others, to *express* oneself mathematically, and to *create* one's own mathematical arguments. In short, one must be able to read, write, and think mathematics. Accordingly, before embarking on our review of mathematics per se, we shall discuss these skills.

Reading Mathematics

Interestingly, many of the same reasons why some people loath mathematics are reasons why other people like it; in particular, mathematics is *objective*, *logical*, and *precise*. Upon proving some mathematical statement, there is no need for a person to seek the confirmation of others; rather, the burden of performing a rigorous proof is rewarded by certainty in the result. For example, based on a little historical knowledge about the modern (Gregorian) calendar, along with an understanding of what the words "century" and "millennium" mean, anyone can perform a simple mathematical calculation and conclude with confidence that January 1, 2001, is the first day of the 21st century, as well as the first day of the third millennium—no matter how many other people (e.g., in the news media) state this date as January 1, 2000.

Because mathematics is so precise, mathematical statements must be carefully read in order to be correctly understood. Furthermore, mathematical discourse is usually very concise, partly because space in books and journals is costly, but also because excess verbiage is distracting to the reader. Hence, since much information is often packed into little space, mathematical writings (including problem statements) should normally be read *slowly* and with *thought*.

As an example, here are a simple mathematical assertion and a justification of it, put forth as a theorem and proof:

> ***Theorem*** For all real numbers x and y such that $x < y$, if $x^2 = y^2$ then $x = -|y|$.
>
> ***Proof*** Since x is real, $\sqrt{x^2} = |x|$; and, likewise for y. Thus, taking the square root of each side of the equation $x^2 = y^2$, and equating the results, we get $|x| = |y|$; therefore, $x = \pm|y|$, since x is real. But, having $x = |y|$ would imply $|y| < y$, which is impossible; hence, it must be that $x = -|y|$. □

As usual, the above mathematical statements utilize various notational devices to enhance readability. In the theorem, arbitrary variables "x" and "y" are introduced to serve as names for particular unknown quantities, thereby enabling these unknowns to be easily referenced; also, special symbols and conventions are used to express the less-than relation, squaring, equality, and so forth. (An inconvenient alternative would be to write the theorem in ordinary language: "For any two real numbers such that the first is less than the second, if their squares are equal then the first value equals the negative magnitude of the second value." And, likewise for the proof.) Of course, it is assumed that the reader is familiar with these notational devices—or, if not, that their meanings will be sought elsewhere, as needed.[1]

The proof of the theorem is expected to be comprehended one logical step at a time. Certain basic facts—e.g., that $\sqrt{x^2} = |x|$ and $x = \pm|x|$ for any real number x—are assumed to be known; moreover, such facts may or may not be explicitly cited when used. Sometimes, though, a logical step that is small, obvious, or mundane in the opinion of the writer might require much greater contemplation on the part of the reader; thus, it is always good to have a pencil and paper handy for performing one's own analyses. Finally, the reader may be alerted to the end of the proof by a symbol such as "□"; alternatively, a proof might be terminated with the abbreviation "Q.E.D." for the Latin *quod erat demonstrandum*, meaning "which was to be demonstrated".

Although any mathematical statement can be reworded, some particular idioms are frequently encountered in mathematics literature. For example, the above theorem might be reworded as follows:

> ***Theorem*** Given real numbers x and y for which $x < y$ and $x^2 = y^2$, we have $x = -|y|$.

Or, it could have begun, "Let x and y be real numbers." or "Suppose x and y are real numbers." Likewise, phrases such as "assuming $x = 4$" and "taking x to be 4" are common. In a professional journal, the theorem might be more succinctly written

> ***Theorem*** Given $x < y$, if $x^2 = y^2$ then $x = -|y|$.

Here it is implicit that x and y are real numbers because the relation $<$ is being applied to them. In informal settings (e.g., a mathematics lecture), various abbreviations may be used to reduce writing and thereby speed up the presentation, besides make taking notes easier; common examples include using "iff" for "if and only if", "∴" for "therefore", and "∵" for "because" or "since".

Often, several similar statements are incorporated into a single statement by incorporating the word "respectively". For example, the compound statement

> The roman numeral "I" represents the number 1; the roman numeral "V" represents the number 5; and, the roman numeral "X" represents the number 10.

can be compactly rewritten

> The roman numerals "I", "V", and "X" represent the numbers 1, 5, and 10, respectively.

---

1. A list of all notation used in this book follows the appendix.

Or, a compound statement such as

> We have $x \leq y$ if and only if $x < y$ or $x = y$, whereas $x \geq y$ if and only if $x > y$ or $x = y$.

might be rewritten by putting respective replacement phrases within parentheses:

> We have $x \leq y$ (respectively, $x \geq y$) if and only if $x < y$ ($x > y$) or $x = y$.

Likewise, two related formulas—e.g.,

$$(x + y)^2 = x^2 + 2xy + y^2,$$
$$(x - y)^2 = x^2 - 2xy + y^2$$

—might be combined:

$$(x \pm y)^2 = x^2 \pm 2xy + y^2 \quad (\pm \text{ respectively}).$$

Similarly,

$$-(x \pm y)^2 = -x^2 \mp 2xy - y^2 \quad (\pm \text{ and } \mp \text{ respectively}).$$

Sometimes the shortening of a mathematical argument is signaled by the phrase "without loss of generality", which is often used to avoid proving several similar cases one by one. For example, for the purpose of proving that $x + 1/x \geq 2$ for all values $x > 0$, one can immediately assume without loss of generality that $x \geq 1$; because, x and $1/x$ each equals the reciprocal of the other, and for $0 < x < 1$ we have $1/x > 1$. Of course, some assumptions *do* incur a loss of generality; e.g., here we could not assume $x \geq 2$ (for which the proof of the original inequality becomes trivial).

The reader must always be prepared to adjust to the style of the writer. For example, whereas in this book the conditions under which a formula is valid are stated as parenthetical remarks—e.g.,

$$x = -\sqrt{x^2} \quad (x \leq 0)$$

—some authors omit the parentheses, inserting a separating comma instead. We will also routinely write, e.g., "$a \in \mathbb{R}$ and $b \in \mathbb{R}$" more briefly as "$a, b \in \mathbb{R}$"; and, rather than saying (as in the above proof), "taking the square root of each side of the equation $x^2 = y^2$, and equating the results", we might simply say, "taking the square root of the equation $x^2 = y^2$". However, although it is common for definitions to be expressed using the word "if" when the phrase "if and only if" is intended—e.g.,

> A real number x is defined to be "positive" if $x > 0$.

—we will avoid this practice, for greater clarity.

Writing Mathematics

Of course, the precision of mathematics requires care on the part of the writer as well. Inadvertently changing the wording of a theorem might, if one is lucky, produce another statement having the same meaning. More likely, though, such a lapse will at best produce a true statement that is unjustified; it will at worst produce a falsehood. Likewise, slightly rewording a valid proof can make it invalid.

Less serious, but still important, the mathematical symbols one chooses should conform with customary usage so as not to be misleading. In particular: "x", "y", and "z" are frequently used as arbitrary variables; "a", "b", and "c" often represent constants; and, "ε" and "δ" are typically reserved for small positive values. For variables restricted to integer values (e.g., indices of sequences and sums), "i", "j", "k", "l", "m", and "n" are typical; "θ" and "ϕ" are often used for angles; and, "f", "g", and "h" usually symbolize functions. In physical applications, other common practices apply (some specific to particular scientific fields), such as using "t" for time.

When a mathematical argument is well written, the intended reader can generally follow along without extreme difficulty (which is not to say that *no* effort is required). The argument is factually accurate, logically sound, and readable. It is also concise and as simple as possible, while at the same time being complete and sufficiently detailed. No step in the argument is too large, and each leads naturally to the next.

In particular, when writing mathematics, you should:

- Precisely state variable ranges. (To say that "the number x is between 5 and 10" is to assert $5 < x < 10$, not $5 \leq x \leq 10$; for the latter case, x lies between 5 and 10, *inclusive*.)
- Be careful when stating the size of a collection. (For either of the sets $S = \{a, b\}$ and $T = \{x, y, z\}$, one can correctly say that it "has two elements"; but, T actually has *more* than two elements. Thus, when there is a chance of confusion, greater clarity can be achieved by saying that S has *exactly* two elements, and that T has *at least* two elements.)
- Distinguish objects as intended. (The statement "Let a and b be points", or even "Let a and b be two points", might be interpreted by the reader to mean that a and b can be the *same* point. If this is not intended, then state that a and b are *distinct* points.)
- Keep in mind that uniqueness is not generally assumed by the reader, and so must be explicitly asserted. (Compare the assertion "$\sqrt{x}$ is a value $y \geq 0$ such that $x = y^2$" with the more descriptive "$\sqrt{x}$ is the unique value $y \geq 0$ such that $x = y^2$", both of which are true for all $x \geq 0$.)
- Apply the word "not" with care. (The logical negation of "x and y are even" is not "x and y are not even" (meaning neither x nor y is even), but rather "It is not true that x and y are even", which is equivalent to "Either x or y is not even".)
- Use the words "always" and "never" with care. (The assertion "At least one of the numbers x, y, and z is always zero" is not equivalent to "it is always the case that at least one of the numbers x, y, and z is zero". Likewise, the assertion "either x or y is never zero" is not equivalent to "it is never the case that either x or y is zero" (meaning neither x nor y is ever zero).)
- Be aware that although the words "all", "each", "every", and "any" are often interchangeable (e.g., "for all $x \in \{-1, 0, 1\}$ we have $x = x^2$"), such is not always the case. (Compare "we do not have $x > 0$ for all $x \in \{-1, 0, 1\}$" (which is true, since $-1 \not> 0$) with "we do not have $x > 0$

for any $x \in \{-1, 0, 1\}$" (which is false, since $1 > 0$). Also, "each roman numeral represents particular number" is preferred to "every roman numeral represents a particular number", since the latter could be misinterpreted as saying that every roman numeral represents the *same* number.)

- Give verbal cues (e.g., "therefore", "however", "conversely", "nevertheless") to help the reader follow the flow of your arguments.

Finally, be aware that, as with all exposition, writing mathematics well generally requires much *re*writing.

Thinking Mathematics

Not surprisingly, the ability to think mathematically is related to how much mathematics a person knows. One reason is that every mathematical theorem can be viewed as a statement of the form, "If such and such conditions hold, then these other conditions also hold."[2] Thus, when analyzing a mathematical problem, one might observe that the "if conditions" (i.e., premises) of a particular theorem hold, and thereby conclude that its "other conditions" (i.e., conclusions) also hold. Then, having determined more about the problem, another such step might be performed, and so forth, until a solution to the problem is obtained.

Learning mathematics, though, is as much about mastering techniques for solving problems as it is about acquiring a bank of theorems. By seeing how others solve mathematical problems (including proving theorems), then practicing the same techniques on similar problems, one becomes more likely to recognize when to apply those techniques to other situations. For example, suppose that for a particular sequence $\{x_i\}_{i=1}^{\infty}$ of numbers $x_i \geq 0$, we suspect $x_i \to 0$ as $i \to \infty$. The expression given for x_i in terms of i could be quite complicated; or, perhaps the dependence of x_i on i is only partially known. But, we might be able to prove an upper bound—say, that $x_i \leq 2^{-i}$ for all i. Then, since $2^{-i} \to 0$ as $i \to \infty$, we can conclude (by invoking a theorem) that $x_i \to 0$ as $i \to \infty$ as well. Thus, a proof of the suspected fact has been obtained by employing a common technique: overbounding one sequence of numbers by another that is both easier to work with and has a helpful property (viz., it approaches 0). Needless to say, it is good to know as many such problem-solving techniques as possible.

In general, although *following* a given solution to some mathematical problem is a systematic step-by-step procedure, *providing* a solution to the problem is more difficult; because, to solve a problem, both insight and creativity are usually required.[3] In the previous paragraph, some insight led us to suspect that the sequence $\{x_i\}_{i=1}^{\infty}$ converged to 0; then, creativity was applied to devise a way to prove it. As a more specific example, suppose

$$x_n = 1 + 2 + 3 + \cdots + n$$

2. Even a theorem that simply states, e.g., "$2 + 3 = 5$" may be interpreted as saying, "If $x = 2$ and $y = 3$, then $x + y = 5$."

3. Accordingly, whenever you are told that the solution to a particular problem is "easy", be sure to clarify whether that means there exists a solution that is easily *understood* or is easily *found*.

($n = 1, 2, \ldots$). Does there exist a simple formula, valid for all values n, expressing x_n as a function of n? (*Try to solve this problem yourself before reading further.*) After looking at this problem from various angles, tumbling it around in your mind while trying to spot some helpful pattern(s) or relationship(s), you might consciously take note of the fact that the terms in the sum are increasing at a steady rate, each being one greater than the preceding. Based on this trivial insight, you might creatively think to counterbalance this effect by reversing the order of the terms, thereby obtaining the sum

$$x_n = n + (n-1) + (n-2) + \cdots + 1,$$

in which the terms decrease at the same steady rate. That is, adding this sum to the original term by term—i.e., getting new terms $1+n$, $2+(n-1)$, $3+(n-2)$, $\ldots$, $n+1$—you obtain another sum,

$$2x_n = (n+1) + (n+1) + (n+1) + \cdots + (n+1),$$

in which the terms are constant. Now, since this sum also has n terms, each being $n+1$, you find that $2x_n = n \cdot (n+1)$, from which the desired formula follows: $x_n = \frac{1}{2}n(n+1)$ ($n = 1, 2, \ldots$). To some extent, then, solving a mathematical problem involves a bit of luck to stumble upon a fruitful line of attack (e.g., reversing a sum, adding sums term by term); however, the more problems one solves, the more such ideas will come to mind thereafter, making future problems easier to solve.

Since this book is more a summary of mathematical facts than a guide to thinking mathematically, the remaining comments on this subject will be limited. Nevertheless, the observations made are representative of others that can be found elsewhere. In particular, more on how to develop one's ability to reason mathematically can be found in several books by mathematician George Polya, cited in the bibliography.[4]

First, when trying to gain some insight into a mathematical situation, especially one pertaining to geometry, it is often helpful to think pictorially—i.e., to *visualize*—perhaps by actually drawing a picture. For example, to determine how many days there are from today to a week from next Tuesday, it's easier for most people to visualize a calendar than it is to reason arithmetically. As another example, the real line (see Figure 3.1) is a picture of the set of real numbers that captures the ordering of this set; namely, for any real numbers x and y, we have $x < y$ if and only if the point on the line corresponding to x lies to the left of the point corresponding to y. From this visualization it is obvious that for any real numbers x, y, and z, if $x < y$ and $y < z$, then $x < z$. All visualization, though, must be performed with care, because one must always be mindful that *specific* pictures are being used to represent the *general* mathematical situation under consideration; that is, one's pictures should not lead one to make unjustified assumptions.

Occasionally, a mathematical problem exhibits some form of *symmetry*, the recognition of which can be beneficial. For example, if it is known for some real numbers x and y that $x^2 + 3xy + y^2 = 0$ and $x + y = 2$, then we might observe that interchanging

4. See [46], [45], and [47], in that order.

the variables "x" and "y" yields essentially the same two equations; therefore, given a solution (X, Y) for the pair (x, y), it must be that (Y, X) is also a solution for (x, y). Hence, upon combining the equations and solving for x to find $x = 1 \pm \sqrt{5}$, we may immediately deduce that $y = 1 \pm \sqrt{5}$, after which applying the fact that $x + y = 2$ leads to the conclusion that (x, y) must equal either $(1+\sqrt{5}, 1-\sqrt{5})$ or $(1-\sqrt{5}, 1+\sqrt{5})$. As another example, for the unit vectors $\hat{\mathbf{i}}$, $\hat{\mathbf{j}}$, and $\hat{\mathbf{k}}$ in three-dimensional Euclidean space, we have the cross product $\hat{\mathbf{i}} \times \hat{\mathbf{j}} = \hat{\mathbf{k}}$. Furthermore, the geometric relationship of $\hat{\mathbf{i}}$, $\hat{\mathbf{j}}$, and $\hat{\mathbf{k}}$ is the same as the geometric relationship of $\hat{\mathbf{j}}$, $\hat{\mathbf{k}}$, and $\hat{\mathbf{i}}$, *respectively*. Now, note the circular permutation that was just performed from the first three vectors to the second three: "$\hat{\mathbf{i}}$" became "$\hat{\mathbf{j}}$", "$\hat{\mathbf{j}}$ became "$\hat{\mathbf{k}}$", and "$\hat{\mathbf{k}}$" became "$\hat{\mathbf{i}}$". Therefore, we immediately conclude that $\hat{\mathbf{j}} \times \hat{\mathbf{k}} = \hat{\mathbf{i}}$; and, likewise (performing the same circular permutation again) $\hat{\mathbf{k}} \times \hat{\mathbf{i}} = \hat{\mathbf{j}}$.

When the answer to a problem is known, it might give a clue to how the problem can be solved. Accordingly, when the answer to a problem is *not* known, it is often helpful to guess the answer and then attempt to justify it. For example, we found earlier that $1 + 2 + 3 + \cdots + n = \frac{1}{2}n(n+1)$ $(n = 1, 2, \ldots)$. Observing that the expression on the right side of this equation has the same form as the usual formula for the area of a triangle—viz., one-half the base of the triangle times its height—one might suspect that the equation can be derived geometrically. Indeed, this is done in Section 4.2 (see Figure 4.1).

Again, much mathematical reasoning involves recognizing *patterns* and *relationships*. For example, upon observing that for all integers $n \geq 1$ we have $1^0+2^0+\cdots+n^0 = n$ (a first-degree polynomial), and $1^1 + 2^1 + \cdots + n^1 = \frac{1}{2}n(n+1)$ (a second-degree polynomial), and $1^2 + 2^2 + \cdots + n^2 = \frac{1}{6}n(2n+1)(n+1)$ (a third-degree polynomial), one might suspect that $1^3+2^3+\cdots+n^3$ can be expressed as a fourth-degree polynomial. Indeed, based on this hypothesis it is not difficult to prove (via mathematical induction) that $1^3 + 2^3 + \cdots + n^3 = \frac{1}{4}n^2(n+1)^2$ $(n \geq 1)$. Then, observing that $\frac{1}{4}n^2(n+1)^2 = \left[\frac{1}{2}n(n+1)\right]^2$, an interesting fact emerges: $1^3 + 2^3 + \cdots + n^3 = (1 + 2 + \cdots + n)^2$ $(n \geq 1)$.

Finally, in addition to being objective, logical, and precise, mathematics is *abstract*. This aspect of the theory can be a conceptual hurdle for scientists and engineers, who are mostly occupied with real-world applications. In fact, however, mathematics is so widely applicable to real-world problems precisely because mathematicians have performed the process of *abstraction*, whereby the essential features common to many diverse situations are extracted for specific attention, while nonessential features are discarded. For example, rather than prove that combining 2 apples with 3 apples yields 5 apples, and then perform a separate proof that combining 2 oranges with 3 oranges yields 5 oranges, the mathematician simply proves once and for all that $2+3 = 5$, all the while knowing that the numbers 2, 3, and 5 may be used to count *any* objects. Accordingly, characterizing pure mathematics, Bertrand Russell said: "Thus mathematics may be defined as the subject in which we never know what we are talking about, nor whether what we are saying is true."[5] Just "what" mathematics is talking about physically is unknown because

5. From "Mathematics and the Metaphysicians", 1901 (reprinted in [52]).

its abstract objects (e.g., numbers, points, sets) can serve as surrogates for many different real objects (e.g., apples, oranges, collections thereof), while capturing some of their properties (e.g., quantity, order, category) having analytical significance. Moreover, it is not for the mathematician to determine whether mathematics is in any way applicable to the real world; that is, the theorems of mathematics may or may not be "true" in some physical sense.[6]

6. For more on this particular point, see the statement by Albert Einstein quoted on p. 227.

CHAPTER 1

Logic

Logic—that is, *deductive* logic (in contrast to other kinds of logic, such as inductive logic and fuzzy logic)—has been called the "science of necessary inference". It provides general **rules of inference** by which rigorous logical deductions can be made. A **(logical) deduction** is essentially a conversion of information from one form to another, with no additional information being created in the process. More specifically, given a collection of **statements** (i.e., declarative sentences) called **premises** that are *assumed* to be true, the truth of another statement—the **conclusion**—is shown to *necessarily* follow (i.e., is deduced) by applying one of the rules of inference.

In mathematics, logic helps us expand our understanding of numbers, geometric shapes, and other abstract objects. Each mathematical discipline—arithmetic, geometry, and so forth—begins with a collection of (usually simple) statements called **axioms** (or **postulates**) that are accepted as being true. A **theorem** is then an assertion that certain premises (implicitly including the axioms) logically entail a particular conclusion. For example, based on the axioms of arithmetic, one can show that if x and y are integers (first premise), and $x < y$ (second premise), then $x + 1 < y + 1$ (conclusion). Every theorem is accompanied by a **proof**, which is a finite series of deductions; thus, a proof consists of a finite sequence of statements, each being logically deduced from the original premises and previously deduced statements in the sequence, with the last statement of the sequence being the conclusion of the theorem. Overall, mathematics is simply a large and ever-growing collection of theorems.[1]

Again, a deduction merely converts given information (the premises) into a new *form* (the conclusion), perhaps with some loss; therefore, the conclusion contains no new information beyond what is already contained in the premises, possibly less. In a another sense, though, the act of performing a deduction typically *does* provide new information; for one then knows that the conclusion indeed follows from the premises. Moreover, proving a theorem provides a practical benefit of enabling further conclusions to be drawn more readily. For example, upon showing that 2 is the only even prime

1. Logically, a theorem is any statement having a proof; but, in typical mathematical exposition, the word "theorem" is reserved for *major* statements. A statement of moderate importance, yet still having fairly broad applicability, is often called a **proposition**, whereas a **lemma** is a narrowly applicable statement used to prove another statement, perhaps another lemma. Finally, a **corollary** is a more-or-less immediate consequence of some statement already proved.

number, then proving that an integer is even when the rightmost digit of its fractionless decimal expansion is even, we can immediately recognize the large number 1029384756 as not being prime. Thus, a theorem represents an *investment*, in that its utility can more than compensate for the time and effort expended to prove it.

1.1 Sentential Logic

The study of logic typically begins with **sentential logic** (also known as the **propositional calculus**), in which progressively more complicated statements are constructed by repeatedly applying **logical connectives**, the most basic of which are defined in Table 1.1 in terms of arbitrary statements A and B.[2] The construction process starts when one is given a collection of **atomic statements**, from which all other statement are built. A statement is atomic if it cannot be decomposed in terms of other statements; in other words, it has no internal structure. For example, given atomic statements A, B, and C—which each can be viewed as representing any declarative sentence we like (e.g., "The ball is red.")—we can form the conjunction $A \wedge B$ and call it statement D (nonatomic); then, we might form the disjunction $C \vee D$, which fully expanded in terms of the atomic statements is $C \vee (A \wedge B)$.

The **truth value**—viz., **true** (T) or **false** (F)—of every nonatomic statement is uniquely determined by the truth values of the statements from which it is constructed. This aspect of the statement is commonly expressed in tabular form as a **truth table**. In particular, the truth table for each of the logical connectives in Table 1.1 is given in Table 1.2 (e.g., it shows that if A is true and B is false, then $A \vee B$ must be true). By using these truth tables, we can find the truth table for any other statement formed from the connectives in Table 1.1. Here, we discuss five simple examples, tabulated together in Table 1.3.

The third column of Table 1.3 shows that the statements A and $\neg(\neg A)$ always have the same truth value, regardless of the truth value of A. In general, two sentential-logic statements are said to be **logically equivalent** if and only if they have the same truth value for *every* assignment of truth values to their constituent atomic statements (irrespective of the *actual* truth values these statements might have). Hence, comparing the fourth

TABLE 1.1 Basic Logical Connectives

Connective	Statement	Meaning
Negation ($\neg$)	$\neg A$	Not A
Conjunction ($\wedge$)	$A \wedge B$	A and B
Disjunction ($\vee$)	$A \vee B$	A or B
Conditional ($\Rightarrow$)	$A \Rightarrow B$	If A then B
Biconditional ($\Leftrightarrow$)	$A \Leftrightarrow B$	A if and only if B

2. Various other symbols are also commonly used for logical purposes—e.g., "$\sim$" for a negation, and "$\rightarrow$" or "$\supset$" for a conditional. The symbols we have chosen, though, allow others (such as those just mentioned) to be used for purely mathematical purposes.

TABLE 1.2 Truth Tables for the Basic Logical Connectives

A	$\neg A$	A	B	$A \wedge B$	$A \vee B$	$A \Rightarrow B$	$A \Leftrightarrow B$
F	T	F	F	F	F	T	T
T	F	F	T	F	T	T	F
		T	F	F	T	F	F
		T	T	T	T	T	T

TABLE 1.3 Some Truth Tables

A	B	$\neg(\neg A)$	$(\neg A) \vee B$	$A \vee (\neg A)$	$(\neg A) \wedge (\neg B)$	$(\neg B) \Rightarrow (\neg A)$
F	F	F	T	T	T	T
F	T	F	T	T	F	T
T	F	T	F	T	F	F
T	T	T	T	T	F	T

column of Table 1.3 with Table 1.2, we also see that the statement $(\neg A) \vee B$ is logically equivalent to $A \Rightarrow B$. It follows that the conditional connective, though convenient, does not provide any expressive capability beyond that provided by the negation and disjunction connectives together.

The fifth column of Table 1.3 shows that a statement of the form $A \vee (\neg A)$ is always true, regardless of the truth value of A. In general, a sentential-logic statement that is true for every assignment of truth values to its constituent atomic statements is called a **tautology**.[3] Two sentential-logic statements A and B are logically equivalent if and only if their biconditional $A \Leftrightarrow B$ is a tautology.

Comparing the next-to-last column of Table 1.3 with Table 1.2, we see that the statements $\neg(A \vee B)$ and $(\neg A) \wedge (\neg B)$ are logically equivalent. This is one of **DeMorgan's laws**. The other is the complementary fact that $\neg(A \wedge B)$ and $(\neg A) \vee (\neg B)$ are logically equivalent.

Finally, Tables 1.2 and 1.3 show that the statements $A \Rightarrow B$ and $(\neg B) \Rightarrow (\neg A)$ are logically equivalent. The latter conditional is called the **contrapositive** of the former. This is in distinction to the **converse** of $A \Rightarrow B$, which is simply the reverse conditional $B \Rightarrow A$. In general, the **antecedent** of a conditional $A \Rightarrow B$ is the statement A, and its **consequent** is the statement B (which can be the same as A). Thus, its converse obtains by interchanging the antecedent and consequent.

A frequently used rule of inference is **modus ponens**:

$$\text{Given that} \quad A \quad \text{and} \quad A \Rightarrow B \quad \text{are true,} \quad \text{conclude} \quad B \quad \text{is true.} \tag{1.1}$$

In sentential logic, though, no explicit rules of inference are necessary; rather, any deduction can be performed directly via truth tables, by a process of elimination. For example, based on the truth table for a conditional, shown separately in Table 1.4, we can justify

3. The particular tautology $A \vee (\neg A)$—asserting that for every statement A, either it or its negation $\neg A$ must be true—is known as the **law of (the) excluded middle**.

TABLE 1.4 The Truth Table for a Conditional

A	B	$A \Rightarrow B$
F	F	T
F	T	T
T	F	F
T	T	T

modus ponens itself: If A and $A \Rightarrow B$ are both true, then the first three rows of the table can be eliminated from being possible; therefore, based on the remaining rows (here, only one), B must also be true.

In general, a finite collection of sentential-logic statements $A_1, \ldots, A_n$ $(n \geq 1)$ **logically implies** another statement B—i.e.,

$$A_1, \ldots, A_n \vDash B \tag{1.2}$$

—if and only if every assignment of truth values to atomic statements that makes A_1, $\ldots$, A_n all true also makes B true. Equivalently, (1.2) asserts that the single statement[4]

$$(A_1 \wedge \cdots \wedge A_n) \Rightarrow B$$

is a tautology. It follows that two statements A and B are logically equivalent if and only if each logically implies the other; that is, both $A \vDash B$ and $B \vDash A$—i.e.,

$$A \vDash \Dashv B$$

—which is the case if and only if the single statement $A \Leftrightarrow B$ is a tautology. As an example, it was just observed that

$$A \quad \vDash \Dashv \quad \neg(\neg A).$$

By contrast, whereas

$$A \wedge B \quad \vDash \quad A \vee B,$$

the reverse implication does not hold; that is,

$$A \vee B \quad \nvDash \quad A \wedge B$$

(consider, e.g., the case of A being true and B being false).

Note that a logical implication $A \vDash B$ is at least as "strong" as the corresponding "conditional implication" $A \Rightarrow B$,[5] perhaps stronger. That is, whenever the former

4. Strictly speaking, the streamlined syntax used in the following statement is ambiguous because it is not clear in what order the conjunctions are to be performed in $A_1 \wedge \cdots \wedge A_n$. If desired, this order can be uniquely specified by inserting additional parentheses to provide grouping; however, the truth value of the resulting statement is independent of the order chosen. A similar comment applies to a multiple disjunction $A_1 \vee \cdots \vee A_n$.

5. The word "imply" and its derivatives are often used with regard to the logical connective $\Rightarrow$. Namely, a conditional $A \Rightarrow B$ is sometimes referred to as an "implication" and may be verbalized as "A implies B"

implication can be asserted, so can the latter; but, not necessarily vice versa. This is because the truth value of $A \Rightarrow B$ is determined only by the *actual* truth values of A and B, whereas finding the truth value of $A \models B$ requires us to consider *all* possible assignments of truth values to the atomic statements making up A and B. Specifically, it follows from Table 1.4 that if $A \models B$ is true (i.e., it is *logically impossible* to simultaneously have A true and B false), then $A \Rightarrow B$ must be true (i.e., in *fact*, we do not have A true and B false); in other words, $A \models B$ asserts that the statement $A \Rightarrow B$ must be true based on *logical structure* alone, irrespective of any actual truth values. Conversely, though, $A \Rightarrow B$ can be true without having $A \models B$. For example, A might be an atomic statement, and B might be $A \wedge C$ for another atomic statement C that *happens* to be true, thereby making $A \Rightarrow B$ true (whatever the truth value of A); however, this conditional is not true for *every* assignment of truth values to A and C (viz., consider A true and C false), so $A \not\models B$. Similarly, a logical equivalence $A \models\!\!\!\dashv B$ is at least as strong as the corresponding "biconditional equivalence" $A \Leftrightarrow B$, perhaps stronger.

1.2 First-Order Logic

Consider the following syllogism:

All humans are mortal.	(premise 1)
Sophie is a human.	(premise 2)
Sophie is mortal.	(conclusion)

This deduction, though clearly correct, cannot be justified using sentential logic alone. Likewise, most mathematical arguments require an enhancement of sentential logic called **first-order logic**, which enables deductions to be performed upon certain statements containing the locution "for all" (possibly expressed by just "all", as in the first premise).[6]

A precise treatment of first-order logic requires much additional syntactical machinery—including variables, names, and predicates—plus several axioms of quantification to fully capture the meaning of "for all", which is symbolized by the **universal quantifier** (∀). As a simple demonstration, our syllogism can be translated into the syntax of first-order logic as follows:

$\forall x,\ H(x) \Rightarrow M(x)$	(premise 1)
$H(\text{Sophie})$	(premise 2)
$M(\text{Sophie})$	(conclusion)

---

rather than "if A then B". As long as it is clear how this conditional implication differs from the assertion that A *logically* implies B, no confusion results.

6. The deductive power of first-order logic is limited as well. Successively greater power in handling statements containing "for all" is provided by logics of higher order (second-order, third-order, . . .); however, first-order logic is sufficiently powerful to express the vast majority of mathematics.

Here, "x" is a variable taking its value from, say, the collection of all physical objects, exactly one of which is a human named "Sophie". Also, "H" and "M" are predicates, with "$H(x)$" meaning "x is a human" and "$M(x)$" meaning "x is mortal". The syllogism can now be proved by first noting that, in light of the meaning of "$\forall$", we are free to substitute $x =$ Sophie in the conditional within premise 1, thereby deducing

$$H(\text{Sophie}) \Rightarrow M(\text{Sophie})$$

—i.e., if Sophie is a human, then Sophie is mortal. Then, applying modus ponens (1.1) to this result along with premise 2, we deduce the stated conclusion.

In general, a **variable** (logical or mathematical) is an arrangement of one or more symbols—e.g., a single symbol such as "x" or a subscripted symbol such as "x_1"—that can be assigned a value taken from the (nonempty) **universe** of things under discussion. Thus, the variable "x" represents the value x (e.g., a number), the variable "x_1" represents the value x_1, and so forth.[7] In a mathematical context, a variable under consideration might be called, e.g., a "real variable", indicating that its value may be any real number, but not anything else; hence, in this case the universe would need to at least contain all real numbers.

In addition to such *generic* referencing of things in the universe by variables, the values of which can vary, we often desire the capability of referring to some things in the universe *specifically*. For this purpose, any particular thing in the universe may be uniquely designated by some arrangement of symbols called a **name** (or **constant**). Unlike a variable, which can represent any one of a *range* of values, a name (e.g., "π") always represents the *same* value (3.14159 ...); that is, its value is *fixed*.

An ***n*-ary** (or ***n*-place**) **predicate**, for some fixed integer $n \geq 0$, is yet another arrangement of symbols; it enables a particular assertion to be made about the things in the universe, taken n at a time. Thus, "H" above is a 1-ary (or **unary**) predicate, with "$H(x)$" being a statement asserting that x is a human. The most frequently occurring 2-ary (or **binary**) predicate is the **identity predicate** "$=$", with "$x = y$" asserting that x and y are equal (i.e., the variables "x" and "y" refer to the same thing). Similarly, there are 3-ary (or **ternary**) predicates; e.g., "$W(x, y, z)$" might assert that x, y, z are collinear points in a particular geometric space, with y lying between x and z. Furthermore, each atomic statement in a sentential logic is essentially a 0-ary predicate, because (to the extent that it conveys any meaning) it makes an assertion without being provided any objects from the universe; hence, since the logical connectives are also part of first-order logic, sentential logic can be viewed as a special case.

To give, and then explain, a collection of axioms sufficient to enable the deductions of first-order logic would require more space than is warranted here.[8] Fortunately, it is rare for a proof in mathematics to be expressed and carried out in the formal syntax of first-order logic. Rather, just as the correctness of the above syllogism was easily determined from its original informal expression, so all that is required to follow the

7. With this relationship of *variables* "x", "x_1", ... to their *values* $x, x_1, \ldots$ understood, it is common in mathematical writings for the quotation marks to be omitted; that is, one might see a reference to variables x, y, and z when what is really meant is variables "x", "y", and "z". Likewise, where no confusion results, quotation marks are routinely omitted elsewhere as well.

8. Moreover, including the identity predicate would require additional axioms.

logic of a typical mathematical proof is a clear understanding of the logical connectives, the universal quantifier, and the **existential quantifier** ($\exists$).

This latter quantifier occurs frequently, since existence is an important consideration throughout logic and mathematics. A statement of the form

$$\exists x,\ P(x), \tag{1.3}$$

for some unary predicate "P", asserts "there exists an x such that $P(x)$"; more briefly, "$\exists$" can be read "for some".[9] Actually, the existential quantifier is not really necessary; because, it is logically correct to view "$\exists x$" as simply an abbreviation for "$\neg\forall x,\ \neg$". Thus, the existential statement in (1.3) may be interpreted as

$$\neg\forall x,\ \neg P(x).$$

This is because to assert that there exists an object x having a property P is equivalent to asserting that it is not the case that property P fails to hold for every object x. Also, note that whereas "some" is often interpreted in ordinary discourse as meaning "some, but not all", such is not the case in formal logic (e.g., to say that *some* of the numbers in a set are even does not the preclude possibility that *all* of the numbers in the set are even). In other words, (1.3) alone does not imply $\neg\forall x,\ P(x)$.[10]

Utilizing the available logical connectives, variables, names, predicates, and quantifiers, the statements of a first-order logic are constructed in accordance with strict grammatical rules (left unstated in this discussion). A statement A is then said to be **logically true** (or **logically valid**)—i.e.,

$$\models A$$

—if and only if A is true by virtue of its *logical structure* alone. Specifically, only the meanings of the logical connectives, the quantifiers, and the identity predicate (if present) are *essential*; that is, the truth of A is *independent* of any meanings associated with the names and other predicates, as well as the truth values assigned to the atomic statements. It follows that although a logically true statement must be true in fact, a factually true statement need not be logically true; in other words, logical truth is generally stronger than factual truth. In sentential logic, the logically true statements are the tautologies.

As an example, let us boil the above syllogism down to its essence—viz.,

If *premise 1* and *premise 2*, then *conclusion*.

---

9. Some authors would write "$\exists x \ni P(x)$" instead of (1.3), with the symbol "$\ni$" being read "such that". Also, the reader should be careful to understand any shorthand notation used. For example, it is not uncommon to write "$\exists x,\ \exists y$" more compactly as "$\exists x, y$", and likewise for successive universal quantifications. Additionally, in this book we use, e.g., "$x, y \geq z$" as shorthand for "$x \geq z$ and $y \geq z$". Therefore, it must be explicitly stated that the expression "$\exists x, y \geq z$" correctly expands to "$\exists x \geq z,\ \exists y \geq z$", not "$\exists x,\ \exists y \geq z$"; that is, variables "bind" more tightly to relations such as "$\geq$" than to quantifiers. Without such binding rules, many logical expressions must be cluttered with parentheses to precisely express them; nevertheless, when confusion might occur, parentheses should be used.

10. On the other hand, since the universe of discourse is always taken to be nonempty, a universal statement $\forall x,\ P(x)$ always implies the corresponding existential statement $\exists x,\ P(x)$.

We then obtain the statement

$$\{[\forall x,\ H(x) \Rightarrow M(x)] \wedge H(\text{Sophie})\} \Rightarrow M(\text{Sophie}), \tag{1.4}$$

which is logically true. That is, this statement is true no matter what meanings are associated with the name "Sophie" and the predicates "H" and "M". Thus, Sophie might be a happy Jersey cow, "$H(x)$" might mean "x is a Holstein cow" and "$M(x)$" might mean "x gives milk"; then, the conditional (1.4) would be true simply because the statement "$M(\text{Sophie})$" is true. But, if the meaning of "$M(x)$" is changed to "x is male", then "$M(\text{Sophie})$" becomes false; yet, (1.4) is still true because the statement "$H(\text{Sophie})$" is false, thereby making false the conjunction within the braces. However, if we also change the meaning of "$H(x)$" to "x is happy", then "$H(\text{Sophie})$" becomes true; nevertheless, (1.4) remains true because the statement "$\forall x,\ H(x) \Rightarrow M(x)$" is now false (as seen by letting $x = \text{Sophie}$)—though this statement was true with the predicate meanings first assumed.

Furthermore, generalizing to first-order logic the definitions given in Section 1.1 for sentential logic, a finite collection of $n \geq 1$ statements $A_1, \ldots, A_n$ **logically implies** another statement B—i.e.,

$$A_1, \ldots, A_n \vDash B$$

—if and only if, by virtue of the logical structure of these statements alone, B is true whenever $A_1, \ldots, A_n$ are all true. Equivalently,

$$\vDash (A_1 \wedge \cdots \wedge A_n) \Rightarrow B.$$

And, two statements A and B are **logically equivalent**—i.e.,

$$A \vDash\dashv B$$

—if and only if each logically implies the other, which is the case if and only if

$$\vDash A \Leftrightarrow B.$$

For example, for any unary predicate "P", we have

$$\exists x,\ \neg P(x) \quad \vDash\dashv \quad \neg\forall x,\ P(x), \tag{1.5}$$

since the statements "$\exists x,\ \neg P(x)$" and "$\neg\forall x,\ P(x)$" must always have the same truth value, regardless of the meaning of "P".

Finally, consider again premise 1 of our syllogism,

$$\forall x,\ H(x) \Rightarrow M(x), \tag{1.6}$$

with the original meanings of "$H(x)$" and "$M(x)$". Rather than assert this universal statement as written, we could just as well say that being a human is a **sufficient condition** for being mortal; equivalently, being mortal is a **necessary condition** for being a human. Also, in the course of analyzing the syllogism, we derived the more specific statement

$$H(\text{Sophie}) \Rightarrow M(\text{Sophie}); \tag{1.7}$$

and, this conditional can be expressed by saying either that Sophie being a human is a sufficient condition for Sophie being mortal, or that Sophie being mortal is a necessary condition for Sophie being a human. Moreover, were we not aware of the meanings of the predicates "H" and "M", we could express (1.6) by saying that property H is a sufficient condition for property M, or that property M is a necessary condition for property H; likewise, we could express (1.7) by saying that Sophie possessing property H is a sufficient condition for Sophie possessing property M, or that Sophie possessing property M is a necessary condition for Sophie possessing property H.

However, these last conditions of sufficiency and necessity might be perplexing once the meanings of the predicates become known. For example, it could happen that "$H(x)$" means "x is happy" and "$M(x)$" means "x is miserable"; then, (1.7) would be a true statement if Sophie were miserable but not happy (a perfectly sensible state of affairs). But, above we would be asserting (1.7) by saying either that Sophie being happy is a sufficient condition for Sophie being miserable, or that Sophie being miserable is a necessary condition for Sophie being happy. Note, though, that because of the absence of a universal quantifier in (1.7), we would *not* be asserting the generalization (1.6)—i.e., that being happy is a sufficient condition for being miserable, or that being miserable is a necessary condition for being happy—nor could we, since being miserable precludes being happy. The source of the perplexity regarding (1.7) is that if this conditional is indeed true with these meanings of "H" and "M", then it must be also be true that Sophie is not happy—i.e., "$H(\text{Sophie})$" is false—thereby robbing the conditional (1.7) of its utility.

1.3 Types of Proof

Whenever a theorem is proved, there are infinitely many logically correct paths from the premises to the final conclusion; moreover, even among proofs that do not contain extraneous steps (e.g., citing an axiom without using it in a subsequent deduction), there can be substantial differences. Qualitatively, a **direct proof** of a conditional $A \Rightarrow B$ is a straightforward sequence of deductions starting by assuming the truth of A and concluding with the truth of B. Alternatively, one might prove $A \Rightarrow B$ indirectly by providing a direct proof of its contrapositive $\neg B \Rightarrow \neg A$—i.e., assuming $\neg B$ and concluding $\neg A$—which might actually be easier to do. Yet another approach to proving $A \Rightarrow B$ is **proof by contradiction** (also called **reductio ad absurdum**, meaning "reduction to absurdity"), whereby simultaneously assuming the truth of A and $\neg B$ is shown to lead to a **contradiction**—i.e., that some statement C and its negation $\neg C$ are both true (which, of course, is impossible).

Proofs of mathematical *existence* receive further classification. To prove an existential statement $\exists x,\ P(x)$, one might actually construct a particular mathematical object x having the property P; or, as well, a step-by-step procedure could be specified by which such a construction could by carried out in principle (though it might not be practical to do so). In contrast to such a **constructive proof**, one might instead give a **pure existence proof**, whereby it is shown that assuming the negation $\neg\exists x,\ P(x)$ leads to a contradiction.[11] Also, per (1.5), to *disprove* a universal statement $\forall x,\ P(x)$—i.e., prove $\neg\forall x,\ P(x)$—we may equivalently prove $\exists x,\ \neg P(x)$. This could be done by constructing an object x that does *not* possess property P; accordingly, such an object is called a **counterexample**.

Finally, consider a statement A of the form $\forall x,\ P(x) \Rightarrow Q(x)$, which states that everything having property P also has property Q. Suppose, though, that *nothing* in the universe under discussion possesses property P; that is, $P(x)$ is false for every x. What, then, of the truth of A? From the truth table for the logical connective $\Rightarrow$ (Table 1.4), we see that $P(x) \Rightarrow Q(x)$ is true whenever $P(x)$ is false, *regardless* of the truth value of $Q(x)$; therefore, we conclude that A must be true. Briefly put, A asserts that property P implies property Q, and here this assertion is **vacuously true** because *no* thing in the universe possesses property P. Should the reader find this reasoning a bit unsettling, observe that the universal statement $\forall x,\ P(x) \Rightarrow Q(x)$ must now be true simply because it cannot be false; no counterexample to it can possibly exist. Specifically, for this statement to be false, there would have to exist some thing x for which the conditional $P(x) \Rightarrow Q(x)$ is false; but, this would require $P(x)$ to be true and $Q(x)$ to be false, whereas it has been stipulated that $P(x)$ is true for no x.

---

11. This is a common way of proving, e.g., that there are infinitely many prime numbers; that is, it is shown that the assumption that there are only finitely many primes leads to a contradiction.

CHAPTER 2

Sets

The basic building blocks of mathematics are *sets*. Intuitively, a **set** is a *collection* of objects, each such object being an **element** (or **member**) *of* (or *in*) the set. To assert that an object a is an element of a set A—i.e., that *a belongs to A*—we write "$a \in A$"; whereas "$a \notin A$" asserts the opposite. Two sets A and B are equal if and only if they have exactly the same elements:

$$A = B \quad \Leftrightarrow \quad \forall x,\, x \in A \Leftrightarrow x \in B.$$

Simple sets can be specified by listing their elements within braces. For example, $\{0, 1\}$ is the set having just the elements 0 and 1; likewise, $\{1, 2, 3, \dots\}$ is the set of positive integers. By definition, a set is an *unordered* collection; thus, $\{0, 1\} = \{1, 0\}$. Furthermore, it is understood that repetitions in a listing have no effect; so, e.g., $\{0, 1, 0\}$ is also the same set as $\{0, 1\}$ and $\{1, 0\}$.

The most fundamental set is the **empty set** $\emptyset$, which is defined as the unique set having *no* elements. (It should not be confused with $\{\emptyset\}$, which is a set having *one* element—namely, $\emptyset$.) In fact, it is possible to view *all* mathematical objects (numbers, functions, etc.) as sets built up from $\emptyset$; for example, a common way to define the natural numbers is to first let $0 \triangleq \emptyset$,[1] and then let $1 \triangleq \{0\}$, $2 \triangleq \{0, 1\}$, $3 \triangleq \{0, 1, 2\}$, and so forth. Such foundational matters, though, are rarely mentioned in most mathematical discourse.

The *number* of elements in a set A—i.e., the *size* of the set—is called its **cardinality** $|A|$. (For example, $|\{0, 1\}| = 2$.) If $|A| < \infty$, then A is referred to as a **finite set**; otherwise, it is an **infinite set**. In particular, $|A| = 0$ if (and only if) $A = \emptyset$; so, $\emptyset$ is finite.

It is tempting to assume that every mathematical property P determines a corresponding set, the elements of which are precisely those mathematical objects possessing the property; that is, P would effectively "define" the set

$$\{x \mid P(x)\}, \tag{2.1}$$

1. The symbol "$\triangleq$" stands for "equals by definition". Other symbols commonly used for this purpose are "$:=$" and "$\equiv$".

where the vertical bar is read "such that". However, suppose we let R be the set of all mathematical objects that are not elements of themselves:

$$R = \{x \mid x \notin x\}.$$

Then, is R an element of itself? Well, if $R \in R$, then since $x \notin x$ for each $x \in R$, we must also have $R \notin R$; but, if $R \notin R$, then since $x \in R$ for each x such that $x \notin x$, we must also have $R \in R$. Hence, we have deduced that the conditions $R \in R$ and $R \notin R$ imply each other; so, the very existence of R leads to a contradiction. Therefore, the only way out of this quandary—known as **Russell's paradox**—is to conclude that no such set R exists.

Although not every mathematical property determines a set, it is always permissible to impose a property P on a *given* set A, thereby obtaining the more restricted set

$$\{x \in A \mid P(x)\}. \tag{2.2}$$

For example, by introducing the symbol "$\neq$" meaning "is not equal to", we have $\{x \in A \mid x \neq x\} = \emptyset$ for every set A. However, it now follows that it is impossible to collect all mathematical objects into a **universal set** U; for then we could form the set $\{x \in U \mid x \notin x\}$, which is precisely the set R shown above not to exist. Thus, we have the profound fact that although every set is a collection of mathematical objects, not every collection of mathematical objects is a set.[2]

As for relating one set A to another B, we say that A is a **subset** of B—i.e., $A \subseteq B$—if and only if every element of A is also an element of B.[3] Thus, $A = B$ if and only if both $A \subseteq B$ and $B \subseteq A$. Also, every set is a subset of itself.

It is important that the difference between the meanings of "$\subseteq$" and "$\in$" be clearly understood. To wit, here is how the former relation can be expressed in terms of the latter:

$$A \subseteq B \quad \Leftrightarrow \quad \forall x,\ x \in A \Rightarrow x \in B.$$

Hence, letting $A = \emptyset$ shows that $\emptyset \subseteq B$ for every set B; because, the empty set has no elements, and thus it is vacuously true that every element of $\emptyset$ is an element of B. Also, it is possible for elements of a set to themselves be sets (as was already observed with $\{\emptyset\}$). In particular, for every set A there exists a corresponding **power set**, defined as the set of all subsets of A—e.g., the power set of $\{0, 1\}$ is $\big\{\emptyset, \{0\}, \{1\}, \{0, 1\}\big\}$—the cardinality of which is $2^{|A|}$ when A is finite.

A set A is a **proper subset** of B—i.e.,[4] $A \subset B$—if and only if $A \subseteq B$ and there exists an element of B that is not in A. In other words, $A \subset B$ if and only if $A \subseteq B$ and $A \neq B$. Thus, for no set A do we have $A \subset \emptyset$, whereas $\emptyset \subset B$ for every nonempty set B. Also, no set is a proper subset of itself.

---

2. In particular, in a first-order logic, the universe under discussion may be a set, but need not be.

3. This is often expressed by saying "set A is *contained in* set B" or "set B *contains* set A". However, when speaking generally about sets, we will avoid such use of the word "contain" because, it can be widely found to mean *either* "$\subseteq$" or "$\in$", as determined by the context. For example, in geometry (see Section 10.1) one might say "plane p contains line l" (i.e., $l \subseteq p$), yet also say "plane p contains point A" (i.e., $A \in p$).

4. Unfortunately, the sensible notation presented here is not universally adopted; rather, some authors use "$\subset$" for our "$\subseteq$", and likewise use "$\supset$" below for our "$\supseteq$".

TABLE 2.1 Basic Operations on Sets

Operation	**Set**	**Definition**
Intersection ($\cap$)	$A \cap B$	$\{x \mid x \in A \text{ and } x \in B\}$
Union ($\cup$)	$A \cup B$	$\{x \mid x \in A \text{ or } x \in B\}$
Difference ($-$)[a]	$A - B$	$\{x \in A \mid x \notin B\}$

[a] *Some authors prefer the symbol "\" for this purpose, and thus write "$A \setminus B$" instead.*

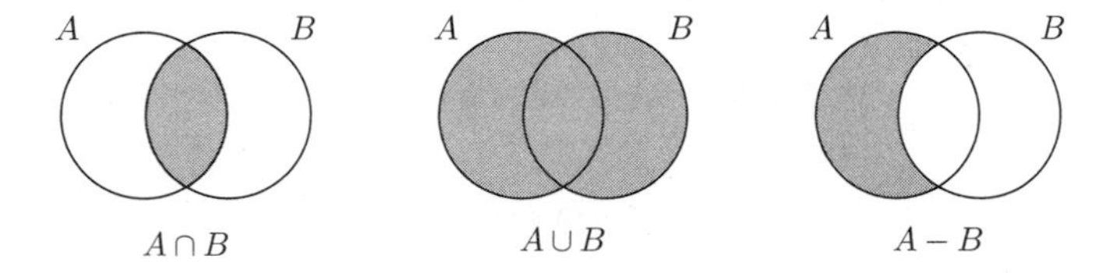

FIGURE 2.1 Venn Diagrams for the Operations in Table 2.1

If $A \subseteq B$, then B is a **superset** of A—i.e., $B \supseteq A$—and, vice versa. Similarly, B is a **proper superset** of A—i.e., $B \supset A$—if and only if $A \subset B$. Also, just as "$\in$" is negated to "$\notin$", and "$=$" to "$\neq$", the logical negations of "$\subseteq$", "$\subset$", "$\supseteq$", and "$\supset$" are compactly written "$\nsubseteq$", "$\not\subset$", "$\nsupseteq$", and "$\not\supset$", respectively.

Some common ways of forming a new set from given sets A and B are shown in Table 2.1. Each of these operations is illustrated in Fig. 2.1 by a **Venn diagram**, with circles (including their interiors) representing A and B, and the shaded region indicating the resulting set. Additionally, in any particular context, one might specify a reference set A that is a superset of all of the sets under discussion; then, for each set B, the difference $A - B$ (read "A minus B") can be simply referred to as the **complement** B^{c} (*relative* to A) of B. For any sets B and C in the same context, we then have

$$B - C = B \cap C^{\mathrm{c}}.$$

The complement, intersection, and union operations for sets have obvious relationships to the logical connectives $\neg$, $\wedge$, and $\vee$ (see Table 1.1), respectively. Also, analogous to the tautology $A \Leftrightarrow \neg(\neg A)$ for any *statement* A, here for any *set* A we have

$$A = (A^{\mathrm{c}})^{\mathrm{c}} \tag{2.3}$$

(the complement being relative to any superset of A). Likewise, **DeMorgan's laws** (see p. 11) can be expressed in terms of sets as

$$(A \cap B)^{\mathrm{c}} = A^{\mathrm{c}} \cup B^{\mathrm{c}}, \quad (A \cup B)^{\mathrm{c}} = A^{\mathrm{c}} \cap B^{\mathrm{c}}.$$

Furthermore, intersections and unions are often performed upon several sets at a time:

$$\bigcap_{i=m}^{n} A_i \triangleq A_m \cap \cdots \cap A_n = \{x \mid x \in A_i \text{ for all } i \in \{m, m+1, \ldots, n\}\}, \tag{2.4a}$$

$$\bigcup_{i=m}^{n} A_i \triangleq A_m \cup \cdots \cup A_n = \{x \mid x \in A_i \text{ for some } i \in \{m, m+1, \ldots, n\}\} \tag{2.4b}$$

($n \geq 1$).[5] Likewise, for any nonempty set $\mathcal{S}$ of sets we define

$$\bigcap_{A \in \mathcal{S}} A \triangleq \{x \mid x \in A \text{ for all } A \in \mathcal{S}\},$$

$$\bigcup_{A \in \mathcal{S}} A \triangleq \{x \mid x \in A \text{ for some } A \in \mathcal{S}\}.$$

Hence, the set operations $\cap$ and $\cup$ are also related to the logical quantifiers $\forall$ and $\exists$, respectively.

Two sets A and B are said to be **disjoint** if and only if they have no elements in common; that is, $A \cap B = \emptyset$. More generally, a *collection* of sets is disjoint if and only if every pair of sets in the collection are disjoint; thus, such a collection may also be referred to as **pairwise disjoint**.

In addition to sets, which are unordered collections, it is common in mathematics to consider *ordered* collections; this is indicated by using parentheses, rather than braces, to enclose the objects in a list. Specifically, an **ordered pair** is a list (a, b) of two objects, a and b, where the order of the list *is* significant; thus, (a, b) and (b, a) are not the same unless $a = b$. Similarly, an **ordered triple** is a list (a, b, c) of three objects a, b, and c, their order again being significant; and, generalizing to an **ordered tuple** having any number of objects, an ***n*-tuple** is an ordered list of $n \geq 1$ objects.[6] In contrast to a listed set, repetitions in the listing of an ordered tuple *do* matter—e.g., $(0, 1)$ and $(0, 1, 0)$ are not the same.

The most common way to obtain a *set* of ordered pairs is to form the **Cartesian product** of two sets A and B:

$$A \times B = \{(x, y) \mid x \in A, y \in B\}.$$

More generally, given $n \geq 1$ sets A_i ($i = 1, \ldots, n$), one can form an n-fold Cartesian product

5. Strictly speaking, the streamlined syntax used in "$A_1 \cap \cdots \cap A_n$" and "$A_1 \cup \cdots \cup A_n$" is ambiguous because it is not clear in what order the intersections and unions are to be performed. If desired, this order can be uniquely specified by inserting parentheses to provide grouping; however, as the right sides of (2.4) show, the resulting sets are independent of the order chosen (cf. fn. 4 in Chapter 1).

6. Although an ordered collection is more structured than an unordered collection (which, again, is what every set is), the previous remark that every mathematical object can be viewed as a set still applies. For example, a common way to define an ordered pair (a, b) in terms of sets is $\{\{a\}, \{a, b\}\}$; and, similarly for any ordered tuple. This, though, is another foundational matter.

$$A_1 \times \cdots \times A_n = \{(x_1, \ldots, x_n) \mid x_i \in A_i \text{ (all } i)\},$$

which is a set of n-tuples.[7] It follows that $A_1 \times \cdots \times A_n = \emptyset$ if and only if $A_i = \emptyset$ for at least one i. Often the sets A_i are the same set A, in which case the above n-fold Cartesian product may be compactly written

$$A \times \cdots \times A = A^n.$$

It is useful to note that $|A_1 \times \cdots \times A_n| = |A_1| \cdot \cdots \cdot |A_n|$, and thus $|A^n| = |A|^n$.

2.1 Relations

Often, one wishes to express the fact that certain elements of a set A possess a particular property P. But, rather than give a direct description of P, one can instead provide the corresponding subset of A given in (2.2), thereby characterizing P with regard to A. Likewise, a particular relationship might hold among the elements of A when taken several at a time. For example, there are certain pairs (x, y) of elements x and y taken from the set $A = \{1, 2, 3\}$ for which $x < y$; and, this fact can also be expressed as a subset, now of the Cartesian product $A \times A = A^2$:

$$\{(x, y) \in A^2 \mid x < y\} = \{(1, 2), (1, 3), (2, 3)\}. \tag{2.5}$$

More generally, for any fixed integer $n \geq 1$, an n-ary **relation** *on a set* A is defined as a subset R of A^n. Implicitly, the elements of R are precisely those n-tuples $(x_1, \ldots, x_n) \in A^n$ that possess a corresponding n-ary property P; that is,

$$\forall x_1, \ldots, x_n \in A,\ (x_1, \ldots, x_n) \in R \Leftrightarrow P(x_1, \ldots, x_n), \tag{2.6a}$$

and thus

$$R = \{(x_1, \ldots, x_n) \in A^n \mid P(x_1, \ldots, x_n)\} \tag{2.6b}$$

(cf. (2.2)). Finally, to be completely general, an n-ary relation is defined as any subset R of a n-fold Cartesian product $A_1 \times \cdots \times A_n$, where the sets $A_1, \ldots, A_n$ may differ; hence, (2.6a) and (2.6b) generalize to

$$\forall x_1 \in A_1, \ldots, x_n \in A_n,\ (x_1, \ldots, x_n) \in R \Leftrightarrow P(x_1, \ldots, x_n)$$

7. In contrast to intersections and unions (see fn. 5), the expression "$A_1 \times \cdots \times A_n$" *does* change in meaning if grouping parentheses are added (which is rarely done). For example, the sets

$$A_1 \times A_2 \times A_3 = \{(x_1, x_2, x_3) \mid x_i \in A_i\ (i = 1, 2, 3)\}$$

and

$$(A_1 \times A_2) \times A_3 = \{((x_1, x_2), x_3) \mid x_i \in A_i\ (i = 1, 2, 3)\}$$

are generally not the same.

and

$$R = \{(x_1, \ldots, x_n) \in A_1 \times \cdots \times A_n \mid P(x_1, \ldots, x_n)\},$$

where P is again the n-ary property implicitly expressed by R.[8]

For the frequently occurring case of $n = 2$—i.e., a **binary relation**—a special notation is commonly used. To wit, let a binary relation symbolized by "R" have corresponding binary predicate "P". Then, just as it is customary to write "$x \, P \, y$" (e.g., "$x < y$") in lieu of "$P(x, y)$", one may also write "$x \, R \, y$" to indicate "$(x, y) \in R$", thereby using the same symbol "R" for both relation and predicate.

The **domain** D of a binary relation $R \subseteq A \times B$ (for some sets A and B) is the set

$$D = \{x \in A \mid x \, R \, y \text{ for some } y \in B\},$$

and its **range** E is the set

$$E = \{y \in B \mid x \, R \, y \text{ for some } x \in A\}.$$

Thus, we also have $R \subseteq D \times E$; moreover, it might happen that $R \subset D \times E$ (e.g., consider the set R on the right of (2.5), for which $D = \{1, 2\}$ and $E = \{2, 3\}$). However, if $D' \subset D$, then $R \not\subseteq D' \times E$; and, if $E' \subset E$, then $R \not\subseteq D \times E'$.

A **one-to-one relation** is a binary relation $R \subseteq A \times B$ such that whenever $x \, R \, y$ and $x' \, R \, y'$ ($x, x' \in A$; $y, y' \in B$), we have $x = x'$ if and only if $y = y'$. In other words, if R has domain D and range E, then for each $x \in D$ there is exactly one $y \in E$ for which $x \, R \, y$, and for each $y \in E$ there is exactly one $x \in D$ for which $x \, R \, y$; thus, we now have $R = D \times E$. (For example, assuming monogamy, the husband–wife relation is one-to-one, since every man can have at most one wife, and every woman can have at most one husband. Here, we can let A be the set of all men, B be the set of all women, D be the set of all married men, and E be the set of all married women.) Such a relation R provides a **one-to-one correspondence** between (the elements of) the two sets D and E; and, from this it follows that $|D| = |E|$.

A **many-to-one relation** is a binary relation R such that having $x \, R \, y$ and $x \, R \, y'$ implies $y = y'$; that is, for each x in the domain of R there is exactly one y in its range for which $x \, R \, y$. (Whereas each person has, e.g., only one birthday, each calendar date can be the birthday of more than one person; hence, the person–birthday relation is many-to-one.) Conversely, R is a **one-to-many relation** when having $x \, R \, y$ and $x' \, R \, y$ implies $x = x'$; that is, for each y in the range of R there is exactly one x in its domain for which $x \, R \, y$. (For example, biologically, whereas a mother can have many children, a child has only one mother; hence, the mother–child relation is one-to-many.) It follows that a relation is one-to-one if and only if it is both many-to-one and one-to-many.

---

8. In informal mathematical discourse, a logical predicate (e.g., "P") may itself be referred to as a "relation", as may the property (e.g., "less than") that the predicate expresses; however, it is always understood that the relation under discussion is really the corresponding set of ordered tuples.

Some important properties that a binary relation R on a set A might have are:

- R is **reflexive**: $\forall x \in A,\ x\,R\,x$.
- R is **symmetric**: $\forall x, y \in A,\ x\,R\,y \Rightarrow y\,R\,x$.
- R is **transitive**: $\forall x, y, z \in A,\ (x\,R\,y \wedge y\,R\,z) \Rightarrow x\,R\,z$.

For example, the relation $<$ on the set $A = \{1, 2, 3\}$ is neither reflexive nor symmetric; but, it is transitive. A binary relation that is reflexive, symmetric, and transitive is called an **equivalence relation**, an obvious example of which is the usual identity relation $=$ (on any set A). Another example of an equivalence relation is the logical-equivalence relation $\models\,\dashv$ on the set of all statements in a particular first-order logic; because, for any statements U, V, and W in the logic, the following hold.

- $U \models\,\dashv U$.
- If $U \models\,\dashv V$, then $V \models\,\dashv U$.
- If $U \models\,\dashv V$ and $V \models\,\dashv W$, then $U \models\,\dashv W$.

The main significance of having an *equivalence* relation, say "$\equiv$", on a nonempty set A is that it induces a *partitioning* of A. In general, a **partition** of a set A is another set $\mathcal{P}$, the elements of which are all disjoint subsets of A, with the union of these subsets being A; that is:

$$\forall B \in \mathcal{P},\ B \subseteq A; \quad \forall B, C \in \mathcal{P},\ B \neq C \Rightarrow B \cap C = \emptyset; \quad \bigcup_{B \in \mathcal{P}} B = A$$

(e.g., $\big\{\{0,4\}, \{1\}, \{2,3,5\}\big\}$ is a partition of $\{0,1,2,3,4,5\}$). For each $x \in A$, the set

$$\{y \in A \mid y \equiv x\}$$

of those elements of A that are "equivalent" to x is called an **equivalence class** under $\equiv$; and, it can be shown that the set comprising these equivalence classes—viz.,

$$\big\{\{y \in A \mid y \equiv x\} \;\big|\; x \in A\big\}$$

—is a partition of A (see Figure 2.2). Two elements $x, y \in A$ are called **equivalent**, with respect to the relation $\equiv$, if and only if they belong to the same equivalence class.

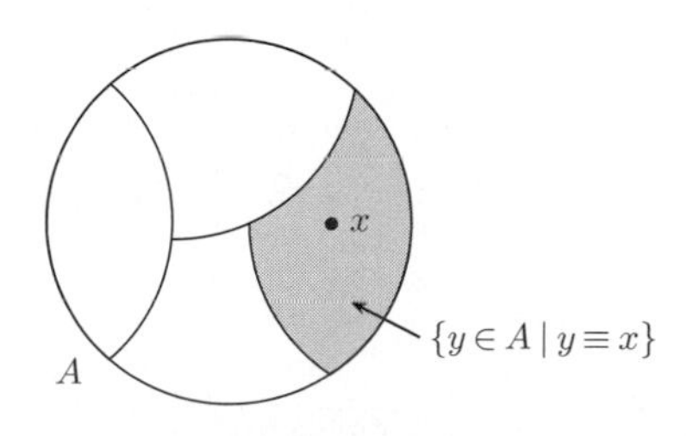

FIGURE 2.2 Partitioning a Set into Equivalence Classes

The idea behind this partitioning is that although the elements of A *can* be distinguished individually, the elements in each equivalence class are, as "seen" by $\equiv$, indistinguishable. Accordingly, rather than simply thinking of A as being composed of individual elements, we may instead think of A as being composed of equivalence classes, which are in turn composed of individual elements. (Such a viewpoint is not valid, though, for every binary relation—only for equivalence relations. Because, only for them is such a partitioning possible.) When desired, *any* element of an equivalence class can be selected as its "representative", since each element of a particular equivalence class has essentially the same significance as any other.

A trivial example obtains when $\equiv$ is the identity relation $=$ on A, in which case the equivalence classes are just the single-element sets $\{x\}$ $(x \in A)$; thus, the identity relation *does* distinguish the elements of A individually. The other extreme of partitioning comes from the binary relation $R = A \times A$, which is also an equivalence relation; in this case, $x\,R\,y$ for all $x, y \in A$, so the one and only equivalence class under R is the entire set A. As a nontrivial example, let A be a set of people, and R be the set of all pairs (x, y) of persons $x, y \in A$ having the same parents; then, R is an equivalence relation, and the equivalence class corresponding to each person comprises that individual and his or her siblings. As a final example, it follows from the equivalence relation mentioned above regarding a first-order logic that the set of all sentences in the logic can be partitioned into equivalence classes, with two sentences belonging to the same equivalence class if and only if they are logically equivalent.

CHAPTER 3

Numbers

3.1 Real Numbers and Infinity

More often than not, mathematics deals with *numbers*. These basic objects are typically introduced into the theory via a succession of number *sets*, each being a proper subset of the next.[1]

The first and simplest of these is the set of **natural numbers**,

$$\mathbb{N} \triangleq \{0, 1, 2, \dots\},$$

which captures the intuitive idea of "counting"—i.e., the act of repeatedly adding one (1)—here indefinitely, starting from zero (0).[2] Applying to each number $x \in \mathbb{N}$ a **sign** of either **plus** $(+)$ or **minus** $(-)$, gives the two numbers $+x = x$ and $-x$, with $+x = -x$ if and only if $x = 0$. Then, collecting all of these numbers together, we obtain the set of **integers**,

$$\mathbb{Z} \triangleq \{\pm x \mid x \in \mathbb{N}\} = \{0, \pm 1, \pm 2, \pm 3, \dots\} = \{\dots, -3, -2, -1, 0, 1, 2, 3, \dots\}$$

—where the symbol "$\pm$" is read "plus or minus".[3] The subsets

$$\mathbb{Z}^+ \triangleq \{1, 2, 3, \dots\}, \quad \mathbb{Z}^- \triangleq \{-1, -2, -3, \dots\} \tag{3.1}$$

are the sets of positive integers and negative integers, respectively.

All of the sets $\mathbb{N}$, $\mathbb{Z}$, $\mathbb{Z}^+$, and $\mathbb{Z}^-$ are "closed" under addition; that is, adding any two numbers from one of these sets yields another number in that set. However, only $\mathbb{Z}$ is closed under subtraction.

Both $\mathbb{N}$ and $\mathbb{Z}$ are also closed under multiplication; however, neither is closed under division, even upon barring division by 0. Accordingly, we next introduce the set of

---

1. For brevity, our definitions of these sets will be a bit lax, since a complete and rigorous treatment (e.g., in terms of the "Peano axioms", "Cauchy sequences", and "Dedekind cuts") would require much space and be rather involved.

2. Some authors define $\mathbb{N}$ with 0 omitted, which for us is the set $\mathbb{Z}^+$ defined in (3.1).

3. Likewise, there is the minus-or-plus symbol "$\mp$", which occasionally is more appropriate (e.g., see p. 3).

rational numbers,

$$\mathbb{Q} \triangleq \{x/y \mid x, y \in \mathbb{Z} \text{ with } y \neq 0\}; \tag{3.2}$$

that is, every rational number can be expressed as a **fraction** $\frac{x}{y}$ in which the **numerator** x and the **denominator** y are integers, with the latter being nonzero.[4] The set $\mathbb{Q}$ is closed under addition, subtraction, multiplication, and division (other than by 0).

Many frequently occurring numbers, though—such as $\sqrt{2}$, π, and e—are not rational; however, each of these can be *approximated*, as closely as desired, by a rational number. (For example, a popular approximation to π is 22/7; even better is 355/113.) Thus, considering all such approximations by rational numbers, we are led to the set of **real numbers,**

$$\mathbb{R} \triangleq \{x \mid \forall n \in \mathbb{Z}^+, \exists y \in \mathbb{Q}, |x - y| \leq 1/n\}; \tag{3.3}$$

that is, no matter how large the positive integer n is chosen (thereby making $1/n$ arbitrarily small), for each real number x there exists a rational number y within $\pm 1/n$ of x. We now have the hierarchy $\mathbb{N} \subset \mathbb{Z} \subset \mathbb{Q} \subset \mathbb{R}$, with $\mathbb{R} - \mathbb{Q}$ being the set of **irrational numbers** (e.g., $\sqrt{2}$, π, and e). The set $\mathbb{R}$ is closed under addition, subtraction, multiplication, and division (other than by 0); but, unlike $\mathbb{Q}$, it is also closed under limits (see Section 4.1).

3.1.1 Numbers versus Numerals

A common way to represent a real number x is as a **decimal expansion**

$$\pm \boldsymbol{x}_n \ldots \boldsymbol{x}_1 \boldsymbol{x}_0 . \boldsymbol{x}_{-1} \boldsymbol{x}_{-2} \boldsymbol{x}_{-3} \ldots \tag{3.4a}$$

where each **digit** $\boldsymbol{x}_n$, $\boldsymbol{x}_{n-1}$, ... $(n \geq 0)$ is one of the ten symbols "0", "1", "2", "3", "4", "5", "6", "7", "8", "9". The dot "." following $\boldsymbol{x}_0$ is called the **decimal point.** The plus-or-minus symbol "±" is taken as "+" (and may be omitted) if $x > 0$, and as "−" if $x < 0$; whereas for $x = 0$, it can be taken as either "+" (perhaps omitted) or "−". Expressing a number x in the decimal notation (3.4a) is to assert

$$x = \pm \sum_{i=-\infty}^{n} x_i \cdot 10^i \tag{3.4b}$$

(± respectively), where each $x_i \in \{0, 1, \ldots, 9\}$ is the numerical value symbolized by the digit $\boldsymbol{x}_i$—e.g., if $\boldsymbol{x}_i$ is the symbol "4", then $x_i = 4$. (For example, $\pi = +3.14\ldots = 3.14\ldots = 3 \times 10^0 + 1 \times 10^{-1} + 4 \times 10^{-2} + \cdots$.) By (3.4b), each decimal expansion (3.4a) represents only one real number; however, as explained in Section 4.1, some real numbers have *two* decimal expansions (e.g., $1.000\ldots = 0.999\ldots$).

It is important to observe that a digit—e.g., "4"—is not itself a number, but a *symbol* representing a number—viz., 4. Likewise, a decimal expansion is not itself a number, but an *arrangement of symbols* (sign, digits, decimal point) that represents a number.

---

4. More generally, for any division $x/y = x \div y$, the number x being divided is the **dividend**, and the number y by which the division is performed is the **divisor**.

A symbolic representation of a number is called a **numeral**. (Another example: "IV" is the roman numeral that also represents the number 4.) It is because of this distinction between numbers and numerals that a special device (bold italic letters) is used in (3.4a) to indicate the digits, rather than letting "x_n", "x_{n-1}", ... represent *both* numbers and numerals by instead writing the expansion as "$\pm x_n \ldots x_1 x_0 . x_{-1} x_{-2} x_{-3} \ldots$". However, once this distinction is clearly understood, performing such an "abuse of notation" can streamline a discussion without causing confusion, and thus is often done.

Since the factor 10^i multiplying x_i in (3.4b) is largest for $i = n$, we refer to $\boldsymbol{x}_n$ as the **most significant digit** of the expansion (3.4a). The portion of the expansion to the left of the decimal point—i.e.,

$$\pm \boldsymbol{x}_n \ldots \boldsymbol{x}_1 \boldsymbol{x}_0 \tag{3.5}$$

—is its **integer part** (or **characteristic**). Usually, n is chosen to be as small as possible; that is, $\boldsymbol{x}_n$ is rarely "0" when $n > 0$. However, it is standard practice to always have at least one digit to the left of the decimal point, thereby making it more visible; hence, it is not uncommon for $\boldsymbol{x}_n$ to be "0" when $n = 0$.

The remainder of the expansion—i.e.,

$$.\boldsymbol{x}_{-1} \boldsymbol{x}_{-2} \boldsymbol{x}_{-3} \ldots$$

—is its **fractional part** (or **mantissa**). For each integer $j \geq 1$, the position of the jth digit $\boldsymbol{x}_{-j}$ to the right of the decimal point is called the jth **decimal place**.

If for some $m \geq 0$ we have $x_i = 0$ for all $i < -m$—i.e., if (3.4a) can be written

$$\pm \boldsymbol{x}_n \ldots \boldsymbol{x}_1 \boldsymbol{x}_0 . \boldsymbol{x}_{-1} \boldsymbol{x}_{-2} \ldots \boldsymbol{x}_{-m} 000 \ldots \tag{3.6}$$

—then we have a **finite decimal expansion**, which if desired can be terminated at $\boldsymbol{x}_{-m}$ to

$$\pm \boldsymbol{x}_n \ldots \boldsymbol{x}_1 \boldsymbol{x}_0 . \boldsymbol{x}_{-1} \boldsymbol{x}_{-2} \ldots \boldsymbol{x}_{-m} \tag{3.7}$$

in which case $\boldsymbol{x}_{-m}$ becomes the **least significant digit**. In particular, for $m = 0$ the fractional part is empty, in which case the decimal point may also be omitted.[5] On the other hand, if (3.4a) cannot be written as (3.6), then we have an **infinite decimal expansion**.

In general, the act of removing, from some decimal place forward, all of the digits of a given decimal expansion (infinite or finite) is called **truncation**. Unless the digits removed are all zeros, the value of the expansion is thereby changed. Notationally, this change in value can be avoided by using an ellipsis (...), as has already been done, to stand for the omitted digits (e.g., $\pi = 3.1415\ldots$).

Every finite decimal expansion represents a rational number; but, some rational numbers such as

$$\tfrac{1}{3} = 0.333\ldots \tag{3.8}$$

5. The number x is then an integer. Accordingly, the integers are sometimes referred to as **whole numbers**, since each can be written as a decimal expansion (3.5) that is fractionless (i.e., has no fractional part).

require infinitely many digits. In general, an infinite decimal expansion represents a rational number if and only if it contains a finite segment of one or more digits that continually repeats thereafter; that is, we have a **recurring decimal expansion**. For example, upon using a grouping overbar called a **vinculum** to indicate the repeating segment, called the **repetend**, we can write

$$\begin{aligned}123456.78967896789\ldots &= 12345\overline{6.789}\ldots = 123450 + \overline{6.789}\ldots \\ &= 123450 + 6.789/0.9999 = 411481480/3333; \qquad (3.9)\end{aligned}$$

because, the value of the entire recurring portion of a recurring decimal expansion can be obtained by dividing the value of the first instance of the repetend by $0.99\ldots 9$, where the number of nines equals the number of digits in the repetend (e.g., $0.333\ldots = 0.\overline{3}\ldots = 0.3/0.9 = \frac{1}{3}$).

Generalizing (3.4) leads to other **numeral systems** (often erroneously called "number systems"), many of which are widely used. Specifically, given a particular **base** (or **radix**) $b \in \{2, 3, 4, \ldots\}$, we restrict the digits $\boldsymbol{x}_i$ in the expansion (3.4a) to have values $x_i \in \{0, 1, \ldots, b-1\}$ (all i); then, the value of the expansion is taken to be

$$x = \pm \sum_{i=-\infty}^{n} x_i \cdot b^i \qquad (3.10)$$

($\pm$ respectively with (3.4a)). Thus, the **decimal numeral system** corresponds to $b = 10$, which is generally assumed when one is reading or writing an expansion of the form (3.4a) unless another base is explicitly stated. Alternatively, taking $b = 2$ yields the **binary numeral system**, for which each digit is called a **bit**—a contraction (sans apostrophe) of "*b*inary dig*it*". This system is particularly suited for expressing numbers within a digital computer, since for each bit the two possibilities ("0" and "1") can be physically represented as a switch in the computer hardware that is off or on. The **octal numeral system**, for which $b = 8 = 2^3$, effectively groups three consecutive bits into each octal digit; because, the octal numerals "0", "1", "2", "3", "4", "5", "6", and "7" represent the same numbers as the binary numerals "000", "001", "010", "011", "100", "101", "110", and "111", respectively. (Thus, octal 12.34 easily converts to binary 001 010.011 100 = 1010.0111, and also equals decimal $1 \cdot 8^1 + 2 \cdot 8^0 + 3 \cdot 8^{-1} + 4 \cdot 8^{-2} = 10.4375$.)

When a numeral system has a base $b > 10$, it must introduce distinct single-symbol representations of the decimal numbers $10, 11, \ldots, b-1$, to be used for the digits $\boldsymbol{x}_i$ in (3.4a). A popular example is the **hexadecimal numeral system**, with $b = 16 = 2^4$, for which the letter digits "A", "B", "C", "D", "E", and "F" represent the decimal numbers 10, 11, 12, 13, 14, and 15, respectively. (Thus, hexadecimal F0.A8 equals decimal $15 \cdot 16^1 + 0 \cdot 16^0 + 10 \cdot 16^{-1} + 8 \cdot 16^{-2} = 240.65625$.) This system effectively groups four consecutive bits into each hexadecimal digit (e.g., hexadecimal F0.A8 easily converts to binary 1111 0000.1010 0100 = 11110000.101001); therefore, a pair of consecutive hexadecimal digits represents eight consecutive bits, which is called a **byte**—a basic unit of computer memory space.

The terms "integer part" ("characteristic"), "fractional part" ("mantissa"), "finite expansion", "infinite expansion", "recurring expansion", "truncation", "most significant digit", and "least significant digit" apply to expansions of any base; also, in the binary

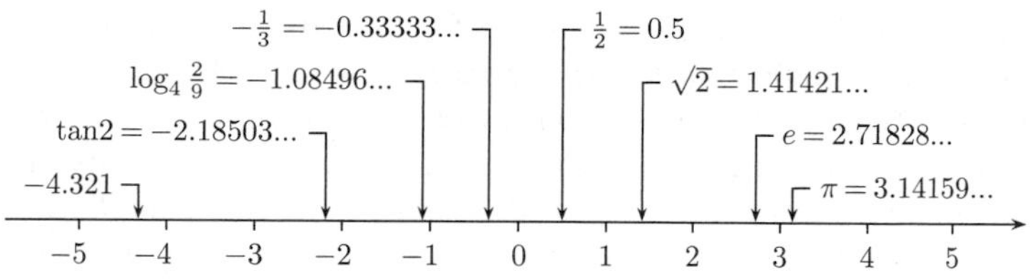

FIGURE 3.1 The Real Line

case, the word "digit" may be replaced with "bit". However, for a nondecimal expansion, the word "decimal" in terms such as "decimal point" and "decimal place" should either be avoided or replaced with the type of the expansion (e.g., "binary point" and "octal place"). Furthermore, when dealing simultaneously with expansions having different bases, it is common to indicate the base as a subscript on the expansion (e.g., $101_2 = 5_{10}$). Other distinguishing notation can also be found; for example, the C programming language uses a prefix "0" to indicate octal numerals (e.g., $0103 = 103_8 = 67_{10}$), and the prefix "0x" to indicate hexadecimal numerals (e.g., $0x2B = 2B_{16} = 43_{10}$). Additionally, for easier viewing, expansions having several digits (e.g.,"10293.84756") are often parsed into 3-digit groups by using either commas ("10,393.847,56") or spaces ("10 393.847 56").[6]

3.1.2 More on Numbers

Regarding the real numbers themselves, the set $\mathbb{R}$ is often depicted geometrically by the **real line**, shown in Figure 3.1 as a horizontal line (which, being infinite, is only partially visible).[7] Each real number corresponds to a unique *point* on the line; conversely, each point on the line represents a unique real number. Once the point representing 0 has been specified, along with the *direction* of the line (indicated in the figure by an arrow pointing to the right) and the *unit distance*,[8] then the position on the line of each real number is determined.

The real line clearly displays the fact that the set $\mathbb{R}$ is *ordered*. Specifically, for any two real numbers x and y, exactly one of the following conditions holds:

- x is **less than** y: $x < y$.
- x **equals** y: $x = y$.
- x is **greater than** y: $x > y$ (which is equivalent to having $y < x$).

Furthermore, if z is also a real number, then

$$x < y \quad \wedge \quad y < z \qquad \Rightarrow \qquad x < z; \tag{3.11}$$

that is, the binary relation $<$ is transitive—as are $=$ and $>$.

6. For international usage, spaces are preferred to commas because in some locales, a comma is used instead of a period for the decimal point.

7. By virtue of the "smoothness" of this line, the set $\mathbb{R}$ is also called *the* **continuum**.

8. The number 1 is commonly referred to as **unity**. Hence, the word **unit**, when used as an adjective, indicates something of size 1.

In distinction to the expression "$x = y$", which asserts an **equality**, the expression "$x < y$" asserts an **inequality**, as does the expression "$x > y$". Also, the assertion that x is less than *or* equal to y—i.e., $x \leq y$—is an inequality as well, as is the assertion that x is greater than *or* equal to y—i.e., $x \geq y$. Moreover, since having $x < y$ implies that $x \neq y$, but not so when $x \leq y$, the relation $<$ is said to express a **strict inequality**; likewise, so does $>$. And, just as the negation of the relation $=$ can be compactly written "$\neq$", so the logical negations of relations $<$, $>$, $\leq$, and $\geq$ can be written "$\nless$", "$\ngtr$", "$\nleq$", and "$\ngeq$", respectively.

Additionally, just as it is often convenient to say that a number x is "*approximately* equal to" another number y—i.e., $x \approx y$—so it can be convenient to say that x is "*much* less than" or "*much* greater than" y—i.e., $x \ll y$ or $x \gg y$, respectively. If $x \ll y$ (respectively, $x \gg y$), then it is certainly true that $x < y$ ($x > y$); however, the precise meanings of the relations $\approx$, $\ll$, and $\gg$ can vary. For example, in some contexts having $x \approx y$ means that $x - y$ is nearly 0, whereas in other contexts it means that x/y is nearly 1; similarly for $\ll$ and $\gg$. Also, it is occasionally useful to write "$x \lesssim y$" to mean "$x < y$ or $x \approx y$", and write "$x \gtrsim y$" to mean "$x > y$ or $x \approx y$".

A real number x is referred to as **positive** (respectively, **negative**) if and only if $x > 0$ ($x < 0$). On the other hand, x is **nonpositive** (respectively, **nonnegative**) if and only if $x \ngtr 0$ ($x \nless 0$)—equivalently, $x \leq 0$ ($x \geq 0$).

When $x \leq y$, we say that x is a **lower bound** on y, and y is an **upper bound** on x. It then follows that $-x \geq -y$; and, if x and y are nonzero and have the same sign, we also have $1/x \geq 1/y$. Additionally, for any $z \in \mathbb{R}$ we have both $x + z \leq y + z$ and $x - z \leq y - z$; also, $x \cdot z \leq y \cdot z$ and $x/z \leq y/z$ if $z > 0$, whereas $x \cdot z \geq y \cdot z$ and $x/z \geq y/z$ if $z < 0$. Furthermore, the statements in this paragraph remain valid upon making *all* of the inequalities strict.

It is often convenient to augment $\mathbb{R}$ with the values **infinity** (∞) and its negative ($-\infty$), thereby obtaining the set of **extended real numbers**,

$$\overline{\mathbb{R}} \triangleq \mathbb{R} \cup \{\pm\infty\}.$$

We then have $-\infty < x < \infty$ for all $x \in \mathbb{R}$, and thus $-\infty < \infty$ for consistency with (3.11). Both of the numbers $\pm\infty$ are described as **infinite**, whereas all other numbers (including 0 and the "complex" numbers—see Section 3.3) are **finite**.

An **interval** is a subset of $\overline{\mathbb{R}}$ consisting of all numbers lying between two fixed **endpoints** $a, b \in \overline{\mathbb{R}}$, perhaps along with either or both endpoints as well. Thus, an interval has four possible forms—

$$(a, b) \triangleq \{x \in \overline{\mathbb{R}} \mid a < x < b\}, \tag{3.12a}$$

$$(a, b] \triangleq \{x \in \overline{\mathbb{R}} \mid a < x \leq b\}, \tag{3.12b}$$

$$[a, b) \triangleq \{x \in \overline{\mathbb{R}} \mid a \leq x < b\}, \tag{3.12c}$$

$$[a, b] \triangleq \{x \in \overline{\mathbb{R}} \mid a \leq x \leq b\} \tag{3.12d}$$

—each, when nonempty, having **length** $b - a$. For example, $(-\infty, \infty) = \mathbb{R}$ and $[-\infty, \infty] = \overline{\mathbb{R}}$, whereas $[0, \infty)$ is the set of nonnegative numbers.

A set $A \subseteq \overline{\mathbb{R}}$ has a **maximum** x (i.e., $x = \max A$) if and only if $x \in A$, and $y \leq x$ for all $y \in A$. Whereas A has a **minimum** x (i.e., $x = \min A$) if and only if $x \in A$, and $y \geq x$ for all $y \in A$. It follows that A can have at most one maximum value, and likewise for its minimum; also, if A is nonempty and finite,[9] then its maximum and minimum must both exist. However, in general, neither of these values need exist. For example, every nonempty subset of $\mathbb{N}$ has a minimum, but not necessarily a maximum; in particular, $\min \mathbb{N} = 0$, whereas $\max \mathbb{N}$ does not exist (since $\infty \notin \mathbb{N}$). Likewise, the set $\{\frac{1}{2}, \frac{1}{3}, \frac{1}{4}, \ldots\}$ has no minimum, and the interval $(0, 1)$ has neither a maximum nor minimum.

Every set $A \subseteq \overline{\mathbb{R}}$ *does*, though, have a **supremum** (or **least upper bound**) $\sup A$, which is the (unique) smallest value $x \in \overline{\mathbb{R}}$ such that $y \leq x$ for all $y \in A$. Also, A has an **infimum** (or **greatest lower bound**) $\inf A$, which is the (unique) largest value $x \in \overline{\mathbb{R}}$ such that $y \geq x$ for all $y \in A$. For example, $\sup \mathbb{N} = \infty$, $\inf \mathbb{N} = 0$, and $\inf\{\frac{1}{2}, \frac{1}{3}, \frac{1}{4}, \ldots\} = 0$; also, each interval in (3.12), when nonempty, has infimum a and supremum b. It follows that a set $A \subseteq \overline{\mathbb{R}}$ has a maximum if and only if $\sup A \in A$, in which case $\max A = \sup A$; likewise, A has a minimum if and only if $\inf A \in A$, in which case $\min A = \inf A$. Finally, just as we have $\min A \leq \max A$ when both of these values exist, so we have $\inf A \leq \sup A$ when $A \neq \emptyset$.

3.2 Arithmetic

Of course, there are the four basic arithmetic operations of **addition** ($+$), **subtraction** ($-$), **multiplication** ($\cdot$, $\times$, or implicitly by juxtaposition), and **division** ($/$, $\div$, or as a fraction).[10] Except for division by 0, which is disallowed, these operations can be applied to any pair of real numbers x and y, thereby yielding a **sum** $(x + y)$, a **difference** $(x - y)$, a **product** $(x \cdot y = x \times y = xy)$, and a **quotient** $(x/y = x \div y = \frac{x}{y})$, respectively. Both addition and multiplication are **commutative**,

$$x + y = y + x, \quad x \cdot y = y \cdot x, \tag{3.13a}$$

as well as **associative**,

$$(x + y) + z = x + (y + z), \quad (x \cdot y) \cdot z = x \cdot (y \cdot z); \tag{3.13b}$$

and, multiplication is **distributive** over addition

$$x \cdot (y + z) = (x \cdot y) + (x \cdot z) \tag{3.13c}$$

$(x, y, z \in \mathbb{R})$. By contrast, subtraction and division are neither commutative nor associative.

---

9. Note, e.g., that the set $\{\infty\}$ is itself finite, though the value within is infinite.
10. The forward-slash symbol "/" is called a **solidus**.

In addition to using these basic properties, other arithmetic manipulations are commonly performed. In particular, for $w, x, y, z \in \mathbb{R}$, we have

$$1 \Big/ \frac{x}{y} = \frac{y}{x} \quad (x, y \neq 0), \tag{3.14a}$$

$$x \cdot \frac{y}{z} = \frac{xy}{z} \quad (z \neq 0), \tag{3.14b}$$

$$\frac{x}{y} \Big/ z = \frac{x/y}{z} = \frac{x}{yz} \quad (y, z \neq 0), \tag{3.14c}$$

$$x \Big/ \frac{y}{z} = \frac{x}{y/z} = \frac{xz}{y} \quad (y, z \neq 0), \tag{3.14d}$$

$$\frac{w}{x} \cdot \frac{y}{z} = \frac{wy}{xz} \quad (x, z \neq 0), \tag{3.14e}$$

$$\frac{w}{x} \Big/ \frac{y}{z} = \frac{w/x}{y/z} = \frac{wz}{xy} \quad (x, y, z \neq 0), \tag{3.14f}$$

$$\frac{x}{z} \pm \frac{y}{z} = \frac{x \pm y}{z} \quad (z \neq 0) \tag{3.14g}$$

($\pm$ respectively). However, we do *not* generally have $\frac{x}{y} + \frac{x}{z} = \frac{x}{y+z}$ or $\frac{x}{y} - \frac{x}{z} = \frac{x}{y-z}$, even when division by 0 does not occur; rather, the above facts yield

$$\begin{aligned} \frac{x}{y} \pm \frac{x}{z} &= \frac{x}{y} \cdot 1 \pm \frac{x}{z} \cdot 1 = \frac{x}{y} \cdot \frac{z}{z} \pm \frac{x}{z} \cdot \frac{y}{y} = \frac{xz}{yz} \pm \frac{xy}{zy} = \frac{xz}{yz} \pm \frac{xy}{yz} = \frac{xz \pm xy}{yz} \\ &= \frac{x(z \pm y)}{yz} \quad (y, z \neq 0) \end{aligned} \tag{3.14h}$$

($\pm$ respectively).

With some restrictions, the four arithmetic operations can be applied to *infinite* values as well. First, for all $x \in \mathbb{R}$, we can safely let

$$x + \infty = \infty, \qquad x \cdot \infty = \begin{cases} \infty & x > 0 \\ -\infty & x < 0 \end{cases}; \tag{3.15a}$$

also,

$$\infty + \infty = \infty, \quad \infty \cdot \infty = \infty. \tag{3.15b}$$

Moreover, depending upon the context (viz., the kind of mathematical analyses anticipated), additional conventions regarding infinity *might* be adopted; for example, it is common to complete the right side of (3.15a) by letting

$$0 \cdot \infty = 0, \tag{3.15c}$$

as is assumed throughout this book.[11] Further identities involving $\pm\infty$ now obtain by performing simple manipulations as one would for finite values; e.g.,

$$x - \infty = -[(-x) + \infty] = -\infty \quad (x \in \mathbb{R}),$$

$$\infty/x = \infty \cdot (1/x) = (1/x) \cdot \infty = \begin{cases} \infty & 0 < x < \infty \\ -\infty & -\infty < x < 0 \end{cases},$$

$$\infty \cdot (-\infty) = -(\infty \cdot \infty) = -\infty.$$

On the other hand, $\infty - \infty$ and ∞/∞ are explicitly *undefined*. Also, $x/0$ is usually left undefined for all $x \in \overline{\mathbb{R}}$; however, in a context where all of the numbers under consideration are nonnegative, it might be convenient to define $x/0$ as ∞ for $x > 0$.

Because $x + 0 = x$ for all $x \in \overline{\mathbb{R}}$, the number 0 is called the **additive identity**; moreover, there is no other number $y \in \overline{\mathbb{R}}$ such that $x + y = x$ for all x. Likewise, because $x \cdot 1 = x$ for all $x \in \overline{\mathbb{R}}$, the number 1 is called the **multiplicative identity**; and, there is no other number $y \in \overline{\mathbb{R}}$ such that $x \cdot y = x$ for all x. (These statements remain valid when x and y are allowed to vary over the set $\mathbb{C}$ of "complex" numbers—see Section 3.3—but, the next paragraph does not.)

Each nonzero number $x \in \overline{\mathbb{R}}$ has a unique sign, which is essentially extracted by the **signum** function:

$$\operatorname{sgn} x \triangleq \begin{cases} 1 & x > 0 \\ 0 & x = 0 \\ -1 & x < 0 \end{cases}. \tag{3.16}$$

Also, each $x \in \overline{\mathbb{R}}$ has a **magnitude**, given by the **absolute-value** function:[12]

$$|x| \triangleq \begin{cases} x & x \geq 0 \\ -x & x < 0 \end{cases}. \tag{3.17}$$

Hence, any extended real number can be expressed in **sign-magnitude form**:

$$x = (\operatorname{sgn} x)|x| \quad (x \in \overline{\mathbb{R}}). \tag{3.18}$$

3.2.1 Notational Conventions

As we have seen, a mathematical expression may have within it explicit groupings—indicated by pairs of parentheses, (square) brackets, or (curly) braces—to specify the order in which certain operations are to be performed. Even when not necessary, the

---

11. Another common convention is to let $x/\infty = 0$ ($x \in \mathbb{R}$). Although there is some leeway in choosing such conventions, the choice is not arbitrary; rather, the choices must be consistent with the theory of limits—see Section 4.1.

12. Occasionally, the same notation is used in mathematics for more than one purpose, in which case its meaning must be determined from context. Here, vertical lines denote the absolute value of a number, not the cardinality of a set as in Chapter 2. Further instances of reusing notation may occur henceforth without comment. (A complete list of the notation used in this book follows the appendix.)

inclusion of such groupings sometimes make an expression more readable. In any event, an explicit grouping is an indication that, when one is evaluating the overall expression, what is inside the grouping is itself to be evaluated as a mathematical expression, with only the resulting value being passed to what is outside the grouping for the remainder of the evaluation. Moreover, the expression inside a grouping might itself contain other groupings—that is, groupings can be *nested*—in which case the evaluation is performed outward starting with the innermost group(s). For example,

$$6 - \{[5 + (4 - 3)]/(2 + 1)\} = 6 - [(5 + 1)/3] = 6 - (6/3) = 6 - 2 = 4, \quad (3.19)$$

where for greater readability a combination of parentheses, brackets, and braces has been used; specifically, as is commonly done, parentheses appear at the deepest levels, followed outward by brackets and then braces (repeating this cycle thereafter as long as necessary).[13]

Note that for the left side of (3.19), the order in which the operations therein are to be performed is not fully specified, since either pair of parentheses can be dealt with first; nevertheless, all allowable ways of evaluating the expression yield the same final value. Similarly, an explicit grouping may be omitted from an expression if its value remains unambiguous. In particular, it follows from (3.13b) that the parentheses there can be omitted—that is, "$x + y + z$" and "$x \cdot y \cdot z$" have unambiguous values—and likewise for any sum or product having finitely many terms.

Furthermore, by observing the following simple conventions, other explicit groupings become unnecessary. To begin, we consider how to interpret an expression in which the only mathematical operations are addition, subtraction, multiplication, and division. Assuming that the expression contains no explicit groupings, the following rules of *operator precedence* determine the order in which the operations are to be performed:

- First, perform all multiplications (in any order).
- Next, perform all divisions as they occur from left to right.
- Finally, perform all additions and subtractions as they occur from left to right.

Thus, e.g.,

$$6/2 \cdot 3 = 6/(2 \cdot 3) = 1, \quad (3.20a)$$

$$8 + 0/4 = 8 + (0/4) = 8, \quad (3.20b)$$

$$9 - 5 + 3 = (9 - 5) + 3 = 7. \quad (3.20c)$$

(Note that in each of these examples, performing the two arithmetic operations on the left side in the order opposite to that shown produces a different result.) With a little practice, no thought at all is required to correctly apply these rules, even for very complicated expressions.[14]

13. Adjustments to this "rule" must be made, though, when these grouping symbols are used for other purposes, such as braces for sets or parentheses for ordered tuples.

14. Nowadays (perhaps owing to the proliferation of calculators and programming languages, each with its own evaluation rules), there is some disagreement about whether multiplication should be given *sole*

A few more comments about notation: First, although an expression containing successive divisions such as "8/4/2" can be evaluated unambiguously according to the above rules, it is prone to be interpreted with puzzlement as

$$\frac{8}{\frac{4}{2}}$$

(which *is* ambiguous); therefore, an additional visual cue is in order, such as the explicit grouping in "(8/4)/2" or a variation in symbol size:

$$\frac{\frac{8}{4}}{2}.$$

Alternatively, one can write "$\frac{8}{4}/2$", since a fraction written with a *horizontal* bar implicitly groups itself; thus, $8/\frac{4}{2} = 8/(4/2)$. Similarly, other functions such as powers and roots also provide implicit groupings. Finally, the effect of a leading minus sign (indicating a change of sign rather than subtraction) can be determined by simply viewing it as multiplication by -1; for example, we interpret "$-2^3 + 4$" as "$(-1) \cdot 2^3 + 4$".[15] However, consecutive arithmetic operators should be eschewed, rewriting expressions such as "$5 + -6$" and "$7 \cdot -8$" as "$5 + (-6)$" and "$7 \cdot (-8)$" instead, for better clarity.

3.2.2 More on Arithmetic

Returning to arithmetic per se, for any two *integers* x and $y \neq 0$ there exist unique integers q and r such that

$$x = qy + r \quad \text{where} \quad 0 \leq r < |y|. \tag{3.21}$$

top precedence among the four basic arithmetic operations. Specifically, another school of thought favors first performing all multiplications *and* divisions as they occur from left to right; thus, the left side of (3.20a) would be evaluated as $(6/2) \cdot 3 = 9$. However, the "old school" of thought put forth here has certain advantages. For one, it treats all multiplications the same, including those implicitly indicated by juxtaposition—e.g., for $a = 6$, $b = 2$, and $c = 3$, we obtain $a/bc = 1$, as in (3.20a)—whereas some in the other school would treat multiplication by juxtaposition as an exception. (The practice of first performing multiplications by juxtaposition is ubiquitous in mathematics literature, as the reader can verify by perusing a few of the journals.) Another advantage is that a ratio of products such as the expanded binomial coefficient

$$\binom{5}{3} = \frac{5 \cdot 4}{2 \cdot 1}$$

can be written "$5 \cdot 4/2 \cdot 1$" rather than "$\frac{5 \cdot 4}{2 \cdot 1}$" or "$(5 \cdot 4)/(2 \cdot 1)$", the first form being a handy option when one is writing in cramped places such as within text (as here) or within an exponent. Also, the old-school convention works well with computations in exponential notation; for example, "$6 \times 10^9/3 \times 10^4$" evaluates to $(6 \times 10^9)/(3 \times 10^4) = 2 \times 10^5$, as generally intended, rather than $[(6 \times 10^9)/3] \times 10^4 = 2 \times 10^{13}$. Moreover, the approach is consistent with how units of measure are expressed in terms of the multiplication and division of other units; for example, the approximate power density of solar radiation falling on the earth can be conveniently written "2.0 cal/cm^2·min", meaning 2.0 cal/(cm^2·min) not 2.0 (cal/cm^2)·min.

15. Likewise, a subtraction "$x - y$" can be viewed as a sign change followed by an addition: "$x + (-y)$". With this approach, successive additions and subtractions can be safely done in *any* order; e.g., $1 - 2 + 3 = (1 - 2) + 3 = 1 + (-2 + 3)$.

Thus, since $x/y = q + r/y$, we say that this **integer division** of x by y has an **(integer) quotient** q with a corresponding **remainder** r. Also, x is said to be **divisible** *by* y if and only if $r = 0$; equivalently, y **divides** x, y is a **divisor** of x, y is a **factor** of x, x is a **multiple** of y, and y is a **submultiple** of x.

When x and y are positive, their **least common multiple (LCM)** is the smallest positive integer z that is a multiple of both, and their **greatest common divisor (GCD)** is the largest integer z that is a divisor of both. (For example, $\text{LCM}(9, 12) = 36$ and $\text{GCD}(9, 12) = 3$.) In particular, if x is a multiple of y, then $\text{LCM}(x, y) = x$ and $\text{GCD}(x, y) = y$. The LCM and GCD of a collection of more than two positive integers are similarly defined.

When $|x| \geq |y|$, the fraction "$\frac{x}{y}$" is called an **improper fraction**;[16] whereas it is a **proper fraction** when $|x| < |y|$. If desired, an improper fraction can be written as a **mixed fraction**, which is a combination of an integer and a proper fraction; specifically, when $x \geq y > 0$ the fraction "$\frac{x}{y}$" may be rewritten "$q\frac{r}{y}$", with q and r satisfying (3.21), from which the cases of x and/or y being negative also obtain.[17] (Thus, e.g., $\frac{-25}{-7} = \frac{25}{7} = 3\frac{4}{7}$ and $\frac{-25}{7} = \frac{25}{-7} = -\frac{25}{7} = -3\frac{4}{7}$.)

An **even number** is an integer that is divisible by 2, whereas an **odd number** is an integer that is not. A **prime (number)** is an integer greater than 1 that is divisible by no positive integer other than itself and 1; thus, the only even prime number is 2, and the first five primes are 2, 3, 5, 7, and 11.

Every integer $x \geq 2$ has a unique **prime factorization**; that is, letting the prime numbers in increasing order be $p_1 = 2, p_2 = 3, p_3 = 5, \ldots$, there is exactly one choice of nonnegative integers $n_1, n_2, n_3, \ldots$ such that

$$x = {p_1}^{n_1} {p_2}^{n_2} {p_3}^{n_3} \cdots$$

—where the ith factor ${p_i}^{n_i}$ may be omitted when $n_i = 0$ (since then ${p_i}^{n_i} = 1$), and thus only finitely many factors need appear (e.g., $17\,578\,750 = 2^1 5^4 7^3 41^1$). Accordingly, an integer greater than 1 that is *not* prime is called a **composite number**, since it can be viewed as the product of several prime factors. Given the prime factorization of a number, all of its positive-integer divisors can be easily found by considering all possible products composed of its prime factors. For example, the positive-integer divisors of $60 = 2^2 3^1 5^1$ are $2^0 3^0 5^0 = 1$, $2^1 3^0 5^0 = 2$, $2^2 3^0 5^0 = 4$, $2^0 3^1 5^0 = 3$, $2^1 3^1 5^0 = 6$, $2^2 3^1 5^0 = 12$, $2^0 3^0 5^1 = 5$, $2^1 3^0 5^1 = 10$, $2^2 3^0 5^1 = 20$, $2^0 3^1 5^1 = 15$, $2^1 3^1 5^1 = 30$, and $2^2 3^1 5^1 = 60$.

Two nonzero integers x and y are **relatively prime** if and only if there is no integer $z > 1$ that divides them both—i.e., $\text{GCD}(|x|, |y|) = 1$. Whereas if such a z *does* exist, then the fraction x/y can be converted into another fraction x'/y' composed of integers $x' = x/z$ and $y' = y/z$ having smaller magnitudes than x and y, respectively (e.g., $6/12 = (6 \div 3)/(12 \div 3) = 2/4$). In particular, when x' and y' are relatively prime—obtained by taking $z = \text{GCD}(|x|, |y|)$—the fraction x'/y' is called **reduced** and is said to be **in lowest terms**.

16. Some authors exclude the case of $|x| = |y|$ from being "improper".

17. In formal mathematical discourse, though, mixed fractions are eschewed.

The remainder $r \in \{0, 1, \ldots, |y| - 1\}$ in (3.21) is also called the value of x **modulo** y:

$$r = x \bmod y. \tag{3.22}$$

Correspondingly, there is the arithmetic operation of **modular addition** with respect to any fixed integer **base** $n \geq 2$; namely, symbolizing this operation by "$\oplus$", we define

$$x \oplus y \triangleq (x + y) \bmod n \quad (x, y \in \mathbb{Z}). \tag{3.23}$$

This operation is particularly useful for analyses of a cyclic nature; for example, if it is presently 5 o'clock, then—taking $n = 12$ (since there are 12 hours in one clock cycle)—we conclude that 10 hours from now it will be $5 \oplus 10 = 15 \bmod 12 = 3$ o'clock. For the simplest case of $n = 2$,

$$0 \oplus 0 = 1 \oplus 1 = 0, \quad 0 \oplus 1 = 1 \oplus 0 = 1 \tag{3.24}$$

—here restricting our attention to $x, y \in \{0, 1, \ldots, n - 1\}$, as is often done. As with ordinary addition, modular addition (any base) is both commutative and associative.

The quotient x/y of two *real* numbers x and $y \neq 0$ is sometimes called a **ratio**. Alternatively, the ratio *of* x *to* y may be written

$$x : y \tag{3.25}$$

where all we now require is that x and y not *both* be zero. This ratio is equivalent to another ratio

$$x' : y' \tag{3.26}$$

if and only if for some constant α we have $x = \alpha x'$ and $y = \alpha y'$ (in which case α is unique and nonzero). That is, we have $x/y = x'/y'$ if $y, y' \neq 0$, and $y/x = y'/x'$ if $x, x' \neq 0$; or, more succinctly, $xy' = yx'$. The equivalence of the ratios in (3.25) and (3.26) may be expressed by writing

$$x : y :: x' : y' \tag{3.27}$$

which is read "x is (in proportion) to y as x' is (in proportion) to y'". Furthermore, the notation in (3.25)–(3.27) extends in the obvious way to arbitrarily many numbers; thus, a multiple ratio $x : y : z$ of three real values x, y, and z (not all zero) is equivalent to another such ratio $x' : y' : z'$—i.e.,

$$x : y : z :: x' : y' : z'$$

—if and only if $x' = \alpha x$, $y' = \alpha y$, and $z' = \alpha z$ for some (unique and nonzero) constant α.

Perhaps the most frequently occurring kind of ratio is **percentage**, which is the comparison of a real number p with the number 100; specifically, it is the ratio $p : 100$, which is commonly written

$$p\% \triangleq p/100$$

—the left side being read "p percent" (e.g., $5\% = 0.05$). To convert a real number x (often a fraction) to a percentage p, we simply multiply by 100%, which (as the above formula shows) is equivalent to a factor of 1: $x = x \cdot 1 = x \cdot 100\% = (100x)\%$ (e.g., $0.05 = 5\%$). To increase (respectively, decrease) a real number x by some percentage p, we add to (subtract from) x the amount $x \cdot p\% = xp/100$. Accordingly, a real number y is $p\%$ larger than a real number x when

$$y = x + x \cdot p\% = (1 + p/100)x;$$

whereas replacing each "+" with "−" expresses that y is $p\%$ smaller than x.

3.3 Complex Numbers

For no real number x do we have $x^2 < 0$; nevertheless, "imaginary numbers" having this property are quite useful in science and engineering. Furthermore, real and imaginary numbers can be combined together. To wit, introducing the **imaginary unit** $i \triangleq \sqrt{-1}$,[18] and thus

$$i^2 = -1, \tag{3.28}$$

we define the set of **complex numbers**

$$\mathbb{C} \triangleq \{x + iy \mid x, y \in \mathbb{R}\}$$

(where $x + iy$ may also be written "$x + yi$"). Note that every real number x is *also* a complex number $x + i0$;[19] hence, $\mathbb{R} \subset \mathbb{C}$, since not every complex number is real. Similarly, an **imaginary number** is any complex number of the form $0 + iy = iy$ for some $y \in \mathbb{R}$; therefore, the only number that is both real and imaginary is $0 + i0 = 0$. In general, a complex number $z = x + iy$ ($x, y \in \mathbb{R}$) has **real part** $\operatorname{Re} z \triangleq x$ and **imaginary part** $\operatorname{Im} z \triangleq y$ (not iy);[20] so, z is real (respectively, imaginary) if and only if $\operatorname{Im} z = 0$ ($\operatorname{Re} z = 0$). Also, if z' is another complex number, then $z = z'$ if and only if $\operatorname{Re} z = \operatorname{Re} z'$ and $\operatorname{Im} z = \operatorname{Im} z'$.

Since a complex number

$$z = \operatorname{Re} z + i \operatorname{Im} z \tag{3.29}$$

is essentially a pair $(\operatorname{Re} z, \operatorname{Im} z)$ of real numbers, the set $\mathbb{C}$ is essentially the same as $\mathbb{R} \times \mathbb{R} = \mathbb{R}^2$; moreover, just as $\mathbb{R}^2$ can be viewed geometrically as a plane (see Section 11.1), so too can $\mathbb{C}$ be viewed as the **complex plane** shown in Figure 3.2.

---

18. In electrical engineering, the symbol "j" is typically used instead for this purpose, since "i" often denotes a current within an electrical circuit.

19. Accordingly, the adjective "nonreal" should be used rather than "complex" whenever one wishes to indicate that a particular number is not real.

20. Thus, the imaginary part of a complex number is actually a *real* value.

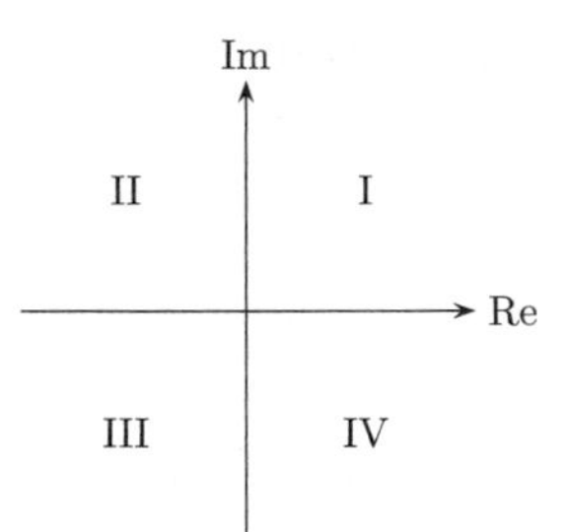

FIGURE 3.2 The Complex Plane

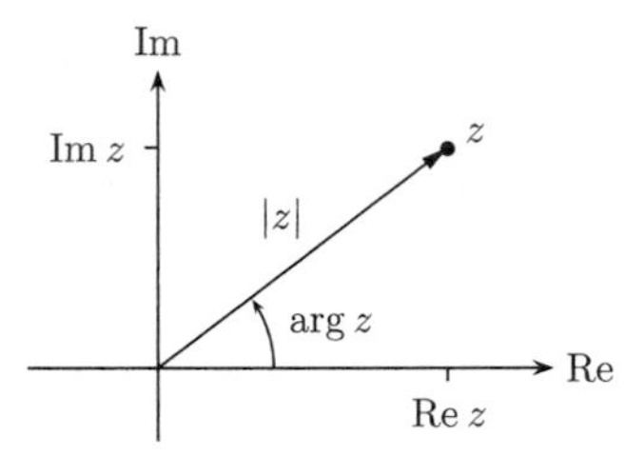

FIGURE 3.3 A Complex Number as a Vector

Two perpendicular axes divide the plane into four **quadrants**, numbered counterclockwise I through IV—viz., $\{z \in \mathbb{C} \mid \operatorname{Re} z > 0, \operatorname{Im} z > 0\}$, $\{z \in \mathbb{C} \mid \operatorname{Re} z < 0, \operatorname{Im} z > 0\}$, $\{z \in \mathbb{C} \mid \operatorname{Re} z < 0, \operatorname{Im} z < 0\}$, and $\{z \in \mathbb{C} \mid \operatorname{Re} z > 0, \operatorname{Im} z < 0\}$, respectively. The horizontal axis is called the **real axis**, because it is used to convert the first element $\operatorname{Re} z$ of the pair $(\operatorname{Re} z, \operatorname{Im} z)$ into a distance; likewise, the second element $\operatorname{Im} z$ is converted into a distance via the vertical **imaginary axis**. These two distances combine rectilinearly (as described in Chapter 5 with regard to Figure 5.1) to assign the complex number z a *point* in the complex plane; conversely, each point in the plane represents a unique complex number. In particular, the number 0 corresponds to the point where the real and imaginary axes intersect, called the **origin** of the plane.

As illustrated by the **Argand diagram** in Figure 3.3, it is common to think of a complex number z as a *vector* (see Section 11.3) in the complex plane, drawn from the origin to the point z. Accordingly, we define the **magnitude** (or **modulus**) $|z|$ of z to be the length of this vector,

$$|z| = \sqrt{(\operatorname{Re} z)^2 + (\operatorname{Im} z)^2}, \tag{3.30a}$$

which is consistent with (3.17) when z is real (since $\sqrt{x^2} = |x|$ for all $x \in \mathbb{R}$). It follows that $|z|$ is unique; and, $|z| \geq 0$, with equality being attained if and only if $z = 0$. Also, although a nonreal complex number has no *sign*, every complex number $z \neq 0$ has a **phase** $\arg z$, defined as the angle of the vector in Figure 3.3 measured counterclockwise from the positive half of the real axis; specifically, we can let[21]

$$\arg z = \begin{cases} \tan^{-1}\left(\frac{\operatorname{Im} z}{\operatorname{Re} z}\right) & \operatorname{Re} z > 0 \\ \tan^{-1}\left(\frac{\operatorname{Im} z}{\operatorname{Re} z}\right) + \pi & \operatorname{Re} z < 0 \\ \pi/2 & \operatorname{Re} z = 0, \operatorname{Im} z > 0 \\ -\pi/2 & \operatorname{Re} z = 0, \operatorname{Im} z < 0 \\ \text{undefined} & \operatorname{Re} z = \operatorname{Im} z = 0 \end{cases}. \tag{3.30b}$$

The value of $\arg z$ is not unique, though, but is determined only up to an integer multiple of 2π (360°); that is, adding $\pm 2\pi$ ($\pm 360°$) to a given value of $\arg z$ yields *another* acceptable value—which is why we refer to any particular value of $\arg z$ as *an* **argument** of z.[22]

From Figure 3.3 we also obtain

$$\operatorname{Re} z = |z| \cos(\arg z), \quad \operatorname{Im} z = |z| \sin(\arg z). \tag{3.31}$$

Therefore, by **Euler's identity**,[23]

$$e^{i\theta} = \cos\theta + i \sin\theta \quad (\theta \in \mathbb{R}) \tag{3.32}$$

(see Chapter 7), it follows from (3.29) and (3.31) that

$$z = |z| e^{i \arg z} \triangleq \underbrace{|z| \angle \arg z}_{\text{shorthand}} \tag{3.33}$$

—cf. (3.18).[24] This is the **polar form** of the complex number z, in contrast to its **rectangular form** in (3.29); the conversion from rectangular to polar form is provided by (3.30), whereas (3.31) provides the opposite conversion.

---

21. Many authors are sloppy here, stating simply that $\arg z = \tan^{-1}\left(\frac{\operatorname{Im} z}{\operatorname{Re} z}\right)$. Clearly, though, this equation cannot always be valid (even if we assume $\operatorname{Re} z \neq 0$), since the right side never yields an angle in quadrant II or III. To deal with this "problem", some computer languages introduce an arctangent-like function "atan2" having *two* real arguments, with $\operatorname{atan2}(y, x) \triangleq \arg(x + iy)$ ($x, y \in \mathbb{R}$ not both 0). For an alternative formula, which breaks up the calculation of $\arg z$ into fewer cases than in (3.30b), see Problem 11.17 (which pertains to Figure 11.3).

22. All assertions involving "arg" must be interpreted in light of this ambiguity. For example, if it is asserted for two complex values z_1 and z_2 that $\arg z_1 \neq \arg z_2$, then one must be careful not to thereby conclude that $z_1 \neq z_2$; for it could be that the values assigned to $\arg z_1$ and $\arg z_2$ differ by a nonzero multiple of 2π. (Perhaps a better notation would be to write "$\arg z_1 \equiv \arg z_2$" to indicate that $\arg z_1 = \arg z_2 + 2\pi n$ for some integer n. Then, having $\arg z_1 = \arg z_2$ would imply $\arg z_1 \equiv \arg z_2$, but not vice versa; and, one *could* conclude from having $\arg z_1 \not\equiv \arg z_2$ that $z_1 \neq z_2$.) One way to achieve definiteness is to instead use the **principal value of the argument**, $\operatorname{Arg} z$, which is defined as the unique value $\arg z$ such that $-\pi < \arg z \leq \pi$.

23. The name "Euler" is pronounced *OY-ler*. Also, (3.32) is often referred to as "Euler's formula"; however, we will use that term for a different purpose—see (10.23).

24. The convention that 0 times any number equals 0 allows us to view (3.31) and (3.33) as valid even when $z = 0$, though $\arg z$ is then undefined; likewise for (3.34) below. Also, it is common to see the notation "$\angle z$" used to denote $\arg z$; however, this usage of the symbol "$\angle$" conflicts with (3.33), where it represents a complex exponential function.

By (3.32), we have

$$e^{-i\theta} = e^{i(-\theta)} = \cos(-\theta) + i\sin(-\theta) = \cos\theta - i\sin\theta \quad (\theta \in \mathbb{R}).$$

Thus, defining the **(complex) conjugate** of z as

$$z^* \triangleq \operatorname{Re} z - i \operatorname{Im} z,$$

it follows from (3.31) that

$$z^* = |z|e^{-i\arg z}. \tag{3.34}$$

We also note that for $\theta_1, \theta_2 \in \mathbb{R}$,

$$\begin{aligned} e^{i\theta_1}e^{i\theta_2} &= (\cos\theta_1 + i\sin\theta_1)(\cos\theta_2 + i\sin\theta_2) \\ &= (\cos\theta_1\cos\theta_2 - \sin\theta_1\sin\theta_2) + i(\sin\theta_1\cos\theta_2 + \cos\theta_1\sin\theta_2) \\ &= \cos(\theta_1 + \theta_2) + i\sin(\theta_1 + \theta_2), \end{aligned}$$

by the trigonometric identities in (11.16); therefore,

$$e^{i\theta_1}e^{i\theta_2} = e^{i(\theta_1+\theta_2)} \quad (\theta_1, \theta_2 \in \mathbb{R}). \tag{3.35}$$

The four basic arithmetic operations can also be applied to complex numbers; and, the same commutative, associative, and distributive properties (3.13), as well as basic rules such as (3.14), continue to hold. Moreover, using (3.28), we get the following identities for $x_1, y_1, x_2, y_2 \in \mathbb{R}$:

$$\begin{aligned} (x_1 + iy_1) + (x_2 + iy_2) &= (x_1 + x_2) + i(y_1 + y_2), \\ (x_1 + iy_1) - (x_2 + iy_2) &= (x_1 - x_2) + i(y_1 - y_2), \\ (x_1 + iy_1) \times (x_2 + iy_2) &= (x_1x_2 - y_1y_2) + i(x_2y_1 + x_1y_2), \\ (x_1 + iy_1) \div (x_2 + iy_2) &= \frac{x_1x_2 + y_1y_2}{x_2{}^2 + y_2{}^2} + i\frac{x_2y_1 - x_1y_2}{x_2{}^2 + y_2{}^2} \quad (x_2, y_2 \text{ not both } 0). \end{aligned}$$

The last identity obtains by multiplying the numerator and denominator of the fraction $(x_1 + iy_1)/(x_2 + iy_2)$ by the complex conjugate of the denominator. This step is unnecessary, though, in the multiplication or division of two complex numbers z_1 and z_2 expressed in *polar* form; for then we simply use the facts

$$\begin{aligned} |z_1z_2| &= |z_1| \cdot |z_2|, \\ |z_1/z_2| &= |z_1|/|z_2| \quad (z_2 \neq 0), \\ \arg(z_1z_2) &= \arg z_1 + \arg z_2 \quad (z_1, z_2 \neq 0), \\ \arg(z_1/z_2) &= \arg z_1 - \arg z_2 \quad (z_1, z_2 \neq 0). \end{aligned}$$

Accordingly, multiplication and division are generally easier when complex numbers are expressed in polar form, whereas addition and subtraction are generally easier via the rectangular form.

Some additional basic facts for $z, z_1, z_2 \in \mathbb{C}$ are:[25]

$$1/i = -i = i^*, \tag{3.36a}$$

$$|i| = 1, \tag{3.36b}$$

$$\arg(\pm i) = \pm\pi/2 \quad (\pm \text{ respectively}), \tag{3.36c}$$

$$\operatorname{Re} az = a \operatorname{Re} z \quad (a \in \mathbb{R}), \tag{3.36d}$$

$$\operatorname{Im} az = a \operatorname{Im} z \quad (a \in \mathbb{R}), \tag{3.36e}$$

$$\operatorname{Re} iz = -\operatorname{Im} z, \tag{3.36f}$$

$$\operatorname{Im} iz = \operatorname{Re} z, \tag{3.36g}$$

$$|-z| = |z|, \tag{3.36h}$$

$$\arg a = \begin{cases} 0 & a > 0 \\ \pi & a < 0 \end{cases}, \tag{3.36i}$$

$$\arg az = \begin{cases} \arg z & a > 0 \\ \arg z + \pi & a < 0 \end{cases} \quad (z \neq 0), \tag{3.36j}$$

$$|1/z| = 1/|z| \quad (z \neq 0), \tag{3.36k}$$

$$\arg(1/z) = -\arg z \quad (z \neq 0), \tag{3.36l}$$

$$\operatorname{Re} z^* = \operatorname{Re} z, \tag{3.36m}$$

$$\operatorname{Im} z^* = -\operatorname{Im} z, \tag{3.36n}$$

$$|z^*| = |z|, \tag{3.36o}$$

$$\arg z^* = -\arg z \quad (z \neq 0), \tag{3.36p}$$

$$z + z^* = 2 \operatorname{Re} z, \tag{3.36q}$$

$$z - z^* = i2 \operatorname{Im} z, \tag{3.36r}$$

$$zz^* = |z|^2, \tag{3.36s}$$

$$z/z^* = 1\angle 2 \arg z \quad (z \neq 0), \tag{3.36t}$$

$$\operatorname{Re}(z_1 \pm z_2) = \operatorname{Re} z_1 \pm \operatorname{Re} z_2 \quad (\pm \text{ respectively}), \tag{3.36u}$$

$$\operatorname{Im}(z_1 \pm z_2) = \operatorname{Im} z_1 \pm \operatorname{Im} z_2 \quad (\pm \text{ respectively}), \tag{3.36v}$$

$$(z_1 \pm z_2)^* = z_1{}^* \pm z_2{}^* \quad (\pm \text{ respectively}), \tag{3.36w}$$

$$(z_1 z_2)^* = z_1{}^* z_2{}^*, \tag{3.36x}$$

$$(z_1/z_2)^* = z_1{}^*/z_2{}^* \quad (z_2 \neq 0). \tag{3.36y}$$

In addition, there is the **triangle inequality**,

$$|z_1 + z_2| \leq |z_1| + |z_2|. \tag{3.37}$$

25. Further properties of complex numbers are given in Chapter 7.

Unlike $\mathbb{R}$, the set $\mathbb{C}$ is not ordered by the less-than relation $<$. In fact, direct inequalities such as "$z_1 < z_2$" ($z_1, z_2 \in \mathbb{C}$) are meaningful only when z_1 and z_2 are both real. In particular, nonreal complex numbers are neither positive nor negative.

Finally, it follows from (3.33) that

$$|e^{i\theta}| = 1, \quad \arg e^{i\theta} = \theta$$

($\theta \in \mathbb{R}$). Also, via Euler's identity (3.32)—or, more easily, from Figure 3.3—we obtain

$$e^{i2n\pi} = 1, \qquad e^{i(2n+1)\pi} = -1, \qquad e^{i(2n\pm1/2)\pi} = \pm i \quad (\pm \text{ respectively}) \tag{3.38}$$

($n \in \mathbb{Z}$).

CHAPTER 4

Sequences

A **finite sequence** $\{x_i\}_{i=m}^{n}$ is a *list* of objects x_i $(i = m, m+1, \ldots, n)$, where m and n are integers such that $m \le n$; thus, $\{x_i\}_{i=m}^{n}$ is essentially the same as the $(n-m+1)$-tuple $(x_m, x_{m+1}, \ldots, x_n)$, now with the elements being ordered with respect to an integer **index** i.[1] Furthermore, one can have a **singly infinite sequence**, $\{x_i\}_{i=m}^{\infty}$ or $\{x_i\}_{i=-\infty}^{n}$, as well as a **doubly infinite sequence** $\{x_i\}_{i=-\infty}^{\infty}$; essentially, these are the same, respectively, as the infinite lists $(x_m, x_{m+1}, x_{m+2}, \ldots)$, $(\ldots, x_{n-2}, x_{n-1}, x_n)$, and $(\ldots, x_{-1}, x_0, x_1, \ldots)$. In all of these cases, the object x_i is called the ith **term** of the sequence. Also, whenever the range of the index i is either understood or unspecified, the sequence may be simply referred to as "$\{x_i\}$".

Similarly, an object $x_{i,j}$ might depend on *two* integer indices, here i and j, in which case we have a **double sequence** $\{x_{i,j}\}$. Specifically, suppose i varies from m to n $(-\infty < m \le n < \infty)$, and j varies from p to q $(-\infty < p \le q < \infty)$; then, fixing $i \in \{m, m+1, \ldots, n\}$ yields a sequence $\{x_{i,j}\}_{j=p}^{q}$ of terms $x_{i,j}$ versus the index j, whereas fixing $j \in \{p, p+1, \ldots, q\}$ yields a sequence $\{x_{i,j}\}_{i=m}^{n}$ of terms $x_{i,j}$ versus the index i. A double sequence can also be singly or doubly infinite versus either (perhaps both) of its indices.

Henceforth, each sequence under discussion is understood to be a **complex sequence**; that is, all of its terms are complex numbers. A special case is a **real sequence**, for which all of the terms are real numbers. Likewise, a sequence is called **positive, negative, nonpositive**, or **nonnegative** when every term is a number of the respective type.

An **arithmetic progression** is any sequence $\{x_i\}_{i=m}^{\infty}$ for which the difference $x_{i+1} - x_i$ of consecutive terms is the same for all i; e.g., $x_i = i$ $(i \ge m)$. Whereas for a **geometric progression**, the ratio x_{i+1}/x_i of consecutive terms is the same for all i (thereby precluding $x_i = 0$ for any i); e.g., $x_i = 2^i$ $(i \ge m)$.

A real sequence $\{x_i\}$ is said to be **increasing** (or **nondecreasing**) if and only if

$$x_i \le x_{i+1} \quad \text{(all } i\text{)}$$

1. In contexts where "i" is used to represent $\sqrt{-1}$, it is wise to use other symbols—e.g., "j", "k", ...—for sequence indices.

—equivalently, $x_i \leq x_j$ $(i < j)$; whereas it is **strictly increasing** if and only if

$$x_i < x_{i+1} \quad \text{(all } i\text{)}$$

—equivalently, $x_i < x_j$ $(i < j)$. Likewise, the sequence is **decreasing** (or **nonincreasing**) if and only if

$$x_i \geq x_{i+1} \quad \text{(all } i\text{)}$$

—equivalently, $x_i \geq x_j$ $(i < j)$; whereas it is **strictly decreasing** if and only if

$$x_i > x_{i+1} \quad \text{(all } i\text{)}$$

—equivalently, $x_i > x_j$ $(i < j)$. Hence, a sequence is both increasing and decreasing if and only if it is a **constant sequence**—i.e., all of its elements x_i have the same value. A sequence that is either increasing or decreasing is called **monotonic**; and, a sequence that is either strictly increasing or strictly decreasing is **strictly monotonic**.[2]

The **maximum** and **minimum** of a real sequence $\{x_i\}$ are, respectively,

$$\max_i x_i \triangleq \max\{x_i\}, \quad \min_i x_i \triangleq \min\{x_i\}$$

(when each exists), where "$\{x_i\}$" is now interpreted simply as the *set* of values x_i. Similarly, the **supremum** and **infimum** of $\{x_i\}$ are, respectively,

$$\sup_i x_i \triangleq \sup\{x_i\}, \quad \inf_i x_i \triangleq \inf\{x_i\}$$

(which always exist). Additionally, for any of these definitions a restriction may be placed on the index i. For example,

$$\max_{3 \leq i < 6} x_i \triangleq \max\{x_3, x_4, x_5\}$$

(assuming x_i is defined for $i \in \{3, 4, 5\}$), and

$$\inf_{i \in I} x_i \triangleq \inf\{x_i \mid i \in I\},$$

where I is any set of possible index values. A complex sequence $\{x_i\}$ is **bounded** if and only if

$$|x_i| \leq M \text{ (all } i\text{)} \quad \text{for some} \quad M > 0,$$

which is the case if and only if $\sup_i x_i$ and $\inf_i x_i$ are both finite.

2. It is not uncommon to see "increasing" and "decreasing" preceded by either "monotone" or "monotonically"; the practice, however, is unnecessary.

4.1 Limits

Consider an infinite sequence $\{x_i\}_{i=1}^{\infty}$.[3] We say that this sequence **converges**, or that it is **convergent**, if and only if there exists a complex value x such that x_i becomes arbitrarily close to x upon taking i sufficiently large. More precisely, given a constant $\varepsilon > 0$ (no matter how small), the magnitude of the difference between x_i and x is no more than ε when i is sufficiently large:

$$\forall \varepsilon > 0,\ \exists n \in \mathbb{Z}^+,\ \forall i \geq n,\ |x_i - x| \leq \varepsilon. \tag{4.1}$$

We then say that the sequence $\{x_i\}_{i=1}^{\infty}$ converges *to* x, which is indicated by writing any one of the following three expressions:

$$x_i \to x \text{ as } i \to \infty, \quad x_i \xrightarrow[i\to\infty]{} x, \quad \lim_{i\to\infty} x_i = x. \tag{4.2}$$

It follows from (4.1) that a necessary and sufficient condition for having $x_i \to x$ as $i \to \infty$ is that $x_i - x \to 0$; moreover, this occurs if and only if $|x_i - x| \to 0$. Verbally, (4.2) is expressed by saying that x_i *approaches* x *as* i *approaches* ∞, or that x is *the* **limit** of the sequence—the word "the" being used because a sequence can have at most one limit (including when infinite limits are allowed—see below).

Two simple examples: If $x_i = -1/\sqrt{i}$ $(i = 1, 2, \ldots)$, then x_i becomes arbitrarily close to $x = 0$ for i sufficiently large; because, for any $\varepsilon > 0$ we can choose a positive integer n (e.g., the smallest integer larger than $1/\varepsilon^2$) such that $|x_i - x| = 1/\sqrt{i} \leq \varepsilon$ for all $i \geq n$. Hence,

$$-1/\sqrt{i} \xrightarrow[i\to\infty]{} 0.$$

Likewise, for any complex constant c such that $|c| < 1$, we have $\lim_{i\to\infty} c^i \to 0$; whereas the only other value of c for which the sequence $\{c^i\}_{i=1}^{\infty}$ converges is 1, in which case $c^i = 1$ (all i), making $\lim_{i\to\infty} c^i \to 1$.

As a more interesting example, comparing the definition of $\mathbb{R}$ in (3.3) with (4.1) shows that for each real number x, there exists a sequence of rational numbers converging to x. Indeed, an infinite decimal expansion (3.4a) provides one such sequence; because, as each digit is annexed to the expansion, a better rational-valued approximation obtains to the ultimate real-valued limit. (For example, per (3.8), $\frac{1}{3}$ is the limit of the sequence $(0.3, 0.33, 0.333, \ldots)$.) We also note here that although only some real numbers have a finite decimal expansion, *every* nonzero real number has an infinite decimal expansion (e.g., $1 = 0.999\ldots$). However, other than zero—which has two finite expansions ("0" and "−0") but no infinite expansion—a real number can have at most one finite expansion, and it always has exactly one infinite expansion (assuming, as usual, that the leading digit $\boldsymbol{x}_n$ in (3.4a) is allowed to be "0" only when $n = 0$).

An infinite sequence that does not converge is said to **diverge** and is called **divergent**. In particular, a sequence $\{x_i\}_{i=1}^{\infty}$ of *real* values x_i diverges *to* ∞ if and only if x_i becomes

3. For convenience, we now focus on singly infinite sequences starting with an index value of 1; however, the following discussion generalizes in obvious ways to other sequences.

arbitrarily large upon taking i sufficiently large; that is, given a constant $\xi > 0$ (no matter how large), the value x_i must be at least as large as ξ when i is sufficiently large:

$$\forall \xi > 0,\ \exists n \in \mathbb{Z}^+,\ \forall i \geq n,\ x_i \geq \xi \tag{4.3}$$

(cf. (4.1)). This may be indicated by any of the expressions in (4.2) with $x = \infty$. Similarly, the sequence $\{x_i\}_{i=1}^{\infty}$ might diverge to $-\infty$, which is the case if and only if we can replace "$x_i \geq \xi$" in (4.3) with "$x_i \leq -\xi$"; and, this is indicated by any of the expressions in (4.2) with $x = -\infty$.

Sometimes infinite limits are allowed; but, usually not. When infinite limits are not allowed, we say for a sequence $\{x_i\}_{i=1}^{\infty}$ that its limit *exists* only if (4.2) holds for some value $x \in \mathbb{C}$; whereas when infinite limits are allowed, the limit of $\{x_i\}_{i=1}^{\infty}$ also exists should (4.2) hold for some value $x \in \mathbb{C} \cup \{\pm\infty\}$. In general, it is understood that infinite limits are allowed only when this is stated explicitly. Consequently, the existence of a limit typically implies that it is *finite*, and that the sequence at hand *converges*.

Examples: For $x_i = -\sqrt{i}$ (all i), the sequence $\{x_i\}_{i=1}^{\infty}$ diverges to $-\infty$; because, for any $\xi > 0$ we can choose a positive integer n (e.g., the smallest integer larger than ξ^2) such that $x_i \leq -\xi$ for all $i \geq n$. Hence, when infinite limits are allowed, we may write

$$-\sqrt{i} \xrightarrow[i\to\infty]{} -\infty. \tag{4.4}$$

Also, for any constant $c > 1$,[4] the sequence $\{c^i\}_{i=1}^{\infty}$ diverges to ∞; whereas for any other complex value $c \neq 1$ such that $|c| \geq 1$, this sequence diverges, but not to $\pm\infty$.

A monotonic sequence that is also bounded must converge; whereas a monotonic sequence that is not bounded must diverge to $\pm\infty$ (as in (4.4)). However, it is possible for a divergent sequence to be real and unbounded but not diverge to $\pm\infty$; e.g., consider $\{(-1)^i i\}_{i=1}^{\infty}$, or $\{c^i\}_{i=1}^{\infty}$ with $c < -1$.

Some basic facts for two convergent sequences $\{x_i\}_{i=1}^{\infty}$ and $\{y_i\}_{i=1}^{\infty}$ are

$$\lim_{i\to\infty} (x_i + y_i) = \lim_{i\to\infty} x_i + \lim_{i\to\infty} y_i, \tag{4.5a}$$

$$\lim_{i\to\infty} (x_i - y_i) = \lim_{i\to\infty} x_i - \lim_{i\to\infty} y_i, \tag{4.5b}$$

$$\lim_{i\to\infty} x_i y_i = \left(\lim_{i\to\infty} x_i\right)\left(\lim_{i\to\infty} y_i\right), \tag{4.5c}$$

$$\lim_{i\to\infty} x_i / y_i = \left(\lim_{i\to\infty} x_i\right) \Big/ \left(\lim_{i\to\infty} y_i\right), \tag{4.5d}$$

assuming for the last equality that $\lim_{i\to\infty} y_i \neq 0$.[5] Hence, since for each constant $a \in \mathbb{C}$ we have

$$\lim_{i\to\infty} a = a,$$

4. This bound on c implies that it is must be real; see p. 45.

5. Note that this assumption implies that $y_i = 0$ at most finitely many times, and so any divisions by 0 on the left side of (4.5d) are not a problem, since x_i/y_i will exist for i sufficiently large.

it follows from (4.5c) and (4.5d) that

$$\lim_{i\to\infty} ax_i = a \lim_{i\to\infty} x_i, \quad \lim_{i\to\infty} a/x_i = a \Big/ \lim_{i\to\infty} x_i, \tag{4.6}$$

assuming for the latter equality that $\lim_{i\to\infty} x_i \neq 0$. Be aware, though, that for each part of (4.5) there exist sequences $\{x_i\}_{i=1}^{\infty}$ and $\{y_i\}_{i=1}^{\infty}$ for which the limit on the left exists yet at least one of those on the right does not (and thus the equality fails to hold); e.g., take $x_i = y_i = (-1)^i$ (all i) in (4.5b).

Equalities are preserved in the limit (when the limits therein exist):

$$x_i = y_i \quad (\text{all } i) \qquad \Rightarrow \qquad \lim_{i\to\infty} x_i = \lim_{i\to\infty} y_i. \tag{4.7a}$$

And, so are inequalities that are not strict:

$$x_i \leq y_i \quad (\text{all } i) \qquad \Rightarrow \qquad \lim_{i\to\infty} x_i \leq \lim_{i\to\infty} y_i. \tag{4.7b}$$

However, strict inequalities need not be preserved. For example, if $x_i = 0$ and $y_i = 1/i$ ($i = 1, 2, \ldots$), then $x_i < y_i$ (all i); but, $\lim_{i\to\infty} x_i = 0 = \lim_{i\to\infty} y_i$, so $\lim_{i\to\infty} x_i \not< \lim_{i\to\infty} y_i$.

When infinite limits are allowed, the basic facts in (4.5) must be applied with care. Specifically, (4.5a) and (4.5b) are each valid when the right side of the equation can be evaluated (viz., "$\infty - \infty$" or such does not occur). The same, though, cannot be said of (4.5c) and (4.5d); e.g., if $x_i = i^2$ and $y_i = 1/i$ ($i = 1, 2, \ldots$), then the left side of (4.5c) equals ∞, whereas the right side equals $\infty \cdot 0 = 0$. On the other hand, (4.6) and (4.7) still hold as before—*if* for the right side of (4.6) we interpret a/∞ as 0 for all $a \in \mathbb{C}$.

Now, suppose we have a double sequence $\{x_{i,j}\}_{i,j=1}^{\infty}$; that is, $x_{i,j} \in \mathbb{C}$ for each pair (i,j) of positive integers i and j. Then, the **iterated limit**

$$\lim_{i\to\infty} \lim_{j\to\infty} x_{i,j} = \lim_{i\to\infty} \left(\lim_{j\to\infty} x_{i,j} \right) \tag{4.8}$$

might exist. As emphasized by the parentheses, we first let $j \to \infty$ to obtain the value $\lim_{j\to\infty} x_{i,j}$ for *each* i; then, assuming that these values exist when i is sufficiently large, we find *their* limit (if it exists) as $i \to \infty$. Thus, the *inner* limit in (4.8) is evaluated *first*, after which the outer limit is evaluated. Indeed, changing the order of the limits can yield different results; for example,

$$\lim_{i\to\infty} \lim_{j\to\infty} \frac{i}{i+j} = \lim_{i\to\infty} 0 = 0, \quad \lim_{j\to\infty} \lim_{i\to\infty} \frac{i}{i+j} = \lim_{j\to\infty} 1 = 1. \tag{4.9}$$

By contrast, we say that the double sequence $\{x_{i,j}\}_{i,j=1}^{\infty}$ *converges* if and only if there exists a value $x \in \mathbb{C}$ such that $x_{i,j}$ becomes arbitrarily close to x upon taking *both* i and j sufficiently large:

$$\forall \varepsilon > 0, \exists n \in \mathbb{Z}^+, \forall i,j \geq n, |x_{i,j} - x| \leq \varepsilon$$

(cf. (4.1)). This may be indicated by writing any one of the following three expressions:

$$x_{i,j} \to x \text{ as } i, j \to \infty, \quad x_{i,j} \xrightarrow[i,j\to\infty]{} x, \quad \lim_{i,j\to\infty} x_{i,j} = x$$

(cf. (4.2)). Also, the unique value x for which the condition holds is called the **multiple limit** of the double sequence $\{x_{i,j}\}_{i,j=1}^{\infty}$. The foregoing remarks regarding infinite limits, along with uses of the words "converge", "diverge", "exist", "approach", and their derivatives, carry over in the obvious way to multiple limits.

If the multiple limit $\lim_{i,j\to\infty} x_{i,j}$ exists, and the ordinary limits $\lim_{j\to\infty} x_{i,j}$ $(i = 1, 2, \ldots)$ exist for i sufficiently large, then

$$\lim_{i\to\infty} \lim_{j\to\infty} x_{i,j} = \lim_{i,j\to\infty} x_{i,j}; \tag{4.10a}$$

likewise,

$$\lim_{j\to\infty} \lim_{i\to\infty} x_{i,j} = \lim_{i,j\to\infty} x_{i,j} \tag{4.10b}$$

when $\lim_{i,j\to\infty} x_{i,j}$ exists, and the limits $\lim_{i\to\infty} x_{i,j}$ $(j = 1, 2, \ldots)$ exist for j sufficiently large. Therefore, whenever all three limits shown in (4.10) exist, they must all be equal; accordingly, it follows from the example in (4.9), for which the two limits on the left of (4.10) exist but are unequal, that it is possible for both iterated limits $\lim_{i\to\infty} \lim_{j\to\infty} x_{i,j}$ and $\lim_{j\to\infty} \lim_{i\to\infty} x_{i,j}$ to exist without the multiple limit $\lim_{i,j\to\infty} x_{i,j}$ existing. Likewise, the existence of a multiple limit does not guarantee the existence of a corresponding iterated limit. For example, if $x_{i,j} = (-1)^i/j$ $(i, j = 1, 2, \ldots)$, then $\lim_{i,j\to\infty} x_{i,j} = 0$; but, the iterated limit $\lim_{j\to\infty} \lim_{i\to\infty} x_{i,j}$ does not exist, because for no j does the limit $\lim_{i\to\infty} x_{i,j}$ exist.

The basic facts (4.5) and (4.6) generalize to multiple limits as one would expect. Namely, given two double sequences $\{x_{i,j}\}_{i,j=1}^{\infty}$ and $\{y_{i,j}\}_{i,j=1}^{\infty}$ that converge, we have:

$$\lim_{i,j\to\infty} (x_{i,j} + y_{i,j}) = \lim_{i,j\to\infty} x_{i,j} + \lim_{i,j\to\infty} y_{i,j},$$

$$\lim_{i,j\to\infty} (x_{i,j} - y_{i,j}) = \lim_{i,j\to\infty} x_{i,j} - \lim_{i,j\to\infty} y_{i,j},$$

$$\lim_{i,j\to\infty} x_{i,j} y_{i,j} = \left(\lim_{i,j\to\infty} x_{i,j}\right)\left(\lim_{i,j\to\infty} y_{i,j}\right),$$

$$\lim_{i,j\to\infty} x_{i,j}/y_{i,j} = \left(\lim_{i,j\to\infty} x_{i,j}\right) \Big/ \left(\lim_{i,j\to\infty} y_{i,j}\right),$$

assuming for the last equality that $\lim_{i,j\to\infty} y_{i,j} \neq 0$; and,

$$\lim_{i,j\to\infty} a x_{i,j} = a \lim_{i,j\to\infty} x_{i,j}, \quad \lim_{i,j\to\infty} a/x_{i,j} = a \Big/ \lim_{i,j\to\infty} x_{i,j},$$

assuming for the latter equality that $\lim_{i,j\to\infty} x_{i,j} \neq 0$. Also, (4.7a) and (4.7b) generalize in the obvious way.

If desired, an ordinary sequence $\{x_i\}_{i=1}^{\infty}$ can be viewed as a double sequence $\{x_{i,j}{}'\}_{i,j=1}^{\infty}$ by simply letting $x_{i,j}{}' = x_i$ (all i, j); accordingly,

$$\lim_{i,j\to\infty} x_i = \lim_{i\to\infty} x_i$$

when either side exists. Furthermore, for two convergent sequences $\{x_i\}_{i=1}^{\infty}$ and $\{y_j\}_{j=1}^{\infty}$, we have

$$\lim_{i,j\to\infty} (x_i + y_j) = \lim_{i\to\infty} x_i + \lim_{j\to\infty} y_j, \tag{4.11a}$$

$$\lim_{i,j\to\infty} (x_i - y_j) = \lim_{i\to\infty} x_i - \lim_{j\to\infty} y_j, \tag{4.11b}$$

$$\lim_{i,j\to\infty} x_i y_j = \left(\lim_{i\to\infty} x_i\right)\left(\lim_{j\to\infty} y_j\right), \tag{4.11c}$$

$$\lim_{i,j\to\infty} x_i / y_j = \left(\lim_{i\to\infty} x_i\right)\Big/\left(\lim_{j\to\infty} y_j\right), \tag{4.11d}$$

assuming for the last equality that $\lim_{j\to\infty} y_j \neq 0$. Moreover, in contrast to (4.5), the following can also be said about (4.11): If the limit on the left side of (4.11a) or (4.11b) exists, then the same is true of the limits on the right side, and thus the equality holds; also, if the limit on the left side of (4.11c) or (4.11d) exists and is nonzero, then the same is true of the limits on the right side, and thus the equality holds.[6]

The conventions given earlier, in (3.15), for the arithmetic manipulation of infinity can now be justified in terms of infinite limits. For instance, it is reasonable to let $0 \cdot \infty = 0$, because this equality can be interpreted as stating

$$\lim_{i\to\infty} x_i = \infty \quad \Rightarrow \quad \lim_{i\to\infty} (0 \cdot x_i) = \lim_{i\to\infty} 0 = 0, \tag{4.12a}$$

which is true. Similarly, we have $\infty + \infty = \infty$ because

$$\lim_{i\to\infty} x_i = \infty \quad \wedge \quad \lim_{j\to\infty} y_j = \infty \quad \Rightarrow \quad \lim_{i,j\to\infty} (x_i + y_j) = \infty; \tag{4.12b}$$

whereas $\infty - \infty$ is undefined, since the premises on the left of (4.12b) do not even guarantee that the multiple limit $\lim_{i,j\to\infty}(x_i - y_j)$ exists—e.g., let $x_i = i$ (all i) and $y_j = j$ (all j). Likewise, $1/0$ is generally undefined, since having $\lim_{i\to\infty} x_i = 0$ does not necessarily yield a value for $\lim_{i\to\infty} 1/x_i$—e.g., let $x_i = (-1)^i/i$ (all i).

---

6. Here, infinite limits cannot be allowed; for example, for $x_i = i$ (all i) and $y_j = (-1)^j$ (all j) we would have $\lim_{i,j\to\infty}(x_i + y_j) = \infty$ if infinite limits were allowed, but $\lim_{j\to\infty} y_j$ does not exist. Also, the assumption that the left sides of (4.11c) and (4.11d) are nonzero cannot be removed; for example, for $x_i = (-1)^i$ (all i) and $y_j = 1/j$ (all j) we have $\lim_{i,j\to\infty} x_i y_j = 0$, but $\lim_{i\to\infty} x_i$ does not exist.

4.2 Sums and Products

A common operation to perform on a sequence $\{x_i\}$ is to add finitely many (perhaps all) of its terms to obtain a **finite sum**

$$\sum_{i=m}^{n} x_i \triangleq x_m + x_{m+1} + \cdots + x_n. \tag{4.13a}$$

(It is assumed, of course, that x_i exists for $i = m, m+1, \ldots, n$; and, likewise below.) Or, the same terms can be multiplied to obtain the **finite product**

$$\prod_{i=m}^{n} x_i \triangleq x_m \cdot x_{m+1} \cdot \cdots \cdot x_n. \tag{4.13b}$$

In each case, a **lower limit** m and **upper limit** n,[7] which are integers such that $m \leq n$,[8] determine the set $\{m, m+1, \ldots, n\}$ over which the **index** i ranges while the operation is performed. In (4.13a), the generic term x_i to which the symbol "Σ" applies is the **summand** of the sum; and, in (4.13b), the generic term x_i to which the symbol "Π" applies is the **multiplicand** of the product.

When the limits of a sum or product as in (4.13) are either understood or unspecified, one may write simply "$\sum_i x_i$" or "$\prod_i x_i$". Likewise, one may write "$\sum_m^n x_i$" or "$\prod_m^n x_i$" when the index of summation or multiplication is understood.[9] Thus, when both the index is understood and the limits are understood or unspecified, one may write "$\sum x_i$" or "$\prod x_i$". Furthermore, additional constraints to be satisfied by the index may be explicitly appended. For example, if it is understood that i ranges over the set $\{1, 2, \ldots, 100\}$ unless further restricted, then one might write

$$\sum_{i \text{ odd}} x_i$$

rather than the completely detailed

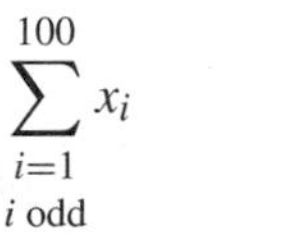

$$\sum_{\substack{i=1 \\ i \text{ odd}}}^{100} x_i \tag{4.14}$$

to indicate the sum $x_1 + x_3 + x_5 + \cdots + x_{99}$.

Infinite sums and products are defined as limits of finite sums and products. (A sum or product is called "finite" or "infinite" based on the number of terms it has, not on its

7. These uses of the word "limit" are different from those in Section 4.1. Often, the meaning of this word must be determined from context. Likewise, the word "term" has more than one meaning in mathematics.

8. Regarding the possibility of having $m > n$, see Subsection 4.2.1.

9. In (4.13a), the index of the sum happens to be the same as the index of the sequence $\{x_i\}$ upon which this operation is being performed; and, likewise for the index of the product in (4.13b). However, this need not be the case; e.g., see the right sides of (4.17) below.

value.) Specifically, we have the **singly infinite sums**

$$\sum_{i=m}^{\infty} x_i \triangleq \lim_{n\to\infty} \sum_{i=m}^{n} x_i \quad (m \in \mathbb{Z}), \qquad \sum_{i=-\infty}^{n} x_i \triangleq \lim_{m\to-\infty} \sum_{i=m}^{n} x_i \quad (n \in \mathbb{Z}), \tag{4.15a}$$

and the **singly infinite products**

$$\prod_{i=m}^{\infty} x_i \triangleq \lim_{n\to\infty} \prod_{i=m}^{n} x_i \quad (m \in \mathbb{Z}), \qquad \prod_{i=-\infty}^{n} x_i \triangleq \lim_{m\to-\infty} \prod_{i=m}^{n} x_i \quad (n \in \mathbb{Z}), \tag{4.15b}$$

as well as the **doubly infinite sum**

$$\sum_{i=-\infty}^{\infty} x_i \triangleq \lim_{m\to-\infty, n\to\infty} \sum_{i=m}^{n} x_i, \tag{4.16a}$$

and the **doubly infinite product**

$$\prod_{i=-\infty}^{\infty} x_i \triangleq \lim_{m\to-\infty, n\to\infty} \prod_{i=m}^{n} x_i. \tag{4.16b}$$

As with sequences generally, there may be situations where the sequence $\{x_i\}$ is real and the above limits are explicitly allowed to take on infinite values; otherwise, a sum or product is said to "exist" only if the corresponding limit is finite. In particular, when infinite limits are allowed, all of the infinite sums above exist whenever either $x_i \geq 0$ (all i) or $x_i \leq 0$ (all i), and all of the infinite products above exist whenever either $x_i \geq 1$ (all i) or $0 \leq x_i \leq 1$ (all i).

The terms "summand", "multiplicand", "lower limit", and "upper limit", as well as the various kinds of notation mentioned above, carry over in the obvious way to infinite sums and products. Additionally, an infinite sum is often called a **series** (a word that serves as both singular and plural); and, occasionally a finite sum is so called as well.

An infinite sum for which the corresponding limit in (4.15a) or (4.16a) is finite is called **convergent**, and is said to **converge**; otherwise, the sum is **divergent**, and is said to **diverge**. Additionally, if the limit equals $\pm\infty$, then the sum is said to diverge *to* $\pm\infty$ ($\pm$ respectively). For example, the **harmonic series** $\sum_{i=1}^{\infty} 1/i$ diverges to ∞. But, regardless of whether infinite limits are allowed, whenever a sum exists its value is unique. Also, if the infinite sum on the left side of (4.15a) converges, then we must have $x_i \to 0$ as $i \to \infty$; if the infinite sum on the right side (4.15a) converges, then $x_i \to 0$ as $i \to -\infty$; whereas if the doubly infinite sum in (4.16a) converges, then $x_i \to 0$ as $i \to \pm\infty$. The converses of these statements, however, are not true (e.g., $1/i \to 0$ as $i \to \infty$, yet the harmonic series does not converge).

The above terminology regarding convergence and divergence carries over from sums to products in the obvious way, *except* for products that equal 0; in that case, a more detailed examination of the product is required.[10] Specifically, an infinite product $\prod_i x_i$ is called **convergent**, and is said to **converge**, when

- only *finitely* many (perhaps none) of the terms x_i equal 0;[11] and,
- upon changing these x_i to 1, the corresponding limit in (4.15b) or (4.16b) converges to a *nonzero* value.

Otherwise, the product is **divergent**, and is said to **diverge**. As examples of convergence to 0 versus divergence to 0, the product $\prod_{i=0}^{\infty} x_i$

- diverges to 0 for $x_i = 1 + (-1)^i$ (all i), simply because $x_i = 0$ infinitely many times;
- diverges to 0 for $x_i = i$ (all i), because when x_0 is changed from 0 to 1 the corresponding limit does not converge;
- diverges to 0 for $x_i = \frac{1}{2}$ (all i), because every x_i is nonzero yet the corresponding limit converges to zero;
- converges to 0 for $x_i = \operatorname{sgn} i$ (all i), because changing x_0 from 0 to 1 makes every $x_i = 1$, thereby causing the corresponding limit to converge to a nonzero value (viz., 1).

Additionally, a product can diverge to $\pm\infty$. But, regardless of whether infinite limits are allowed, whenever a product exists its value is unique. Also, if the infinite product on the left side of (4.15b) converges, then we must have $x_i \to 1$ as $i \to \infty$; if the infinite product on the right side (4.15b) converges, then $x_i \to 1$ as $i \to -\infty$; whereas if the doubly infinite product in (4.16b) converges, then $x_i \to 1$ as $i \to \pm\infty$. The converses of these statements, however, are not true.

For each of the sums and products above, the variable "i" serving as the index can be described as a **dummy variable**; because, it has no intrinsic significance. The sole purpose of a dummy variable is to enable one to express some mathematical relationship or operation—here, a sum or a product. Thus, e.g., we can choose to replace the summation index in (4.13a) with "j", and multiplication index in (4.13b) with "k", and write

$$\sum_{j=m}^{n} x_j = \sum_{i=m}^{n} x_i, \quad \prod_{k=m}^{n} x_k = \prod_{i=m}^{n} x_i.$$

In general, the choice of a dummy variable is arbitrary except that it cannot conflict with other variables already introduced. For example, in the equation

$$\sum_{i=m}^{n} a x_i = y,$$

we are free to replace "i" with any variable other than "m", "n", "a", or "y", as long as the new variable has not been attributed a special significance elsewhere in the same

10. A reason for the special treatment of this case is given later (see fn. 16).

11. Of course, when *any* $x_i = 0$, the product equals 0. The issue to be settled, then, is whether the product *converges* to 0. This is done by evaluating the following limit, which has no other significance.

context. (Even an unsubscripted "x" might be used, though that would normally be eschewed to avoid confusion.) Moreover, with care, dummy variables can be reused *within* an expression; the key is that the index of a sum or product cannot be reused within its own *scope*. For example, if $x_i = 0$ (all i), then it is correct to write

$$\sum_{i=m}^{n} x_i = \prod_{i=m}^{n} x_i,$$

since the scope of the sum ends at the identity symbol "=".

More complicated index changes can also be performed. For example, if $x_i = 0$ for odd values of i, then

$$\sum_{i=1}^{100} x_i = \sum_{j=1}^{50} x_{2j}; \tag{4.17a}$$

whereas if $x_i = 0$ for even values of i, then

$$\sum_{i=1}^{100} x_i = \sum_{j=1}^{50} x_{2j-1}. \tag{4.17b}$$

Similarly, under any conditions, the right side of (4.17b) equals the sum in (4.14).

As another example, let a sequence $\{x_i\}_{i=-\infty}^{\infty}$ and limits $m, n \in \mathbb{Z} \cup \{\pm\infty\}$ with $m \leq n$ be given. Then, for any fixed integer c, we can change indices from i to $j = i - c$ and write

$$\sum_{i=m}^{n} x_i = \sum_{j=m-c}^{n-c} x_{j+c}.$$

(If $n = \infty$, then $n - c$ simplifies to ∞; and, likewise if $m = -\infty$.) If desired, this result can be derived in two simple steps: first substitute $i = j + c$ throughout the left side to get

$$\sum_{j+c=m}^{n} x_{j+c};$$

then, subtract c from the new "index" $j + c$, as well as from the limits m and n. Similarly, letting $i = -j + c$ instead gives

$$\sum_{i=m}^{n} x_i = \sum_{-j+c=m}^{n} x_{c-j} = \sum_{-j=m-c}^{n-c} x_{c-j} = \sum_{j=c-n}^{c-m} x_{c-j}, \tag{4.18}$$

where we have reordered the last sum so that the lower limit does not exceed the upper limit.[12] Of course, manipulations such as these can be performed on products as well.

12. Reversing the order of a sum or product is discussed in Subsection 4.2.1.

(Analogous to (4.17), the corresponding cases for products would be to have $x_i = 1$ for either i odd or i even.)

Observe that the limits in (4.16a) and (4.16b) are *multiple*, not iterated. It follows that each of the equalities

$$\sum_{i=-\infty}^{\infty} x_i = \lim_{n\to\infty} \sum_{i=-n}^{n} x_i, \qquad \prod_{i=-\infty}^{\infty} x_i = \lim_{n\to\infty} \prod_{i=-n}^{n} x_i$$

holds when the left side exists; however, the same cannot be said about the right sides. For example, if $x_i = i$ (all i), then the doubly infinite sum defined in (4.16a) does not exist, even if infinite limits are allowed; nevertheless, we now have

$$\lim_{n\to\infty} \sum_{i=-n}^{n} x_i = \lim_{n\to\infty} 0 = 0.$$

Based on (4.11), though, is always possible to split a *convergent* doubly infinite sum or product into two convergent singly infinite parts:

$$\sum_{i=-\infty}^{\infty} x_i = \left(\sum_{i=-\infty}^{l} x_i \right) + \left(\sum_{i=l+1}^{\infty} x_i \right), \tag{4.19a}$$

$$\prod_{i=-\infty}^{\infty} x_i = \left(\prod_{i=-\infty}^{l} x_i \right) \left(\prod_{i=l+1}^{\infty} x_i \right), \tag{4.19b}$$

for any integer l.[13] Furthermore, (4.19a) holds whenever both singly infinite sums exist, including when infinite limits are allowed—except if one of these sums equals ∞ and the other equals $-\infty$, in which case the addition "+" is not allowed and the doubly infinite sum does not exist. Whereas (4.19b) holds whenever both singly infinite products exist, including when infinite limits are allowed—except if one of the products equals $\pm\infty$, the other equals 0, and $x_i \neq 0$ (all i), in which case the doubly infinite product does not exist.

Given a double sequence $\{x_{i,j}\}$, we can perform an iterated sum

$$\sum_{i=m}^{n} \sum_{j=p}^{q} x_{i,j} = \sum_{i=m}^{n} \left(\sum_{j=p}^{q} x_{i,j} \right) \tag{4.20a}$$

or an iterated product

$$\prod_{i=m}^{n} \prod_{j=p}^{q} x_{i,j} = \prod_{i=m}^{n} \left(\prod_{j=p}^{q} x_{i,j} \right) \tag{4.20b}$$

13. The parentheses in (4.19a) are optional, as explained shortly in fn. 15.

for any allowable limits m, n, p, and q. As emphasized by the parentheses in (4.20a), we first obtain the value $\sum_{j=p}^{q} x_{i,j}$ for *each* i; then, these values are summed over the range of i. That is, the *inner* sum is evaluated *first*, after which the outer sum is evaluated; and, likewise for the iterated product in (4.20b).

By the commutative and associative properties of addition and multiplication, two finite sums comprising an iterated sum can be interchanged without affecting the resulting value, as can two finite products comprising an iterated product; that is, when m, n, p, and q are finite, we have

$$\sum_{i=m}^{n}\sum_{j=p}^{q} x_{i,j} = \sum_{j=p}^{q}\sum_{i=m}^{n} x_{i,j}, \tag{4.21a}$$

$$\prod_{i=m}^{n}\prod_{j=p}^{q} x_{i,j} = \prod_{j=p}^{q}\prod_{i=m}^{n} x_{i,j}. \tag{4.21b}$$

It then follows from (4.5a) that (4.21a) holds when the sum over i is finite, even if the sum over j is infinite, as long as the left side of the equation exists; likewise, it follows from (4.5c) that (4.21b) holds when the product over i is finite, even if the product over j is infinite, as long as the left side of the equation exists. However, when the sum or product over i is infinite, the rearrangement above might not be allowed; for example,

$$\sum_{i=1}^{\infty}\sum_{j=0}^{1}(-1)^j = \sum_{i=1}^{\infty} 0 = 0, \tag{4.22a}$$

whereas even upon allowing infinite limits we obtain

$$\sum_{j=0}^{1}\sum_{i=1}^{\infty}(-1)^j = \sum_{j=0}^{1}\begin{Bmatrix} \infty & j=0 \\ -\infty & j=1 \end{Bmatrix} = \infty + (-\infty), \tag{4.22b}$$

which is undefined.

It is also possible for both sides of (4.21a) or (4.21b) to converge, yet the equality not hold. For example, for the double sequence $\{x_{i,j}\}_{i,j=1}^{\infty}$ with

$$x_{i,j} = \begin{cases} 1 & i = j+1 \\ -1 & i = j-1 \\ 0 & \text{otherwise} \end{cases},$$

we have

$$\sum_{i=1}^{\infty}\sum_{j=1}^{\infty} x_{i,j} = \sum_{i=1}^{\infty}\begin{Bmatrix} -1 & i=1 \\ 0 & \text{otherwise} \end{Bmatrix} = -1,$$

$$\sum_{j=1}^{\infty}\sum_{i=1}^{\infty} x_{i,j} = \sum_{j=1}^{\infty}\begin{Bmatrix} 1 & j=1 \\ 0 & \text{otherwise} \end{Bmatrix} = 1.$$

On the other hand, should infinite limits be allowed, (4.21a) holds whenever either $x_i \geq 0$ (all i) or $x_i \leq 0$ (all i), and (4.21b) holds whenever either $x_i \geq 1$ (all i) or $0 \leq x_i \leq 1$ (all i).

If either of the inner limits p and q in (4.20a) is replaced with a function of the outer index i, then interchanging (if possible) the two sums over i and j typically requires a bit of work; and, likewise for the products in (4.20b). A common approach is to guide the calculation by drawing a map in the ij-plane of the pairs (i,j) that occur in the given iterated sum or product. Alternatively, here is a purely analytical technique, which will be explained with an example:

$$\sum_{i=7}^{\infty} \sum_{j=i-1}^{i+4} x_{i,j} = \sum_{j=?}^{?} \sum_{i=?}^{?} x_{i,j}. \tag{4.23}$$

First, introducing the appropriate **indicator function** $\mathbf{1}(\cdot,\cdot)$—viz., $\mathbf{1}(i,j) = 1$ for the pairs (i,j) occurring on the left, and $\mathbf{1}(i,j) = 0$ otherwise—we have

$$\mathbf{1}(i,j) = \begin{cases} 1 & 7 \leq i < \infty,\, i-1 \leq j \leq i+4 \\ 0 & \text{otherwise} \end{cases} \tag{4.24a}$$

$$= \begin{cases} 1 & 6 \leq j < \infty,\, \max(7, j-4) \leq i \leq j+1 \\ 0 & \text{otherwise} \end{cases}. \tag{4.24b}$$

Observe that in (4.24a), j depends on i, as on the left side of (4.23); and, this form is converted into (4.24b), where i depends on j, which facilitates interchanging the two sums:

$$\sum_{i=7}^{\infty} \sum_{j=i-1}^{i+4} x_{i,j} = \sum_{i=-\infty}^{\infty} \sum_{j=-\infty}^{\infty} x_{i,j}\mathbf{1}(i,j) = \sum_{j=-\infty}^{\infty} \sum_{i=-\infty}^{\infty} x_{i,j}\mathbf{1}(i,j)$$
$$= \sum_{j=6}^{\infty} \sum_{i=\max(7,j-4)}^{j+1} x_{i,j} \tag{4.25}$$

—*if* the interchanging of infinite sums to obtain the second equality can be justified.

To draw a more definite conclusion, we begin instead by expressing the infinite sum in (4.23), from $i = 7$ to ∞, as the limit as $n \to \infty$ of the corresponding sum from $i = 7$ to n:

$$\sum_{i=7}^{\infty} \sum_{j=i-1}^{i+4} x_{i,j} = \lim_{n\to\infty} \sum_{i=7}^{n} \sum_{j=i-1}^{i+4} x_{i,j}.$$

Then, the appropriate indicator functions for the right side are

$$\mathbf{1}_n(i,j) = \begin{cases} 1 & 7 \le i \le n,\, i-1 \le j \le i+4 \\ 0 & \text{otherwise} \end{cases}$$

$$= \begin{cases} 1 & 6 \le j \le n+4,\, \max(7,j-4) \le i \le \min(n,j+1) \\ 0 & \text{otherwise} \end{cases}$$

($n = 7, 8, \ldots$), yielding

$$\sum_{i=7}^{n} \sum_{j=i-1}^{i+4} x_{i,j} = \sum_{i=-\infty}^{\infty} \sum_{j=-\infty}^{\infty} x_{i,j}\mathbf{1}_n(i,j) = \sum_{j=-\infty}^{\infty} \sum_{i=-\infty}^{\infty} x_{i,j}\mathbf{1}_n(i,j)$$

$$= \sum_{j=6}^{n+4} \sum_{i=\max(7,j-4)}^{\min(n,j+1)} x_{i,j} \quad (n = 7, 8, \ldots). \tag{4.26}$$

Here, interchanging the two infinite sums is certainly allowed, because for each n the summand $x_{i,j}\mathbf{1}_n(i,j)$ is nonzero for only finitely many pairs (i,j). Hence, letting $n \to \infty$ in (4.26), we obtain

$$\sum_{i=7}^{\infty} \sum_{j=i-1}^{i+4} x_{i,j} = \lim_{n\to\infty} \sum_{j=6}^{n+4} \sum_{i=\max(7,j-4)}^{\min(n,j+1)} x_{i,j} \tag{4.27}$$

when either side of this equation exists (regardless of whether infinite limits are allowed).[14]

We now present some basic facts about sums and products for any sequences $\{x_i\}_{i=m}^n$ and $\{y_i\}_{i=m}^n$ (m, $n \in \mathbb{Z} \cup \{\pm\infty\}$ with $m \le n$). First, if m and n are finite, then for any

14. It might be thought that since for each j we have $\min(n,j+1) \to j+1$ as $n \to \infty$, the right side of (4.27) can always be simplified to be the same as right side of (4.25); however, this is not so. Because, the sequence of numbers

$$\sum_{j=6}^{n} \sum_{i=\max(7,j-4)}^{j+1} x_{i,j} \quad (n = 7, 8, \ldots)$$

need not be the same as the sequence of numbers

$$\sum_{j=6}^{n} \sum_{i=\max(7,j-4)}^{\min(n,j+1)} x_{i,j} \quad (n = 7, 8, \ldots),$$

nor have the same limit (as a counterexample can show).

constant $a \in \mathbb{C}$ we have [15]

$$\sum_{i=m}^{n} a x_i = a \sum_{i=m}^{n} x_i, \tag{4.28}$$

$$\sum_{i=m}^{n} (x_i \pm y_i) = \sum_{i=m}^{n} x_i \pm \sum_{i=m}^{n} y_i \quad (\pm \text{ respectively}), \tag{4.29}$$

$$\prod_{i=m}^{n} x_i y_i = \left(\prod_{i=m}^{n} x_i \right) \left(\prod_{i=m}^{n} y_i \right), \tag{4.30}$$

$$\prod_{i=m}^{n} x_i / y_i = \left(\prod_{i=m}^{n} x_i \right) \Big/ \left(\prod_{i=m}^{n} y_i \right), \tag{4.31}$$

assuming in (4.31) that the last product is nonzero. Thus, suppose $m = -\infty$ and/or $n = \infty$; and, let be it understood whether infinite limits are allowed (the statements to follow are valid in either case). Then, in (4.28) the existence of the sum on one side of the equation guarantees the existence of the sum on the other side, and the equality holds. In (4.29), though, it is possible for the left side of the equality to exist without the equality itself being valid, as is essentially demonstrated by (4.22); but, whenever the two sums on the right side of (4.29) exist, and the addition or subtraction there is allowed (viz., the process does not amount to an evaluation of "$\infty - \infty$"), the equality holds. Likewise, (4.30) holds if the two products on the right side exist, whereas (4.31) holds if the two products on the right side exist and the division on the right is allowed (e.g., no attempt is made to evaluate "∞/∞").

Generalizing the triangle inequality (3.37), we have

$$\left| \sum_{i=m}^{n} x_i \right| \le \sum_{i=m}^{n} |x_i| \tag{4.32a}$$

whenever both sums exist. By contrast,

$$\left| \prod_{i=m}^{n} x_i \right| = \prod_{i=m}^{n} |x_i| \tag{4.32b}$$

---

15. Parentheses are not required around either of the two sums on the right side of (4.29), because the scope of the first of these sums is understood to end at the succeeding addition or subtraction. (Accordingly, we are able to reuse the index "i" in the other sum.) For the same reason, parentheses *are* required on the left side of (4.29). On the left side of (4.30), though, no parentheses are required around the product "$x_i y_i$", which is understood to lie completely within the scope of the product over i; and, likewise for the quotient "x_i/y_i" in (4.31). For the same reason, the parentheses on the right side of (4.30) are inserted to avoid it from being incorrectly interpreted as

$$\prod_{i=m}^{n} \left(x_i \prod_{j=m}^{n} y_j \right)$$

(a change of indices being introduced here for clarity); and, likewise for the right side of (4.31).

whenever the product on the left exists. Moreover, these statements remain valid when infinite limits are allowed (in which case the sum on the right of (4.32a) always exists).

By using exponential and logarithm functions, a sum of *real* numbers x_i can be expressed in terms of a related product:

$$\exp\left(\sum_{i=m}^{n} x_i\right) = \prod_{i=m}^{n} e^{x_i} \quad \Rightarrow \quad \sum_{i=m}^{n} x_i = \ln\left(\prod_{i=m}^{n} e^{x_i}\right). \tag{4.33a}$$

Conversely, a product of *positive* numbers x_i can be expressed in terms of a related sum:

$$\ln\left(\prod_{i=m}^{n} x_i\right) = \sum_{i=m}^{n} \ln x_i \quad \Rightarrow \quad \prod_{i=m}^{n} x_i = \exp\left(\sum_{i=m}^{n} \ln x_i\right). \tag{4.33b}$$

These four equations always hold when m and n are finite. As for having $m = -\infty$ and/or $n = \infty$, let infinite limits be allowed, and let $e^{-\infty} = 0$, $e^{\infty} = \infty$, $\ln 0 = -\infty$, and $\ln \infty = \infty$ (per the limiting behavior of the exponential and logarithm functions). Then (owing to the continuity of these functions), when either the sum or the product exists in one of the four equations in (4.33), that equation is valid.[16]

A *sum of differences* will commonly **telescope**, whereby each term in the sum partially cancels the next, with the net result being that all but the first and last terms are totally canceled. Specifically, for any finite sequence $\{x_i\}_{i=m}^{n+1}$ $(m \le n+1)$ we have

$$\sum_{i=m}^{n} (x_{i+1} - x_i) = x_{n+1} - x_m. \tag{4.34a}$$

Analogously, a *product of quotients* can also telescope:

$$\prod_{i=m}^{n} x_{i+1}/x_i = x_{n+1}/x_m, \tag{4.34b}$$

assuming no divisions by 0 occur.

Here are some frequently occurring sums, where $m, n \in \mathbb{Z}$ with $m \le n$, and $c \in \mathbb{C}$.

- Sum of a constant:

$$\sum_{i=m}^{n} c = (n - m + 1)c. \tag{4.35}$$

16. This correspondence between sums and products explains why they have different conditions for convergence; for we can now say that an infinite sum of real numbers converges if and only if the corresponding infinite product of positive numbers converges. A snag occurs, though, upon allowing $x_i = 0$ for some values of i, since then the product $\prod_{i=m}^{n} x_i$ has no corresponding sum $\sum_{i=m}^{n} y_i$; because, there is no number y_i for which $e^{y_i} = x_i$ when $x_i = 0$. This problem is handled in the convergence definition for products by allowing *finitely* many zeros x_i to be replaced by ones; for then, unless infinitely many $x_i = 0$, a corresponding sum can be made to exist without changing the asymptotic behavior of the sequence $\{x_i\}_{i=m}^{n}$.

- Sum of an arithmetic progression:

$$\sum_{i=m}^{n} i = \tfrac{1}{2}[n(n+1) - (m-1)m]. \tag{4.36}$$

- Sum of a geometric progression:

$$\sum_{i=m}^{n} c^i = \begin{cases} n-m+1 & c = 1 \\ \dfrac{c^{n+1} - c^m}{c-1} & c \neq 1, \text{ with } c \neq 0 \text{ for } m < 0 \end{cases} \tag{4.37}$$

(where $0^0 = 1$).

- **Geometric series**:

$$\sum_{i=0}^{\infty} c^i = \frac{1}{1-c} \quad (|c| < 1) \tag{4.38}$$

(where $0^0 = 1$).[17]

We now demonstrate some of the preceding material by proving (4.35)–(4.38).

Equation (4.35) obtains simply because the number of terms in the sum is $n-m+1$ (not $n-m$). By contrast, we will justify (4.36) by reasoning geometrically.[18] In Figure 4.1, each value $i \in \{1, 2, \ldots, n\}$ is represented by the area of a rectangle of width 1 and height i, positioned on the horizontal axis from $i-1$ and i. The sum of these areas, $\sum_{i=1}^{n} i$, equals the area of the solid-line polygon shaped like a staircase; but, that area also equals the area of the large triangle lying below the dashed line, plus the areas of the n small triangles lying above the dashed line. Hence,

$$\sum_{i=1}^{n} i = \tfrac{1}{2} \cdot n \cdot n + n \cdot \tfrac{1}{2} = \tfrac{1}{2}n(n+1) \quad (n = 1, 2, \ldots),$$

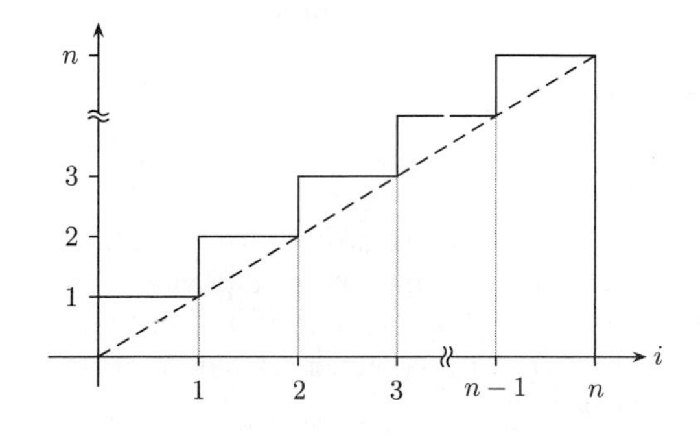

FIGURE 4.1 A Geometric Interpretation of Summing an Arithmetic Progression

17. More generally, a geometric series is any sum $\sum_{i=m}^{\infty} x_i$ such that $x_{i+1}/x_i = c$ (all i) for some constant c with $|c| < 1$; and, this sum equals x_m times that in (4.38).

18. A nongeometric proof of (4.36) for $m = 1$ is given in the prologue.

giving (4.36) for $m = 1$; therefore, for an arbitrary m, it follows from a change of indices that

$$\sum_{i=m}^{n} i = \sum_{j=1}^{n-m+1} (j + m - 1) = \sum_{j=1}^{n-m+1} j + \sum_{j=1}^{n-m+1} (m-1)$$
$$= \tfrac{1}{2}(n-m+1)(n-m+2) + (n-m+1)(m-1),$$

which simplifies to the right side of (4.36).

As for (4.37), the case of $c = 1$ follows from (4.35). Therefore, assume $c \neq 1$; furthermore, require $c \neq 0$ for $m < 0$, to avoid division by 0. Then, one way to proceed is to multiply the sum by c, obtaining

$$c \sum_{i=m}^{n} c^i = \sum_{i=m}^{n} c^{i+1} = \sum_{i=m+1}^{n+1} c^i = \sum_{i=m}^{n} c^i + c^{n+1} - c^m,$$

and thus

$$(c-1) \sum_{i=m}^{n} c^i = c^{n+1} - c^m, \tag{4.39}$$

which upon dividing by $c - 1 \neq 0$ yields the remainder of (4.37). Alternatively, (4.39) obtains by bringing the factor $c - 1$ inside the sum to produce a telescoping sum:

$$(c-1) \sum_{i=m}^{n} c^i = \sum_{i=m}^{n} (c-1)c^i = \sum_{i=m}^{n} (c^{i+1} - c^i) = c^{n+1} - c^m,$$

by (4.34a). Finally, (4.38) obtains from (4.37) by taking $m = 0$ and letting $n \to \infty$; for then $c^{n+1} \to 0$, having assumed $|c| < 1$ (otherwise, the series does not converge).

4.2.1 Summation and Multiplication in Reverse

Heretofore we have only considered sums (4.13a) and products (4.13b) for which the lower limit m is no greater than the upper limit n. As for allowing $m > n$, there are various possible conventions, with the one chosen in a particular context being largely a matter of personal preference. To discuss these options, let a sequence $\{x_i\}_{i=-\infty}^{\infty}$ be given, along with limits $l, m, n \in \mathbb{Z} \cup \{\pm\infty\}$; furthermore, to minimize distraction by convergence issues, assume that all of the sums and products introduced below exist.

The simplest approach regarding order is to not distinguish between the "forward" and "backward" summation of a sequence; and, likewise for its multiplication. Accordingly, we would always have

$$\sum_{i=m}^{n} x_i = \sum_{i=n}^{m} x_i, \quad \prod_{i=m}^{n} x_i = \prod_{i=n}^{m} x_i \quad \text{(all } m, n\text{)}. \tag{4.40}$$

This convention reflects the fact that reversing the order in which the terms of a finite sequence are summed or multiplied has no effect on the result; moreover, in light of how infinite sums and products are defined as limits of finite sums and products, the same is true for them too.

Another reasonable approach, though, is to interpret the left sides of (4.13a) and (4.13b) as

$$\sum_{i=m}^{n} x_i = \sum_{m \le i \le n} x_i, \quad \prod_{i=m}^{n} x_i = \prod_{m \le i \le n} x_i \quad \text{(all } m, n\text{);}$$

or, using set notation,

$$\sum_{i=m}^{n} x_i = \sum_{i \in I} x_i, \quad \prod_{i=m}^{n} x_i = \prod_{i \in I} x_i,$$

where $I = \{i \in \mathbb{Z} \mid m \le i \le n\}$.[19] In particular, when $I = \emptyset$ we have either an **empty sum** or an **empty product**; and, a common convention is to evaluate every empty sum as 0 (the additive identity), and every empty product as 1 (the multiplicative identity)—i.e.,

$$\sum_{i \in \emptyset} x_i = 0, \quad \prod_{i \in \emptyset} x_i = 1.$$

Thus, in contrast to (4.40), we now have

$$\sum_{i=m}^{n} x_i = 0, \quad \prod_{i=m}^{n} x_i = 1 \quad (m > n), \tag{4.41}$$

since $I = \emptyset$ when $m > n$. A benefit of this convention is that it enables the usual splitting of sums and products as

$$\sum_{i=l}^{n} x_i = \sum_{i=l}^{m-1} x_i + \sum_{i=m}^{n} x_i = \sum_{i=l}^{m} x_i + \sum_{i=m+1}^{n} x_i, \tag{4.42a}$$

$$\prod_{i=l}^{n} x_i = \left(\prod_{i=l}^{m-1} x_i\right)\left(\prod_{i=m}^{n} x_i\right) = \left(\prod_{i=l}^{m} x_i\right)\left(\prod_{i=m+1}^{n} x_i\right) \tag{4.42b}$$

$(l \le m \le n)$, *including* when $m = l$ or $m = n$. Other benefits also accrue; for instance, we can now define $n!$ simply as $\prod_{i=1}^{n} i$ for all $n \ge 0$, rather than treating $n = 0$ as a separate case as shown later in (8.1).

---

19. This notation for sums and products in terms of an index set I is more flexible. (For example, consider taking $I = \{\text{prime numbers}\}$, or letting the elements of I be nonnumerical.) However, additional information might be required when I is infinite; because, for a given infinite sequence $\{x_i\}_{i \in I}$, the value of the sum $\sum_{i \in I} x_i$ or the product $\prod_{i \in I} x_i$—indeed, its very existence—can be altered by changing the order in which the values i are taken from I.

The convention (4.41), however, does not make (4.42) always valid for *all* integers l, m, n. Such is the case if and only if we let

$$\sum_{i=m}^{n} x_i = \begin{cases} 0 & m = n+1 \\ -\sum_{i=n+1}^{m-1} x_i & m > n+1 \end{cases}, \tag{4.43a}$$

$$\prod_{i=m}^{n} x_i = \begin{cases} 1 & m = n+1 \\ \left(\prod_{i=n+1}^{m-1} x_i\right)^{-1} & m > n+1 \end{cases}, \tag{4.43b}$$

assuming in (4.43b) that the product in parentheses is nonzero. This convention yields the same results as (4.41) when $m \leq n+1$, but not (in general) otherwise.

For each of the three conventions (4.40), (4.41), and (4.43), *some* basic properties—e.g., (4.28)–(4.31), (4.32b), and (4.33)—remain valid when the lower limits of sums and products are allowed to exceed their upper limits; but, such might not be the case for other properties. In particular, (4.32a) applies generally only for conventions (4.40) and (4.41), whereas (4.34) applies generally only for convention (4.43). Furthermore, when making a change of indices with conventions (4.41) and (4.43), one must be careful that the resulting sum or product has the proper direction; for example, by both of these conventions the leftmost and rightmost sums in (4.18) are always equal; but not so if the limits of the latter sum are reversed.

4.3 Mathematical Induction

Often, an infinite sequence exhibits a *pattern*. More generally, suppose that the terms y_n of a sequence $\{y_n\}_{n=0}^{\infty}$ appear to possess a particular property P vis-à-vis the index $n \in \{0, 1, \ldots\} = \mathbb{N}$; that is,

$$\forall n \in \mathbb{N},\ P(n). \tag{4.44}$$

One way to show that this conjecture is, in fact, true is to perform a **proof by (mathematical) induction**;[20] this is a two-step process:

20. Despite its name, "mathematical induction" is a method of *deductive* logic. Broadly speaking, "inductive logic" (often referred to as "induction" in nonmathematical contexts) is the process of concluding, based on numerous observed individual cases, that a general law is *likely* to be true. By contrast, deductive logic derives, from given premises, conclusions that are *necessarily* true. (For example, upon flipping a coin 100 times and observing all of the outcomes to be heads, one might come to believe—inductively—that the coin is biased; but, this need not be so in fact.) Thus, inductive reasoning is typically subjective and intuitive, making appeals to one's personal experiences; whereas deductive reasoning is objective, rational, and independent of experience. Regarding the present context, a universal statement $\forall x,\ P(x)$ cannot be proved by merely producing finitely many examples x of $P(x)$ being true—*unless* the universe of things under discussion is finite, in which case an *exhaustive* verification is required. (On the other hand, as discussed in Section 1.3, a universal statement can be *disproved* by producing a single counterexample.)

1. Prove $P(0)$.
2. Prove that whenever $P(n)$ for a given $n \in \mathbb{N}$, then $P(n+1)$ as well.

Upon completing these two steps, we may conclude (4.44); because, it is an axiom of mathematical logic that

$$P(0) \wedge [\, \forall n \in \mathbb{N},\, P(n) \Rightarrow P(n+1)] \quad \Rightarrow \quad \forall n \in \mathbb{N},\, P(n) \tag{4.45}$$

(the reverse implication being obvious). The assertion that (4.45) holds for every unary predicate "P" is known as the **principle of mathematical induction**; more briefly, it is referred to as the **(weak) induction principle**. Also, while performing step 2, it is common to *assume* the truth of $P(n)$ for an arbitrary $n \in \mathbb{N}$, in order to show that $P(n)$ implies $P(n+1)$ for each such n; this assumption is called the **induction hypothesis**.

As an example, suppose

$$y_n = \sum_{i=0}^{n} i \quad (n \in \mathbb{N}). \tag{4.46}$$

Thus, we know from (4.36) that

$$y_n = \tfrac{1}{2}n(n+1) \quad (n \in \mathbb{N}). \tag{4.47}$$

To prove this fact by mathematical induction, we identify (4.47) as the property P in (4.44); that is, having $P(n)$ for a given $n \in \mathbb{N}$ simply means that the equality in (4.47) holds for that n. Next, we verify from (4.46) that (4.47) holds for $n = 0$, which is immediately evident. Finally, we assume (induction hypothesis) that the equality in (4.47) holds for a given $n \in \mathbb{N}$; then, it follows that

$$\begin{aligned} y_{n+1} &= \sum_{i=0}^{n+1} i = \sum_{i=0}^{n} i + (n+1) = y_n + (n+1) = \tfrac{1}{2}n(n+1) + (n+1) \\ &= \tfrac{1}{2}(n+1)(n+2), \end{aligned}$$

which also obtains from (4.47) upon replacing each "n" there with "$n+1$". Hence, for this property P, we have both $P(0)$ and $\forall n \in \mathbb{N},\, P(n) \Rightarrow P(n+1)$; therefore, we conclude from the induction principle (4.45) that (4.44) holds—i.e., as stated in (4.47), the equality there holds for *all* $n \in \mathbb{N}$.

Our two-step process makes intuitive sense because it can be viewed as proving the statements $P(n)$ ($n = 0, 1, 2, \ldots$) sequentially: first, we prove $P(0)$; then, we prove $P(1)$ by showing $P(0) \Rightarrow P(1)$; then, we prove $P(2)$ by showing $P(1) \Rightarrow P(2)$; and so on. As a matter of practicality, though, performing these infinitely many steps individually is not feasible; so instead, those after the first are effectively performed all at once via step 2 above. However, note that in the sequential interpretation, at the time we attempt to prove $P(n+1)$ for a particular n, we have not only proved $P(n)$ already, but $P(m)$ for all $m \in \mathbb{N}$ such that $m \leq n$; therefore, it is reasonable to suspect that we can strengthen the induction hypothesis, and modify step 2 thus:

2′. Prove that whenever $P(m)$ for all $m \in \mathbb{N}$ less than or equal to a given $n \in \mathbb{N}$, then $P(n+1)$ as well.

That is, the induction hypothesis is now the assumption that, for an arbitrary $n \in \mathbb{N}$, we have $P(n)$ for all $m \in \mathbb{N}$ such that $m \leq n$; and, based on this assumption, we prove that $P(n+1)$ as well, thereby satisfying step $2'$. This modification of the mathematical induction process is indeed justified, since it can be shown that an implication of the weak induction principle (4.45) is the **strong induction principle**:

$$P(0) \wedge \{\forall n \in \mathbb{N}, [\forall m \in \mathbb{N}, m \leq n \Rightarrow P(m)] \Rightarrow P(n+1)\} \quad \Rightarrow \quad \forall n \in \mathbb{N}, P(n) \tag{4.48}$$

(the reverse implication again holding). Conversely, replacing "$\leq$" in (4.48) with "$=$" yields a statement equivalent to (4.45); because, for each $n \in \mathbb{N}$ it is true that $[\forall m \in \mathbb{N}, m = n \Rightarrow P(m)] \Leftrightarrow P(n)$.

As is customary, the statement (4.44) to be proved, and the corresponding induction principles (4.45) and (4.48), have been expressed in terms of a property P defined on the set $\mathbb{N} = \{0, 1, \dots\}$ of natural numbers; but, the method of mathematical induction can be applied to any property defined on a set that has a "first" element (0, for $\mathbb{N}$) from which the remaining elements obtain by repeatedly performing some "successor" operation (adding 1, for $\mathbb{N}$). Examples: the set $\mathbb{Z}^+ = \{1, 2, \dots\}$, for which the first element is 1 and the successor operation is adding 1; the set $\{2, 4, 8, 16, \dots\}$, for which the first element is 2 and the successor operation is multiplying by 2; and, the set $\{-1, -2, -3, \dots\}$, for which the first element is -1 and the successor operation is subtracting 1. All that is required to prove via mathematical induction that $P(n)$ holds for all n in a set such as one of these is to make the obvious minor modifications in the two-step processes above.

As an example of applying the strong induction principle, consider the **Fibonacci numbers** f_n ($n = 1, 2, \dots$), which are generated sequentially by the equations[21]

$$f_1 = 1, \qquad f_2 = 1, \tag{4.49a}$$

$$f_n = f_{n-1} + f_{n-2} \quad (n \geq 3). \tag{4.49b}$$

Thus, $f_1 = 1, f_2 = 1, f_3 = 2, f_4 = 3, f_5 = 5, f_6 = 8$, etc. We now prove by mathematical induction that

$$f_n \leq 2^{n-1} \quad (n \geq 1). \tag{4.50}$$

First, note that $f_1 = 1 = 2^{1-1}$ and $f_2 < 2 = 2^{2-1}$; hence, the inequality in (4.50) holds for $n = 1, 2$. Now, arbitrarily fixing $n \geq 2$, assume (induction hypothesis) that

$$f_m \leq 2^{m-1} \quad (1 \leq m \leq n); \tag{4.51}$$

then, since $n + 1 \geq 3$, we can use (4.49b) with "n" replaced with "$n + 1$" to obtain

$$f_{n+1} = f_n + f_{n-1} \leq 2^{n-1} + 2^{n-2} = \tfrac{1}{2} \cdot 2^n + \tfrac{1}{4} \cdot 2^n = \tfrac{3}{4} \cdot 2^n \leq 2^n,$$

21. This is an example of a **recursive definition** (or **inductive definition**). That is, the sequence $\{f_n\}_{n=1}^{\infty}$ is determined by first specifying one or more **initial values**, as in (4.49a); then, the remaining terms f_n are found by performing a **recursion**, as in (4.49b), whereby each new value f_n is calculated using the previously calculated values $f_1, f_2, \dots, f_{n-1}$.

and thus the inequality in (4.51) also holds for $m = n + 1$. Therefore, using the strong induction principle (4.48)—here, modified for the set $\{2, 3, \dots\}$ in lieu of $\mathbb{N}$—we conclude that (4.50) holds.

Finally, when trying to prove the universal statement (4.44) for some sequence $\{y_n\}_{n=0}^{\infty}$ and property P, a direct application of mathematical induction might be difficult; sometimes, though, an indirect route via a *more general* property Q is easier. For example, suppose that we wish to prove that no prime number other than 3 appears in the sequence $\{y_n\}_{n=0}^{\infty}$ generated by (4.46); accordingly, we would define $P(n)$ ($n \in \mathbb{N}$) as

$$y_n = 3 \quad \vee \quad \neg(y_n \text{ is prime}). \tag{4.52}$$

However, although (4.44) is now true, proving it directly via (4.45) or (4.48) is difficult. But, suppose we can find another property Q on $\mathbb{N}$ that implies P—i.e., $\forall n \in \mathbb{N},\ Q(n) \Rightarrow P(n)$—where Q *can* be readily proved by mathematical induction; then, we obtain an indirect proof of (4.44) for P, since

$$[\forall n \in \mathbb{N},\ Q(n)] \wedge [\forall n \in \mathbb{N},\ Q(n) \Rightarrow P(n)] \quad \Rightarrow \quad \forall n \in \mathbb{N},\ P(n). \tag{4.53}$$

For the present example, one such property Q is provided by (4.47), which we have already proved by mathematical induction; because: From (4.47) we get $y_0 = 0$, $y_1 = 1$, and $y_2 = 3$; so, (4.52) holds for $n = 0, 1, 2$. Whereas for $n \geq 3$, either n or $n + 1$ is an even integer greater than or equal to 4; therefore, by (4.47), either $\frac{1}{2}n$ or $\frac{1}{2}(n + 1)$ is an integer greater than or equal to 2 that divides y_n; so, y_n is not prime, and we again have (4.52). Hence, Q implies P, so by (4.53) we may conclude (4.44).

CHAPTER 5

Functions

A **function** (or **mapping**) $f(\cdot)$ is a *rule* for converting a given mathematical object x—the **argument** of the function—into another object $f(x)$—the **value** of the function.[1] When we introduce another variable "y" and let $y = f(x)$, the value y then depends (potentially) on the value x; thus, we say that y is a function *of* x. Additionally, we say that x **maps to** y under the function $f(\cdot)$, which is commonly indicated by writing

$$x \overset{f}{\longmapsto} y$$

—or, simply

$$x \mapsto y$$

when the function performing the mapping is either understood or unnamed.

The rule that defines the function $f(\cdot)$ is applicable to every element x in a nonempty set D called the **domain** of the function; accordingly, we say that $f(\cdot)$ is a function *on* D. This is indicated by writing

$$f : D \to S,$$

where S is any set such that $f(x) \in S$ for all $x \in D$; hence, we say that the function $f(\cdot)$ **maps** D **into** S. The smallest such set S—viz., $\{f(x) \mid x \in D\}$—is called the **range** of the function; and, for this set E (but no other) we may also say that $f(\cdot)$ **maps** D **onto** E, since for each $y \in E$ there exists at least one $x \in D$ such that $y = f(x)$. By contrast, for each $x \in D$ there is *exactly* one $y \in E$ such that $y = f(x)$. Thus, a function may be defined as a nonempty many-to-one relation (see Section 2.1), the domain and range of which become the domain and range of the function. Two functions $f(\cdot)$ and $g(\cdot)$ are equal—i.e., $f(\cdot) = g(\cdot)$—if and only if they express the same relation; equivalently, the domains of $f(\cdot)$ and $g(\cdot)$ are the same set D, and $f(x) = g(x)$ $(x \in D)$.

1. As shown here, the notation "$f(\cdot)$" can be used to denote a function f having a single argument, with the *place* for that argument being indicated by a dot. This device avoids giving the argument a name—e.g., "x". On the other hand, naming the argument and then referring to the function as, e.g., "$f(x)$" is often convenient and usually harmless; however, it does risk confusing a *function* f with its *value* $f(x)$ for a *particular* argument value x.

For a **constant function**, the value $f(x)$ is the same for all $x \in D$; that is, the range E contains exactly one point. For an **identity function**, we have $f(x) = x$ $(x \in D)$, and therefore $E = D$. Also, although not necessary mathematically, in practice the domains and ranges of most functions are sets of numbers. In particular, for a **real-valued function** (or **real function**) we have $E \subseteq \mathbb{R}$, whereas for a **complex-valued function** (or **complex function**) $E \subseteq \mathbb{C}$. (Hence, every real-valued function is also a complex-valued function, but not vice versa.)

There are many ways by which a function can be specified. Often, one can simply state the mapping all at once for the entire domain; e.g., for a function $f\colon \mathbb{R} \to \mathbb{R}$ we might have

$$f(x) = 2\sin 3x - 4 \quad (x \in \mathbb{R}).$$

But, sometimes it is more convenient for a function to be **piecewise defined**, whereby the domain is partitioned into subsets, and a separate mapping is given for each; e.g.,

$$f(x) = \begin{cases} 0 & x \leq 1 \\ x^5 & x > 1 \end{cases}.$$

Occasionally, though, a mapping $x \mapsto y = f(x)$ is not *explicitly* provided, perhaps because it is difficult to express y directly in terms of x. Instead, the function $f(\cdot)$ might be **implicitly defined** by giving its domain D and range E, along with a condition to be satisfied by pairs $(x, y) \in D \times E$; then, it is proved that the corresponding relation on $D \times E$ is both nonempty and many-to-one. For example, it is easily shown that for each $x > 0$ there is exactly one $y \in \mathbb{R}$ such that

$$x = e^y + e^{6y}; \tag{5.1}$$

hence, there exists a unique function $f\colon (0, \infty) \to \mathbb{R}$ such that $x = \exp[f(x)] + \exp[6f(x)]$ for each $x \in (0, \infty)$.

Sometimes, given two functions $f(\cdot)$ and $g(\cdot)$, we let

$$h(x) = g[f(x)] \quad (x \in A)$$

to form a **composite function** $h(\cdot)$, the domain A of which can be any nonempty subset of the domain of $f(\cdot)$. That is, $h(\cdot)$ is *a* **composition** of the functions $f(\cdot)$ and $g(\cdot)$—here with $f(\cdot)$ being applied first to the argument x of $h(\cdot)$, then $g(\cdot)$ being applied to the result $f(x)$ to produce $h(x)$.[2] This composition is possible only if the **image** of A under $f(\cdot)$—i.e., the set[3]

$$f(A) \triangleq \{f(x) \mid x \in A\} \tag{5.2a}$$

2. It is generally not correct to speak of *the* composition of two given functions, since the order in which they are applied might be significant (and usually is).

3. Note the abuse of the notation "$f(x)$" being performed here upon writing "$f(A)$", which is commonly done; for whereas x is an argument value, A is a *set* of such values. A similar abuse occurs below with regard to the "inverse image" of a function versus its "inverse".

—is a subset of the domain of $g(\cdot)$. In particular, note that the range E of any function $f(\cdot)$ is the image of its domain D: $E = f(D)$.

Comparable to (5.2a), but in the opposite direction, the **inverse image** (or **pre-image**) of a set B, under a function $f(\cdot)$, is the set

$$f^{-1}(B) \triangleq \{x \in D \mid f(x) \in B\}, \tag{5.2b}$$

where D is the domain of $f(\cdot)$. That is, $f^{-1}(B)$ is the set of elements in the domain D of $f(\cdot)$ that map into B; equivalently, $f^{-1}(B)$ is the largest set $A \subseteq D$ for which $f(A) \subseteq B$. In particular, the inverse image under $f(\cdot)$ of its range E is its domain D: $f^{-1}(E) = D$.

Although for each $x \in D$ there is exactly one $y \in E$ such that $y = f(x)$, there may be more than one $x \in D$ yielding this equality for a given $y \in E$; because, the mapping performed by a function is many-to-one. If, however, for each $y \in E$ there is only *one* such $x \in D$ for which $y = f(x)$, then the function $f(\cdot)$ is called **one-to-one**. Only such functions are **invertible**; that is, only for a one-to-one function $f(\cdot)$ does there exist a corresponding **inverse function** $f^{-1}(\cdot)$, defined as follows:

$$\text{For each } y \in E, \quad f^{-1}(y) \quad \text{is the unique} \quad x \in D \quad \text{for which} \quad y = f(x).$$

(Note that although inverse functions sometimes do not exist, inverse images always do.) More briefly,

$$y = f(x) \quad \Rightarrow \quad f^{-1}(y) = x$$

$(x \in D, y \in E)$; therefore, $f^{-1}[f(x)] = x\ (x \in D)$. It then follows that the function $f^{-1}(\cdot)$ has domain E and range D; that is, $f^{-1}(\cdot)$ maps E *onto* D. Moreover, $f^{-1}(\cdot)$ is also one-to-one, and thus is itself invertible; in fact, *its* inverse is $f(\cdot)$, since $f[f^{-1}(y)] = y$ $(y \in E)$.

If a function $f(\cdot)$ has a domain D that can be expressed as a Cartesian product—i.e., $D = D_1 \times \cdots \times D_n$ for $n \geq 2$ nonempty sets D_i $(i = 1, \ldots, n)$—then $f(\cdot)$ can be viewed as a function $f(\cdot, \ldots, \cdot)$ of *several* variables. Namely, we let

$$f(x_1, \ldots, x_n) = f(x) \quad (x_1 \in D_1, \ldots, x_n \in D_n),$$

where $x = (x_1, \ldots, x_n)$. For example, the **maximum function** $\max(\cdot, \ldots, \cdot)$ and **minimum function** $\min(\cdot, \ldots, \cdot)$ are defined on $\overline{\mathbb{R}}^n$ by

$$\max(x_1, \ldots, x_n) \triangleq \max\{x_1, \ldots, x_n\}, \quad \min(x_1, \ldots, x_n) \triangleq \min\{x_1, \ldots, x_n\}$$

$(x_1, \ldots, x_n \in \overline{\mathbb{R}})$, where $\{x_1, \ldots, x_n\}$ is the set comprising $x_1, \ldots, x_n$.

Many of the properties defined for sequences in Chapter 4 extend to functions.[4] A real-valued function $f(\cdot)$ having a domain D is called **positive**, **negative**, **nonpositive**, or **nonnegative** if and only if all of its values $f(x)$ $(x \in D)$ are numbers of that type. If $f(\cdot)$ is real-valued and $D \subseteq \mathbb{R}$, then it is **increasing** (or **nondecreasing**) if and only if

$$f(x_1) \leq f(x_2) \quad (x_1, x_2 \in D \text{ with } x_1 < x_2);$$

4. Indeed, by letting $f(i) = x_i$ (all i), any sequence $\{x_i\}$ becomes a *function* $f(\cdot)$ on a subset of $\mathbb{Z}$; e.g., a sequence $\{x_i\}_{i=1}^{\infty}$ is a essentially function on $\mathbb{Z}^+$.

whereas it is **strictly increasing** if and only if this inequality can be made strict. Likewise, $f(\cdot)$ is **decreasing** (or **nonincreasing**) if and only if

$$f(x_1) \geq f(x_2) \quad (x_1, x_2 \in D \text{ with } x_1 < x_2);$$

whereas it is **strictly decreasing** if and only if this inequality can be made strict. Hence, a function that is both increasing and decreasing must be constant. A function that is either increasing or decreasing is called **monotonic**; and, a function that is either strictly increasing or strictly decreasing is **strictly monotonic**.[5] Every strictly increasing (respectively, strictly decreasing) function is invertible, and the inverse function is also strictly increasing (strictly decreasing).

Given a real-valued function $f(\cdot)$ having a domain D, the **maximum** and **minimum** of $f(\cdot)$ over a set $S \subseteq D$ are, respectively,

$$\max_{x \in S} f(x) \triangleq \max\{f(x) \mid x \in S\}, \quad \min_{x \in S} f(x) \triangleq \min\{f(x) \mid x \in S\}$$

(when each exists); similarly, the **supremum** and **infimum** of $f(\cdot)$ over S are, respectively,

$$\sup_{x \in S} f(x) \triangleq \sup\{f(x) \mid x \in S\}, \quad \inf_{x \in S} f(x) \triangleq \inf\{f(x) \mid x \in S\}$$

(which always exist). For the frequently occurring case of $S = D$, the left sides of these definitions may be more succinctly written

$$\max_x f(x), \quad \min_x f(x), \quad \sup_x f(x), \quad \inf_x f(x),$$

respectively. When $\max_x f(x)$ exists, it is the unique **(global) maximum** of the function $f(\cdot)$; likewise, $\min_x f(x)$ is the unique **(global) minimum**, should it exist. (Of course, although a function can have at most one global maximum and at most one global minimum, either of these values can occur at multiple points in the function's domain.) A complex-valued function $f(\cdot)$ having a domain D is **bounded** if and only if

$$|f(x)| \leq M \ (x \in D) \quad \text{for some} \quad M > 0,$$

which is the case if and only if $\sup_x f(x)$ and $\inf_x f(x)$ are both finite.

Given a function $f : D \to S$ and a nonempty set $D' \subseteq D$, we may define another function $g : D' \to S$ via the *same* rule; namely,

$$g(x) = f(x) \quad (x \in D').$$

We then refer to $g(\cdot)$ as the **restriction** of $f(\cdot)$ to D'; conversely, $f(\cdot)$ is an **extension** of $g(\cdot)$. Many, though not all, function properties are preserved upon restriction. For example, a noninvertible function might become invertible; e.g., the function $f(x) =$

5. As with sequences, it is not uncommon here to see the words "increasing" and "decreasing" preceded by either "monotone" or "monotonically", though this is unnecessary.

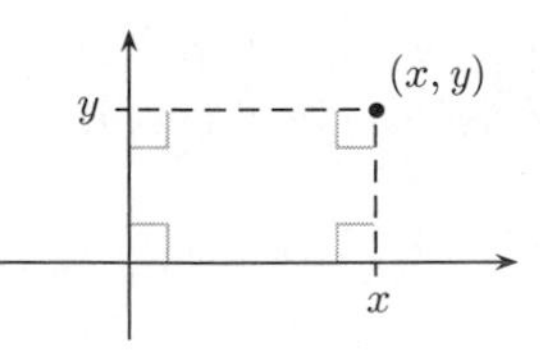

FIGURE 5.1 The Plot of a Point

$\sin x$ defined for $x \in \mathbb{R}$ is not invertible, whereas the function $g(x) = \sin x$ defined for $x \in [-\pi/2, \pi/2]$ is.

5.1 Plots

It is often helpful to display a function—more generally, a relation—*visually* as a **plot** (or **graph**). Here, we focus on two-dimensional plots—i.e., plots in the plane $\mathbb{R}^2$—though similar remarks apply to plots in the three-dimensional space $\mathbb{R}^3$.

To begin, Figure 5.1 shows the *rectangular* plot of a single point, as given by an ordered pair $(x, y) \in \mathbb{R}^2$. By using two perpendicular **axes**—a **horizontal axis** and a **vertical axis**—each acting as a real line, the real values x and y are converted into unique positions on these respective axes; then, moving orthogonally to the axes, these two positions combine to locate the point (x, y). Conversely, each point in the plane represents a unique element of $\mathbb{R}^2$. Accordingly, the values x and y are referred to as the **rectangular** (or **Cartesian**) **coordinates** of the point,[6] with the horizontal position x being the **abscissa** and vertical position y being the **ordinate**.[7] By convention, the horizontal axis is usually directed to the right and the vertical axis directed upward; moreover, though it is sometimes convenient to do otherwise, the axes are normally drawn to intersect at the **origin** of the plane—i.e., the point $(0, 0)$. The plane is divided by the axes into four **quadrants**, numbered counterclockwise I through IV, as done in Figure 3.2 for the complex plane.

To plot a real-valued function $f(\cdot)$ having domain $D \subseteq \mathbb{R}$, we simply plot the point $\big(x, f(x)\big)$ for each value $x \in D$; and, the resulting plane figure is generally called a **curve** (whether or not it has a curved shape). Stated another way, let $y = f(x)$, thereby making the value y uniquely determined by the value x; then, as indicated in Figure 5.2, we plot a **dependent variable** "y" on the vertical axis *versus* an **independent variable** "x" on the horizontal axis (not x versus y). Also, with the axes so labeled, one may appropriately

---

6. For a more thorough discussion of coordinate systems, including "polar coordinates" and some three-dimensional systems, see Section 11.1.

7. It is not uncommon to see the horizontal and vertical *axes* themselves referred to as the "abscissa" and "ordinate", respectively.

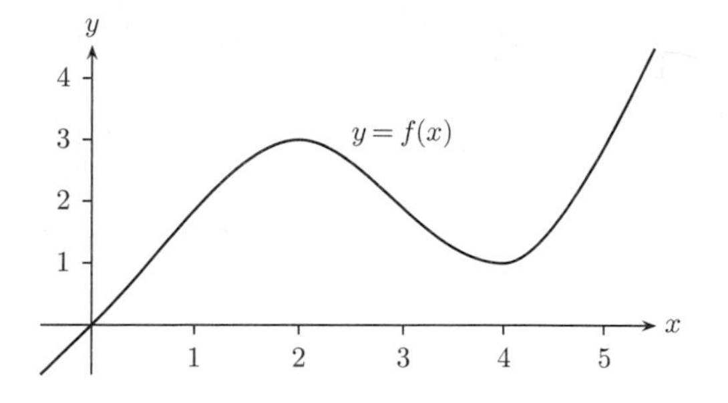

FIGURE 5.2 The Plot of a Function

refer to the horizontal axis as the ***x*-axis**, and the vertical axis as the ***y*-axis**, with the plane on which the plot is drawn now being the ***xy*-plane**.[8]

The function $f(\cdot)$ is said to have a **local maximum** of value $f(X)$ at $X \in D$ if and only if for all $x \in D$ sufficiently close to X, the value $f(x)$ does not exceed $f(X)$; that is,

$$\exists \varepsilon > 0,\ \forall x \in D,\ |x - X| \leq \varepsilon \Rightarrow f(x) \leq f(X).$$

Likewise, $f(\cdot)$ has a **local minimum** of value $f(X)$ at $X \in D$ if and only if

$$\exists \varepsilon > 0,\ \forall x \in D,\ |x - X| \leq \varepsilon \Rightarrow f(x) \geq f(X).$$

Thus, the function plotted in Figure 5.2 has a local maximum of 3 at 2, and a local minimum of 1 at 4, neither of which is global. Of course, a global maximum (respectively, global minimum) must also be a local maximum (local minimum); and, a function may have several local maxima (or local minima), and possibly none. Collectively, maxima and minima (global or local) are referred to as **extrema**.[9]

As shown in the upper right of Figure 5.3, to translate a plot of $f(x)$ versus x (indicated in gray) upward by an amount Y, we instead plot $f(x) + Y$ versus x (indicated in black); whereas plotting $f(x) - Y$ produces a downward translation by Y. Similarly (but perhaps less obviously), to translate a plot of $f(x)$ versus x to the right by an amount X, we instead plot $f(x - X)$ versus x; whereas plotting $f(x + X)$ produces a translation to the left by X.

If a translation to the left or right causes the original plot to reappear, then that plot must be repetitive horizontally. Accordingly, a function $f(\cdot)$ having domain $D \subseteq \mathbb{R}$ is called a **periodic function** if and only if there is a value $X > 0$— called *a* **period** of the function—such that

$$f(x) = f(x + X) \quad (x \in D) \tag{5.3}$$

---

8. Of course, other variables can be used as well for the coordinates, in which case the axes and plane are renamed accordingly.

9. Everything said in this paragraph also applies to any real function $f(\cdot)$ having a domain $D \subseteq \mathbb{C}$.

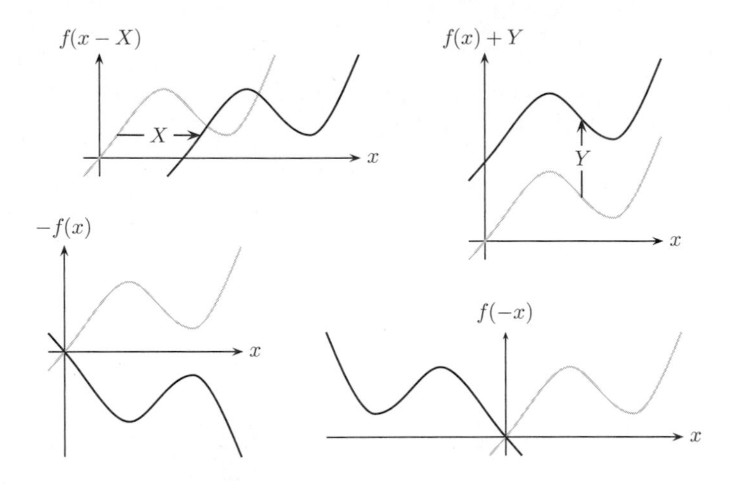

FIGURE 5.3 Plot Translations and Axis Inversions

(replacing the "+" with "−" yields an equivalent condition).[10] Stated succinctly, $f(\cdot)$ is an "X-periodic" function (e.g., as is evident from Figure 11.8, sine and cosine are 2π-periodic functions). Furthermore, the smallest such value X, if there is one, is commonly referred to as *the* period of $f(\cdot)$, or *its* period.[11]

As also shown in Figure 5.3, a plot of $-f(x)$ versus x obtains from a plot of $f(x)$ versus x by inverting the plot along the vertical axis, thereby yielding a mirror image with respect to the horizontal axis; whereas a plot of $f(-x)$ versus x obtains from a plot of $f(x)$ versus x by inverting along the horizontal axis, thereby yielding a mirror image with respect to the vertical axis. When

$$f(x) = f(-x) \quad (x \in D), \tag{5.4a}$$

$f(\cdot)$ is said to possess **even symmetry**, and is called an **even(-symmetric) function**; whereas when

$$f(x) = -f(-x) \quad (x \in D), \tag{5.4b}$$

$f(\cdot)$ has **odd symmetry** and is an **odd(-symmetric) function.**[12] (For example, sine and cosine are odd and even functions, respectively.)

To contract a plot of $f(x)$ versus x vertically, we instead plot $af(x)$ versus x for some constant a such that $0 < a < 1$, with greater contraction being obtained for smaller a; whereas for $a > 1$ a vertical expansion is obtained, the more so for larger a. Similarly

---

10. Implicitly, (5.3) requires D and X to be such that $x + X \in D$ for each $x \in D$, in order for the right side of the equation to exist; it then follows that $x + kX \in D$ $(k = 0, \pm 1, \pm 2, \ldots)$ for each $x \in D$. For example, having a domain $D = \mathbb{R}$ is allowed, in which case any period $X \in \mathbb{R}$ is possible; whereas for $D = \mathbb{Z}$, only $X \in \mathbb{Z}$ is allowed.

11. Every periodic function $f(\cdot)$ has, in fact, *infinitely* many periods; because, (5.3) implies that $f(x) = f(x + X) = f(x + 2X) = f(x + 3X) = \cdots$ (all x), making $X, 2X, 3X, \ldots$ all periods of $f(\cdot)$.

12. Implicitly, (5.4a) and (5.4b) each require the domain $D \subseteq \mathbb{R}$ to be such that $-x \in D$ for each $x \in D$.

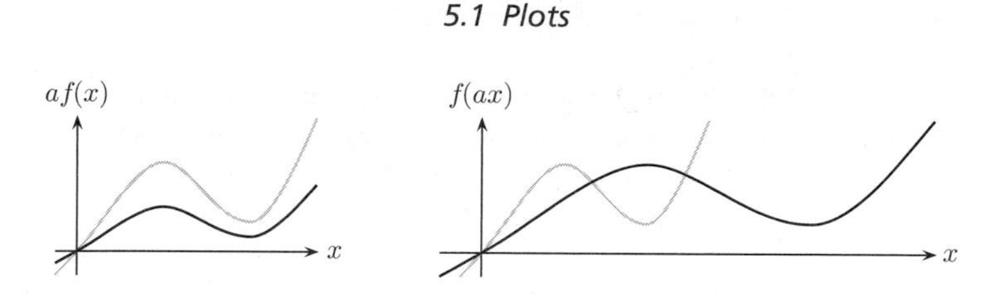

FIGURE 5.4 Plot Contraction (Vertically) and Expansion (Horizontally)

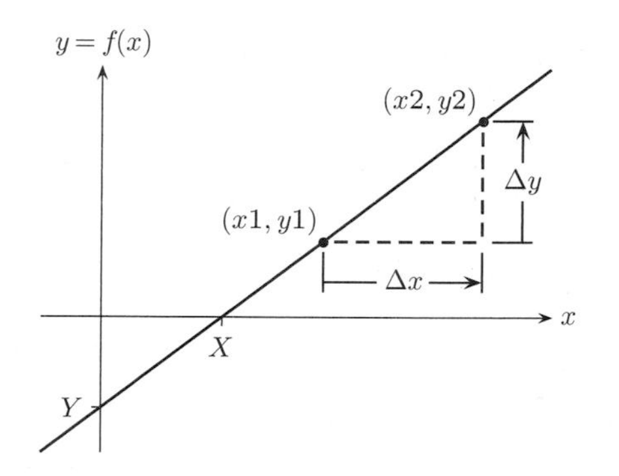

FIGURE 5.5 The Plot of a Line

(but with the cases reversed vis-à-vis a), to horizontally expand a plot of $f(x)$ versus x, we instead plot $f(ax)$ versus x for $0 < a < 1$, with greater expansion being obtained for smaller a; whereas for $a > 1$ a horizontal contraction is obtained, the more so for larger a. These transformations are shown in Figure 5.4 (with the original plots in gray and the resulting plots in black) for $a = \frac{1}{2}$.

The kind of function $f : \mathbb{R} \to \mathbb{R}$ most frequently plotted is that producing a straight line. As shown in Figure 5.5 for a plot of $y = f(x)$ versus x, the line has **slope**

$$m = \frac{\Delta y}{\Delta x} = \frac{y_2 - y_1}{x_2 - x_1}, \tag{5.5}$$

for any two distinct points (x_1, y_1) and (x_2, y_2) on the line.[13] A positive slope indicates that $f(\cdot)$ is strictly increasing, and a negative slope that $f(\cdot)$ is strictly decreasing. For a horizontal line—i.e., $\Delta y = 0$ for any $\Delta x \neq 0$—y is a constant function of x, and $m = 0$. However, for a vertical line—i.e., $\Delta x = 0$ for any $\Delta y \neq 0$—y cannot be expressed as a function of x, and the slope m of the line is undefined because of division by 0 in (5.5).[14]

13. As just done, the uppercase Greek letter delta (Δ) is often used to indicate a variation (i.e., change) in the value of the variable that follows.

14. Qualitatively, the slope of a vertical line may be safely described as "infinite"; however, a problem arises when such a slope is expressed numerically, since then one must somehow choose appropriately between $m = \infty$ and $m = -\infty$. The *general* equation for a line (possibly vertical) in the xy-plane is discussed Subsection 11.5.1; see (11.88).

Using the slope of the line, along with its **intercept** of the y-axis—i.e., the value Y on the vertical axis where the line crosses—we can express the function $f(\cdot)$ in the form

$$f(x) = mx + Y \quad (\text{all } x). \tag{5.6a}$$

Since $Y = f(0)$, the y-intercept Y always exists and is unique. However, when $m = 0$, there might not be a value X for which $f(X) = 0$; thus, an x-intercept X—i.e., a value on the horizontal axis where the line crosses—need not exist. But, if $m \neq 0$, then there *is* a unique x-intercept X, and we also have

$$f(x) = m(x - X) \quad (\text{all } x). \tag{5.6b}$$

Whenever an x-intercept $X \neq 0$ exists, $m = -Y/X$. Finally, if $f(\cdot)$ expresses a line of some slope m passing through a point (x_1, y_1), then

$$f(x) = m(x - x_1) + y_1 \quad (\text{all } x). \tag{5.6c}$$

In particular, (5.6c) becomes (5.6a) when $(x_1, y_1) = (0, Y)$, and (5.6b) when $(x_1, y_1) = (X, 0)$.

In general, when one variable quantity y is related to another variable quantity x by the equation $y = mx$ for some constant $m > 0$, we say that y is **proportional** to x. Thus, this relationship—written[15]

$$y \propto x,$$

thereby avoiding mention of the **proportionality constant** (or **constant of proportionality**) m, which might be unknown—is represented in the xy-plane by a line of positive slope passing through the origin. Likewise, we say that y is **inversely proportional** to x when $y \propto 1/x$ (which assumes x never equals zero).

Given a rectangular plot of a function $f(\cdot)$, one can easily obtain a plot of its inverse $f^{-1}(\cdot)$, when it exists. On each pair of axes in Figure 5.6, a plot of $y = f(x)$ versus x is shown for a particular function $f(\cdot)$, with the function on the left being taken from Figure 5.2. As indicated on the left plot, since the same unit distance has been chosen for the horizontal and vertical axes, reflecting the curve for $f(\cdot)$ about the line described by the equation $y = x$ yields a plot of all points (x, y) for which $f(y) = x$; moreover, should the inverse $f^{-1}(\cdot)$ exist, as is the case for the function $f(\cdot)$ on the right, then this new plot is also a plot of $f^{-1}(x)$ versus x.[16] (More simply, upon plotting $y = f(x)$ versus x on a transparent sheet, a plot of $f^{-1}(y)$ versus y obtains by flipping the sheet over and rotating it to make the y-axis point to the right; furthermore, this method works regardless of whether the axes have the same unit distance.) That $f^{-1}(\cdot)$ does not exist on the left side of Figure 5.6 is evident because a vertical line can be drawn on the plot that intersects the curve for the equation $f(y) = x$ at more than one point; equivalently,

15. Note the difference between the proportional-to symbol "$\propto$" and the lowercase Greek letter alpha, "α".
16. Several examples of applying this method appear in this book, including: converting the plots in Figure 6.1 of x^n (restricted to $x \geq 0$) to the plots in Figure 6.3 of $\sqrt[n]{x}$ ($n = 1, 2, 3$); converting in Figure 7.1 the plot of $\exp x$ to the plot of $\ln x$; and, converting the plots in Figure 11.8 of $\sin\theta$ (restricted to $-\pi/2 \leq \theta \leq \pi/2$) and $\cos\theta$ (restricted to $0 \leq \theta \leq \pi$) to the plots in Figure 11.11 of $\sin^{-1} x$ and $\cos^{-1} x$.

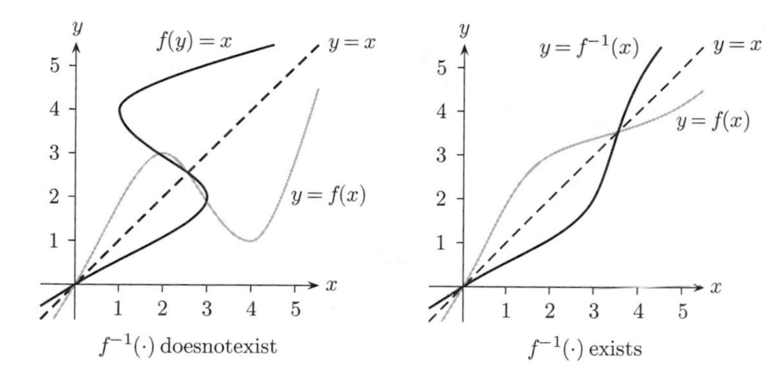

FIGURE 5.6 Inverting a Function Graphically

a horizontal line can be drawn on the plot in Figure 5.2 that intersects the curve for $f(\cdot)$ at more than one point.

Plotting a real-valued function $f(\cdot)$ having domain $D \subseteq \mathbb{R}$ is actually a special case of plotting a binary relation $R \subseteq \mathbb{R}^2$; because, upon letting $R = \{(x, y) \in \mathbb{R}^2 \mid y = f(x)\}$, the points (x, y) in the xy-plane for which $y = f(x)$ are precisely the points $(x, y) \in R$. A simple example of plotting a relation that is not a function appears on the left side of Figure 5.6; for although the curve representing the equation $f(y) = x$ is a plot of the relation $R = \{(x, y) \in \mathbb{R}^2 \mid f(y) = x\}$, it is not a plot of a function of x, since $f^{-1}(\cdot)$ does not exist. (Other such examples are the plots in Figures 11.23–11.26 in the discussion of the conic sections, Subsection 11.5.2.) But, unlike the curves obtained when one is plotting functions, plots of relations (especially those expressed in terms of inequalities) are often *regions*. For example, Figure 5.7 shows a plot of the relation

$$R = \{(x, y) \in \mathbb{R}^2 \mid x < 10,\ 120y < 11x^3 - 179x^2 + 758x,\ 3x + 5y > 40\}, \tag{5.7}$$

which we obtain by intersecting the regions to the left of the line for the equation $x = 10$, below the curve for the cubic equation $120y = 11x^3 - 179x^2 + 758x$, and above the line for the equation $3x + 5y = 40$. Hence, for this relation we see that the plotted region happens to have two disconnected parts.

In all of the plots considered thus far (as well as for the real line in Figure 3.1), each axis has had a **linear scale** by which to convert real values into corresponding positions on the axis; such axes are generally assumed except when explicitly indicated otherwise. Specifically, letting $d(x_1, x_2)$ be the distance by which a point $x_2 \in \mathbb{R}$ lies to the right of a point $x_1 \in \mathbb{R}$—therefore, $d(x_1, x_2)$ being *negative* means that x_2 lies a distance $-d(x_1, x_2)$ to the *left* of x_1—the distance by which a point $x \in \mathbb{R}$ lies to the right of the point corresponding to 0 is

$$d(0, x) = \kappa x \quad (x \in \mathbb{R}), \tag{5.8}$$

for some fixed **scale factor** $\kappa > 0$; and, likewise for distances measured vertically along the y-axis. (As illustrated by Figure 5.2, the same scale factor need not be used for all axes of a plot.) Alternatively, when an axis is used to convert only *positive* numbers to

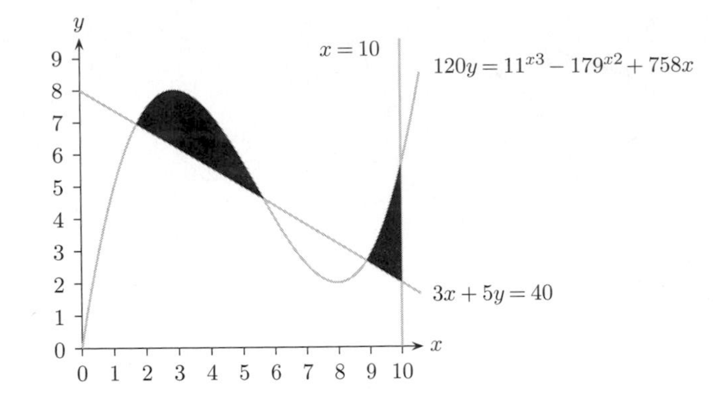

FIGURE 5.7 The Plot of a Relation

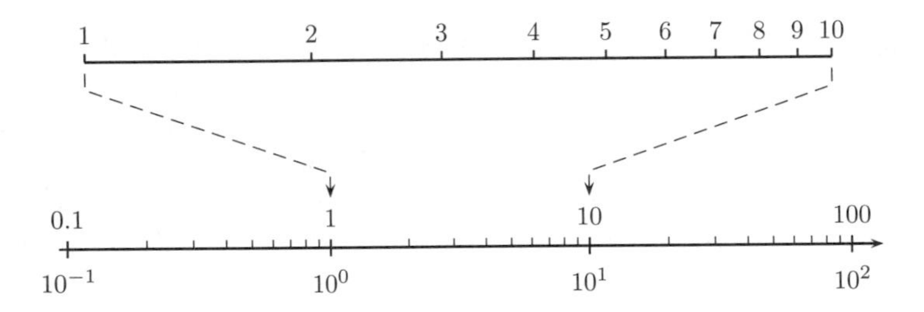

FIGURE 5.8 A Logarithmic Scale

positions, it may be more appropriate to perform the conversion via a **logarithmic scale**; specifically, for a logarithmic x-axis, the distance by which a point $x > 0$ lies to the right of the point corresponding to 1 (there is no point corresponding to 0) is

$$d(1,x) = \kappa \log_b x \quad (x > 0), \tag{5.9}$$

for some fixed constant $\kappa > 0$ and logarithm base $b > 1$.[17] As shown in Figure 5.8, a logarithmic scale can be viewed as a concatenation of **cycles**, each extending from 10^i to 10^{i+1} for some integer i, with the numbers $j \cdot 10^i$ ($j = 2, 3, \ldots, 9$) appearing at the next level of detail.

For a linear scale, the distance between any two numbers x_1 and x_2 depends only on their *difference* $x_2 - x_1$; because, $d(x_1,x_2) = d(0,x_2) - d(0,x_1)$, so (5.8) yields

$$d(x_1,x_2) = \kappa(x_2 - x_1) \quad (x_1, x_2 \in \mathbb{R}). \tag{5.10}$$

By contrast, for a logarithmic scale, the distance between any two numbers x_1 and x_2 depends only on their *ratio* x_2/x_1; because, $d(x_1,x_2) = d(1,x_2) - d(1,x_1)$,

17. The pair (κ, b) is not unique; rather, it follows from (7.8) that any pair (κ', b') for which $\kappa' = \kappa \log_b b'$ produces the same values $d(1,x)$ in (5.9). Accordingly, for a given logarithmic scale, one may choose $b > 1$ as desired (a popular choice being 10), from which the value of κ is determined.

so (5.9) yields

$$d(x_1, x_2) = \kappa \log_b(x_2/x_1) \quad (x_1, x_2 > 0). \tag{5.11}$$

In particular, as can be seen in Figure 5.8, two numbers differing by a factor of 10—i.e., a **decade**—are separated by the same distance as any other such pair of numbers. Likewise, a factor of 2—i.e., an **octave**—corresponds to the same separation distance anywhere along the scale.

In general, for any two numbers $x_1, x_2 > 0$, we say that x_2 is $\log_{10}(x_2/x_1)$ *decades larger* than x_1—even if x_2/x_1 is not an integer power of 10—because,

$$x_2 = x_1 \cdot 10^{\log_{10}(x_2/x_1)}.$$

(Thus, e.g., 987 is $\log_{10}(987/65) \approx 1.18$ decades larger than 65.) Likewise, x_2 is $\log_2(x_2/x_1)$ *octaves larger* than x_1, because

$$x_2 = x_1 \cdot 2^{\log_2(x_2/x_1)}.$$

(Thus, e.g., 987 is $\log_2(987/65) \approx 3.92$ octaves larger than 65.) For quick conversions between these two measures, it is useful to observe that there are $\log_{10} 2 \approx 0.3$ decades per octave, and $\log_2 10 \approx 3.3$ octaves per decade (thus, e.g., $2^{40} \approx 10^{0.3 \cdot 40} = 10^{12}$).

Finally, it is often convenient to plot *several* functions $f_1(\cdot), \ldots, f_n(\cdot)$ on a single pair of axes; that is, a **family** of $n \geq 2$ functions $f_i(\cdot)$ $(i = 1, \ldots, n)$ is plotted *versus* a **parameter** i—which is itself a variable, in addition to those associated with the axes. Later, for example, Figure 6.1 shows a plot versus i of the family of functions $f_i(\cdot)$ $(i = 1, 2, 3)$, given $f_i(x) = x^i$ $(i = 1, 2, 3; x \in \mathbb{R})$, and where for each i the value $f_i(x)$ is plotted versus $x \in \mathbb{R}$. Such a plot, though, should not be confused with a **parametric plot**, which again is a plot of a *single* curve. In this case, the point (x, y) is determined by a pair of **parametric equations**,

$$x = f(t), \quad y = g(t), \tag{5.12}$$

for some real-valued functions $f(\cdot)$ and $g(\cdot)$; specifically, the point is plotted versus a parameter $t \in D$, where D is a subset of the domains of both functions. For example, with $f(t) = \cos t$ and $g(t) = \sin t$, letting t vary over $D = [0, 2\pi)$ causes the point (x, y) to trace out a circle of unit radius centered at the origin (as in Figure 11.24 with $a = 1$)—a curve that is impossible to obtain by plotting $y = h(x)$ versus x for a *single* function $h(\cdot)$.

5.2 Limits and Continuity

The notion of a "limit", discussed in Section 4.1 for an infinite sequence of numbers, extends naturally to any function $f : D \to \mathbb{C}$ having a domain $D \subseteq \mathbb{C}$. First, we say that $f(x)$ **converges** as $x \to X$, for a particular $X \in D$, if and only if there exists a value $Y \in \mathbb{C}$ such that $f(x)$ becomes arbitrarily close to Y by taking $x \in D - \{X\}$ sufficiently close to X; that is,

$$\forall \varepsilon > 0, \exists \delta > 0, \forall x \in D, 0 < |x - X| \leq \delta \Rightarrow |f(x) - Y| \leq \varepsilon. \tag{5.13}$$

We then say that $f(x)$ converges *to* Y, or that $f(x)$ *approaches* Y, or that the **limit** Y of $f(x)$ *exists*, as $x \to X$; and, this is indicated by writing one of the following three expressions:

$$f(x) \to Y \text{ as } x \to X, \quad f(x) \xrightarrow[x \to X]{} Y, \quad \lim_{x \to X} f(x) = Y. \tag{5.14}$$

Otherwise, $f(x)$ **diverges** as $x \to X$.

We can also have $X = \pm\infty$. Namely, $f(x) \to Y$ as $x \to \infty$ if and only if $f(x)$ becomes arbitrarily close to Y by taking $x \in D$ to be a sufficiently large real value—i.e.,

$$\forall \varepsilon > 0, \exists \eta > 0, \forall x \in D, x \geq \eta \Rightarrow |f(x) - Y| \leq \varepsilon \tag{5.15}$$

(cf. (4.1))—and similarly for $x \to -\infty$. Or, if $f(\cdot)$ is real valued, then $Y = \pm\infty$ is possible; namely, we have $f(x) \to \infty$—i.e., $f(x)$ diverges *to* ∞—as $x \to X$ (X now finite) if and only if

$$\forall \xi > 0, \exists \delta > 0, \forall x \in D, 0 < |x - X| \leq \delta \Rightarrow f(x) \geq \xi; \tag{5.16}$$

and, similarly for $f(x)$ diverging to $-\infty$. In contexts where infinite limits have been explicitly allowed, we may again say that the limit *exists*. Finally, combining the two situations in the obvious way, we can have *both* $X = \pm\infty$ and $Y = \pm\infty$.

In each of the above cases, it is typical for the domain D and the value $X \in \mathbb{C} \cup \{\pm\infty\}$ to be such that X can be approached by *other* values in D; that is, X is a **limit point** of D:

$$\text{There exists a sequence} \quad \{x_i\}_{i=1}^{\infty} \quad \text{such that} \tag{5.17a}$$

$$x_i \in D - \{X\} \text{ (all } i), \quad \lim_{i \to \infty} x_i = X. \tag{5.17b}$$

(We need not have $X \in D$, though.) In particular, for $D = \mathbb{C}$ this condition holds for all $X \in \mathbb{C} \cup \{\pm\infty\}$, and for $D = \mathbb{R}$ it holds for all $X \in \overline{\mathbb{R}}$; but, e.g., for $D = \mathbb{Z}$ it holds only for $X = \pm\infty$. A benefit of (5.17) is that when it holds, the value of the limit $\lim_{x \to X} f(x)$ is unique, should that limit exist (infinite limits being allowed throughout this paragraph);[18] moreover, we then have

$$\lim_{x \to X} f(x) = \lim_{i \to \infty} f(x_i) \tag{5.18}$$

for *every* sequence $\{x_i\}_{i=1}^{\infty}$ satisfying (5.17b). Conversely, if the limit on the right side of (5.18) exists and has the same value for all sequences $\{x_i\}_{i=1}^{\infty}$ satisfying (5.17b)—again, assuming that there is at least one—then the limit on the left also exists and (5.18) obtains. It follows that when (5.17) holds, the limit properties of sequences given in (4.5)–(4.7) generalize in the obvious way to analogous properties for functions; for example, given

18. By contrast, if $X \in D = \mathbb{Z}$, then by (5.13) we have $\lim_{x \to X} f(x) = Y$ for *any* $Y \in \mathbb{C}$ (i.e., the limit is not unique); because, given $\varepsilon > 0$ we can take $\delta = \frac{1}{2}$, and then it is vacuously true that $|f(x) - Y| \leq \varepsilon$ for all $x \in D$ such that $0 < |x - X| \leq \delta$, since no such values x exist.

functions $f(\cdot)$ and $g(\cdot)$ having the same domain, of which X is a limit point, it follows from (4.5a) that

$$\lim_{x\to X}[f(x)+g(x)] = \lim_{x\to X} f(x) + \lim_{x\to X} g(x)$$

when the two limits on the right both exist, unless one equals ∞ and the other equals $-\infty$. Additionally, the definitions given in Section 4.1 of **iterated limit** and **multiple limit**, along with the corresponding notation, generalize naturally to apply to functions.

Of particular interest is the case of $f(\cdot)$ having a domain $D \subseteq \mathbb{R}$. Here, for each $X \in D$ we also define the **limit from the right** of $f(\cdot)$ at X—denoted

$$f(X+) \quad \text{or} \quad \lim_{x\to X+} f(x) \quad \text{or} \quad \lim_{x\downarrow X} f(x)$$

—as the limit Y of $f(x)$ as $x \to X$ when only values $x > X$ are allowed. Thus, e.g., if Y is finite, then

$$\forall\varepsilon > 0,\ \exists\delta > 0,\ \forall x \in D,\ X < x \leq X + \delta \Rightarrow |f(x) - Y| \leq \varepsilon \tag{5.19}$$

(cf. (5.13)). Similarly, the **limit from the left** of $f(\cdot)$ at X—denoted

$$f(X-) \quad \text{or} \quad \lim_{x\to X-} f(x) \quad \text{or} \quad \lim_{x\uparrow X} f(x)$$

—is defined to be the limit of $f(x)$ as $x \to X$ when only values $x < X$ are allowed. (For example, for the function signum defined in (3.16) we have $\operatorname{sgn} 0+ = 1$ and $\operatorname{sgn} 0- = -1$.) Equivalently, $f(X+) = \lim_{x\to X} g(x)$ where $g(\cdot)$ is the restriction of $f(\cdot)$ to $D \cap [X, \infty)$, and $f(X-) = \lim_{x\to X} g(x)$ where $g(\cdot)$ is the restriction of $f(\cdot)$ to $D \cap (-\infty, X]$. Accordingly, $f(X+)$ is unique if (5.17) holds with $D - \{X\}$ replaced with $D \cap (X, \infty)$, and $f(X-)$ is unique if (5.17) holds with $D - \{X\}$ replaced with $D \cap (-\infty, X)$.

Now, let $f(\cdot)$ and $g(\cdot)$ be complex-valued functions having a common domain $D \subseteq \mathbb{C}$; and, suppose that either $X \in \mathbb{C}$ or $X = \pm\infty$. Then, irrespective of whether $f(x)$ and $g(x)$ converge themselves, one might ask whether they have the same *asymptotic behavior* as $x \to X$. In this regard, we write

$$f(x) \sim g(x) \quad \text{as} \quad x \to X \tag{5.20a}$$

if and only if

$$\lim_{x\to X} \frac{f(x)}{g(x)} = 1. \tag{5.20b}$$

The status of this condition (i.e., whether or not it holds) is especially significant when $\lim_{x\to X} f(x)$ and $\lim_{x\to X} g(x)$ both equal either 0 or $\pm\infty$.

For example, if $f(x) = x^2 + x$ and $g(x) = x^2$ ($x \in \mathbb{R}$), then $f(x) \sim g(x)$ as $x \to \infty$, which shows that it is not necessary to have $f(x) - g(x) \to 0$ as $x \to X$ in order for (5.20) to hold. On the other hand, for these same functions, we do not have $f(x) \sim g(x)$ as $x \to 0$—i.e., $f(x) \nsim g(x)$ as $x \to 0$—which shows that it is not sufficient to have

TABLE 5.1 Asymptotic Behavior of Stirling's Approximation

n	$\dfrac{n!}{\sqrt{2\pi n}(n/e)^n}$
1	1.08443...
2	1.04220...
5	1.01678...
10	1.00836...
20	1.00417...
50	1.00166...
100	1.00083...

$\lim_{x\to X} f(x) = \lim_{x\to X} g(x)$ for (5.20) to hold. A more interesting example is **Stirling's approximation** for the factorial function defined on $\mathbb{N}$:

$$n! \sim \sqrt{2\pi n}(n/e)^n \quad \text{as} \quad n \to \infty, \tag{5.21}$$

for which the convergence of the limit in (5.20b) is evident in Table 5.1.

It is easily shown that the binary relation $\sim$ is symmetric and transitive; that is, for any complex-valued functions $f(\cdot)$, $g(\cdot)$, and $h(\cdot)$ having a common domain $D \subseteq \mathbb{C}$, and any point $X \in \mathbb{C} \cup \{\pm\infty\}$, we have the following as $x \to X$:

- if $f(x) \sim g(x)$, then $g(x) \sim f(x)$;
- if $f(x) \sim g(x)$ and $g(x) \sim h(x)$, then $f(x) \sim h(x)$.

However, unless the set of functions considered is restricted, $\sim$ is not reflexive; because, e.g., we might have $f(x) = 0$ (all x), thereby precluding $f(x) \sim f(x)$ owing to division by 0 in (5.20). Additionally, let $f_1(\cdot), f_2(\cdot)$, $g_1(\cdot)$, and $g_2(\cdot)$ be complex-valued functions on a common domain $D \subseteq \mathbb{C}$ such that $f_1(x) \sim f_2(x)$ and $g_1(x) \sim g_2(x)$ as $x \to X$, for some point $X \in \mathbb{C} \cup \{\pm\infty\}$; then, $f_1(x)g_1(x) \sim f_2(x)g_2(x)$ and $f_1(x)/g_1(x) \sim f_2(x)/g_2(x)$ as $x \to X$.

Another way of relating the asymptotic behavior of one function to that of another is via the **asymptotic order symbols** "O" and "o". Again, let $f(\cdot)$ and $g(\cdot)$ be complex-valued functions having a common domain $D \subseteq \mathbb{C}$; and, suppose that either $X \in \mathbb{C}$ or $X = \pm\infty$. Then, we write

$$f(x) = O[g(x)] \quad \text{as} \quad x \to X$$

if and only if there exists a constant $M > 0$ such that $|f(x)| \leq M|g(x)|$ for all $x \in D - \{X\}$ sufficiently close to X. That is, if $X \in \mathbb{C}$, then

$$\exists M > 0,\ \exists \delta > 0,\ \forall x \in D,\ 0 < |x - X| \leq \delta \Rightarrow |f(x)| \leq M|g(x)|;$$

whereas if $X = \infty$, then

$$\exists M > 0,\ \exists \eta > 0,\ \forall x \in D,\ x \geq \eta \Rightarrow |f(x)| \leq M|g(x)|,$$

and similarly for $X = -\infty$. For example, with $D = \mathbb{R}$ we have $4 \sin x = O(x)$ and $x^2/(x^3 + 1) = O(x^2)$ as $x \to 0$; also, $4 \sin x = O(1)$ and $x^2/(x^3 + 1) = O(x^{-1})$ as $x \to \infty$. Likewise, we write

$$f(x) = o[g(x)] \quad \text{as} \quad x \to X$$

if and only if for any constant $\varepsilon > 0$ we have $|f(x)| \leq \varepsilon |g(x)|$ for all $x \in D - \{X\}$ sufficiently close to X. That is, if $X \in \mathbb{C}$, then

$$\forall \varepsilon > 0, \exists \delta > 0, \forall x \in D, 0 < |x - X| \leq \delta \Rightarrow |f(x)| \leq \varepsilon |g(x)|;$$

whereas if $X = \infty$, then

$$\forall \varepsilon > 0, \exists \eta > 0, \forall x \in D, x \geq \eta \Rightarrow |f(x)| \leq \varepsilon |g(x)|,$$

and similarly for $X = -\infty$. For example, with $D = \mathbb{R}$ we have $4 \sin x = o(1)$ and $x^2/(x^3 + 1) = o(x)$ as $x \to 0$; also, $4 \sin x = o(x)$ and $x^2/(x^3 + 1) = o(1)$ as $x \to \infty$. Often (as in the above examples), $g(x) \neq 0$ for $x \in D - \{X\}$ sufficiently close to X, in which case we can simply say that $f(x) = O[g(x)]$ as $x \to X$ if and only if $f(x)/g(x)$ is bounded for $x \in D - \{X\}$ sufficiently close to X, and $f(x) = o[g(x)]$ as $x \to X$ if and only if $f(x)/g(x) \to 0$ as $x \to X$.

A complex-valued function $f(\cdot)$ having a domain $D \subseteq \mathbb{C}$ is said to be **continuous** *at a point* $X \in D$ if and only if $f(x)$ becomes arbitrarily close to $f(X)$ by taking $x \in D$ sufficiently close to X (here $x = X$ is allowed, since an equivalent definition results when $x = X$ is disallowed):

$$\forall \varepsilon > 0, \exists \delta > 0, \forall x \in D, |x - X| \leq \delta \Rightarrow |f(x) - f(X)| \leq \varepsilon \tag{5.22}$$

(cf. (5.13)). When (5.17) holds—i.e., X is a limit point of D—an equivalent condition for continuity at X is

$$\lim_{x \to X} f(x) = f(X); \tag{5.23}$$

whereas when (5.17) does not hold, continuity at X automatically obtains. A function that is continuous at *every* point in its domain—i.e.,

$$\forall X \in D, \forall \varepsilon > 0, \exists \delta > 0, \forall x \in D, |x - X| \leq \delta \Rightarrow |f(x) - f(X)| \leq \varepsilon \tag{5.24}$$

—is a **continuous function**. Qualitatively, continuity is one of many properties by which a function may be judged "smooth" or "well behaved".

For any function $f: D \to \mathbb{C}$ with $D \subseteq \mathbb{C}$ that is continuous at $X \in D$: If $g(\cdot)$ is also a complex-valued function on D that is continuous at X, then so are $f(x) + g(x)$, $f(x) - g(x)$, $f(x) \cdot g(x)$, and $f(x)/g(x)$ as functions of $x \in D$ (assuming for the last case that division by 0 does not occur). Or, if the range of $f(\cdot)$ is a subset of the domain of another complex-valued function $g(\cdot)$, and $g(\cdot)$ is continuous at the point $f(X)$, then the composite function $g[f(\cdot)]$ on D is also continuous at X. Or, if $g(\cdot)$ is the restriction of $f(\cdot)$ to a set $D' \subseteq D$, and $X \in D'$, then $g(\cdot)$ is also continuous at X.

For any continuous *real*-valued function $f(\cdot)$ defined on an *interval* D: If $a, b \in D$ with $a \leq b$ and $f(a) \leq f(b)$, then for each value $y \in [[f(a), f(b)]$ there exists a value $x \in [a, b]$ such that $y = f(x)$; and, similarly if $f(a) \geq f(b)$ instead. In other words, on the interval $[a, b]$, the function $f(\cdot)$ takes on every value from $f(a)$ to $f(b)$. (For example, if $f(a) < 0$ and $f(b) > 0$, or vice versa, then $f(x) = 0$ for some $x \in [a, b]$.) This is known as the **intermediate-value theorem**. It follows that such a function $f(\cdot)$ is invertible if and only if it is strictly monotonic, in which case $f^{-1}(\cdot)$ is also continuous.

When a function $f(\cdot)$ is not continuous at some point X, we say that it is **discontinuous** *at* X, or that it has a **discontinuity** *at* X (e.g., the function signum is discontinuous at 0). An example of a function on $\mathbb{R}$ that is discontinuous at *every* point is

$$f(x) = \begin{cases} 0 & x \text{ rational} \\ 1 & x \text{ irrational} \end{cases}. \tag{5.25}$$

Furthermore, suppose a complex-valued function $f(\cdot)$ on $D \subseteq \mathbb{R}$ is given for which a particular $X \in D$ can be approached from *either* the right or the left by other values in D; that is, (5.17) holds with $D - \{X\}$ replaced with $D \cap (X, \infty)$, as well as $D \cap (-\infty, X)$. Then, $f(\cdot)$ is continuous at X if and only if $f(X+) = f(X) = f(X-)$. Should $f(X+)$ and $f(X-)$ exist without both equaling $f(X)$, then the function $f(\cdot)$ is said to have a **jump discontinuity** at X (e.g., signum has a jump discontinuity at 0). However, not every discontinuity is of this kind; for example, the function

$$f(x) = \begin{cases} 0 & x \leq 0 \\ \sin(1/x) & x > 0 \end{cases}$$

is continuous at every point other than 0, where it is discontinuous because $f(0+)$ does not exist.

A complex-valued function $f(\cdot)$ having domain $D \subseteq \mathbb{C}$ is called **uniformly continuous** if and only if, given $\varepsilon > 0$, a value $\delta > 0$ can be chosen such that for *any* point $X \in D$, we have $|f(x) - f(X)| \leq \varepsilon$ for every $x \in D$ such that $|x - X| \leq \delta$; that is,

$$\forall \varepsilon > 0, \exists \delta > 0, \forall X \in D, \forall x \in D, |x - X| \leq \delta \Rightarrow |f(x) - f(X)| \leq \varepsilon. \tag{5.26}$$

Observe that here, in contrast to condition (5.24) for the mere continuity of $f(\cdot)$, the quantification over X is performed *after* the quantification over δ; accordingly, whereas in (5.24) the choice of δ might depend not only on ε but on X as well, here for each ε a single value of δ can be chosen that "works" for *all* $X \in D$. Accordingly, uniform continuity is a stronger condition than continuity alone; for example, the function x^2 ($x \in \mathbb{R}$) is continuous, but not uniformly. It can be shown, though, that every continuous function $f: [a, b] \to \mathbb{C}$ ($-\infty < a < b < \infty$) is uniformly continuous. (Note that the domain of the function is a *finite* interval and *includes* both endpoints.)

Finally, a sequence $\{f_i(\cdot)\}_{i=1}^{\infty}$ of functions $f_i: D \to \mathbb{C}$ having a common domain $D \subseteq \mathbb{C}$ is said to **converge** *to a function* $f: D \to \mathbb{C}$—i.e.,

$$f_i(\cdot) \to f(\cdot) \text{ as } i \to \infty, \quad f_i(\cdot) \xrightarrow[i\to\infty]{} f(\cdot), \quad \lim_{i\to\infty} f_i(\cdot) = f(\cdot)$$

(cf. (4.2) and (5.14))—if and only if

$$\lim_{i\to\infty} f_i(x) = f(x) \quad (x \in D);$$

that is,

$$\forall x \in D,\ \forall \varepsilon > 0,\ \exists n \in \mathbb{Z}^+,\ \forall i \geq n,\ |f_i(x) - f(x)| \leq \varepsilon$$

(cf. (4.1) and (5.13)). Moreover, this sequence **converges uniformly** when

$$\forall \varepsilon > 0,\ \exists n \in \mathbb{Z}^+,\ \forall i \geq n,\ \forall x \in D,\ |f_i(x) - f(x)| \leq \varepsilon$$

(cf. (5.26)); that is, for each $\varepsilon > 0$ a single integer $n \geq 1$ can be chosen such that when $i \geq n$, we have $|f_i(x) - f(x)| \leq \varepsilon$ for *all* $x \in D$. Accordingly, uniform convergence is a stronger condition than convergence alone; for example, the functions

$$f_i(x) = \begin{cases} 0 & x < i \\ 1 & x \geq i \end{cases} \qquad (i = 1, 2, \dots) \tag{5.27}$$

on $D = \mathbb{R}$ converge to $f(x) = 0$ $(x \in \mathbb{R})$ as $i \to \infty$, but not uniformly.

CHAPTER 6

Powers and Roots

The operation of multiplication is often performed repetitively; accordingly, the ***n*th power** of a complex number x is defined as

$$x^n \triangleq \underbrace{x \cdot x \cdot \cdots \cdot x}_{n \text{ occurrences of "}x\text{"}} \quad (n = 1, 2, \ldots) \tag{6.1}$$

—i.e., x multiplied by itself $n - 1$ times. Hence, for the trivial case of $n = 1$ we obtain $x^1 = x$; also, $x^2 = x \cdot x$ is referred to as the **square** of x, and $x^3 = x \cdot x \cdot x$ is its **cube**. These functions of $x \in \mathbb{C}$ are plotted versus $x \in \mathbb{R}$ in Figure 6.1. It follows from (6.1) that for $x, y \in \mathbb{C}$ and $m, n \in \mathbb{Z}^+$:

$$x^m x^n = x^{m+n}, \tag{6.2a}$$

$$(x^m)^n = x^{mn}, \tag{6.2b}$$

$$(xy)^n = x^n y^n, \tag{6.2c}$$

$$(x/y)^n = x^n / y^n \quad (y \neq 0). \tag{6.2d}$$

In addition to (6.1) we define

$$x^0 \triangleq 1, \quad x^{-n} \triangleq 1/x^n \quad (n = 1, 2, \ldots), \tag{6.3}$$

for all $x \in \mathbb{C}$ other than 0;[1] see Figure 6.2. Hence, x^{-1} is the same as the **reciprocal** $1/x$ of x.[2] It now follows that for $x, y \neq 0$, the properties in (6.2) hold for *all* integers m, n; therefore, by (6.2a) and the right side of (6.3),

$$x^m / x^n = x^{m-n} \quad (x \neq 0) \tag{6.4}$$

1. Negative powers of 0 are undefined to avoid division by zero. As for 0^0, it is generally left undefined too; however, for a particular purpose (e.g., a polynomial (6.16) or the binomial expansion (8.4)) it may be conveniently assigned a value of 0 or 1 (usually the latter)—see Problem 6.1.

2. The nth power of a function $f(\cdot)$ is customarily indicated as "$f^n(\cdot)$", not "$f(\cdot)^n$"; and, likewise for noninteger powers (see discussion following (6.10)). However, an exception is the notation "$f^{-1}(\cdot)$", which (see definition, p. 72) denotes the *inverse* of the function $f(\cdot)$, not its reciprocal (unless such is explicitly stated).

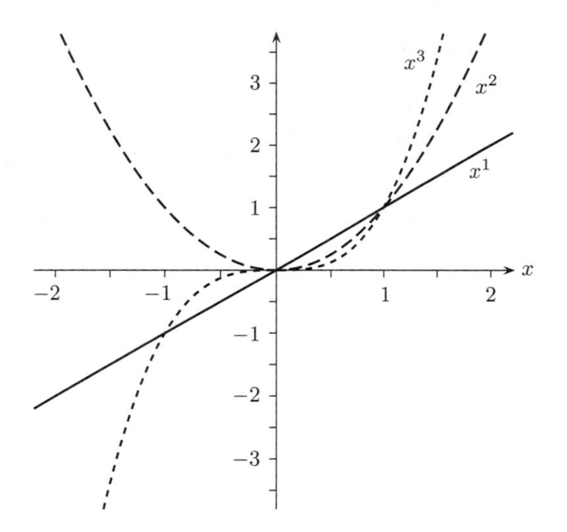

FIGURE 6.1 The nth-Power Function for $n = 1, 2, 3$

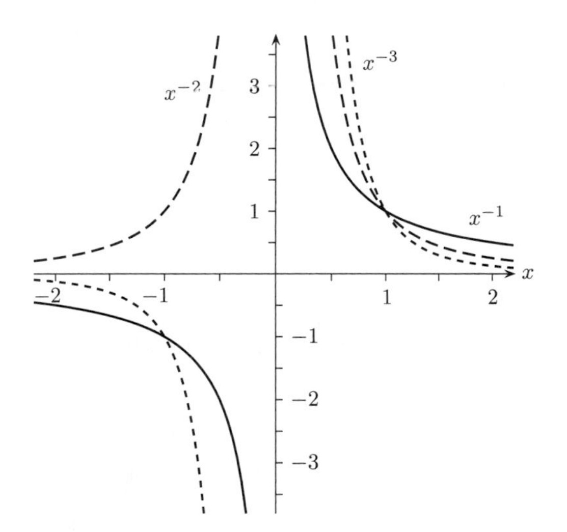

FIGURE 6.2 The nth-Power Function for $n = -1, -2, -3$

for all $m, n \in \mathbb{Z}$. We also note that

$$0^n = 0 \quad (n = 1, 2, \dots), \qquad 1^n = 1 \quad (n = 0, \pm 1, \pm 2, \dots). \tag{6.5}$$

Besides the basic properties of complex numbers given in Section 3.3, repeated multiplication and application of (3.35) yields

$$\left(re^{i\theta}\right)^n = r^n e^{in\theta} \quad (r > 0, \theta \in \mathbb{R}, n \in \mathbb{Z}). \tag{6.6}$$

Thus, letting $x = re^{i\theta}$, we obtain

$$x^n = |x|^n e^{in \arg x} \quad (x \in \mathbb{C}, n \in \mathbb{Z}, \text{ with } x \neq 0 \text{ for } n \leq 0), \tag{6.7}$$

having included the possibility of $x = 0$.[3] It follows that for $x \in \mathbb{C}$ and $n \in \mathbb{Z}$:

$$\operatorname{Re} x^n = |x|^n \cos(n \arg x) \quad (x \neq 0 \text{ for } n \leq 0), \tag{6.8a}$$

$$\operatorname{Im} x^n = |x|^n \sin(n \arg x) \quad (x \neq 0 \text{ for } n \leq 0), \tag{6.8b}$$

$$|x^n| = |x|^n \quad (x \neq 0 \text{ for } n \leq 0), \tag{6.8c}$$

$$\arg x^n = n \arg x \quad (x \neq 0), \tag{6.8d}$$

$$(x^n)^* = (x^*)^n \quad (x \neq 0 \text{ for } n \leq 0), \tag{6.8e}$$

$$(i^n)^* = i^{-n}. \tag{6.8f}$$

For each $n \geq 1$, the value x^n is a continuous function of $x \in \mathbb{C}$; but, when $n > 1$, this function is not one-to-one—e.g., $(-1)^2 = 1^2$ and $[(-1 \pm i\sqrt{3})/2]^3 = 1^3$. If we restrict x to $\mathbb{R}$, though, then for each *odd* $n \geq 1$, x^n is strictly increasing versus x, and therefore is a one-to-one function of x; but, such is not the case for n even. However, let $f(x) = x^n$ with x restricted to $[0, \infty)$; then, for *every* integer $n \geq 1$ we have a function $f(\cdot)$ on $[0, \infty)$ that is strictly increasing, and therefore one-to-one. Hence, $f(\cdot)$ has an inverse $f^{-1}(\cdot)$; moreover, the domain of $f^{-1}(\cdot)$ is also $[0, \infty)$, since this is the range of $f(\cdot)$. It follows that given $y \in [0, \infty)$, there exists a unique $x \in [0, \infty)$ such that $f(x) = y$; namely, $x = f^{-1}(y)$. In other words, for each integer $n \geq 1$ and number $y \geq 0$, there exists a unique *nonnegative* real value $\sqrt[n]{y}$—called the **nth root** of y—for which

$$(\sqrt[n]{y})^n = y. \tag{6.9}$$

In particular, $\sqrt[1]{y} = y$; also $\sqrt{y} \triangleq \sqrt[2]{y}$ is the **square root** of y, and $\sqrt[3]{y}$ is its **cube root**.[4] These functions are plotted in Figure 6.3.

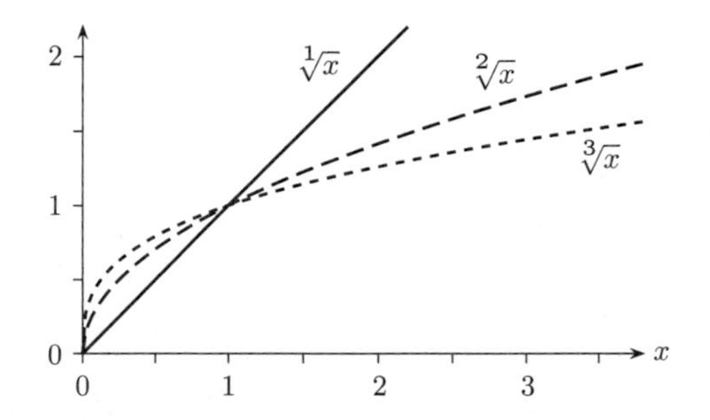

FIGURE 6.3 The nth-Root Function for $n = 1, 2, 3$

3. Regarding this inclusion, see fn. 24 in Chapter 3.

4. The symbol "$\sqrt{\ }$" (often followed by a vinculum, as here) is called a **radical** (or **surd**). Thus, "$\sqrt[n]{y}$" may be read "y radical n".

For each $n \in \mathbb{Z}^+$, the value $\sqrt[n]{x}$ is a continuous, strictly increasing, nonnegative function of $x \in [0, \infty)$, with $\sqrt[n]{0} = 0$ and $\sqrt[n]{1} = 1$. Also, if $n > 1$, then $\sqrt[n]{x} > x$ for $0 < x < 1$ and $\sqrt[n]{x} < x$ for $x > 1$, as compared with having $x^n < x$ for $0 < x < 1$ and $x^n > x$ for $x > 1$. Additionally, for $m, n \in \mathbb{Z}^+$ and $x, y \geq 0$:

$$\sqrt[n]{x^n} = x, \tag{6.10a}$$

$$\sqrt[m]{\sqrt[n]{x}} = \sqrt[mn]{x}, \tag{6.10b}$$

$$\sqrt[n]{xy} = \sqrt[n]{x}\sqrt[n]{y}, \tag{6.10c}$$

$$\sqrt[n]{x/y} = \sqrt[n]{x}/\sqrt[n]{y} \quad (y \neq 0). \tag{6.10d}$$

We now extend the nth-power function defined in (6.1) and (6.3) to allow noninteger powers. Specifically, we shall define x^p—i.e., *x to the power p*—for all powers $p \in \mathbb{R}$. As with the nth-root function, though, the variable x will be required to be nonnegative.

First, using (6.9) with $y = x^m$, and wishing to extend property (6.2b) to noninteger powers, we define

$$x^{m/n} \triangleq \sqrt[n]{x^m} \quad (x \geq 0;\ m, n = 1, 2, \ldots). \tag{6.11}$$

Hence, since taking $m = 1$ gives

$$x^{1/n} = \sqrt[n]{x} \quad (x \geq 0;\ n = 1, 2, \ldots),$$

each nth-root function can be viewed as a reciprocal-integer power. Although there are infinitely many pairs (m, n) for which m/n has any particular value, the definition (6.11) of $x^{m/n}$ is unambiguous. Because, if $M, N \in \mathbb{Z}^+$ are such that $M/N = m/n$, thereby making $Mn = mN$, then applying (6.2b) and (6.9) yields

$$\left(\sqrt[N]{x^M}\right)^{Nn} = \left[\left(\sqrt[N]{x^M}\right)^N\right]^n = (x^M)^n = (x^m)^N = \left[\left(\sqrt[n]{x^m}\right)^n\right]^N = \left(\sqrt[n]{x^m}\right)^{Nn}; \tag{6.12}$$

so, using (6.10a) to take the (Nn)th roots of the left and right sides, we obtain $\sqrt[N]{x^M} = \sqrt[n]{x^m}$.

Effectively, (6.11) defines x^q for $x \geq 0$ and *rational* powers $q > 0$—e.g., see Figure 6.4 for $x^{0.456}$. Moreover, for a fixed $x \in [0, 1)$, the value x^q is a decreasing function of such q; because, were we to assume above that $M/N > m/n$, thereby making $Mn > mN$, then the only change to (6.12) would be the middle "=" becoming "$\leq$" (with equality holding only if $x = 0$), and so we would now obtain $x^{M/N} \leq x^{m/n}$. Similarly, for a fixed $x \in [1, \infty)$, x^q is an increasing function of such q. Accordingly, to extend from positive rational powers to positive *real* powers, we recall the fact that any real number p can be approximated arbitrarily well by a rational number q, and thus define

$$x^p \triangleq \begin{cases} \inf\{x^q \mid q \in \mathbb{Q}, 0 < q < p\} & 0 \leq x < 1 \\ \sup\{x^q \mid q \in \mathbb{Q}, 0 < q < p\} & 1 \leq x < \infty \end{cases} \qquad (p > 0)$$

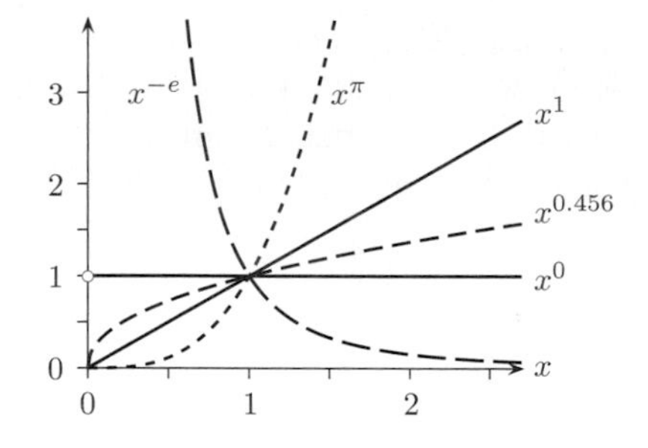

FIGURE 6.4 The pth-Power Function for $p = -e, 0, 0.456, 1, \pi$

—e.g., see Figure 6.4 for x^{π}. In particular, since $0^q = 0$ and $1^q = 1$ for all positive $q \in \mathbb{Q}$, we have

$$0^p = 0 \quad (p > 0), \qquad 1^p = 1 \quad (p \geq 0) \tag{6.13}$$

—using (6.5) to include here the assertion that $1^0 = 1$. The result is that for $x > 0$ fixed, the value x^p is a continuous positive function of $p \in [0, \infty)$, which is strictly decreasing if $0 < x < 1$, and strictly increasing if $x > 1$. Alternatively, for $p > 0$ fixed (as in Figures 6.1 and 6.3), x^p is a continuous, strictly increasing, nonnegative function of $x \in [0, \infty)$ that is positive when $x \neq 0$.

As for *negative* real powers, we simply proceed analogously to the second part of (6.3) and define

$$x^{-p} \triangleq 1/x^p \quad (x > 0, p > 0)$$

—e.g., see Figure 6.4 for x^{-e}. Hence, the second part of (6.13) now generalizes to

$$1^p = 1 \quad (p \in \mathbb{R}).$$

Based on the end of the previous paragraph, we conclude that for $x > 0$ fixed, x^{-p} is a continuous positive function of $p \in [0, \infty)$, which is strictly increasing if $0 < x < 1$, and strictly decreasing if $x > 1$; whereas for $p > 0$ fixed (as in Figure 6.2), x^{-p} is a continuous, strictly decreasing, positive function of $x \in (0, \infty)$. Combining the cases of $p > 0$, $p < 0$, and $p = 0$, it follows that for $x > 0$ fixed, x^p is a continuous positive function of $p \in \mathbb{R}$, which is strictly decreasing if $0 < x < 1$, and strictly increasing if $x > 1$. Furthermore, for $x, y > 0$ and $p, q \in \mathbb{R}$:

$$x^p x^q = x^{p+q}, \tag{6.14a}$$

$$x^p / x^q = x^{p-q}, \tag{6.14b}$$

$$(x^p)^q = x^{pq}, \tag{6.14c}$$

$$(xy)^p = x^p y^p, \tag{6.14d}$$

$$(x/y)^p = x^p / y^p \tag{6.14e}$$

—akin to (6.2) and (6.4).

In Section 5.2, with regard to (5.23), it was explained how the continuity of a function at a point may be expressed in terms of a limit. Furthermore, any real number can be expressed as the limit of a sequence of rational numbers. Applied to the above development, these facts along with (5.18) produce a simple view of real-valued powers: Given $x > 0$ and $p \in \mathbb{R}$,

$$x^p = \lim_{j \to \infty} x^{q_j}$$

for any sequence $\{q_j\}_{j=1}^{\infty}$ of rational numbers q_j such that $q_j \to p$ as $j \to \infty$, where for each j the value x^{q_j} obtains from (6.11). For example, the value $2^\pi \approx 8.825$ can be viewed as the limit of the sequence $(2^3, 2^{3.1}, 2^{3.14}, \dots) = (2^{3/1}, 2^{31/10}, 2^{314/100}, \dots) = (\sqrt[1]{2^3}, \sqrt[10]{2^{31}}, \sqrt[100]{2^{314}}, \dots)$.

Conspicuously absent from the above discussion is a definition of the nth root of a *negative* number, even though in Section 3.3 we identified the imaginary unit i as $\sqrt{-1}$. A complete treatment of this topic is beyond the scope of this book;[5] however, one should not assume that square roots of negative numbers behave like square roots of positive numbers. Indeed, if this were so, then by using (6.10c) with $x = y = -1$ and $n = 2$ we could prove that $1 = -1$:

$$1 = \sqrt{1} = \sqrt{(-1)(-1)} = \sqrt{-1}\sqrt{-1} = i \cdot i = -1.$$

For our purposes—e.g., when using the quadratic formula (6.23c) below—it suffices to know

$$\sqrt{-x} = i\sqrt{x} \quad (x \geq 0). \tag{6.15}$$

6.1 Polynomials

A **polynomial** is a function $f : \mathbb{C} \to \mathbb{C}$ that can be expressed in the form

$$f(x) = a_n x^n + a_{n-1}x^{n-1} + \cdots + a_1 x + a_0 = \sum_{j=0}^{n} a_j x^j \quad (x \in \mathbb{C}) \tag{6.16}$$

(where $0^0 = 1$), for some integer $n \geq 0$ and **coefficients** $a_j \in \mathbb{C}$ $(j = 0, 1, \dots, n)$.[6] If $a_n \neq 0$, then the coefficients are unique, and n is called the **degree** of the polynomial—i.e.,

$$\deg f(\cdot) = n.$$

5. Essentially, what is covered in this book—primarily in Section 3.3, here in the beginning of Chapter 6, and later in Chapter 7—are integer powers of complex numbers, and complex powers of positive numbers.

6. Strictly speaking, a polynomial is not a function itself, but a way (if possible) of *expressing* a function in a particular form—viz., the middle of (6.16), assuming the variable "x" is used. Nevertheless, it is verbally convenient, and usually harmless, to say that some function *is* a polynomial, rather than to say, more correctly, that it can be expressed *as* a polynomial.

TABLE 6.1 Types of Polynomial

Degree	Polynomial
1	**Linear**
2	**Quadratic**
3	**Cubic**
4	**Quartic**
5	**Quintic**

In particular, a nonzero constant is a polynomial of degree 0; whereas the degree of the **zero polynomial**—i.e., $f(x) = 0$ (all x)—is undefined. For the next few degrees, polynomials are often referred to using the adjectives given in Table 6.1. Additionally, for any degree, $f(\cdot)$ is called a **monic polynomial** when $a_n = 1$.

If the polynomial $f(\cdot)$ in (6.16) has degree $n \geq 1$ (which is assumed in the next few paragraphs, so $a_n \neq 0$), then it can be *factored* into the form

$$f(x) = a_n(x - x_1)(x - x_2) \cdots (x - x_n) = a_n \prod_{k=1}^{n} (x - x_k) \quad (x \in \mathbb{C}) \tag{6.17}$$

for some complex numbers x_k ($k = 1, 2, \ldots, n$). A simple and useful example is factoring a difference of two squares:

$$x^2 - y^2 = (x + y)(x - y) \quad (x, y \in \mathbb{C}). \tag{6.18}$$

Except for the ordering of the values x_k (which need not be distinct), the factorization (6.17) is unique. Of course, if two polynomials yield the same factorization, then they must represent the same function.

A **root** of a polynomial $f(\cdot)$ is a complex number x for which $f(x) = 0$. Thus, in terms of (6.17), x is a root if and only if $x = x_k$ for some value of the index k, with the *number* of such index values being called the **multiplicity** of that root. (For example, the third-degree polynomial $5x^3 + 10x^2 - 75x - 180 = 5(x-4)(x+3)(x+3)$ has a root at 4 of multiplicity 1, and a root at -3 of multiplicity 2.) If a root has multiplicity 1, then it is called a **simple root**; otherwise, its multiplicity is greater than 1 and it is a **multiple root**. By collecting the factors corresponding to each root (e.g., $5(x-4)(x+3)(x+3) = 5(x-4)(x+3)^2$), (6.17) can be rewritten

$$f(x) = a_n \prod_{k=1}^{\hat{n}} (x - \hat{x}_k)^{m_k} \quad (x \in \mathbb{C}), \tag{6.19}$$

where $\hat{x}_1, \ldots, \hat{x}_{\hat{n}}$ ($1 \leq \hat{n} \leq n$) are the $\hat{n}$ *distinct* roots of $f(\cdot)$—i.e.,

$$\{\hat{x}_1, \ldots, \hat{x}_{\hat{n}}\} = \{x_1, \ldots, x_n\} \qquad \text{with} \qquad \hat{x}_j \neq \hat{x}_k \quad (j \neq k)$$

—having respective multiplicities $m_1, \ldots, m_{\hat{n}}$, thereby making

$$\sum_{k=1}^{\hat{n}} m_k = n. \tag{6.20}$$

Hence, the total number of roots (including repetitions) of a polynomial equals its degree. Furthermore, if the coefficients a_j in (6.16) are all real, then for each nonreal root $\hat{x}_k$ there is a $j \neq k$ such that both $\hat{x}_j = \hat{x}_k{}^*$ and $m_j = m_k$. Conversely, if this latter condition holds, then *expanding* the factorization (6.19)—i.e., combining the factors to get the form in (6.16)—will produce a polynomial having real coefficients.

Dividing $f(\cdot)$ in (6.16) by a_n yields the unique monic polynomial having exactly the same roots and respective multiplicities. Also, expanding the factorization (6.17) shows that

$$\sum_{k=1}^{n} x_k = -\frac{a_{n-1}}{a_n}, \tag{6.21a}$$

$$\prod_{k=1}^{n} x_k = (-1)^n \frac{a_0}{a_n}. \tag{6.21b}$$

It follows that for a monic polynomial, if it is known that the roots are all *integers*, then they must all be divisors of a_0; therefore, when $a_0 \neq 0$, the roots can be found systematically via the prime factorization of a_0, even when the degree n is large.

Using (6.7), it can be verified that for any integer $n \geq 1$ and complex constant $c \neq 0$, the polynomial

$$x^n - c \quad (x \in \mathbb{C}) \tag{6.22a}$$

has n distinct roots:

$$|c|^{1/n} \exp[i(\arg c + 2\pi k)/n] \quad (k = 0, 1, \ldots, n-1) \tag{6.22b}$$

—which, if desired, can be put into rectangular form via Euler's identity (3.32). In particular, for $c = 1$ the above values—i.e., $e^{i2\pi k/n}$ $(k = 0, 1, \ldots, n-1)$—are called the ***n*th roots of unity**, since these are precisely the values $x \in \mathbb{C}$ for which $x^n = 1$. When viewed in the complex plane (see Figure 6.5), these n values are uniformly spaced on the **unit circle**—i.e., the circle of unit radius centered at the origin—with one of the values being 1 (for $k = 0$).

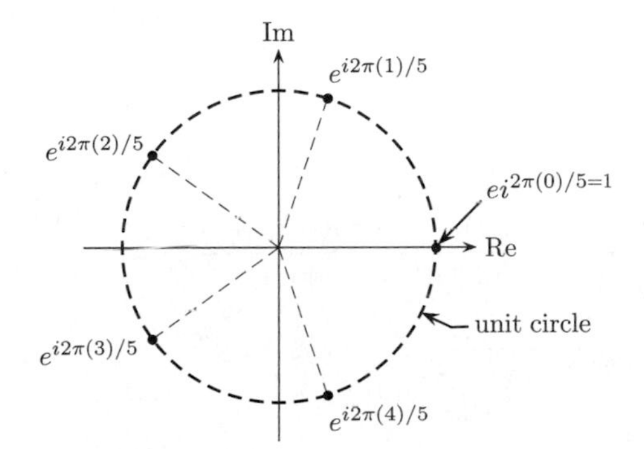

FIGURE 6.5 The nth-Roots of Unity for $n = 5$

For a second-degree (i.e., quadratic) polynomial

$$f(x) = ax^2 + bx + c \quad (x \in \mathbb{C}) \tag{6.23a}$$

with complex coefficients $a \neq 0$, b, and c, we have the factorization

$$f(x) = a(x - x_1)(x - x_2) \quad (x \in \mathbb{C}), \tag{6.23b}$$

where x_1 and x_2 are given by the **quadratic formula**:

$$x_1, x_2 = \frac{-b \pm \sqrt{b^2 - 4ac}}{2a} = \frac{-(b/2) \pm \sqrt{(b/2)^2 - ac}}{a}. \tag{6.23c}$$

In particular, when the coefficients are real, there are three possibilities in terms of the **discriminant** $d = b^2 - 4ac$:

- $d > 0$: x_1 and x_2 are real with $x_1 \neq x_2$.
- $d = 0$: x_1 and x_2 are real with $x_1 = x_2$.
- $d < 0$: x_1 and x_2 are nonreal with $x_1 = x_2^*$.

Regardless of whether all of the coefficients a, b, and c are real, (6.23c) may be used to find the roots x_1 and x_2 of the polynomial (6.23a), simply by interpreting the square root $\sqrt{\alpha}$ of a complex number α as any complex number β for which $\alpha = \beta^2$. To wit, suppose

$$\alpha_r + i\alpha_i = (\beta_r + i\beta_i)^2 \quad (\alpha_r, \alpha_i, \beta_r, \beta_i \in \mathbb{R}), \tag{6.24a}$$

where α_r and $\alpha_i \neq 0$ are given. Expanding the right side, then equating the real and imaginary parts of the result with those on the left, we have $\alpha_r = {\beta_r}^2 - {\beta_i}^2$ and $\alpha_i = 2\beta_r\beta_i$; and, combining these two equations to eliminate either β_i or β_r, then solving for the remaining variable, we obtain

$$\beta_r = \pm\sqrt{\frac{\sqrt{{\alpha_r}^2 + {\alpha_i}^2} + \alpha_r}{2}}, \quad \beta_i = \pm\sqrt{\frac{\sqrt{{\alpha_r}^2 + {\alpha_i}^2} - \alpha_r}{2}}, \tag{6.24b}$$

with the signs of β_r and β_i being chosen to be the same (respectively, opposite) when $\alpha_i > 0$ ($\alpha_i < 0$).[7] For example, applying (6.23) and (6.24), we find that the polynomial

$$x^2 + (2 + i1)x + (-23 + i43)$$

has roots

$$x_1, x_2 = \frac{-(2 + i1) \pm \sqrt{(2 + i1)^2 - 4(1)(-23 + i43)}}{2(1)} = \frac{(-2 - i1) \pm \sqrt{95 - i168}}{2}$$
$$= \frac{(-2 - i1) \pm (12 - i7)}{2} = 5 - i4, \; -7 + i3.$$

7. Alternatively, using (6.22) with $n = 2$ yields $\sqrt{\alpha} = \pm\sqrt{|\alpha|}e^{i(\arg\alpha)/2}$ ($\alpha \in \mathbb{C}$), which can then be converted to rectangular form.

Given finitely many points $(x_j, y_j) \in \mathbb{C}^2$ $(j = 1, \ldots, n)$, where the values x_j are distinct, an easy way to obtain a polynomial $f(\cdot)$ that passes through (i.e., **interpolates**) all of these points—i.e., $f(x_j) = y_j$ (all j)—is to use the **Lagrange interpolation formula**:

$$f(x) = \sum_{j=1}^{n} \left(\prod_{k \neq j} \frac{x - x_k}{x_j - x_k} \right) y_j \quad (x \in \mathbb{C}). \tag{6.25}$$

(In the trivial case of $n = 1$, the product is empty and is understood to equal 1.) For example, for $n = 3$ we obtain

$$f(x) = \frac{(x - x_2)(x - x_3)}{(x_1 - x_2)(x_1 - x_3)} y_1 + \frac{(x - x_1)(x - x_3)}{(x_2 - x_1)(x_2 - x_3)} y_2 + \frac{(x - x_1)(x - x_2)}{(x_3 - x_1)(x_3 - x_2)} y_3$$

$(x \in \mathbb{C})$—which can be expanded into the form in (6.16), if desired. The formula (6.25) works because for each j, the function of x inside the parentheses is a polynomial that equals 1 for $x = x_j$ and 0 for $x = x_l$ $(l \neq j)$.

Along with the polynomial $f(\cdot)$ in (6.16)—now allowing $n \geq 0$ and no longer assuming $a_n \neq 0$—suppose

$$g(x) = \sum_{j=0}^{n} b_j x^j \quad (x \in \mathbb{C}) \tag{6.26}$$

for some coefficients $b_j \in \mathbb{C}$ $(j = 0, 1, \ldots, n)$.[8] Of course, if $a_j = b_j$ for all j, then $f(x) = g(x)$ for all $x \in \mathbb{C}$. But, the converse statement is also true; that is, the *only* way to have $f(x) = g(x)$ for all $x \in \mathbb{C}$ is to have $a_j = b_j$ for all j. In fact, it can be shown that since the degrees of the polynomials in (6.16) and (6.26) are at most n, if $f(x) = g(x)$ at any $n + 1$ distinct points $x \in \mathbb{C}$, then $a_j = b_j$ (all j). It follows, e.g., that given any polynomials $f(\cdot)$ and $g(\cdot)$ for which $f(x) = g(x)$ $(x \in \mathbb{R})$, we must have $f(x) = g(x)$ $(x \in \mathbb{C})$.

For any polynomials $f(\cdot)$ and $g(\cdot)$ in (6.16) and (6.26), we have

$$f(x) \pm g(x) = \sum_{j=0}^{n} (a_j \pm b_j) x^j \quad (x \in \mathbb{C})$$

($\pm$ respectively); and,

$$f(x)g(x) = \sum_{j=0}^{2n} c_j x^j \quad (x \in \mathbb{C}),$$

---

8. No loss of generality incurs by assuming the same number of coefficients for both polynomials $f(\cdot)$ and $g(\cdot)$, since any of these coefficients may be 0.

where each c_j equals the sum of those products $a_k b_l$ $(k,\ l = 0, 1, \ldots, n)$ for which $k + l = j$:

$$c_j = \sum_{k=\max(0,j-n)}^{\min(n,j)} a_k b_{j-k} \quad (j = 0, 1, \ldots, 2n).$$

However, a *ratio* of two polynomials—i.e., a **rational function**—need not be expressible itself as a polynomial. Rather, assuming that neither $f(\cdot)$ nor $g(\cdot)$ is the zero polynomial, the function $f(x)/g(x)$ of x equals a polynomial versus x if and only if each root of $g(\cdot)$ is also a root of $f(\cdot)$, with the multiplicity of the former being no greater than the multiplicity of the latter. For example, if

$$f(x) = 8x^7 - 56x^6 + 8x^5 + 392x^4 + 88x^3 - 1000x^2 - 1128x - 360,$$
$$g(x) = -2x^4 + 4x^3 + 24x^2 + 28x + 10$$

$(x \in \mathbb{C})$, then the factorizations of $f(\cdot)$ and $g(\cdot)$ yield

$$\frac{f(x)}{g(x)} = \frac{8(x+1)^4(x-3)^2(x-5)}{-2(x+1)^3(x-5)} = -4(x+1)(x-3)^2$$
$$= -4x^3 + 20x^2 - 12x - 36 \quad (x \neq -1, 5).$$

Moreover, in such cases it is often convenient to let $f(x)/g(x)$ equal the resulting polynomial for *all* $x \in \mathbb{C}$—even when $g(x) = 0$—thereby obtaining a continuous function defined on all of $\mathbb{C}$.[9] Regardless, since 0 times any number equals 0, it is always safe to write

$$f(x) = g(x)\frac{f(x)}{g(x)} \quad (x \in \mathbb{C})$$

for any complex-valued functions $g(\cdot)$ and $f(\cdot)$.

In general, the division of a polynomial $f(\cdot)$ by a polynomial $g(\cdot)$ is analogous to the division of one integer by another. Namely—assuming $g(\cdot)$ is not the zero polynomial—we have

$$f(x) = q(x)g(x) + r(x) \quad (x \in \mathbb{C}) \tag{6.27}$$

(cf. (3.21)) for some polynomial $q(\cdot)$ called the **quotient** and some polynomial $r(\cdot)$ called the **remainder**, where $r(\cdot)$ either is the zero polynomial or has a degree less than that of $g(\cdot)$; moreover, $q(\cdot)$ and $r(\cdot)$ are unique. For example, if

$$f(x) = 8x^3 - 22x^2 + 27x - 25, \quad g(x) = 2x - 3 \quad (x \in \mathbb{C}), \tag{6.28a}$$

then

$$q(x) = 4x^2 - 5x + 6, \quad r(x) = -7 \quad (x \in \mathbb{C}), \tag{6.28b}$$

9. Accordingly, the points $x = -1, 5$ at which the function $f(x)/g(x)$ is undefined are called **removable singularities**.

$$\begin{array}{r}
4x^2 - 5x + 6 \\
2x - 3 \overline{\smash{\big)} 8x^3 - 22x^2 + 27x - 25} \\
\underline{8x^3 - 12x^2 } \\
-10x^2 + 27x \\
\underline{-10x^2 + 15x } \\
12x - 25 \\
\underline{12x - 18} \\
-7
\end{array}$$

FIGURE 6.6 A Long Division

as obtained from the **long division** shown in Figure 6.6.

We say that $f(\cdot)$ is **divisible** *by* $g(\cdot)$ if and only if $r(\cdot)$ is the zero polynomial (as is the case in the preceding paragraph); equivalently, $g(\cdot)$ **divides** $f(\cdot)$, $g(\cdot)$ is a **divisor** of $f(\cdot)$, $g(\cdot)$ is a **factor** of $f(\cdot)$, and $f(\cdot)$ is a **(polynomial) multiple** of $g(\cdot)$. A useful fact is that given a polynomial $f(\cdot)$ and constants $a, b \in \mathbb{C}$ with $a \neq 0$, the remainder of dividing $f(x)$ by $ax - b$ is simply $f(b/a)$; e.g., in (6.28), $r(x) = f(3/2) = -7$ (all x).

CHAPTER 7

Exponentials and Logarithms

In Chapter 6 we considered x^p mainly as a function of x, for some fixed power p. We now consider b^x as function of $x \in \mathbb{R}$ for a fixed $b > 0$; this is called the **exponential function** having **base** b, with the power x being the **exponent**. From our previous discussion, we know that this function is positive and continuous; also, it is strictly decreasing if $b < 1$, constant (1) if $b = 1$, and strictly increasing if $b > 1$. Furthermore,

$$b^0 = 1, \quad b^1 = b,$$
$$b^{-x} = 1/b^x = (1/b)^x \quad (x \in \mathbb{R}).$$

Comparing the asymptotic behavior of an exponential with that of a fixed power, we have

$$\lim_{x\to\infty} \frac{b^x}{x^p} = \begin{cases} 0 & 0 < b < 1 \\ \infty & b > 1 \end{cases} \qquad (p \in \mathbb{R}). \tag{7.1}$$

Thus, for all powers $p \in \mathbb{R}$, and every base $b > 0$ other than the uninteresting case of $b = 1$, the limiting behavior of b^x/x^p as $x \to \infty$ is exactly the same as that of b^x alone.

The "preferred" choice for the base b is **Napier's number**,

$$e \triangleq \lim_{\alpha\to\infty} (1 + 1/\alpha)^\alpha = 2.71828\ldots \tag{7.2}$$

(where α is a real variable). Accordingly, the function e^x of $x \in \mathbb{R}$ is commonly referred to as *the* **exponential function**, exp, plotted on the left side of Figure 7.1. As indicated in the figure by a line drawn tangent to the plot, one reason why the number e is special is that it is the only base $b > 0$ for which the function b^x of $x \in \mathbb{R}$ has a slope of 1 at $x = 0$.[1]

The exponential function can be generalized to allow *complex*-valued exponents. Namely, we define

$$e^{x+iy} \triangleq e^x(\cos y + i \sin y) \quad (x, y \in \mathbb{R}), \tag{7.3}$$

1. Additionally, see Problem 7.3(a) for an elementary interpretation of (7.2).

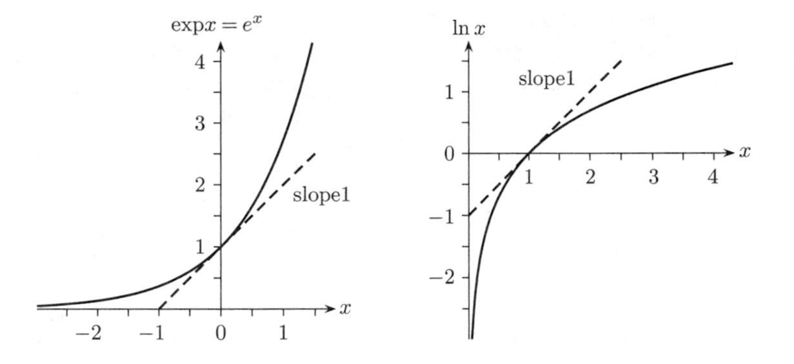

FIGURE 7.1 The Exponential and Natural Logarithm Functions

which yields Euler's identity (3.32) by taking $x = 0$.[2] It follows that e^z is a continuous function of $z \in \mathbb{C}$, with $e^z \neq 0$ for all z. Also, for $z, z_1, z_2 \in \mathbb{C}$:

$$\operatorname{Re} e^z = e^{\operatorname{Re} z} \cos(\operatorname{Im} z), \tag{7.4a}$$

$$\operatorname{Im} e^z = e^{\operatorname{Re} z} \sin(\operatorname{Im} z), \tag{7.4b}$$

$$|e^z| = e^{\operatorname{Re} z}, \tag{7.4c}$$

$$\arg e^z = \operatorname{Im} z, \tag{7.4d}$$

$$(e^z)^* = e^{z^*}, \tag{7.4e}$$

$$(e^z)^n = e^{nz} \quad (n \in \mathbb{Z}), \tag{7.4f}$$

$$1/e^z = e^{-z}, \tag{7.4g}$$

$$e^{z_1} e^{z_2} = e^{z_1+z_2}, \tag{7.4h}$$

$$e^{z_1}/e^{z_2} = e^{z_1-z_2}. \tag{7.4i}$$

2. To motivate the definition (7.3) requires mathematics more advanced than what we shall discuss. Briefly, one way is to first prove the three infinite series

$$e^x = \sum_{n=0}^{\infty} \frac{x^n}{n!}, \quad \cos x = \sum_{n=0}^{\infty} (-1)^n \frac{x^{2n}}{(2n)!}, \quad \sin x = \sum_{n=0}^{\infty} (-1)^n \frac{x^{2n+1}}{(2n+1)!} \quad (x \in \mathbb{R})$$

(where $0^0 = 1$). Then, by *formally* substituting $x = i\theta$ ($\theta \in \mathbb{R}$) in the first series—a step that cannot be *rigorously* justified at this point since $i\theta \notin \mathbb{R}$ when $\theta \neq 0$—we can "derive" Euler's identity:

$$e^{i\theta} = \sum_{n=0}^{\infty} \frac{(i\theta)^n}{n!} = \sum_{n=0}^{\infty} \frac{(i\theta)^{2n}}{(2n)!} + \sum_{n=0}^{\infty} \frac{(i\theta)^{2n+1}}{(2n+1)!} = \sum_{n=0}^{\infty} (i^2)^n \frac{\theta^{2n}}{(2n)!} + i \sum_{n=0}^{\infty} (i^2)^n \frac{\theta^{2n+1}}{(2n+1)!}$$

$$= \cos\theta + i \sin\theta \quad (\theta \in \mathbb{R}).$$

From here, (7.3) obtains by requiring $e^{x+iy} = e^x e^{iy}$ ($x, y \in \mathbb{R}$), analogous to (6.14a).

Having observed for all $b > 0$ other than $b = 1$ that b^x is a strictly monotonic function of $x \in \mathbb{R}$, we conclude that there must exist a corresponding inverse function for each such b. Namely, since the range of each of these exponential functions is $(0, \infty)$, there exists on $(0, \infty)$ a unique function $\log_b$—called the **logarithm** having **base** b—for which

$$\log_b b^x = x \quad (x \in \mathbb{R}). \tag{7.5a}$$

Conversely, the inverse of $\log_b$ is the exponential function with base b—i.e.,

$$b^{\log_b x} = x \quad (x > 0). \tag{7.5b}$$

—and thus an exponential function is sometimes called an **antilogarithm**.

The function $\log_b$ is continuous; also, it is strictly decreasing if $0 < b < 1$, and strictly increasing if $b > 1$. Also,

$$\log_b 1 = 0, \quad \log_b b = 1,$$
$$\log_b x = -\log_{1/b} x \quad (x > 0).$$

(The last equation explains why bases $b < 1$ are rarely used, since changing the base to $1/b > 1$ merely causes the logarithm to change sign.) Furthermore, for any base $b \neq 1$ and numbers $x, x_1, x_2 > 0$:

$$\log_b x^p = p \log_b x \quad (p \in \mathbb{R}), \tag{7.6a}$$
$$\log_b(1/x) = -\log_b x, \tag{7.6b}$$
$$\log_b(x_1 x_2) = \log_b x_1 + \log_b x_2, \tag{7.6c}$$
$$\log_b(x_1/x_2) = \log_b x_1 - \log_b x_2. \tag{7.6d}$$

Comparing the asymptotic behavior of a logarithm with that of a fixed power, we have

$$\lim_{x \to 0+} \frac{\log_b x}{x^p} = \begin{cases} 0 & p < 0 \\ -\infty & p \geq 0 \end{cases}, \qquad \lim_{x \to \infty} \frac{\log_b x}{x^p} = \begin{cases} \infty & p \leq 0 \\ 0 & p > 0 \end{cases} \tag{7.7}$$

$(b > 1)$. Thus, for all bases $b > 1$, and every nonzero power $p \in \mathbb{R}$, the limiting behavior of $(\log_b x)/x^p$ as either $x \to 0+$ or $x \to \infty$ is exactly the same as that of $1/x^p$, except for a sign change when $x \to 0+$ for $p > 0$.

For any positive constants a, $b \neq 1$, taking the base-a logarithm of (7.5b) and applying (7.6a), we obtain the first equality of

$$\log_a x = (\log_a b) \log_b x = \frac{\log_b x}{\log_b a} \quad (x > 0); \tag{7.8}$$

then, interchanging "a" and "b" yields the second equality. These equalities show how to convert a logarithm of one base a into another of base b. Also, the second equality for $x = b$ gives

$$\log_a b = 1/\log_b a.$$

For a base of 10, we have the **common logarithm** $\log_{10}$. This base is popular because, by (7.6c), multiplying any number $x > 0$ by 10 corresponds to increasing the value $\log_{10} x$ by 1. However, the most widely used logarithm base is e, because of its unique properties; accordingly, $\log_e$ is called the **natural logarithm**. This function is often written "ln"; hence,

$$\ln(e^x) = x \quad (x \in \mathbb{R}), \qquad \ln e = 1, \qquad e^{\ln x} = x \quad (x > 0).$$

The natural logarithm function is plotted in Figure 7.1; as shown, it is the only logarithm for which the plot crosses the horizontal axis with a slope of 1. In general, when the base b is understood, one may indicate $\log_b$ by simply writing "log"; for example, in purely mathematical literature, "log" typically means the natural logarithm.

Finally, it follows from (7.5b) that for any positive constants a and $b \neq 1$, we have $a = b^{\log_b a}$, and thus $a^x = \left(b^{\log_b a}\right)^x$ for any real value x; therefore, using (6.14c) we obtain

$$a^x = b^{(\log_b a)x} \quad (x \in \mathbb{R}), \tag{7.9}$$

which shows how to convert an exponential of one base to another. Proceeding analogously for a *complex* variable z, for each base $a > 0$ we take $b = e$ in (7.9) and define

$$a^z \triangleq e^{(\ln a)z} \quad (z \in \mathbb{C}) \tag{7.10}$$

—the right side being evaluated by (7.3).[3] It can then be shown via (7.4) that for $a > 0$ and $z, z_1, z_2 \in \mathbb{C}$:

$$\operatorname{Re} a^z = a^{\operatorname{Re} z} \cos[(\ln a) \operatorname{Im} z], \tag{7.11a}$$

$$\operatorname{Im} a^z = a^{\operatorname{Re} z} \sin[(\ln a) \operatorname{Im} z], \tag{7.11b}$$

$$|a^z| = a^{\operatorname{Re} z}, \tag{7.11c}$$

$$\arg a^z = (\ln a) \operatorname{Im} z, \tag{7.11d}$$

$$\left(a^z\right)^* = a^{z^*}, \tag{7.11e}$$

$$(a^z)^n = a^{nz} \quad (n \in \mathbb{Z}), \tag{7.11f}$$

$$1/a^z = a^{-z}, \tag{7.11g}$$

$$a^{z_1} a^{z_2} = a^{z_1 + z_2}, \tag{7.11h}$$

$$a^{z_1}/a^{z_2} = a^{z_1 - z_2}. \tag{7.11i}$$

3. For all $a > 0$ and $n \in \mathbb{Z}$, (7.10) yields the same value for a^n as does (7.9); therefore, since (7.9) is consistent with the definition given in (6.1) and (6.3) for a^n, so is (7.10).

CHAPTER 8

Possibility and Probability

8.1 Permutations and Combinations

Occasionally, one considers the possible ways to *arrange* the elements of a finite set A. To begin, a **permutation** of A is a particular *ordering* of its elements.[1] Specifically, for a set of n objects—i.e., $|A| = n$—one of the objects is designated as the first, then another is designated the second, then another the third, and so forth, until the last object is finally designated the nth. The number of possible permutations of n objects is n **factorial**:

$$n! \triangleq \begin{cases} 1 & n = 0 \\ 1 \cdot 2 \cdot 3 \cdot \cdots \cdot n & n > 0 \end{cases} \qquad (n \in \mathbb{N}). \tag{8.1}$$

Because, for $n > 0$ there are exactly n ways to designate the first object, after which there are exactly $n - 1$ ways to designate the second object, and so forth, until finally there is exactly 1 way to designate the last object, thereby making $n \cdot (n-1) \cdot \cdots \cdot 1 = n!$ possible permutations. Whereas for the empty case of $n = 0$—i.e., $A = \emptyset$—experience shows that it is mathematically convenient to say that the number of permutations of A is 1, which is how $0!$ is defined. (For example, letting $N(n)$ be the number of permutations of n objects, we can now succinctly say that $N(n) = nN(n-1)$ for all $n \geq 1$—including $n = 1$.)

A **combination** is a *selection* of a particular number of elements from a finite set A, *without* respect to order; thus, it is simply a *subset* of A. The total number of combinations of k objects chosen from a set of n objects is

$$\binom{n}{k} \triangleq \frac{n!}{k!(n-k)!} \qquad (n = 0, 1, \ldots; k = 0, 1, \ldots, n) \tag{8.2}$$

—the left side being read "n choose k". Because, for $k = 0$, the chosen subset of k objects must be $\emptyset$; hence, $\binom{n}{0} = 1$ for all $n \geq 0$. Whereas for $0 < k \leq n$, there are exactly

1. In addition to denoting an ordered set itself, the word "permutation" is commonly used to express the *action* of *re*ordering such a set; that is, this term may also denote a one-to-one mapping of a set onto itself. For example, one might perform a reordering of the sequence $(1, 2, 3, 4)$ to obtain $(1, 3, 4, 2)$, for which applying the same reordering yields $(1, 4, 2, 3)$. To avoid confusion, we will not be using the word "permutation" is this latter sense, though the verb "permute" may be so used.

n ways to select a first object, after which there are exactly $n-1$ ways to select a second object, and so forth, until finally there are exactly $n-k+1$ ways to select a kth object, thereby making

$$n \cdot (n-1) \cdot \cdots \cdot (n-k+1) = \frac{n!}{(n-k)!} \tag{8.3}$$

permutations of k objects that can be chosen from $n \geq k$ objects; then, to disregard the order in which these objects are chosen, we divide (8.3) by the number $k!$ of permutations of k objects, getting (8.2).

A common application of (8.2) is the **binomial expansion**

$$(x+y)^n = \sum_{k=0}^{n} \binom{n}{k} x^k y^{n-k} \quad (x, y \in \mathbb{C}; n \in \mathbb{N}) \tag{8.4}$$

(where $0^0 = 1$).[2] This identity holds because by rewriting the left side as n factors of $(x+y)$, we see that for each $k \in \{0, 1, \ldots, n\}$ the term $x^k y^{n-k}$ obtains only by having k of the factors $(x+y)$ contribute a factor of x, with the remaining $n-k$ factors $(x+y)$ contributing a factor of y; and, there are $\binom{n}{k}$ ways of choosing the first k factors. In light of (8.4), the expression on the right side of (8.2), symbolized on the left, is called the **binomial coefficient**.

8.2 Probability and Statistics

Suppose an object x is *randomly* selected from a nonempty set A; thus, the variable "x" is a **random variable**. Moreover, suppose A is finite, and that the selection is done in such a way that no element of A receives preference over any other.[3] Then, according to the **principle of indifference** (or **principle of insufficient reason**), we assign the same likelihood to *each* particular element being selected; that is, for each $a \in A$, the **probability** of a being the selected object x is

$$\Pr(x = a) = 1/|A| \tag{8.5}$$

—the abbreviation "Pr" being read "the probability of having". For example, flipping a fair coin corresponds to a set $A = \{\text{HEAD}, \text{TAIL}\}$ of $|A| = 2$ possibilities, with $\Pr(\text{flip} = \text{HEAD}) = \Pr(\text{flip} = \text{TAIL}) = \frac{1}{2}$.

More generally, the probability of the object x being selected from a particular subset $S \subseteq A$ is the proportion of A that is in S:

$$\Pr(x \in S) = |S|/|A| = \sum_{a \in S} \Pr(x = a). \tag{8.6}$$

---

2. The statement that (8.4) is true is known as the **binomial theorem**.

3. The reader is alerted that often some random outcomes *do* receive preference over others (e.g., a coin being flipped might be biased rather than fair), and for such cases, some of the following discussion would need to be modified. The treatment of such cases, though touched on briefly in Section E.4, is beyond the scope of this book.

(Hence, for each $a \in A$ we have $\Pr(x = a) = \Pr(x \in S)$ for $S = \{a\}$.) Clearly, $0 \leq \Pr(x \in S) \leq 1$ (all S). In particular, $\Pr(x \in A) = 1$, since it is certain that the selected object will be an element of A; and, at the other extreme, $\Pr(x \in \emptyset) = 0$, since having $x \in \emptyset$ is impossible.

It follows from (8.6) that the probabilities of *disjoint* sets add to give the probability of their union. Because, if $S_1, S_2 \subseteq A$ with $S_1 \cap S_2 = \emptyset$, then

$$\sum_{a \in S_1 \cup S_2} \Pr(x = a) = \sum_{a \in S_1} \Pr(x = a) + \sum_{a \in S_2} \Pr(x = a),$$

making

$$\Pr(x \in S_1 \text{ or } x \in S_2) = \Pr(x \in S_1 \cup S_2) = \Pr(x \in S_1) + \Pr(x \in S_2); \tag{8.7}$$

and, likewise for any finite number of disjoint sets. For example, upon throwing an ordinary six-sided die (for which $A = \{1, 2, 3, 4, 5, 6\}$), the probability of the resulting number being either even ($S_1 = \{2, 4, 6\}$) or five ($S_2 = \{5\}$) is $3/6 + 1/6 = 2/3$.

There are also times when probabilities multiply. In particular, suppose two objects x_1 and x_2 are *independently* selected from, respectively, two finite nonempty sets A_1 and A_2 (which need not be disjoint); then, for any subsets $S_1 \subseteq A_1$ and $S_2 \subseteq A_2$, we have

$$\Pr(x_1 \in S_1 \text{ and } x_2 \in S_2) = \Pr[(x_1, x_2) \in S_1 \times S_2] = \Pr(x \in S_1) \cdot \Pr(x \in S_2) \tag{8.8}$$

—and, likewise for more than two objects. This is because

$$\begin{aligned}\Pr[(x_1, x_2) \in S_1 \times S_2] &= |S_1 \times S_2|/|A_1 \times A_2| = (|S_1| \cdot |S_2|)/(|A_1| \cdot |A_2|) \\ &= (|S_1|/|A_1|) \cdot (|S_2|/|A_2|) = \Pr(x \in S_1) \cdot \Pr(x \in S_2).\end{aligned}$$

For example, upon throwing a die twice (i.e., $A_1 = A_2 = \{1, 2, 3, 4, 5, 6\}$), the probability of the first number being even ($S_1 = \{2, 4, 6\}$) and the second number being five ($S_2 = \{5\}$) is $(3/6) \cdot (1/6) = 1/12$.[4]

As an example of calculating probabilities in terms of combinations, the 52 cards in a standard deck are divided into four suits (♠, ♡, ◇, ♣), each having 13 cards of different ranks (ACE, KING, QUEEN, JACK, 10, 9, . . . , 2). A poker hand comprises any five cards from the deck; therefore, there are $\binom{52}{5} = 2\,598\,960$ possible hands. Of these, $4\binom{13}{5} = 5\,148$ are flushes—i.e., every card in the hand has the same suit (here, including *straight* flushes, in which the five cards are consecutively ranked). Thus, the probability of being dealt a flush (of some kind) from a well-shuffled deck is $5\,148/2\,598\,960 \approx 0.00198 = 0.198\%$.

In other words, the **odds** of being dealt a flush are 5 148 in 2 598 960; because, in general, the odds of a randomly selected element $a \in A$ falling within a subset $S \subseteq A$

4. Note, though, that the probability of throwing *two* dice and having one of the numbers be even and the other be 5 is

$$\begin{aligned}&\Pr[(x_1 \in \{2, 4, 6\} \text{ and } x_2 \in \{5\}) \text{ or } (x_1 \in \{5\} \text{ and } x_2 \in \{2, 4, 6\})] \\ &\quad = \Pr(x_1 \in \{2, 4, 6\} \text{ and } x_2 \in \{5\}) + \Pr(x_1 \in \{5\} \text{ and } x_2 \in \{2, 4, 6\}) = 1/12 + 1/12 = 1/6;\end{aligned}$$

because, since the dice were not designated in a particular order, we must consider *all* their permutations.

may be expressed as "$|S|$ in $|A|$". Alternatively, these odds may also be expressed as either "$|S|$ to $|S^c|$ *in favor* of S" (here, 5 148 to 2 593 812), or "$|S^c|$ to $|S|$ *against* S" (here, 2 593 812 to 5 148), where the complement S^c of S is relative to A (hence, $|S^c| = |A| - |S|$). Furthermore, dividing $|A|$, $|S|$, and $|S^c|$ all by any positive value yields equivalent expressions of these same odds. For example, noting that the greatest common divisor of 5 148 and 2 598 960 is 156, we might divide throughout by 156 and state that the odds of a flush are 33 in 16 660, or 33 to 16 627 in favor, or 16 627 to 33 against. It then follows from (8.6) that for a random event having probability p, the corresponding odds are p in 1, as well as p to $1-p$ in favor, as well as $1-p$ to p against; conversely, given odds of either α in β ($0 \le \alpha \le \beta > 0$), or α to β in favor ($\alpha, \beta \ge 0$ not both 0), or α to β against ($\alpha, \beta \ge 0$ not both 0), the corresponding probability is α/β, or $\alpha/(\alpha+\beta)$, or $\beta/(\alpha+\beta)$, respectively.[5]

Any odds of α to β may be written as the ratio "$\alpha : \beta$". Moreover, it is common to scale the ratio to make $\alpha + \beta = 100$. Then, odds of $\alpha : \beta$ in favor correspond to a probability of $\alpha\%$, and odds of $\alpha : \beta$ against correspond to a probability of $\beta\%$. For example, "even odds" of $1:1$ (in favor or against) are often stated as "fifty-fifty", meaning "$50:50$", and correspond to a probability of $50\% = \frac{1}{2}$.

Now, suppose an object x is randomly selected n successive times ($1 \le n < \infty$) from a set $A \ne \emptyset$; and, let the outcomes be x_i ($i = 1, 2, \ldots, n$). That is, in the parlance of statistics, a **population** A is **sampled** n times, thereby yielding n **samples** x_i. Furthermore, in contrast to **sampling without replacement**, whereby each sample is removed from the population before the next sample is taken (thus causing the sampled set to decrease, and forcing the sample values x_i to be distinct), suppose we perform **sampling with replacement**, whereby each sample is "put back" into the population before the next sample is taken (thus keeping the sampled set fixed as A, and allowing any of the values x_i to be the same). Then, if the set A is finite and n is very large (more precisely, $n \gg |A|$), we expect that for each element $a \in A$, the *fraction* of the n samples that equal a will be about $1/|A|$. In other words, define the **frequency** of a as the *number* k of samples x_i that equal a; then, for n sufficiently large it is highly likely that the **relative frequency** of a, defined as k/n, will approximately equal $\Pr(x = a) = 1/|A|$. (For example, when flipping a fair coin, we expect heads and tails to each occur about 50% of the time.) This phenomenon, which relates a relative frequency to a corresponding probability, is known as the **law of large numbers**.

8.2.1 Statistical Measures

Often, the elements of a sampled set A (no longer assumed to be finite) are numbers, in which case various statistical averages can be calculated upon the samples (regardless of whether taken with or without replacement). First, averaging the samples x_i

5. Betting odds are usually expressed as odds *against*. (Thus, a random event with odds such as 100-to-1 is called a "long shot" since it is very improbable.) This way, a winning bet on an event having odds of α to β returns α dollars for every β dollars bet; that is, the bet "pays" α to β. Accordingly, here it is convenient to scale the odds to have $\beta = 1$; and, since most events bet upon have probabilities less that $\frac{1}{2}$, their odds will be α to 1 for some $\alpha > 1$. (Of course, betting odds set by a gambling house are rarely intended to exactly reflect the corresponding probabilities, but rather to give the house a payoff advantage.)

($i = 1, 2, \ldots, n < \infty$) themselves, we obtain their **(arithmetic) mean**:

$$\overline{x} \triangleq \frac{1}{n} \sum_{i=1}^{n} x_i. \tag{8.9}$$

For real-valued samples (assumed hereafter), the mean is a point on the real line that is roughly centered within the sample points.

Occasionally, one might wish to attribute more significance to some samples than others (e.g., particular samples might be more recent, or deemed more "reliable"). Accordingly, the **weighted (arithmetic) mean** of the samples x_i is defined as

$$\sum_{i=1}^{n} w_i x_i, \tag{8.10a}$$

for any specified **weights** w_i ($i = 1, 2, \ldots, n$) such that

$$w_i \geq 0 \quad (\text{all } i), \qquad \sum_{i=1}^{n} w_i = 1. \tag{8.10b}$$

Thus, the larger the value of w_i for a particular i, the more "weight" is given in the average to the corresponding sample x_i; whereas uniform weighting—i.e., $w_i = 1/n$ (all i)—yields the ordinary arithmetic mean (8.9).

When $x_i > 0$ (all i), it is sometimes more appropriate to use the **geometric mean**,[6]

$$\exp\left(\overline{\ln x}\right) = \exp\left(\frac{1}{n}\sum_{i=1}^{n} \ln x_i\right) = \exp\left[\ln\left(\prod_{i=1}^{n} x_i\right)^{1/n}\right] = \sqrt[n]{\prod_{i=1}^{n} x_i}. \tag{8.11}$$

This value is never greater than $\overline{x}$; moreover, it equals $\overline{x}$ if and only if the samples x_i are all equal. Alternatively, one might compute

$$\frac{1}{\left(\overline{1/x}\right)} = 1 \Bigg/ \left(\frac{1}{n}\sum_{i=1}^{n} \frac{1}{x_i}\right) = n \Bigg/ \sum_{i=1}^{n} \frac{1}{x_i}, \tag{8.12}$$

which is called the **harmonic mean** of the samples. It too is never greater than $\overline{x}$, and equals $\overline{x}$ if and only if the samples are all equal. In fact, the harmonic mean is never greater than the geometric mean, with the two being equal if and only if the samples are all equal.

An alternative to *averaging* samples in order to obtain a "middle estimate" is to calculate their **median**, which is roughly defined as the real number for which half of the samples have lesser value and half have greater value. More precisely, let the n samples $x_1, x_2, \ldots, x_n$ be reordered (if necessary) as values $X_1, X_2, \ldots, X_n$ such that $X_1 \leq X_2 \leq \cdots \leq X_n$. Then, for n odd, the median of the samples x_i is the simply middle

---

6. Per definition (8.9), we again use an overbar to indicate the operation of summing a collection of values—here, $\ln x_1, \ln x_2, \ldots, \ln x_n$—then dividing by their number. This shorthand notation is likewise used hereafter.

value $X_{(n+1)/2}$; whereas for n even, there is no unique middle value, so the median is instead $(X_{n/2} + X_{n/2+1})/2$. (For example, the $n = 3$ samples $-3, 2, 4$ have mean $\overline{x} = 1$ and median $X_2 = 2$, whereas the $n = 4$ samples 2, 2, 4, 8 have mean $\overline{x} = 4$ and median $(X_2 + X_3)/2 = 3$.) A notable feature of the median is that it is possible (when $n \geq 3$) for a sample x_i to change value yet the median remain the same; whereas changing *any* sample causes changes in the arithmetic, geometric, and harmonic means, as well as in any weighted mean for which that sample is assigned a nonzero weight.

A **mode** of a finite collection of samples is any sample value that occurs with the greatest frequency (e.g., the samples 5, 6, 6, 6, 7, 7 have mode 6). Of course, more than one value can occur with the maximum frequency, and thus a collection can have more than one mode (e.g., the samples 5, 6, 6, 7, 7 have modes 6 and 7). As exception is made, though, for the case of every sample value having frequency 1 (i.e., the samples are distinct), in which case we say that no mode exists. The modes of a collection give a loose indication of where the samples are concentrated.

The mth **moment** of n samples x_i $(i = 1, 2, \ldots, n < \infty)$ is

$$\overline{x^m} = \frac{1}{n}\sum_{i=1}^{n} x_i{}^m \quad (m = 1, 2, \ldots).$$

Hence, per (8.9), the first $(m = 1)$ moment is the mean. Additionally, the mth moment about the mean—i.e., the mth **central moment**—is

$$\overline{(x - \overline{x})^m} = \frac{1}{n}\sum_{i=1}^{n} (x_i - \overline{x})^m \quad (m = 1, 2, \ldots).$$

It follows that the first central moment always equals 0. The second central moment,

$$\sigma_x{}^2 \triangleq \overline{(x - \overline{x})^2} = \frac{1}{n}\sum_{i=1}^{n} (x_i - \overline{x})^2, \tag{8.13}$$

is called the **variance** of x.[7]

The square root $\sigma_x \geq 0$ of the variance is called the **standard deviation** of x, which is a measure of how widely the samples x_i are dispersed about the mean. Indeed, $\sigma_x = 0$ if and only if the samples are all equal (to the mean). Additionally, the **root mean square (RMS)** of x is the square root of the second moment—i.e.,

$$\sqrt{\overline{x^2}} = \sqrt{\frac{1}{n}\sum_{i=1}^{n} x_i{}^2} \tag{8.14}$$

7. For reasons (having to do with "estimator bias") that we will not discuss, the division by n in (8.13) is often replaced with division by $n - 1$ (thereby requiring $n \geq 2$). When n is large, though, this amounts to introducing a factor of $n/(n-1) \approx 1$, which has little effect.

—which is also called the **quadratic mean**. It measures how widely the samples are dispersed about 0. Also, since expanding (8.13) yields

$$\sigma_x{}^2 = \frac{1}{n}\sum_{i=1}^{n}(x_i{}^2 - 2\bar{x}x_i + \bar{x}^2) = \frac{1}{n}\sum_{i=1}^{n}x_i{}^2 - 2\bar{x}\frac{1}{n}\sum_{i=1}^{n}x_i + \frac{1}{n}\sum_{i=1}^{n}\bar{x}^2$$
$$= \overline{x^2} - 2\bar{x}\cdot\bar{x} + \bar{x}^2 = \overline{x^2} - \bar{x}^2, \qquad (8.15)$$

the RMS of x equals its standard deviation σ_x if and only if $\bar{x} = 0$; furthermore, $\overline{x^2} \geq \bar{x}^2$, with equality if and only if $\sigma_x = 0$.

The value M of each of the means defined in (8.9)–(8.12), always satisfies the bounds $\min_i x_i \leq M \leq \max_i x_i$. Moreover, both inequalities can be made strict unless the samples x_i are all equal (assuming for the weighted arithmetic mean (8.10a) that none of weights w_i equals 0), in which case both inequalities attain equality. Also, if all of the samples are multiplied by a real constant α, then M is multiplied by α as well; whereas for each $m \geq 1$, the mth moment and the mth central moment are both multiplied by α^m. Therefore, the variance is multiplied by α^2, with the standard deviation and RMS being multiplied by α. Furthermore, if a real constant β is added to all of samples, then β is also added to the arithmetic mean, regardless of any weighting; whereas the central moments (including the variance) and standard deviation do not change.

8.2.2 Statistical Plots

A common way to visually display real-valued statistical data is to plot a **(frequency) histogram**. To make such a two-dimensional plot, an interval on the horizontal axis is divided into contiguous subintervals, called **bins**, the "widths" of which (i.e., the lengths of subintervals) are typically all the same; then, over each bin is plotted a rectangle of the same width, with its height equal to the number of samples "falling" in that bin (i.e., its frequency). For example, Figure 8.1 shows two histograms for the randomly generated numbers x given in Table 8.1. As shown in the table, six dice were thrown together 40 times, and for each throw x is the sum of the numbers on the first three dice; hence, $x \in \{3, 4, 5, \ldots, 18\}$. Correspondingly, the histogram on the left of Figure 8.1 has been drawn with 16 bins of width 1, each centered on a possible value of

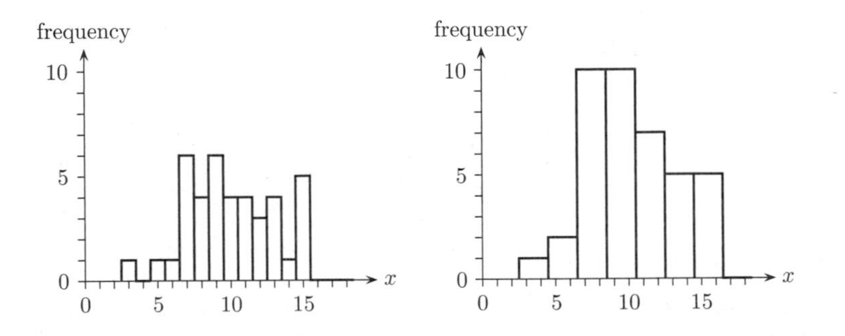

FIGURE 8.1 Two Histograms of the Same Samples

TABLE 8.1 Some Randomly Generated Numbers

Thrown Dice	x	y	y'	y''	Thrown Dice	x	y	y'	y''
	11	9	9	9		15	14	10	10
	12	11	14	13		9	9	10	14
	14	15	12	13		6	8	11	10
	12	7	11	9		9	9	11	10
	10	11	10	5		10	14	12	12
	12	13	10	6		8	9	8	9
	9	12	9	10		9	10	11	11
	11	11	10	11		15	11	10	7
	13	13	10	12		5	7	8	13
	11	12	7	7		7	9	11	10
	13	11	9	9		7	9	9	5
	8	8	11	12		9	12	14	15
	10	10	6	7		9	9	13	14
	7	10	12	15		7	11	16	16
	15	13	12	9		8	8	6	6
	3	3	4	5		15	13	15	10
	7	7	9	7		13	12	11	7
	15	17	12	9		8	8	9	6
	10	13	11	13		11	9	8	9
	13	14	12	13		7	6	5	5

x; that is, the bins are the intervals $[c-\frac{1}{2}, c+\frac{1}{2})$ $(c = 3, 4, 5, \ldots, 18)$.[8] (For example, the rectangle plotted over the bin $[11.5, 12.5)$ has height 3, since there were 3 occurrences of $x \in [11.5, 12.5)$—i.e., 3 occurrences of 12.) For comparison, the histogram on the right side of the figure has 8 bins of width 2—viz., $[c-1, c+1)$ $(c = 3.5, 5.5, 7.5, \ldots, 17.5)$. As can be seen, increasing the bin width typically has the effect of smoothing the plot, because random fluctuations in the frequencies of adjacent bins are lessened; if, the bin width is chosen too large, however, then desirable detail will be lost.

For a frequency histogram of n samples, with all bins having the same width w, the total area under the plot is nw; thus, in Figure 8.1 the area (80) of the right histogram is twice that (40) of the left, since n remained fixed while w was doubled. Dividing vertically by n converts a frequency histogram into a **relative-frequency histogram**, for then the height of each rectangle equals the *fraction* of samples falling into the corresponding bin; now, the total area is w. Further dividing vertically by w, we obtain a **probability histogram**, for which the *area* of each bin equals the fraction of samples falling in it, and the total area is 1. Generalizing these definitions to allow bins of variable width: for a relative-frequency histogram (respectively, probability histogram), the height (area) of

8. The association of interval endpoints with bins is usually arbitrary; thus, here the bins could just as well have been taken to be $(c-\frac{1}{2}, c+\frac{1}{2}]$ $(c = 3, 4, 5, \ldots, 18)$.

each rectangle equals the relative frequency of the corresponding bin. Consequently, a *probability* histogram can have varying bin widths without being visually misleading; also, for it the height of each rectangle equals the relative frequency of the corresponding bin divided by its width, and the total area under the plot is 1.

To display the statistical relationship between *two* random variables—say, x and y—one could sample both several times simultaneously, then make a three-dimensional histogram of the sample pairs (x, y). That is, a portion of the xy-plane would be divided into contiguous two-dimensional bins of some shape (e.g., squares or regular hexagons), and the number of samples falling in each bin would be plotted along a third dimension perpendicular to the plane. A similar two-dimensional approach, though, is to plot a **scattergram** (or **scatter diagram**, or **scatter plot**). Here, each sample pair (x, y) is simply plotted as a dot at the corresponding point in the xy-plane; then, rather than have a third dimension to numerically indicate how many dots fall within selected regions of the plane, one visually estimates this value by observing the relative *density* of dots in each region.

For examples, we again use the data in Table 8.1. For each throw of the six dice, each of the tabulated values x, y, y'', and y'' is the sum of the numbers showing on three of the dice; specifically, letting d_i be the number shown on the ith die ($i = 1, 2, \ldots, 6$), we have

$$x = d_1 + d_2 + d_3, \quad y = d_2 + d_3 + d_4, \quad y' = d_3 + d_4 + d_5, \quad y'' = d_4 + d_5 + d_6.$$

(Thus, e.g., for first throw, Table 8.1 states that $(d_1, d_2, d_3, d_4, d_5, d_6) = (6, 2, 3, 4, 2, 3)$, so $x = 6+2+3 = 11$, $y = 2+3+4 = 9$, $y' = 3+4+2 = 9$, and $y'' = 4+2+3 = 9$.) Figure 8.2 shows the scattergram for each of these four variables versus x, using black circular dots. (For none of the four plots, though, do 40 distinct dots appear, since some of the dots are plotted over others; this is a general problem with scattergrams.) As expected, the dots for the plot of x versus itself all fall on a line of slope 1 passing through the origin. For the plot of y versus x, the dots disperse somewhat, but there is a strong statistical trend for y to increase when x does; this is also as expected, since $d_2 + d_3$ contributes to both values. For y' versus x, the dots disperse even more, but there is a loose trend of y' increasing versus x, because of the contribution of d_3 to both. Finally, for y'' versus x, the plot shows no statistical dependence of y'' on x; this too is as expected, since none of the dice contributes to both values.

Often, one desires the ability to *predict* the value of one random variable y based on the observed value of another x. One way to do this is to sample the pair (x, y) a large number of times, plot the samples as a scattergram, and draw a curve through the plot that roughly captures the dependence of y on x; then, the function $f(\cdot)$ expressed by the curve is used to perform future predictions. That is, upon observing each additional value of x, we predict $f(x)$ to be the corresponding value of y. The difference $f(x) - y$ is called the **error** in the prediction.

The process of generating such a function $f(\cdot)$ from a given collection of $n \geq 1$ sample pairs (x_i, y_i) $(i = 1, 2, \ldots, n)$ is called **regression**, the most common kind being **linear regression**, whereby $f(\cdot)$ plots to a line:

$$f(x) = a_0 + a_1 x \quad (x \in \mathbb{R}), \tag{8.16}$$

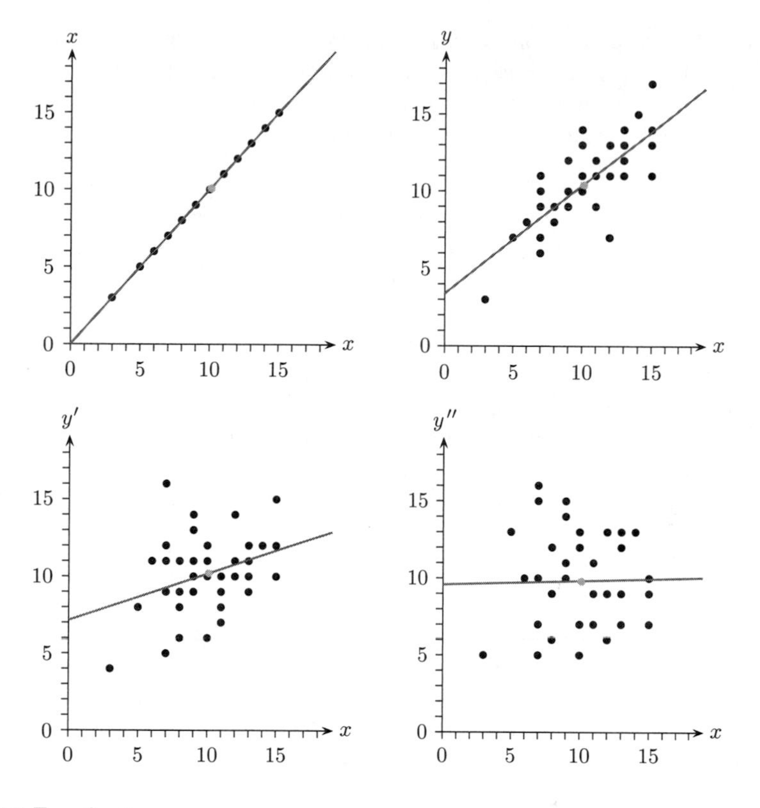

FIGURE 8.2 Four Scattergrams

for some real coefficients a_0 and a_1. A popular way to choose a_0 and a_1 is to minimize the **mean square error** of the samples:

$$\overline{[f(x)-y]^2} = \frac{1}{n}\sum_{i=1}^{n}[f(x_i)-y_i]^2 = \frac{1}{n}\sum_{i=1}^{n}(a_0+a_1x_i-y_i)^2. \tag{8.17}$$

For this criterion, it can be shown that the coefficients are unique when $n \geq 2$, with

$$a_0 = \frac{\overline{x^2}\cdot\overline{y}-\overline{x}\cdot\overline{xy}}{\overline{x^2}-\overline{x}^2} = \frac{\left(\sum_{i=1}^{n}x_i{}^2\right)\left(\sum_{i=1}^{n}y_i\right)-\left(\sum_{i=1}^{n}x_i\right)\left(\sum_{i=1}^{n}x_iy_i\right)}{n\sum_{i=1}^{n}x_i{}^2-\left(\sum_{i=1}^{n}x_i\right)^2}, \tag{8.18a}$$

$$a_1 = \frac{\overline{xy}-\overline{x}\cdot\overline{y}}{\overline{x^2}-\overline{x}^2} = \frac{n\sum_{i=1}^{n}x_iy_i-\left(\sum_{i=1}^{n}x_i\right)\left(\sum_{i=1}^{n}y_i\right)}{n\sum_{i=1}^{n}x_i{}^2-\left(\sum_{i=1}^{n}x_i\right)^2} \tag{8.18b}$$

—assuming at least two values x_i differ, to avoid division by 0.[9] This regression line is shown on each of the scattergrams in Figure 8.2. Also plotted, with a gray dot, is

9. This particular linear regression is a special case of a "least-squares fit", which is further discussed in Subsection 12.2.3.

the **centroid** $\overline{(x,y)} = (\overline{x},\overline{y})$ of the samples. As can be seen, the regression line passes through the centroid in each case; this always happens, since $a_0 + a_1\,\overline{x} = \overline{y}$.

One way to quantify the "strength" of the linear correlation between two random variables x and y is to calculate their **correlation coefficient**

$$r \triangleq \frac{\overline{(x-\overline{x})(y-\overline{y})}}{\sqrt{\overline{(x-\overline{x})^2}\cdot\overline{(y-\overline{y})^2}}} = \frac{\sum_{i=1}^{n}(x_i-\overline{x})(y_i-\overline{y})}{\sqrt{\sum_{i=1}^{n}(x_i-\overline{x})^2\sum_{i=1}^{n}(y_i-\overline{y})^2}} \tag{8.19}$$

—assuming at least two values x_i differ, and at least two values y_i differ, to avoid division by 0. In general, $-1 \leq r \leq 1$. If $r = 0$, then x and y are said to be **uncorrelated**; whereas they are **positively correlated** (respectively, **negatively correlated**) if $r > 0$ ($r < 0$), with the correlation being stronger the closer r is to 1 (-1). We have $r = 0$ if and only if $a_1 = 0$ in (8.18b); otherwise, the signs of r and a_1 agree. (Hence, the sign of the correlation coefficient r can be determined from the slope a_1 of the above regression line.) Also, notice from (8.19) that the value of r does not change if "x" and "y" are interchanged throughout; however, it is usually *not* the case that if a scattergram for y versus x is replotted with its axis variables interchanged, then the regression line (given by the equation $y = a_1x + a_0$) simply inverts (as given by the equation $x = y/a_1 - a_0/a_1$) to become the regression line for x versus y. (In other words, if a scattergram and a regression line for y versus x are plotted on a transparent sheet, then although flipping the sheet over and rotating it to make the y-axis point to the right *does* produce the scattergram for x versus y, the corresponding regression line is usually not obtained.)

For the data in Table 8.1, the values of the correlation coefficient r for x, y, y', and y'' versus x are, respectively, 1, 0.767, 0.360, and 0.023 (the last three values being rounded). In general, as can be seen in Figure 8.2, when $|r|$ is nearer to 1, the dots of a scattergram are more concentrated near the regression line; whereas greater dispersion of the dots occurs when $|r|$ is nearer to 0.

CHAPTER 9

Matrices

It is sometimes useful to arrange the elements of a finite nonempty set into an *array*. In general, a **matrix** having **elements** (or **components**, or **entries**) $a_{i,j}$ ($i = 1, \ldots, m$; $j = 1, \ldots, n$) is the *rectangular* array

$$\mathbf{A} = [a_{i,j}]_{i,j=1}^{m,n} = \begin{bmatrix} a_{1,1} & a_{1,2} & \cdots & a_{1,n} \\ a_{2,1} & a_{2,2} & \cdots & a_{2,n} \\ \vdots & \vdots & & \vdots \\ a_{m,1} & a_{m,2} & \cdots & a_{m,n} \end{bmatrix} \tag{9.1a}$$

—which may be briefly written

$$\mathbf{A} = [a_{i,j}] \tag{9.1b}$$

when the ranges of the indices i (vertical, downward) and j (horizontal, to the right) are understood. The integers $m, n \geq 1$ are, respectively, the vertical and horizontal **dimensions** of this $m \times n$ (read "m by n") matrix; thus, the number of elements in $\mathbf{A}$ is mn.

For each i, the n horizontal elements $a_{i,j}$ ($j = 1, \ldots, n$) constitute the ith **row** of $\mathbf{A}$; and, for each j, the m vertical elements $a_{i,j}$ ($i = 1, \ldots, m$) constitute its jth **column**. A matrix having only one row or one column is called a **vector**; specifically, for $m = 1$ the matrix $\mathbf{A}$ is an n-dimensional **row vector**, whereas for $n = 1$ it is an m-dimensional **column vector**.

When $m = n$, $\mathbf{A}$ is a **square matrix** of **order** n, and we may write

$$\mathbf{A} = [a_{i,j}]_{i,j=1}^{n}.$$

For such a matrix, the n diagonal elements $a_{i,j}$ with $i = j$—i.e., $a_{i,i}$ ($i = 1, \ldots, n$)—are collectively referred to as the **main diagonal**.

The **transpose** of the $m \times n$ matrix $\mathbf{A}$ in (9.1) is the $n \times m$ matrix $\mathbf{A}^{\mathrm{T}}$ obtained by converting the rows of $\mathbf{A}$ into respective columns, and columns into respective rows; that is,

$$\mathbf{A}^{\mathrm{T}} \triangleq [c_{i,j}]_{i,j=1}^{n,m} \qquad \text{where} \qquad c_{i,j} = a_{j,i} \quad (\text{all } i, j).$$[1]

For example,

$$\begin{bmatrix} 1 & 2 & 3 \\ 4 & 5 & 6 \end{bmatrix}^{\mathrm{T}} = \begin{bmatrix} 1 & 4 \\ 2 & 5 \\ 3 & 6 \end{bmatrix}.$$

It follows that

$$\left(\mathbf{A}^{\mathrm{T}}\right)^{\mathrm{T}} = \mathbf{A}.$$

The transpose of a row vector is a column vector, and vice versa. A **symmetric matrix** is a matrix $\mathbf{A}$ for which $\mathbf{A} = \mathbf{A}^{\mathrm{T}}$; thus, a symmetric matrix is necessarily square.

Henceforth, each matrix under discussion is understood to be a **complex matrix**; that is, all of its elements are complex numbers. A special case is a **real matrix**, for which all of the elements are real numbers. Similarly, a matrix is called **positive**, **negative**, **nonpositive**, or **nonnegative** if and only if every element is a number of that type.

The **adjoint** of the $m \times n$ matrix $\mathbf{A}$ in (9.1) is the $n \times m$ matrix $\mathbf{A}^*$ obtained by complex conjugating the elements of $\mathbf{A}$, then taking the transpose of the result (or, transposing first, then conjugating); that is,

$$\mathbf{A}^* \triangleq [c_{i,j}]_{i,j=1}^{n,m} \qquad \text{where} \qquad c_{i,j} = a_{j,i}{}^* \quad (\text{all } i, j).$$

It follows that

$$(\mathbf{A}^*)^* = \mathbf{A}.$$

The adjoint of a row vector is a column vector, and vice versa. A **Hermitian** (or **self-adjoint) matrix** is a matrix $\mathbf{A}$ for which $\mathbf{A} = \mathbf{A}^*$; thus, a Hermitian matrix is necessarily square.

Note that for a *matrix*, a superscript asterisk ($*$) not only indicates complex conjugation, but includes transposition; thus, using the abbreviated notation of (9.1b), we have $\mathbf{A}^* = [a_{i,j}{}^*]^{\mathrm{T}}$. On occasion, though, one may wish to conjugate the elements of a matrix $\mathbf{A}$ without also taking its transpose; for this purpose, we introduce the following notation for the **complex conjugate** of $\mathbf{A}$:

$$\overline{\mathbf{A}} \triangleq [c_{i,j}]_{i,j=1}^{m,n} \qquad \text{where} \qquad c_{i,j} = a_{i,j}{}^* \quad (\text{all } i, j)$$

Thus, $\mathbf{A}^* = \overline{\mathbf{A}}^{\mathrm{T}} = \overline{\mathbf{A}^{\mathrm{T}}}$. (For convenience and consistency, one may also define $\overline{\alpha} \triangleq \alpha^*$ for all complex numbers α.)

Two matrices $\mathbf{A} = [a_{i,j}]$ and $\mathbf{B} = [b_{i,j}]$ may be added or subtracted if and only if they have the same respective dimensions—i.e., same number of rows and same number

---

1. Many authors use a prime symbol to write the transpose of a matrix $\mathbf{A}$ as "$\mathbf{A}'$". However, we will not be using this alternative notation, thereby allowing "$\mathbf{A}'$" to denote *any* matrix.

of columns—which become the respective dimensions of the resulting sum or difference matrix. Specifically, matrix addition and subtraction are performed componentwise:

$$\mathbf{A} \pm \mathbf{B} \triangleq [c_{i,j}] \qquad \text{where} \qquad c_{i,j} = a_{i,j} \pm b_{i,j} \quad (\text{all } i, j) \tag{9.2}$$

($\pm$ respectively). Matrix addition is both commutative and associative, whereas matrix subtraction is neither. Also,

$$(\mathbf{A} \pm \mathbf{B})^{\mathrm{T}} = \mathbf{A}^{\mathrm{T}} \pm \mathbf{B}^{\mathrm{T}},$$
$$(\mathbf{A} \pm \mathbf{B})^{*} = \mathbf{A}^{*} \pm \mathbf{B}^{*},$$
$$\overline{\mathbf{A} \pm \mathbf{B}} = \overline{\mathbf{A}} \pm \overline{\mathbf{B}}$$

($\pm$ respectively in each case).

If $\mathbf{A}$ is $m \times n$ and $\mathbf{B}$ is $m' \times n'$, then we may multiply $\mathbf{A}$ *on the right* by $\mathbf{B}$ if and only if $n = m'$, thereby obtaining the $m \times n'$ matrix

$$\mathbf{AB} \triangleq [c_{i,j}] \qquad \text{where} \qquad c_{i,j} = \sum_{k=1}^{n} a_{i,k} b_{k,j} \quad (\text{all } i, j); \tag{9.3}$$

similarly, we may multiply $\mathbf{A}$ *on the left* by $\mathbf{B}$ to obtain the $m' \times n$ matrix $\mathbf{BA}$ if and only if $n' = m$. It follows that each element $c_{i,j}$ of $\mathbf{AB}$ can be viewed as the matrix product of the ith row of $\mathbf{A}$ and the jth column of $\mathbf{B}$:

$$c_{i,j} = \begin{bmatrix} a_{i,1} & a_{i,2} & \cdots & a_{i,n} \end{bmatrix} \begin{bmatrix} b_{1,j} \\ b_{2,j} \\ \vdots \\ b_{n,j} \end{bmatrix} \qquad (\text{all } i, j). \tag{9.4}$$

Matrix multiplication is associative, but not commutative (indeed, reversing the order of two matrices might cause their product to no longer exist). Also,

$$(\mathbf{AB})^{\mathrm{T}} = \mathbf{B}^{\mathrm{T}}\mathbf{A}^{\mathrm{T}},$$
$$(\mathbf{AB})^{*} = \mathbf{B}^{*}\mathbf{A}^{*},$$
$$\overline{\mathbf{AB}} = \overline{\mathbf{A}}\,\overline{\mathbf{B}}.$$

Matrix division, though, is not generally defined; however, it is sometimes possible to achieve the same effect by multiplying one matrix (on one side or the other, as appropriate) by another's *inverse* (see Section 9.1).

A matrix $\mathbf{A}$ may always be multiplied, on either side, by a complex number α; such a multiplier α is called a **scalar**.[2] The effect is that each element of $\mathbf{A}$ is multiplied by α to obtain a new matrix having the same respective dimensions:

$$\alpha\mathbf{A} \triangleq [c_{i,j}] \triangleq \mathbf{A}\alpha \qquad \text{where} \qquad c_{i,j} = \alpha a_{i,j} \quad (\text{all } i, j). \tag{9.5}$$

2. It follows that a scalar is not equivalent to a 1×1 matrix; because, not every matrix can be left- or right-multiplied by a 1×1 matrix. Nevertheless, as done in (9.4), it is sometimes convenient to treat a 1×1 matrix as simply a number.

The negative of the matrix $\mathbf{A}$ is then defined by taking $\alpha = -1$:

$$-\mathbf{A} \triangleq [c_{i,j}] \qquad \text{where} \qquad c_{i,j} = -a_{i,j} \quad (\text{all } i, j).$$

Scalar-matrix multiplication is distributive over matrix addition—i.e.,

$$\alpha(\mathbf{A} + \mathbf{B}) = (\alpha\mathbf{A}) + (\alpha\mathbf{B}),$$

where $\mathbf{A}$ and $\mathbf{B}$ have the same dimensions. As for dividing a matrix $\mathbf{A}$ by a complex number $\beta \neq 0$, we define

$$\mathbf{A}/\beta \triangleq \mathbf{A}(1/\beta).$$

Then,

$$\begin{aligned} (\alpha\mathbf{A})^{\mathrm{T}} &= \alpha\mathbf{A}^{\mathrm{T}}, & (\mathbf{A}/\beta)^{\mathrm{T}} &= \mathbf{A}^{\mathrm{T}}/\beta, \\ (\alpha\mathbf{A})^{*} &= \alpha^{*}\mathbf{A}^{*}, & (\mathbf{A}/\beta)^{*} &= \mathbf{A}^{*}/\beta^{*}, \\ \overline{\alpha\mathbf{A}} &= \alpha^{*}\overline{\mathbf{A}}, & \overline{\mathbf{A}/\beta} &= \overline{\mathbf{A}}/\beta^{*}. \end{aligned}$$

In addition to its interpretation (9.4) in terms of vector multiplication, the matrix multiplication defined in (9.3) can also be interpreted via scalar-vector multiplication in a way that is often helpful. Namely, each row of $\mathbf{AB}$ can be viewed as a *linear combination* of the rows of $\mathbf{B}$; and, each column of $\mathbf{AB}$ can be viewed as a linear combination of the columns of $\mathbf{A}$.[3] Specifically, the ith row of $\mathbf{AB}$ is

$$\begin{bmatrix} c_{i,1} & c_{i,2} & \cdots & c_{i,n'} \end{bmatrix} = \sum_{k=1}^{n} a_{i,k} \begin{bmatrix} b_{k,1} & b_{k,2} & \cdots & b_{k,n'} \end{bmatrix}, \tag{9.6a}$$

where the scalars $a_{i,k}$ obtain from the ith row of $\mathbf{A}$; likewise, the jth column of $\mathbf{AB}$ is

$$\begin{bmatrix} c_{1,j} \\ c_{2,j} \\ \vdots \\ c_{m,j} \end{bmatrix} = \sum_{k=1}^{n} \begin{bmatrix} a_{1,k} \\ a_{2,k} \\ \vdots \\ a_{m,k} \end{bmatrix} b_{k,j}, \tag{9.6b}$$

where the scalars $b_{k,j}$ obtain from the jth column of $\mathbf{B}$.

Given an $m \times n$ matrix $\mathbf{A}$, and letting $\mathbf{0}$ be the $m \times n$ **zero matrix**—i.e., the $m \times n$ matrix for which every element is 0—it follows from (9.2) that

$$\mathbf{A} + \mathbf{0} = \mathbf{A}, \quad \mathbf{0} + \mathbf{A} = \mathbf{A}.$$

Conversely, neither of these equalities holds when another matrix is substituted for $\mathbf{0}$. Similarly, by (9.3) we have

$$\mathbf{AI} = \mathbf{A} \tag{9.7a}$$

---

3. In general, a **linear combination** of mathematical objects $x_1, x_2, \ldots, x_n$ (e.g., numbers, or matrices, or geometric vectors) is a sum $a_1x_1 + a_2x_2 + \cdots + a_nx_n$ for some choice of scalars $a_1, a_2, \ldots, a_n$, the result being another object of the same kind.

when $\mathbf{I}$ is the $n \times n$ **identity matrix**—i.e., the square matrix

$$\mathbf{I} \triangleq \begin{bmatrix} 1 & 0 & \cdots & 0 \\ 0 & 1 & \cdots & 0 \\ \vdots & \vdots & \ddots & \vdots \\ 0 & 0 & \cdots & 1 \end{bmatrix}$$

of order n, where the main-diagonal elements are all 1 and the other elements are all 0. Thus, an identity matrix is an example of a **diagonal matrix**—i.e., a square matrix for which all the elements not on the main diagonal are 0. Additionally, if $\mathbf{I}$ is the $m \times m$ identity matrix, then

$$\mathbf{IA} = \mathbf{A}. \tag{9.7b}$$

Conversely, upon substituting any other matrix for $\mathbf{I}$, neither (9.7a) nor (9.7b) will hold for *all* $m \times n$ matrices $\mathbf{A}$ (but might hold for some).

The ***n*th power** of a *square* matrix $\mathbf{A}$ is defined as the matrix

$$\mathbf{A}^n \triangleq \underbrace{\mathbf{AA}\cdots\mathbf{A}}_{n \text{ occurrences of "}\mathbf{A}\text{"}} \quad (n = 1, 2, \ldots)$$

—i.e., $\mathbf{A}$ multiplied by itself $n - 1$ times. Also, for $\mathbf{A} \neq \mathbf{0}$ we define $\mathbf{A}^0 \triangleq \mathbf{I}$, where $\mathbf{I}$ is the identity matrix having the same order as $\mathbf{A}$.[4]

A finite collection of n-dimensional vectors $\mathbf{x}_k$ $(k = 1, \ldots, m)$—either all row vectors or all column vectors—is said to be **linearly independent** if and only if for all scalars α_k $(k = 1, \ldots, m)$, we have

$$\sum_{k=1}^{m} \alpha_k \mathbf{x}_k = \mathbf{0} \qquad \Rightarrow \qquad \alpha_k = 0 \quad (k = 1, \ldots, m);$$

otherwise, the collection is **linearly dependent**. (Of course, the reverse implication always holds.) Equivalently, the collection is linearly dependent if and only if at least one of the vectors $\mathbf{x}_l$ $(l = 1, \ldots, m)$ can be expressed as a linear combination of the *other* vectors in the collection—i.e.,

$$\mathbf{x}_l = \sum_{k \neq l} \alpha_k \mathbf{x}_k$$

for some scalars α_k $(k \neq l)$. It follows that the collection must be linearly independent when $m = 1$, and linearly dependent when $m > n$. Finally, every subcollection of a collection of linearly independent vectors is also linearly independent (but, the corresponding assertion cannot be made about linear *de*pendence).

---

4. As with the 0th power of 0 (see fn. 1 in Chapter 6), it is sometimes convenient for $\mathbf{0}^0$ to be defined (usually as $\mathbf{I}$).

9.1 Determinants and Inverses

Consider now the matrix $\mathbf{A}$ in (9.1) with $m = n$. Let P be the set of all $n!$ permutations of the n-tuple $(1, 2, \ldots, n)$. For each such tuple $(j_1, j_2, \ldots, j_n)$, let $s(j_1, j_2, \ldots, j_n) = 1$ if it is possible, by successively interchanging elements *two* at a time, to transform $(j_1, j_2, \ldots, j_n)$ back into $(1, 2, \ldots, n)$ by performing an *even* number of interchanges; otherwise, let $s(j_1, j_2, \ldots, j_n) = -1$.[5] Using this permutation set P and sign function $s(\cdot\ \ldots, \cdot)$, the **determinant** of the *square* matrix $\mathbf{A}$ is defined as the number

$$\det \mathbf{A} \triangleq \sum_{(j_1, j_2, \ldots, j_n) \in P} s(j_1, j_2, \ldots, j_n)\, a_{1,j_1} a_{2,j_2} \cdots a_{n,j_n} \tag{9.8}$$

—often written "$|\mathbf{A}|$".[6] In particular, for the first two orders n we have

$$\det \mathbf{A} = \begin{cases} a_{1,1} & n = 1 \\ a_{1,1}a_{2,2} - a_{1,2}a_{2,1} & n = 2 \end{cases}. \tag{9.9}$$

Also, the determinant of a diagonal matrix simply equals the product of the elements on the main diagonal; thus, the determinant of every identity matrix is 1. If $\det \mathbf{A} = 0$, then $\mathbf{A}$ is called **singular**; otherwise, it is **nonsingular**.

When $n \geq 2$, the determinant of $\mathbf{A}$ can be calculated systematically by expanding $\det \mathbf{A}$ in terms of the determinants of successively smaller submatrices. First, we introduce the **cofactors** of $\mathbf{A}$,

$$c_{i,j} = (-1)^{i+j} \det \mathbf{A}_{i,j} \quad (i, j = 1, \ldots, n), \tag{9.10}$$

where each $\mathbf{A}_{i,j}$ is the $(n-1) \times (n-1)$ matrix obtained by deleting the ith row and jth column of $\mathbf{A}$. Then, for any $j \in \{1, \ldots, n\}$, a **row expansion** about the jth row of $\mathbf{A}$ can be performed, yielding

$$\det \mathbf{A} = \sum_{i=1}^{n} a_{i,j} c_{i,j}; \tag{9.11a}$$

alternatively, for any $i \in \{1, \ldots, n\}$, a **column expansion** about the ith column of $\mathbf{A}$ can be performed, yielding

$$\det \mathbf{A} = \sum_{j=1}^{n} a_{i,j} c_{i,j}. \tag{9.11b}$$

Repeating these steps as necessary—i.e., likewise expanding each $\det \mathbf{A}_{i,j}$ in (9.10) in terms of the determinants of $(n-2) \times (n-2)$ matrices, and so forth—we eventually

5. For example, if $n = 3$, then $P = \{(1,2,3), (1,3,2), (2,1,3), (2,3,1), (3,1,2), (3,2,1)\}$. Also, $s(2,3,1) = 1$, since we can permute $(2,3,1)$ into $(1,3,2)$ and then into $(1,2,3)$; whereas $s(3,2,1) = -1$. In general, for $n \geq 2$ we have $s(j_1, j_2, \ldots, j_n) = 1$ for half of the tuples $(j_1, j_2, \ldots, j_n)$, with $s(j_1, j_2, \ldots, j_n) = -1$ for the other half.

6. We will not be using this alternative notation, though, for by avoiding it, we can write the absolute value of a determinant without causing confusion.

express $\det \mathbf{A}$ completely in terms of the determinants of either 2×2 or 1×1 matrices, which can be evaluated by (9.9).

Here are some basic facts for any square matrices $\mathbf{A}$ and $\mathbf{B}$ of the same order $n \geq 1$, and any complex number α:

$$\det \mathbf{A}^{\mathrm{T}} = \det \mathbf{A}, \tag{9.12a}$$

$$\det \mathbf{A}^{*} = (\det \mathbf{A})^{*}, \tag{9.12b}$$

$$\det \overline{\mathbf{A}} = (\det \mathbf{A})^{*}, \tag{9.12c}$$

$$\det \mathbf{AB} = \det \mathbf{A} \cdot \det \mathbf{B}, \tag{9.12d}$$

$$\det(\alpha \mathbf{A}) = \alpha^{n} \det \mathbf{A}. \tag{9.12e}$$

Also, it follows immediately from (9.11) that if any row or column of $\mathbf{A}$ is multiplied by some complex number α, then $\det \mathbf{A}$ is multiplied by α as well; hence (choosing $\alpha = 0$), if every element of some row or column of $\mathbf{A}$ is 0, then $\det \mathbf{A} = 0$. Additionally, if any two rows (or columns) of $\mathbf{A}$ are interchanged, then $\det \mathbf{A}$ changes sign. It can also be shown that if a complex multiple α of a particular row of $\mathbf{A}$ is added to any *other* row, then the value of $\det \mathbf{A}$ is unchanged; and, likewise for the columns of $\mathbf{A}$. Therefore (choosing $\alpha = -1$), if any two rows (or columns) of $\mathbf{A}$ are the same, then $\det \mathbf{A} = 0$. In fact, $\det \mathbf{A} = 0$ if and only if the rows of $\mathbf{A}$ are linearly dependent; and, likewise for its columns.

Given a square matrix $\mathbf{A}$ of order n, the corresponding **inverse matrix** is the unique square matrix $\mathbf{A}^{-1}$ of order n such that

$$\mathbf{A}\mathbf{A}^{-1} = \mathbf{I}, \quad \mathbf{A}^{-1}\mathbf{A} = \mathbf{I}, \tag{9.13}$$

where $\mathbf{I}$ is the nth-order identity matrix. Such a matrix $\mathbf{A}^{-1}$ exists if and only if $\mathbf{A}$ is nonsingular; thus, for matrices, a synonym for "nonsingular" is **invertible**, and a synonym for "singular" is **noninvertible**. Taking $m = n$ in (9.1), the inverse of the matrix $\mathbf{A}$ for $n = 1$ is

$$\left[a_{1,1}\right]^{-1} = \left[{a_{1,1}}^{-1}\right];$$

whereas for $n > 1$ we have

$$\mathbf{A}^{-1} = \frac{\mathbf{C}^{\mathrm{T}}}{\det \mathbf{A}}, \tag{9.14}$$

where $\mathbf{C}$ is the $n \times n$ **cofactor matrix** $[c_{i,j}]$ formed from the cofactors (9.10) of $\mathbf{A}$. In particular, for $n = 2$,

$$\begin{bmatrix} a_{1,1} & a_{1,2} \\ a_{2,1} & a_{2,2} \end{bmatrix}^{-1} = \frac{1}{a_{1,1}a_{2,2} - a_{1,2}a_{2,1}} \begin{bmatrix} a_{2,2} & -a_{1,2} \\ -a_{2,1} & a_{1,1} \end{bmatrix}. \tag{9.15}$$

The inverse of a nonsingular diagonal matrix obtains by simply replacing each element on the main diagonal with its reciprocal; thus, the inverse of an identity matrix is itself.

Here are some basic facts for any square matrices **A** and **B** of the same order, and any complex number $\alpha \neq 0$:

$$\begin{aligned}
(\mathbf{A}^{-1})^{-1} &= \mathbf{A}, \\
(\mathbf{A}^{-1})^{\mathrm{T}} &= (\mathbf{A}^{\mathrm{T}})^{-1}, \\
(\mathbf{A}^{-1})^{*} &= (\mathbf{A}^{*})^{-1}, \\
\overline{\mathbf{A}^{-1}} &= (\overline{\mathbf{A}})^{-1}, \\
(\mathbf{AB})^{-1} &= \mathbf{B}^{-1}\mathbf{A}^{-1}, \\
(\alpha\mathbf{A})^{-1} &= \mathbf{A}^{-1}/\alpha, \\
\det \mathbf{A}^{-1} &= 1/\det \mathbf{A}
\end{aligned}$$

—assuming for the first equality that $\mathbf{A}^{-1}$ exists, and for each of the remaining equalities that one side or the other exists. Furthermore, if either $\mathbf{AB} = \mathbf{I}$ or $\mathbf{BA} = \mathbf{I}$, then we must have $\mathbf{B} = \mathbf{A}^{-1}$; therefore, by (9.13), the other equality of this pair holds as well.

9.2 Simultaneous Linear Equations

Many problems in mathematics are expressed in terms of constraints placed on $n \geq 1$ variables "x_j" ($j = 1, \ldots, n$), the values of which are initially unknown. A solution to the problem then consists of an n-tuple $(x_1, x_2, \ldots, x_n)$ of values x_j for which the constraints are simultaneously satisfied. For example, we might want to know for what pairs (x_1, x_2) of real values x_1 and x_2 do we have both $e^{x_1} > x_2$ and ${x_1}^3 \leq |x_2|$, each such pair (if any exist) being a solution. The general approach of assigning variables to unknown values, then manipulating the variables to perform analyses upon these values, is called **algebra**.

A frequently occurring case is where the given constraints are in the form of $m \geq 1$ **simultaneous linear equations**. Namely, what is required is

$$\begin{aligned}
a_{1,1}x_1 + a_{1,2}x_2 + \cdots + a_{1,n}x_n &= b_1, \\
a_{2,1}x_1 + a_{2,2}x_2 + \cdots + a_{2,n}x_n &= b_2, \\
&\vdots \\
a_{m,1}x_1 + a_{m,2}x_2 + \cdots + a_{m,n}x_n &= b_m,
\end{aligned} \tag{9.16}$$

for some given **coefficients** $a_{i,j} \in \mathbb{C}$ and $b_i \in \mathbb{C}$ ($i = 1, \ldots, m; j = 1, \ldots, n$). The equations (9.16) are said to be **consistent** if there exists at least one solution; otherwise, they are **inconsistent**.

Taking $\mathbf{A}$ to be the $m \times n$ matrix in (9.1), and introducing the vectors

$$\mathbf{x} = \begin{bmatrix} x_1 \\ x_2 \\ \vdots \\ x_n \end{bmatrix}, \quad \mathbf{b} = \begin{bmatrix} b_1 \\ b_2 \\ \vdots \\ b_m \end{bmatrix},$$

the m equations in (9.16) can be succinctly expressed by the single matrix equation

$$\mathbf{Ax} = \mathbf{b}; \tag{9.17}$$

equivalently,

$$\mathbf{x}^{\mathrm{T}}\mathbf{A}^{\mathrm{T}} = \mathbf{b}^{\mathrm{T}},$$

should row vectors be preferred to column vectors. For the special case of $m = n$, if the square matrix $\mathbf{A}$ is nonsingular, then for each vector $\mathbf{b}$ there exists a *unique* solution $\mathbf{x}$ to (9.17); namely, left-multiplying (9.17) by $\mathbf{A}^{-1}$ yields

$$\mathbf{x} = \mathbf{A}^{-1}\mathbf{b}.$$

Alternatively, each component x_j of $\mathbf{x}$ can be solved for individually by applying **Cramer's rule**:

$$x_j = \frac{\det \mathbf{A}_j}{\det \mathbf{A}} \quad (j = 1, \ldots, n), \tag{9.18}$$

where $\mathbf{A}_j$ is the matrix that results from replacing the jth column of $\mathbf{A}$ with $\mathbf{b}$. By contrast, if $\mathbf{A}$ is singular, then for each vector $\mathbf{b}$ there are either infinitely many solutions $\mathbf{x}$ to (9.17) or none at all.

Finally, when simultaneous equations are manipulated directly (i.e., by hand, without using matrices), it is usually best to *postpone division* as long as possible. For example, suppose

$$2x_1 + 5x_2 = 8, \tag{9.19a}$$

$$9x_1 + 7x_2 = 6. \tag{9.19b}$$

To obtain the value of x_1 without using matrices, one might first solve (9.19a) for x_2 in terms of x_1, getting

$$x_2 = (8 - 2x_1)/5; \tag{9.20}$$

then, substituting this result into (9.19b) yields the equation

$$9x_1 + 7 \cdot (8 - 2x_1)/5 = 6,$$

which can be solved by further manipulation to obtain $x_1 = -26/31$. This last step, though, is made more arduous by the division introduced in (9.20). Alternatively, observe how much easier it is to *multiply* (9.19a) by 7—i.e., the coefficient of x_2 in (9.19b)—to get

$$14x_1 + 35x_2 = 56, \tag{9.21a}$$

and multiply (9.19b) by 5—i.e., the coefficient of x_2 in (9.19a)—to get

$$45x_1 + 35x_2 = 30, \tag{9.21b}$$

which has the same coefficient for x_2 as in (9.21a); then, subtracting (9.21b) from (9.21a) eliminates x_2, giving

$$-31x_1 = 26.$$

Now, performing a division as the *last* step, we again obtain $x_1 = -26/31$.

9.3 Eigenvectors and Eigenvalues

Given an $n \times n$ matrix $\mathbf{A}$, we say that an $n \times 1$ vector $\mathbf{x}$ is an **eigenvector** (or **characteristic vector**) of $\mathbf{A}$ if and only if $\mathbf{x}$ is nonzero (i.e., at least one of its elements is nonzero) and

$$\mathbf{Ax} = \lambda\mathbf{x} \tag{9.22}$$

for some value $\lambda \in \mathbb{C}$; accordingly, λ is the **eigenvalue** (or **characteristic value**) of $\mathbf{A}$ associated with $\mathbf{x}$.[7] Note that (9.22) holds if and only if

$$\mathbf{0} = \lambda\mathbf{x} - \mathbf{Ax} = \lambda\mathbf{Ix} - \mathbf{Ax} = (\lambda\mathbf{I} - \mathbf{A})\mathbf{x},$$

where $\mathbf{0}$ is the $n \times 1$ zero matrix and $\mathbf{I}$ is the $n \times n$ identity matrix. It follows that the $n \times n$ matrix $\lambda\mathbf{I} - \mathbf{A}$ must be singular; for otherwise we would have $(\lambda\mathbf{I} - \mathbf{A})^{-1}\mathbf{0} = \mathbf{x}$, which is impossible since $\mathbf{x} \neq \mathbf{0}$. Hence, *if* a constant $\lambda \in \mathbb{C}$ is an eigenvalue of $\mathbf{A}$, then

$$\det(\lambda\mathbf{I} - \mathbf{A}) = 0.$$

Conversely, when this **characteristic equation** holds for some $\lambda \in \mathbb{C}$, the columns of $\lambda\mathbf{I} - \mathbf{A}$ are linearly dependent, so $(\lambda\mathbf{I} - \mathbf{A})\mathbf{x} = \mathbf{0}$ for some nonzero $\mathbf{x}$; therefore, (9.22) holds, and λ *must* be an eigenvalue of $\mathbf{A}$.

Letting λ be variable makes $\det(\lambda\mathbf{I} - \mathbf{A})$ an nth-degree polynomial in λ; thus, the n roots (including repetitions) of this **characteristic polynomial** are the *unique* eigenvalues of $\mathbf{A}$. Accordingly, borrowing some terminology regarding polynomials, each eigenvalue is either "simple" or "multiple", with its "multiplicity" being the number of times that it occurs. Multiple eigenvalues are also called **degenerate**; hence, simple eigenvalues are **nondegenerate**. Collectively, the eigenvalues of a (square) matrix are called its **spectrum**.

The eigen*vectors* of a matrix $\mathbf{A}$, though, are never unique. For one thing, it is clear from (9.22) that multiplying any eigenvector of $\mathbf{A}$ by a nonzero constant yields another eigenvector of $\mathbf{A}$ having the same associated eigenvalue. More generally, if

---

7. Here we are discussing *left*-multiplication by $\mathbf{A}$, and therefore $\mathbf{x}$ is a column vector. The treatment of right-multiplication by $\mathbf{A}$—for which the corresponding "eigenequation" is $\mathbf{xA} = \lambda\mathbf{x}$, with $\mathbf{x}$ being a row vector—is similar.

$\mathbf{x}_1, \mathbf{x}_2, \ldots, \mathbf{x}_m$ $(m \geq 1)$ are all eigenvectors of $\mathbf{A}$ having the same associated eigenvalue λ, then so is any linear combination

$$\alpha_1\mathbf{x}_1 + \alpha_2\mathbf{x}_2 + \cdots + \alpha_m\mathbf{x}_m \quad (\alpha_1, \alpha_2, \ldots, \alpha_m \in \mathbb{C})$$

which does not equal $\mathbf{0}$. As an extreme example, *every* nonzero $n \times 1$ vector is an eigenvector of the $n \times n$ identity matrix $\mathbf{I}$, with associated eigenvalue 1.

If a nonzero $n \times 1$ vector $\mathbf{x}$ is not an eigenvector of an $n \times n$ matrix $\mathbf{A}$, then multiplying $\mathbf{x}$ by $\mathbf{A}$ is a more complicated operation than merely scaling $\mathbf{x}$ by some factor λ as in (9.22); however, for some matrices $\mathbf{A}$ we can view any matrix-vector multiplication $\mathbf{Ax}$ as a *sum* of scalar-vector multiplications. Specifically, let the n eigenvalues of $\mathbf{A}$ be λ_k $(k = 1, 2, \ldots, n)$, and suppose there exist n corresponding eigenvectors

$$\mathbf{x}_k = \begin{bmatrix} x_{k,1} \\ x_{k,2} \\ \vdots \\ x_{k,n} \end{bmatrix} \quad (k = 1, 2, \ldots, n)$$

that are *linearly independent*. (In particular, this is necessarily the case when the eigenvalues are nondegenerate.) Then, the related matrix

$$\mathbf{X} = \begin{bmatrix} \mathbf{x}_1 & \mathbf{x}_2 & \cdots & \mathbf{x}_n \end{bmatrix} = \begin{bmatrix} x_{1,1} & x_{2,1} & \cdots & x_{n,1} \\ x_{1,2} & x_{2,2} & \cdots & x_{n,2} \\ \vdots & \vdots & & \vdots \\ x_{1,n} & x_{2,n} & \cdots & x_{n,n} \end{bmatrix} \tag{9.23a}$$

is nonsingular; therefore, we can let

$$\begin{bmatrix} \alpha_1 \\ \alpha_2 \\ \vdots \\ \alpha_n \end{bmatrix} = \mathbf{X}^{-1}\mathbf{x}, \tag{9.23b}$$

thereby making

$$\mathbf{x} = \mathbf{X}\begin{bmatrix} \alpha_1 \\ \alpha_2 \\ \vdots \\ \alpha_n \end{bmatrix} = \sum_{k=1}^{n} \alpha_k\mathbf{x}_k.$$

Now, as a generalization of (9.22), we have

$$\mathbf{Ax} = \mathbf{A}\left(\sum_{k=1}^{n} \alpha_k\mathbf{x}_k\right) = \sum_{k=1}^{n} \alpha_k\mathbf{Ax}_k = \sum_{k=1}^{n} \alpha_k \cdot \lambda_k\mathbf{x}_k = \sum_{k=1}^{n} \lambda_k \cdot \alpha_k\mathbf{x}_k.$$

Thus, we can view the multiplication of $\mathbf{x}$ by $\mathbf{A}$ as a two-step process: first, express $\mathbf{x}$ as a linear combination of the eigenvectors $\mathbf{x}_k$; then, multiply each term in this combination by the corresponding eigenvalue λ_k. Moreover, the vectors $\mathbf{x}_k$ and values λ_k can be predetermined once and for all from the matrix $\mathbf{A}$; only the scalars α_k in the linear combination, which obtain from (9.23), depend on the vector $\mathbf{x}$.

Unfortunately, not every $n \times n$ matrix has n linearly independent eigenvectors. For example, the matrix

$$\mathbf{A} = \begin{bmatrix} 1 & 0 \\ 1 & 1 \end{bmatrix}$$

has the characteristic polynomial

$$\det(\lambda\mathbf{I} - \mathbf{A}) = \det\left(\begin{bmatrix} \lambda & 0 \\ 0 & \lambda \end{bmatrix} - \begin{bmatrix} 1 & 0 \\ 1 & 1 \end{bmatrix}\right) = \det\begin{bmatrix} \lambda - 1 & 0 \\ -1 & \lambda - 1 \end{bmatrix} = (\lambda - 1)^2,$$

making both of its eigenvalues 1. Hence, taking $\lambda = 1$ yields

$$\lambda\mathbf{I} - \mathbf{A} = \begin{bmatrix} 0 & 0 \\ -1 & 0 \end{bmatrix}.$$

Therefore, $(\lambda\mathbf{I} - \mathbf{A})\mathbf{x} = \mathbf{0}$—i.e., (9.22) holds—if and only if

$$\mathbf{x} = \begin{bmatrix} 0 \\ \beta \end{bmatrix} \qquad (\beta \in \mathbb{C});$$

but, no two such vectors that are nonzero (i.e., $\beta \neq 0$) are linearly independent.

The product of the eigenvalues λ_k ($k = 1, 2, \ldots, n$) of any square matrix $\mathbf{A}$ equals its determinant:

$$\det \mathbf{A} = \prod_{k=1}^{n} \lambda_k. \tag{9.24}$$

Because, the characteristic polynomial of $\mathbf{A}$ is

$$\det(\lambda\mathbf{I} - \mathbf{A}) = \prod_{k=1}^{n} (\lambda - \lambda_k);$$

therefore, taking $\lambda = 0$ yields

$$\det(-\mathbf{A}) = \prod_{k=1}^{n} (-\lambda_k) = (-1)^n \prod_{k=1}^{n} \lambda_k,$$

whereas by (9.12e) we have $\det(-\mathbf{A}) = (-1)^n \det \mathbf{A}$. It can also be shown that the sum of the eigenvalues λ_k, which is called the **trace** of $\mathbf{A}$ (tr $\mathbf{A}$) equals the sum of main-diagonal elements of $\mathbf{A}$; that is, if $\mathbf{A} = [a_{i,j}]_{i,j=1}^{n}$, then

$$\operatorname{tr} \mathbf{A} \triangleq \sum_{k=1}^{n} \lambda_k = \sum_{i=1}^{n} a_{i,i}.$$

It follows from (9.24) that an $n \times n$ matrix $\mathbf{A}$ is singular if and only if it has zero as an eigenvalue; that is, $\mathbf{Ax} = \mathbf{0}$ for some nonzero $n \times 1$ matrix $\mathbf{x}$. This fact also obtains directly from the interpretation (9.6b) of the general matrix multiplication in (9.3); namely, multiplying $\mathbf{A}$ on the right by $\mathbf{x}$ performs a linear combination of the columns of $\mathbf{A}$, which are linearly dependent if and only if $\det \mathbf{A} = 0$. Likewise, interpreting (9.3) by (9.6a) shows that $\mathbf{A}$ is singular if and only if $\mathbf{xA} = \mathbf{0}$ for some nonzero $1 \times n$ matrix $\mathbf{x}$ (with $\mathbf{0}$ now being $1 \times n$), since the rows of $\mathbf{A}$ are linearly dependent if and only if $\det \mathbf{A} = 0$.

CHAPTER 10

Euclidean Geometry

Most people view mathematics as dealing primarily with numbers. However, it also is concerned with the fundamental properties of *space*. The study of spatial figures is called **geometry**.

Although our treatment of this topic will be somewhat informal, geometry is often cited as an exemplar of a theory developed strictly according to the **axiomatic method**: First, conceptual objects such as "point", "line", and "plane" are introduced, along with relations that relate them, such as "lying on", "between", and "congruent". Though often based on intuition, these initial **primitive** items by which the theory is fundamentally expressed are, in fact, *undefined*; that is, the existence of each is simply posited without explanation in terms of anything else.[1] Additional objects and relations such as "line segment", "collinear", and "triangle" are then defined in terms of the primitive objects and relations; then, in turn, any objects and relations that have been defined may be used to define further objects and relations, and so on. Next, all *assumed* geometric facts are explicitly stated as axioms. Only through these accepted assertions (as opposed to anyone's subjective intuition) do the objects and relations (primitive and defined) in the theory become imbued with any theoretical interpretation (i.e., "meaning"). In particular, in this chapter we concentrate on **plane geometry** and **solid geometry**—the two- and three-dimensional forms, respectively, of **Euclidean geometry**—in which the axioms are intended to capture the basic properties of ordinary space.[2] Finally, additional geometric facts are derived by proving theorems from the axioms; then, in turn, these theorems may be used to prove further theorems, and so on.

Even when one is adhering to the axiomatic method, there are various ways by which plane and solid geometry can be developed. Here, in part for greater continuity with the next chapter, all geometric objects other than individual points will be expressed

1. Examples of nongeometrical primitives are "set" and "member of" in set theory.

2. Other axiom systems for geometry can be useful as well. For example, a *non*-Euclidean geometry—in which space can be "curved" rather than "flat"—was used by Albert Einstein to express his theory of gravitation, the general theory of relativity. Also, although difficult to visualize, it is actually a straightforward mathematical matter to generalize Euclidean geometry to spaces of *higher* than three dimensions; see Subsection 11.3.2.

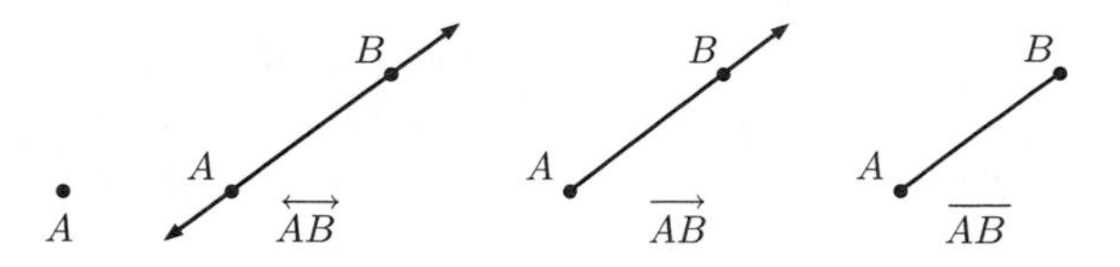

FIGURE 10.1 A Point, a Line, a Ray, and a Segment

as *sets* of points.[3] It should also be mentioned that in a more step-by-step development than ours, there would be the option of proceeding by either the "metric" approach, in which numerical measures of distance and angle appear at the outset and are used to *define* relations such as betweenness and congruence, or by the classical approach, in which such relations are *primitive*. For ease of exposition, though, no care is taken here to specify which concepts are primitive and which are defined; furthermore, to motivate the concepts, heuristic remarks are freely made.

10.1 Basic Geometric Components

To begin, a **point** is an indivisible, zero-dimensional geometric object. As illustrated at the left of Figure 10.1 by a dot, any particular point A may be visualized as specifying a fixed position in the **space** under consideration, which is the set of *all* points.

A **(geometric) figure** is any particular set of points; that is, it is a subset of the space. Two or more figures are said to **coincide**, or be **coincident**, if and only if they are all the same set of points.[4] Two figures **intersect** if and only if they have at least one point in common; that is, their intersection (as a set) is nonempty. Thus, two figures are nonintersecting if and only if they are disjoint (as sets). A point A **lies on** a figure f if and only if $A \in f$, whereas a figure f **lies in** another figure g if and only if $f \subseteq g$;[5] equivalently, one may say that (in terms of sets) f *contains* A, and g *contains* f.[6]

A **(straight) line** is a straight, unbounded, complete, one-dimensional geometric figure. It comprises infinitely many points, each of which is also contained in infinitely many other lines lying in the space. Two or more figures are said to be **collinear** if and

3. An alternative approach would be to avoid set theory and continually state, e.g., "point A *lies on* line l" and "line l *lies in* plane p", instead of simply writing "$A \in l$" and "$l \subseteq p$". Also, rather than refer to a specific geometric figure as the "set" of points satisfying a particular condition, we would refer to it as the **locus** of points satisfying that condition.

4. Strictly speaking, this definition leaves open what it means for individual *points* to "coincide"; because, a point A is not the same as the set $\{A\}$ containing it, and thus is not a figure. Nevertheless, to streamline the discussion, all obvious extensions of our definitions (for "coincide" here, and below for, e.g., "collinear" and "coplanar") from geometric figures to individual points, and combinations thereof, will be understood. Thus, two or more *points* coincide if and only if they are all the same point; and, a *figure* f coincides with a *point* A if and only if $f = \{A\}$.

5. Often, the word "lies" in these expressions is replaced with "is". Also, it is common to hear that a *point* "lies *in*" some figure—e.g., "point A lies in plane p"—as if treating the point A as the corresponding figure $\{A\}$ (see preceding footnote).

6. See fn. 3 in Chapter 2.

only if they all lie in a single line. A figure for which all the points are collinear is a **line figure**; hence, any collinear figures must all be line figures.

Through any two distinct points A and B there passes exactly one line, $\overleftrightarrow{AB}$ (or $\overleftrightarrow{BA}$); see Figure 10.1. The point A (and, similarly, the point B) separates this line into two parts: the **half-line** comprising those points on $\overleftrightarrow{AB}$ lying on the same side of A as B, and the half-line comprising those points on $\overleftrightarrow{AB}$ lying on the opposite side of A as B. (Since A does not belong to either half-line, it together with the two half-lines comprise a partition of $\overleftrightarrow{AB}$.) Combining the first of these half-lines with the point A itself, we obtain the **ray** $\overrightarrow{AB}$ (or $\overleftarrow{BA}$), of which A is the unique **endpoint** (which is also the "endpoint" of the corresponding half-line, though not part of it); see Figure 10.1. Any half-line (respectively, ray) can be extended to form a line, and the extension is unique; that is, exactly one line contains the half-line (ray). In fact, $\overleftrightarrow{AB} = \overrightarrow{AB} \cup \overleftarrow{AB}$.

A point C lies **between** two points A and B if and only if these three points are distinct and collinear, with A and B lying on opposite sides of C (on the line passing though them all); thus, in Figure 10.2, C is not between A and B, whereas the point P is. For any two distinct points A and B, the **(line) segment** $\overline{AB}$ (or $\overline{BA}$)[7] is the set composed of both A and B—called the **endpoints** of the segment—along with all the points between, which constitute its **interior**; see Figure 10.1. It follows that $\overline{AB} = \overrightarrow{AB} \cap \overleftarrow{AB}$. Furthermore, any segment can be extended in either direction to form a ray, and for each direction the extension is unique; that is, for each of its endpoints, the segment lies in exactly one ray for which that endpoint is also an endpoint of the ray. And, by extending either of these rays, the segment can be extended to form the unique line containing it.

A **plane** is a flat, unbounded, complete, two-dimensional geometric figure. It comprises infinitely many points, each of which is contained in infinitely many lines lying in the plane; and, in solid geometry, each of these lines is also contained in infinitely many other planes lying in the space. Two or more figures are said to be **coplanar** if and only if they all lie in a single plane. A figure for which all the points are coplanar is a **plane figure**; hence, any coplanar figures must all be plane figures.

Through any three noncollinear points (which, therefore, must be distinct), there passes exactly one plane; similarly, given a line and a point not on it, there is a exactly one plane containing both (hence, every line figure is a plane figure). It follows that for any two distinct points lying on a plane, the line passing through those points must lie in that plane (consider a third point on the plane that is noncollinear with the other

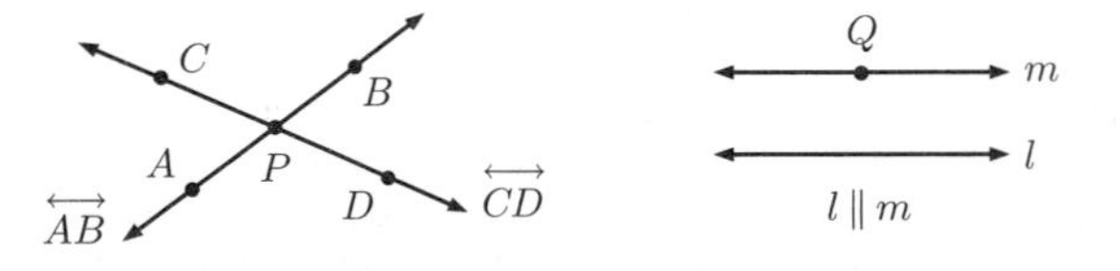

FIGURE 10.2 Crossing Lines and Parallel Lines

7. Notation in geometry differs somewhat from author to author. Here, another common way to indicate this segment is simply "AB".

two). Furthermore, just as a point on a line separates the line into two half-lines, so a line in a plane separates the plane into the **half-plane** on one side of the line and the half-plane on the other side; and, the unique **edge** of each of these half-planes is this line. (The line and the two half-planes are disjoint, and together comprise a partition of the plane.) A **closed half-plane** is the union of a half-plane and its edge (which is also the "edge" of the closed half-plane). Any half-plane (respectively, closed half-plane) can be extended to form a plane, and the extension is unique; that is, exactly one plane contains the half-plane (closed half-plane).

If two lines $\overleftrightarrow{AB}$ and $\overleftrightarrow{CD}$ intersect, then either they share exactly one point P—i.e., they **cross** at P, as shown on the left side of Figure 10.2—or they coincide.[8] In other words, two distinct lines cannot intersect at more than one point. Given two crossing lines, there is exactly one plane containing both.

In plane geometry, the space *is* a plane; hence, in this case there are no other planes to speak of, and all figures are coplanar. By contrast, in solid geometry the space is visualized as a unbounded, complete, three-dimensional expanse; and, this particular space is often referred to simply as **space**.[9] A **solid figure** (or **solid**) is a figure in space that is not a plane figure.

Every plane in space separates it into the **half-space** on one side of the plane and the half-space on the other side; and, the unique **face** of each of these half-spaces is this plane. (The plane and the two half-spaces are disjoint, and together comprise a partition of space.) A **closed half-space** is the union of a half-space and its face (which is also the "face" of the closed half-space).

If two planes intersect, then either the points they share constitute a line l—i.e., they **cross** at l—or they coincide. Whereas if a line and a plane intersect, then either they share exactly one point or the line lies in the plane.

Two lines l and m that are coplanar but do not cross are called **parallel lines**—i.e., $l \parallel m$ (or $m \parallel l$);[10] see the right side of Figure 10.2. It follows that two parallel lines coincide if and only if they intersect. More than two lines are said to be "parallel" if and only if each pair are parallel; in solid geometry, though, such lines need not be coplanar. Two lines that are noncoplanar—i.e., neither cross nor are parallel—are called **skew lines**.

Two planes p and q are said to be **parallel planes**—i.e., $p \parallel q$ (or $q \parallel p$)—if and only if they are either coincident or nonintersecting. It follows that two parallel planes coincide if and only if they intersect. More than two planes are said to be "parallel" if and only if each pair are parallel.

---

8. As with notation, terminology may vary slightly among authors. For example, some define "intersecting lines" as crossing only (and thus they do not formally introduce the term "cross"); whereas for us, every line intersects (but does not cross) itself.

9. When greater clarity is desired, the three-dimensional space of solid geometry may be referred to as **3-space**, as distinct from the **2-space** of plane geometry. *All statements in this chapter, except where explicitly indicated otherwise, are valid for both plane geometry and solid geometry (though perhaps not for higher-dimensional Euclidean spaces, or for 1-space).*

10. By this definition, every line is parallel to itself; likewise below for planes. Also, as with some of the terms to follow (e.g., "perpendicular"), the definition of "parallel" can be generalized in a natural way to apply to other line figures as well (e.g., a ray and a segment can be parallel); however, to avoid cluttering the discussion, such obvious extensions will not be explicitly stated.

A *line* l and a *plane* p are **parallel**—i.e., $l \parallel p$ (or $p \parallel l$)—if and only if either the line lies in the plane or the line and the plane are nonintersecting. It follows that if a line and a plane are parallel, then the line lies in the plane if and only if the two intersect; whereas if a line and a plane are not parallel, then they share exactly one point.

The primary axiom that distinguishes Euclidean geometry from most other geometries is the **parallel postulate**. One version of the postulate states that for any line l and point Q, there is exactly one line m parallel to l that contains Q; see Figure 10.2. A similar fact is that for any plane p and point Q, there is exactly one plane parallel to p that contains Q. Furthermore, for any plane p and parallel line l, there is exactly one plane parallel to p that contains l.

For each line l in a plane, there are infinitely many lines in the plane that are parallel to l; and, for each plane p in space, there are infinitely many planes in space that are parallel to p. If l_1 and l_2 are lines that are both parallel to another line, then $l_1 \parallel l_2$; but, it is possible for two lines l_1 and l_2 to both be parallel to a plane, yet not have $l_1 \parallel l_2$. Likewise, if p_1 and p_2 are planes that are both parallel to another plane, then $p_1 \parallel p_2$; but, it is possible for two planes p_1 and p_2 to both be parallel to a line, yet not have $p_1 \parallel p_2$. Also, if a line l and a plane p are both parallel to another line or plane, then $l \parallel p$.

As shown in Figure 10.3, any two rays $\overleftarrow{AB}$ and $\overrightarrow{BC}$ sharing an endpoint (thus, the point B is distinct from the points A and C, which may be the same) combine to give a **(plane) angle** $\angle ABC \triangleq \overleftarrow{AB} \cup \overrightarrow{BC}$. (Hence, $\angle CBA$ is the same angle.) The **sides** of this angle are the rays, and its **vertex** is the common endpoint B. (In many contexts, an angle $\angle ABC$ may be unambiguously indicated by its vertex alone as "$\angle B$".) If $\overleftarrow{AB}$ and $\overrightarrow{BC}$ are collinear, then either $\overleftarrow{AB} = \overrightarrow{BC}$—in which case $\angle ABC$ is itself this same ray and is called a **zero angle**—or $\angle ABC$ is the line $\overleftrightarrow{AC}$ and is called a **straight angle** (see the leftmost and rightmost parts of Figure 10.4). Otherwise, we will refer to $\angle ABC$ as a **proper angle**.

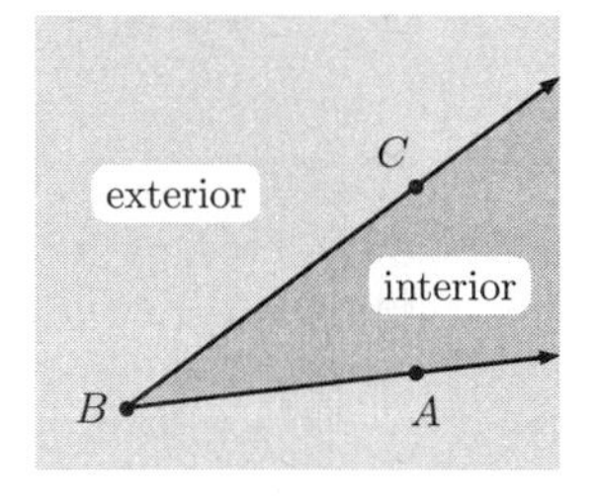

FIGURE 10.3 An Angle $\angle ABC$

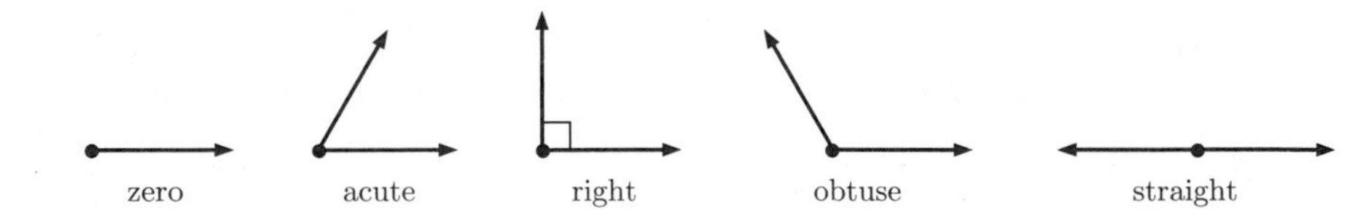

FIGURE 10.4 Types of Angles

Each proper angle $\angle ABC$ lies in exactly one plane. Its **interior** is the intersection of two half-planes lying in this plane (see Figure 10.3): the half-plane of points lying on the same side of $\overleftrightarrow{AB}$ as C, and the half-plane of points lying on the same side of $\overleftrightarrow{BC}$ as A. The **exterior** of $\angle ABC$ is then the portion of the plane other than the angle and its interior. (The angle, its interior, and its exterior are disjoint, and together they comprise a partition of the plane.)

10.2 Shape, Size, and Measure

Having introduced the basic objects and relations of Euclidean geometry, we now turn to matters of *shape* and *size*, beginning our discussion of these intuitive notions without the benefit of any measures of distance or angle. To wit, two geometric figures f and g are said to be **similar**—i.e., $f \sim g$ (or $g \sim f$)—if and only if they have the same shape; whereas f and g are **congruent**—i.e., $f \cong g$ (or $g \cong f$)—if and only if they have *both* the same shape and the same size.[11] Hence, congruence implies similarity, but not vice versa. Clearly, both the similarity relation $\sim$ and the congruence relation $\cong$ are reflexive, symmetric, and transitive; thus, each is an equivalence relation (on the set of all geometric figures in the space under consideration).

All points (individually) are congruent, as are all lines, all half-lines, all rays, all planes, all half-planes, and all closed half-planes. For any two segments $\overline{AB}$ and $\overline{CD}$, we have $\overline{AB} \sim \overline{CD}$; therefore, $\overline{AB} \cong \overline{CD}$ if (and only if) the segments have the same size. Whereas for two angles $\angle ABC$ and $\angle DEF$, we have $\angle ABC \cong \angle DEF$ if (and only if) $\angle ABC \sim \angle DEF$, since angles having the same shape also have the same size.[12]

Regarding the four proper angles formed by the two crossing lines in Figure 10.2—viz.,

$$\angle APC, \quad \angle APD, \quad \angle BPC, \quad \angle BPD$$

—any pair of these angles that do not share a side—e.g., $\angle APC$ and $\angle BPD$—are called **vertical angles**; such angles are necessarily congruent. Whereas any pair of these angles that do share a side—e.g., $\angle APC$ and $\angle APD$—(and thus the remaining two sides form a line) are called **supplementary angles**, with each angle being a **supplement** of the other. A **right angle**—commonly indicated by a corner symbol "$\neg$", as in Figure 10.4—is any proper angle having a congruent supplement; moreover, all right angles are congruent, even when not formed by the same pair of crossing lines. An **oblique angle** is a proper angle that is not a right angle.

11. If two figures do not have the same shape, then there may be more than one way to assess whether they have the same size; therefore, a general same-size relation is not introduced.

12. For *segments* and *angles*, no particular correspondence is implied between the geometric points explicitly appearing on one side of the binary relation $\sim$ and those appearing on the other side; and, likewise for the binary relation $\cong$. Thus, we may just as well express the similarity of $\overline{AB}$ and $\overline{CD}$ by writing "$\overline{AB} \sim \overline{DC}$", and the congruence $\angle ABC$ and $\angle DEF$ by writing "$\angle ABC \cong \angle FED$". However, for some other figures—e.g., polygons (see Subsection 10.3.1)—the order of the points explicitly appearing in such expressions *is* significant.

Two lines l and m are **perpendicular lines**—i.e., $l \perp m$ (or $m \perp l$)—if and only if they cross and one of the resulting proper angles is a right angle,[13] in which case all are right angles. A *line* l and a *plane* p are **perpendicular**—i.e., $l \perp p$ (or $p \perp l$)—if and only if they intersect at a single point P, and for all other points Q on p we have $\overleftrightarrow{PQ} \perp l$. Whereas if a line l and a plane p intersect at a single point but are not perpendicular, then there is exactly one line lying in p that is perpendicular to l.

In plane geometry, given a line l and a point P on it, there is exactly one line that is perpendicular to l at P. In solid geometry, given a plane p and a point P on it, there is exactly one line that is perpendicular to p at P.

In solid geometry, for any crossing lines l_1 and l_2, there is exactly one line l such that $l \perp l_1$ and $l \perp l_2$; moreover, l crosses l_1 and l_2 at the same point, and is perpendicular to the plane containing l_1 and l_2. For any skew lines l_1 and l_2, there is exactly one line l such that $l \perp l_1$ and $l \perp l_2$.

Two planes p and q are **perpendicular planes**—i.e., $p \perp q$ (or $q \perp p$)—if and only if:

(i) The planes p and q cross at some line l.
(ii) For every line $l_p \subseteq p$ such that $l_p \perp l$, we have $l_p \perp q$; and, for every line $l_q \subseteq q$ such that $l_q \perp l$, we have $l_q \perp p$.

When condition (i) holds, a sufficient (and necessary) condition for (ii) to hold is that there exist lines $l_p \subseteq p$ and $l_q \subseteq q$ such that $l_p \perp l$, $l_q \perp l$, and $l_p \perp l_q$ (i.e., the lines l, l_p, and l_q are *pairwise* perpendicular).

In solid geometry: For any crossing planes p_1 and p_2, there are infinitely many planes p such that $p \perp p_1$ and $p \perp p_2$, with all being parallel; moreover, if we are also given a point P in space, then exactly one such p contains P. Furthermore, the intersection of p_1, p_2, and each such p is a single point. Similarly, given a plane p and a parallel line l, there are infinitely many planes perpendicular to both p and l, with all being parallel; moreover, if we are also given a point P, then exactly one such plane contains P. By contrast, given a plane p and a line l that intersect but are not perpendicular, there is exactly one plane perpendicular to p that contains l.

Figure 10.5 shows the situation of two distinct coplanar lines $\overleftrightarrow{AB}$ and $\overleftrightarrow{CD}$ that are crossed at two distinct points P and Q by third line $\overleftrightarrow{EF}$—called a **transversal**—thereby causing eight proper angles to be formed. Four of these are the **interior angles**

$$\angle APF, \quad \angle BPF, \quad \angle CQE, \quad \angle DQE;$$

and, the other four are the **exterior angles**

$$\angle APE, \quad \angle BPE, \quad \angle CQF, \quad \angle DQF.$$

Moreover, these angles associate into various pairs. First, $\angle APF$ and $\angle DQE$ are called **alternate interior angles**, as are $\angle BPF$ and $\angle CQE$; likewise, $\angle APE$ and $\angle DQF$ are called **alternate exterior angles**, as are $\angle BPE$ and $\angle CQF$. Additionally, $\angle APF$ and

13. But, see fn. 30 in Chapter 11.

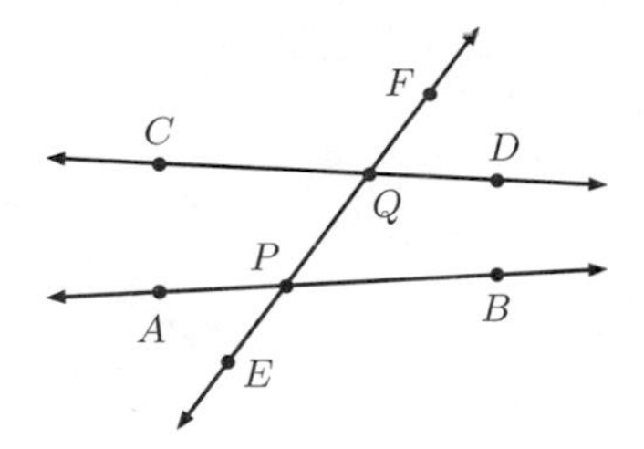

FIGURE 10.5 A Transversal across Two Lines

$\angle CQF$ are called **corresponding angles**, as are $\angle BPF$ and $\angle DQF$, $\angle APE$ and $\angle CQE$, and $\angle BPE$ and $\angle DQE$. If the two angles in *any* one of the above pairs are congruent, then the lines $\overleftrightarrow{AB}$ and $\overleftrightarrow{CD}$ are parallel; and, vice versa.

It follows that if l_1 and l_2 are coplanar lines that are both perpendicular to some other line, then $l_1 \parallel l_2$. It can also be shown that for any lines l_1 and l_2 that are both perpendicular to some plane, we have $l_1 \parallel l_2$. Likewise, if p_1 and p_2 are planes that are both perpendicular to some line, then $p_1 \parallel p_2$; but, it is possible for two planes p_1 and p_2 to both be perpendicular to another plane, yet not have $p_1 \parallel p_2$. Also, if a line l and a plane p are both perpendicular to another line or plane, then $l \parallel p$.

Furthermore, let f be either a line or a plane. If l_1 and l_2 are coplanar lines such that $l_1 \parallel f$ and $l_2 \perp f$, then $l_1 \perp l_2$. Likewise, if p_1 and p_2 are planes such that $p_1 \parallel f$ and $p_2 \perp f$, then $p_1 \perp p_2$. Also, if l is a line and p is a plane such that $l \parallel f$ and $p \perp f$, then $l \perp p$ when f is a line, but not necessarily when f is a plane. Similarly, if l is a line and p is a plane such that $l \perp f$ and $p \parallel f$, then $l \perp p$ when f is a plane, but not necessarily when f is a line.

In the classical approach to Euclidean geometry, the advancement from congruence relations to *numerical measures* of distance and angle (which are taken as given in the metric approach) mirrors the development of the real numbers via the integers and the rational numbers. Summarizing briefly, a measure $d(A,B)$ of the **distance** *between* two arbitrary points A and B (or, *from* one of these points *to* the other)—which, when $A \neq B$, is identical to the **length** $|\overline{AB}|$ of the segment $\overline{AB}$—can be developed as follows. First, for all points A and B we must have $d(A,B) \geq 0$, with equality attaining if and only if $A = B$. Also, for all segments $\overline{AB}$ and $\overline{CD}$ we require that $d(A,B) = d(C,D)$ if and only if $\overline{AB} \cong \overline{CD}$; hence, $d(A,B) = d(B,A)$. Next, one particular segment $\overline{OI}$ is specified as providing the **unit length** upon which the distance measure is based; that is, we set $d(O,I) = 1$, thereby determining the "scale" of the measure. Furthermore, we require for all points A, B, and C that

$$d(A,C) \leq d(A,B) + d(B,C); \tag{10.1}$$

this is called the **triangle inequality**, since it reflects the fact that for any triangle, the sum of the lengths of any two sides must be greater than the length of the remaining side. Additionally, when A and C are distinct, equality is attained in (10.1) if and only if $B \in \overline{AC}$; that is, the distance measure is additive, as shown on the left most side of Figure 10.6. Therefore, if $\overline{AB} \cong \overline{OI}$, $\overline{BC} \cong \overline{OI}$, and B lies between A and C, then

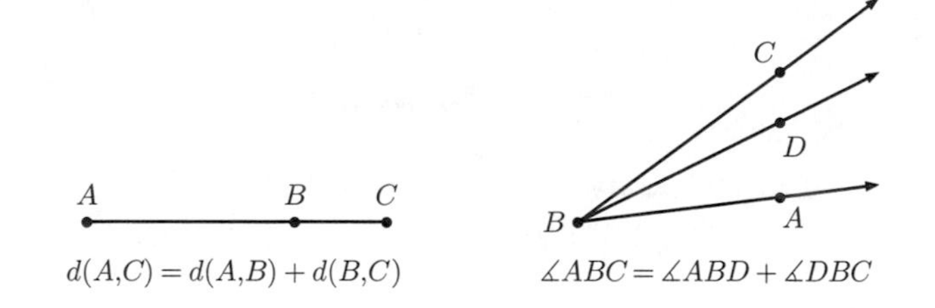

FIGURE 10.6 Additivity of Distance and Angle Measures

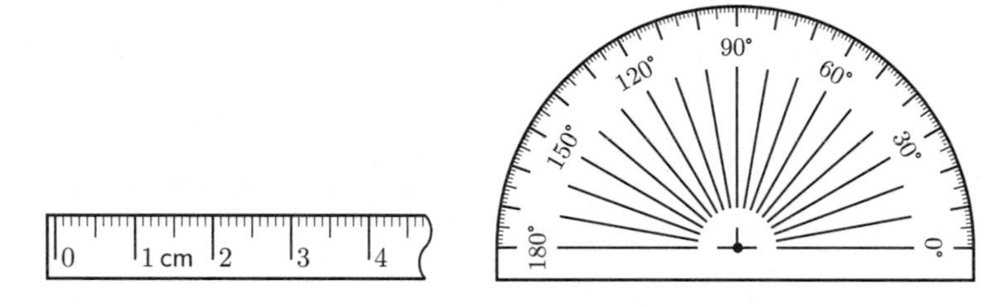

FIGURE 10.7 A Ruler and a Protractor

$d(A, C) = 1 + 1 = 2$; likewise, a distance of any integer value $m \geq 1$ obtains by concatenating m collinear segments of unit length. Moreover, given another integer $n \geq 1$ and a segment $\overline{EF}$, if $\overline{OI}$ obtains by concatenating n collinear segments that are all congruent to $\overline{EF}$, then additivity implies that $n \cdot d(E, F) = 1$, making $d(E, F) = 1/n$; and, a length of the rational value m/n obtains by concatenating m collinear segments of length $1/n$. Finally, by imposing an appropriate continuity condition on the measure, a distance having any nonnegative real value (perhaps irrational) may be obtained as a limit of rational-valued distances. The end result—upon showing that the process just described is self-consistent and succeeds in assigning a unique distance to each pair of points—is a distance measure that is applied in practice by using a **ruler**, such as that shown on the left side of Figure 10.7 (cf. the real line in Figure 3.1).[14]

A similar process yields a measure $\measuredangle ABC$ (respectively, $\measuredangle B$) of an arbitrary angle $\angle ABC$ ($\angle B$).[15] For all angles $\angle ABC$ and $\angle DEF$, we have $\measuredangle ABC = \measuredangle DEF$ if and only if $\angle ABC \cong \angle DEF$ (hence, $\measuredangle ABC = \measuredangle CBA$). Taking the unit angle to be one **degree**—i.e., 1°—we have $0° \leq \measuredangle ABC \leq 180°$, with the left inequality attaining equality if and only if $\angle ABC$ is a zero angle, and the right inequality attaining equality if and only if $\angle ABC$ is a straight angle.[16] The angle measure is additive in that for any point D lying

14. Although a typical ruler possesses a straight edge along its ruled side, its primary purpose is to measure distances; whereas the primary purpose of a **straightedge** is to help one to draw straight lines. Though often done by nonprofessionals, it is considered poor draftsmanship to use a ruler as a straightedge.

15. The word "angle" is commonly used not only to refer to an angle itself, but also its measure. However, for clarity, such ambiguous usage (which in practice is convenient and rarely confusing) is avoided in this chapter.

16. A widely used subdivision of the degree is the **minute**, with one minute—i.e., $1'$—being one-sixtieth of a degree. Subdividing further, one **second**—i.e., $1''$—is one-sixtieth of a minute.

on the interior of a proper angle $\angle ABC$, we have $\measuredangle ABC = \measuredangle ABD + \measuredangle DBC$, as shown on the right side of Figure 10.6. Moreover, the sum of the measures of two supplementary angles is always 180°, from which it follows that the measure of every right angle is 90°.[17] An angle $\angle ABC$ is **acute** if and only if $0° < \measuredangle ABC < 90°$, whereas it is **obtuse** if and only if $90° < \measuredangle ABC < 180°$ (see Figure 10.4); hence, an angle is oblique if and only if it is either acute or obtuse. In practice, an angle is measured by using a **protractor**, such as that shown on the right side of Figure 10.7.

Once one has a distance measure (primitive or defined), the relations of similarity and congruence can be expressed in terms of that measure. Namely, for any two geometric figures f and g, we have $f \sim g$ if and only if there exists a one-to-one function $\varphi(\cdot)$ mapping f onto g such that

$$d[\varphi(P), \varphi(Q)] = \kappa \cdot d(P, Q) \quad (P, Q \in f), \tag{10.2}$$

for some **scale factor** $\kappa > 0$. In other words, the function $\varphi(\cdot)$ provides a one-to-one correspondence between the points of f and the points of g, with the distance between any two points $\varphi(P)$ and $\varphi(Q)$ on g being κ times the distance between the corresponding two points P and Q on f. Furthermore, $f \cong g$ if and only if we can take $\kappa = 1$.

A geometric object or operation is said to **bisect** a segment if and only if the object or operation somehow splits the segment into two segments of equal length. For example, if on the left side of Figure 10.6 we were to have $d(A, B) = d(B, C)$, then the point B would bisect—or, be a **bisector** of—the segment $\overline{AC}$. In this particular case, B is the unique **midpoint** of $\overline{AC}$. Similarly, a geometric object or operation might split a segment into three segments of equal length, and thus **trisect** the segment. Likewise, an angle can be bisected or trisected.

Given a line l and a point P not on it, there is exactly one point Q on l such that $l \perp \overline{PQ}$ (see Figure 10.8); the **distance** from P to l is then defined as $d(P, Q)$. Whereas for any point P on l, the distance from P to l is defined as 0. Also, given two parallel lines l and m, the distance from P to l is the same for every point P on m; accordingly, this is defined to be the **distance** between l and m.

Similarly, given a plane p and a point P not on it, there is exactly one point Q on p such that $p \perp \overline{PQ}$; the **distance** from P to p is then defined as $d(P, Q)$. Whereas for any point P on p, the distance from P to p is defined as 0. Also, given two parallel planes p and q, the distance from P to p is the same for every point P on q; accordingly, this is

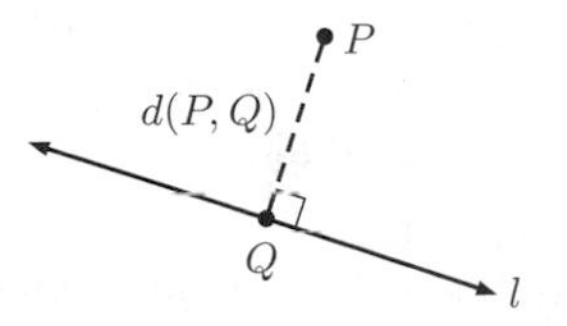

FIGURE 10.8 The Distance from a Point to a Line

17. Often, **supplementary angles** are defined as *any* two angles with measures summing to 180°, regardless of whether they share a side. Likewise, two angles with measures summing to 90° are called **complementary angles**, each angle being a **complement** of the other.

defined to be the **distance** between p and q. Moreover, for any plane p and line l that are parallel, the distance from P to p is the same for every point P on l, which we define to be the **distance** between p and l; but, the distance from P to l is not the same for every point P on p.

As is evident from Figure 10.8, given a line l and a point P (perhaps on l), and allowing a point Q to be chosen anywhere on l, there is exactly one point Q that minimizes the distance $d(P, Q)$; moreover, this minimum distance is the distance from P to l. Likewise, given a plane p and a point P, there is exactly one point Q on p that minimizes $d(P, Q)$; moreover, this minimum distance is the distance from P to p. It follows that if f and g are each either a line or a plane (hence, there are four possible cases), with $f \parallel g$, then except in the case where f is a plane and g is a line, for each point P on f the minimum of $d(P, Q)$ over all points Q on g equals the distance between f and g. Moreover, in all four cases, the distance between f and g equals the minimum of $d(P, Q)$ over all points P on f and Q on g.

Finally, given skew lines l_1 and l_2, let a point P_1 be chosen anywhere on l_1, and a point P_2 be chosen anywhere on l_2; then, there is exactly one pair (P_1, P_2) that minimizes $d(P_1, P_2)$. In fact, if l is the one line for which $l \perp l_1$ and $l \perp l_2$, then P_1 is the point at which l and l_1 cross, and P_2 is the point at which l and l_2 cross.

10.3 Plane Figures

10.3.1 Polygons

A **polygon** is a plane figure f consisting of finitely many segments forming a cycle; that is,

$$f = \overline{A_1A_2} \cup \overline{A_2A_3} \cup \cdots \cup \overline{A_nA_1}, \tag{10.3}$$

for some coplanar points

$$A_1, \quad A_2, \quad \ldots, \quad A_n \tag{10.4}$$

($n \geq 2$) such that $A_1 \neq A_2, A_2 \neq A_3, \ldots, A_{n-1} \neq A_n$, and $A_n \neq A_1$.[18] This figure f may be succinctly referred to as "polygon $A_1A_2 \cdots A_n$". Observe that—for the same points (10.4)—polygon $A_2A_3 \cdots A_nA_1$ is the identical figure, as is polygon $A_nA_{n-1} \cdots A_1$; that is, it follows from (10.3) that if we rearrange the points (10.4) such that their *cyclic order* is either unchanged or reversed, then the resulting polygon is unchanged. Moreover, in *some* instances, other rearrangements are also allowed; for example, the left most polygon in Figure 10.9 could just as well be expressed as "polygon $A_1A_2A_3A_4A_6A_5$".

Often, as in Figure 10.10, a polygon f is such that the points (10.4) used to express it via (10.3) can be chosen to be distinct, with none being an interior point of one of the segments

$$\overline{A_1A_2}, \quad \overline{A_2A_3}, \quad \ldots, \quad \overline{A_nA_1}, \tag{10.5}$$

18. For greater generality, one might allow these points (and, thus, the resulting polygon) to be noncoplanar; then, what we have just defined would be referred to as a "planar polygon".

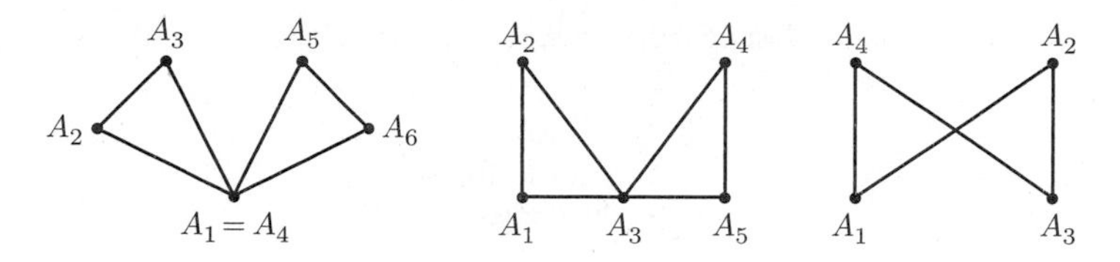

FIGURE 10.9 Three Nonsimple Polygons

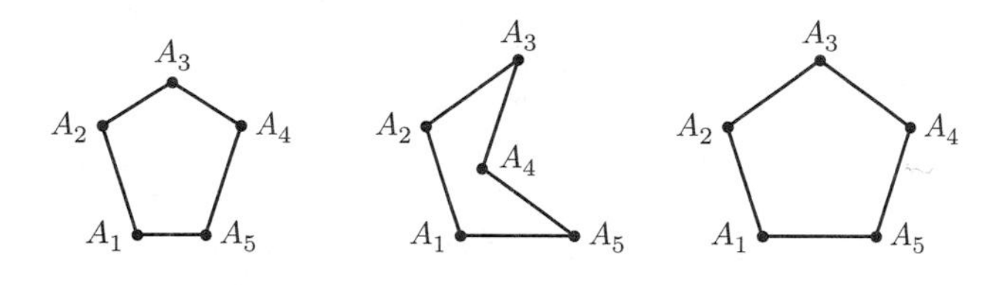

FIGURE 10.10 Three Simple Polygons (Pentagons)

and with none of these segments intersecting at another point;[19] then, the figure f in (10.3) has the appearance of a single "loop". In this case, f is called a **simple polygon**, with the region enclosed by the loop being the **interior** of f, and the region outside being its **exterior**. (The polygon, its interior, and its exterior are disjoint, and they together comprise a partition of the plane in which the polygon lies.)

Given a simple polygon f, we must have $n \geq 3$ in (10.3). If n has the minimum value possible for that f, then the points (10.4) are distinct; also, the segments (10.5) are distinct, as are the angles

$$\angle A_nA_1A_2, \quad \angle A_1A_2A_3, \quad \ldots, \quad \angle A_{n-1}A_nA_1. \tag{10.6}$$

Moreover, for each minimum-n expression (10.3) of a simple polygon f, the points (10.4) taken together must constitute the same *set* as for any other such expression; and, likewise for the segments (10.5) and the angles (10.6). Therefore, using any one of these expressions, we can unambiguously define *the* **vertices** of f as the points (10.4), *the* **sides** of f as the segments (10.5), and *the* **angles** of f as the angles (10.6), with n being the number of each.[20]

Two vertices of a simple polygon are **adjacent vertices** if and only if both are endpoints of the same side; two sides of a simple polygon are **adjacent sides** if and only if they share an endpoint; and, two angles of a simple polygon are **adjacent angles** if and only if the vertices of the angles are adjacent vertices of the polygon. The **perimeter**

19. Notice that each of the three conditions just stated is violated in Figure 10.9.

20. The identification of vertices, sides, and angles of a nonsimple polygon is not such a straightforward matter. (For example, the reader might ponder whether—irrespective of the vertex labelings shown—the polygons in Figure 10.9 should be viewed as having the same number of vertices, or sides, or angles.) Usually, this identification is made easier by first constraining the above definition of a "polygon"; for as it stands, any segment alone is a polygon (e.g., take $n = 2$), as are some other figures not normally thought of as "polygonal". However, even seemingly reasonable constraints upon the points (10.4) and segments (10.5) can lead to some surprising results; e.g., see Problem 10.7.

TABLE 10.1 Polygon Names

n	n-gon
3	**Triangle, trigon**
4	**Quadrangle, tetragon**
5	**Pentagon**
6	**Hexagon**
7	**Heptagon**
8	**Octagon**
9	**Nonagon, enneagon**
10	**Decagon**
11	**Hendecagon**
12	**Dodecagon**

of a simple polygon is defined as the sum of the lengths of its sides,[21] which is the total distance traveled by traversing the cycle (10.3) exactly once. The union of a simple polygon and its interior is the **polygonal region** *bounded by* the polygon; and, it has the same "interior", "exterior", "vertices", "sides", "angles", and "perimeter".

For each integer $n \geq 3$, an ***n*-gon** is defined as a polygon having *n angles*. The names of the first several n-gons are given in Table 10.1. A triangle is always a simple polygon; and, a simple quadrangle is also called a **quadrilateral** (i.e., a four-*sided* figure).[22] Since triangles and quadrilaterals are the polygons receiving the most attention, it is convenient to abbreviate the phrases "triangle $A_1A_2A_3$" and "quadrilateral $A_1A_2A_3A_4$" as "$\triangle A_1A_2A_3$" and "$\square A_1A_2A_3A_4$", respectively. Regardless of whether an abbreviation is used, though, the order of the points (10.4) appearing in phrases such as "n-gon $A_1A_2 \cdots A_n$" and "pentagon $A_1A_2A_3A_4A_5$" *is* significant in the same way as it is for the phrase "polygon $A_1A_2 \cdots A_n$"; namely, it is implicit that (10.3) holds for the cited figure f.

In general, a geometric figure f is called a **convex set** when for any two distinct points A and B on f, the segment $\overline{AB}$ lies in f (i.e., for any $A, B \in f$, we have $\overline{AB} \subseteq f$); otherwise, f is a **concave set**. Correspondingly, a **convex polygon** is defined to be a simple polygon for which its *interior* is a convex set, which is the case if and only if the corresponding polygonal region (though not the polygon itself) is a convex set; any other polygon is a **concave polygon**. Thus, in Figure 10.10, the left- and rightmost polygons are convex, whereas the middle polygon is concave. Given a simple polygon f and a segment having one endpoint in the interior of f and the other endpoint in the exterior, the segment and f must intersect; moreover, f is convex if and only if for each such segment the intersection contains exactly one point.

---

21. The word "side" is commonly used to refer not only to a side itself, but also to its measure (length). However, for clarity, such ambiguous usage (which in practice is convenient and rarely confusing) is avoided in this chapter.

22. Some authors would also classify, e.g., the polygon on the right of Figure 10.9 as a "quadrilateral". All agree, though, that every triangle is a **trilateral** (i.e., a three-sided figure), and vice versa; so this term gets little use.

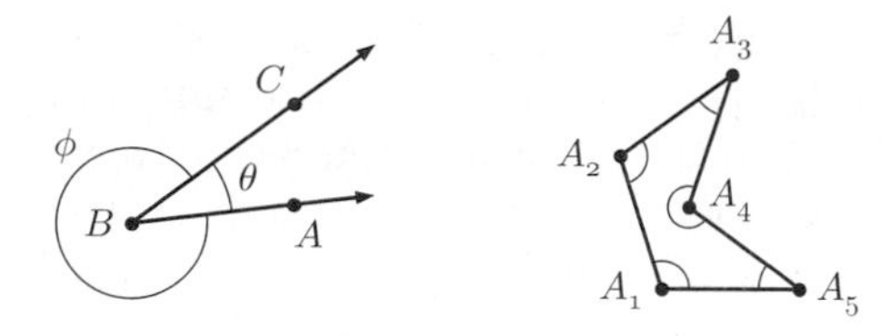

FIGURE 10.11 Angle Indication and the Interior Angles of a Polygon

A **diagonal** of a simple polygon is any segment, other than a side of the polygon, for which the endpoints are vertices of the polygon. Thus, for each of the polygons in Figure 10.10, the diagonals are $\overline{A_1A_3}, \overline{A_1A_4}, \overline{A_2A_4}, \overline{A_2A_5}$, and $\overline{A_3A_5}$. A sufficient condition for a simple polygon to be convex is that none of its diagonals pass through its exterior.

We now turn to some basic definitions and facts pertaining to polygon angles. In the preceding section we defined an "angle" $\angle B$ as the union of two rays having a common endpoint B, and then we defined the "measure" $\measuredangle B$ of this angle as a corresponding number in the range of 0° to 180°. An advantage of this classical approach is that each angle is a definite geometric figure, and its measure is *unique*. A disadvantage, though, is that it can be too restrictive. For as it stands, $\measuredangle B$ is the amount by which we "go around" the vertex B to get from one side of $\angle B$ to the other, *if* our path is through the interior of the angle (recall Figure 10.3); however, we might also like to consider the less direct route of passing through the angle's exterior. Indeed, we presently would like to measure all of the angles of a given simple polygon via paths passing through the *polygon*'s interior; but, the interior of the polygon need not be within the interior of each of its angles (e.g., consider $\angle A_4$ for the middle polygon in Figure 10.10).

Hence, we are led to extend the terms "angle" and "angle measure": As shown on the left side of Figure 10.11, given two rays $\overleftarrow{AB}$ and $\overrightarrow{BC}$ sharing an endpoint B, a circular arc centered at B may be drawn to indicate either of the two paths around B from one ray to the other; these two **angles** have corresponding *measures*—say, θ and ϕ, which are written nearby. (Thus, although the word "angle" may still refer to a geometric figure $\overleftarrow{AB} \cup \overrightarrow{BC}$, it now may also refer to a particular circular path around a vertex B; accordingly, without some additional cue, we can no longer unambiguously speak of *the* "measure" $\measuredangle ABC$ of an angle $\angle ABC$.[23]) The value of θ (corresponding to the *interior* of $\angle ABC$), is provided by the previously defined measure $\measuredangle ABC$; and, the value of ϕ (corresponding to the *exterior* of $\angle ABC$) obtains from the requirement that

$$\theta + \phi = 360°.$$

It follows that every angle now has a measure in the range from 0° to 360°, and there are *two* possible values—except when the angle is a straight angle, which has the one possible value of 180°.[24]

23. An example of such a cue would be to refer to the "interior" of an "angle", or to a "proper angle", both of which imply the classical meaning of "angle" (i.e., the union of two rays).

24. In Chapter 11 (with regard to Figure 11.4), the definition of angle measure is extended still further to allow *any* real value (perhaps negative). For one thing, this will allow the term "supplementary angles" to be

As for the angles formed at the vertices of a simple polygon, each angle corresponding to a path through the polygon's interior is called an **interior angle** of the polygon;[25] thus, e.g., the right side of Figure 10.11 indicates the interior angles of the middle polygon in Figure 10.10. Moreover, when one is referring to a simple polygon, it is customary to understand the unqualified term "angle" to mean "interior angle". The fundamental fact regarding the angles of a simple n-gon ($n \geq 3$) is that the sum of their measures equals $(n-2) \cdot 180°$.

An **equiangular polygon** is defined as a simple polygon for which all of the angles are congruent, whereas an **equilateral polygon** is a simple polygon for which all of the sides are congruent (i.e., have the same length). A simple polygon that is both equiangular and equilateral is a **regular polygon**. (Thus, the first polygon in Figure 10.10 is equiangular but not equilateral, the second is equilateral but not equiangular, and the third is regular.) Two regular polygons are similar if and only if they have the same number of sides, in which case they are congruent if and only if their sides have the same length.

A simple polygon is convex if and only if each of its angles has a measure less than 180°. Hence, since the angles of a triangle sum to 180°, every triangle is convex; and, as seen by dividing by n the above angle-sum formula for a simple n-gon, every equiangular polygon is convex. A simple polygon is regular if and only if it is convex and equilateral.

A simple polygon cannot be similar, or congruent, to a nonsimple polygon; and, two simple polygons that are similar, or congruent, must have the same number of vertices. Therefore, suppose that for some $n \geq 3$ we have two simple polygons $A_1A_2 \cdots A_n$ and $B_1B_2 \cdots B_n$. By convention, upon writing

$$(\text{polygon } A_1A_2 \cdots A_n) \sim (\text{polygon } B_1B_2 \cdots B_n),^{26} \tag{10.7}$$

we assert not only that these two polygons are similar; in addition, this use of the symbol "$\sim$" implies the similarity of the corresponding angles:

$$\angle A_1 \sim \angle B_1, \quad \angle A_2 \sim \angle B_2, \quad \ldots, \quad \angle A_n \sim \angle B_n. \tag{10.8a}$$

Also, (10.7) implies that the ratios of the lengths of corresponding sides are equal:

$$\frac{|\overline{B_1B_2}|}{|\overline{A_1A_2}|} = \frac{|\overline{B_2B_3}|}{|\overline{A_2A_3}|} = \cdots = \frac{|\overline{B_nB_1}|}{|\overline{A_nA_1}|}. \tag{10.8b}$$

Likewise, upon writing

$$(\text{polygon } A_1A_2 \cdots A_n) \cong (\text{polygon } B_1B_2 \cdots B_n), \tag{10.9}$$

---

redefined as any pair of angles having measures summing to 180°, whereas at present it is unclear what the supplement of an angle having a measure greater than 180° is.

25. Contrary to what one might expect, though, the angles corresponding to paths through the polygon's exterior are *not* commonly referred to as "exterior angles". Rather, for a polygon, an **exterior angle** is the *supplement* of an interior angle (as is the case for the angles defined in regard to Figure 10.5).

26. In particular, for a triangle we might write "$\triangle A_1A_2A_3 \sim \triangle B_1B_2B_3$", and likewise regarding congruence in (10.9) below.

we assert not only that these two polygons are congruent; in addition, this use of the symbol "≅" implies the congruence of the corresponding angles, and of the corresponding sides:

$$\angle A_1 \cong \angle B_1, \quad \angle A_2 \cong \angle B_2, \quad \ldots, \quad \angle A_n \cong \angle B_n; \tag{10.10a}$$

$$\overline{A_1A_2} \cong \overline{B_1B_2}, \quad \overline{A_2A_3} \cong \overline{B_2B_3}, \quad \ldots, \quad \overline{A_nA_1} \cong \overline{B_nB_1}. \tag{10.10b}$$

If preferred, one can equivalently express (10.8a) and (10.10a) each as

$$\measuredangle A_1 = \measuredangle B_1, \quad \measuredangle A_2 = \measuredangle B_2, \quad \ldots, \quad \measuredangle A_n = \measuredangle B_n, \tag{10.11a}$$

and (10.10b) as

$$\left|\overline{A_1A_2}\right| = \left|\overline{B_1B_2}\right|, \quad \left|\overline{A_2A_3}\right| = \left|\overline{B_2B_3}\right|, \quad \ldots, \quad \left|\overline{A_nA_1}\right| = \left|\overline{B_nB_1}\right|. \tag{10.11b}$$

Also, in contrast to (10.7) and (10.9), writing

$$(\text{polygon } A_1A_2 \cdots A_n) = (\text{polygon } B_1B_2 \cdots B_n)$$

asserts only that these two polygons are the same geometric figure; it does not imply a particular one-to-one correspondence between their vertices.

A sufficient condition for two simple n-gons to be similar is that they be expressible as "polygon $A_1A_2 \cdots A_n$" and "polygon $B_1B_2 \cdots B_n$" such that (10.8) holds; for then, letting these polygons be f and g respectively, (10.2) holds for a function $\varphi(\cdot)$ such that $\varphi(A_i) = B_i$ ($i = 1, 2, \ldots, n$), with κ being the common value of the ratios in (10.8b). Likewise, a sufficient condition for these n-gons to be congruent is to be able to express them in such a way that (10.10) holds.

As shown in Table 10.2, triangles are classified both by their angles and by their sides. We thus see that a given triangle is either right or oblique, but not both; in the latter case, it is either acute (possibly equiangular) or obtuse, but not both. Also, it must be either isosceles (possibly equilateral) or scalene, but not both.

For a right triangle $\triangle ABC$ having right angle $\angle B$, as on the left side of Figure 10.12, the side $\overline{AC}$ opposite to $\angle B$ is called the **hypotenuse** of the triangle, with the other two sides $\overline{AB}$ and $\overline{BC}$ being its **legs**; furthermore,

$$\left|\overline{AB}\right|^2 + \left|\overline{BC}\right|^2 = \left|\overline{AC}\right|^2, \tag{10.12}$$

TABLE 10.2 Types of Triangles

Triangle	Defining Property
Right	One right angle
Oblique	No right angles
Acute	All angles acute
Obtuse	One angle obtuse
Isosceles	At least two sides congruent
Scalene	No two sides congruent

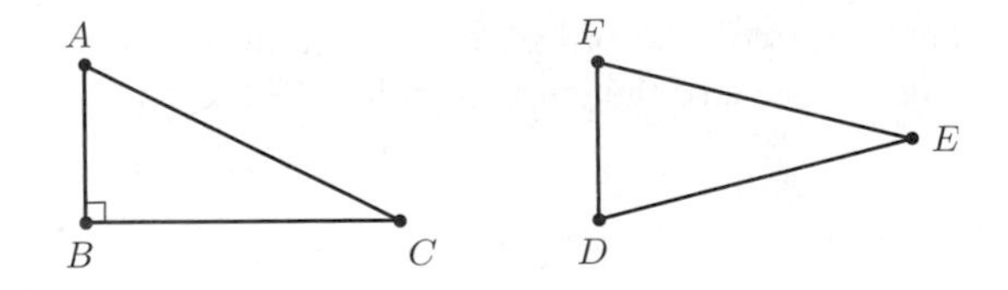

FIGURE 10.12 A Right Triangle and an Isosceles Triangle

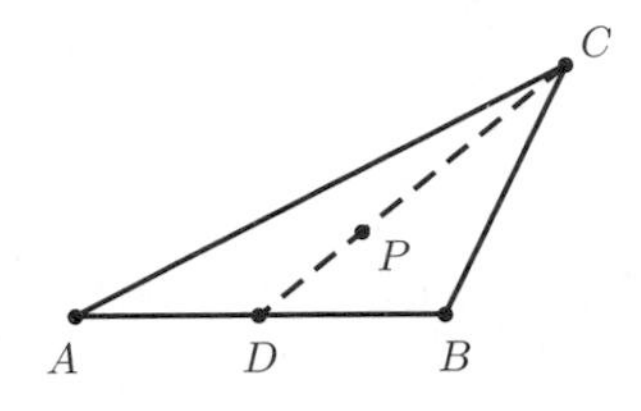

FIGURE 10.13 A Median and the Centroid of a Triangle

by the **Pythagorean theorem**.[27] For an isosceles triangle $\triangle DEF$ that is not equilateral, as on the right side of Figure 10.12, its unique **base** is the side $\overline{DF}$ not congruent to the other two sides; and, the corresponding angles $\angle D$ and $\angle F$ are its **base angles**.

For any particular side $\overline{AB}$ of a triangle $\triangle ABC$, the corresponding **median** of the triangle is the segment $\overline{CD}$ connecting the midpoint D of this side and the opposite vertex C; see Figure 10.13. The three medians of $\triangle ABC$ intersect at a single point P called the **centroid** of the triangle. Moreover, $d(C,P) = 2d(P,D)$.

For any two triangles $\triangle A_1A_2A_3$ and $\triangle B_1B_2B_3$, each of the following is a sufficient (and necessary) condition to have $\triangle A_1A_2A_3 \sim \triangle B_1B_2B_3$:

- **AA**: $\measuredangle A_1 = \measuredangle B_1$ and $\measuredangle A_2 = \measuredangle B_2$.
- **SAS**: $|\overline{A_1A_2}|/|\overline{B_1B_2}| = |\overline{A_2A_3}|/|\overline{B_2B_3}|$ and $\measuredangle A_2 = \measuredangle B_2$.
- **SSS**: $|\overline{A_1A_2}|/|\overline{B_1B_2}| = |\overline{A_2A_3}|/|\overline{B_2B_3}| = |\overline{A_3A_1}|/|\overline{B_3B_1}|$.

(Each condition is given a name in which the letters "A" and "S" indicate whether angles or sides appear in the condition; and, if both appear, then their relationship is reflected in the order of these letters.) Likewise, each of the following is a sufficient (and necessary) condition to have $\triangle A_1A_2A_3 \cong \triangle B_1B_2B_3$:

- **ASA**: $\measuredangle A_1 = \measuredangle B_1$, $|\overline{A_1A_2}| = |\overline{B_1B_2}|$, and $\measuredangle A_2 = \measuredangle B_2$.
- **SAS**: $|\overline{A_1A_2}| = |\overline{B_1B_2}|$, $\measuredangle A_2 = \measuredangle B_2$, and $|\overline{A_2A_3}| = |\overline{B_2B_3}|$.
- **SSS**: $|\overline{A_1A_2}| = |\overline{B_1B_2}|$, $|\overline{A_2A_3}| = |\overline{B_2B_3}|$, and $|\overline{A_3A_1}| = |\overline{B_3B_1}|$.
- **SAA**: $|\overline{A_1A_2}| = |\overline{B_1B_2}|$, $\measuredangle A_2 = \measuredangle B_2$, and $\measuredangle A_3 = \measuredangle B_3$.

27. Accordingly, any triple (x,y,z) of *integers* $x, y, z > 0$ for which $x^2 + y^2 = z^2$—e.g., $(3,4,5)$—is called a **Pythagorean triple**.

TABLE 10.3 Types of Quadrilaterals

Quadrilateral	Defining Property
Square	Regular
Rectangle	Equiangular
Rhombus	Equilateral
Parallelogram	Each pair of opposite sides parallel
Trapezoid	At least one pair of opposite sides parallel

(Thus, when referring to either of the names "SAS" and "SSS", one must explicitly state whether it is the similarity or the congruence condition that is intended.) It follows, e.g., from the above congruence conditions (taking $\Delta A_1A_2A_3 = \Delta B_1B_2B_3$, then varying how A_1, A_2, and A_3 correspond to B_1, B_2, and B_3) that a triangle is equiangular if and only if it is equilateral.

For a right isoceles triangle, the Pythagorean theorem implies that the length of the hypotenuse equals $\sqrt{2}$ times the length of either leg. Another useful fact is that for a right triangle having a 30° angle, the length of the opposite side is half that of the hypotenuse (e.g., in Figure 10.12, where $\measuredangle C = 30°$, we have $|\overline{AB}| = |\overline{AC}|/2$). This fact obtains by reflecting the triangle about the other leg to form an equilateral triangle; specifically: Extending the left side of Figure 10.12, imagine A' is the point on $\overleftrightarrow{AB}$ that lies on the opposite side of B than A such that $|\overline{A'B}| = |\overline{AB}|$. Then, since $\angle A'BC$ and $\angle ABC$ are supplementary angles, the former must also be a right angle; therefore, it follows from the SAS condition for triangle congruence that $\Delta A'BC \cong \Delta ABC$. Hence, by (10.11a), $\measuredangle A' = \measuredangle A = 60°$, so $\measuredangle ACA' = 180° - \measuredangle A' - \measuredangle A = 60°$. Thus, $\Delta ACA'$ is equiangular, and therefore equilateral, thereby making $|\overline{AC}| = |\overline{AA'}| = |\overline{AB}| + |\overline{BA'}| = |\overline{AB}| + |\overline{A'B}| = 2|\overline{AB}|$.

Table 10.3 defines the most common quadrilaterals, all of which are convex. We thus see that every square is both a rectangle and a rhombus; and, every parallelogram is a trapezoid. It can also be shown that every rectangle is a parallelogram, as is every rhombus.

Any two squares are similar; whereas they are congruent if and only if their sides are of equal length. Two arbitrary rectangles, though, need not be similar; and, likewise for rhombi, parallelograms, and trapezoids.

10.3.2 Some Nonpolygonal Figures

A **circle** is a geometric figure f consisting of all points Q on a plane p that are the same distance $r > 0$ from a particular point P on p; that is,

$$f = \{Q \in p \mid d(P, Q) = r\}$$

(see Figure 10.14). The distance r is called *the* **radius** of f, or *its* radius, and the point P is its **center**; whereas each of the infinitely many segments $\overline{PQ}$ for which $Q \in f$ is

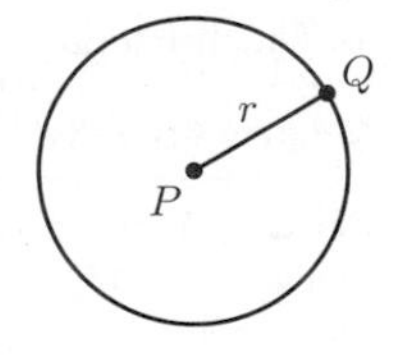

FIGURE 10.14 A Circle, Its Center, and a Radius

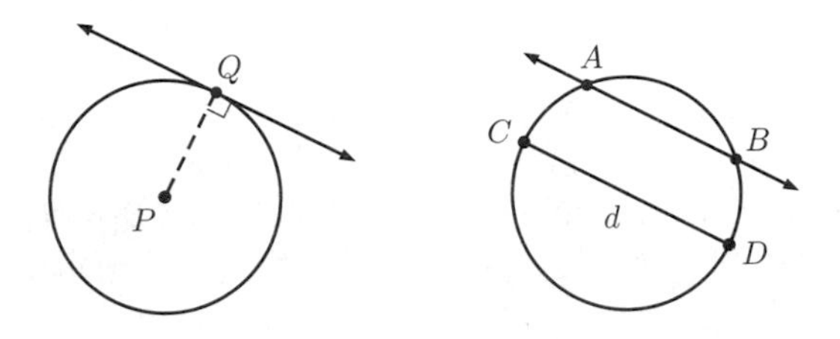

FIGURE 10.15 A Tangent Line, a Secant Line, and a Diameter

referred to as "*a* radius" of f.[28] The region enclosed by the circle is its **interior**, and the region outside is its **exterior**. (The circle, its interior, and its exterior are disjoint, and together comprise a partition of the plane p.) The interior of a circle is also called an **(open) disk**; and, the union of a circle and its interior is called a **closed disk**. In either case, the disk is *bounded by* the circle, is a convex set (though the circle itself is not), and has the same "center", "radii", "interior", and "exterior" (as well as other features of circles defined below).

Through any three noncollinear points, there passes exactly one circle. Two or more circles are said to be **concentric** if and only they are coplanar and have the same center. Also, any two circles are similar; whereas they are congruent if and only if the radius of one equals the radius of the other.

A line that intersects a coplanar circle at a single point is said to be a **tangent line** of (or be "tangent to") the circle; in this case, as well as the case of a line and a coplanar circle that do not intersect, none of the line lies within the interior of the circle. For any point Q on a circle having center P, the tangent line to the circle at Q is the unique coplanar line passing through Q that is perpendicular to the segment $\overline{PQ}$ (see left side of Figure 10.15). A line and a circle cannot intersect at more than two points; whereas a line that intersects a (necessarily coplanar) circle at exactly two points A and B is a **secant line** of the circle. The segment $\overline{AB}$ (see right side of Figure 10.15) is then a **chord** of the circle, and the interior of the chord is the portion of the secant line that lies within the interior of the circle. As also shown on the right side of Figure 10.15, infinitely many chords $\overline{CD}$ pass through the center of a circle; and, each such chord is called *a* **diameter** of the circle. Moreover, for a circle of radius r, a chord is a diameter if and only if it has

28. This usage of the articles "the" and "a" is occasionally violated, appropriately, without confusion; and, likewise below when these articles are applied to the term "diameter". For example, given a circle f having center P, one might speak of "*the* radius $\overline{PQ}$" for any point Q on f.

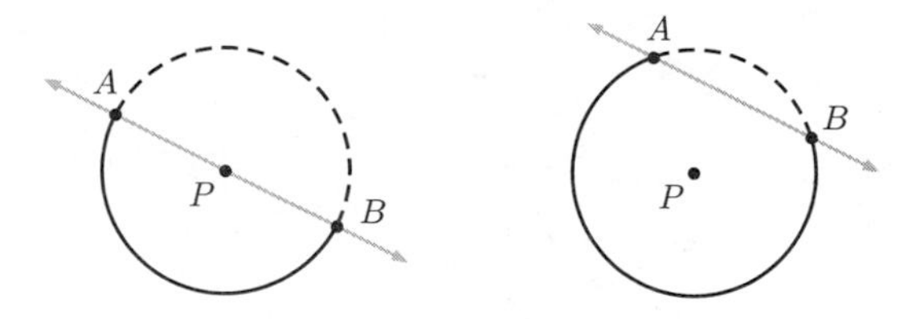

FIGURE 10.16 Two Semicircles, a Major Arc, and a Minor Arc

length $d = 2r$—called *the* diameter of the circle, or *its* diameter—the longest a chord can be.

Roughly stated, a **(circular) arc** $\widehat{AB}$ of a circle f is any "continuous" portion of f that has two **endpoints** A and B. More precisely, $\widehat{AB}$ is the figure composed of two distinct points $A, B \in f$ along with all points in f lying on one particular side of the secant line $\overleftrightarrow{AB}$;[29] hence, there are actually *two* arcs $\widehat{AB}$. If the secant line passes through the center P of the circle f, then the two arcs are congruent, each being called a **semicircle** of f (see the solid and dashed curves on the left side Figure 10.16); otherwise, the arcs are not congruent, with the one lying on the same side of the secant line as P being the **major arc** (the solid curve on the right side of the figure), and the other (the dashed curvc) being the **minor arc**. Thus, for given points A and B, the notation "$\widehat{AB}$" is ambiguous in two ways: the circle in which the arc lies is not determined; and, even when the circle is specified, there are two possible arcs to consider. Both of these ambiguities, though, can be removed by providing another point on the arc; for example, the major arc in Figure 10.16 corresponds to the arc $\widehat{ACB}$ in Figure 10.15, which is the same as $\widehat{ADB}$. Nevertheless, when used in context, the former notation is often convenient and usually harmless.

Because each arc contains three noncoplanar points, there is only one circle in which it lies. Hence, we may define the **radius** of the arc to be the radius of this circle.

A proper angle $\angle P$ is said to **intercept** an arc $\widehat{AB}$ of a given circle if and only if both points A and B lie on $\angle P$, with each side of $\angle P$ containing at least one of these points (perhaps as the vertex P), and the remainder of $\widehat{AB}$ lies in the interior of $\angle P$. Thus, e.g., the circle on the left side of Figure 10.17 has one of its arcs intercepted by the angle $\angle P$, whereas $\angle Q$ intercepts two of its arcs (the most possible).

A **central angle** of a circle f is an angle $\angle P$ for which the vertex P is the center of f.[30] Moreover, as shown on the right side of Figure 10.17, for each such angle there are unique points A and B (arbitrarily ordered and not necessarily distinct) on f such that $\angle P = \angle APB$; therefore, if $\angle P$ is a proper angle, then it intercepts exactly one arc of f—viz., the minor arc $\widehat{AB}$. Conversely, given two points A and B on a circle f having center P, there is a unique corresponding central angle of f—viz., $\angle APB$. Accordingly,

---

29. Even more precisely, if p is the plane containing f, then $\widehat{AB} = f \cap h$ where h is one of the two closed half-planes of p having $\overleftrightarrow{AB}$ as its edge.

30. In this paragraph and the next (as in the preceding paragraph), the term "angle" is used in the classical sense—i.e., the union of two rays having a measure from 0° to 180°. Later, though, it will be convenient to consider "central angles" more generally, allowing their measure to exceed 180° (see Figure 10.19 below).

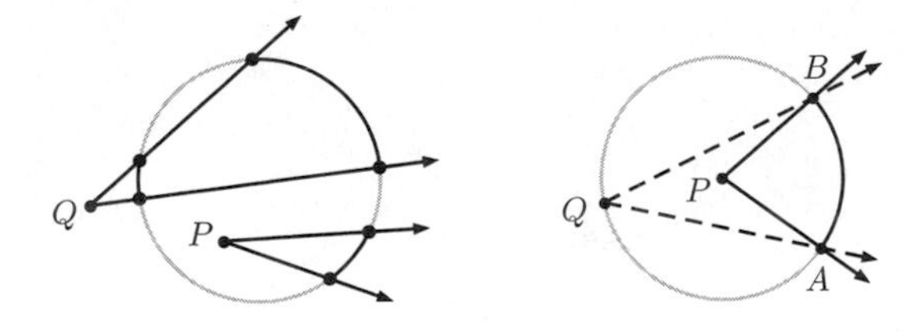

FIGURE 10.17 Intercepted Arcs, a Central Angle, and an Inscribed Angle

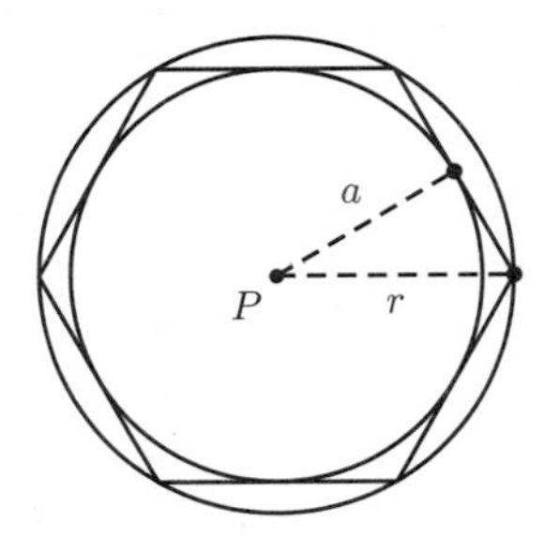

FIGURE 10.18 Circles Circumscribing and Inscribing a Polygon

if (as indicated in the figure) $\widehat{AB}$ is a minor arc of f, then we say that the **angle measure** of this *arc* is $\measuredangle APB < 180°$; if it is a semicircle, then its angle measure is $\measuredangle APB = 180°$; whereas if it is a major arc, then its angle measure is $360° - \measuredangle APB > 180°$. Two arcs are similar if and only if they have the same angle measure, in which case they are congruent if and only if the radius of one equals the radius of the other.

An **inscribed angle** of a circle f is an angle $\angle AQB$ for which the points A, B, and Q all lie on f (thus, the point Q is distinct from the others); see the dashed angle in Figure 10.17. If A and B are distinct, then this angle intercepts a unique (and not necessarily minor) arc $\widehat{AB}$ of f; moreover, the measure $\measuredangle AQB$ of this angle always equals one-half the measure of the corresponding central angle (i.e., one-half the angle measure of $\widehat{AB}$).

If the vertices of a simple polygon all lie on a particular circle (i.e., if the sides of the polygon are all chords of the circle), then that circle is said to **circumscribe** (or "to be circumscribed about") the polygon; whereas if all of the sides of the polygon are tangent (at interior points) to the circle, then the circle is said to **inscribe** (or "to be inscribed in") the polygon. Equivalently, in the former (respectively, latter) case—as illustrated by the larger (smaller) circle in Figure 10.18—the *polygon* inscribes (circumscribes) the circle. As shown in the figure, any regular polygon can be both circumscribed and inscribed by circles; moreover, the two circles are concentric, and their common center P is called the **center** of the polygon. Additionally, the **radius** of the polygon is defined as the radius r of the circumscribed circle, and its **apothem** is the radius a of the inscribed circle.

Having defined "length" for *line* segments, we are now in a position to extend this term to *circular* segments—i.e., arcs—and to full circles as well. Intuitively, the total distance traveled by traversing either of the circular paths in Figure 10.18 exactly once is approximately equal to the perimeter of the polygon shown. Also, it is apparent that for any particular circle, this approximation of its "circular perimeter" by the perimeter of a regular n-gon, either circumscribed or inscribed, becomes successively better as n is increased. In fact, the perimeters $p_{i,n}$ and $p_{c,n}$ of the inscribed and circumscribed n-gons ($n \geq 3$), respectively, are strictly decreasing and strictly decreasing versus n, with $p_{i,n} < p_{c,n}$ (all n); moreover,

$$\lim_{n\to\infty} p_{i,n} = \lim_{n\to\infty} p_{c,n}. \tag{10.13}$$

Thus, this common limit is defined as the **circumference** c of the circle. Furthermore, applying this analysis to concentric circles shows that c is proportional to the diameter d of the circle; that is, naming the proportionality constant with the Greek letter **pi** (π), we have

$$c = \pi d = 2\pi r,$$

where r is the radius of the circle. In fact, $\pi = 3.14159\ldots$, as found by further analysis.

A similar approach yields the **length** l of a given arc, the result being that l is proportional to both the radius r of the circle containing the arc *and* the angle measure θ of the arc itself:

$$l = 2\pi r \cdot (\theta \text{ in degrees})/360°. \tag{10.14}$$

(Note that for $\theta = 360°$ we have $l = c$, as required.) Conversely, we can use this proportionality to define angle measure in terms of arc length and radius, and thereby introduce an alternative unit of angle measure—the **radian**:

$$(\theta \text{ in radians}) \triangleq l/r. \tag{10.15}$$

It then follows from (10.14) that 2π radians equals 360°, and therefore 1 radian $= 180°/\pi \approx 57.3°$. Also, with the understanding that *angles are henceforth expressed in radians unless stated otherwise*, we can rewrite (10.15) simply as

$$l = r\theta \tag{10.16}$$

(cf. (10.14)), as illustrated in Figure 10.19. In particular, taking $r = 1$ shows that the radian measure of an angle equals the length of the corresponding arc for a circle having unit radius.

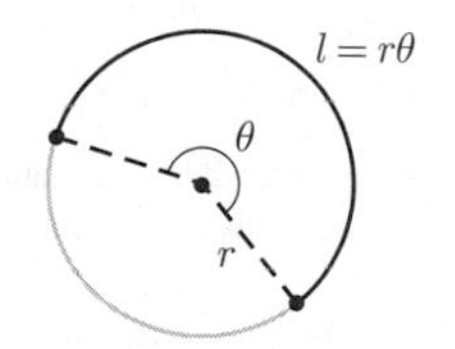

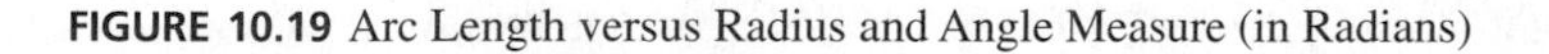

FIGURE 10.19 Arc Length versus Radius and Angle Measure (in Radians)

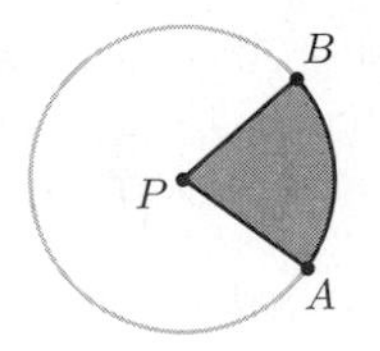

FIGURE 10.20 A Sector

Both arc length and angle measure are additive.[31] Namely, let A, B, and C be distinct points on a circle having center P, such that the arcs $\widehat{AB}$ and $\widehat{BC}$ intersect only at B (i.e., they are contiguous). Then, the sum of the lengths of $\widehat{AB}$ and $\widehat{BC}$ equals the length of the arc $\widehat{AC}$; and, for the corresponding angles we have $\measuredangle APB + \measuredangle BPC = \measuredangle APC$.

A **(circular) sector** is the region in a plane bounded by an arc of a circle and the two radii sharing the endpoints of the arc; see Figure 10.20. Equivalently, the sector corresponding to an arc $\widehat{AB}$ of a circle having center P is the figure f composed of all segments (radii) $\overline{PQ}$ such that the endpoint Q is on $\widehat{AB}$; i.e.,

$$f = \{\overline{PQ} \mid Q \in \widehat{AB}\}.$$

Hence, f includes the arc $\widehat{AB}$ and the corresponding two radii $\overline{PA}$ and $\overline{PB}$, which together form the boundary of the sector. The **radius** of the sector is defined as the radius of the circle, and its **angle measure** is that of the arc.

A **parabola** is a geometric figure f consisting of all points P in a plane q that are equidistant from a particular point $F \in q$ and a particular line $d \subseteq q$, where $F \notin d$; thus, as shown in Figure 10.21,[32]

$$f = \{P \in q \mid d(P,F) = d(P,Q)\}, \tag{10.17}$$

where Q is the unique point on d such that $\overline{PQ} \perp d$.[33] For each parabola f in a plane q, there is only one pair (F,d) of a point $F \in q$—called the **focus** of f—and a line $d \subseteq q$—called the **directrix** of f—that generates f this way. The line containing F that is perpendicular to d is the **axis** of f, and the **vertex** of f is the one point V on f that is also on the axis. It follows that the axis is $\overleftrightarrow{FV}$, and the distance from F to V equals the distance from V to d. Every two parabolas are similar; however, they are congruent if and only if the distance from focus to directrix—i.e., the **focal parameter** p—is the same for both.

Generalizing (10.17), consider the geometric figure

$$f = \{P \in q \mid d(P,F) = e\,d(P,Q)\} \tag{10.18}$$

31. Angle additivity in the classical sense has already been mentioned (see the right side of Figure 10.6), whereas here we are treating angles more generally (as was discussed with regard to the left side of Figure 10.11).

32. Since the line d extends indefinitely, so does the parabola, though only a portion of it appears in the figure; likewise for the hyperbola in Figure 10.23 below.

33. Note that if $P \in d$, then no such point Q exists; therefore, it is implicit that $P \notin d$.

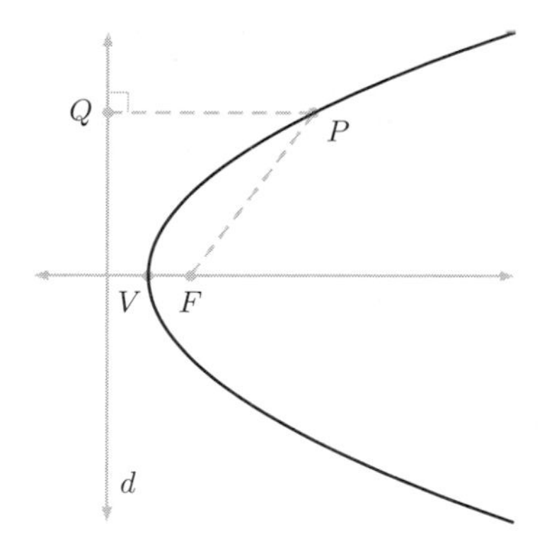

FIGURE 10.21 A Parabola

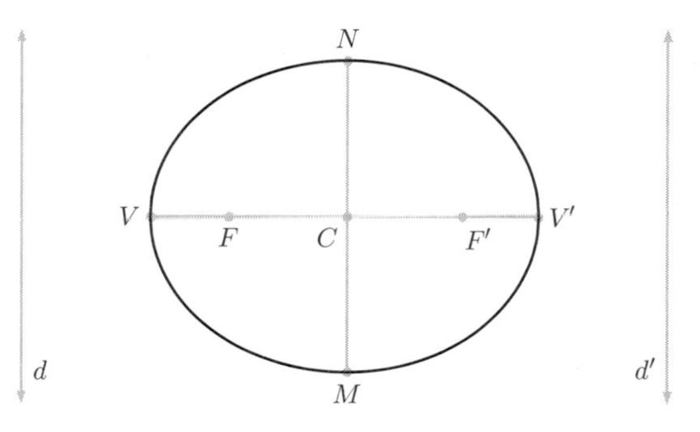

FIGURE 10.22 An Ellipse

—given the same plane q, point F, and line d, and with the point Q being determined by the variable point P as before—for some scale factor $e > 0$ called the **eccentricity** of f.[34] Thus, setting $e = 1$, we obtain a parabola; whereas each value $e < 1$ yields an **ellipse**, one of which is shown in Figure 10.22 (for $e = 0.6$). Unlike a parabola, though, any ellipse f is likewise generated (for the same value of e) by a *another* pair (F', d') of a point $F' \in q$ and a line $d' \subseteq q$, where $F' \notin d'$; that is, as shown in the figure, f has *two* distinct **foci**, F and F', with corresponding distinct **directrices**, d and d'. The **center** of f is the midpoint C of the segment $\overline{FF'}$, which is perpendicular to both d and d' (hence, $d \parallel d'$); and, its **vertices** are the two distinct points $V, V' \in f$ for which the segment $\overline{VV'}$ contains both F and F'. This segment is called the **major axis** of f; its **minor axis** is the unique segment $\overline{MN}$, with $M, N \in f$, that contains C (as its midpoint) and is perpendicular to the major axis. Introducing the **semimajor axis**

34. This parameter e is not to be confused with Napier's number defined in (7.2), though it *may* take on that value.

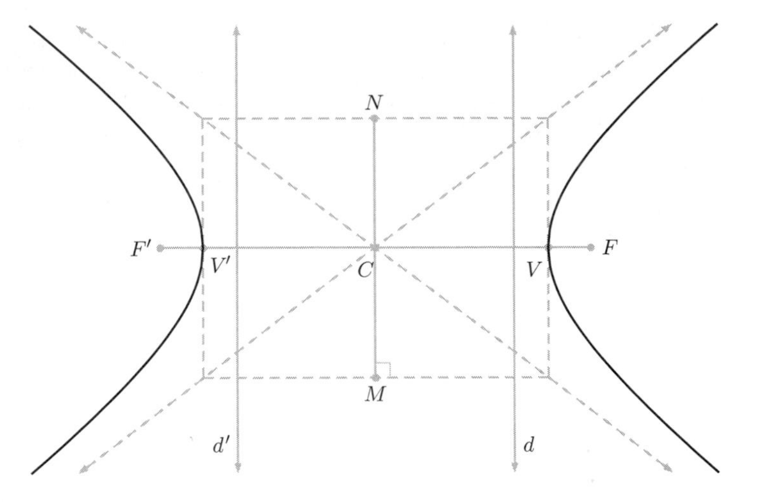

FIGURE 10.23 A Hyperbola

$a = d(V, C) = d(V', C)$ and the **semiminor axis** $b = d(M, C) = d(N, C)$,[35] we have $a > b$. Also, letting $c = \sqrt{a^2 - b^2}$, it can be shown that $d(F, C) = d(F', C) = c$ and $e = c/a$; moreover, for each directrix, its distance from C is a^2/c, and its distance from the corresponding focus—i.e., the **focal parameter**—is $p = b^2/c$. It follows that $a = ep/(1 - e^2)$, $b = ep/\sqrt{1 - e^2}$, and $c = e^2p/(1 - e^2)$. Furthermore, for each point $P \in f$ we have

$$d(P, F) + d(P, F') = 2a; \tag{10.19}$$

conversely, given a constant $a > 0$ and two distinct points F and F' in a plane q, the set of all points $P \in q$ for which (10.19) holds is an ellipse, with a being the semimajor axis and F and F' being the foci. Two ellipses are similar if and only if they have the same eccentricity e, in which case they are congruent if and only if they have the same focal parameter p; moreover, they are similar if and only if they have the same value for the ratio b/a, whereas they are congruent if and only if the values for a and b are respectively the same.

Each value $e > 1$ in (10.18) yields a **hyperbola**, one of which is shown in Figure 10.23 (for $e = 1.25$). Like ellipses, each hyperbola f has two distinct **foci,** F and F', with corresponding distinct **directrices**, d and d'; that is, both of the pairs (F, d) and (F', d') generate f, per (10.18). The **center** of f is the midpoint C of the segment $\overline{FF'}$, which is perpendicular to both d and d' (hence, $d \parallel d'$); its **vertices** are the two distinct points V, $V' \in f$ that are on this segment; and, the segment $\overline{VV'}$ is the **transverse axis** of f. A feature of hyperbolas not shared by parabolas or ellipses is the possession of **asymptotes**; for our hyperbola f, these are two lines (shown dashed

35. Like the word "radius", the terms "major axis" and "minor axis" are commonly used to refer to both a segment and its length; thus, the terms "semimajor axis" and "semiminor axis" are used here for the respective half-lengths.

in Figure 10.23) crossing at C, such that if a point P is moved continuously along f to make $d(P,C) \to \infty$, then the distance from P to one of these lines approaches 0. The **conjugate axis** of f is the unique segment $\overline{MN}$ that contains C (as its midpoint), is perpendicular to the transverse axis, and has endpoints M and N on the rectangle shown in Figure 10.23—all four vertices of which are on the asymptotes, with V and V' being midpoints of two opposite sides. (Then, M and N are the midpoints of the remaining two sides.) The hyperbola is separated into two **branches**, one on each side of the line $\overleftrightarrow{MN}$. Introducing the **semitransverse axis** $a = d(V,C) = d(V',C)$ and the **semiconjugate axis** $b = d(M,C) = d(N,C)$,[36] and letting $c = \sqrt{a^2+b^2}$, it can be shown that $d(F,C) = d(F',C) = c$ and $e = c/a$; moreover, for each directrix, its distance from C is a^2/c, and its distance from the corresponding focus—i.e., the **focal parameter**—is $p = b^2/c$. It follows that $a = ep/(e^2-1)$, $b = ep/\sqrt{e^2-1}$, and $c = e^2p/(e^2-1)$. Furthermore, for each point $P \in f$ we have

$$|d(P,F) - d(P,F')| = 2a; \tag{10.20}$$

conversely, given a constant $a > 0$ and two distinct points F and F' in a plane q, the set of all points $P \in q$ for which (10.20) holds is a hyperbola, with a being the semitransverse axis and F and F' being the foci. Two hyperbolas are similar if and only if they have the same eccentricity e, in which case they are congruent if and only if they have the same focal parameter p; moreover, they are similar if and only if they have the same value for the ratio b/a, whereas they are congruent if and only if the values for a and b are respectively the same.

10.4 Solid Figures

10.4.1 Polyhedra

A **polyhedron** is a *surface* consisting of *polygonal regions*, which are typically connected, with the surface enclosing a region of space.[37] Thus, a polyhedron is a solid figure analogous to a polygon (a planar figure composed of segments). Rather than attempt to provide a precise general definition of the term "polyhedron", though, we will proceed as we did for polygons, restricting our attention to the "simple" case. Roughly put, a **simple polyhedron**, such as those shown in Figure 10.24, has a surface that can be converted into a sphere by stretching alone (i.e., without tearing). By contrast, this conversion is not possible for, e.g., the donut-shaped polyhedron in Figure 10.25 (constructed from 32 square regions).

One way to precisely define a simple polyhedron is as a geometric figure

$$f = f_1 \cup f_2 \cup \cdots \cup f_n \tag{10.21}$$

36. Unlike the case of an ellipse (see preceding paragraph), we need not have $a > b$.

37. Some authors include in a "polyhedron" the enclosed region. Our limitation to the surface alone, though, is consistent with the definitions of, e.g., "simple polygon" and "sphere" (see Subsection 10.4.2), which do not include interiors.

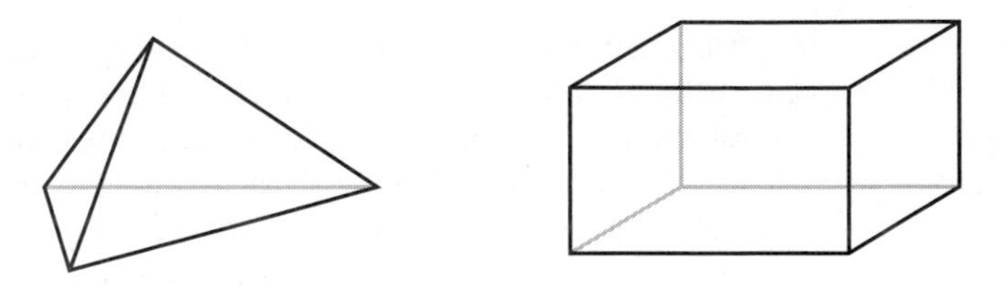

FIGURE 10.24 Two Simple Polyhedra (a Tetrahedron and a Hexahedron)

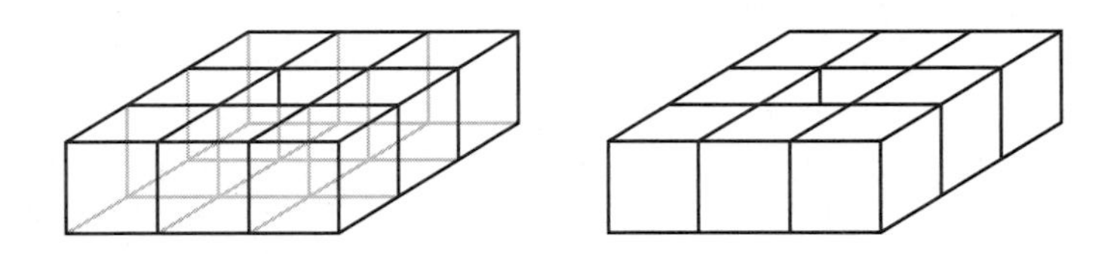

FIGURE 10.25 A Nonsimple Polyhedron (Transparent and Opaque Views)

($n \geq 4$),[38] for some polygonal regions f_i ($i = 1, \ldots, n$), such that:

(i) For each i, no interior point of f_i is a point (interior or otherwise) on f_j for $j \neq i$.
(ii) For each i, each side of f_i is also a side of f_j for exactly one $j \neq i$; also, f_i and f_j are noncoplanar.
(iii) For each $i > 1$, the sides that f_i has in common with the f_j for $j < i$ are connected (i.e., together form a single path).

The requirement in (ii) that f_i and f_j be noncoplanar causes the constituent figures f_i of f to be unique (except for their order);[39] therefore, we can unambiguously define *the* **faces** of the simple polyhedron f to be the polygonal regions f_i, the sides of which are *the* **edges** of f, and the vertices of which are *the* **vertices** of f.

Two vertices of a simple polyhedron are **adjacent vertices** if and only if both are endpoints of the same edge; two edges of a simple polyhedron are **adjacent edges** if and only if they share an endpoint; and, two faces of a simple polyhedron are **adjacent faces** if and only if they share a side. The region enclosed by a simple polyhedron is its **interior**, and the region outside is its **exterior**. (The polyhedron, its interior, and its exterior are disjoint, and together comprise a partition of space.) The union of a simple polyhedron and its interior is the **solid polyhedron** *bounded by* the polyhedron; and, it has the same "vertices", "edges", "faces", "interior", and "exterior".

In general, a polyhedron having n faces is called an ***n*-hedron**. Also, the same prefixes (i.e., "tetra-", "penta-", etc.) used to form the names of polygons (see Table 10.1) can be used to form the names of polyhedra (viz., a 4-gon is a "tetrahedron", a 5-gon is

38. It is impossible to satisfy the following conditions—in particular, (ii)—for $n < 4$.

39. If f_i and f_j were coplanar and shared a single side (e.g., which is the case for some pairs of the square regions in Figure 10.25), then they could be replaced with $f_i \cup f_j$—itself a polygonal region. By contrast, the definition does allow a simple polyhedron to have multiple coplanar faces that intersect at *finitely* many points (perhaps none).

"pentahedron", etc.). A **convex polyhedron** is defined to be a simple polyhedron for which its *interior* is a convex set, which is the case if and only if the corresponding solid polyhedron (though not the polyhedron itself) is a convex set; any other polyhedron is a **concave polyhedron**. Given a simple polyhedron f and a segment having one endpoint in the interior of f and the other endpoint in the exterior, the segment and f must intersect; moreover, f is convex if and only if for each such segment the intersection contains exactly one point.

A **diagonal** of a simple polyhedron is any segment for which the endpoints are vertices of the polyhedron, but do not belong to the same face.[40] A sufficient condition for a simple polyhedron to be convex is that none of its diagonals pass through the exterior of the polyhedron.

For any simple polyhedron f with V vertices, E edges, and F faces, we have $V \geq 4$, $E \geq 6$, and $F \geq 4$ (the minimum values being achieved by every tetrahedron). Furthermore, V, E, and F must satisfy **Euler's formula** (or **Euler's theorem**):

$$V - E + F = 2. \tag{10.22}$$

This fact is easily proved by constructing f one face f_i at a time, per (10.21) and the conditions (i)–(iii) that follow. Briefly, for the first face f_1 alone, we begin with $V = E$ and $F = 1$, and thus $V - E + F = 1$. Then, upon adding each face $f_2, \ldots, f_{n-1}$, we increase V by some amount $m \geq 1$, with E correspondingly increasing by $m + 1$ owing to condition (iii) on (10.21), while F increases by 1; therefore, we continue to have $V - E + F = 1$ after each such addition. Finally, adding the last face f_n—the vertices and edges of which are all already part of the construction—leaves V and E unchanged, whereas F increases by 1, thereby giving (10.22).

One way to form a simple polyhedron f is to take a simple polygon g and a non-coplanar point P, then let f be the set comprising the interior of g and all segments having P as one endpoint with the other endpoint Q being on g; i.e.,

$$f = (\text{interior of } g) \cup \left\{ \overline{PQ} \;\middle|\; Q \in g \right\} \tag{10.23}$$

(see Figure 10.26). Such a figure f is called a **pyramid**, of which P is the **apex** and the one face bounded by g is the **base**. (Thus, e.g., every tetrahedron is a pyramid.) The

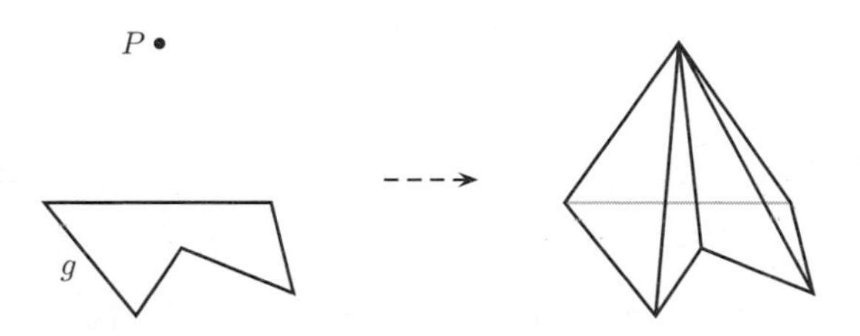

FIGURE 10.26 The Generation of a Pyramid

40. Some authors allow the endpoints to belong to the same face, while disallowing the segment from being an edge; that is, they define a "diagonal" of a polyhedron to be any segment for which the endpoints are vertices of the polyhedron, but are not adjacent.

other faces—i.e., those meeting at P—are the **lateral faces** of f (the boundary of each being a triangle); and, the edges of f meeting at P (i.e., the edges that are not sides of the base) are its **lateral edges.**

A **regular pyramid** is pyramid for which

(i) The boundary of the base is a regular polygon.

(ii) The segment having the center of the base as one endpoint, and the apex of the pyramid as the other, is perpendicular to (the plane containing) the base.

For any pyramid satisfying condition (i), each of the following conditions both implies and is implied by (ii):

- The lateral edges have the same length.
- The boundaries of the lateral faces are congruent isosceles triangles.

Also, a pyramid is a convex polyhedron if and only if its base is a convex set (which is the case if and only if the boundary of the base is a convex polygon); hence, every regular pyramid is convex.

In general, given two distinct parallel planes p and q, along with a line l intersecting p at a single point, then for each point P on p there is exactly one point Q on q such that the segment $\overline{PQ}$ is parallel to l. Moreover, for any figure g lying in p, the corresponding figure

$$g' = \{Q \in q \mid \overline{PQ} \parallel l \text{ for some } P \in g\} \tag{10.24a}$$

lying in q is congruent to g (see Figure 10.27); in particular, if g is a simple polygon, which we henceforth assume, then so is g'. Taking together the interior points of g and g', along with all of the points on the segments $\overline{PQ}$ occurring in (10.24a)—i.e., letting

$$f = (\text{interior of } g) \cup (\text{interior of } g') \cup \{\overline{PQ} \mid P \in g, Q \in g', \overline{PQ} \parallel l\} \tag{10.24b}$$

—we obtain a simple polyhedron f called a **prism**, of which the faces bounded by g and g' are the two **bases**. The other faces of f are its **lateral faces** (the boundary of each being a parallelogram); and, the edges of f that are not sides of either base are its **lateral edges** (which are all parallel).

A **right prism** is a prism for which any (and therefore all) of the lateral edges are perpendicular to either (and therefore both) of the bases, which is the case if and only if

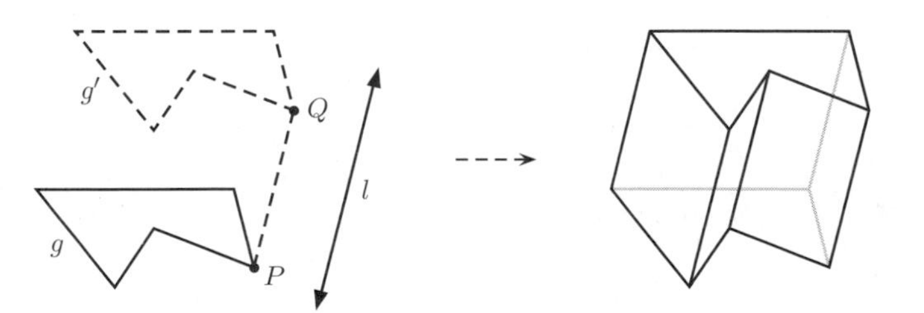

FIGURE 10.27 The Generation of a Prism

TABLE 10.4 The Regular Polyhedra

Polyhedron	V	E	F	Face Boundary	M
Regular tetrahedron	4	6	4	Equilateral triangle	3
Regular hexahedron (cube)	8	12	6	Square	3
Regular octahedron	6	12	8	Equilateral triangle	4
Regular dodecahedron	20	30	12	Regular pentagon	3
Regular icosahedron	12	30	20	Equilateral triangle	5

the boundaries of the lateral faces are all rectangles; any other prism is an **oblique prism**. A **regular prism** is a right prism for which the boundary of either (and therefore both) of the bases is a regular polygon (thereby making the lateral faces congruent). Also, a prism is a convex polyhedron if and only if either (and therefore both) of the bases is a convex set; hence, every regular prism is convex.

A **parallelepiped** is a prism for which the boundary of either base is a parallelogram; hence, the boundary of *every* face is then a parallelogram. Equivalently, a parallelepiped is a simple hexahedron for which each pair of opposite faces are parallel. A **rectangular parallelepiped** is a parallelepiped for which the boundary of each face is a rectangle; in particular, a **cube** is a parallelepiped for which the boundary of each face is a square.

Pyramids and prisms can be classified by their bases. Thus, e.g., we may speak of a "triangular pyramid" or a "quadrilateral prism" (examples of which are shown in Figure 10.24).

A **regular polyhedron** is a convex polyhedron having congruent faces, each bounded by a regular polygon, with the same number of faces meeting at every vertex. Remarkably, there are only five types of regular polyhedra (also called the **Platonic solids**); these are are given in Table 10.4, where V, E, and F are as defined for Euler's formula (10.22), and M is the number of faces (and edges) meeting at each vertex. (Again, the names of these polyhedra derive from F.) Two regular polyhedra are similar if and only if they have the same number of faces, in which case they are congruent if and only if their edges have the same length.

10.4.2 Some Nonpolyhedral Figures

Analogous to a pyramid, a **circular cone** is a figure f that results from (10.23) given a circle g and a noncoplanar point P. The closed disk bounded by g is the **base** of f, the point P is its **apex**, and the portion of f other than the interior of g is its **lateral surface**. If the segment that has as its endpoints the center of g and P is perpendicular to (the plane containing) g, then f is a **right circular cone**, which is ordinarily referred to simply as a "cone";[41] otherwise, f is an **oblique circular cone** (see Figure 10.28).

Analogous to a prism, a **circular cylinder** is a figure f that results from (10.24) given two distinct parallel planes p and q, a circle g lying in p, and a line l intersecting

41. In more general contexts, a "cone" may be defined as the figure f resulting from (10.23) for *some* simple closed curve g (i.e., a single "loop" of arbitrary shape) lying in a plane p, and a point P not on p. Likewise, in the next paragraph, the figure f resulting from (10.24) for some such g may be referred to as a "cylinder".

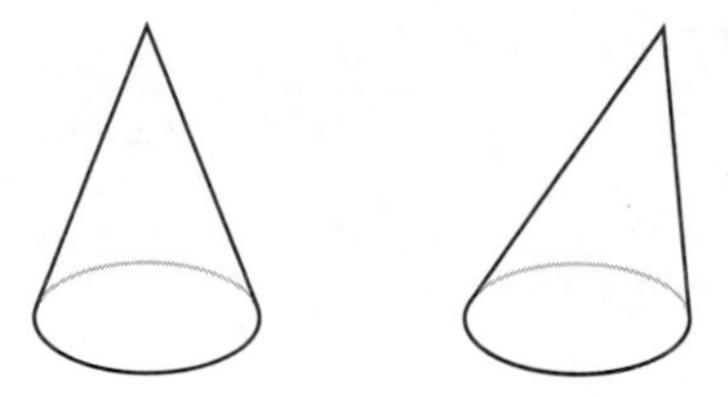

FIGURE 10.28 A Right Circular Cone and an Oblique Circular Cone

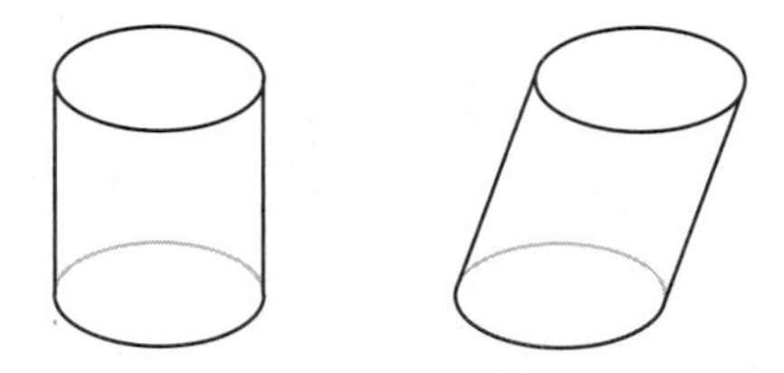

FIGURE 10.29 A Right Circular Cylinder and an Oblique Circular Cylinder

p at a single point. The closed disk bounded by g is one **base** of f, and the closed disk bounded by g' in (10.24a) is the other base (lying in q); also, the portion of f other than the interiors of g and g' is its **lateral surface**. If for each point $P \in g$ there is a point $Q \in g'$ such that $\overline{PQ} \perp p$, then f is a **right circular cylinder**, which is ordinarily referred to simply as a "cylinder"; otherwise, f is an **oblique circular cylinder** (see Figure 10.29). Letting C and C' be the centers of its two bases, f is a right circular cylinder if and only if $\overline{CC'} \perp p$.

A **sphere** is a figure f consisting of all points Q in space that are the same distance $r > 0$ from a particular point P. (Thus, a sphere is a solid figure analogous to a circle; indeed, a simple drawing of a sphere would look like one of a circle—e.g., Figure 10.14.) The distance r is called *the* **radius** of f, or *its* radius, and the point P is its **center**; whereas each of the infinitely many segments $\overline{PQ}$ for which $Q \in f$ is referred to as "*a* radius" of f. The region enclosed by the sphere is its **interior**, and the region outside is its **exterior**. (The sphere, its interior, and its exterior are disjoint, and together they comprise a partition of space.) The interior of a sphere is also called an **(open) ball**; and, the union of a sphere and its interior is called a **closed ball** or **solid sphere**. In either case, the ball (or solid sphere) is *bounded by* the sphere, is a convex set (though the sphere itself is not), and has the same "center", "radii", "interior", and "exterior" (as well as other features of spheres defined below).

Through any four noncoplanar points, such that no three of them are collinear, there passes exactly one sphere. Two or more spheres are said to be **concentric** if and only if they have the same center. Any two spheres are similar; whereas they are congruent if and only if the radius of one equals the radius of the other.

A line that intersects a sphere at a single point is said to be a **tangent line** of (or to be "tangent to") the sphere. A line and a sphere cannot intersect at more than two points. A segment having both of its endpoints on a sphere is a **chord** of the sphere, and its interior lies completely within the interior of the sphere. Each of the infinitely many chords that passes through the center of a sphere is called *a* **diameter** of the sphere. Moreover, a chord is a diameter if and only if its length equals twice the radius—called *the* diameter of the sphere, or *its* diameter—the longest a chord can be.

A plane that intersects a sphere at a single point is said to be a **tangent plane** of (or to be "tangent to") the sphere; in this case, as well as the case of a plane and a sphere that do not intersect, none of the plane lies within the interior of the sphere. For any point Q on a sphere having center P, the tangent plane to the sphere at Q is the unique plane p passing through Q that is perpendicular to the segment $\overline{PQ}$; furthermore, a line passing through Q is tangent to the sphere if and only if it lies in p.

A plane that intersects a sphere at more than one point is a **secant plane** of the sphere; and, the intersection is a circle. When the plane passes through the center of the sphere, the radius of the circle is as large as possible (viz., the radius of the sphere), and thus such a circle is called a **great circle** of the sphere; also, the union of such a circle and the portion of the sphere on one side of the plane is called a **hemisphere**, of which the circle is its **edge**. (Hence, a hemisphere is a solid figure analogous to a semicircle, with its edge corresponding to the endpoints of the semicircle.) The **circumference** of a sphere is defined as the circumference of any of its great circles.

If the vertices of a simple polyhedron all lie on a particular sphere (i.e., if the edges of the polyhedron are all chords of the sphere), then that sphere is said to **circumscribe** (or to be "circumscribed about") the polyhedron; whereas if all of the faces of the polyhedron are tangent (at interior points) to the sphere, then the sphere is said to **inscribe** (or to be "inscribed in") the polyhedron. Equivalently, in the former (respectively, latter) case, the *polyhedron* inscribes (circumscribes) the sphere. Any regular polyhedron can be both circumscribed and inscribed by spheres; moreover, the two spheres are concentric, and their common center is called the **center** the polyhedron.

The nonempty intersection of a solid figure and a plane is called a **section** of the figure. (Thus, e.g., great circles are sections of spheres.) Every section of a pyramid made by a plane parallel to its base, but not passing through the apex or the base, is similar though not congruent to the boundary of the base; whereas every section of a prism made by a plane parallel to either (and therefore both) of the bases, but not passing through either, is congruent to the boundary of each base. Likewise, every section of a circular cone made by a plane parallel to its base, but not passing through the apex or the base, is a circle having a smaller radius than the base; whereas every section of a circular cylinder made by a plane parallel to either (and therefore both) of the bases, but not passing through either, is a circle having the same radius as each base. Circles, ellipses, parabolas, and hyperbolas are often referred to as the **conic sections** (or **conics**) because they obtain from the sections of two infinite right circular cones placed apex-to-apex and oppositely aligned.[42] A solid figure is a convex set if and only if all of its sections are convex sets.

---

42. For example, the cones can be taken to be the set of all points in *xyz*-space (see Section 11.1) such that $x^2 + y^2 = z^2$. For more on the conic sections, see Subsection 11.5.2.

10.5 Area and Volume

Just as *length* is a one-dimensional measure of geometric size, so "area" and "volume" are, respectively, two- and three-dimensional measures of geometric size. As with other aspects of geometry, these measures can be developed in various ways, which differ by what properties are assumed (via axioms) and what properties are proved (via theorems) thereafter.

Intuitively, any **area** measure $\alpha(\cdot)$ for plane figures should be such that[43]

- If f is a plane figure and its area $\alpha(f)$ exists, then $0 \leq \alpha(f) \leq \infty$.
- If f and g are similar plane figures, and f has an area, then so does g; also, $\alpha(f) = \alpha(g)$ if and only if either $\alpha(f) = 0$ or $f \cong g$.
- If f and g are coplanar figures that each has an area, then the planar figures $f \cup g$, $f \cap g$, and $f - g$ have areas too. Moreover, if f and g are also nonintersecting, then $\alpha(f \cup g) = \alpha(f) + \alpha(g)$; that is, the measure $\alpha(\cdot)$ is additive.
- If f is the union of a unit square (i.e., a square having sides of length 1) and its interior—that is, f is a unit square region—then $\alpha(f) = 1$.
- If f is a line figure (perhaps $\emptyset$), then $\alpha(f) = 0$.

By dividing square regions into smaller square regions (having overlapping boundaries), such a measure can be consistently extended (if necessary) to provide the area of *any* square region. After that, further dissection arguments can be used to obtain the area of any rectangular region, then the area of any right triangular region, then the area of *any* triangular region, and then the area of any polygonal region.

Furthermore, some nonpolygonal plane regions can be assigned areas by employing a successive-approximation method akin to that used to define the circumference of a circle (recall Figure 10.18 and (10.13)). Namely, if for a given plane figure f there exist polygonal regions g_n and h_n $(n = 1, 2, \ldots)$ such that $g_n \subseteq f \subseteq h_n$ (all n) and

$$\lim_{n\to\infty} \alpha(g_n) = \lim_{n\to\infty} \alpha(h_n),$$

then this common limit is defined as the area $\alpha(f)$ of f; or, if there exist polygonal regions h_n $(n = 1, 2, \ldots)$ such that $f \subseteq h_n$ (all n) and

$$\lim_{n\to\infty} \alpha(h_n) = 0,$$

then we define $\alpha(f)$ as 0. (Since the value $\alpha(f)$ thus obtained is otherwise independent of how the polygonal regions g_n and h_n are chosen, this definition is unambiguous.) A common way of choosing the inner approximations g_n and the outer approximations h_n is by using an unbounded two-dimensional array of squares that is successively refined to

43. Note that in what follows we are careful not assume that every plane figure has an area; nor do we assume that every solid figure has a volume. In fact, modern mathematics (specifically, "measure theory") concludes that these assumptions are erroneous; that is, there exist plane figures and solid figures that are *nonmeasurable*. Such figures, though, are rather intricate, whereas all geometric figures normally encountered in science and engineering are sufficiently "smooth" to be measurable. (Be aware that a statement such as "f has an area" merely asserts the *existence* of $\alpha(f)$, not that this area is nonzero; nor does a statement such as "f does not have an area" mean that $\alpha(f) = 0$. And, likewise for such statements regarding volume.)

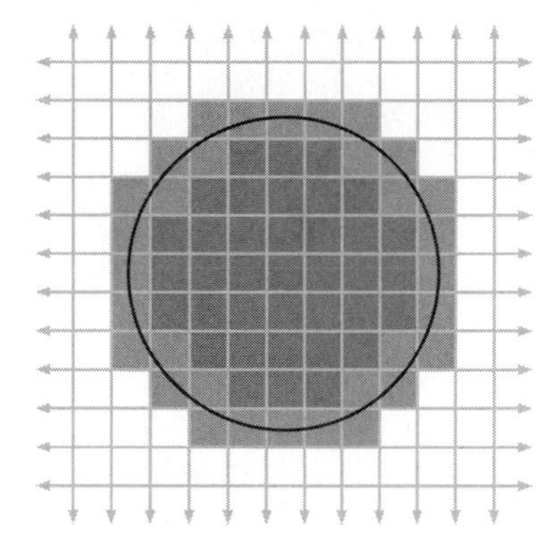

FIGURE 10.30 Inner and Outer Approximations to the Area of a Circle

TABLE 10.5 The Areas of Some Plane Figures

Plane Figure	Area
Square	s^2
Rectangle	$s_1 s_2$
Right triangle	$\frac{1}{2} l_1 l_2$
Triangle	$\frac{1}{2} bh$
Parallelogram	bh
Trapezoid	$\frac{1}{2}(b_1 + b_2)h$
Circle	πr^2
Sector	$\frac{1}{2}\theta r^2$
Ellipse	πab

make the squares smaller. For example, for a circle f, such inner and outer approximations are respectively indicated in Figure 10.30 by the dark shaded region and by that region combined with the lighter shaded region.

The usual formulas for the areas of some common plane figures are given in Table 10.5.[44] Here, s is the length of each side of the square; s_1 and s_2 are the lengths of any two adjacent sides of the rectangle; and, l_1 and l_2 are the lengths of the legs of the right triangle. As for an arbitrary triangle $\triangle ABC$, b is the length of the chosen **base**—which in the present context can be *any* side of the triangle, say $\overline{AB}$—and h is the corresponding **height**; namely (see Figure 10.31), h is the length of the corresponding **altitude**, which is the unique segment $\overline{CD}$ such that $D \in \overleftrightarrow{AB}$ and $\overline{CD} \perp \overleftrightarrow{AB}$. (Note that, as in the figure, we need not have $D \in \overline{AB}$.) Likewise, any side of an arbitrary parallelogram $\square ABCD$, say $\overline{AB}$, may be chosen to be its **base**, the length of which is b; then, the corresponding

44. Contrary to the table, the area of any polygon—being the union of finitely many segments—is actually 0; and, so is the area of a circle. However, we henceforth follow the convention of implicitly including the interiors of such figures when speaking of their areas.

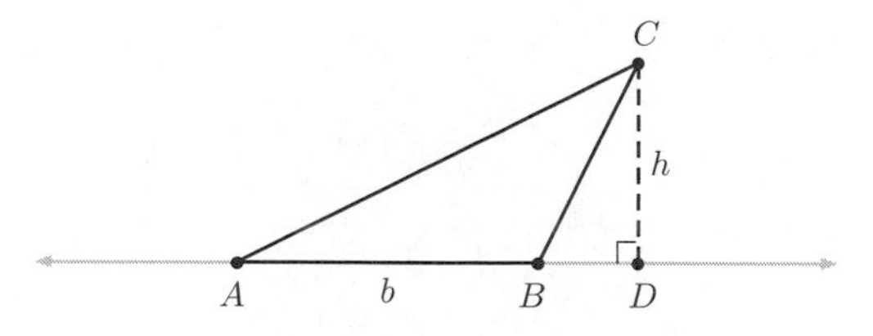

FIGURE 10.31 The Altitude and Height of a Triangle for a Particular Base

height h is the distance between the line $\overleftrightarrow{AB}$ and the parallel line $\overleftrightarrow{CD}$ containing the opposite side $\overline{CD}$. However, a trapezoid $\square ABCD$ that is not a parallelogram has *two unique* **bases**, which are the two parallel sides—say $\overline{AB}$ and $\overline{CD}$, having lengths b_1 and b_2—and, the *unique* **height** h of the trapezoid is the distance between the corresponding parallel lines $\overleftrightarrow{AB}$ and $\overleftrightarrow{CD}$. The circle in Table 10.5 has radius r; and, likewise for the sector, with θ being the angle measure of its boundary arc (as in Figure 10.19). Finally, for the ellipse, a is the semimajor axis and b is the semiminor axis.

There are other area formulas as well. For instance, if the three sides of a triangle have lengths a, b, and c, then its area equals

$$\sqrt{s(s-a)(s-b)(s-c)}, \tag{10.25}$$

where $s = (a+b+c)/2$ is the **semiperimeter** (i.e., half the perimeter) of the triangle (and similarly for any polygon). This is known as **Heron's formula**.

Intuitively, any **volume** measure $\nu(\cdot)$ should be such that:

- If f is a figure and its volume $\nu(f)$ exists, then $0 \leq \nu(f) \leq \infty$.
- If f and g are similar figures, and f has a volume, then so does g; also, $\nu(f) = \nu(g)$ if and only if either $\nu(f) = 0$ or $f \cong g$.
- If f and g are figures that each has a volume, then the figures $f \cup g, f \cap g$, and $f - g$ have volumes too. Moreover, if f and g are also nonintersecting, then $\nu(f \cup g) = \nu(f) + \nu(g)$; that is, the measure $\nu(\cdot)$ is additive.
- If f is the union of a unit cube (i.e., a cube having edges of length 1) and its interior—that is, f is a unit solid cube—then $\nu(f) = 1$.
- If f is a plane figure (perhaps $\emptyset$), then $\nu(f) = 0$.

By dividing solid cubes into smaller solid cubes (having overlapping boundaries), such a measure can be consistently extended (if necessary) to provide the volume of any solid rectangular parallelepiped. Then, all solid simple polyhedra, as well as some nonpolyhedral solid regions, can be assigned volumes by employing a successive-approximation method analogous to that used earlier for areas; namely, inner and outer volume approximations are obtained via an unbounded three-dimensional array of cubes that is successively refined to make the cubes smaller.

Additionally, every simple polyhedron f has a **surface area** $\sigma(f)$, which is simply the sum of the areas of its faces. Also, some nonpolyhedral solid figures can be assigned surface areas. In particular, by using the formula for the volume of a sphere (see Table 10.6 and accompanying discussion below), the formula for its surface area can be derived

TABLE 10.6 The Volumes and Surfaces Areas of Some Solid Figures

Solid Figure	Volume	Surface Area
Cube	e^3	$6e^2$
Rectangular parallelepiped	$e_1e_2e_3$	$2(e_1e_2 + e_1e_3 + e_2e_3)$
Regular pyramid	$\frac{1}{3}bh$	$b + \frac{1}{2}ph_s$
Pyramid	$\frac{1}{3}bh$	—
Right prism	bh	$2b + ph$
Prism	bh	$2b + ph_s$
Right circular cone	$\frac{1}{3}bh = \frac{1}{3}\pi r^2h$	$b + \frac{1}{2}ch_s = \pi r(r + h_s)$
Circular cone	$\frac{1}{3}bh = \frac{1}{3}\pi r^2h$	—
Right circular cylinder	$bh = \pi r^2h$	$2b + ch = 2\pi r(r + h)$
Circular cylinder	$bh = \pi r^2h$	$2b + ch_s = 2\pi r(r + h_s)$
Sphere	$\frac{4}{3}\pi r^3$	$4\pi r^2$

as follows. Consider the "spherical shell" g bounded by two concentric spheres having radii r and $r + \delta$, where $r > 0$ and $\delta > 0$; that is, letting P be the center of the spheres, we have

$$g = \{Q \mid r \leq d(P, Q) < r + \delta\}.$$

If $\delta \ll r$, then it is reasonable to say that surface area of the radius-r sphere is approximately equal to the volume of this shell divided by its "thickness" δ; moreover, fixing r, this approximation appears to become better as δ is made smaller. Therefore, since the volume $\nu(g)$ of the shell equals the difference of the volumes of the two spheres, we obtain

$$\frac{\nu(g)}{\delta} = \frac{\frac{4}{3}\pi(r+\delta)^3 - \frac{4}{3}\pi r^3}{\delta} = \frac{\frac{4}{3}\pi(3r^2\delta + 3r\delta^2 + \delta^3)}{\delta} = 4\pi(r^2 + r\delta + \delta^2/3)$$
$$\xrightarrow[\delta \to 0]{} 4\pi r^2, \tag{10.26}$$

which is identified as the surface area $\sigma(f)$ of a sphere f having radius r.

The usual formulas for the volumes and surface areas of some common solid figures are given in Table 10.6.[45] Here, e is the length of each edge of the cube. For the rectangular parallelepiped, e_1, e_2, and e_3 are the lengths of the three edges that meet at any vertex. For the pyramids, prisms, circular cones, and circular cylinders, b is the area of each base; and, in each case h is the unique **height**, which for a pyramid or circular cone is the distance from the apex to the plane containing the base, and for a prism or circular cylinder is the distance between the parallel planes containing the two bases. For the pyramids and prisms, p is the perimeter of each base; whereas for the circular cones and circular cylinders, r is the radius and c is the circumference of each base. (Thus, for the latter two figures, $b = \pi r^2$ and $c = 2\pi r$.) Finally, r is the radius of the sphere.

45. Contrary to the table, the volume of any polyhedron—being the union of finitely many faces—is actually 0; and, so is the volume of a sphere. However, we henceforth follow the convention of implicitly including the interiors of such figures when speaking of their volumes.

Additionally, each regular pyramid, prism, right circular cone, and circular cylinder has a unique **slant height** h_s. For a regular pyramid, this is height of the triangle bounding any lateral face, as measured from the base of the pyramid (to the apex); equivalently, it is the smallest distance from the apex to the boundary of the base. Whereas for a right circular cone, h_s is the distance from the apex to *any* point on the boundary of the base; moreover, it follows from the Pythagorean theorem that $h_s = \sqrt{h^2 + r^2}$. For a prism f generated by (10.24b), h_s is the length of all the segments $\overline{PQ}$; and, likewise for a circular cylinder. Therefore, we always have $h_s \geq h$, with equality being attained for a right prism and a right circular cylinder.

Regarding the effects of scaling on area and volume, suppose f and g are similar figures, and let κ be the scale factor from f to g, per (10.2).

- If f and g are plane figures having respective areas $\alpha(f)$ and $\alpha(g)$, then $\alpha(g) = \kappa^2\alpha(f)$.
- If f and g have respective volumes $\nu(f)$ and $\nu(g)$, then $\nu(g) = \kappa^3\nu(f)$.
- If f and g are solid figures having respective surface areas $\sigma(f)$ and $\sigma(g)$, then $\sigma(g) = \kappa^2\sigma(f)$.

Finally, the concept "angle", as well as its measure, can be generalized from two to three dimensions. First, recall that a plane angle $\angle ABC$ is simply the union of two rays $\overrightarrow{BA}$ and $\overrightarrow{BC}$ (and thus is plane figure). For the "proper" case of the rays being noncollinear (see Figure 10.3), our original measure $\measuredangle ABC$ of this angle was respect to its interior (i.e., θ in Figure 10.11), thereby giving $\measuredangle ABC$ a value between 0 and π radians (180°); also, we defined $\measuredangle ABC = 0$ radians for every zero angle (see Figure 10.4), and $\measuredangle ABC = \pi$ radians for every straight angle. We then considered measuring the angle $\angle ABC$ with respect to its exterior as well (ϕ in Figure 10.11), thereby allowing $\measuredangle ABC$ to range anywhere from 0 to 2π radians; accordingly, with the exception of straight angles, each plane angle has *two* possible measures. The intended measure of a plane angle is made unambiguous, though, by instead referring to the angle measure of a corresponding *arc*, lying in a circle centered at the angle's vertex; namely (see Figure 10.19), the arc has its endpoints where the rays of the angle intersect the circle, and is chosen such that its interior lies in either the interior or the exterior of the angle, as intended for the measurement. The angle measure of the arc, and thus the intended measure of the corresponding plane angle, equals the length of the arc divided by the radius of the circle; hence, for a unit (radius 1) circle, the angle measure of the arc *is* its length.

Proceeding analogously in three dimensions, let a circle c lie in (the surface of) a unit sphere; and, to begin, suppose c is not a great circle (see the left side of Figure 10.32 where c, shown as a dashed segment, is viewed from the side). Letting point P be the center of sphere, consider the set of rays

$$f = \{\overrightarrow{PQ} \mid Q \in c\} \tag{10.27}$$

(only two of the infinitely many rays being shown in the figure). This **solid angle** f is an infinite right circular cone that "cuts" the sphere into two regions, separated by the circle c—much as, given distinct points A and C on a circle having center B, the rays $\overrightarrow{BA}$ and $\overrightarrow{BC}$ cut the circle into two arcs $\widehat{AC}$ (sans endpoints), separated by A and C. Moreover, f separates *space* into two parts, which are analogous to the interior and exterior of the

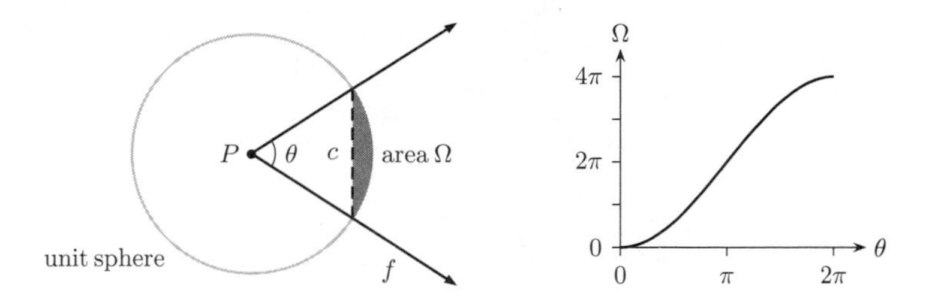

FIGURE 10.32 The Solid-Angle Measure of a Right Circular Cone

plane angle $\angle ABC$; namely, the **interior** of f is the set of all half-lines having endpoint P that pass through the interior of c, and the **exterior** of f is the set of all half-lines having endpoint P that pass through neither c nor its interior. (The solid angle, its interior, and its exterior are disjoint, and together they comprise a partition of space.)

The measure Ω of the solid angle f now has two possible values, depending on whether the measurement is made relative to the angle's interior or exterior. Namely, analogous to how a plane angle equals the length of an arc on a unit circle, here we define Ω to be the *surface area* of the portion of the unit sphere lying within the interior or the exterior of f, as intended (see Figure 10.32, where Ω is relative to the interior of f); also, the unit of solid-angle measure is the **steradian** (which, like the radian, is understood when no unit is explicitly given). Furthermore, if c were a great circle of the unit sphere, then it would be analogous to a straight angle in two dimensions; in this case, f cuts the sphere (having surface area 4π) into two hemispheres, and thus we now define Ω to be $\frac{1}{2} \cdot 4\pi = 2\pi$ steradians.

To treat all of these cases at once, let $\theta \in [0, 2\pi]$ be the maximum value of $\measuredangle QPQ'$ $(Q, Q' \in c)$, with each plane angle $\angle QPQ'$ being measured with respect to an arc lying in the chosen region of the sphere (which will be the larger of the two regions when $\theta > \pi$). It can then be shown that the solid angle f has measure

$$\Omega = 4\pi \sin^2 \theta/4 = 2\pi(1 - \cos \theta/2), \tag{10.28}$$

as plotted on the right side of Figure 10.32. Hence, Ω can range anywhere from 0 to 4π steradians; and, 1 steradian corresponds to $\theta \approx 1.144$ radians (i.e., about 65.5°), which is the angle θ shown on the left side of the figure.

In general, a solid angle is *any* figure f of the form in (10.27), where c is a simple closed curve lying in (the surface of) a sphere having center P; and, the point P is then the **vertex** of the solid angle. The measure Ω of f is then defined as a/r^2 steradians, where a is the area of one of the two regions on the sphere separated by c, chosen as desired, and r is the radius of the sphere. (If c is not a circle, though, then f might not have an obvious "interior" and "exterior".) Given a solid angle f having vertex P, the same value of Ω obtains for *every* sphere centered at P, whatever its radius r may be; because, the area a bounded by the resulting curve c is proportional to r^2. Again, though, a unit sphere (i.e., $r = 1$) is most convenient, since the division by r^2 is then unnecessary, and Ω simply equals the area of the chosen region bounded by c.

CHAPTER 11

Analytic Geometry

Analytic geometry is not itself a *kind* of geometry, but rather a modern *approach* to geometry; that is, it is an analytical method for *expressing* a geometry, be it Euclidean or otherwise. Essentially, this method uses numbers to designate geometric points; then, geometric relations are defined and analyzed via these numbers. Accordingly, theorems are proved *algebraically*, which is often easier than performing a purely geometrical proof.[1] Conversely, analytic geometry often enables one to better understand *numerical* relations by displaying them geometrically (e.g., via function plots, as discussed in Section 5.1).

11.1 Coordinate Systems

Although there are many ways to assign numbers to the points in a plane (e.g., for plane geometry), or to the points in space (e.g., for solid geometry), the most common method is to represent each point in a plane by an ordered *pair* (x_1, x_2) of real numbers x_1 and x_2, and to represent each point in space by an ordered *triple* (x_1, x_2, x_3) of real numbers x_1, x_2, and x_3. Moreover, however these representations are made, no particular pair or triple is assigned to more than one geometric point (though a particular point might be assigned multiple pairs or triples).[2] Thus, in the former case a one-to-many (perhaps one-to-one) correspondence is set up between the points in the plane and the elements of some subset of $\mathbb{R}^2$ (viz., the set of all pairs $(x_1, x_2) \in \mathbb{R}^2$ that have been assigned to points in the plane); likewise, in the latter case such a correspondence is set up between the points in space and the elements of some subset of $\mathbb{R}^3$. The values x_1, x_2, and x_3 are then called the **coordinates** of the corresponding geometric points; accordingly, analytic geometry is often referred to as **coordinate geometry**.

The most straightforward way to express plane geometry analytically is to simply *identify* the plane of interest as the set $\mathbb{R}^2$; then, the desired geometric structure is obtained by viewing the coordinate variables "x_1" and "x_2" as expressing two orthogonal

1. Indeed, many of the facts stated in the preceding chapter are much more easily proved algebraically than geometrically.

2. Hence, the numeric pairs and triples effectively act as *names* of the geometric points, with each point possibly having more than one name.

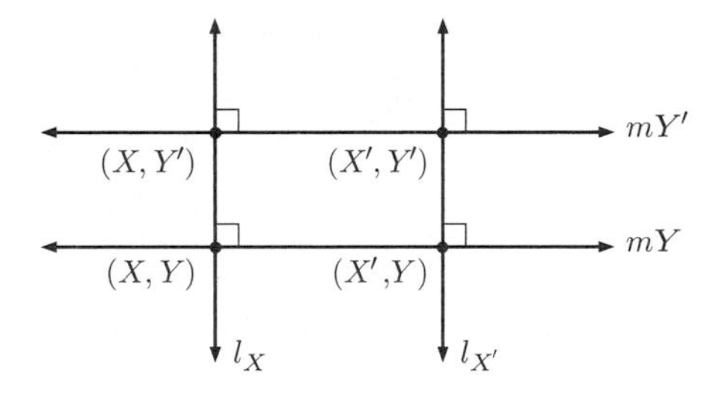

FIGURE 11.1 Basic Lines in the xy-Plane

dimensions of the plane. To wit, letting (as is customary) the variables "x_1" and "x_2" in this context be instead "x" and "y", respectively, each geometric point now *is* a pair $(x, y) \in \mathbb{R}^2$; and, vice versa. (Thus, here the correspondence between geometric points and ordered pairs of numbers is one-to-one.) Furthermore, for each $X \in \mathbb{R}$, the set

$$l_X = \{(x, y) \in \mathbb{R}^2 \mid x = X\}$$

is essentially a "real line" (recall Figure 3.1) lying in the plane $\mathbb{R}^2$; likewise, so is the set

$$m_Y = \{(x, y) \in \mathbb{R}^2 \mid y = Y\}$$

for each $Y \in \mathbb{R}$. Moreover (see Figure 11.1), the lines l_X ($X \in \mathbb{R}$) are viewed as being parallel (since they are coplanar and nonintersecting), as are the lines m_Y ($Y \in \mathbb{R}$), whereas each line l_X is viewed as being perpendicular to each line m_Y. Finally, it is natural to take the distance between any two points (X, Y) and (X, Y') on a line l_X to be $|Y - Y'|$, and the distance between any two points (X, Y) and (X', Y) on a line m_Y to be $|X - X'|$; then, the distance $d(P, P')$ between any two points $P = (X, Y)$ and $P' = (X', Y')$ in $\mathbb{R}^2$ must be

$$d(P, P') = \sqrt{(X - X')^2 + (Y - Y')^2}, \tag{11.1}$$

as seen by letting $Q = (X', Y)$ and applying the Pythagorean theorem (10.12) to the right triangle $\Delta PQP'$—assuming $X \neq X'$ and $Y \neq Y'$. Whereas if $X = X'$ or $Y = Y'$, then $\Delta PQP'$ does not exist, since the points P, P', and Q are not distinct; nevertheless, (11.1) still holds, as can be directly verified. The set $\mathbb{R}^2$ (which alone has no geometric structure) along with the distance measure $d\colon \mathbb{R}^2 \times \mathbb{R}^2 \to [0, \infty)$ just defined combine to give $\mathbb{E}^2$, the **two-dimensional Euclidean space.**[3]

Points in the space $\mathbb{E}^2$ are located via the **two-dimensional rectangular** (or **Cartesian**) **coordinate system.** As shown on the left side of Figure 11.2 (and already

3. In common discourse, a Euclidean space $\mathbb{E}^n$ ($n = 1, 2, \ldots$) is often treated like the corresponding set $\mathbb{R}^n$ of real n-tuples; e.g., one might write "$(x, y) \in \mathbb{E}^2$" to mean "$(x, y) \in \mathbb{R}^2$". However, by using $\mathbb{E}^n$ rather than $\mathbb{R}^n$, it is implied that a particular distance measure is to be used.

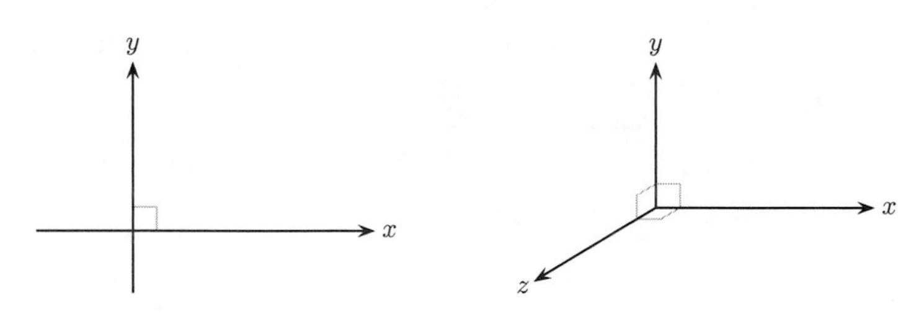

FIGURE 11.2 Two- and Three-Dimensional Rectangular Coordinate Systems

discussed in Section 5.1 with regard to Figure 5.1), this system is indicated by two perpendicular **axes**, one for each of the coordinates x and y. Thus, the axes are often called the $\boldsymbol{x}$**-axis** and the $\boldsymbol{y}$**-axis**; and, the corresponding plane is the $\boldsymbol{xy}$**-plane**.[4] Typically, the x-axis is drawn horizontally and pointing to the right, and the y-axis is drawn vertically and pointing upward, with the direction of each axis corresponding to its coordinate increasing. Unless explicitly indicated otherwise, the axes are understood to intersect at the **origin** of the plane—i.e., the point $(0,0)$—in which case the x-axis is the set $\{(x,y) \in \mathbb{R}^2 \mid y = 0\}$ and the y-axis is the set $\{(x,y) \in \mathbb{R}^2 \mid x = 0\}$. As in Figure 3.2 for the rectangular coordinate system representing the complex plane, the two axes divide the plane into four **quadrants**, numbered I, II, III, and IV—viz., $\{(x,y) \in \mathbb{R}^2 \mid x > 0, y > 0\}$, $\{(x,y) \in \mathbb{R}^2 \mid x < 0, y > 0\}$, $\{(x,y) \in \mathbb{R}^2 \mid x < 0, y < 0\}$, and $\{(x,y) \in \mathbb{R}^2 \mid x > 0, y < 0\}$, respectively.

In the above development of the two-dimensional Euclidean space $\mathbb{E}^2$, the set $\mathbb{R}^2$ is viewed as a union of parallel lines m_Y (the x-axis being m_0)—i.e.,

$$\mathbb{R}^2 = \bigcup_{Y \in \mathbb{R}} m_Y$$

—with the lines m_Y being perpendicular to the parallel lines l_X (the y-axis being l_0). Analogously—now replacing the variables "x_1", "x_2", and "x_3" with "x", "y", and "z", respectively—the **three-dimensional Euclidean space** $\mathbb{E}^3$ obtains by viewing the set $\mathbb{R}^3$ as a union of parallel planes

$$p_Z = \{(x,y,z) \in \mathbb{R}^3 \mid z = Z\} \quad (Z \in \mathbb{R})$$

(each being essentially an xy-plane)—i.e.,

$$\mathbb{R}^3 = \bigcup_{Z \in \mathbb{R}} p_Z$$

—with the planes p_Z being perpendicular to the parallel lines

$$l_{X,Y} = \{(x,y,z) \in \mathbb{R}^3 \mid x = X, y = Y\} \quad (X \in \mathbb{R}, Y \in \mathbb{R}).$$

4. Of course, other variables can be used as well for the coordinates, in which case the axes and plane are renamed accordingly.

Moreover, here the distance $d(P, P')$ between any two points $P = (X, Y, Z)$ and $P' = (X', Y', Z')$ in $\mathbb{R}^3$ can be defined via the right triangle $\Delta PQP'$ formed with $Q = (X', Y', Z)$—assuming P, P', and Q are distinct—where $\angle Q$ is the right angle. Namely, it follows from the Pythagorean theorem that

$$d(P, P') = \sqrt{d^2(P, Q) + d^2(Q, P')};$$

therefore, using (11.1) for $d(P, Q)$ (since P and Q have the same third coordinate) and letting $d(Q, P') = |Z - Z'|$ (since Q and P' have the same first and second coordinates), we obtain

$$d(P, P') = \sqrt{(X - X')^2 + (Y - Y')^2 + (Z - Z')^2} \qquad (11.2)$$

—which also holds when P, P', and Q are not distinct. The set $\mathbb{R}^3$ along with this distance measure $d\colon \mathbb{R}^3 \times \mathbb{R}^3 \to [0, \infty)$ constitute the space $\mathbb{E}^3$.

Points in this space are located by the **three-dimensional rectangular** (or **Cartesian**) **coordinate system**, as indicated on the right side of Figure 11.2; and, since the coordinates have been chosen to be x, y, and z, the corresponding space is referred to as ***xyz*-space**. Assuming again that the axes intersect at the **origin** of the space—now the point $(0, 0, 0)$—the plane of the paper on which the system is drawn is understood to be the set of all points (x, y, z) for which $z = 0$ (i.e., the plane p_0 above); and, the ***z*-axis** (i.e., the line $l_{0,0}$) is visualized as pointing out of the page.[5] We could have, though, directed the z-axis *into* the page; however, a standard axis convention is the **right-hand rule**, whereby pointing the thumb of one's right hand in the direction of the z-axis corresponds to the remaining fingers curling from the direction of the x-axis to the direction of the y-axis. (Accordingly, when the z-axis is drawn pointing upward—e.g., see Figure 11.6 later—the x-axis points out of the page and the y-axis points to the right.) Finally, besides the just-mentioned plane $\{(x, y, z) \in \mathbb{R}^3 \mid z = 0\}$ formed by the x- and y-axes, there is the plane $\{(x, y, z) \in \mathbb{R}^3 \mid y = 0\}$ formed by the x- and z-axes, as well as the plane $\{(x, y, z) \in \mathbb{R}^3 \mid x = 0\}$ formed by the y- and z-axes. These three planes, which are pairwise perpendicular, divide the space into eight **octants**—viz., $\{(x, y, z) \in \mathbb{R}^3 \mid x > 0, y > 0, z > 0\}$, $\{(x, y, z) \in \mathbb{R}^3 \mid x < 0, y > 0, z > 0\}$, $\{(x, y, z) \in \mathbb{R}^3 \mid x > 0, y < 0, z > 0\}, \ldots, \{(x, y, z) \in \mathbb{R}^3 \mid x < 0, y < 0, z < 0\}$.

Using the rectangular coordinate systems, other coordinate systems are readily defined. For example, in two dimensions there is the **polar coordinate system**, shown in Figure 11.3. In this system, each point P in the xy-plane is specified by a pair $(r, \theta) \in [0, \infty) \times \mathbb{R}$, where $r \geq 0$ is the distance from the origin O to P, and $\theta \in \mathbb{R}$ is the counterclockwise angle of the segment $\overline{OP}$ relative to the positive portion of the x-axis. Moreover, if desired, one can also allow r to be negative by understanding that for $r < 0$, a pair (r, θ) is equivalent to $(-r, \theta + n\pi)$ for any odd integer n; then, for each point $P = (r, \theta) \in \mathbb{R} \times \mathbb{R}$, the distance from O to P is $|r|$. In general, the distance

5. For clarity, the negative portions of the three axes are not shown in the figure.

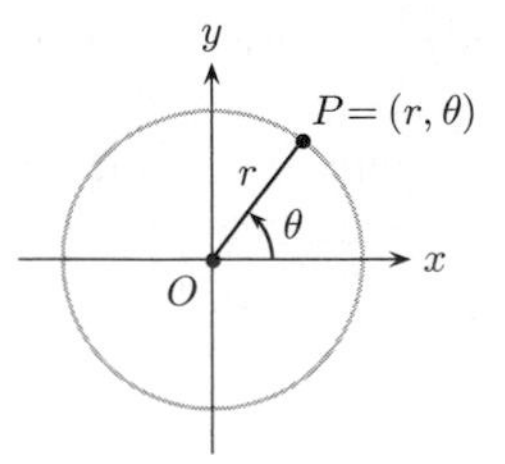

FIGURE 11.3 The Polar Coordinate System

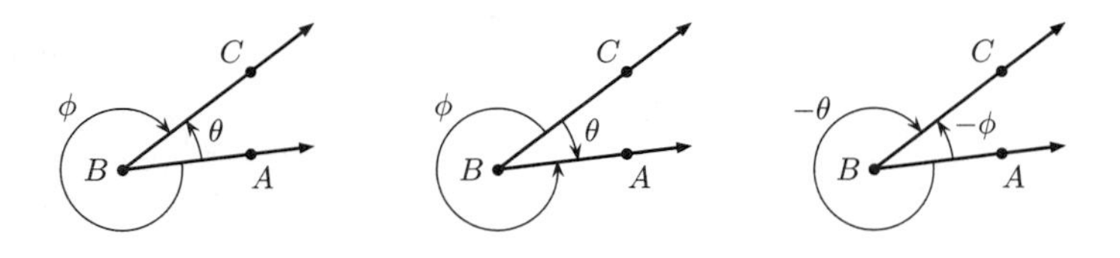

FIGURE 11.4 Directed Angle Measurement

$d(P, P')$ between any two points $P = (R, \Theta)$ and $P' = (R', \Theta')$ expressed in polar coordinates is

$$d(P, P') = \sqrt{R^2 + R'^2 - 2RR' \cos(\Theta - \Theta')}, \tag{11.3}$$

in contrast to (11.1). Geometric figures having a rotational quality about the origin, such as circles and spirals, are typically easier to plot and analyze by using polar coordinates rather than rectangular coordinates.

Note that the definition of "angle"—more precisely, "angle measure"—has now been extended from that given in Chapter 10 (recall Figure 10.19), where all angles were directionless, nonnegative, and no greater than 2π radians (360°). For comparison with measuring directed distances, recall that the value x of a point on the x-axis indicates its distance from the origin in the direction of the axis; hence, negative values are assigned to points arrived at by moving away from the origin in the opposite direction. Similarly, given two rays having a common endpoint, we can measure the angle of either ray relative to the other, in a specified *direction—clockwise* or *counterclockwise*—around the endpoint, with negative angles corresponding to moving in the opposite direction; also, the direction chosen for the measurement is typically indicated by an *arced arrow*.[6] Thus, on the left side of Figure 11.4, the arced arrows next to ϕ and θ indicate that $\overrightarrow{BC}$ is at a counterclockwise angle of θ relative to $\overrightarrow{BA}$ and that $\overrightarrow{BC}$ is at a clockwise angle of ϕ relative to $\overrightarrow{BA}$; here, $\theta = 30°$ and $\phi = 330°$. Or, reversing the reference directions as done in the middle of the figure, we might instead say—for the *same* values of θ and ϕ—that $\overrightarrow{BA}$ is at a clockwise angle of θ relative to $\overrightarrow{BC}$ and that $\overrightarrow{BA}$ is at a counterclockwise angle of ϕ relative to $\overrightarrow{BC}$. Additionally, as shown on the right of the figure, which has the same reference directions as the left side,

6. The direction of an angle (measurement) is sometimes referred to as its "sense".

we can just as well say that $\overrightarrow{BC}$ is at a counterclockwise angle of $-\phi$ relative to $\overrightarrow{BA}$, and that $\overrightarrow{BC}$ is at a clockwise angle of $-\theta$ relative to $\overrightarrow{BA}$—again, for the *same* values of θ and ϕ.

Furthermore, angles of *any* real value α are now allowed by interpreting each multiple of 2π radians as a full **rotation** (or **revolution**, or **turn**) in the specified direction; because, each $\alpha \in \mathbb{R}$ can be uniquely expressed in the form

$$\alpha = n \cdot 2\pi + \alpha',$$

where n is an integer (perhaps negative) and $0 \leq \alpha' < 2\pi$, and thus α can be viewed as n full rotations plus an additional angle of α' (both in the specified direction). For example, comparing the interior angles on the left and right sides of Figure 11.4 implies that we must have $\theta = -\phi + n \cdot 360°$ for some integer n; indeed, for $\theta = 30°$ and $\phi = 330°$, this equation holds with $n = 1$. Also, as first discussed in Section 10.2 with regard to the right side of Figure 10.6, and then in Subsection 10.3.2 with regard to Figure 10.19, angle measurement is again additive—assuming that all angles are measured in the same direction.

The above arrow convention for indicating directed angles, which is consistent with how axis arrows indicate directed distances, tells us how the arced arrow in Figure 11.3 is to be interpreted for the indicated angle θ: namely, when $\theta \geq 0$ (respectively, $\theta < 0$), the point P is at an angle of $|\theta|$ counterclockwise (clockwise) relative to the positive portion of the x-axis. However, once this definition of the polar coordinate θ is understood, it is common practice for arced arrows to be used differently when polar plots are drawn. For example, Figure 11.5 shows two angle indications that would typically appear on a polar plot. In *both* cases, the value of θ is placed next to the arrow, regardless of its direction, even though our arrow convention is thereby violated for the angle indicated as $-300°$. Thus, except for the definition of polar coordinates given in Figure 11.3, arced arrows on a polar plot often do not specify the direction by which the indicated value θ is *measured* (which is already understood to be counterclockwise); rather, as in Figure 11.5, they might instead be used indicate the direction of *motion* that the angle sweeps out as θ varies continuously from 0 to the stated value. Indeed, when $|\theta| > 2\pi$, it is not uncommon to see the angle θ indicated by an arced arrow encircling the origin one or more times, with its direction chosen according to the sign of θ.

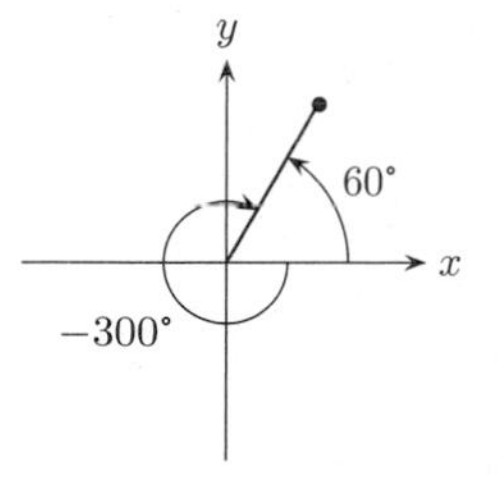

FIGURE 11.5 Typical Angle Indications on a Polar Plot

Unlike the two-dimensional rectangular coordinate system, in which there is exactly one pair $(x, y) \in \mathbb{R} \times \mathbb{R}$ for each point in the xy-plane (and likewise for the three-dimensional rectangular coordinate system), there is not a one-to-one correspondence between the pairs $(r, \theta) \in [0, \infty) \times \mathbb{R}$ and the points in the plane. For this reason, it is sometimes convenient to restrict the angle θ to a particular 2π-interval such as $[0, 2\pi)$ or $(-\pi, \pi]$; then, assuming $r \geq 0$, each point *does* have a unique representation (r, θ) (with the origin corresponding to having $r = 0$ with θ *undefined*).

It follows from the definitions of "sine" and "cosine" to be given in the next section—viz., (11.7a) vis-à-vis Figure 11.7—that the conversion from polar to rectangular coordinates is provided by the formulas

$$x = r\cos\theta, \quad y = r\sin\theta \tag{11.4a}$$

(including when $r \leq 0$). Whereas to convert from rectangular to polar coordinates, we have

$$r = \sqrt{x^2 + y^2}, \qquad \theta = \begin{cases} \tan^{-1}(y/x) & x > 0 \\ \tan^{-1}(y/x) + \pi & x < 0 \\ \pi/2 & x = 0, y > 0 \\ -\pi/2 & x = 0, y < 0 \\ \text{undefined} & x = y = 0 \end{cases} \tag{11.4b}$$

(assuming $r \geq 0$ and $-\pi/2 \leq \theta < 3\pi/2$).[7] The equation for r obtains from the Pythagorean theorem. The equation for θ obtains by applying the definitions of "tangent" and "arctangent"—see (11.7b) and (11.12c) below—while separately considering the cases where the point P in Figure 11.7 lies within each of the four quadrants, as well as on the x- and y-axes.

An alternative coordinate system in three dimensions is the **cylindrical coordinate system** shown in Figure 11.6, which essentially amounts to appending the rectangular

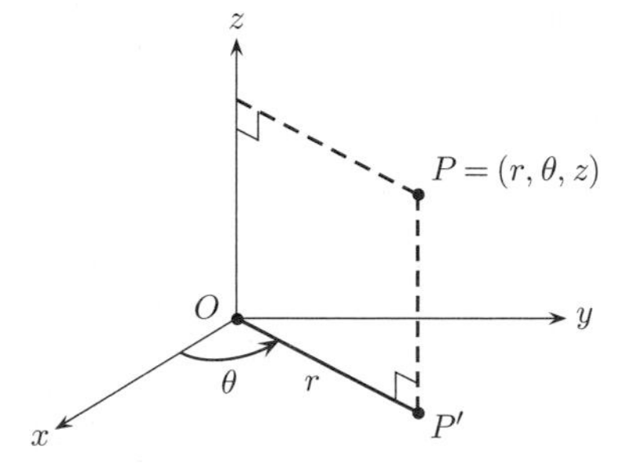

FIGURE 11.6 The Cylindrical Coordinate System

7. The corresponding conversions for complex numbers appear in Section 3.3 as (3.31) and (3.30), respectively. Also, fn. 21 in Chapter 3 likewise applies here.

coordinate z onto the polar coordinate system, thereby designating each point P in xyz-space by a triple $(r, \theta, z) \in [0, \infty) \times \mathbb{R} \times \mathbb{R}$.[8] Specifically, letting P' be the point in this space having the same x and y coordinates as P but a z coordinate of 0, then $r \geq 0$ is the distance from the origin O to P', and $\theta \in \mathbb{R}$ is (for $r \neq 0$) the counterclockwise angle of the segment $\overline{OP'}$ relative to the positive portion of the x-axis; hence, r is also the distance from P to the z-axis. Although there is not a one-to-one correspondence between the triples (r, θ, z) and the points in the space, this can be achieved by restricting the angle θ to a particular 2π-interval. Or, one might allow r to be negative by interpreting (r, θ, z) for $r < 0$ as $(-r, \theta + n\pi, z)$ for some odd integer n; then, for each point $P = (r, \theta, z) \in \mathbb{R} \times \mathbb{R} \times \mathbb{R}$, the distance from O to P is $\sqrt{r^2 + z^2}$. In general, it follows from (11.3) that the distance $d(P, P')$ between two arbitrary points $P = (R, \Theta, Z)$ and $P' = (R', \Theta', Z')$ expressed in cylindrical coordinates is

$$d(P, P') = \sqrt{R^2 + R'^2 - 2RR' \cos(\Theta - \Theta') + (Z - Z')^2}, \qquad (11.5)$$

in contrast to (11.2). Also, the conversions between cylindrical and rectangular coordinates are provided by (11.4), since the z coordinates of these systems are identical.

11.2 Trigonometry

The general study of the metric properties of triangles is called **trigonometry**. As is commonly done, we begin our discussion of this subject without reference to any coordinate system, after which analytic geometry is applied to extend the theory further.

Starting with *right* triangles, recall that the side opposite the right angle of a given right triangle is called the "hypotenuse"; thus, for the triangle $\triangle ABC$ in Figure 11.7 having a right angle $\angle B$, the hypotenuse is $\overline{AC}$. Additionally, it is convenient to specify the remaining two sides of $\triangle ABC$ (i.e., its "legs") relative to the other (acute) angles. Namely, with respect to $\angle A$, the **adjacent side** of $\triangle ABC$ is $\overline{AB}$, and the **opposite side** is $\overline{BC}$; whereas the reverse is the case with respect to $\angle C$.

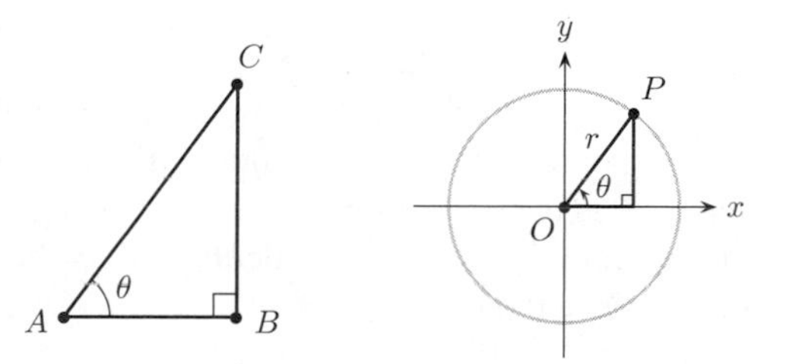

FIGURE 11.7 A Right Triangle, Itself and within a Coordinate System

8. Yet another three-dimensional system—viz., "spherical coordinates"—is given in Problem 11.4.

Since a triangle has three sides, there are $3 \cdot 2 = 6$ ratios that can be formed from their lengths. Upon choosing one of the two acute angles of a given right triangle $\triangle ABC$ as the reference angle—say $\angle A$, as indicated in Figure 11.7—the names of these ratios relative to that angle are:

$$\textbf{sine} \triangleq \frac{|\text{opposite side}|}{|\text{hypotenuse}|}, \qquad \textbf{cosine} \triangleq \frac{|\text{adjacent side}|}{|\text{hypotenuse}|}, \tag{11.6a}$$

$$\textbf{tangent} \triangleq \frac{|\text{opposite side}|}{|\text{adjacent side}|}, \qquad \textbf{cotangent} \triangleq \frac{|\text{adjacent side}|}{|\text{opposite side}|}, \tag{11.6b}$$

$$\textbf{secant} \triangleq \frac{|\text{hypotenuse}|}{|\text{adjacent side}|}, \qquad \textbf{cosecant} \triangleq \frac{|\text{hypotenuse}|}{|\text{opposite side}|} \tag{11.6c}$$

(where, as in Chapter 10, the notation "$|\cdot|$" is used to indicate the length of a segment). It follows from the AA condition for triangle similarity that if $\triangle A'B'C'$ has right angle $\angle B'$, and $\measuredangle A' = \measuredangle A$, then $\triangle A'B'C' \sim \triangle ABC$; so, by the SSS condition for triangle similarity, the ratios (11.6) for $\triangle A'B'C'$ with respect to $\angle A'$ have the same respective values as for $\triangle ABC$ with respect to $\angle A$. Therefore, the value of each ratio depends only on the measure θ of the reference angle $\angle A$, not on the size of the triangle $\triangle ABC$; that is, each ratio is a *function* of θ alone. Thus, the relations in (11.6) effectively specify the six basic **trigonometric functions**.

Since $\angle A$ is acute, these functions are presently defined only for $0 < \theta < \pi/2$. To extend them to all $\theta \in \mathbb{R}$ (with some exceptions to avoid division by 0), we consider a triangle similar to $\triangle ABC$ within a two-dimensional coordinate system, as shown on the right side of Figure 11.7. Letting the indicated point P have rectangular coordinates x and y, as well as polar coordinates $r \geq 0$ and $\theta \in \mathbb{R}$, it follows from (11.6) that for $\theta \in (0, \pi/2)$ we have

$$\sin\theta \triangleq \frac{y}{r}, \qquad \cos\theta \triangleq \frac{x}{r}, \tag{11.7a}$$

$$\tan\theta \triangleq \frac{y}{x}, \qquad \cot\theta \triangleq \frac{x}{y}, \tag{11.7b}$$

$$\sec\theta \triangleq \frac{r}{x}, \qquad \csc\theta \triangleq \frac{r}{y} \tag{11.7c}$$

—here using the abbreviated names of the functions, which are arranged as in (11.6). Now, to allow virtually any value $\theta \in \mathbb{R}$, we simply ignore the triangle and impose (11.7) for every point P in the xy-plane other than the origin O (i.e., we require $r > 0$).[9] The only exceptions are that $\tan\theta$ and $\sec\theta$ are undefined when θ is an odd multiple of $\pi/2$ (since then $x = 0$), whereas $\cot\theta$ and $\csc\theta$ are undefined when θ is an even multiple of $\pi/2$ (i.e., a multiple of π—since then $y = 0$).

9. Just as the size of the triangle $\triangle ABC$ is unimportant above, so for each value θ the ratios (11.7) are independent of the value of r. Therefore, if desired, one may fix r to be 1—thereby restricting P to the unit circle centered at O—and simply define $\sin\theta$ as the resulting value y, $\cos\theta$ as the resulting value x, etc. Accordingly, the six trigonometric functions are often referred to as the **circular functions**.

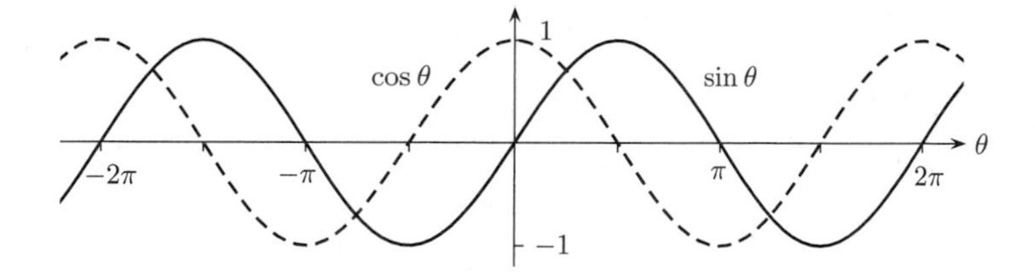

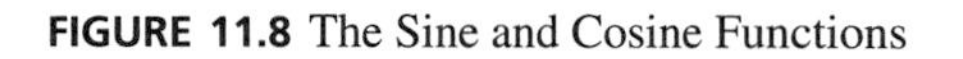
FIGURE 11.8 The Sine and Cosine Functions

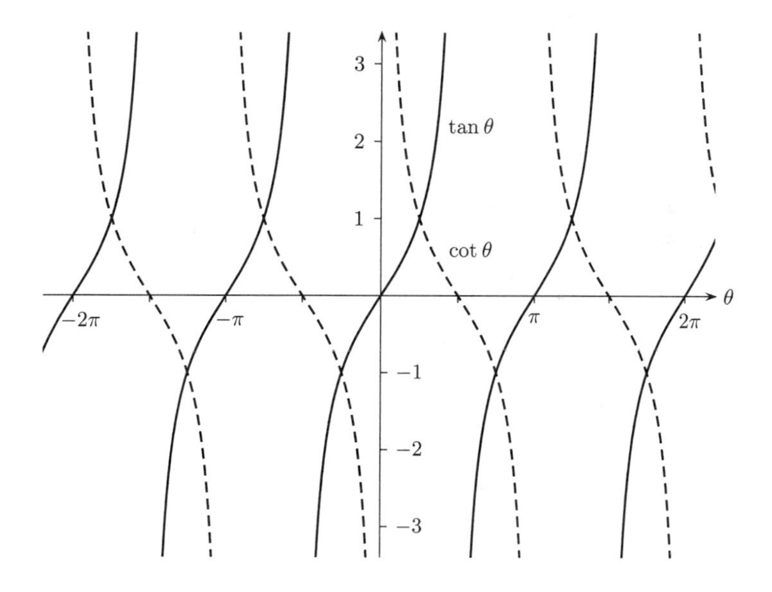

FIGURE 11.9 The Tangent and Cotangent Functions

The six trigonometric functions are plotted in Figures 11.8–11.10, each being continuous at every point θ for which it is defined. In practice (as evidenced by the keypad of a typical scientific calculator), the three functions sine, cosine, and tangent receive the most use; because, their reciprocals provide the remaining three functions:

$$\cot\theta = 1/\tan\theta, \quad \sec\theta = 1/\cos\theta, \quad \csc\theta = 1/\sin\theta \tag{11.8}$$

(with, in each equality, $\theta \in \mathbb{R}$ for which both sides exist). Furthermore,

$$\tan\theta = \frac{\sin\theta}{\cos\theta} \tag{11.9}$$

($\theta \in \mathbb{R}$ not an odd multiple of $\pi/2$).

The values of sine, cosine, and tangent for some frequently occurring angles are given in Table 11.1, with the values at related angles—i.e., 120°, 135°, 150°, . . . , including the corresponding negative angles—being easily calculated from the properties of these functions discussed below. Also, it is useful to note that the signs of these functions

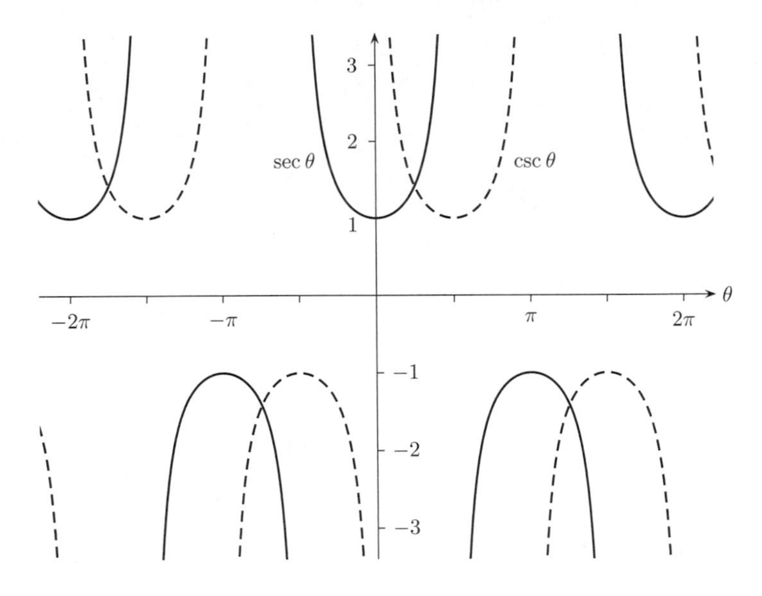

FIGURE 11.10 The Secant and Cosecant Functions

TABLE 11.1 Sine, Cosine, and Tangent of Some Frequently Occurring Angles

θ	$\sin\theta$	$\cos\theta$	$\tan\theta$
$0\ (0°)$	0	1	0
$\pi/6\ (30°)$	$1/2$	$\sqrt{3}/2$	$\sqrt{3}/3$
$\pi/4\ (45°)$	$\sqrt{2}/2$	$\sqrt{2}/2$	1
$\pi/3\ (60°)$	$\sqrt{3}/2$	$1/2$	$\sqrt{3}$
$\pi/2\ (90°)$	1	0	—

TABLE 11.2 Signs of Sine, Cosine, and Tangent

Quadrant of θ	$\sin\theta$	$\cos\theta$	$\tan\theta$
I	+	+	+
II	+	−	−
III	−	−	+
IV	−	+	−

are uniquely determined by the quadrant of the angle θ (i.e., the quadrant of the point P in Figure 11.7); see Table 11.2.

It clearly follows from Figure 11.7 and (11.7) that all six trigonometric functions are periodic with period 2π; in fact, as is evident from Figures 11.8–11.10, the periods of sine, cosine, secant, and cosecant are 2π, whereas the periods of tangent and cotangent are π. Additionally,

$$\sin\theta = -\sin(\theta \pm \pi) = \cos(\theta - \pi/2), \tag{11.10a}$$

$$\cos\theta = -\cos(\theta \pm \pi) = \sin(\theta + \pi/2) \tag{11.10b}$$

($\theta \in \mathbb{R}$); thus, it is said that sine **lags** cosine by $\pi/2$ radians, and cosine **leads** sine by $\pi/2$ radians (see Figure 11.8). Also, cosine and secant are even functions, whereas sine, tangent, cotangent, and cosecant are odd; hence, it follows from (11.10) that

$$\sin\theta = \sin(\pi - \theta) = \cos(\pi/2 - \theta), \tag{11.11a}$$

$$\cos\theta = -\cos(\pi - \theta) = \sin(\pi/2 - \theta) \tag{11.11b}$$

($\theta \in \mathbb{R}$).

As for inverse functions, observe that none of the six trigonometric functions is one-to-one. However, the functions sine, cosine, and tangent are strictly monotonic on the intervals $[-\pi/2, \pi/2]$, $[0, \pi]$, and $(-\pi/2, \pi/2)$, respectively; therefore, each of these functions becomes one-to-one when restricted to that interval. Moreover, each of these restricted functions has the same range—viz., $[-1, 1]$, $[-1, 1]$, and $(-\infty, \infty)$, respectively—as before the restriction. Hence, we can implicitly define the **inverse trigonometric functions**—viz., **arcsine** ($\sin^{-1}$) on $[-1, 1]$, **arccosine** ($\cos^{-1}$) on $[-1, 1]$, and **arctangent** ($\tan^{-1}$) on $(-\infty, \infty)$—thus:

$$x = \sin\theta \quad \Rightarrow \quad \sin^{-1} x = \theta \qquad (-\pi/2 \le \theta \le \pi/2,\ -1 \le x \le 1), \tag{11.12a}$$

$$x = \cos\theta \quad \Rightarrow \quad \cos^{-1} x = \theta \qquad (0 \le \theta \le \pi,\ -1 \le x \le 1), \tag{11.12b}$$

$$x = \tan\theta \quad \Rightarrow \quad \tan^{-1} x = \theta \qquad (-\pi/2 < \theta < \pi/2,\ -\infty < x < \infty) \tag{11.12c}$$

(see Figure 11.11).[10] It follows that

$$\sin(\sin^{-1} x) = x \quad (-1 \le x \le 1), \tag{11.13a}$$

$$\cos(\cos^{-1} x) = x \quad (-1 \le x \le 1), \tag{11.13b}$$

$$\tan(\tan^{-1} x) = x \quad (-\infty < x < \infty), \tag{11.13c}$$

and

$$\sin^{-1}(\sin\theta) = \theta \quad (-\pi/2 \le \theta \le \pi/2), \tag{11.14a}$$

$$\cos^{-1}(\cos\theta) = \theta \quad (0 \le \theta \le \pi), \tag{11.14b}$$

$$\tan^{-1}(\tan\theta) = \theta \quad (-\pi/2 < \theta < \pi/2). \tag{11.14c}$$

(Note that although $\sin(\sin^{-1} x) = x$ for all x in the domain $[-1, 1]$ of arcsine, we do not have $\sin^{-1}(\sin\theta) = \theta$ for all θ in the domain $(-\infty, \infty)$ of sine; likewise for cosine

10. Alternative abbreviations for these functions are "arcsin", "arccos", and "arctan".

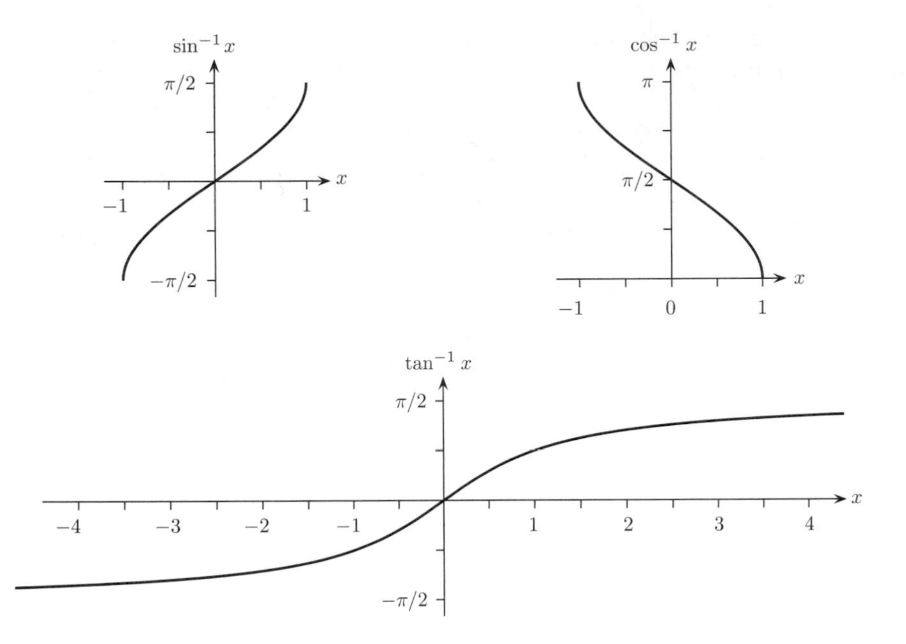

FIGURE 11.11 The Arcsine, Arccosine, and Arctangent Functions

composed with arccosine, and for tangent composed with arctangent.) All three of the inverse trigonometric functions are continuous; also, both arcsine and arctangent are strictly increasing and possess odd symmetry, whereas arccosine is strictly decreasing and possesses neither even nor odd symmetry.

Regarding notational conventions: In addition to "$\sin(\theta)$", e.g., being commonly rewritten "$\sin\theta$" (as above), one may unambiguously rewrite "$\sin(a\theta)$" as "$\sin a\theta$", and "$\sin(\theta/a)$" as "$\sin\theta/a$", and "$(\sin\theta)(\cos\phi)$" as "$\sin\theta\cos\phi$", and even "$(\sin a\theta)(\cos b\phi)$" as "$\sin a\theta \cos b\phi$". However, potentially confusing expressions such as "$\sin\theta/\cos\phi$" and "$\sin\theta/(\cos\phi)$" should probably be avoided, instead writing "$(\sin\theta)/(\cos\phi)$" or "$(\sin\theta)/\cos\phi$" if these are intended, while otherwise writing "$\sin[\theta/(\cos\phi)]$" or "$\sin(\theta/\cos\phi)$". Similar conventions apply to the other trigonometric functions, as well as to the inverse trigonometric functions and the "hyperbolic functions" defined below.

The six trigonometric functions and their inverses are interrelated by various equations—e.g., (11.8), (11.9), (11.10), (11.11), (11.13), and (11.14)—commonly referred to as "trigonometric identities". We now discuss further the most frequently used of these identities, briefly indicating how they can be derived. For example, applying the Pythagorean theorem (10.12) to the triangle on the right of Figure 11.7 yields $x^2 + y^2 = r^2$; therefore, by (11.7),

$$\sin^2\theta + \cos^2\theta = 1, \tag{11.15a}$$

$$\tan^2\theta + 1 = \sec^2\theta, \tag{11.15b}$$

$$\cot^2\theta + 1 = \csc^2\theta. \tag{11.15c}$$

Here and for the remainder of this section, it is understood that each expression in terms of "θ" and/or "ϕ" is asserted only for those values θ, $\phi \in \mathbb{R}$ for which all of the operations performed in the expression are defined.

Using Euler's identity (3.32) to expand each of the complex exponentials in the two equations $e^{i(\theta \pm \phi)} = e^{i\theta} e^{\pm i\phi}$ ($\pm$ respectively), then equating the real and imaginary parts respectively within each resulting equation, we obtain the sum and difference identities

$$\sin(\theta \pm \phi) = \sin\theta \cos\phi \pm \cos\theta \sin\phi, \tag{11.16a}$$

$$\cos(\theta \pm \phi) = \cos\theta \cos\phi \mp \sin\theta \sin\phi, \tag{11.16b}$$

$$\tan(\theta \pm \phi) = \frac{\tan\theta \pm \tan\phi}{1 \mp \tan\theta \tan\phi} \tag{11.16c}$$

($\pm$ and $\mp$ respectively in each case)—the last being found from the other two via (11.9). From (11.16a) and (11.16b) we then obtain the product identities

$$\sin\theta \sin\phi = [\cos(\theta - \phi) - \cos(\theta + \phi)]/2, \tag{11.17a}$$

$$\cos\theta \cos\phi = [\cos(\theta - \phi) + \cos(\theta + \phi)]/2, \tag{11.17b}$$

$$\sin\theta \cos\phi = [\sin(\theta - \phi) + \sin(\theta + \phi)]/2; \tag{11.17c}$$

and these, after appropriate substitutions have been made for θ and ϕ, yield additional sum and difference identities:

$$\sin\theta + \sin\phi = 2\cos\tfrac{1}{2}(\theta - \phi)\sin\tfrac{1}{2}(\theta + \phi), \tag{11.18a}$$

$$\sin\theta - \sin\phi = 2\sin\tfrac{1}{2}(\theta - \phi)\cos\tfrac{1}{2}(\theta + \phi), \tag{11.18b}$$

$$\cos\theta + \cos\phi = 2\cos\tfrac{1}{2}(\theta - \phi)\cos\tfrac{1}{2}(\theta + \phi), \tag{11.18c}$$

$$\cos\theta - \cos\phi = -2\sin\tfrac{1}{2}(\theta - \phi)\sin\tfrac{1}{2}(\theta + \phi). \tag{11.18d}$$

Taking $\phi = \theta$ in (11.16) yields the double-angle identities

$$\sin 2\theta = 2\sin\theta\cos\theta, \tag{11.19a}$$

$$\cos 2\theta = \cos^2\theta - \sin^2\theta = 1 - 2\sin^2\theta = 2\cos^2\theta - 1, \tag{11.19b}$$

$$\tan 2\theta = \frac{2\tan\theta}{1 - \tan^2\theta} \tag{11.19c}$$

—having used (11.15a) to obtain the last two equalities in (11.19b). Substituting "$\theta/2$" for each occurrence of "θ" in (11.19b), then manipulating the results using (11.9) and (11.15a), we obtain the half-angle identities

$$\sin\theta/2 = \pm\sqrt{\frac{1 - \cos\theta}{2}}, \tag{11.20a}$$

$$\cos\theta/2 = \pm\sqrt{\frac{1 + \cos\theta}{2}}, \tag{11.20b}$$

$$\tan\theta/2 = \pm\sqrt{\frac{1 - \cos\theta}{1 + \cos\theta}} = \frac{\sin\theta}{1 + \cos\theta} = \frac{1 - \cos\theta}{\sin\theta} \tag{11.20c}$$

—where the actual sign of each "±" is determined by the quadrant of the angle $\theta/2$ (see Table 11.2).

Just as the trigonometric identities in (11.16) were derived with the aid of complex exponentials, so it is often helpful to use complex exponentials to express the trigonometric functions:

$$\sin\theta = \operatorname{Im} e^{i\theta} = \frac{e^{i\theta} - e^{-i\theta}}{i2}, \tag{11.21a}$$

$$\cos\theta = \operatorname{Re} e^{i\theta} = \frac{e^{i\theta} + e^{-i\theta}}{2}, \tag{11.21b}$$

$$\tan\theta = \frac{\operatorname{Im} e^{i\theta}}{\operatorname{Re} e^{i\theta}} = -i\frac{e^{i\theta} - e^{-i\theta}}{e^{i\theta} + e^{-i\theta}}. \tag{11.21c}$$

For example, we can now obtain

$$\sin^2\theta = \left(\frac{e^{i\theta} - e^{-i\theta}}{i2}\right)^2 = \frac{e^{i2\theta} - 2 + e^{-i2\theta}}{-4} = \frac{1 - (e^{i2\theta} + e^{-i2\theta})/2}{2}$$
$$= (1 - \cos 2\theta)/2$$

(which also readily follows from (11.19b)); or, by a slightly different derivation,

$$\sin^2\theta = \operatorname{Re}(\sin^2\theta) = \operatorname{Re}\left(\frac{e^{i\theta} - e^{-i\theta}}{i2}\right)^2 = \operatorname{Re}\frac{e^{i2\theta} - 2 + e^{-i2\theta}}{-4}$$
$$= \frac{\cos 2\theta - 2 + \cos 2\theta}{-4} = (1 - \cos 2\theta)/2.$$

Thus, we have derived the first of the square identities

$$\sin^2\theta = (1 - \cos 2\theta)/2, \tag{11.22a}$$

$$\cos^2\theta = (1 + \cos 2\theta)/2, \tag{11.22b}$$

$$\tan^2\theta = \frac{1 - \cos 2\theta}{1 + \cos 2\theta}; \tag{11.22c}$$

and, (11.22b) similarly obtains, after which (11.22c) results from (11.9). Likewise, other power identities can be derived, such as

$$\sin^3\theta = (3\sin\theta - \sin 3\theta)/4, \tag{11.23a}$$

$$\cos^3\theta = (3\cos\theta + \cos 3\theta)/4, \tag{11.23b}$$

$$\tan^3\theta = \frac{3\sin\theta - \sin 3\theta}{3\cos\theta + \cos 3\theta}. \tag{11.23c}$$

In the same vein, taking $r = 1$ in (6.6) or $z = i\theta$ in (7.4f) yields **DeMoivre's theorem**,

$$(e^{i\theta})^n = e^{in\theta} \quad (\theta \in \mathbb{R}, n \in \mathbb{Z}), \tag{11.24a}$$

which by Euler's identity can be written

$$(\cos\theta + i\sin\theta)^n = \cos n\theta + i\sin n\theta \quad (\theta \in \mathbb{R}, n \in \mathbb{Z}). \tag{11.24b}$$

Expanding the left side for each $n \geq 1$, then equating the real and imaginary parts respectively within each resulting equation, one can obtain multiple-angle identities for all positive multiples. For $n = 2$, this approach again yields (11.19a) and the leftmost equality in (11.19b); whereas for $n = 3$ it yields

$$\sin 3\theta = 3\sin\theta\cos^2\theta - \sin^3\theta,$$

$$\cos 3\theta = \cos^3\theta - 3\sin^2\theta\cos\theta,$$

which upon using (11.15a) become the triple-angle identities

$$\sin 3\theta = 3\sin\theta - 4\sin^3\theta, \tag{11.25a}$$

$$\cos 3\theta = 4\cos^3\theta - 3\cos\theta. \tag{11.25b}$$

(which also follow from (11.23)). Whereas taking $\phi = 2\theta$ in (11.16c) then substituting (11.19c) yields

$$\tan 3\theta = \frac{3\tan\theta - \tan^3\theta}{1 - 3\tan^2\theta}. \tag{11.25c}$$

Additionally, from (11.24a) it quickly follows that

$$e^{in\pi} = (-1)^n \quad (n \in \mathbb{Z}), \tag{11.26}$$

which upon taking real and imaginary parts per (11.24b) yields

$$\cos n\pi = (-1)^n, \quad \sin n\pi = 0 \quad (n \in \mathbb{Z}).$$

Likewise,

$$e^{in\pi/2} = i^n \quad (n \in \mathbb{Z}), \tag{11.27}$$

which yields

$$\cos n\pi/2 = \begin{cases} (-1)^{n/2} & n \text{ even} \\ 0 & n \text{ odd} \end{cases}, \qquad \sin n\pi/2 = \begin{cases} 0 & n \text{ even} \\ (-1)^{(n-1)/2} & n \text{ odd} \end{cases}.$$

(Compare (11.26) and (11.27) with (3.38).)

As for identities involving the inverse trigonometric functions, we have

$$\sin^{-1}x = \pi/2 - \cos^{-1}x \quad (-1 \leq x \leq 1), \tag{11.28a}$$

$$\cos^{-1}(-x) = \pi - \cos^{-1}x \quad (-1 \leq x \leq 1), \tag{11.28b}$$

$$\tan^{-1}1/x = \pi/2 - \tan^{-1}x \quad (x > 0). \tag{11.28c}$$

Also,

$$\sin^{-1} x = \begin{cases} -\cos^{-1}\sqrt{1-x^2} & -1 \le x \le 0 \\ \cos^{-1}\sqrt{1-x^2} & 0 \le x \le 1 \end{cases}, \tag{11.29a}$$

$$\sin^{-1} x = \tan^{-1} x/\sqrt{1-x^2} \quad (-1 < x < 1), \tag{11.29b}$$

$$\cos^{-1} x = \begin{cases} \pi - \sin^{-1}\sqrt{1-x^2} & -1 \le x \le 0 \\ \sin^{-1}\sqrt{1-x^2} & 0 \le x \le 1 \end{cases}, \tag{11.29c}$$

$$\cos^{-1} x = \begin{cases} \pi + \tan^{-1}\sqrt{1-x^2}/x & -1 \le x < 0 \\ \tan^{-1}\sqrt{1-x^2}/x & 0 < x \le 1 \end{cases}, \tag{11.29d}$$

$$\tan^{-1} x = \sin^{-1} x/\sqrt{1+x^2} \quad (-\infty < x < \infty), \tag{11.29e}$$

$$\tan^{-1} x = \begin{cases} -\cos^{-1} 1/\sqrt{1+x^2} & x \le 0 \\ \cos^{-1} 1/\sqrt{1+x^2} & x \ge 0 \end{cases}. \tag{11.29f}$$

(For example, (11.29b) obtains for $0 < x < 1$ by letting $\left|\overline{BC}\right| = x$ and $\left|\overline{AC}\right| = 1$ in Figure 11.7, thereby making $\left|\overline{AB}\right| = \sqrt{1-x^2}$, with $\sin^{-1}\left|\overline{BC}\right|/\left|\overline{AC}\right| = \theta = \tan^{-1}\left|\overline{BC}\right|/\left|\overline{AB}\right|$; and, we also conclude that (11.29b) holds for $-1 < x \le 0$, since arcsine and arctangent are odd functions.) It then follows that

$$\cos(\sin^{-1} x) = \sqrt{1-x^2} \quad (-1 \le x \le 1), \tag{11.30a}$$

$$\tan(\sin^{-1} x) = x/\sqrt{1-x^2} \quad (-1 < x < 1), \tag{11.30b}$$

$$\sin(\cos^{-1} x) = \sqrt{1-x^2} \quad (-1 \le x \le 1), \tag{11.30c}$$

$$\tan(\cos^{-1} x) = \sqrt{1-x^2}/x \quad (0 < |x| \le 1), \tag{11.30d}$$

$$\sin(\tan^{-1} x) = x/\sqrt{1+x^2} \quad (-\infty < x < \infty), \tag{11.30e}$$

$$\cos(\tan^{-1} x) = 1/\sqrt{1+x^2} \quad (-\infty < x < \infty). \tag{11.30f}$$

With some care, other inverse trigonometric identities can be derived from the preceding facts. For example, if $x, y \ge 0$, then we can apply (11.16c) and (11.13c) to get

$$\tan(\tan^{-1} x - \tan^{-1} y) = \frac{\tan(\tan^{-1} x) - \tan(\tan^{-1} y)}{1 + \tan(\tan^{-1} x) \cdot \tan(\tan^{-1} y)} = \frac{x-y}{1+xy}$$

(where division by 0 is impossible, since $xy \ge 0$); therefore, since $-\pi/2 < \tan^{-1} x - \tan^{-1} y < \pi/2$, it follows from (11.14c) that we can take the arctangent of the above equation to obtain

$$\tan^{-1} x - \tan^{-1} y = \tan^{-1}\frac{x-y}{1+xy} \quad (x, y \ge 0).$$

(Notice, though, that this equation fails, e.g., for $x = 1$ and $y = -2$.)

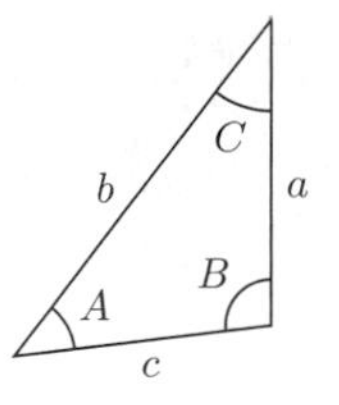

FIGURE 11.12 A Side-versus-Angle Notation for an Arbitrary Triangle

Regarding the general case of an arbitrary (i.e., not necessarily right) triangle, Figure 11.12 shows the frequently used notation whereby the measure of each angle is indicated by an uppercase letter, and the length of the opposite side is indicated by the corresponding lowercase letter. The **law of sines** then states that

$$a/\sin A = b/\sin B = c/\sin C; \tag{11.31}$$

in fact, dividing this value by 2 gives the radius of the circumscribed circle.[11] It follows that a triangle is equilateral if and only if it is equiangular (as already stated in Subsection 10.3.1); and, a triangle is isosceles (respectively, scalene) if and only if at least two (no two) of its *angles* are congruent (cf. Table 10.2). Also, there is the **law of cosines**,

$$a^2 = b^2 + c^2 - 2bc\cos A \tag{11.32}$$

(and similarly vis-à-vis the angles B and C), which can be viewed as a generalization of the Pythagorean theorem (10.12).

Finally, analogous to the "circular functions" as expressed in (11.21), we have the **hyperbolic functions**, the most frequently occurring being **hyperbolic sine**, **hyperbolic cosine**, and **hyperbolic tangent**, respectively defined as

$$\sinh x \triangleq \frac{e^x - e^{-x}}{2}, \tag{11.33a}$$

$$\cosh x \triangleq \frac{e^x + e^{-x}}{2}, \tag{11.33b}$$

$$\tanh x \triangleq \frac{e^x - e^{-x}}{e^x + e^{-x}} \tag{11.33c}$$

($x \in \mathbb{R}$); see Figure 11.13.[12] Then, analogous to (11.8), we may also define the **hyperbolic cotangent**, **hyperbolic secant**, and **hyperbolic cosecant** functions as, respectively,

$$\coth x \triangleq 1/\tanh x, \quad \operatorname{sech} x \triangleq 1/\cosh x, \quad \operatorname{csch} x \triangleq 1/\sinh x$$

11. As for the radius of the inscribed circle, see Problem 10.37.

12. The "hyperbolic functions" are so named because they can be straightforwardly defined by replacing the circle in Figure 11.7 with a hyperbola.

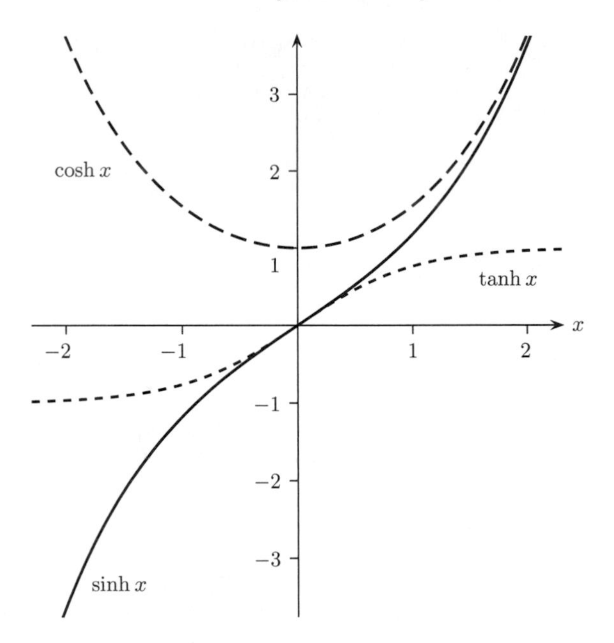

FIGURE 11.13 The Main Hyperbolic Functions

($x \in \mathbb{R}$, with $x \neq 0$ for the first and third equalities). Each of these six functions is continuous at every point x for which it is defined; but, none are periodic. Additionally, cosh and sech are even functions, whereas sinh, tanh, coth, and csch are odd.

The functions sinh and tanh, being strictly increasing, are one-to-one; and, their ranges are $(-\infty, \infty)$ and $(-1, 1)$, respectively. Whereas cosh, though not one-to-one, is strictly increasing on $[0, \infty)$; and, when restricted to this interval, it has the same range—viz., $[1, \infty)$—as before the restriction. Hence, we can implicitly define the **inverse hyperbolic functions**—viz., $\sinh^{-1}$ on $(-\infty, \infty)$, $\cosh^{-1}$ on $[1, \infty)$, and $\tanh^{-1}$ on $(-1, 1)$—thus:

$$y = \sinh x \quad \Rightarrow \quad \sinh^{-1} y = x \qquad (-\infty < x < \infty, -\infty < y < \infty),$$
$$y = \cosh x \quad \Rightarrow \quad \cosh^{-1} y = x \qquad (0 \leq x < \infty, 1 \leq y < \infty),$$
$$y = \tanh x \quad \Rightarrow \quad \tanh^{-1} y = x \qquad (-\infty < x < \infty, -1 < y < 1).$$

It follows that

$$\sinh(\sinh^{-1} y) = y \quad (-\infty < y < \infty),$$
$$\cosh(\cosh^{-1} y) = y \quad (1 \leq y < \infty),$$
$$\tanh(\tanh^{-1} y) = y \quad (-1 < y < 1),$$

and

$$\sinh^{-1}(\sinh x) = x \quad (-\infty < x < \infty), \tag{11.34a}$$

$$\cosh^{-1}(\cosh x) = x \quad (0 \le x < \infty), \tag{11.34b}$$

$$\tanh^{-1}(\tanh x) = x \quad (-\infty < x < \infty). \tag{11.34c}$$

(Note that although $\cosh(\cosh^{-1} y) = y$ for all y in the domain $[1, \infty)$ of $\cosh^{-1}$, we do not have $\cosh^{-1}(\cosh x) = x$ for all x in the domain $(-\infty, \infty)$ of the function cosh; the functions sinh and tanh, though, encounter no such discrepancies when composed with their respective inverses.) In fact, each of the equations (11.33) can be solved for x, thereby providing explicit expressions for the inverse hyperbolic functions:

$$\sinh^{-1} y = \ln\left(y + \sqrt{y^2 + 1}\right) \quad (y \in \mathbb{R}), \tag{11.35a}$$

$$\cosh^{-1} y = \ln\left(y + \sqrt{y^2 - 1}\right) \quad (y \ge 1), \tag{11.35b}$$

$$\tanh^{-1} y = \tfrac{1}{2} \ln \frac{1 + y}{1 - y} \quad (-1 < y < 1). \tag{11.35c}$$

All three of these functions are continuous and strictly increasing; also, both $\sinh^{-1}$ and $\tanh^{-1}$ possess odd symmetry, whereas $\cosh^{-1}$ possesses neither even nor odd symmetry.

11.3 Vectors

Conceptually, a **vector** is a *directed length*; that is, it is a geometric object characterized by a particular length pointing in a particular direction.

11.3.1 Abstract Vector Spaces

We begin our discussion of these objects by viewing them abstractly; that is, we focus on their *essential* characteristics without restricting our attention to specific cases. Thus, a **vector space** is a set V of elements called **vectors**. Accompanying V is a **scalar field** F, which is a set of elements called **scalars**; for us, F will be either the set $\mathbb{R}$ of real numbers or the set $\mathbb{C}$ of complex numbers. (Thus, although the term "vector space" is commonly used to refer to V alone, it actually means the pair (V, F).) If $F = \mathbb{R}$, then V is called a **real vector space**; whereas for $F = \mathbb{C}$, it is a **complex vector space**.

Besides the usual arithmetic operations on $\mathbb{R}$ and $\mathbb{C}$ of addition and multiplication, there are similar operations of vector addition and scalar-vector multiplication. First, for any vectors $\vec{x}$ and $\vec{y}$, their sum $\vec{x} + \vec{y}$ is also a vector; and, for any scalar a and vector $\vec{x}$, their product $a\vec{x}$ is a vector.[13] Additionally, these operations possess the following basic properties for $\vec{x}, \vec{y}, \vec{z} \in V$ and $a, b \in F$:

$$\vec{x} + \vec{y} = \vec{y} + \vec{x} \quad \text{(commutative property);} \tag{11.36a}$$

$$\vec{x} + (\vec{y} + \vec{z}) = (\vec{x} + \vec{y}) + \vec{z}, \quad a(b\vec{x}) = (ab)\vec{x} \quad \text{(associative properties);}^{14} \tag{11.36b}$$

$$a(\vec{x} + \vec{y}) = a\vec{x} + a\vec{y}, \quad (a + b)\vec{x} = a\vec{x} + b\vec{x} \quad \text{(distributive properties);} \tag{11.36c}$$

and,

$$1\vec{x} = \vec{x}, \tag{11.36d}$$

thus making the scalar 1 the multiplicative identity not only for scalar multiplication, but also for scalar-vector multiplication. Furthermore, there exists a unique element $\vec{0} \in V$—called the **zero vector** of the space—such that

$$\vec{x} + \vec{0} = \vec{x} \tag{11.36e}$$

for all $\vec{x} \in V$; that is, $\vec{0}$ is the additive identity for vector addition.[15] Finally, for each $\vec{x} \in V$, there exists a unique element $-\vec{x} \in V$—the **(additive) inverse** of $\vec{x}$—for which

$$\vec{x} + (-\vec{x}) = \vec{0}. \tag{11.36f}$$

From the above properties, others follow. In particular, here are some "obvious" facts for $\vec{x} \in V$ and $a \in F$:

$$-\vec{0} = \vec{0}, \tag{11.37a}$$

$$0\vec{x} = \vec{0}, \tag{11.37b}$$

$$a\vec{0} = \vec{0}, \tag{11.37c}$$

$$(-1)\vec{x} = -\vec{x}, \tag{11.37d}$$

$$-(-\vec{x}) = \vec{x}, \tag{11.37e}$$

$$a(-\vec{x}) = (-a)\vec{x} = -(a\vec{x}). \tag{11.37f}$$

Also, unlike vector addition, vector subtraction is not a primitive operation; rather, the difference $\vec{x} - \vec{y}$ of two vectors $\vec{x}$ and $\vec{y}$ is defined as $\vec{x} + (-\vec{y})$.

An advantage of approaching vector spaces abstractly is that it enables one to treat as vectors mathematical objects that are not normally thought of as such. For example,

---

13. As is commonly done, we are using the same symbol "+" to indicate the addition of two scalars, as well as the addition of two vectors, though in a stricter presentation each of these *different* operations would be assigned its own symbol. Likewise, concatenation will be used to indicate the two different operations of multiplying two scalars and multiplying a vector by a scalar.

14. It follows that the parentheses in these two equations can be omitted, since the expressions "$\vec{x} + \vec{y} + \vec{z}$" and "$ab\vec{x}$" are unambiguous with regard to their results.

15. Observe the distinction between the *scalar* 0 (the additive identity for scalar addition) and the *vector* $\vec{0}$.

let V be the set of all polynomials having complex coefficients such that none of the polynomials has a degree greater than some fixed integer $n \geq 0$; that is,

$$V = \{\text{functions } a_n x^n + a_{n-1}x^{n-1} + \cdots + a_1 x + a_0 \text{ of } x \mid a_j \in \mathbb{C}\ (j = 0, 1, \ldots, n)\},$$

where we have arbitrarily chosen the variable "x" to express the polynomials. Also, let $F = \mathbb{C}$. Then, assuming the usual arithmetic operations, we observe that the addition of any two polynomials in V is also a polynomial in V, as is the multiplication of any polynomial in V by a number a in F. Moreover, upon interpreting $\vec{0}$ as the element of V for which $a_j = 0$ (all j)—i.e., the zero polynomial—all of the properties (11.36) hold. Therefore, this set V (accompanied by F) is a vector space, with the vectors being polynomials.

A **norm** on a vector space V is a function $\|\cdot\|$ on V such that for all $x, y \in V$ and $a \in F$:

$$\|\vec{x}\,\| \geq 0, \quad \text{with equality if and only if } \vec{x} = \vec{0}; \tag{11.38a}$$

$$\|a\vec{x}\,\| = |a| \cdot \|\vec{x}\,\|; \tag{11.38b}$$

$$\|\vec{x} + \vec{y}\,\| \leq \|\vec{x}\,\| + \|\vec{y}\,\| \tag{11.38c}$$

—the last expression being called the **triangle inequality**. The norm $\|\vec{x}\,\|$ of a vector $\vec{x}$ is often referred to as its **length** (or **magnitude**); in particular, a vector having of length 1 is called a **unit vector**. A vector space on which a norm has been defined is a **normed space**.

As an example, suppose we take $V = \mathbb{C}$ with $F = \mathbb{R}$, thereby viewing each complex number $z \in \mathbb{C}$ as a vector $\vec{z} \in V$; in particular, assuming the usual arithmetic operations, it follows from (11.37b) that the complex number 0 must the vector $\vec{0}$. The usual norm for this space, consistent with the Argand diagram in Figure 3.3, is

$$\|\vec{z}\,\| = |\vec{z}\,| = \sqrt{(\text{Re}\,\vec{z}\,)^2 + (\text{Im}\,\vec{z}\,)^2} \quad (\vec{z} \in V), \tag{11.39}$$

per (3.30a). However, another valid norm on this space is

$$\|\vec{z}\,\| = |\,\text{Re}\,\vec{z}\,| + |\,\text{Im}\,\vec{z}\,| \quad (\vec{z} \in V); \tag{11.40}$$

because, this function $\|\cdot\|$ also satisfies (11.38).

An **inner product** (or **scalar product**) on a vector space V is a function $[\cdot,\cdot]$ on $V \times V$, taking on values in F, such that for all $\vec{x}, \vec{y}, \vec{z} \in V$ and $a, b \in F$:

$$[\vec{x}, \vec{x}\,] \geq 0, \quad \text{with equality if and only if } \vec{x} = \vec{0}; \tag{11.41a}$$

$$[\vec{x}, \vec{y}\,] = [\vec{y}, \vec{x}\,]^*; \tag{11.41b}$$

$$[a\vec{x} + b\vec{y}, \vec{z}\,] = a[\vec{x}, \vec{z}\,] + b[\vec{y}, \vec{z}\,]; \tag{11.41c}$$

and, it then follows that

$$[\vec{x}, a\vec{y} + b\vec{z}\,] = a^*[\vec{x}, \vec{y}\,] + b^*[\vec{x}, \vec{z}\,]. \tag{11.42}$$

Thus, an inner product is linear in the first argument, and "conjugate linear" in the second; moreover, if $F = \mathbb{R}$, then all of the complex conjugations above are superfluous, yielding ordinary linearity in the second argument as well. A vector space on which an inner product has been defined is an **inner product space**.

Every inner product on a space V induces a corresponding norm, according to the equation

$$\|\vec{x}\,\| = \sqrt{[\vec{x},\vec{x}\,]} \quad (\vec{x} \in V); \tag{11.43}$$

that is, this function $\|\cdot\|$ on V automatically satisfies the requisite conditions (11.38) of a norm. Conversely, the following **polarization laws** (which are easily verified by using (11.43) and expanding the resulting inner products) show how any inner product can be expressed in terms of its induced norm:

$$F = \mathbb{R}: \quad [\vec{x},\vec{y}\,] = \left\|\frac{\vec{x}+\vec{y}}{2}\right\|^2 - \left\|\frac{\vec{x}-\vec{y}}{2}\right\|^2, \tag{11.44a}$$

$$F = \mathbb{C}: \quad [\vec{x},\vec{y}\,] = \left(\left\|\frac{\vec{x}+\vec{y}}{2}\right\|^2 - \left\|\frac{\vec{x}-\vec{y}}{2}\right\|^2\right) + i\left(\left\|\frac{\vec{x}+i\vec{y}}{2}\right\|^2 - \left\|\frac{\vec{x}-i\vec{y}}{2}\right\|^2\right) \tag{11.44b}$$

$(\vec{x},\vec{y} \in V)$. However, it is not true that for *any* norm $\|\cdot\|$ on V, the function $[\cdot\,,\cdot]$ given by the appropriate part of (11.44) is a valid inner product. For example, again considering $V = \mathbb{C}$ with $F = \mathbb{R}$, the norm (11.39) substituted into (11.44a) yields

$$[\vec{x},\vec{y}\,] = \mathrm{Re}(\vec{x}\vec{y}^{\,*}) \quad (\vec{x},\vec{y} \in V), \tag{11.45}$$

which satisfies (11.41) and thus is indeed an inner product on V; hence, (11.45) expresses the one and only inner product on this space that induces the norm (11.39). But, for this same pair (V, F), a valid inner product does not obtain when (11.44a) is applied to the norm (11.40); therefore, *no* inner product on V induces this norm. Also, note that for $V = \mathbb{C}$ but now with $F = \mathbb{C}$, the norm (11.39) substituted into (11.44b) yields

$$[\vec{x},\vec{y}\,] = \vec{x}\vec{y}^{\,*} \quad (\vec{x},\vec{y} \in V), \tag{11.46}$$

which is a valid inner product on *this* space, but not on the previous space; because, this function sometimes takes on nonreal values, which are outside of the previous scalar field F. *Hereafter, when an inner product is mentioned, it is understood that its induced norm is also to be used.*

A fundamental property of an inner product on a vector space V is the **Schwarz inequality**,

$$\big|[\vec{x},\vec{y}\,]\big| \le \|\vec{x}\,\| \cdot \|\vec{y}\,\| \quad (\vec{x},\vec{y} \in V). \tag{11.47}$$

Furthermore, equality is attained if and only if either $\vec{y} = \vec{0}$ or $\vec{x} = a\vec{y}$ for some $a \in F$. It follows, either directly or via (11.41b), that an equivalent condition for equality is that either $\vec{x} = \vec{0}$ or $\vec{y} = a\vec{x}$ for some $a \in F$.

Given an inner product space V, two vectors $\vec{x}, \vec{y} \in V$ are said to be **orthogonal** (or **perpendicular**)—i.e., $\vec{x} \perp \vec{y}$—if and only if $[\vec{x}, \vec{y}\,] = 0$. (Hence, if $\vec{x} \perp \vec{y}$ then $\vec{y} \perp \vec{x}$.) For one thing, this definition yields an abstract form of the Pythagorean theorem:

$$\vec{x} \perp \vec{y} \quad \Rightarrow \quad \|\vec{x}\,\|^2 + \|\vec{y}\,\|^2 = \|\vec{x} + \vec{y}\,\|^2 \qquad (\vec{x}, \vec{y} \in V). \tag{11.48}$$

It is also worth noting that the only vector that is orthogonal to all vectors in the space is $\vec{0}$; indeed, $\vec{0}$ is the only vector that is orthogonal to itself.

As an example, consider again the space $V = \mathbb{C}$ with $F = \mathbb{R}$ and the inner product in (11.45). Given vectors $\vec{x}, \vec{y} \neq \vec{0}$, let these complex numbers be expressed in polar form:

$$\vec{x} = Ae^{i\alpha}, \quad \vec{y} = Be^{i\beta} \quad (A, B > 0; \alpha, \beta \in \mathbb{R}). \tag{11.49a}$$

Then,

$$[\vec{x}, \vec{y}\,] = \mathrm{Re}(Ae^{i\alpha} \cdot Be^{-i\beta}) = AB\cos(\alpha - \beta). \tag{11.49b}$$

Therefore, if we restrict the angle $\alpha - \beta$ to the interval $(-\pi, \pi]$, then

$$\vec{x} \perp \vec{y} \quad \Leftrightarrow \quad |\alpha - \beta| = \pi/2,$$

as one would expect from the Argand diagram of the vectors.

Given an inner product space V and two vectors $\vec{x}, \vec{y} \in V$, there is exactly one way express $\vec{x}$ in the form

$$\vec{x} = \vec{x}_y + \vec{z} \quad (\vec{x}_y, \vec{z} \in V) \tag{11.50a}$$

such that

$$\vec{x}_y = a\vec{y} \quad \text{for some} \quad a \in F \tag{11.50b}$$

and

$$\vec{z} \perp \vec{y} \tag{11.50c}$$

(see Figure 11.14). The vector $\vec{x}_y$ is called the **(orthogonal) projection** of $\vec{x}$ onto the vector $\vec{y}$. If $\vec{y} \neq \vec{0}$, then

$$a = [\vec{x}, \vec{y}\,]/\|\vec{y}\,\|^2; \tag{11.50d}$$

otherwise, a is arbitrary.[16] It follows from (11.50b) and (11.50c) that $\vec{z} \perp \vec{x}_y$ as well; also, we have $\vec{x} \perp \vec{y}$ if and only if $\vec{x}_y = \vec{0}$—i.e., $\vec{z} = \vec{x}$. Furthermore, the scalar multiple of $\vec{y}$ that is nearest to $\vec{x}$ is $\vec{x}_y$; that is, for all $a' \in F$,

$$a'\vec{y} \neq \vec{x}_y \quad \Rightarrow \quad \|\vec{x} - a'\vec{y}\,\| > \|\vec{x} - \vec{x}_y\|. \tag{11.51}$$

16. The scalar a may be referred to as the "component" of $\vec{x}$ in the direction of $\vec{y}$.

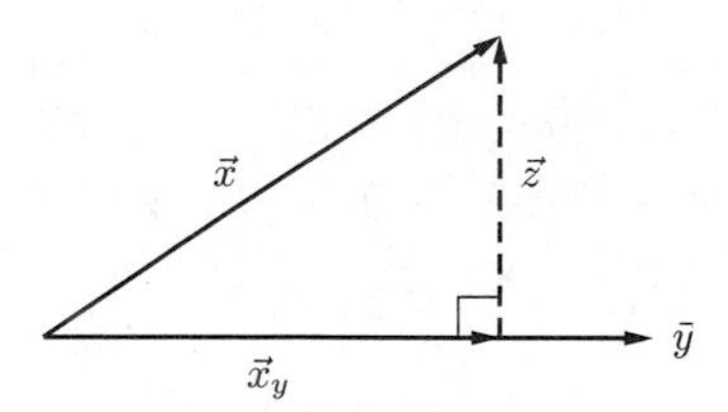

FIGURE 11.14 The Orthogonal Projection of One Vector onto Another

By substituting (11.50d) into (11.50b) and manipulating the result, we can express the projection $\vec{x}_y$ as a scalar multiple of the unit vector $\vec{y}/\|\vec{y}\,\|$ pointing in the direction of $\vec{y} \neq \vec{0}$:

$$\vec{x}_y = [\vec{x}, \vec{y}/\|\vec{y}\,\|] \cdot \vec{y}/\|\vec{y}\,\|. \tag{11.52}$$

In this form, the coefficient $[\vec{x}, \vec{y}/\|\vec{y}\,\|] \in F$ acts as a "scalar magnitude"—which may be negative if $F = \mathbb{R}$, or even nonreal if $F = \mathbb{C}$.

Finally, the operation of projecting one vector onto another provides a heuristic interpretation of the Schwarz inequality (11.47). For it follows from (11.52)—again, assuming $\vec{y} \neq \vec{0}$—that

$$\|\vec{x}_y\| = \big|[\vec{x}, \vec{y}/\|\vec{y}\,\|]\big| = \big|[\vec{x}, \vec{y}\,]\big| \big/ \|\vec{y}\,\|;$$

whereas (11.47) can be rewritten

$$\big|[\vec{x}, \vec{y}\,]\big| \big/ \|\vec{y}\,\| \leq \|\vec{x}\,\|.$$

Hence, the Schwarz inequality can be interpreted as stating that (as is evident from Figure 11.14) if a vector $\vec{x}$ is projected onto a vector $\vec{y} \neq \vec{0}$, then the length of the projection $\vec{x}_y$ can be no greater than the length of $\vec{x}$—i.e.,

$$\|\vec{x}_y\,\| \leq \|\vec{x}\,\| \tag{11.53}$$

—and, this bound also holds when $\vec{y} = \vec{0}$, since then $\vec{x}_y = \vec{0}$. Indeed, because we have $\vec{x}_y \perp \vec{z}$ in (11.50a), it follows from (11.48) that

$$\|\vec{x}_y\|^2 + \|\vec{z}\,\|^2 = \|\vec{x}\,\|^2,$$

which yields (11.53) directly.

11.3.2 Euclidean Vector Spaces

The fundamental properties (11.36) of vector spaces in general are largely motivated by the analytical properties of Euclidean geometric spaces—in particular, $\mathbb{E}^2$ (plane geometry) and $\mathbb{E}^3$ (solid geometry).[17] Consequently, all of the abstract theory presented

17. Briefly, the distinction between a *vector* space and a *geometric* space is that the former is a set of vectors, whereas the latter is a set of points.

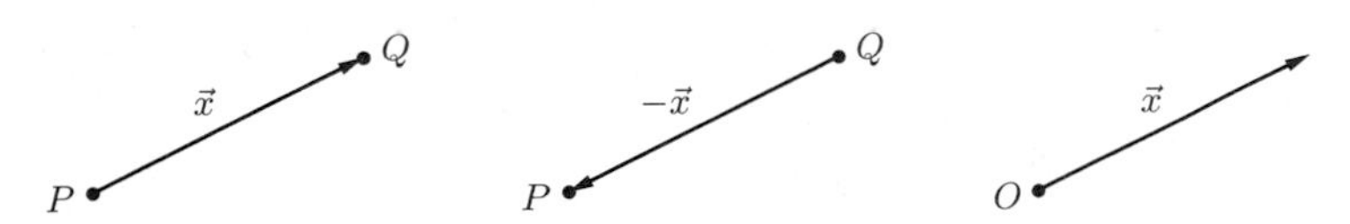

FIGURE 11.15 Vectors in a Euclidean Space

above applies to the special case of Euclidean vector spaces, which we address here. Additionally, Euclidean vector spaces can be expressed, in a natural way, using matrices.

For a Euclidean geometric space E—henceforth referred to simply as a "Euclidean space" (as done, e.g., in Section 11.1)—a vector, say $\vec{x}$, can be specified by giving an ordered pair (P, Q) of points on E, composed of an **initial point** P and a **terminal point** Q. The length (i.e., norm) of $\vec{x}$ is then the distance $d(P, Q)$, and its direction is that from P to Q. As shown on the left side of Figure 11.15, the vector $\vec{x}$ is typically indicated by an arrow, with its *tail* being at the initial point and its *head* being at the terminal point. If $P = Q$, then $\vec{x} = \vec{0}$; and, conversely. Also, as shown in the middle of the figure, interchanging the roles of the initial and terminal points—i.e., considering the pair (Q, P)—converts $\vec{x}$ into its inverse $-\vec{x}$.

Although vectors and rays are similar—indeed, both are displayed pictorially by arrows (recall Figure 10.1)—there are significant differences between these two geometric objects. First, a ray in a Euclidean space E is a *subset* of E—viz., the set of all points lying in a particular direction away from the endpoint of the ray, along with the endpoint itself. A vector, though, is not a set of points, but a particular length (magnitude) paired with a particular direction.[18] Therefore, although a given ray has a fixed *position* in the space E, a given vector $\vec{x}$ does not; rather, if the initial point P and terminal point Q of $\vec{x}$ are moved within E in any way that preserves both the distance and the direction from P to Q, then the new initial point P' and terminal point Q' taken together *also* represent $\vec{x}$. This explains why, e.g., the vector $\vec{0}$ is represented by the pair (P, P) for *any* point $P \in E$.

For $E = \mathbb{E}^2$, if $\vec{x}$ is expressed by $P = (x_1, y_1)$ and $Q = (x_2, y_2)$, then the same vector $\vec{x}$ is also expressed by every pair (P', Q') of points $P' = (x_1', y_1')$ and $Q' = (x_2', y_2')$ for which

$$x_1' - x_1 = x_2' - x_2, \quad y_1' - y_1 = y_2' - y_2 \tag{11.54}$$

—i.e., the shift in the first coordinate from P to P' is the same as the shift in the first coordinate from Q to Q', and likewise for the second coordinates. Similarly, for $E = \mathbb{E}^3$, if $\vec{x}$ is expressed by $P = (x_1, y_1, z_1)$ and $Q = (x_2, y_2, z_2)$, then $\vec{x}$ is also expressed by every pair (P', Q') of points $P' = (x_1', y_1', z_1')$ and $Q' = (x_2', y_2', z_2')$ for which

$$x_1' - x_1 = x_2' - x_2, \quad y_1' - y_1 = y_2' - y_2, \quad z_1' - z_1 = z_2' - z_2. \tag{11.55}$$

Hence, strictly speaking, a particular vector $\vec{x}$ of E is not just a single ordered pair (P, Q) of points $P, Q \in E$, but a *collection* of such pairs, with each pair in the collection being

18. Accordingly, whereas we refer to the rays *in* E, we refer to the vectors *of* E.

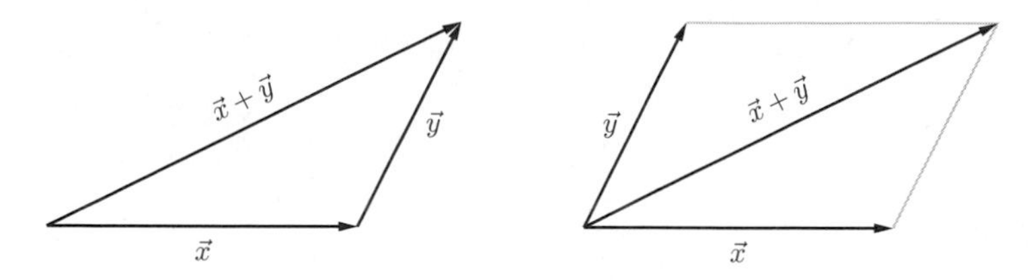

FIGURE 11.16 Vector Addition in a Euclidean Space

a *representative* of $\vec{x}$; but, although each vector $\vec{x}$ is represented by infinitely many pairs (P, Q), each pair (P, Q) of points $P, Q \in E$ represents only one vector.[19] Accordingly, it is common to refer to an individual pair (P, Q) as a "vector".[20]

For a Euclidean space, one way to perform the addition of two vectors $\vec{x}$ and $\vec{y}$ is to place them so that the head (terminal point) of $\vec{x}$ coincides with the tail (initial point) of $\vec{y}$,[21] as shown on the left side of Figure 11.16; then, the vector extending from the tail of $\vec{x}$ to the head of $\vec{y}$ is $\vec{x} + \vec{y}$. Alternatively, we can place $\vec{x}$ and $\vec{y}$ tail-to-tail; then, these vectors (assuming that neither is $\vec{0}$ and that they do not point in the same direction or in opposite directions) form adjacent sides of a parallelogram, as shown on the right side of the figure, for which the vector extending from the vertex at the tails of $\vec{x}$ and $\vec{y}$ to the corresponding diagonal vertex is $\vec{x} + \vec{y}$. As for vector subtraction, it can be performed likewise by using the fact that $\vec{x} - \vec{y} = \vec{x} + (-\vec{y})$.

For each point Q on a Euclidean space E, the pair (O, Q) formed with the origin O of E is called the **position vector** corresponding to Q; because, it effectively locates the point Q within E. Conversely, of the infinitely many pairs (P, Q) of points in E that represent a particular vector $\vec{x}$ of E, there is exactly one for which the initial point P is O (see the right side of Figure 11.15); that is, each vector of the space E has a unique corresponding position vector. Hence, via position vectors we have a one-to-one correspondence between the vectors $\vec{x}$ of E and the points Q on E. In particular, the zero vector $\vec{0}$ corresponds to the origin O—viz., $(0, 0)$ for $\mathbb{E}^2$, and $(0, 0, 0)$ for $\mathbb{E}^3$. Also, if a vector $\vec{x}$ corresponds to either a point $(x, y) \in \mathbb{E}^2$ or $(x, y, z) \in \mathbb{E}^3$, then its inverse $-\vec{x}$ corresponds to the point $(-x, -y)$ or $(-x, -y, -z)$, respectively. Furthermore, for any scalar $a \in \mathbb{R}$ (scalars for Euclidean vector spaces always being real), the vector $a\vec{x}$ corresponds to the point (ax, ay) or (ax, ay, az). By (11.1) and (11.2), the length of $\vec{x}$ is either $\sqrt{x^2 + y^2}$ or $\sqrt{x^2 + y^2 + z^2}$. Finally, if another vector $\vec{y}$ corresponds to a point

---

19. Let a pair (P', Q') be "equivalent to" a pair (P, Q) if and only if the former can be obtained by shifting the latter—per (11.54) for $E = \mathbb{E}^2$, (11.55) for $E = \mathbb{E}^3$, and likewise for higher-dimensional Euclidean spaces E (see below). Then (as is easily verified), (P, Q) is also equivalent to (P', Q'), as well as to itself; moreover, if another pair (P'', Q'') of points $P'', Q'' \in E$ is equivalent to (P', Q'), then it is also equivalent to (P, Q). Therefore, this binary relation of "equivalent to" between pairs of points taken from E is an equivalence relation (see Section 2.1); and, the collections of pairs we have just identified as "vectors" are the *equivalence classes* of this relation. It follows that every vector is represented by at least one pair (P, Q), and each pair (P, Q) represents one and only one vector.

20. Indeed, this is done in the definition of "position vector" below.

21. More precisely, *representatives* of $\vec{x}$ and $\vec{y}$ are chosen that are, respectively, head-to-tail. Henceforth, such precise wording will be invoked only when necessary to avoid confusion.

$(x',y') \in \mathbb{E}^2$ or $(x',y',z') \in \mathbb{E}^3$, then the sum $\vec{x}+\vec{y}$ corresponds to the point $(x+x',y+y')$ or $(x+x',y+y',z+z')$.

In light of these facts, it is natural to represent and manipulate the vectors of $\mathbb{E}^2$ and $\mathbb{E}^3$ as matrices (see Chapter 9) rather than as ordered tuples of points; moreover, this is the case for a Euclidean space of *any* dimension. First, for each integer $n \geq 1$, the ***n*-dimensional Euclidean (geometric) space** (or ***n*-space**) $\mathbb{E}^n$ is defined as the set $\mathbb{R}^n$ of geometric points, along with the distance measure $d\colon \mathbb{R}^n \times \mathbb{R}^n \to [0,\infty)$ given by

$$d(P,P') = \sqrt{\sum_{j=1}^{n}(x_j - x_j{}')^2}, \tag{11.56}$$

for any two points $P = (x_1,x_2,\ldots,x_n)$ and $P' = (x_1{}',x_2{}',\ldots,x_n{}')$ on $\mathbb{R}^n$—thus generalizing (11.1) and (11.2) to $n > 3$.[22] Now, fixing the dimension n of the space, let $\mathbb{V}^n$ be the set of all real $n \times 1$ matrices

$$\mathbf{x} = \begin{bmatrix} x_1 \\ x_2 \\ \vdots \\ x_n \end{bmatrix};^{23} \tag{11.57}$$

and let $\mathbb{R}$ be the scalar field associated with $\mathbb{V}^n$. Each such matrix $\mathbf{x}$ uniquely represents the vector $\vec{x}$ of $\mathbb{E}^n$ that corresponds to the point $(x_1,x_2,\ldots,x_n) \in \mathbb{E}^n$; in particular, the zero vector $\vec{0}$ is represented by the $n \times 1$ matrix

$$\mathbf{0} = \begin{bmatrix} 0 \\ 0 \\ \vdots \\ 0 \end{bmatrix}.$$

Furthermore: the inverse vector $-\vec{x}$ is represented by the matrix

$$-\mathbf{x} = \begin{bmatrix} -x_1 \\ -x_2 \\ \vdots \\ -x_n \end{bmatrix};$$

22. A space having more than three dimensions is often referred to as a **hyperspace**.

23. Here we are using "column vectors" (in the matrix sense) to constitute $\mathbb{V}^n$, whereas $1 \times n$ "row vectors" could be used just as well by making minor adjustments to what follows. An advantage of the latter representation is conservation of space, which is why it is common to see a *column* vector $\mathbf{x}$ specified as $\mathbf{x} = \begin{bmatrix} x_1 & x_2 & \cdots & x_n \end{bmatrix}^{\mathrm{T}}$.

for each $a \in \mathbb{R}$, the vector $a\vec{x}$ is represented by the matrix

$$a\mathbf{x} = \begin{bmatrix} ax_1 \\ ax_2 \\ \vdots \\ ax_n \end{bmatrix};$$

and, if another vector $\vec{y}$ is represented by the matrix

$$\mathbf{y} = \begin{bmatrix} y_1 \\ y_2 \\ \vdots \\ y_n \end{bmatrix} \tag{11.58}$$

in $\mathbb{V}^n$, then the sum $\vec{x} + \vec{y}$ is represented by the matrix

$$\mathbf{x} + \mathbf{y} = \begin{bmatrix} x_1 + y_1 \\ x_2 + y_2 \\ \vdots \\ x_n + y_n \end{bmatrix}.$$

Moreover, note that

$$\mathbf{x} - \mathbf{y} = \begin{bmatrix} x_1 - y_1 \\ x_2 - y_2 \\ \vdots \\ x_n - y_n \end{bmatrix},$$

and thus $\mathbf{x} - \mathbf{y} = \mathbf{x} + (-\mathbf{y})$, as desired. Based on these facts, it is readily verified that the conditions (11.36) for $\mathbb{V}^n$ to be a vector space itself are satisfied; therefore, not only are the elements $\mathbf{x} \in \mathbb{V}^n$ vectors in the *matrix* sense (i.e., arrays), they are also vectors in the sense discussed in Subsection 11.3.1.

Thus, for each $n \geq 1$, we have established a one-to-one correspondence between each pair of the following four sets:

- the set of all vectors $\vec{x}$ of $\mathbb{E}^n$ (each being a collection of ordered pairs (P, Q) of points P, $Q \in \mathbb{E}^n$);
- the set of position vectors of $\mathbb{E}^n$ (each being an ordered pair (O, Q) for a point $Q \in \mathbb{E}^n$);
- the set $\mathbb{E}^n$ of geometric points Q (each being the terminal point of a position vector (O, Q));
- the set of real $n \times 1$ matrices $\mathbf{x}$ (each being the matrix representation (11.57) of a point $Q = (x_1, x_2, \ldots, x_n) \in \mathbb{E}^n$).

(Figure 11.17 illustrates for $n = 2$ the correspondences between the last three sets.) Hence, for conciseness, it is not uncommon for an $n \times 1$ matrix $\mathbf{x}$ to be discussed as if it *were* a vector $\vec{x}$ of $\mathbb{E}^n$ (usually drawn as a position vector, but sometimes shifted); also, $\mathbf{x}$ might be discussed (perhaps in the same context) as if it *were* a point Q in $\mathbb{E}^n$.

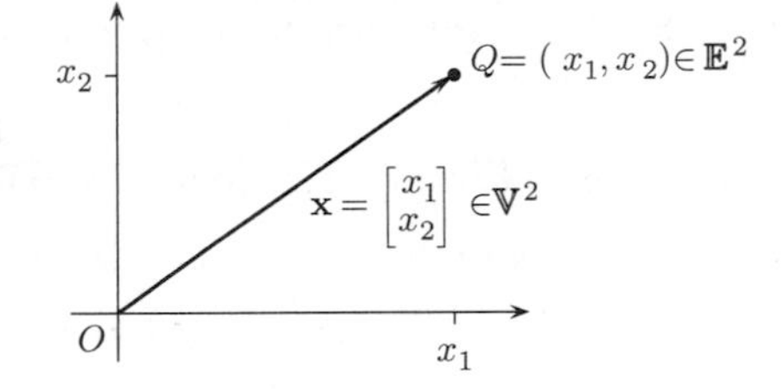

FIGURE 11.17 A Point and the Matrix Representation of Its Position Vector

For example, a reference to "the direction of **x**" implicitly indicates that **x** is be viewed as either a vector $\vec{x}$ or a position vector, whereas a phrase such as "the distance from **x** to" indicates that **x** is to be viewed as a geometric point (viz., the terminal point of the corresponding position vector). Rarely, though, does any confusion result from this abuse of terminology, since the intended meaning is usually obvious.

Per (11.56) with $P' = O$, the norm $\|\cdot\|$ on the space $\mathbb{V}^n$ is defined by

$$\|\mathbf{x}\| \triangleq \sqrt{\sum_{j=1}^{n} x_j{}^2} = \sqrt{\mathbf{x}^{\mathrm{T}}\mathbf{x}} \quad (\mathbf{x} \in \mathbb{V}^n), \tag{11.59}$$

where the second equality follows from the definitions of matrix transposition and multiplication.[24] Hence, the following versions of the properties (11.38) hold for all $\mathbf{x}, \mathbf{y} \in \mathbb{V}^n$ and $a \in \mathbb{R}$:

$$\|\mathbf{x}\| \geq 0, \quad \text{with equality if and only if } \mathbf{x} = \mathbf{0};$$
$$\|a\mathbf{x}\| = |a| \cdot \|\mathbf{x}\|;$$
$$\|\mathbf{x} + \mathbf{y}\| \leq \|\mathbf{x}\| + \|\mathbf{y}\|$$

—the last property being the triangle inequality. Moreover, the associated inner product on $\mathbb{V}^n$ is defined by

$$[\mathbf{x}, \mathbf{y}] \triangleq \sum_{j=1}^{n} x_j y_j = \mathbf{x}^{\mathrm{T}}\mathbf{y} \quad (\mathbf{x}, \mathbf{y} \in \mathbb{V}^n); \tag{11.60a}$$

because, this formula obtains from using (11.59) in the polarization law (11.44a), and the resulting function $[\cdot, \cdot]$ on $\mathbb{V}^n \times \mathbb{V}^n$ satisfies the conditions (11.41) for being an inner product. Using this norm and inner product, the space $\mathbb{V}^n$ is the ***n*-dimensional Euclidean vector space**.

In the context of a *Euclidean* vector space, it is common to refer to an inner product as a **dot product**, which has its own notation:

$$\mathbf{x} \bullet \mathbf{y} \triangleq [\mathbf{x}, \mathbf{y}] \quad (\mathbf{x}, \mathbf{y} \in \mathbb{V}^n). \tag{11.60b}$$

24. Also, recall fn. 2 in Chapter 9.

Hence, by (11.41), the dot product possesses the following properties for all $\mathbf{x}, \mathbf{y}, \mathbf{z} \in \mathbb{V}^n$ and $a, b \in \mathbb{R}$:

$$\mathbf{x} \bullet \mathbf{x} \geq 0, \quad \text{with equality if and only if } \mathbf{x} = \mathbf{0}; \tag{11.61a}$$

$$\mathbf{x} \bullet \mathbf{y} = \mathbf{y} \bullet \mathbf{x}; \tag{11.61b}$$

$$(a\mathbf{x} + b\mathbf{y}) \bullet \mathbf{z} = a(\mathbf{x} \bullet \mathbf{z}) + b(\mathbf{y} \bullet \mathbf{z}); \tag{11.61c}$$

and therefore

$$\mathbf{x} \bullet (a\mathbf{y} + b\mathbf{z}) = a(\mathbf{x} \bullet \mathbf{y}) + b(\mathbf{x} \bullet \mathbf{z}). \tag{11.62}$$

Also,

$$\|\mathbf{x}\| = \sqrt{\mathbf{x} \bullet \mathbf{x}}, \tag{11.63}$$

by (11.43). Moreover, by (11.47), we have the Schwarz inequality,

$$|\mathbf{x} \bullet \mathbf{y}| \leq \|\mathbf{x}\| \cdot \|\mathbf{y}\|, \tag{11.64}$$

with equality being attained if and only if either $\mathbf{y} = \mathbf{0}$ or $\mathbf{x} = a\mathbf{y}$ for some $a \in \mathbb{R}$.

Two vectors $\mathbf{x}, \mathbf{y} \in \mathbb{V}^n$ are orthogonal (or perpendicular)—i.e., $\mathbf{x} \perp \mathbf{y}$—if and only if $\mathbf{x} \bullet \mathbf{y} = 0$. Moreover, as illustrated in Figure 11.18, the **angle (measure)** *between two vectors* **x** and **y** is

$$\measuredangle(\mathbf{x}, \mathbf{y}) \triangleq \cos^{-1} \frac{\mathbf{x} \bullet \mathbf{y}}{\|\mathbf{x}\| \cdot \|\mathbf{y}\|} \quad (\text{nonzero } \mathbf{x}, \mathbf{y} \in \mathbb{V}^n) \tag{11.65}$$

(thus, $0 \leq \measuredangle(\mathbf{x}, \mathbf{y}) \leq \pi$); that is, referring to (11.57) and (11.58), if $Q = (x_1, x_2, \ldots, x_n)$ and $Q' = (y_1, y_2, \ldots, y_n)$ are points on $\mathbb{E}^n$ other than the origin O, then the angle $\angle QOQ'$ has measure $\measuredangle QOQ' = \measuredangle(\mathbf{x}, \mathbf{y})$. It follows that

$$\mathbf{x} \bullet \mathbf{y} = \|\mathbf{x}\| \cdot \|\mathbf{y}\| \cdot \cos \measuredangle(\mathbf{x}, \mathbf{y}) \quad (\mathbf{x}, \mathbf{y} \in \mathbb{V}^n),^{25} \tag{11.66}$$

from which the Schwarz inequality (11.64) obtains immediately. As expected, $\mathbf{x} \perp \mathbf{y}$ if and only if either $\mathbf{x} = \mathbf{0}$, $\mathbf{y} = \mathbf{0}$, or $\measuredangle(\mathbf{x}, \mathbf{y}) = \pi/2$.

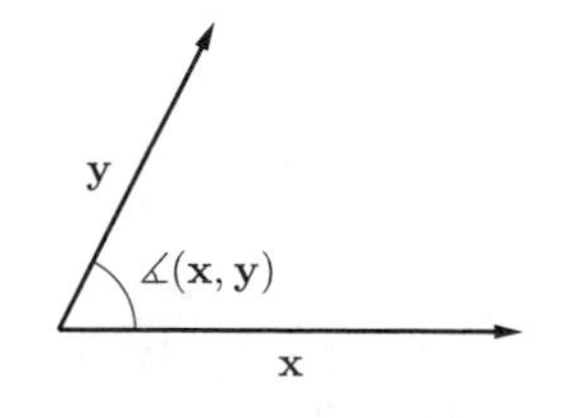

FIGURE 11.18 The Angle between Two Vectors

25. Compare with (11.49), regarding the angle between two nonzero complex numbers—the complex plane with inner product (11.45) being essentially a two-dimensional Euclidean vector space.

The n vectors

$$\mathbf{e}_j = \begin{bmatrix} e_{j,1} \\ e_{j,2} \\ \vdots \\ e_{j,n} \end{bmatrix} \qquad (j = 1, 2, \ldots, n) \tag{11.67a}$$

with

$$e_{j,k} = \begin{cases} 1 & j = k \\ 0 & j \neq k \end{cases} \qquad (j, k = 1, 2, \ldots, n) \tag{11.67b}$$

—i.e.,

$$\mathbf{e}_1 = \begin{bmatrix} 1 \\ 0 \\ \vdots \\ 0 \end{bmatrix}, \quad \mathbf{e}_2 = \begin{bmatrix} 0 \\ 1 \\ \vdots \\ 0 \end{bmatrix}, \quad \ldots, \quad \mathbf{e}_n = \begin{bmatrix} 0 \\ 0 \\ \vdots \\ 1 \end{bmatrix} \tag{11.67c}$$

—are called the **standard basis vectors** for the n-dimensional space $\mathbb{V}^n$. By using them, any vector $\mathbf{x}$ in (11.57) can be expanded in terms of its components:

$$\mathbf{x} = \sum_{j=1}^{n} x_j \mathbf{e}_j. \tag{11.68}$$

For $n = 2$, a commonly used alternative notation is

$$\hat{\mathbf{i}} = \begin{bmatrix} 1 \\ 0 \end{bmatrix}, \quad \hat{\mathbf{j}} = \begin{bmatrix} 0 \\ 1 \end{bmatrix}, \tag{11.69}$$

and thus

$$\mathbf{x} = x_1 \hat{\mathbf{i}} + x_2 \hat{\mathbf{j}};$$

whereas for $n = 3$ we have

$$\hat{\mathbf{i}} = \begin{bmatrix} 1 \\ 0 \\ 0 \end{bmatrix}, \quad \hat{\mathbf{j}} = \begin{bmatrix} 0 \\ 1 \\ 0 \end{bmatrix}, \quad \hat{\mathbf{k}} = \begin{bmatrix} 0 \\ 0 \\ 1 \end{bmatrix}, \tag{11.70}$$

and thus

$$\mathbf{x} = x_1 \hat{\mathbf{i}} + x_2 \hat{\mathbf{j}} + x_3 \hat{\mathbf{k}}$$

in this case.[26]

---

26. The vectors in (11.69) and (11.70)—for which the overset caret (^), or "hat", explicitly indicates that they are *unit* vectors—are typically read "i-hat", "j-hat", and "k-hat".

For two vectors $\mathbf{x}, \mathbf{y} \in \mathbb{V}^3$ expressed by (11.57) and (11.58) with $n = 3$, their **cross product** (or **vector product**) is defined as the *vector*

$$\mathbf{x} \times \mathbf{y} \triangleq \det \begin{bmatrix} \hat{\mathbf{i}} & \hat{\mathbf{j}} & \hat{\mathbf{k}} \\ x_1 & x_2 & x_3 \\ y_1 & y_2 & y_3 \end{bmatrix} \triangleq \det \begin{bmatrix} x_2 & x_3 \\ y_2 & y_3 \end{bmatrix} \hat{\mathbf{i}} + \det \begin{bmatrix} x_3 & x_1 \\ y_3 & y_1 \end{bmatrix} \hat{\mathbf{j}} + \det \begin{bmatrix} x_1 & x_2 \\ y_1 & y_2 \end{bmatrix} \hat{\mathbf{k}} \tag{11.71}$$

(i.e., the 3×3 determinant is evaluated as if the vectors there were numbers). Equivalently, $\mathbf{x} \times \mathbf{y}$ is a vector of length

$$\|\mathbf{x} \times \mathbf{y}\| = \|\mathbf{x}\| \cdot \|\mathbf{y}\| \cdot \sin \measuredangle(\mathbf{x}, \mathbf{y}) \tag{11.72}$$

that is orthogonal to both $\mathbf{x}$ and $\mathbf{y}$; and, if $\|\mathbf{x} \times \mathbf{y}\| \neq 0$, then the direction of $\mathbf{x} \times \mathbf{y}$ is determined by the **right-hand rule**: pointing the thumb of one's right hand in the direction of $\mathbf{x} \times \mathbf{y}$ corresponds to the remaining fingers curling from the direction of $\mathbf{x}$ to the direction of $\mathbf{y}$. (In particular, $\hat{\mathbf{i}} \times \hat{\mathbf{j}} = \hat{\mathbf{k}}$, $\hat{\mathbf{j}} \times \hat{\mathbf{k}} = \hat{\mathbf{i}}$, and $\hat{\mathbf{k}} \times \hat{\mathbf{i}} = \hat{\mathbf{j}}$.) Some basic facts for $\mathbf{x}, \mathbf{y}, \mathbf{z} \in \mathbb{V}^3$ and $a, b \in \mathbb{R}$ are:

$$\mathbf{x} \times \mathbf{x} = \mathbf{0}, \tag{11.73a}$$

$$\mathbf{x} \times \mathbf{y} = -(\mathbf{y} \times \mathbf{x}), \tag{11.73b}$$

$$(a\mathbf{x} + b\mathbf{y}) \times \mathbf{z} = a(\mathbf{x} \times \mathbf{z}) + b(\mathbf{y} \times \mathbf{z}). \tag{11.73c}$$

Thus, since $\mathbf{x} \times \mathbf{y} \neq \mathbf{0}$ is possible, it follows from (11.73b) that the operation $\times$ is not commutative; nor is it associative. Also, (11.66) and (11.72) yield

$$\|\mathbf{x}\|^2 \|\mathbf{y}\|^2 = (\mathbf{x} \bullet \mathbf{y})^2 + \|\mathbf{x} \times \mathbf{y}\|^2.$$

Hence, $\mathbf{x} \perp \mathbf{y}$ if and only if $\|\mathbf{x} \times \mathbf{y}\| = \|\mathbf{x}\| \cdot \|\mathbf{y}\|$.

As shown in Figure 11.19, the length $\|\mathbf{x} \times \mathbf{y}\|$ of a cross product $\mathbf{x} \times \mathbf{y}$ equals the area of the parallelogram formed from the vectors $\mathbf{x}$ and $\mathbf{y}$, assuming that neither is a scalar multiple of the other (so the parallelogram exists); otherwise, $\mathbf{x} \times \mathbf{y} = \mathbf{0}$, making $\|\mathbf{x} \times \mathbf{y}\| = 0$. Likewise, regardless of the dimensionality of the space, if $\mathbf{x} \not\perp \mathbf{y}$ (otherwise, $\mathbf{x} \bullet \mathbf{y} = 0$), then the magnitude $|\mathbf{x} \bullet \mathbf{y}|$ of a dot product $\mathbf{x} \bullet \mathbf{y}$ equals the area

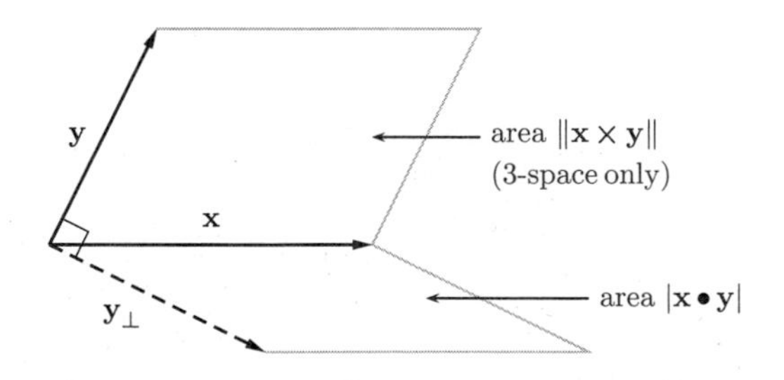

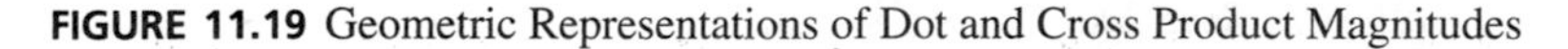

FIGURE 11.19 Geometric Representations of Dot and Cross Product Magnitudes

of the parallelogram formed from $\mathbf{x}$ and $\mathbf{y}_\perp$, where $\mathbf{y}_\perp$ is the vector obtained by rotating $\mathbf{y}$ either toward or away from $\mathbf{x}$ by 90°. More precisely, $\mathbf{y}_\perp$ is chosen such that the vectors $\mathbf{x}$, $\mathbf{y}$, and $\mathbf{y}_\perp$ are coplanar (see next paragraph), with $\mathbf{y}_\perp \perp \mathbf{y}$ and $\|\mathbf{y}_\perp\| = \|\mathbf{y}\|$. (Though there are multiple possibilities for $\mathbf{y}_\perp$, the area of the parallelogram is the same for all.)

Finally, it is sometimes useful to extend certain geometric concepts to vectors.[27] Thus, suppose vectors $\mathbf{x}_1, \ldots, \mathbf{x}_m \in \mathbb{V}^n$ ($m \geq 2$, $n \geq 1$) correspond, respectively, to position vectors $(O, Q_1), \ldots, (O, Q_m)$ of $\mathbb{E}^n$, for points $Q_1, \ldots, Q_m \in \mathbb{E}^n$. Then, we say $\mathbf{x}_1, \ldots, \mathbf{x}_m$ are **collinear** (respectively, **coplanar**) if and only if the points $O, Q_1, \ldots, Q_m$ are collinear (coplanar).[28] It follows that *two* vectors are collinear if and only if they are linearly dependent (i.e., one of the vectors is a scalar multiple of the other), whereas they must be coplanar; and, *three* vectors are coplanar if and only if they are linearly dependent.

Specifically regarding $n \in \{2, 3\}$ (ignoring the trivial case of $n = 1$, for which $\mathbb{E}^n$ is a line), suppose $\mathbf{x} \in \mathbb{V}^n$ is a nonzero vector corresponding to position vector (O, Q) for a point $Q \in \mathbb{E}^n$; and, let a line $l \subseteq \mathbb{E}^n$ and a plane $p \subseteq \mathbb{E}^n$ be given. Then, we say $\mathbf{x}$ is **parallel** to l if and only if $\overleftrightarrow{OQ} \parallel l$; and, $\mathbf{x}$ is parallel to p if and only if $\overleftrightarrow{OQ} \parallel p$. Furthermore, $\mathbf{x}$ is **perpendicular** to p if and only if $\overleftrightarrow{OQ} \perp p$; however, the perpendicularity condition for a vector and a *line* is slightly more complicated. For both $n = 2$ and $n = 3$, $\mathbf{x}$ is perpendicular to l if and only if $\mathbf{x}$ is represented by *some* ordered pair (P', Q') of points $P', Q' \in \mathbb{E}^n$ such that $\overleftrightarrow{P'Q'} \perp l$; and, for $n = 2$, this condition can be simplified to $\overleftrightarrow{OQ} \perp l$ (i.e., we can take $P' = O$ and $Q' = Q$).[29] The reason for the complication when $n = 3$ is that, since we are concerned only with the *direction* of $\mathbf{x}$ (because, translating the point-pair representation (P, Q) of a vector $\vec{x}$ of $\mathbb{E}^n$ does not change the vector), $\overleftrightarrow{OQ}$ and l need not intersect for $\mathbf{x}$ and l to be considered perpendicular (whereas recall from Section 10.2 that, by definition, perpendicular *lines* must intersect).[30]

Now, still with $n \in \{2, 3\}$, let figures $f, g \subseteq \mathbb{E}^n$ each be either a line or a plane; and, let a nonzero vector $\mathbf{x} \in \mathbb{V}^n$ be given. If f and g are both parallel to $\mathbf{x}$, and either $n = 2$ or f and g are not both planes, then $f \parallel g$.[31] Similarly, if f and g are both perpendicular to $\mathbf{x}$, and either $n = 2$ or f and g are not both lines, then $f \parallel g$.[32] Finally, suppose f is

---

27. Although the following definitions and facts are expressed in terms of the matrix vectors $\mathbf{x} \in \mathbb{V}^n$, they also apply to the corresponding vectors $\vec{x}$ of $\mathbb{E}^n$, as well as to the corresponding position vectors.

28. It is important to note that the origin O is included among the points. For example, the vectors $\hat{\mathbf{i}}$ and $\hat{\mathbf{j}}$ of $\mathbb{E}^2$ are coplanar, but noncollinear, because the same can be said of the points $(0, 0)$, $(1, 0)$, and $(0, 1)$. However, were we to abuse terminology and speak of $\hat{\mathbf{i}}$ and $\hat{\mathbf{j}}$ as *points*—meaning $(1, 0)$ and $(0, 1)$—then we would say that the points $\hat{\mathbf{i}}$ and $\hat{\mathbf{j}}$ are collinear.

29. Equivalently, for both $n = 2$ and $n = 3$, $\mathbf{x}$ is perpendicular to l if and only if $\mathbf{x}$ is parallel to a line $l' \subseteq \mathbb{E}^n$ such that $l' \perp l$; and, for $n = 2$, this condition can be simplified to $\overleftrightarrow{OQ} \perp l$ (i.e., we can take $l' = \overleftrightarrow{OQ}$).

30. Some authors, though, perhaps reasoning from vectors back to lines, do not require lines to cross in order to be perpendicular. For them, two lines $\overleftrightarrow{PQ}$ and $\overleftrightarrow{P'Q'}$ in $\mathbb{E}^n$ ($n \geq 1$) are perpendicular if and only if the vectors in $\mathbb{V}^n$ represented by the pairs (P, Q) and (P', Q') are orthogonal (e.g., consider $P = (0, 0, 0)$, $Q = (1, 0, 0)$, $P' = (0, 0, 1)$, and $Q' = (0, 1, 1)$ in $\mathbb{E}^3$).

31. For $n = 3$, if f and g are both planes, then they might cross.

32. For $n = 3$, if f and g are both lines, then they need not intersect.

parallel to $\mathbf{x}$ and g is perpendicular to $\mathbf{x}$; then, $f \perp g$ if either $n = 2$ and f is a line, or $n = 3$ and g is a plane.[33]

11.4 Transformations of a Plane

A **geometric transformation** is a one-to-one mapping $\varphi(\cdot)$ of one geometric figure f onto another g;[34] moreover, f and g are often the entire space under consideration. In this section, we specifically consider transformations of the xy-plane to itself; thus, $f = g = \mathbb{E}^2$ and the function $\varphi(\cdot)$ assigns to each point $P = (x, y) \in \mathbb{E}^2$ a particular point $P' = (x', y') \in \mathbb{E}^2$:

$$P' = \varphi(P). \tag{11.74}$$

Furthermore, since $\varphi(\cdot)$ is one-to-one and maps $\mathbb{E}^2$ *onto* itself, for each $P' \in \mathbb{E}^2$ there is exactly one $P \in \mathbb{E}^2$ such that (11.74) holds; that is, an inverse transformation $\varphi^{-1}(\cdot)$ mapping $\mathbb{E}^2$ onto itself also exists.

In terms of the 2×1 vectors

$$\mathbf{x} = \begin{bmatrix} x \\ y \end{bmatrix}, \quad \mathbf{x}' = \begin{bmatrix} x' \\ y' \end{bmatrix}, \tag{11.75a}$$

each of the mappings $\varphi(\cdot)$ from (x, y) to (x', y')—equivalently, from $\mathbf{x}$ to $\mathbf{x}'$—that we will discuss can be expressed in matrix form as

$$\mathbf{x}' = \mathbf{A}\mathbf{x} + \mathbf{b}, \tag{11.75b}$$

for some 2×2 matrix $\mathbf{A}$ and 2×1 matrix $\mathbf{b}$. In general, given an $m \times n$ matrix $\mathbf{A}$ and an $m \times 1$ matrix $\mathbf{b}$ ($m, n \geq 1$), the mapping (11.75b) of an $n \times 1$ vector $\mathbf{x}$ to an $m \times 1$ vector $\mathbf{x}'$ is called an **affine transformation**. When $\mathbf{b} = \mathbf{0}$, it is also called a **linear transformation**;[35] for then, and only then, do we have

$$\mathbf{x} \mapsto \mathbf{x}', \quad \mathbf{y} \mapsto \mathbf{y}' \qquad \Rightarrow \qquad a\mathbf{x} + b\mathbf{y} \mapsto a\mathbf{x}' + b\mathbf{y}'$$

for all $n \times 1$ vectors $\mathbf{x}$, $\mathbf{y}$, and $m \times 1$ vectors $\mathbf{x}'$, $\mathbf{y}'$, and scalars a, b. In the present context, the 2×1 vectors $\mathbf{x}$ and $\mathbf{x}'$ in (11.75a) are required to be real, and thus likewise for the matrices $\mathbf{A}$ and $\mathbf{b}$; furthermore, for the mapping from $\mathbf{x}$ to $\mathbf{x}'$ to be one-to-one, $\mathbf{A}$ must be invertible.

As shown on the left side of Figure 11.20, the **translation** of $\mathbb{E}^2$ by a given vector

$$\mathbf{v} = \begin{bmatrix} X \\ Y \end{bmatrix} \tag{11.76}$$

33. For $n = 2$, if f is a plane then f must be the space itself, and therefore it is parallel to g. For $n = 3$, if f and g are both lines, then they need not intersect; whereas if f is a plane and g is a line, then they need not be perpendicular.

34. The word "transformation" is also frequently used in mathematics for referring to functions (mappings) *generally*—i.e., without necessarily imposing restrictions such as one-to-oneness.

35. Unfortunately, some authors refer to *any* transformation of the form (11.75b) as "linear", regardless of whether $\mathbf{b} = \mathbf{0}$.

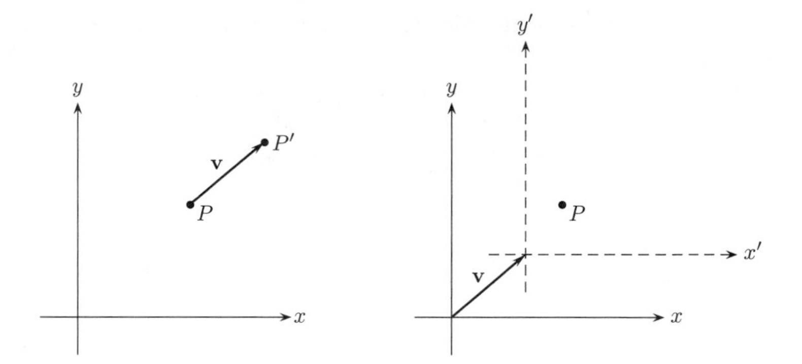

FIGURE 11.20 Translation within, and of, a Coordinate System

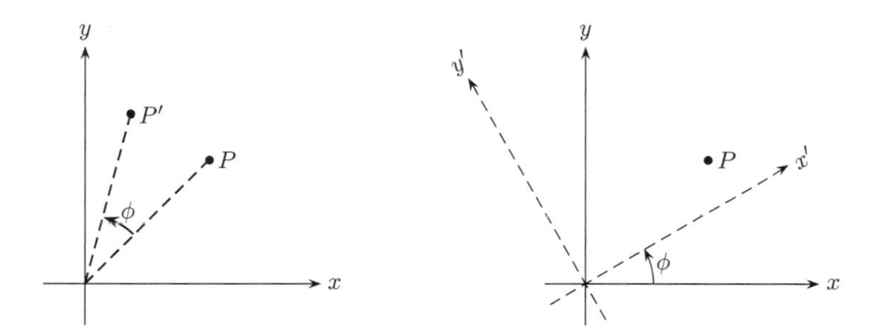

FIGURE 11.21 Rotation within, and of, a Coordinate System

(X, Y real) obtains by adding $\mathbf{v}$ to each point $P \in \mathbb{E}^2$ to get the corresponding translated point $P' \in \mathbb{E}^2$.[36] Hence, for each $\mathbf{x}$ we have $\mathbf{x}' = \mathbf{x} + \mathbf{v}$, which attains via (11.75b) by taking

$$\mathbf{A} = \mathbf{I} = \begin{bmatrix} 1 & 0 \\ 0 & 1 \end{bmatrix}, \quad \mathbf{b} = \mathbf{v} = \begin{bmatrix} X \\ Y \end{bmatrix}. \tag{11.77}$$

In the special case of $X = Y = 0$—i.e., $\mathbf{v} = \mathbf{0}$—we have $\mathbf{A} = \mathbf{I}$ and $\mathbf{b} = \mathbf{0}$, which yields the **identity transformation**; that is, $\mathbf{x}' = \mathbf{x}$ for each $\mathbf{x}$, and thus $P' = P$ for each P.

To obtain a counterclockwise **rotation** of $\mathbb{E}^2$ about the origin $(0, 0)$ by a given angle ϕ, as shown on the left side of Figure 11.21 for an individual point P rotating to get a point P', we take

$$\mathbf{A} = \begin{bmatrix} \cos\phi & -\sin\phi \\ \sin\phi & \cos\phi \end{bmatrix}, \quad \mathbf{b} = \begin{bmatrix} 0 \\ 0 \end{bmatrix} \tag{11.78}$$

36. More accurately, but less concisely, $\mathbf{v}$ is added to the 2×1 matrix representation of the position vector corresponding to P, thereby yielding the 2×1 matrix representation of the position vector corresponding to P'.

in (11.75b); and, this becomes the identity transformation when ϕ equals 0 (or any multiple of 2π).

More generally, a counterclockwise rotation by ϕ about the point (X, Y) specified by a given position vector $\mathbf{v}$, as in (11.76), can be obtained by first translating $\mathbb{E}^2$ by $-\mathbf{v}$ to move the point of rotation to the origin, performing the rotation there, then translating the result back by $\mathbf{v}$; i.e.,

$$\begin{bmatrix} x' \\ y' \end{bmatrix} = \begin{bmatrix} \cos\phi & -\sin\phi \\ \sin\phi & \cos\phi \end{bmatrix} \left(\begin{bmatrix} x \\ y \end{bmatrix} - \begin{bmatrix} X \\ Y \end{bmatrix} \right) + \begin{bmatrix} X \\ Y \end{bmatrix},$$

which is equivalent to taking

$$\mathbf{A} = \begin{bmatrix} \cos\phi & -\sin\phi \\ \sin\phi & \cos\phi \end{bmatrix}, \quad \mathbf{b} = (\mathbf{I} - \mathbf{A})\mathbf{v} = \begin{bmatrix} 1 - \cos\phi & \sin\phi \\ -\sin\phi & 1 - \cos\phi \end{bmatrix} \begin{bmatrix} X \\ Y \end{bmatrix}. \tag{11.79}$$

Conversely, if

$$\mathbf{A} = \begin{bmatrix} a & -b \\ b & a \end{bmatrix} \tag{11.80}$$

for real values a and b such that $a^2 + b^2 = 1$—as is the case in (11.79)—then when $\mathbf{A} \neq \mathbf{I}$ (i.e., $a \neq 1$), the transformation (11.75) performs a counterclockwise rotation of $\mathbb{E}^2$ by $\arg(a + ib)$ about the point specified by the position vector $(\mathbf{I} - \mathbf{A})^{-1}\mathbf{b}$.

A **dilation** of $\mathbb{E}^2$ about the origin, by a given scale factor $\kappa > 0$, is defined as letting $\mathbf{x}' = \kappa\mathbf{x}$ for each $\mathbf{x}$. Thus, we take

$$\mathbf{A} = \kappa\mathbf{I} = \begin{bmatrix} \kappa & 0 \\ 0 & \kappa \end{bmatrix}, \quad \mathbf{b} = \begin{bmatrix} 0 \\ 0 \end{bmatrix} \tag{11.81}$$

in (11.75b); and, this becomes the identity transformation when $\kappa = 1$. More generally, a dilation by κ about the point (X, Y) specified by a given position vector $\mathbf{v}$, as in (11.76), can be obtained by first translating $\mathbb{E}^2$ by $-\mathbf{v}$ to move the point of dilation to the origin, performing the dilation there, then translating the result back by $\mathbf{v}$; i.e.,

$$\begin{bmatrix} x' \\ y' \end{bmatrix} = \begin{bmatrix} \kappa & 0 \\ 0 & \kappa \end{bmatrix} \left(\begin{bmatrix} x \\ y \end{bmatrix} - \begin{bmatrix} X \\ Y \end{bmatrix} \right) + \begin{bmatrix} X \\ Y \end{bmatrix},$$

which is equivalent to taking

$$\mathbf{A} = \begin{bmatrix} \kappa & 0 \\ 0 & \kappa \end{bmatrix}, \quad \mathbf{b} = (\mathbf{I} - \mathbf{A})\mathbf{v} = \begin{bmatrix} 1 - \kappa & 0 \\ 0 & 1 - \kappa \end{bmatrix} \begin{bmatrix} X \\ Y \end{bmatrix}. \tag{11.82}$$

Conversely, if $\mathbf{A} = \kappa\mathbf{I}$ for some constant $\kappa > 0$, then when $\mathbf{A} \neq \mathbf{I}$ (i.e., $\kappa \neq 1$), the transformation (11.75) performs a dilation of $\mathbb{E}^2$ by κ about the point specified by the position vector $(\mathbf{I} - \mathbf{A})^{-1}\mathbf{b} = \mathbf{b}/(1 - \kappa)$. For a scale factor $\kappa > 1$, a dilation is called an **expansion** (or **enlargement**); whereas for $\kappa < 1$, it is a **contraction** (or **reduction**).

As presented above, each transformation is a kind of **alibi transformation** in which the points of the xy-plane $\mathbb{E}^2$ are moved while the coordinate system—in particular, the

x- and y-axes—is held fixed. Alternatively, each of these transformations can instead be performed as an **alias transformation**, whereby the points of the plane remain fixed while the coordinate system is moved, thereby converting the original coordinate variables "x" and "y" into two new coordinate variables—say, "x'" and "y'", respectively.[37]

To wit, as shown on the right side of Figure 11.20, a **translation** of the xy-coordinate system by a position vector $\mathbf{v}$, as given by (11.76), obtains by taking

$$\mathbf{A} = \mathbf{I} = \begin{bmatrix} 1 & 0 \\ 0 & 1 \end{bmatrix}, \quad \mathbf{b} = -\mathbf{v} = \begin{bmatrix} -X \\ -Y \end{bmatrix}$$

in (11.75b)—which amounts to simply changing the sign of $\mathbf{b}$ in the corresponding alibi transformation (11.77). Hence, $\mathbf{x}' = \mathbf{x} - \mathbf{v}$, and thus $\mathbf{x} = \mathbf{x}' + \mathbf{v}$—i.e.,

$$\begin{bmatrix} x \\ y \end{bmatrix} = \begin{bmatrix} x' + X \\ y' + Y \end{bmatrix} \tag{11.83}$$

—showing that a translation by $-\mathbf{v}$ of the $x'y'$-coordinate system returns us to the xy-coordinate system. A counterclockwise **rotation** of the xy-coordinate system about the origin by an angle ϕ, as shown on the right side of Figure 11.21, obtains by taking

$$\mathbf{A} = \begin{bmatrix} \cos\phi & \sin\phi \\ -\sin\phi & \cos\phi \end{bmatrix}, \quad \mathbf{b} = \begin{bmatrix} 0 \\ 0 \end{bmatrix} \tag{11.84}$$

—which amounts to changing the sign of ϕ in the corresponding alibi transformation (11.78). Hence, $\mathbf{x}' = \mathbf{A}\mathbf{x}$, and thus $\mathbf{x} = \mathbf{A}^{-1}\mathbf{x}'$—i.e.,

$$\begin{bmatrix} x \\ y \end{bmatrix} = \begin{bmatrix} \cos\phi & -\sin\phi \\ \sin\phi & \cos\phi \end{bmatrix} \begin{bmatrix} x' \\ y' \end{bmatrix} \tag{11.85}$$

—showing that a counterclockwise rotation by $-\phi$ (i.e., a clockwise rotation by ϕ) of the $x'y'$-coordinate system returns us to the xy-coordinate system. Finally, a **dilation** of the xy-coordinate system about the origin by a scale factor $\kappa > 0$ obtains by taking

$$\mathbf{A} = \mathbf{I}/\kappa = \begin{bmatrix} 1/\kappa & 0 \\ 0 & 1/\kappa \end{bmatrix}, \quad \mathbf{b} = \begin{bmatrix} 0 \\ 0 \end{bmatrix}$$

—which amounts to replacing κ by its reciprocal in the corresponding alibi transformation (11.81). Hence, $\mathbf{x}' = \mathbf{x}/\kappa$, and thus $\mathbf{x} = \kappa\mathbf{x}'$—i.e.,

$$\begin{bmatrix} x \\ y \end{bmatrix} = \begin{bmatrix} \kappa x' \\ \kappa y' \end{bmatrix} \tag{11.86}$$

—showing that a dilation by $1/\kappa$ of the $x'y'$-coordinate system returns us to the xy-coordinate system.

The inverse equations (11.83), (11.85), and (11.86) are particularly useful for determining the effects of these transformations on a given geometric figure, expressed as

---

37. By contrast, when (11.75) is used for an alibi transformation, x' and y' are interpreted as new values of x and y.

a set of points (x, y). Specifically, for each occurrence of the variables "x" and "y" in the defining property of the figure, one substitutes an expression in terms of the new variables "x'" and "y'", as determined by the appropriate inverse equation. For example, given the figure $f = \{(x, y) \in \mathbb{E}^2 \mid x = y\}$—i.e., a line of unit slope passing through the origin—it follows from (11.85) that rotating the xy-coordinate system by $\phi = 45°$ produces the figure

$$\begin{aligned} f' &= \{(x', y') \in \mathbb{E}^2 \mid x' \cos\phi - y' \sin\phi = x' \sin\phi + y' \cos\phi\} \\ &= \left\{(x', y') \in \mathbb{E}^2 \mid x'/\sqrt{2} - y'/\sqrt{2} = x'/\sqrt{2} + y'/\sqrt{2}\right\} = \{(x', y') \mid y' = 0\} \end{aligned}$$

—i.e., a horizontal line through the origin in the $x'y'$-coordinate system.

For the identity transformation, the *same* matrices $\mathbf{A}$ and $\mathbf{b}$—viz., $\mathbf{A} = \mathbf{I}$ and $\mathbf{b} = \mathbf{0}$—are used in (11.75b) whether the transformation is performed as an alibi transformation or an alias transformation. This is also the case for the **axis inversion** (or **coordinate inversion**) transformations: To invert the x-axis alone (e.g., as done in the bottom right of Figure 5.3), we take

$$\mathbf{A} = \begin{bmatrix} -1 & 0 \\ 0 & 1 \end{bmatrix}, \quad \mathbf{b} = \begin{bmatrix} 0 \\ 0 \end{bmatrix},$$

thereby obtaining new coordinates $x' = -x$ and $y' = y$. Likewise, to invert the y-axis alone (e.g., as done in the bottom left of Figure 5.3), we take

$$\mathbf{A} = \begin{bmatrix} 1 & 0 \\ 0 & -1 \end{bmatrix}, \quad \mathbf{b} = \begin{bmatrix} 0 \\ 0 \end{bmatrix},$$

obtaining $x' = x$ and $y' = -y$. Whereas both axes are inverted by taking $\mathbf{A} = -\mathbf{I}$ and $\mathbf{b} = \mathbf{0}$—which, by (11.78) or (11.84), is equivalent to a rotation about the origin by $\phi = \pi$.

In general, a geometric transformation $\varphi(\cdot)$ of a figure f to a figure g is called an **isometry** if and only if it preserves distances between points; that is, for all points P, $Q \in f$ and P', $Q' \in g$,

$$P' = \varphi(P), \quad Q' = \varphi(Q) \qquad \Rightarrow \qquad d(P', Q') = d(P, Q)$$

—in which case the figures f and g are said to be **isometric**. In plane and solid geometry, it follows from the remarks in Section 10.2 about (10.2) that two figures f and g are isometric if and only if they are congruent. Also, an "isometry" is often referred to as a **rigid motion**, which means some combination of translations and rotations; however, this term, which is intended to suggest heuristically how f might be converted into g, is potentially misleading. For example, in the xy-plane, the triangle f having vertices $(0, 0)$, $(1, 0)$, and $(0, 2)$ is congruent to the triangle g having vertices $(0, 0)$, $(1, 0)$, and $(0, -2)$; yet, it is impossible by performing only translations and rotations—i.e., rigid motion *within* the plane—to make f coincide with g. Rather, we must "flip over" one of the triangles, as can be obtained by performing a coordinate inversion. Similarly, in xyz-space, the tetrahedron f having vertices $(0, 0, 0)$, $(1, 0, 0)$, $(0, 2, 0)$, and $(0, 0, 3)$ is congruent to the tetrahedron g having vertices $(0, 0, 0)$, $(1, 0, 0)$, $(0, 2, 0)$, and $(0, 0, -3)$, even though it is impossible by performing translations and rotations alone to make f coincide with g. Thus, although every rigid motion produces an isometry, not every isometry can be accomplished by a rigid motion within the space of interest.

Returning to our discussion of geometric transformations $\varphi(\cdot)$ of the xy-plane $\mathbb{E}^2$ to itself, (11.75) performs an isometry if and only we have either (11.80) or

$$\mathbf{A} = \begin{bmatrix} a & b \\ b & -a \end{bmatrix} \tag{11.87}$$

for some real values a and b such that $a^2 + b^2 = 1$. It follows that the identity transformation, translations, rotations (about any point), and axis inversions are all isometries; whereas, except for a scale factor of 1, dilations (about any point) are not. Also, all of these transformations preserve geometric similarity; that is, given f, f' $g, g' \subseteq \mathbb{E}^2$ such that $g = \varphi(f)$ and $g' = \varphi(f')$, if $f \sim f'$ then $g \sim g'$ (e.g., if f is a segment, then so is g). But, congruence is always preserved only if $\varphi(\cdot)$ is an isometry.

11.5 Basic Geometric Figures

We now use the tools of analytic geometry to discuss some basic geometric figures. First, to say that a figure f lying in a Euclidean space $\mathbb{E}^n$ ($n = 1, 2, \ldots$) is "described" by an equation

$$F(x_1, \ldots, x_n) = 0,$$

for some function $F(\cdot, \ldots, \cdot)$ on $\mathbb{E}^n$, is to assert

$$f = \{(x_1, \ldots, x_n) \in \mathbb{E}^n \mid F(x_1, \ldots, x_n) = 0\}.$$

In particular, a figure f in the xy-plane (respectively, xyz-space) may be so described in terms of the rectangular coordinates x and y (x, y, and z); or, f may instead be described in terms of the polar coordinates r and θ (cylindrical coordinates r, θ, and z). Likewise, multiple equations as well as other conditions (e.g., inequalities) can be used to describe a figure f; that is, f is then the set of points in $\mathbb{E}^n$ satisfying all of the stated conditions.[38]

Alternatively, a *parametric* approach may be used whereby the coordinates x_i ($i = 1, \ldots, n$) of $\mathbb{E}^n$ are themselves expressed as functions $F_i(t_1, \ldots, t_m)$ of $m \geq 1$ external parameters $t_1, \ldots, t_m$, which are allowed to vary over respective sets $D_1, \ldots, D_m$. Specifically, to say that a figure f is "described" by the parametric equations

$$x_i = F_i(t_1, \ldots, t_m) \quad (i = 1, \ldots, n)$$

($t_1 \in D_1, \ldots, t_m \in D_m$) is to assert

$$f = \{(x_1, x_2, \ldots, x_n) \in \mathbb{E}^n \mid x_1 = F_1(t_1, \ldots, t_m),\ x_2 = F_2(t_1, \ldots, t_m),\ \ldots,$$
$$x_n = F_n(t_1, \ldots, t_m)\ (t_1 \in D_1, \ldots, t_m \in D_m)\}$$

38. For example, see the plot in Figure 5.7 of the relation R in (5.7).

—i.e.,

$$f = \{(F_1(t_1, \ldots, t_m), F_2(t_1, \ldots, t_m), \ldots, F_n(t_1, \ldots, t_m)) \in \mathbb{E}^n \mid t_1 \in D_1, \ldots, t_m \in D_m\}.$$

In this section, all parameters are real-valued; moreover, for the figures discussed, at most two parameters are needed, which are taken to be "s" and "t".

11.5.1 Lines and Planes

In Section 5.1, various equations (5.6) were given describing a line in the xy-plane for the case where the y-coordinate can be expressed as a function $f(\cdot)$ of the x-coordinate (see Figure 5.5). Specifically, (5.6a) describes the line in terms its slope m, defined in (5.5), and its y-intercept Y; (5.6b) describes the line in terms of $m \neq 0$ and its x-intercept X; and, (5.6c) describes the line in terms of m and any point (x_1, y_1) on the line. For a vertical line, however, m does not exist, and none of these equations apply.

Any line, though, in the xy-plane can be described by a first-degree equation in the variables "x" and "y":

$$Ax + By + C = 0 \tag{11.88}$$

(A, B, C real), where A and B are not both 0.[39] Conversely, for any real constants A, B, and C such that A and B are not both 0, (11.88) describes a particular line l; specifically, as shown in Figure 11.22, l is the unique line perpendicular to the vector $\mathbf{n} = A\hat{\imath} + B\hat{\jmath} \neq \mathbf{0}$ that is a directed distance $d = -C/\sqrt{A^2 + B^2}$ away from the origin in the direction of $\mathbf{n}$.[40] Since the vector $\mathbf{n}$ is perpendicular—i.e., **normal**—to l, it is called a **normal vector** with respect to this line.

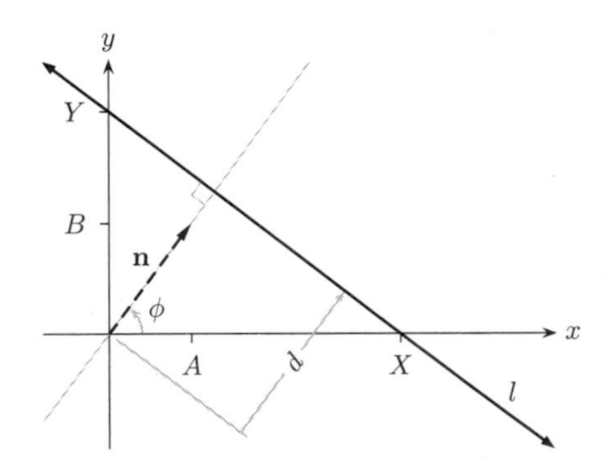

FIGURE 11.22 A Line in the xy-Plane

39. If $A = B = 0$, then (11.88) describes the entire xy-plane when $C = 0$, and the empty set otherwise.

40. Note that d (being a *directed* distance) can be negative, in which case the line is a distance $|d|$ away from the origin in the direction *opposite* to that of the vector $\mathbf{n}$. If desired, one can restrict d to be nonnegative by choosing $\mathbf{n}$ accordingly; however, not making this restriction is usually more convenient.

The half-plane on one side of l is the set of points (x, y) satisfying the inequality

$$Ax + By + C < 0, \tag{11.89a}$$

whereas the half-plane on the other side of l is the set of points (x, y) satisfying the opposite inequality,

$$Ax + By + C > 0; \tag{11.89b}$$

and, the direction of $\mathbf{n}$ is from the former half-plane to the latter. The distance from any point $P = (X, Y)$ to l is $|AX + BY + C|/\sqrt{A^2 + B^2}$; and, the only point $Q = (X', Y')$ on l for which $d(P, Q)$ equals this distance is given by

$$\begin{bmatrix} X' \\ Y' \end{bmatrix} = \frac{1}{A^2 + B^2} \begin{bmatrix} B^2 X - A(BY + C) \\ A^2 Y - B(AX + C) \end{bmatrix}.$$

Also, when $AX + BY + C < 0$ (respectively, > 0), the vector $\mathbf{n}$ $(-\mathbf{n})$ points from P to l.

The line l has an x-intercept X if and only if $A \neq 0$, in which case $X = -C/A$; whereas for $A = 0$, the line is horizontal (i.e., y is constant)—in particular, it is the x-axis when $A = C = 0$. Similarly, l has a y-intercept Y if and only if $B \neq 0$, in which case $Y = -C/B$; whereas for $B = 0$, the line is vertical (i.e., x is constant)—in particular, it is the y-axis when $B = C = 0$. The slope m of l is $-A/B$ when $B \neq 0$; otherwise, it is undefined.

There are many ways of rewriting (11.88), though some are applicable only under special circumstances. For example, the unique line that is perpendicular to a given vector $A\hat{\imath} + B\hat{\jmath} \neq \mathbf{0}$, and that passes through a given point (x_1, y_1), is described by the equation

$$A(x - x_1) + B(y - y_1) = 0. \tag{11.90}$$

Alternatively, given an x-intercept X and a y-intercept Y, the unique line having these intercepts is described by the equation

$$\frac{x}{X} + \frac{y}{Y} = 1 \tag{11.91}$$

—if $X, Y \neq 0$. When either $X = 0$ and $Y \neq 0$, or $X \neq 0$ and $Y = 0$, no line having these intercepts exists;[41] whereas there are infinitely many lines having intercepts $X = Y = 0$.

The unique line l that is parallel to a given vector $a\hat{\imath} + b\hat{\jmath} \neq \mathbf{0}$, and that passes through a given point (x_1, y_1), is described by the parametric equations

$$x = x_1 + ta,$$
$$y = y_1 + tb$$

---

41. Recall that we require a line to *cross* the axis at its intercept (i.e., intersect it at exactly one point). Thus, the line described by the equation $x = 0$ (respectively, $y = 0$) has no x-intercept (y-intercept).

($t \in \mathbb{R}$)—equivalently,

$$\begin{bmatrix} x \\ y \end{bmatrix} = \begin{bmatrix} x_1 \\ y_1 \end{bmatrix} + t \begin{bmatrix} a \\ b \end{bmatrix} \tag{11.92}$$

($t \in \mathbb{R}$). Eliminating the parameter t shows that l is also described by the nonparametric equation

$$\frac{x - x_1}{a} = \frac{y - y_1}{b} \tag{11.93}$$

—if $a, b \neq 0$. When $a = 0$ (and thus $b \neq 0$), l is described by the equation $x = x_1$; whereas when $b = 0$ (and thus $a \neq 0$), l is described by the equation $y = y_1$. Hence, an alternative to (11.93) that works under all circumstances is

$$b(x - x_1) = a(y - y_1)$$

(e.g., we can take $A = b$ and $B = -a$ in (11.90), or take $A = b$, $B = -a$, and $C = ay_1 - bx_1$ in (11.88)), which is expressible in matrix form as

$$\det \begin{bmatrix} x & y & 1 \\ x_1 & y_1 & 1 \\ a & b & 0 \end{bmatrix} = 0. \tag{11.94}$$

Similarly, the unique line l that passes through two distinct points (x_1, y_1) and (x_2, y_2) is described by the parametric equation

$$\begin{bmatrix} x \\ y \end{bmatrix} = \begin{bmatrix} x_1 \\ y_1 \end{bmatrix} + t \begin{bmatrix} x_2 - x_1 \\ y_2 - y_1 \end{bmatrix} \tag{11.95}$$

($t \in \mathbb{R}$). Akin to (11.93), this line is also described by the nonparametric equation

$$\frac{x - x_1}{x_2 - x_1} = \frac{y - y_1}{y_2 - y_1}$$

—if $x_1 \neq x_2$ and $y_1 \neq y_2$. In general, l is described by the equation

$$\det \begin{bmatrix} x & y & 1 \\ x_1 & y_1 & 1 \\ x_2 & y_2 & 1 \end{bmatrix} = 0. \tag{11.96}$$

It follows that three points (x_1, y_1), (x_2, y_2), and (x_3, y_3) in the xy-plane are collinear if and only if

$$\det \begin{bmatrix} x_1 & y_1 & 1 \\ x_2 & y_2 & 1 \\ x_3 & y_3 & 1 \end{bmatrix} = 0. \tag{11.97}$$

Equations (11.94) and (11.96) essentially convert, respectively, the parametric descriptions (11.92) and (11.95) of a line in the xy-plane into a nonparametric equation

(11.88) describing the same line. Conversely, given an equation (11.88) describing a line l, one might somehow find a solution $(x, y) = (x_1, y_1)$; then, upon taking $(a, b) = (-B, A)$, (11.92) also describes l. Or, one might find another solution $(x, y) = (x_2, y_2) \neq (x_1, y_1)$ to (11.88); then, (11.95) describes l.

In terms of the polar coordinates r and θ (see Figure 11.3), the unique line l that is a directed distance d away from the origin, in the direction of a given angle ϕ (see Figure 11.22), is described by the equation

$$r \cos(\theta - \phi) = d \tag{11.98}$$

—regardless of whether r is allowed to be negative. In particular, if $d = 0$—i.e., the line passes through the origin—then l is the set of polar points (r, θ) such that either $r = 0$ (for which θ is undefined), or $r > 0$ and $\theta = \phi \pm \pi/2$. The distance from any polar point (R, Θ) to l is $|R \cos(\Theta - \phi) - d|$, including when R is negative. It follows that if (X, Y) is the equivalent rectangular point, then this distance is also $|X \cos \phi + Y \sin \phi - d|$.

Given two lines in the xy-plane: If their slopes m_1 and m_2 exist, then the lines are parallel if and only if $m_1 = m_2$, whereas they are perpendicular if and only if $m_1 m_2 = -1$. Otherwise, the lines are parallel if and only if both are vertical (i.e., neither slope exists), whereas they are perpendicular if and only if one is horizontal (i.e., its slope is zero) and the other is vertical.

Per (11.88), let two lines in the xy-plane be described by the equations

$$A_1 x + B_1 y + C_1 = 0, \tag{11.99a}$$

$$A_2 x + B_2 y + C_2 = 0, \tag{11.99b}$$

having real coefficients such that A_1 and B_1 are not both 0, and neither are A_2 and B_2. Then, the lines are parallel if and only if $A_2 = \alpha A_1$ and $B_2 = \alpha B_1$ for some constant α (i.e., the normal vectors $\mathbf{n}_1 = A_1 \hat{\mathbf{i}} + B_1 \hat{\mathbf{j}}$ and $\mathbf{n}_2 = A_2 \hat{\mathbf{i}} + B_2 \hat{\mathbf{j}}$ are collinear)—equivalently, $A_1 B_2 = A_2 B_1$—in which case the lines coincide if and only if $C_2 = \alpha C_1$ as well. In fact, the distance between such parallel lines is $\left| C_1 \big/ \sqrt{A_1{}^2 + B_1{}^2} - C_2 \big/ \sqrt{A_2{}^2 + B_2{}^2} \right|$. On the other hand, the lines are perpendicular if and only if $A_1 A_2 + B_1 B_2 = 0$ (i.e., $\mathbf{n}_1 \bullet \mathbf{n}_2 = 0$).

If any two lines given by (11.99) are not parallel, then the point (X, Y) at which they cross is given by

$$\begin{bmatrix} X \\ Y \end{bmatrix} = \begin{bmatrix} A_1 & B_1 \\ A_2 & B_2 \end{bmatrix}^{-1} \begin{bmatrix} -C_1 \\ -C_2 \end{bmatrix};$$

whereas the above 2×2 matrix is singular when the lines are parallel. Also, if the lines given by (11.99) cross but are not perpendicular, then the acute angles formed have measure

$$\cos^{-1} \frac{|A_1 A_2 + B_1 B_2|}{\sqrt{(A_1{}^2 + B_1{}^2)(A_2{}^2 + B_2{}^2)}} = \sin^{-1} \frac{|A_1 B_2 - B_1 A_2|}{\sqrt{(A_1{}^2 + B_1{}^2)(A_2{}^2 + B_2{}^2)}}$$

$$= \tan^{-1} \left| \frac{A_1 B_2 - B_1 A_2}{A_1 A_2 + B_1 B_2} \right| \tag{11.100}$$

(obtained from $\mathbf{n}_1 \bullet \mathbf{n}_2$ and $\mathbf{n}_1 \times \mathbf{n}_2$ via (11.66) and (11.72), upon introducing a third dimension to perform the cross product). Furthermore, the arccosine and arcsine expressions produce 0 when the lines are parallel, and $\pi/2$ when they are perpendicular; thus, either expression may be said to provide the **angle (measure)** *between two lines* described by (11.99). The arctangent expression, though, is invalid when the lines are perpendicular, because of division by 0.

Per (11.98), let two lines be described by the equations $r_1 \cos(\theta_1 - \phi_1) = d_1$ and $r_2 \cos(\theta_2 - \phi_2) = d_2$. Then, the lines are parallel if and only if $\phi_1 = \phi_2 + n\pi$ for some integer n, in which case they coincide if and only if either $d_1 = d_2$ for n even or $d_1 = -d_2$ for n odd. In fact, the distance between such parallel lines is $|d_1 - d_2|$ when n is even, and $|d_1 + d_2|$ when n is odd. Furthermore, the lines are perpendicular if and only if $\phi_1 = \phi_2 + (n + \frac{1}{2})\pi$ for some integer n. Also, if the lines are neither parallel nor perpendicular, then the two acute angles they form have measure $|\phi_1 - \phi_2 - n\pi|$, upon choosing the integer n to make this value less than $\pi/2$.

Turning from two to three dimensions, any plane in xyz-space can be described by a first-degree equation in the variables "x", "y", and "z":

$$Ax + By + Cz + D = 0 \tag{11.101}$$

(A, B, C, D real), where A, B, and C are not all 0.[42] Conversely, for any real constants A, B, C, and D such that A, B, and C are not all 0, (11.101) describes a particular plane p; specifically, analogous to Figure 11.22 for a line l described by (11.88), p is the unique plane perpendicular to the vector $\mathbf{n} = A\hat{\imath} + B\hat{\jmath} + C\hat{\mathbf{k}} \neq \mathbf{0}$ that is a directed distance $d = -D/\sqrt{A^2 + B^2 + C^2}$ away from the origin in the direction of the vector.[43] Thus, $\mathbf{n}$ is a **normal (vector)** with respect to this plane.

The half-space on one side of p is the set of points (x, y, z) satisfying the inequality

$$Ax + By + Cz + D < 0, \tag{11.102a}$$

whereas the half-space on the other side of p is the set of points (x, y, z) satisfying the opposite inequality,

$$Ax + By + Cz + D > 0; \tag{11.102b}$$

and, the direction of $\mathbf{n}$ is from the former half-space to the latter. The distance from any point $P = (X, Y, Z)$ to p is $|AX + BY + CZ + D|/\sqrt{A^2 + B^2 + C^2}$; and, the only point $Q = (X', Y', Z')$ on p for which $d(P, Q)$ equals this distance is given by

$$\begin{bmatrix} X' \\ Y' \\ Z' \end{bmatrix} = \frac{1}{A^2 + B^2 + C^2} \begin{bmatrix} (B^2 + C^2)X - A(BY + CZ + D) \\ (A^2 + C^2)Y - B(AX + CZ + D) \\ (A^2 + B^2)Z - C(AX + BY + D) \end{bmatrix}.$$

Also, when $AX + BY + CZ + D < 0$ (respectively, > 0), the vector $\mathbf{n}$ ($-\mathbf{n}$) points from P to p.

---

42. If $A = B = C = 0$, then (11.101) describes the entire xyz-space when $D = 0$, and the empty set otherwise.

43. For example, whereas the equation $x = y$ describes a line in the xy-plane, it describes a plane in xyz-space.

There are many ways of rewriting (11.101), though some are applicable only under special circumstances. For example, analogous to (11.90), the unique plane that is perpendicular to a given vector $A\hat{\mathbf{\imath}} + B\hat{\mathbf{\jmath}} + C\hat{\mathbf{k}} \neq \mathbf{0}$, and that passes through a given point (x_1, y_1, z_1), is described by the equation

$$A(x - x_1) + B(y - y_1) + C(y - z_1) = 0. \tag{11.103}$$

Alternatively, analogous to (11.91), given an x-intercept X, a y-intercept Y, and a z-intercept Z, the unique plane having these intercepts—i.e., its intersects the x-, y-, and z-axes at, and only at, these respective values—is described by the equation

$$\frac{x}{X} + \frac{y}{Y} + \frac{z}{Z} = 1 \tag{11.104}$$

—if $X, Y, Z \neq 0$. When some, but not all, of the values X, Y, and Z are 0, no plane having these intercepts exists; whereas there are infinitely many planes having intercepts $X = Y = Z = 0$.

Analogous to (11.92), the unique plane p in xyz-space that is parallel to two given noncollinear vectors $a\hat{\mathbf{\imath}} + b\hat{\mathbf{\jmath}} + c\hat{\mathbf{k}} \neq \mathbf{0}$ and $a'\hat{\mathbf{\imath}} + b'\hat{\mathbf{\jmath}} + c'\hat{\mathbf{k}} \neq \mathbf{0}$, and that passes through a given point (x_1, y_1, z_1), is described by the parametric equation

$$\begin{bmatrix} x \\ y \\ z \end{bmatrix} = \begin{bmatrix} x_1 \\ y_1 \\ z_1 \end{bmatrix} + s \begin{bmatrix} a \\ b \\ c \end{bmatrix} + t \begin{bmatrix} a' \\ b' \\ c' \end{bmatrix} \tag{11.105}$$

($s, t \in \mathbb{R}$). Analogous to (11.94), p is also described by the nonparametric equation

$$\det \begin{bmatrix} x & y & z & 1 \\ x_1 & y_1 & z_1 & 1 \\ a & b & c & 0 \\ a' & b' & c' & 0 \end{bmatrix} = 0. \tag{11.106}$$

Similarly, suppose a vector $\mathbf{v} = a\hat{\mathbf{\imath}} + b\hat{\mathbf{\jmath}} + c\hat{\mathbf{k}} \neq \mathbf{0}$ is given, along with two distinct points (x_1, y_1, z_1) and (x_2, y_2, z_2) such that $\mathbf{v}$ and the vector $(x_2 - x_1)\hat{\mathbf{\imath}} + (y_2 - y_1)\hat{\mathbf{\jmath}} + (z_2 - z_1)\hat{\mathbf{k}} \neq \mathbf{0}$ are noncollinear; then, the unique plane that is parallel to $\mathbf{v}$, and passes through both points, is described by the parametric equation

$$\begin{bmatrix} x \\ y \\ z \end{bmatrix} = \begin{bmatrix} x_1 \\ y_1 \\ z_1 \end{bmatrix} + s \begin{bmatrix} x_2 - x_1 \\ y_2 - y_1 \\ z_2 - z_1 \end{bmatrix} + t \begin{bmatrix} a \\ b \\ c \end{bmatrix} \tag{11.107}$$

($s, t \in \mathbb{R}$). This plane is also described by the equation

$$\det \begin{bmatrix} x & y & z & 1 \\ x_1 & y_1 & z_1 & 1 \\ x_2 & y_2 & z_2 & 1 \\ a & b & c & 0 \end{bmatrix} = 0. \tag{11.108}$$

Analogous to (11.95), the unique plane that passes through three noncollinear points (x_1, y_1, z_1), (x_2, y_2, z_2), and (x_3, y_3, z_3) is

$$\begin{bmatrix} x \\ y \\ z \end{bmatrix} = \begin{bmatrix} x_1 \\ y_1 \\ z_1 \end{bmatrix} + s \begin{bmatrix} x_2 - x_1 \\ y_2 - y_1 \\ z_2 - z_1 \end{bmatrix} + t \begin{bmatrix} x_3 - x_1 \\ y_3 - y_1 \\ z_3 - z_1 \end{bmatrix} \tag{11.109}$$

$(s, t \in \mathbb{R})$. Analogous to (11.96), this plane is also described by the equation

$$\det \begin{bmatrix} x & y & z & 1 \\ x_1 & y_1 & z_1 & 1 \\ x_2 & y_2 & z_2 & 1 \\ x_3 & y_3 & z_3 & 1 \end{bmatrix} = 0. \tag{11.110}$$

It follows that four points (x_1, y_1, z_1), (x_2, y_2, z_2), (x_3, y_3, z_3), and (x_4, y_4, z_4) in xyz-space are coplanar if and only if

$$\det \begin{bmatrix} x_1 & y_1 & z_1 & 1 \\ x_2 & y_2 & z_2 & 1 \\ x_3 & y_3 & z_3 & 1 \\ x_4 & y_4 & z_4 & 1 \end{bmatrix} = 0,$$

analogous to (11.97).

Equations (11.106), (11.108), and (11.110) essentially convert, respectively, the parametric expressions (11.105), (11.107), and (11.109) of a plane in xyz-space into a nonparametric equation (11.101) describing the same plane. Conversely, given an equation (11.101) describing a plane p, one might somehow find a solution $(x, y, z) = (x_1, y_1, z_1)$; then, upon taking for (a, b, c) and (a', b', c') two of the three triples $(-B, A, 0)$, $(C, 0, -A)$, and $(0, -C, B)$—chosen so that the vectors $a\hat{\imath} + b\hat{\jmath} + c\hat{\mathbf{k}}$ and $a'\hat{\imath} + b'\hat{\jmath} + c'\hat{\mathbf{k}}$ are noncollinear—(11.105) also describes p. Or, one might find another solution $(x, y) = (x_2, y_2, z_2) \neq (x_1, y_1, z_1)$ to (11.88); then, upon taking for (a, b, c) one of the three triples $(-B, A, 0)$, $(C, 0, -A)$, and $(0, -C, B)$—chosen so that the vectors $a\hat{\imath} + b\hat{\jmath} + c\hat{\mathbf{k}}$ and $(x_1 - x_2)\hat{\imath} + (y_1 - y_2)\hat{\jmath} + (z_1 - z_2)\hat{\mathbf{k}}$ are noncollinear—(11.107) also describes p. Or, one might find yet another solution $(x, y) = (x_3, y_3, z_3)$ to (11.88) such that the points (x_1, y_1, z_1), (x_2, y_2, z_2), and (x_3, y_3, z_3) are noncollinear; then, (11.109) also describes p.

Per (11.101), let two planes in xyz-space be described by the equations

$$A_1x + B_1y + C_1z + D_1 = 0, \tag{11.111a}$$

$$A_2x + B_2y + C_2z + D_2 = 0, \tag{11.111b}$$

having real coefficients such that A_1, B_1, and C_1 are not all 0, and neither are A_2, B_2, and C_2. Then, the planes are parallel if and only if $A_2 = \alpha A_1$, $B_2 = \alpha B_1$, and $C_2 = \alpha C_1$ for some constant α (i.e., the normal vectors $\mathbf{n}_1 = A_1\hat{\imath} + B_1\hat{\jmath} + C_1\hat{\mathbf{k}}$ and $\mathbf{n}_2 = A_2\hat{\imath} + B_2\hat{\jmath} + C_2\hat{\mathbf{k}}$ are collinear), in which case they coincide if and only if $D_2 = \alpha D_1$ as well. In fact, the distance between such parallel planes is $\left|D_1 / \sqrt{A_1{}^2 + B_1{}^2 + C_1{}^2} - D_2 / \sqrt{A_2{}^2 + B_2{}^2 + C_2{}^2}\right|$. On the other hand, the planes are perpendicular if and only if $A_1A_2 + B_1B_2 + C_1C_2 = 0$ (i.e., $\mathbf{n}_1 \bullet \mathbf{n}_2 = 0$).

If the planes given by (11.111) are not parallel, then their intersection is a line; conversely, since any line in three-dimensional Euclidean space can be expressed as the intersection of two planes, each line in xyz-space can be described by a *pair* of nonparametric equations (11.111) for two crossing planes. If these planes are not perpendicular, then intersecting them with another plane that is perpendicular to both forms two nonperpendicular crossing lines, the acute angles of which have measure

$$\cos^{-1}\frac{|A_1A_2+B_1B_2+C_1C_2|}{\sqrt{(A_1{}^2+B_1{}^2+C_1{}^2)(A_2{}^2+B_2{}^2+C_2{}^2)}}$$
$$=\sin^{-1}\sqrt{\frac{(A_1B_2-B_1A_2)^2+(A_1C_2-C_1A_2)^2+(B_1C_2-C_1B_2)^2}{(A_1{}^2+B_1{}^2+C_1{}^2)(A_2{}^2+B_2{}^2+C_2{}^2)}}$$
$$=\tan^{-1}\frac{\sqrt{(A_1B_2-B_1A_2)^2+(A_1C_2-C_1A_2)^2+(B_1C_2-C_1B_2)^2}}{|A_1A_2+B_1B_2+C_1C_2|} \tag{11.112}$$

(obtained from $\mathbf{n}_1 \bullet \mathbf{n}_2$ and $\mathbf{n}_1 \times \mathbf{n}_2$ via (11.66) and (11.72)); and, this formula simplifies to (11.100) when $C_1 = C_2 = 0$ (i.e., both planes are perpendicular to the xy-plane). Furthermore, the arccosine and arcsine expressions produce 0 when the planes are parallel, and $\pi/2$ when they are perpendicular; thus, either expression may be said to provide the **angle (measure)** *between two planes* described by (11.111). The arctangent expression, though, is invalid when the lines are perpendicular, because of division by 0.

If the intersection of any two planes described by (11.111) and a third plane described by the equation

$$A_3x + B_3y + C_3z + D_3 = 0 \tag{11.113}$$

is a single point (X, Y, Z), then this point is given by

$$\begin{bmatrix} X \\ Y \\ Z \end{bmatrix} = \begin{bmatrix} A_1 & B_1 & C_1 \\ A_2 & B_2 & C_2 \\ A_3 & B_3 & C_3 \end{bmatrix}^{-1} \begin{bmatrix} -D_1 \\ -D_2 \\ -D_3 \end{bmatrix}; \tag{11.114}$$

otherwise, the above 3×3 matrix is singular. Accordingly, if the intersection of a line described by (11.111) and a plane described by (11.113) is a single point (X, Y, Z), then this point is given by (11.114); whereas if the above 3×3 matrix is singular, then the line and the plane are parallel, in which case the line lies in the plane if and only if there is at least one solution (x, y, z) to the three simultaneous equations (11.111) and (11.113). Also, a line described by (11.111) and a plane described by (11.113) are perpendicular if and only if the plane is perpendicular to each of the two planes described by (11.111a) and (11.111b).

Analogous to (11.90), the unique line that is perpendicular to two given noncollinear vectors $A\hat{\mathbf{i}} + B\hat{\mathbf{j}} + C\hat{\mathbf{k}} \neq \mathbf{0}$ and $A'\hat{\mathbf{i}} + B'\hat{\mathbf{j}} + C'\hat{\mathbf{k}} \neq \mathbf{0}$, and that passes through a given point (x_1, y_1, z_1), is described by the equations

$$A(x - x_1) + B(y - y_1) + C(z - z_1) = 0, \tag{11.115a}$$
$$A'(x - x_1) + B'(y - y_1) + C'(z - z_1) = 0, \tag{11.115b}$$

per (11.103). Furthermore, it might be possible to manipulate these equations into the form

$$\alpha(x - x_1) = \beta(y - y_1) = \gamma(z - z_1) \tag{11.116}$$

for some real constants α, β, and γ; namely, the first equality obtains by combining (if necessary) equations (11.115) to eliminate the variable z, and the second equality likewise obtains by eliminating the variable x. Conversely, given real constants α, β, and γ of which at most one is 0, (11.116) describes the unique line passing through a given point (x_1, y_1, z_1) that is perpendicular to each of the three vectors $\alpha\hat{\mathbf{i}} - \beta\hat{\mathbf{j}}$, $\beta\hat{\mathbf{j}} - \gamma\hat{\mathbf{k}}$, and $-\alpha\hat{\mathbf{i}} + \gamma\hat{\mathbf{k}}$. (These three vectors are coplanar, but no two are collinear; therefore, any line that is perpendicular to two of them is also perpendicular to the third.)

Analogous to (11.92), the unique line that is parallel to a given vector $a\hat{\mathbf{i}}+b\hat{\mathbf{j}}+c\hat{\mathbf{k}} \neq \mathbf{0}$, and passes through a given point (x_1, y_1, z_1), is described by the parametric equation

$$\begin{bmatrix} x \\ y \\ z \end{bmatrix} = \begin{bmatrix} x_1 \\ y_1 \\ z_1 \end{bmatrix} + t \begin{bmatrix} a \\ b \\ c \end{bmatrix} \tag{11.117}$$

($t \in \mathbb{R}$). Given this description, a pair of nonparametric equations (11.111) describing l obtains by taking for (A_1, B_1, C_1) and (A_2, B_2, C_2) two of the three triples $(-b, a, 0)$, $(c, 0, -a)$, and $(0, -c, b)$—chosen so that the vectors $A_1\hat{\mathbf{i}}+B_1\hat{\mathbf{j}}+C_1\hat{\mathbf{k}}$ and $A_2\hat{\mathbf{i}}+B_2\hat{\mathbf{j}}+C_2\hat{\mathbf{k}}$ are noncollinear—then taking $D_1 = -A_1x_1 - B_1y_1 - C_1z_1$ and $D_2 = -A_2x_1 - B_2y_1 - C_2z_1$. Conversely, given a pair of equations (11.111) describing a line l, one might somehow find a solution $(x, y, z) = (x_1, y_1, z_1)$ to both; then, upon taking any triple $(a, b, c) \neq (0, 0, 0)$ such that

$$\begin{bmatrix} A_1 & B_1 & C_1 \\ A_2 & B_2 & C_2 \end{bmatrix} \begin{bmatrix} a \\ b \\ c \end{bmatrix} = \begin{bmatrix} 0 \\ 0 \\ 0 \end{bmatrix},$$

(11.117) also describes l.

Analogous to (11.93), the line l described by (11.117) is also described by the nonparametric equations

$$\frac{x - x_1}{a} = \frac{y - y_1}{b} = \frac{z - z_1}{c} \tag{11.118}$$

—if a, b, $c \neq 0$. If not, then a pair of equations describing l can be obtained as follows: When only one of the constants a, b, and c equals 0, remove the corresponding term from (11.118) to leave one equation; then, set the numerator of the removed term equal to 0 to get a second equation. Whereas when two of the constants equal 0, the desired pair of equations results by setting the numerators of the two corresponding terms in (11.118) equal to 0.

Analogous to (11.95), the unique line l that passes through two given distinct points (x_1, y_1, z_1) and (x_2, y_2, z_2) is described by the parametric equation

$$\begin{bmatrix} x \\ y \\ z \end{bmatrix} = \begin{bmatrix} x_1 \\ y_1 \\ z_1 \end{bmatrix} + t \begin{bmatrix} x_2 - x_1 \\ y_2 - y_1 \\ z_2 - z_1 \end{bmatrix} \tag{11.119}$$

($t \in \mathbb{R}$). Akin to (11.118), this line is also described by the nonparametric equations

$$\frac{x - x_1}{x_2 - x_1} = \frac{y - y_1}{y_2 - y_1} = \frac{z - z_1}{z_2 - z_1} \tag{11.120}$$

—if $x_1 \neq x_2$, $y_1 \neq y_2$, and $z_1 \neq z_2$. If not, then the divisions by 0 can be dealt with as stated above for (11.118); or, these problems can be avoided altogether by describing l by the equations

$$\det \begin{bmatrix} x & y & 1 \\ x_1 & y_1 & 1 \\ x_2 & y_2 & 1 \end{bmatrix} = \det \begin{bmatrix} x & 1 & z \\ x_1 & 1 & z_1 \\ x_2 & 1 & z_2 \end{bmatrix} = \det \begin{bmatrix} 1 & y & z \\ 1 & y_1 & z_1 \\ 1 & y_2 & z_2 \end{bmatrix} = 0. \tag{11.121}$$

(Although setting these three determinants equal to 0 yields three equations rather than two, one of them can be derived from the other two and therefore discarded; however, *which* of the equations can be discarded is not always the same.)

It follows from (11.121) that three points (x_1, y_1, z_1), (x_2, y_2, z_2), and (x_3, y_3, z_3) in xyz-space are collinear if and only if

$$\det \begin{bmatrix} x_1 & y_1 & 1 \\ x_2 & y_2 & 1 \\ x_3 & y_3 & 1 \end{bmatrix} = \det \begin{bmatrix} x_1 & 1 & z_1 \\ x_2 & 1 & z_2 \\ x_3 & 1 & z_3 \end{bmatrix} = \det \begin{bmatrix} 1 & y_1 & z_1 \\ 1 & y_2 & z_2 \\ 1 & y_3 & z_3 \end{bmatrix} = 0.$$

Moreover, if $x_1 \neq x_2$, $y_1 \neq y_2$, and $z_1 \neq z_2$, then an equivalent condition is

$$\frac{x_3 - x_1}{x_2 - x_1} = \frac{y_3 - y_1}{y_2 - y_1} = \frac{z_3 - z_1}{z_2 - z_1}, \tag{11.122}$$

per (11.120).

The preceding parametric descriptions of lines and planes easily generalize to apply to Euclidean spaces $\mathbb{E}^n$ of *all* dimensions. Moreover, by simply placing restrictions on the parameters, the resulting parametric equations can also be used to describe other basic line and plane figures in $\mathbb{E}^n$. In what follows, then, let the dimension $n \geq 1$ of the space be arbitrarily fixed;[44] and, let the $n \times 1$ matrix

$$\mathbf{x} = \begin{bmatrix} x_1 \\ \vdots \\ x_n \end{bmatrix}$$

in the vector space $\mathbb{V}^n$ correspond to the generic point $(x_1, \ldots, x_n)$ on $\mathbb{E}^n$.

44. For $n = 1$, some of the statements below will only be true vacuously since, e.g., no planes lie in $\mathbb{E}^1$.

Given the matrix representation $\mathbf{p} \in \mathbb{V}^n$ of a point $P = (p_1, \ldots, p_n)$ on $\mathbb{E}^n$, along with the matrix representation $\mathbf{v} \in \mathbb{V}^n$ of a nonzero vector $\vec{v}$ of $\mathbb{E}^n$—i.e.,

$$\mathbf{p} = \begin{bmatrix} p_1 \\ \vdots \\ p_n \end{bmatrix}, \quad \mathbf{v} = \begin{bmatrix} v_1 \\ \vdots \\ v_n \end{bmatrix}$$

—the unique line in $\mathbb{E}^n$ that is parallel to $\mathbf{v}$ and passes through P is described by the parametric equation

$$\mathbf{x} = \mathbf{p} + t\mathbf{v} \tag{11.123}$$

($t \in \mathbb{R}$)—thus generalizing (11.92) and (11.117). Stated more concisely:[45] Given a vector $\mathbf{v}$ of $\mathbb{E}^n$, along with a point $\mathbf{p}$ on $\mathbb{E}^n$, the unique line in $\mathbb{E}^n$ that is parallel to $\mathbf{v}$ and passes through $\mathbf{p}$ is described by (11.123). Likewise, given two distinct points $\mathbf{p}$ and $\mathbf{q}$ on $\mathbb{E}^n$, the unique line $\overleftrightarrow{\mathbf{pq}}$ in $\mathbb{E}^n$ that passes through both points is described by the parametric equation

$$\mathbf{x} = \mathbf{p} + t(\mathbf{q} - \mathbf{p}) \tag{11.124}$$

($t \in \mathbb{R}$)—generalizing (11.95) and (11.119).

For both (11.123) and (11.124), each point $\mathbf{x}$ on the described line obtains for exactly one value of the parameter t. Also, upon imposing the restriction that $t > 0$ (respectively, $t \geq 0$), equation (11.123) describes a half-line (ray) that is parallel to the vector $\mathbf{v}$ and has endpoint $\mathbf{p}$; moreover, the only other such half-line (ray) obtains from the restriction that $t < 0$ ($t \leq 0$). Likewise, the restriction $0 \leq t \leq 1$ causes (11.124) to describe the segment $\overline{\mathbf{pq}}$. Its endpoints $\mathbf{p}$ and $\mathbf{q}$ correspond to the values $t = 0$ and $t = 1$, respectively, and its interior corresponds to $0 < t < 1$; also, the midpoint $\frac{1}{2}(\mathbf{p} + \mathbf{q})$ of the segment obtains for $t = \frac{1}{2}$. Hence, a figure $f \subseteq \mathbb{E}^n$ is often defined to be convex if and only if for any points $\mathbf{p}$ and $\mathbf{q}$ on f, the point $(1 - t)\mathbf{p} + t\mathbf{q}$ is on f for all $t \in [0, 1]$.

Analogous to (11.123), given two noncollinear vectors $\mathbf{u}$ and $\mathbf{v}$ of $\mathbb{E}^n$, along with a point $\mathbf{p}$ on $\mathbb{E}^n$, the unique plane in $\mathbb{E}^n$ that is parallel to both $\mathbf{u}$ and $\mathbf{v}$, and passes through $\mathbf{p}$, is described by the parametric equation

$$\mathbf{x} = \mathbf{p} + s\mathbf{u} + t\mathbf{v} \tag{11.125}$$

(s, $t \in \mathbb{R}$)—generalizing (11.105). It follows that given a vector $\mathbf{v}$ of $\mathbb{E}^n$, along with points $\mathbf{p}$ and $\mathbf{q}$ on $\mathbb{E}^n$, if the vectors $\mathbf{v}$ and $\mathbf{p} - \mathbf{q}$ are noncollinear, then the unique plane in $\mathbb{E}^n$ parallel to $\mathbf{v}$ that passes through both $\mathbf{p}$ and $\mathbf{q}$ is described by the parametric equation

$$\mathbf{x} = \mathbf{p} + s(\mathbf{q} - \mathbf{p}) + t\mathbf{v} \tag{11.126}$$

45. As discussed in Subsection 11.3.2, here is an instance of referring to one matrix ($\mathbf{v}$) as a *vector* of $\mathbb{E}^n$ while at the same time referring to another matrix ($\mathbf{p}$) as a *point* on $\mathbb{E}^n$. Such abuses of terminology will be made henceforth without comment.

($s, t \in \mathbb{R}$)—generalizing (11.107). Likewise, the unique plane in $\mathbb{E}^n$ that passes through three noncollinear points $\mathbf{p}$, $\mathbf{q}$, and $\mathbf{r}$ on $\mathbb{E}^n$ is described by the parametric equation

$$\mathbf{x} = \mathbf{p} + s(\mathbf{q} - \mathbf{p}) + t(\mathbf{r} - \mathbf{p}) \tag{11.127}$$

($s, t \in \mathbb{R}$)—generalizing (11.109).

For (11.125)–(11.127), each point $\mathbf{x}$ on the described plane obtains for exactly one value of the parameter pair (s, t). Also, imposing the restriction that $s > 0$ (respectively, $s \geq 0$) causes (11.125) to describe a half-plane (closed half-plane) parallel to both $\mathbf{u}$ and $\mathbf{v}$, for which the edge of the half-plane (closed half-plane) is described by (11.123); moreover, the only other such half-plane (closed half-plane) obtains from the restriction $s < 0$ ($s \leq 0$).

11.5.2 The Conic Sections

In Subsection 10.3.2, several ways were given for characterizing the four kinds of conic sections—circles, parabolas, ellipses, and hyperbolas—irrespective of any coordinate system. In the xy-plane, every conic section can be described by a second-degree (i.e., quadratic) equation in the variables "x" and "y":

$$Ax^2 + Bxy + Cy^2 + Dx + Ey + F = 0 \tag{11.128}$$

(A, B, C, D, E, F real), where A, B, and C are not all 0. However—in contrast to the first-degree equation (11.88), which *always* describes a line (assuming that the coefficients are real, and A and B are not both 0)—an equation of the form (11.128) need not describe a conic section. For example, consider these similar second-degree equations:

$$x^2 - y^2 = 0, \tag{11.129a}$$

$$x^2 - 2xy + y^2 - 1 = 0, \tag{11.129b}$$

$$x^2 + 2xy + y^2 = 0, \tag{11.129c}$$

$$x^2 + y^2 = 0, \tag{11.129d}$$

$$x^2 + y^2 + 1 = 0. \tag{11.129e}$$

For the first three, the left side can be factored into two first-degree factors (possibly the same)—i.e.,

$$Ax^2 + Bxy + Cy^2 + Dx + Ey + F = (A'x + B'y + C')(A''x + B''y + C'')$$

—using real coefficients. Specifically, (11.129a) is equivalent to the equation $(x-y)(x+y) = 0$, which is satisfied if and only if either $y = x$ or $y = -x$, and thus describes two crossing lines; (11.129b) is equivalent to the equation $(x-y-1)(x-y+1) = 0$, which is satisfied if and only if either $y = x - 1$ or $y = x + 1$, and thus describes two parallel lines; and, (11.129c) is equivalent to the equation $(x + y)^2 = 0$, which is satisfied if and only if $y = -x$, and thus describes a single line. Furthermore, (11.129d) is satisfied if and only if $x = y = 0$, and thus describes the single point $(x, y) = (0, 0)$; whereas (11.129e) describes the empty set, since $x^2 + y^2 \geq 0$ for all points (x, y).

TABLE 11.3 Figures Described by a Second-Degree Equation

Δ	J	I/Δ	K	Figure
$\neq 0$	< 0	—	—	Hyperbola
$\neq 0$	0	—	—	Parabola
$\neq 0$	> 0	< 0	—	Circle ($I^2 = 4J$) or ellipse ($I^2 \neq 4J$)
$\neq 0$	> 0	> 0	—	Empty set
0	< 0	—	—	Two crossing lines
0	0	—	< 0	Two parallel lines
0	0	—	0	One line
0	0	—	> 0	Empty set
0	> 0	—	—	One point

Upon calculating the quantities

$$\Delta = \det \begin{bmatrix} A & B/2 & D/2 \\ B/2 & C & E/2 \\ D/2 & E/2 & F \end{bmatrix}, \tag{11.130a}$$

$$I = A + C, \tag{11.130b}$$

$$J = \det \begin{bmatrix} A & B/2 \\ B/2 & C \end{bmatrix}, \tag{11.130c}$$

$$K = \det \begin{bmatrix} A & D/2 \\ D/2 & F \end{bmatrix} + \det \begin{bmatrix} C & E/2 \\ E/2 & F \end{bmatrix}, \tag{11.130d}$$

Table 11.3 can be used to predict the figure resulting from equation (11.128)—the conic sections being shown above the dashed line.[46] The quantities Δ, I, and J are all invariant (i.e., their values do not change) under both translations and rotations of the coordinate system; and, whereas K is, in general, invariant only under rotations, it is also invariant under translations when $\Delta = J = 0$ (the cases for which K is significant).[47]

As for special cases, the equation

$$y^2 = 4cx \tag{11.131}$$

($c > 0$) describes a parabola having its focus at $(c, 0)$, with its directrix being the vertical line described by the equation $x = -c$; therefore, as shown in Figure 11.23, its axis is the x-axis. Such a parabola is said to "open" to the right. Alternatively, it can be rotated counterclockwise either by $\pi/2$ to open upward ($x^2 = 4cy$), or by π to open to the left ($y^2 = -4cx$), or by $3\pi/2$ to open downward ($x^2 = -4cy$). For each of these parabolas,

46. Note that observing the sign of I/Δ is equivalent to observing whether I and Δ have the same sign. Accordingly, I/Δ need not be evaluated to use the table.

47. These properties of Δ, I, J, and K are not surprising, since all of the figures described in Table 11.3 are invariant in *kind* under translation and rotation (e.g., translating or rotating a hyperbola always yields another hyperbola).

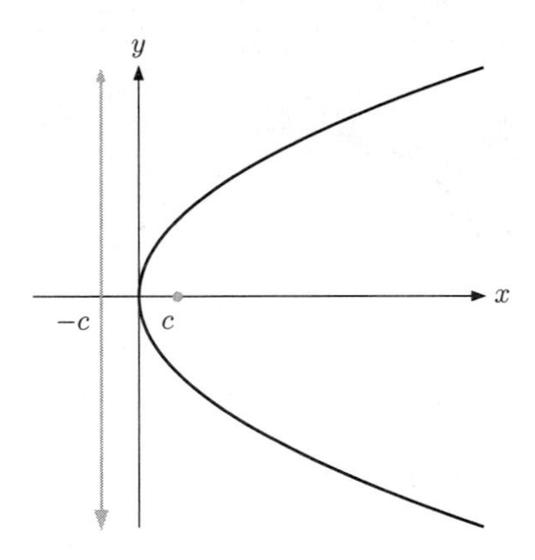

FIGURE 11.23 A Parabola in the xy-Plane

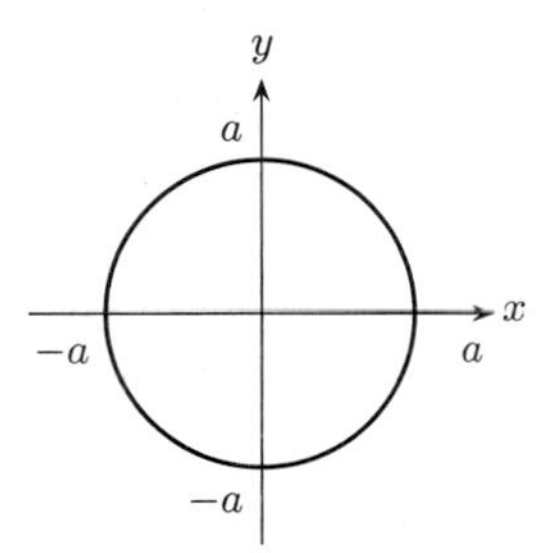

FIGURE 11.24 A Circle in the xy-Plane

the vertex is the point $(0, 0)$, and the focal parameter $p = 2c$; thus, e.g., (11.131) can be rewritten

$$y^2 = 2px. \tag{11.132}$$

The equation

$$x^2/a^2 + y^2/a^2 = 1 \tag{11.133}$$

$(a > 0)$ describes a circle of radius a centered at the origin $(0, 0)$; see Figure 11.24. Whereas the similar equation

$$x^2/a^2 + y^2/b^2 = 1 \tag{11.134}$$

$(a > b > 0)$ describes an ellipse centered at $(0, 0)$; see Figure 11.25. Its major axis is the segment having endpoints $(\pm a, 0)$, which are also the vertices of the ellipse; and, its minor axis is the segment having endpoints $(0, \pm b)$. The foci are the points $(\pm c, 0)$ with $c = \sqrt{a^2 - b^2}$; also the directrices are the vertical lines described by the equations $x = \pm a^2/c$, the eccentricity $e = c/a$, and the focal parameter $p = b^2/c$. Alternatively,

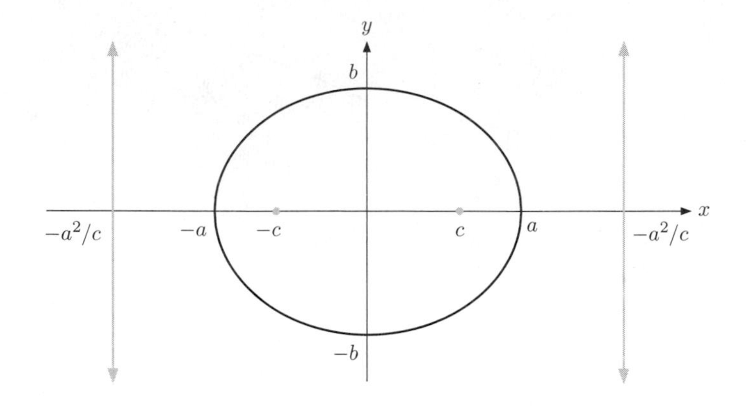

FIGURE 11.25 An Ellipse in the xy-Plane

interchanging the variables "x" and "y" in (11.134) causes the ellipse to rotate $\pm\pi/2$ about the origin, thereby putting its major axis on the y-axis, and so forth.

Finally, the equation

$$x^2/a^2 - y^2/b^2 = 1 \tag{11.135}$$

($a, b > 0$) describes a hyperbola centered at $(0, 0)$; see Figure 11.26. Its transverse axis is the segment having endpoints $(\pm a, 0)$, which are also the vertices of the hyperbola; and, its conjugate axis is the segment having endpoints $(0, \pm b)$. The asymptotes are the dashed lines passing through $(0, 0)$ with slopes $\pm b/a$, and described by the equations $x/a \pm y/b = 0$. The foci are the points $(\pm c, 0)$ with $c = \sqrt{a^2 + b^2}$; also, the directrices are the vertical lines described by the equations $x = \pm a^2/c$, the eccentricity $e = c/a$, and the focal parameter $p = b^2/c$. Alternatively, interchanging the variables "x" and "y" in (11.135) causes the hyperbola to rotate $\pm\pi/2$ about the origin, thereby putting its transverse axis on the y-axis, and so forth.

In general, any conic section lying in the xy-plane can be obtained from one of the above special cases by performing a rotation followed by a translation (or vice versa), thereby yielding another equation of the form (11.128). Conversely, any conic section described by (11.128) can be rotated and then translated to be put into the form of one of these special cases; then, the various geometric features (foci, directrices, ...) of the original conic section obtain by performing the inverse translation, and then inverse rotation, upon the corresponding features of the special case.

To wit, the xy-term in (11.128) vanishes—i.e., the coefficient B is converted to 0—upon rotating the xy-coordinate system counterclockwise by a particular angle ϕ to obtain an $x'y'$-coordinate system.[48] Specifically, using (11.85), the left side

48. Or, rather than perform a rotation and translation upon the *coordinate system* (i.e., perform alias transformations), the same goal can be accomplished by performing a rotation and translation of the *geometric figure* itself (i.e., perform alibi transformations).

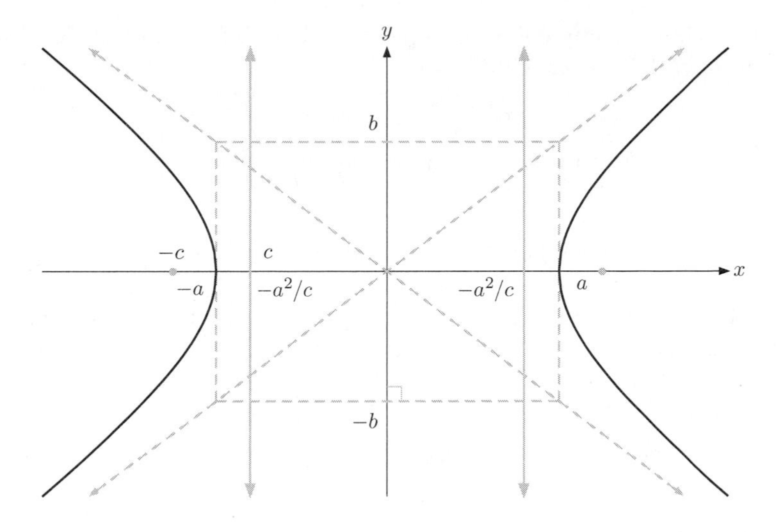

FIGURE 11.26 A Hyperbola in the *xy*-Plane

of (11.128) becomes

$$Ax^2 + Bxy + Cy^2 + Dx + Ey + F$$
$$= A'x'^2 + B'x'y' + C'y'^2 + D'x' + E'y' + F', \quad (11.136)$$

where

$$A' = A\cos^2\phi + B\sin\phi\cos\phi + C\sin^2\phi$$
$$= [(A+C) + (A-C)\cos 2\phi + B\sin 2\phi]/2, \quad (11.137a)$$

$$B' = -2A\sin\phi\cos\phi + B(\cos^2\phi - \sin^2\phi) + 2C\sin\phi\cos\phi$$
$$= -(A-C)\sin 2\phi + B\cos 2\phi, \quad (11.137b)$$

$$C' = A\sin^2\phi - B\sin\phi\cos\phi + C\cos^2\phi,$$
$$= [(A+C) - (A-C)\cos 2\phi - B\sin 2\phi]/2, \quad (11.137c)$$

$$D' = D\cos\phi + E\sin\phi, \quad (11.137d)$$

$$E' = -D\sin\phi + E\cos\phi, \quad (11.137e)$$

$$F' = F. \quad (11.137f)$$

Hence, assuming $B \neq 0$ (if not, simply let $\phi = 0$, in which case none of the coefficients change), we obtain $B' = 0$ if and only if

$$\cot 2\phi = \frac{A-C}{B}. \quad (11.138)$$

Specifically, if $A = C$, then we can take $\phi = \pi/4$; otherwise, we can take $\phi = \frac{1}{2}\tan^{-1}[B/(A-C)]$. Also, although no restrictions have been placed on ϕ, it can be shown that there is exactly one value $\phi' \in (0, \pi/2)$—i.e., an acute angle—such that $\phi = \phi'$ satisfies (11.138); namely,

$$\phi' = \tan^{-1}\left[-\frac{A-C}{B} + \sqrt{\left(\frac{A-C}{B}\right)^2 + 1}\right].$$

Moreover, the solutions to (11.138) are the values $\phi = \phi' + n\pi/2$ for which n is an integer.

Next, we translate the $x'y'$-coordinate system by the position vector corresponding to a particular point $(x', y') = (X, Y)$, thereby obtaining an $x''y''$-coordinate system. Specifically, letting

$$\begin{bmatrix} x' \\ y' \end{bmatrix} = \begin{bmatrix} x'' + X \\ y'' + Y \end{bmatrix},$$

per (11.83), the right side of (11.136) becomes

$$A'x'^2 + B'x'y' + C'y'^2 + D'x' + E'y' + F' \\ = A''x''^2 + C''y''^2 + D''x'' + E''y'' + F', \quad (11.139)$$

where

$$\begin{aligned} A'' &= A', \\ C'' &= C', \\ D'' &= 2A'X + D', \\ E'' &= 2C'Y + E', \\ F'' &= A'X^2 + C'Y^2 + D'X + E'Y + F'. \end{aligned}$$

If $A' = 0$, then we must have $C', D' \neq 0$; for otherwise the given equation (11.128) would not describe a conic section (since $\Delta = 0$). By taking $X = (E'^2/4C' - F')/D'$ and $Y = -E'/2C'$, we now obtain

$$A'' = 0, \quad C'' = C', \quad D'' = D', \quad E'' = 0, \quad F'' = 0,$$

for which the equation

$$A''x''^2 + C''y''^2 + D''x'' + E''y'' + F'' = 0 \quad (11.140)$$

describes a parabola having the x-axis as its axis, and that opens either to the right (if $\operatorname{sgn} C'' \neq \operatorname{sgn} D''$) or to the left (if $\operatorname{sgn} C'' = \operatorname{sgn} D''$). Similarly, if $C' = 0$, then we must have $A', E' \neq 0$; and, taking $X = -D'/2A'$ and $Y = (D'^2/4A' - F')/E'$ yields

$$A'' = A', \quad C'' = 0, \quad D'' = 0, \quad E'' = E', \quad F'' = 0,$$

for which equation (11.140) describes a parabola that has the y-axis as its axis, and that opens either upward (if $\operatorname{sgn} A'' \neq \operatorname{sgn} E''$) or downward (if $\operatorname{sgn} A'' = \operatorname{sgn} E''$). In all cases, the vertex of the parabola is at $(x'', y'') = (0, 0)$.

If A', $C' \neq 0$, then taking $X = -D'/2A'$ and $Y = -E'/2C'$ yields

$$A'' = A', \quad C'' = C', \quad D'' = 0, \quad E'' = 0, \quad F'' = F' - D'^2/4A' - E'^2/4C'.$$

Now, for (11.140) to describe a conic section, we must have $F'' \neq 0$ (otherwise $\Delta = 0$); additionally, we cannot have $\operatorname{sgn} A'' = \operatorname{sgn} C'' = \operatorname{sgn} F''$ (otherwise $J > 0$ and $I/\Delta > 0$). If $\operatorname{sgn} A'' = \operatorname{sgn} C''$, then (11.140) describes either a circle (if $A'' = C''$), or an ellipse with its major axis on either the x-axis (if $|A''| < |C''|$) or the y-axis (if $|A''| > |C''|$). Whereas if $\operatorname{sgn} A'' \neq \operatorname{sgn} C''$, then (11.140) describes a hyperbola with its transverse axis on either the x-axis (if $\operatorname{sgn} C'' = \operatorname{sgn} F''$) or the y-axis (if $\operatorname{sgn} A'' = \operatorname{sgn} F''$). In all cases, the conic section is centered at $(x'', y'') = (0, 0)$.

In terms of the polar coordinates r and θ (see Figure 11.3), first suppose r is allowed to be negative. Then, a conic section in the xy-plane that has eccentricity $e > 0$ and focal parameter $p > 0$—specifically, with a focus at the origin and the corresponding directrix being the vertical line described by the equation $x = p$—is described by the equation

$$r(1 + e\cos\theta) = ep. \tag{11.141a}$$

Furthermore, when r is restricted to be nonnegative, this equation still describes the entire conic section if $e \leq 1$ (i.e., a parabola or an ellipse); however, for $e > 1$, only one branch of a hyperbola is described by (11.141a), the other branch being described by the equation

$$r(-1 + e\cos\theta) = ep. \tag{11.141b}$$

Alternatively, the equation

$$r = e|p - r\cos\theta| \tag{11.141c}$$

describes the entire conic section for any value of e, regardless of whether r is allowed to be negative.

None of the equations (11.141), though, can describe a circle, since the eccentricity e and focal parameter p of a circle are undefined.[49] Instead, regardless of whether r is allowed to be negative, a circle centered at the origin having radius $a > 0$ is described by the simple equation

$$r = a.$$

Moreover, for $a < 0$ this equation describes a circle centered at the origin having radius $|a|$, assuming r may be negative.

Parametrically: The parabola described by (11.132) is also described by the equations

$$x = 2pt^2, \quad y = 2pt$$

49. For more on this matter, see Problem 10.27.

$(-\infty < t < \infty)$. The circle described by (11.133) is also described by the equations

$$x = a\cos t, \quad y = a\sin t$$

$(0 \leq t < 2\pi)$. The ellipse described by (11.134) is also described by the equations

$$x = a\cos t, \quad y = b\sin t$$

$(0 \leq t < 2\pi)$. And, the hyperbola described by (11.135) is also described by the equations

$$x = \pm a\sec t, \quad y = b\tan t$$

$(-\pi/2 < t < \pi/2)$, as well as by the equations

$$x = \pm a\cosh t, \quad y = b\sinh t$$

$(-\infty < t < \infty)$, with each choice of "$\pm$" yielding one branch of the hyperbola.

CHAPTER 12

Mathematics in Practice

For the purposes of science and engineering, mathematics is important not because of its intrinsic value, but because it enables the precise formulation and solution of problems pertaining to the real world. In this modern age, it is largely taken for granted that many physical phenomena can be explained in mathematical terms. In fact, though, it is quite remarkable that physical measurements can be correlated mathematically, and that predictions made in light of these correlations are repeatedly confirmed. Furthermore, it is often the case that one physical quantity can be expressed as a function of only a few others. For example, classical mechanics asserts the fundamental law

$$F = ma$$

relating the force F applied to an object of mass m and the resulting acceleration a in the direction of the force; here, not only are these three quantities related mathematically, the relationship could hardly be simpler.

In science, the general process by which physical phenomena come to be explained mathematically can be outlined as follows. That is, this is what scientists do:

1. Specify the physical phenomena to be explained. In particular, identify which physical quantities, the **observables**, are to be directly measured.
2. Based on present knowledge, experience, and intuition, hypothesize how the observables are correlated; that is, propose a theory to explain the phenomena.
3. Apply logic and mathematics to analyze the proposed theory, thereby formulating predictions about the observables.
4. Conduct experiments to test the predictions.
5. If the predictions are borne out by the experiments, then return to steps 3 and 4 for further analysis and testing. Otherwise, the proposed theory has been contradicted by experiment, and so it must either be modified or discarded; hence, return to step 2.

This endless process of hypothesizing, predicting, testing, and revising constitutes what is commonly referred to as the **scientific method**.[1]

1. Though different authors may express the individual steps somewhat differently, the essential features of the method are the same for all: a theory about some phenomena is proposed, then repeatedly tested experimentally and modified in accordance with the results.

A scientific hypothesis receives greater "confirmation" the more tests it successfully passes. If the predictions made by a proposed theory are consistently observed (within acceptably small experimental uncertainty), then it might eventually gain sufficient credibility among scientists to be accepted as a **scientific theory**. By contrast, in mathematics, although individual cases may lead a mathematician to *suspect* the truth of a universal statement (i.e., "Such and such is *always* the case"), that statement does not become a theorem until it is accompanied by a rigorous deductive proof. Thus, whereas inductive reasoning serves mathematics only as an aid to discovery, it is a cornerstone of science.[2] Accordingly, because, the validity of any scientific theory is a matter of subjective judgment, a wide range of opinions may exist among scientists (and, the lay public).[3] Moreover, no amount of experimentation can confirm a scientific theory beyond all doubt. Conversely, though, a single false prediction is sufficient to overthrow even the most firmly established theory, or at least limit the range of its applicability; for example, in modern physics, quantum mechanics and relativity theory have superseded classical mechanics, which is now known to be applicable (approximately) only to large-scale phenomena involving speeds much less than that of light.

A widely accepted principle by which competing scientific theories are assessed is **Ockham's razor**. For scientific purposes, it is the assertion that, all other things being equal, the simplest theory is preferred. (Of course, simplicity is itself subjective.) Similarly, **ad hoc hypotheses**—i.e., make shift assumptions introduced to patch a flawed theory—are disdained. For example, if a particular theory happens to fail during a test performed on a rainy Thursday morning, then it would be an ad hoc hypothesis to assume that the theory is valid *except* on rainy Thursday mornings.

Furthermore, a scientific theory that includes a metaphysical[4] interpretation of the associated phenomena does so at a risk; for although such an interpretation is likely intended to be conceptually helpful, it might actually make further theoretical advances more difficult. As a classic example, someone who tries to visualize the continuous paths by which photons pass through the slits in a diffraction grating, probably will be perplexed by the pattern of light intensity emitted from the grating; because, the same pattern does not obtain by forcing the photons to pass through each slit individually (e.g., by covering the other slits), then summing the results. Yet, the observed intensity pattern *is* accurately predicted by quantum mechanics, which in its basic form (the "Copenhagen interpretation") eschews such heuristic visualization, being concerned only with *correlating observables*.[5] Here, the primary observables are the frequency of

---

2. See fn. 20 in chapter 4 for a comparison of inductive and deductive reasoning.

3. The reader might think that, in contrast to science, logic and mathematics are objective, and so indisputable. Indeed, as stated by the philosopher Gottfried Leibniz, the "truths of reason" must be "true in all possible worlds". In fact, however, the *philosophies* behind logic and mathematics are not without controversy. Specifically, mathematicians may disagree on what abstract objects "exist", on what axioms (logical and mathematical) should be assumed, on what rules of inference are acceptable (e.g., some argue against the law of excluded middle), and on the significance of the resulting theorems. Nevertheless, once the axioms and rules of inference have been fixed, the validity of any proof can be verified mechanically (i.e., by a machine). (For a brief introduction to the philosophy of mathematics, see Section E.7)

4. By definition, anything "metaphysical" is beyond physical experience, and thus cannot be measured or tested experimentally.

5. For more on the interpretation of quantum mechanics, [42] is recommended.

the light source and the spatial distribution of light intensity coming from the grating (the physical structure of which is assumed to be known); therefore, it is metaphysical to visualize photons traveling along continuous paths through the diffraction grating, because such paths are not measured in *this* experiment. Moreover, any modification of the experiment to determine which slit each photon passes through—assuming that this visualization is indeed meaningful—will disrupt the intensity pattern.

It is important that the relationship between science and mathematics be clearly understood. For one thing, since the real world is quite intricate, the mathematical representation—i.e., **model**—of reality that a scientist uses to analyze a real-world phenomenon is often only an approximation. For example, a botanist wanting to quantify the size of a daisy might do so by measuring the diameter of its yellow disk. In fact, though, the disk is not perfectly circular; therefore, approximating it as such immediately introduces an error. This kind of error, in which an assumed mathematical model does not accurately reflect physical reality, is a **modeling error**. (By contrast, it is not an "experimental error", which is discussed in Subsection 12.1.3; because, it is still possible for the botanist to accurately measure *something*, which is then interpreted via the model as being the diameter of a circular disk.)

Additionally, as Albert Einstein once said, "As far as the propositions of mathematics refer to reality, they are not certain; and as far as they are certain, they do not refer to reality."[6] Because, when mathematical propositions refer to reality, they become predictions that might be refuted by experiment, and thus cannot be certain. On the other hand no amount of empirical evidence can disprove a theorem in mathematics; hence, mathematical propositions that are certain cannot be refuted by experiment, and therefore must not refer to reality, but instead to abstract mathematical objects (e.g., numbers or geometric shapes). Accordingly, if one were to find that the angles of some physical approximation to a triangle do not sum to 180° (which is indeed the case for sufficiently large "triangles" drawn on the surface of the globe), then the appropriate response would not to be to conclude that Euclidean geometry is wrong, but rather that it is *inapplicable* to the phenomenon being analyzed, at least in the manner assumed.

In the practice of engineering, the current state of science is generally accepted without question, and no attempt is made to generate new physical laws; instead, the goal is to design physical systems that can be built to perform useful functions. To analyze such a system (given or proposed), it is first necessary to describe it by a mathematical model; then, logic and mathematics are applied to analyze the model, thereby yielding predictions about the behavior of the system. All theoretical conclusions about the system, then, depend on the choice of the model, which might not fully or accurately capture the behavior of the actual system. Rewording Einstein: Insofar as the conclusions drawn from a model refer to the physical system, they are not certain; and insofar as they are certain, they do not refer to the physical system. Moreover, this lack of certainty is greater in engineering than in science; because, not only is engineering built upon scientific theory, but many systems are so complicated that the models used to describe them

6. From "Geometry and Experience", 1921 (reprinted in [18] pp. 227–40). Einstein then goes on to explain how this understanding helped in his development of the general theory of relativity (by using non-Euclidean geometry) and includes a readable explanation of how physical space can be finite yet unbounded.

are *deliberately* simplified to make their mathematical analyses tractable. Alternatively, more sophisticated models may be used and "tested" by computer simulation. In any case, all theoretical conclusions about a physical system are actually about the mathematical model used to analyze it, and so there is always some doubt about how well they pertain to reality.

12.1 Data Measurement

A physical feature (temperature, velocity, shape, relative size, category, ...) is converted into a corresponding mathematical object or relationship (number, vector, geometric figure, inequality, set, ...) by making an observation. In particular, an observation that quantifies the *amount* of something is called a **measurement**.[7] To make a good measurement is to carefully perform some clearly defined experimental process, then faithfully record both the process and its numerical result—the latter being commonly referred to itself as the "measurement". Also, the process is typically repeatable by others.

In general, any facts (be they direct observations or previously calculated results) upon which analyses may be performed, and conclusions thereby drawn, are collectively referred to as **data**.[8] Hence, when an experiment is performed, its measurements provide us with data. Moreover, measured values are often referred to as **raw data**, to emphasize that the information they provide has not yet been manipulated in any way.

12.1.1 Exponential Notation

An experimental value x_{exp}, be it a direct measurement or a number calculated from given data, is usually expressed in some form of a finite decimal expansion. Specifically, as was discussed in Subsection 3.1.1 with regard to (3.4) and (3.7), we might write[9]

$$x_{\text{exp}} = \pm x_n \ldots x_1 x_0 . x_{-1} x_{-2} \ldots x_{-m} \tag{12.1a}$$

$(m, n \geq 0)$, using the appropriate sign and digits $x_i \in \{0, 1, \ldots, 9\}$ $(i = -m, -m+1, \ldots, n)$; in particular, we will allow $x_n = 0$ only when $n = 0$. By writing (12.1a), one asserts

$$x_{\text{exp}} = \pm \sum_{i=-m}^{n} x_i \cdot 10^i \tag{12.1b}$$

---

7. As is normal, we will assume measurements that produce *real*-valued numbers.

8. Although the noun "data" (preferably pronounced *DAY-ta*) is plural, being derived from the singular **datum**, it is also widely used as a "mass noun"—which, like the word "information", is singular. Thus, although someone might say "This is good data" rather than (properly) "These are good data", one would never say "This is *a* good data" anymore than one would say "This is a good information" (as contrasted with, e.g., "This is a good apple"). Ironically, the word "mathematics" is a plural noun that is properly used as if singular—e.g., "Mathematics *is* helpful to scientists and engineers."

9. Here, per the remarks following (3.4), we streamline our discussion by allowing the symbols "x_n", "x_{n-1}", ... to represent *both* digits (e.g., "4") and the corresponding values (4). Such will be done hereafter without further comment.

($\pm$ respectively). If, however, $|x_{\text{exp}}| \ll 1$ or $|x_{\text{exp}}| \gg 1$, then several digits will be required in the expansion, perhaps including many leading or trailing zeros; therefore, we might instead express x_{exp} in **exponential notation** as

$$x_{\text{exp}} = \pm x_n \ldots x_1 x_0 . x_{-1} x_{-2} \ldots x_{-m} \times 10^p \tag{12.2a}$$

($m, n \geq 0$), for some conveniently chosen integer exponent p, again allowing $x_n = 0$ only when $n = 0$. We thereby assert

$$x_{\text{exp}} = \pm \left(\sum_{i=-m}^{n} x_i \cdot 10^i \right) \times 10^p = \pm \sum_{i=-m}^{n} x_i \cdot 10^{i+p} \tag{12.2b}$$

($\pm$ respectively).[10] (Thus, e.g., $0.000\,000\,123\,45 = 12.345 \times 10^{-8}$.)

To more easily discuss exponential notation, we will refer to the portion to the left of the multiplication symbol "$\times$" as the "decimal part", which is comprised of an "integer part" and a "fractional part" (as defined in Subsection 3.1.1); and, the remainder "$\times\ 10^p$" will be referred to as the "exponential part". Also, the "digits" of the numeral will be understood to be the digits $x_{-m}, x_{-m+1}, \ldots, x_n$ of the decimal part, not any digits used to express the exponent p.

Clearly, an ordinary finite decimal expansion (12.1) can be viewed as a special case of exponential notation (12.2); namely, choosing $p = 0$ makes $10^p = 1$, and thus the exponential part in (12.2a) can be omitted. Another special case is **scientific notation**, for which $n = 0$, and $x_n \neq 0$ when $x_{\text{exp}} \neq 0$:

$$x_{\text{exp}} = \pm x_0 . x_{-1} x_{-2} \ldots x_{-m} \times 10^p. \tag{12.3}$$

(Thus, e.g., $12.345 \times 10^{-8} = 1.2345 \times 10^{-7}$.) The value of the exponent p is now uniquely determined for each number $x_{\text{exp}} \neq 0$; moreover, this value is often referred to as the **order of magnitude** of x_{exp}, since it provides a rough indication of $|x_{\text{exp}}|$.

Yet another special case of exponential notation is **engineering notation**, for which p must be a multiple of 3, $n \in \{0, 1, 2\}$, and $x_n \neq 0$ when $x_{\text{exp}} \neq 0$ (e.g., $12.345 \times 10^{-8} = 123.45 \times 10^{-9}$). These constraints again uniquely determine p. Also, for any number $x_{\text{exp}} \neq 0$ expressed in scientific notation, the magnitude of the integer part ranges from 1 to 9; whereas in engineering notation, it ranges from 1 to 999. The main benefit of engineering notation is that it allows the exponential part to be directly replaced by a standard numeric prefix (see Table 12.3 on p. 233), such as "kilo" (for $p = 3$) and "milli" (for $p = -3$).

12.1.2 Dimensions, Units, and Numeric Prefixes

Usually, an experimental value is associated with a physical **dimension**—e.g., length, mass, time—that specifies the *kind* of physical feature being quantified. Somewhat arbitrarily, some dimensions are taken as fundamental, from which numerous others can be derived by multiplication and/or division. For example, volume = length $\cdot$ length $\cdot$

10. Thus, exponential notation is yet another numeral system.

length = length^3 (read "length cubed"), and speed = length/time (read "length divided by time"); thus, we also have length = speed · time (read "speed times time").

To associate a number with a dimension, and thereby specify its amount, a corresponding **(physical) unit**—e.g., meter, kilogram, second—must be chosen to provide a *scale* for that dimension. Specifically, every physical quantity is expressible as the *product* of a "pure number" and a physical unit. For example, if a length l is measured as 3 inches, then l is the same as the combined length of 3 one-inch rods, laid one immediately after another along a straight line; accordingly, using the abbreviation "in" for "inch", we may say that $l = 3 \times 1$ in—or, more briefly, $l = 3$ in. Thus, the dimension of length may be quantified using the unit inch (symbolized by the abbreviation "in"), which acts as a nonzero constant (just as π is a constant symbolized by the Greek letter "π"); and, likewise for any physical unit. Moreover, as explained in Section 10.2 regarding how length and angle measures can be defined in geometry, we may extend this product of number and unit to allow for rational numbers, and then real numbers. In this way, every physical quantity is essentially a comparison of some physical feature to a unit appropriate to measure the dimension of that feature; namely, the numerical value of a physical quantity equals the *ratio* of the quantity to the chosen unit (e.g., if $l = 3$ in, then $l/(1 \text{ in}) = l/\text{in} = 3$).

Many unit systems exist, including the English system (inch, foot, yard, mile, ounce, pound, second, minute, hour, ...) and the CGS system (centimeter, gram, second, ...). However, the standard system of units for use in science and engineering is the **International System of Units (SI)**, which evolved from the MKS system (meter, kilogram, second, ...). To wit, Table 12.1 lists the seven SI "base" (i.e., fundamental) units, from which all other SI units are derived by multiplication and/or division; also, accompanying each unit is its standard abbreviation. Then, in Table 12.2, several additional SI units are defined in terms of the base units; in particular, above the dashed line are all SI-derived units that have standard names. Thus, e.g., 5 newtons, or 5 N, is defined as 5 kg·m/s^2 (read "five kilogram-meters per second squared"), which means $5 \cdot (1 \text{ kg})(1 \text{ m})/(1 \text{ s})^2$; equivalently, 5 N = 5 kg·m·s^{-2}. As an example of combining physical units with exponential notation, the speed of light is about 3×10^8 m/s = 3×10^8 m·s^{-1}.

TABLE 12.1 The SI Base Units

Dimension	SI Unit	Abbreviation
Length, distance	**meter**	**m**
Mass	**kilogram**	**kg**
Time	**second**	**s**
Electric current	**ampere**	**A**
Thermodynamic temperature	**kelvin**[a]	**K**
Amount of substance[b]	**mole**	**mol**
Luminous intensity	**candela**	**cd**

[a] *This unit was formerly "degree Kelvin" (abbreviated "°K"), which is no longer used.*
[b] *This dimension is relative to a* count *of "elementary entities" (e.g., atoms, molecules), chosen as desired; specifically, the number of elementary entities in the substance is its amount in moles times* ***Avogadro's number*** $N_A \approx 6.022 \times 10^{23}$, *which by definition is the number of atoms in 0.012 kg of the isotope carbon-12.*

TABLE 12.2 Some SI Derived Units

Dimension	SI Unit	Abbreviation
Plane angle	**radian**	**rad** (m/m)
Solid angle	**steradian**	**sr** (m^2/m^2)
Force	**newton**	**N** ($kg{\cdot}m/s^2$)
Pressure	**pascal**	**Pa** (N/m^2)
Energy	**joule**	**J** (N·m)
Power	**watt**	**W** (J/s)
Electric potential	**volt**	**V** (W/A)
Resistance	**ohm**	**Ω** (V/A)
Conductance	**siemens**[a]	**S** (1/Ω)
Electric charge	**coulomb**	**C** (A·s)
Capacitance	**farad**	**F** (C/V)
Magnetic flux	**weber**	**Wb** (V·s)
Magnetic flux density	**tesla**	**T** (Wb/m^2)
Inductance	**henry**	**H** (Wb/A)
Frequency	**hertz**	**Hz** (1/s)
Luminous flux	**lumen**	**lm** (cd·sr)
Illuminance	**lux**	**lx** (lm/m^2)
Activity of a radionuclide	**becquerel**	**Bq** (1/s)
Absorbed dose	**gray**	**Gy** (J/kg)
Dose equivalent	**sievert**	**Sv** (J/kg)
Celsius temperature[b]	**degree Celsius**	**°C** (K)
Area	square meter	m^2
Volume	cubic meter	m^3
Velocity, speed	meter per second	m/s
Acceleration	meter per second squared	m/s^2
Angular velocity, angular speed	radian per second	rad/s
Angular acceleration	radian per second squared	rad/s^2

[a] *The equivalent unit* ***mho*** *(℧), indicating reciprocal–ohms, is not recognized by the SI.*
[b] *The Celsius temperature scale is a* shift *of the Kelvin temperature scale; namely,* $t°C \triangleq (t+273.15)$ K *for* $t \geq -273.15$. *It is convenient because water freezes at about* 0°C *and boils at about* 100°C. *By contrast, the Kelvin scale is an* ***absolute temperature*** *scale, since 0 K corresponds to* ***absolute zero****, the lowest temperature possible. The* ***centigrade*** *temperature scale, which is approximately the same as the Celsius scale, is obsolete.*

The preceding paragraph demonstrates some of the many style conventions commonly followed in the handling of SI units. A unit abbreviation (e.g., "m", "kg", "s") is not terminated with a period unless it occurs at the end of a sentence. When a unit is named after a person (e.g., Isaac Newton, Blaise Pascal, Alessandro Volta), its full name ("newton", "pascal", "volt") begins with a lowercase letter, except at the start of a sentence; whereas its abbreviation ("N", "Pa", "V") always begins with an uppercase letter.[11] For all other units, both full names and abbreviations begin with lowercase letters,

11. In particular, this convention applies to the unit ohm, which is abbreviated with the uppercase *Greek* letter "Ω"; and, it applies to the unit "degree Celsius", in which the word "degree" begins with a lowercase letter.

except at the start of a sentence (where abbreviation should generally be avoided).[12] Also, it is improper to mix full names and abbreviations in a single expression (e.g., one would not write "9 pascals = 9 N/m^2" or "2 grays equals 2 joules per kg").

Additionally, when a nonzero number having a magnitude less than or equal to 1 (respectively, greater than 1) is applied to a unit, the singular (plural) form of its full name is used (e.g., "1 meter", "$\frac{1}{2}$ meter", "−0.2 meter", "8×10^{-3} meter", "−6 meters");[13] but, abbreviated names are never modified (e.g., use "4 kg", not "4 kgs" or "4 kg's"). Multiplication of units can be indicated by a hyphen or a space when using full names (e.g., 1 watt equals 1 volt-ampere, or 1 volt ampere), and by a raised dot when using abbreviations ($1\ \text{W} = 1\ \text{V}{\cdot}\text{A}$). Division can be indicated by the word "per" when the unit names are spelled out (e.g., "meters per second"), and by a solidus (/) when abbreviations are used ("m/s"). Also, the word "reciprocal" can be used with full names (e.g., 7 siemens = 7 reciprocal ohms), whereas a superscript "-1" can be used for the same purpose with abbreviations ($7\ \text{S} = 7/\Omega = 7\ \Omega^{-1}$). As in ordinary arithmetic, multiplication takes precedence over division (e.g., $3\ \text{N}^{-1} = 3\ \text{s}^2/\text{kg}{\cdot}\text{m}$); and, unless parentheses are used (which is rare), at most one solidus should appear.

Whenever a ratio is formed from two quantities having the same physical dimension, their dimensions "cancel" (more on this in Subsection 12.2.1), and the result is a pure number—i.e., a **dimensionless quantity**. Moreover, applying any mathematical function to a dimensionless quantity yields another. For example, the radian measure of an angle is a dimensionless quantity, because (as discussed in Subsection 10.3.2 with regard to Figure 10.19) it is obtained by dividing the length l of a circular arc by its radius r, which is also a length; and, applying any trigonometric function to this ratio l/r yields another dimensionless quantity. However, not every ratio of lengths is naturally viewed as an angle (e.g., sine, cosine, and tangent are all defined as length ratios which, though related to angles, are not themselves angles). More generally, although dimensionless quantities having the same numerical value (e.g., 8 kg/kg and $8\ \text{s}^2/\text{s}^2$) are merely different ways of expressing the same pure number (8), the contexts in which they appear may differentiate how they are interpreted physically. Accordingly, though not necessary mathematically, it is sometimes conceptually helpful to append onto a dimensionless quantity an appropriate **dimensionless unit** indicating the physical nature of the quantity. Specifically (see Table 12.2), the SI includes two dimensionless units: the **radian (rad)** for quantifying plane angles, and the **steradian (sr)** for quantifying solid angles.[14] Furthermore, being both dimensionless and a unit, each is treated by the SI as being another name for the number 1 (unity); accordingly, each has essentially no effect when appearing in a physical quantity (e.g., $6\ \text{rad} = 6 \times 1\ \text{rad} = 6 \times 1 = 6$). But again, this is not to say that it is appropriate to multiply or divide by these units at any

12. This convention is not always followed, though, for units defined outside of the SI. In particular, an uppercase letter is commonly used to abbreviate **liter (L)** (defined as a cubic decimeter), because the lowercase letter "l" might be misread as the numeral "1".

13. The words "siemens", "hertz", and "lux" are both singular and plural.

14. Additional dimensionless units (e.g., the bel—see fn. 29, below) may also be introduced, as long as the overall system of units remains consistent, as discussed in Subsection 12.2.1.

TABLE 12.3 The SI Numeric Prefixes

Prefix (abbreviation)	Multiplier	Prefix (abbreviation)	Multiplier
deka (da)	10^1	**deci (d)**	10^{-1}
hecto (h)	10^2	**centi (c)**	10^{-2}
kilo (k)	10^3	**milli (m)**	10^{-3}
mega (M)	10^6	**micro (μ)**	10^{-6}
giga (G)	10^9	**nano (n)**	10^{-9}
tera (T)	10^{12}	**pico (p)**	10^{-12}
peta (P)	10^{15}	**femto (f)**	10^{-15}
exa (E)	10^{18}	**atto (a)**	10^{-18}
zetta (Z)	10^{21}	**zepto (z)**	10^{-21}
yotta (Y)	10^{24}	**yocto (y)**	10^{-24}

time; rather, dimensionless units should be used only as *labels* to indicate how physical quantities are to be interpreted.[15]

In the English system, there are several acceptable units for each dimension; however, it is often inconvenient to convert from one to another. For example, although it is a straightforward matter to calculate the number of inches in a mile, a calculation is required nonetheless. By contrast, a **metric system**—e.g., CGS, MKS, SI—has but one unit for each dimension, from which a larger or smaller version is easily obtained by applying a **numeric prefix** to represent multiplication by a chosen power of 10; see Table 12.3. Then, since we are using a decimal (i.e., base-10) notation for our numbers, it is a trivial matter to change from one prefix to another (e.g., 12.345×10^{-8} A $=$ 0.12345×10^{-6} A $=$ 0.12345 μA $=$ 123.45×10^{-3} μA $=$ 123.45 nA).

As can be seen, most SI prefixes correspond to factors 10^i for which i is a multiple of 3; the others (above the dashed line) get little use, the main exceptions being "decibel" and "centimeter". Since every prefix–unit combination is effectively another unit (and is sometimes to referred to as such), it is understood that any power that is applied to the unit includes the prefix as well. (Thus, e.g., "cm^3" means $(\text{cm})^3$, not $\text{c}(\text{m})^3$; likewise, $9\,\text{cm}^{-1} = 9(1\,\text{cm})^{-1} = 9(10^{-2}\,\text{m})^{-1} = 9 \times 10^2\,\text{m}^{-1}$, not $9 \times 10^{-2}\,\text{m}^{-1}$.[16]) Of course, applying a numeric prefix to a dimensionless unit produces a dimensionless quantity; however, by convention, such a quantity should not be expressed by a prefix standing

15. Examples of this practice occur in Table 12.2. The lumen (lm) is defined as a candela-steradian (cd·sr); but, since 1 sr $=$ 1, and thus 1 lm $=$ 1 cd·sr $=$ 1 cd $\times$ 1 sr $=$ 1 cd $\times$ 1 $=$ 1 cd, the lumen could just as well have been defined as a candela. However, the original definition is suggestive of how the lumen is used in photometry; namely, the lumen is the luminous flux emitted *within a solid angle of 1 steradian* by a uniform point-source of light having a luminous intensity of 1 candela. Similarly, the units of angular velocity and angular acceleration are given in the table as radian per second (rad/s) and radian per second squared (rad/s^2), respectively, rather than simply per second (1/s) and per second squared (1/s^2).

16. When full names are used, though, care must be taken to avoid confusion. For example, the expression "10 centimeters cubed" will be interpreted by some as $10\,\text{cm}^3$, but by others as $(10\,\text{cm})^3 = 1000\,\text{cm}^3$; whereas there is no ambiguity in "10 cubic centimeters".

alone, or by a prefix applied to the numeral "1" (e.g., the pure number 10^3 should not be written "k" or "k1" or "10 h" or "10 h1").[17]

Some other style conventions: Regardless of whether full names or abbreviations are used, no space or hyphen is placed between a prefix and a unit (e.g., 8 millimoles—i.e., 8 mmol—would not be written "8 milli moles" or "8 milli-moles" or "8 m mol" or "8 m-mol"). There are a few instances, though, in which the full name of the prefix is slightly modified; namely, write **megohm** rather than "megaohm", and **kilohm** rather than "kiloohm". Again, full names and abbreviations are never mixed (e.g., 3 picowatts—i.e., 3 pW—would never be written "3 pwatts" or "3 picoW"). Nor should one use multiple prefixes (e.g., using "mmg"—for "millimilligrams"—instead of "μg"). In particular, since the kilogram is the one SI base unit having a prefix in its name, the appropriate unit of mass to which *any* numeric prefix can be applied is actually the **gram (g)** (e.g., 10^9 kg = 1 Tg, not "1 Gkg"). On the other hand, the unit gram should not appear *without* a prefix (e.g., the approximate mass of an electron can be correctly stated in SI units as "9.1×10^{-31} kg", but not "9.1×10^{-28} g"). Hence, although numeric prefixes are normally avoided in the denominators of unit compounds (e.g., "5 kΩ/H" is generally preferred to "5 Ω/mH"), an exception is made for the kilogram (e.g., "2kcd/kg" is generally preferred to "2 cd/g").

From our introduction to prefixes, we now move to appropriate letter case, which is even more important (e.g., 1 Mm is 10^9 times 1 mm). Full names of prefixes begin with lowercase letters, except at the start of a sentence; whereas prefix abbreviations should always be written as shown in Table 12.3. Also, avoid nonstandard abbreviations, such as "sec" for "s" and "cc" (meaning "cubic centimeter") for "cm^3". Furthermore, although the terms "thousand" (10^3), "million" (10^6), "thousandth" (10^{-3}), and "millionth" (10^{-6}) are unambiguous, avoid the succeeding terms "billion", "trillion", "quadrillion", "quintillion", ..., and "billionth", "trillionth", "quadrillionth", "quintillionth", ..., because their meanings differ between the United States and most other countries.[18] Finally, be aware that outside of the SI, some of the prefixes in Table 12.3 may be defined differently; in particular, with regard to computer memory (usually measured in bits or bytes), "kilo", "mega", "giga", ..., typically represent the multipliers $1024 = 2^{10}$, $1024^2 = 2^{20}$, $1024^3 = 2^{30}$, ..., respectively.

12.1.3 Experimental Error

In the ideal Platonic realm of mathematical objects, there is perfect clarity. Geometric points having no extension can be individually identified; numbers that differ by any small amount can be distinguished; and so forth. But, making such precise identifications and distinctions in the real world is fraught with practical problems; because, the means by which physical measurements can be made are limited, as are the ways by which the results of those measurements can be efficiently recorded. Furthermore, the quality of the laboratory work performed can vary from one experimenter to another.

---

17. This convention is routinely violated in electronic circuit diagrams, though, where it has long been customary to drop the symbol "Ω" from all resistor values—e.g., writing "4.7 k" for 4.7 kΩ.

18. For example, "trillion" in the United States means the same as "billion" elsewhere, whereas "billion" in the United States means the same as "milliard" elsewhere.

The discrepancy between experimental data and physical reality is called **experimental error**. In this section, we specifically consider errors in measurement; that is, the data are raw. Later, in Subsection 12.2.2, we address the additional error that might result from processing data.

The experimental error in the recorded value of a measurement has several possible sources. First, the equipment used to make the measurement might lack **precision**; that is, if several measurements of the same physical quantity were made using a given set of equipment, then the resulting values would likely be somewhat diffused, rather than all being exactly the same.[19] Second, whether or not it is precise, the equipment might lack **accuracy**; that is, the average of these several values will likely differ somewhat from the true real-world value.[20] Third, the method used to record the measurement might lack **resolution**; that is, some equipment readings that are close, but different, might not be distinguishable after they have been recorded.[21] (An extreme example would be the recording of a length measurement, made with a typical ruler, to only the nearest inch.) Finally, **human error** can enter in many ways, including misoperation of equipment, poor reading of equipment outputs, and erroneous recording of equipment readings.

Equipment precision and accuracy, recording resolution, and human error all affect the accuracy of the *measurement*, which is how close the *one* recorded value is to the true value of the physical quantity being measured. (This is in contrast to *equipment* accuracy which, as stated above, is assessed by calculating an average of *several* measurements.) In other words, an accurate measurement is a measurement for which the experimental error is small.

As an example, consider making a paper map of a city. Since the surface of the earth in the vicinity of the city is only approximately flat, mapping it onto a plane is necessarily inaccurate; but, this mismatch between mathematical representation and physical reality is a modeling error, not experimental error. The equipment used to measure the city may include a handheld magnetic compass, which we suppose has not been calibrated to account for the difference between magnetic north and true north; thus, the compass is only approximately accurate. Furthermore, vibrations transmitted to the compass by the person holding it will degrade its precision. Also, this person might commit the human error of not holding the compass level, which could lessen the accuracy of the device. Additional experimental error might then be introduced by simply recording each compass reading as one of the sixteen standard directions—i.e., north (N), north-northeast (NNE), northeast (NE), east-northeast (ENE), east (E), . . . , south (S), . . . , west (W), . . . , north-northwest (NNW)—which effectively constitute a numeral system

19. A common contribution to equipment imprecision is its internal randomness, called "noise".

20. Thus, the accuracy of a piece of equipment having a numerical output, can be assessed by making several measurements of the same physical quantity, then comparing the *mean* value of the measurements to the true physical value being measured; whereas its precision can be assessed by calculating the *standard deviation* of these measurements. This process would be repeated for several values of the physical feature being measured, over the full range of the instrument.

21. The word "precision" is often used in place of "resolution", especially in purely mathematical (i.e., nonexperimental) contexts. This alternate wording could have been used here as well, in which case we would distinguish between equipment precision and "recording precision". However now, in addition to equipment precision, we are able to speak of "equipment resolution" (see Subsection 12.1.4), as distinguished from the recording resolution just described.

for angles (viz., 0°, 22.5°, 45.0°, 67.5°, 90.0°, ..., 180.0°, ..., 270.0°, ..., 337.5°, clockwise from vertical). This method of expressing the compass output, being rather coarse, lacks resolution.

In particular, note that the output of the compass (presumably indicated by a freely rotating magnetic pointer) is a continuously variable angle, whereas the compass measurements are recorded in a form that allows for variation in discrete steps only. In general, the process of converting a continuous variable into a discrete variable is called **quantization** (or **discretization**). Such a conversion necessarily occurs whenever the value of a continuous variable is expressed in an ordinary language (e.g., English); because, such a language—in which each sentence comprises finitely many words taken from a finite vocabulary—is itself a discrete medium. The significance of quantization is that, regardless of how finely it is done, the resulting discrete variable is unable to take on all the possible values of the original continuous variable; therefore, there is the potential (indeed, likelihood) for the introduction of an error, called **quantization error**. In the above example, the quantization error introduced upon recording the compass output contributes to the experimental error of each compass measurement.

In general, the particular physical quantity being measured is called the **measurand**. Thus, the goal of a measurement is to record, as closely as possible, the true value of a specific measurand. However, as we have seen, the act of measuring effectively adds experimental error to this value:

$$\text{(experimental value)} = \text{(true value)} + \text{(experimental error)}. \tag{12.4}$$

Furthermore, experimental error can be split into two components:

$$\text{(experimental error)} = \text{(systematic error)} + \text{(random error)}. \tag{12.5}$$

Conceptually, the **systematic error** in a measurement is the experimental error that would remain if all of the randomness in the experiment were somehow "averaged out"; whereas the **random error** in a measurement is the portion of the experimental error due to the randomness alone. To calculate systematic error, the same experiment is performed infinitely many times (under the same conditions); then, the average value of these measurements is calculated (as a limit), from which the true value of the measurand is subtracted, since averaging (12.4) yields

$$\text{(average experimental value)} = \text{(true value)} + \text{(systematic error)}.$$

The resulting value is then the systematic error for *all* of the infinitely many measurements performed, as well as for any other measurement that might be performed thereafter via another instance of the same experiment.[22] Also, per (12.5), subtracting this *constant* value from the experimental error of each such measurement yields the random error for *that* measurement. Thus, in contrast to systematic error, random error can *vary* from one measurement to another.

22. Of course, the procedure just outlined for determining systematic error is purely theoretical, since it is impractical to perform infinitely many experiments. Moreover, the subtraction requires that the true value of the measurand be known; but, if that were the case, then the measurement would not be necessary!

Upon relating these two types of experimental error back to the above sources of error, it is clear that lack of equipment precision contributes to random error, and lack of equipment accuracy contributes to systematic error. Categorizing recording resolution, though, is a bit trickier. Assuming that equipment outputs are always recorded exactly the same way every time, there is no randomness involved in recording each measurement; rather, the quantization error that is introduced *depends* on the output value.[23] However, when many different measurands are measured, the quantization error usually *appears* to vary randomly from one measurement to the next. Hence, although quantization is actually a deterministic process, it is natural to view it as being random.[24] Finally, human error can be either systematic (e.g., the experimenter sloppily writes the numeric prefix "µ" so that it is consistently read later as "m") or random (e.g., the experimenter errs as a result of an interruption).[25]

As for quantifying experimental error, let x_{exp} be the experimental value recorded from a measurement, and let x_{phy} be the true physical value of the corresponding measurand. The **(absolute) error**[26] of the measurement is then defined as

$$x_{\text{err}} \triangleq x_{\text{exp}} - x_{\text{phy}} \tag{12.6}$$

—hence, $x_{\text{exp}} = x_{\text{phy}} + x_{\text{err}}$, per (12.4). Also, it is sometimes appropriate to compare this error with the true physical value; for this purpose, we define the **relative error**

$$x_{\text{rel-err}} \triangleq \frac{x_{\text{err}}}{x_{\text{phy}}} = \frac{x_{\text{exp}} - x_{\text{phy}}}{x_{\text{phy}}} = \frac{x_{\text{exp}}}{x_{\text{phy}}} - 1 \tag{12.7}$$

(for $x_{\text{phy}} \neq 0$), perhaps multiplied by 100% to convert to a percentage. (Thus, e.g., if a true value $x_{\text{phy}} = \pi$ were measured as $x_{\text{exp}} = 3.14$, then the absolute error would be $x_{\text{err}} = 3.14 - \pi = -0.00159\ldots$, which relative to x_{phy} differs by $x_{\text{rel-err}} = (x_{\text{err}}/x_{\text{phy}}) \cdot 100\% \approx -0.0507\%$.[27]) In fact, however, we do not have direct access to the value x_{phy}; therefore, we can never knowingly perform the calculations (12.6) and (12.7). Rather, in practice, experimental error is calculated with respect to an *assumed* value for x_{phy}, such as a reference value generated by other experiments, or a theoretical value generated by inserting experimental values of *other* measurands into a particular scientific theory.[28]

---

23. As an example, a simple quantizer is provided by the floor function (see Problem 5.11); namely, $\lfloor x \rfloor$ can be viewed as a quantized version of a real variable x. Here, the quantization error is $\lfloor x \rfloor - x$, which is completely determined by the quantizer input x.

24. For example, in digital signal processing (DSP), quantization is deliberately performed in order to digitally process an analog signal; and, the resulting quantization error is typically analyzed as a random process called "quantization noise".

25. Some authors prefer to treat human error (which should be rare) separately, categorizing it as neither systematic nor random.

26. Note that the optional modifier "absolute" does not imply that an absolute value is taken; rather, its purpose is to emphasize that this is not a "relative" error (see below).

27. Note that the signs of errors are not discarded.

28. The latter situation reflects a chicken-and-egg quandary from which there is no escape: sometimes a scientific theory is used to judge the quality of an experiment, and sometimes an experiment is used to judge the quality of a scientific theory.

Measurements of energy and power are often expressed in **decibel (dB)** form; namely, for such a quantity $x > 0$, the value

$$x|_{\mathrm{dB}} \triangleq 10\log_{10}\frac{x}{x_{\mathrm{ref}}} \quad (\mathrm{dB}) \tag{12.8}$$

expresses x in decibels relative to a chosen reference value $x_{\mathrm{ref}} > 0$ (which must have the same physical dimension as x so the argument of the logarithm is dimensionless).[29] So, rather than measuring $x_{\mathrm{exp}} > 0$ directly, suppose we measure its decibel value $x_{\mathrm{exp}}|_{\mathrm{dB}}$; then, motivated by (12.6), we might calculate

$$x_{\mathrm{exp}}|_{\mathrm{dB}} - x_{\mathrm{phy}}|_{\mathrm{dB}} = 10\log_{10}\frac{x_{\mathrm{exp}}}{x_{\mathrm{ref}}} - 10\log_{10}\frac{x_{\mathrm{phy}}}{x_{\mathrm{ref}}} = 10\log_{10}\frac{x_{\mathrm{exp}}}{x_{\mathrm{phy}}} \quad (\mathrm{dB}) \tag{12.9}$$

(again, assuming some value for x_{phy}, now positive). Thus, this difference is independent of the reference x_{ref}, and effectively expresses x_{exp} relative to x_{phy} (cf. (12.8)). Moreover, $x_{\mathrm{exp}}|_{\mathrm{dB}} - x_{\mathrm{phy}}|_{\mathrm{dB}} = 0$ dB if and only if $x_{\mathrm{exp}} = x_{\mathrm{phy}}$.

However, we do *not* proceed to mirror (12.7) and attempt to convert this difference into a relative error by calculating

$$\frac{x_{\mathrm{exp}}|_{\mathrm{dB}} - x_{\mathrm{phy}}|_{\mathrm{dB}}}{x_{\mathrm{phy}}|_{\mathrm{dB}}}.$$

Rather, it follows from (12.7) and (12.9) that

$$x_{\mathrm{exp}}|_{\mathrm{dB}} - x_{\mathrm{phy}}|_{\mathrm{dB}} = 10\log_{10}(1 + x_{\mathrm{rel\text{-}err}}) \quad (\mathrm{dB}), \tag{12.10}$$

which shows how to convert between the decibel-difference $x_{\mathrm{exp}}|_{\mathrm{dB}} - x_{\mathrm{phy}}|_{\mathrm{dB}}$ and the relative error $x_{\mathrm{rel\text{-}err}}$ (with the conversion from the latter to the former requiring $x_{\mathrm{rel\text{-}err}} > -1$, so the logarithm exists). Hence, the difference $x_{\mathrm{exp}}|_{\mathrm{dB}} - x_{\mathrm{phy}}|_{\mathrm{dB}}$ is *itself* a way of expressing the relative error $x_{\mathrm{rel\text{-}err}}$, but now in decibels.[30] For example, $x_{\mathrm{rel\text{-}err}} = 1\%$ corresponds to $x_{\mathrm{exp}}|_{\mathrm{dB}} - x_{\mathrm{phy}}|_{\mathrm{dB}} = 10\log_{10}(1+1\%) \approx 0.043$ dB; whereas $x_{\mathrm{exp}}|_{\mathrm{dB}} - x_{\mathrm{phy}}|_{\mathrm{dB}} = 1$ dB corresponds to $x_{\mathrm{rel\text{-}err}} = 10^{1/10} - 1 \approx 25.9\%$.

---

29. As its name implies, a *deci*bel is one-tenth of a **bel (B)**; thus, the factor of 10 in (12.8) converts from x in bels—i.e., $x|_{\mathrm{B}} = \log_{10}(x/x_{\mathrm{ref}})$ (B)—to decibels. (For consistency, decibel measures of electrical voltage and current are instead defined with a factor of 20 in (12.8), since the power dissipated by a resistor is proportional to the *square* of either the voltage across it or the current through it.) Though discouraged by the SI, the reference value x_{ref}—which is the value of x corresponding to 0 dB—is sometimes indicated by appending appropriate symbols to "dB" (e.g., "dBm" means "decibels relative to 1 milliwatt").

30. Note that, in contrast to (12.8), the relative error $x_{\mathrm{rel\text{-}err}}$ is not converted into decibels by calculating a logarithm of $x_{\mathrm{rel\text{-}err}}$ divided by some reference value. Also, there are many other ways (e.g., a percentage) to express relative quantities. For example, impurities (which are usually very small) are sometimes expressed in **parts per million (ppm)**: x ppm $\triangleq x/10^6 = x \times 10^{-4}\,\%$. This relative measure, though, is not recognized by the SI.

12.1.4 Experimental Uncertainty

For a measurement to be deemed noteworthy, its quality must be assessed. This is usually done by calculating a quantitative bound on the corresponding experimental error. Specifically, upon carefully analyzing how the various sources of error affect the measurement, the experimenter asserts a lower bound x_{lb} and an upper bound x_{ub} on the true physical value x_{phy}. That is, what is actually obtained from a complete measurement is an *interval* of possible values for x_{phy}—viz.,

$$x_{\text{lb}} \leq x_{\text{phy}} \leq x_{\text{ub}} \tag{12.11}$$

—rather than just a single experimental value x_{exp}.[31] This interval is commonly recorded in the form

$$x_{\text{phy}} = x_{\text{exp}} \pm x_{\text{unc}}, \tag{12.12}$$

where $x_{\text{exp}} = (x_{\text{ub}} + x_{\text{lb}})/2$ is the midpoint of the interval, with the $\pm$ variation $x_{\text{unc}} = (x_{\text{ub}} - x_{\text{lb}})/2 > 0$ being the **(absolute) uncertainty** of the measurement. Hence, in an experimental context, equation (12.12) is understood to mean

$$x_{\text{exp}} - x_{\text{unc}} \leq x_{\text{phy}} \leq x_{\text{exp}} + x_{\text{unc}}$$

—not "$x_{\text{phy}} = x_{\text{exp}} - x_{\text{unc}}$ or $x_{\text{phy}} = x_{\text{exp}} + x_{\text{unc}}$", as would be the normal interpretation in a purely mathematical context. It then follows that the experimental error $x_{\text{err}} = x_{\text{exp}} - x_{\text{phy}}$ satisfies the bound

$$|x_{\text{err}}| \leq x_{\text{unc}}, \tag{12.13}$$

with equality being possible; thus, x_{unc} may also be called the **worst-case error**.

Usually, the uncertainty x_{unc} is significant only in comparison with the magnitude of the measured value x_{exp} itself. Accordingly, we define the **relative uncertainty** of the measurement as

$$x_{\text{rel-unc}} \triangleq \frac{x_{\text{unc}}}{|x_{\text{exp}}|} \tag{12.14}$$

(for $x_{\text{exp}} \neq 0$), perhaps multiplied by 100% to convert to a percentage.[32] Then, equivalent to (12.12), one can instead write

$$x_{\text{phy}} = x_{\text{exp}} \pm P\% \tag{12.15}$$

where $P = 100 x_{\text{rel-unc}}$.

---

31. In fact, a thorough assessment of experimental error is often quite involved, owing to the inherent randomness associated with any measurement. So, in addition to providing an interval of values as in (12.11), the experimenter calculates a *confidence* (e.g., 95% likely) that x_{phy} lies within the interval. A discussion of the statistical aspects of experimentation, however, is beyond the scope of this book.

32. In contrast to (12.7), here "relative" means in comparison to $|x_{\text{exp}}|$, not x_{phy} or $|x_{\text{phy}}|$. Also, note that relative error and relative uncertainty are both dimensionless quantities.

For example, suppose that the distance from the earth to the moon is measured as 3.8×10^8 meters, with a relative uncertainty of 5%. Then, this measurement can be recorded as either

$$3.8 \times 10^8 \text{ m} \pm 5\% \tag{12.16a}$$

or

$$(3.80 \pm 0.19) \times 10^8 \text{ m} \tag{12.16b}$$

—the latter form being preferable to

$$3.8 \times 10^8 \pm 1.9 \times 10^7 \text{ m},$$

which, though mathematically correct, is less readable. More concisely, (12.16b) can be written

$$3.80(19) \times 10^8 \text{ m}. \tag{12.16c}$$

That is, for any $x_{\text{exp}} \geq 0$, the exponential notation (12.2a) may be modified to

$$x_{\text{exp}} = x_n \ldots x_1 x_0.x_{-1}x_{-2} \ldots x_{-m}(u_k \ldots u_1 u_0) \times 10^p \tag{12.17a}$$

—for some additional digits $u_0, u_1, \ldots, u_k$ $(k \geq 0)$—thereby asserting

$$x_{\text{exp}} = \left[\sum_{i=-m}^{n} x_i \cdot 10^i \pm \left(\sum_{i=0}^{k} u_i \cdot 10^i \right) \times 10^{-m} \right] \times 10^p; \tag{12.17b}$$

and, for $x_{\text{exp}} < 0$, a minus sign would be applied to both the right side of (12.17a) and the first sum in (12.17b).[33]

Owing to quantization, the mere act of recording a measurement, even if done well, can contribute to the experimental uncertainty x_{unc}. In general, when a physical quantity x_{phy} is measured, an equipment output x_{out} is read by the experimenter and recorded as an experimental value x_{exp}, which is typically expressed in exponential notation (12.2a). But, from (12.2b) we see that given the exponent p and the number m of decimal places, this representation of x_{exp} allows x_{out} to be expressed only to the nearest multiple of 10^{-m+p}.[34] It follows that if the unknown quantity x_{phy} has a continuum of possible values, then somewhere in the chain of converting x_{phy} to x_{out} to x_{exp}, a quantization must occur. If the measurement is made with an **analog meter**—i.e., the displayed value can vary continuously—then recording the output in exponential notation causes a quantization to occur in converting x_{out} to x_{exp}. Whereas for a **digital meter**—i.e., the display (typically a finite decimal expansion) varies in discrete steps—the meter itself

33. Note that if the zero digit in "3.80" is removed, then the meaning of (12.16c) changes; but, not so for (12.16b). Subsection 12.1.5 explains why this digit is appropriate in (12.16b), and might be included in (12.16a) as well.

34. When m and p are allowed to freely vary, though, any rational number can be exactly represented in the form (12.2a), and thus any real number can be approximated arbitrarily well.

has limited resolution; thus, a quantization occurs in converting x_{phy} to x_{out}, the error of which is subsumed into the stated accuracy the meter. Additionally, the output x_{out} of a digital meter might be recorded as x_{exp} using a larger step size, thereby incurring a *coarser* quantization; that is, the recording resolution would be poorer than the equipment resolution.

In any event, the quantization error $x_{\text{q-err}} = x_{\text{exp}} - x_{\text{out}}$ that results from recording x_{out} as x_{exp} must lie somewhere in the interval

$$-10^{-m+p}/2 \leq x_{\text{q-err}} \leq 10^{-m+p}/2;$$

that is,

$$|x_{\text{q-err}}| \leq 10^{-m+p}/2.$$

In other words, the measurement has recording resolution

$$x_{\text{res}} = 10^{-m+p}/2, \tag{12.18}$$

and

$$|x_{\text{q-err}}| \leq x_{\text{res}} \leq x_{\text{unc}}, \tag{12.19}$$

the first inequality being analogous to (12.13).[35] Also, akin to (12.14), we define the **relative resolution** of the measurement as

$$x_{\text{rel-res}} \triangleq \frac{x_{\text{res}}}{|x_{\text{exp}}|} \tag{12.20}$$

(for $x_{\text{exp}} \neq 0$), perhaps multiplied by 100% to convert to a percentage.

Of course, a numerical datum need not be expressed in a decimal form for the recording resolution to be quantifiable; rather, any uniform quantization of a real variable x using steps of size $x_{\text{step}} > 0$ corresponds to a resolution of $x_{\text{step}}/2$. For example, the compass measurements considered in Subsection 12.1.3, which were recorded using the sixteen standard directions (N, NNE, NE, ..., NNW), effectively quantize a full revolution of 360° using steps of size $360°/16 = 22.5°$; therefore, each such measurement has an uncertainty of at least $22.5°/2 = 11.25°$.

12.1.5 Significant Digits

When experimental data are written in exponential notation, it is customary not to use any more digits than necessary; for not only does this conserve on writing and improve readability, but often the very way that a quantity is expressed implicitly conveys an estimate of its uncertainty. Broadly stated, a generally accepted "rule" for expressing a measurement x_{exp} by a numeral in a chosen numeral system is this:

---

35. Actually, it would be more appropriate (but less convenient) to refer to x_{res} as the *lack* of resolution—i.e., the *ir*resolution—since a larger value x_{res} corresponds to a poorer resolution. Hence, when speaking of resolution, we will avoid using quantitative words such "higher", "lower", "increase", and "decrease", instead using qualitative words such as "better", "poorer", "improve", and "degrade".

The numeral should be as compact as possible such that the recording resolution x_{res} does not substantially contribute to the uncertainty x_{unc} of the measurement; that is, $x_{\text{res}} \ll x_{\text{unc}}$.[36]

In this way, the numeral recorded for x_{exp} efficiently expresses a good approximation of the corresponding output value x_{out} read from the equipment, while also providing a rough idea of the experimental uncertainty x_{unc}; because, not only do we know $x_{\text{res}} \ll x_{\text{unc}}$, we also know that this relation fails for any less-compact numeral (in the same numeral system) representing approximately the same value. As for recording x_{exp} in exponential notation, one way of assessing the compactness of the numeral, as well as quickly estimating its resolution, is to count the number of **significant digits** (or **significant figures**) it has, as will be explained forthwith.

Unfortunately (though not surprisingly), not everyone agrees on how to precisely interpret the much-less-than symbol "$\ll$" in the above rule.[37] Part of the disagreement merely reflects differences in subjective judgment; however, it is also true that no single convention regarding significant digits is appropriate for all circumstances. Accordingly, while everyone should always invoke an appropriate convention for the situation at hand, they should also recognize that other experimenters might use different conventions to report *their* results.

One thing on which there *is* general agreement is how to *read* an experimental value $x_{\text{exp}} \neq 0$ expressed in exponential notation (12.2a) and determine which of its digits are "significant".[38] Namely, reading the numeral from left to right:

- All digits (including any zeros) from the first nonzero digit x_j through the last nonzero digit x_k $(-m \leq k \leq j \leq n)$ are significant.
- None of the digits before x_j—i.e., leading zeros—are significant.
- If a decimal point is present, then all digits after x_k—i.e., trailing zeros—are significant.[39]
- If a decimal point is not present, then the first trailing zero *might* be significant, in which case the next digit might be significant, and so forth; however, if any of these digits is deemed insignificant, then so are all the digits that follow.

---

36. Besides this relation between x_{res} and x_{unc}, there may be other considerations regarding the "compact as possible" criterion, especially when several numbers are written together. For instance, a column of numbers expressed in a decimal notation might be neatly written with all of the numerals having the same number of decimal places, which could cause some of them to be less compact than would have been true if the numbers had been written individually. For example, the column might contain "1.234×10^7" and "0.089×10^7" (perhaps with the exponential part "$\times 10^7$" asserted at the top of the column for all of its entries), even though the latter number could be rewritten with two fewer digits as "8.9×10^5".

37. As evidence of this fact, just look at the *title* of [9]. (The content is telling as well.)

38. Not much is written, though, about which digits are significant in a numeral representing zero—i.e.,"0" or "0.00 . . . 0", perhaps with "$\times 10^p$" appended. It is safe to assume, though, that the rightmost digit of such a numeral is significant; moreover, for any digit that is insignificant (by whatever criteria), so are all the digits to its left.

39. This statement remains valid should one allow a decimal point to be present with *no* digits following; e.g., "123.", "120.", and "100." all have three significant digits. However, this practice of ending a numeral with a decimal point does not seem to be widespread, perhaps because of the risk of misinterpreting the period at the end of a sentence.

Thus, these criteria are somewhat indefinite regarding the significance of trailing zeros when the numeral has no decimal point; accordingly, such a numeral must be accompanied by additional information if the reader is to know which of the trailing zeros are significant. On the other hand, when the numeral has a decimal point, the significant digits are simply those from the first nonzero digit forward; therefore, for *scientific* notation (12.3), *all* digits are significant.

Some examples: The numerals "0.0102", "0.102", "1.02", "10.2", and "102" all have three significant digits (in each case, from "1" forward). Whereas the numerals "0.01020", "0.1020", "1.020", "10.20", and "102.0" all have four significant digits (again, from "1" forward). Also, prefixing a minus sign to any of these numerals, and/or appending "$\times 10^p$", leaves the significant digits unchanged. However, all we can say about "1020", "10200", and "102000" is that each numeral has *at least* 3 significant digits (from the portion "102"), perhaps 4 (from "1020"), or 5 in "10200" and "102000" (from "10200"), or all 6 in "102000"; and, likewise upon prefixing a minus sign and/or appending "$\times 10^p$". Therefore, it might be better to express these numbers in another form—e.g., rewriting "1020" in scientific notation as either "1.02×10^3" or "1.020×10^3"—to make definite the intended significant digits.

When one is discussing a numeral representing an experimental value, the terms "most significant digit" and "least significant digit" are not used as defined in Subsection 3.1.1. Rather, the **most significant digit** is the leftmost significant digit (not just the leftmost digit), and the **least significant digit** is the rightmost significant digit (not just the rightmost digit). The reason is that from a purely mathematical standpoint, all digits are significant; whereas in the present context, "significance" is a threshold that some digits achieve, and only for those do we consider degrees thereof.

To see how significant digits relate to resolution, suppose that a value $x_{\text{exp}} \neq 0$ is expressed in exponential notation (12.2a) using $d \geq 1$ significant digits, which we assume includes any trailing zeros (e.g., a decimal point is present). Then, $|x_{\text{exp}}|$ can be as small as $10^{-m+d-1+p}$ (when $x_{-m+d-1} = 1$ and all other digits are zero) or as large as $10^{-m+p}(10^d - 1) = 10^{-m+d+p}(1 - 10^{-d})$ (when $x_i = 9$ for $-m \leq i \leq -m+d-1$, and all other digits are zero). Therefore, it follows from (12.18) and (12.20) that the relative resolution $x_{\text{rel-res}}$ of this numeral is bounded as

$$\frac{0.5 \cdot 10^{-d}}{1 - 10^{-d}} \leq x_{\text{rel-res}} \leq 5 \cdot 10^{-d} \tag{12.21a}$$

(the left side being slightly larger than its numerator); moreover, if all we are told about the numeral is the value of d (in particular, the value of x_{exp} is not provided), then no tighter bounds on $x_{\text{rel-res}}$ can be asserted. Roughly, then,

$$x_{\text{rel-res}} \approx 10^{-d}, \tag{12.21b}$$

and so we may say that x_{exp} is specified to d **digits of resolution**.

There is also general agreement about the process of **rounding** a numeral in exponential notation to fewer significant digits. To wit, assume that the numeral in (12.2a) expressing $x_{\text{exp}} \neq 0$ has $d \geq 2$ significant digits; and, suppose we wish to round it to d' significant digits with $1 \leq d' < d$. First, let x_j be the leftmost nonzero digit, and consider the d'th digit $x_{j-d'}$ to its right; then, if

(i) $x_{j-d'} > 5$, or
(ii) $x_{j-d'} = 5$ and either at least one nonzero digit follows or the preceding digit $x_{j-d'+1}$ is odd,

add $10^{j-d'+1+p}$ to $|x_{\text{exp}}|$ (loosely put, add 1 to $x_{j-d'+1}$). Next, regarding the digits that are d' or more positions to the right of the *present* leftmost nonzero digit,[40] remove those lying to the right of the decimal point (if there is one), and replace each of the others with "0". Finally, the decimal point may be removed if there are now no digits to its right, and must be removed if any digits to its left are insignificant.

Some examples: Successively rounding to one fewer significant digit, we convert the numeral "246.850" to "246.85" to "246.8" to "247" to "250" to "200". And, we convert "0.097531" to "0.09753" to "0.0975" to "0.098" to "0.1"—the last conversion being in contrast to rounding "0.098" to one fewer *decimal place* as "0.10". Also, prefixing a minus sign to any of these numerals, or appending "$\times 10^p$", does the same to its rounded result.

If x_{exp}' is the value that results from rounding a numeral for x_{exp}, then the corresponding **roundoff error** is the difference $x_{\text{exp}}' - x_{\text{exp}}$. We say that x_{exp} has been **rounded up** (respectively, **rounded down**) to x_{exp}' when the latter is larger (smaller); equivalently, the roundoff error is positive (negative). The rationale behind condition (ii) in the above rounding procedure is that when $x_{j-d'} = 5$ is followed by only zeros (or no digits at all), the number x_{exp} lies exactly midway between two multiples of $10^{j-d'+1+p}$; therefore, by always rounding this borderline case to make the new digit for $x_{j-d'}$ even, it is expected that over several roundings, numerals will be rounded up and down about equally often.[41] It is for this borderline case that the magnitude of the roundoff error takes on its worst-case value of $10^{j-d'+1+p}/2$.

Note that performing successive roundings on a given numeral to obtain a particular number of significant digits can produce different results depending on how many digits are rounded at a time. For example, rounding to one fewer significant digit each time converts "0.851" to "0.85" to "0.8", the last numeral having one significant digit; whereas rounding "0.851" immediately to one significant digit yields "0.9". To minimize roundoff error, a *single* rounding to the desired number of significant digits should be performed whenever possible.

Regarding how to *write* an experimental value with an appropriate number of significant digits, consider the measurements depicted in Figure 12.1. There, a ruler is used to measure the distance from the center of a circle to the center of each of three polygons; thus, the measuring device has a continuous output that must be quantized upon recording its value in a decimal form. As shown, the scale of the ruler is divided down to tenths of centimeters, but no further; nevertheless, it is reasonable for the person making the measurement to try to read the ruler to the nearest tenth of a division, whereas attempting

40. As shown by the last example in the next paragraph, if the addition just mentioned is performed, then the position of the leftmost nonzero might move. This is why the first step in the rounding process is not simply to truncate the numeral to the desired number of significant digits; because, another truncation might be required afterward.

41. An alternative would to decide in a truly random way—e.g., by flipping a coin—whether to add $10^{j-d'+1+p}$ to $|x_{\text{exp}}|$.

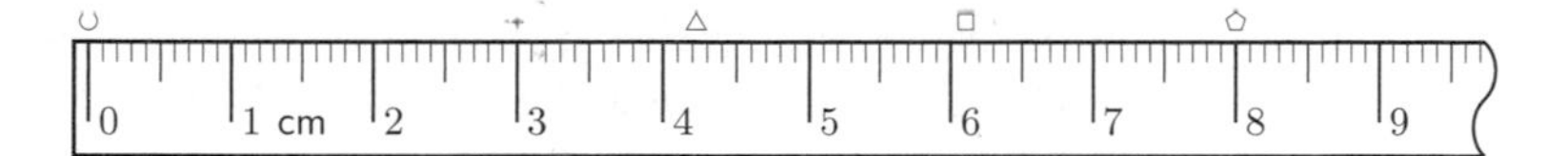

FIGURE 12.1 Some Distance Measurements

any finer resolution would be folly. Accordingly, the following measurements are made, each recorded to the nearest 0.01 cm:

From circle to ...	Distance (cm)
Triangle	4.23
Square	6.10
Pentagon	8.00

All things considered (e.g., the width of the ruler tick marks) though, the person probably cannot say with confidence that the ruler was correctly read to the nearest 0.01 cm—i.e., to within $\pm(0.01 \text{ cm})/2 = \pm 0.005$ cm—but rather within some wider range such as ± 0.01 cm; that is, the uncertainty in each ruler reading is more likely 0.01 cm.

In general, when a continuously variable equipment output is recorded in a decimal form, it is often recommended that all but the least significant digit be certain; and, *that* digit (the "uncertain digit") should be estimated. (Hence, a digit need not be certain to be significant.) So stated, though, this rule is slightly flawed. For example, since the last measurement was read within ± 0.01 cm of 8.00 cm—i.e., the ruler output was determined to lie somewhere from 7.99 to 8.01 cm—*none* of the digits in this measurement are certain. Thus, what is really intended is that the uncertainty of a reading be on the order of a variation in the least significant digit (which, as just shown, might carry over to affect more significant digits).[42]

We now see the purpose for recording trailing zeros beyond the decimal point in a finite decimal expansion—and, more generally, in exponential notation. For although writing an experimental value as, e.g., "6.10" rather than "6.1", or "8.00" rather than "8.0" or "8", has no effect on the value conveyed, the better resolution obtained from having more significant digits implies a smaller uncertainty in the stated value. Thus, two numerals that express exactly the same number in mathematical theory need not be equivalent in experimental practice.

It should be emphasized that at this point in the measurement, the uncertainty under consideration is only that in *reading* the equipment; it does not include other sources of experimental error such as equipment inaccuracy and imprecision, which also contribute to the overall uncertainty x_{unc} in a measured value x_{exp}. For example, a complete

42. Even if the criterion for "on the order of" were made completely unambiguous (a pedantic exercise to many), some subjective judgment would still be required. For instance, given a ruler having a scale divided down to sixteenths of an inch, a distance might be read to the nearest thirty-second of an inch as $4\frac{7}{32}$ inches. Therefore, expressing this value as a decimal expansion, it could be written "4.21875 inches"; but, since $\frac{1}{32} \approx 0.03$, this record conveys much greater certainty in the measurement than is the case. Yet, rounding the numeral to "4.2188" or "4.219" or "4.22" would deliberately introduce additional error, which may or may not be judged acceptable.

evaluation of the uncertainty in our ruler measurements would include determining the ruler's accuracy, estimating the error resulting from not placing the ruler with its origin (0) perfectly aligned with the center of the circle, as well as estimating the error from not placing the ruler parallel with the line that passes through the centers of the circles and the polygons. Suppose, then, that the cumulative effect of all such errors is determined to be, in magnitude, at most 2% of the reading plus 0.03 cm; then, this uncertainty is added to the reading uncertainty of 0.01 cm to obtain the measurement uncertainty x_{unc}. Thus, e.g., the ruler measurement of $x_{\text{exp}} = 4.23$ cm has uncertainty

$$x_{\text{unc}} = 0.01\ \text{cm} + 2\% \times 4.23\ \text{cm} + 0.03\ \text{cm} = 0.1246\ \text{cm}.$$

Accordingly, we *could* now write

$$x_{\text{exp}} = (4.23 \pm 0.1246)\ \text{cm}, \tag{12.22}$$

thereby expressing our confidence that the true physical value x_{phy} of the distance from the circle to the triangle lies in the interval $4.1054 \leq x_{\text{phy}} \leq 4.3546$. However, although (12.22) is numerically correct, writing an uncertainty with more than one or two significant digits is usually viewed as excessive. Some experimenters prefer to express all uncertainties to one significant digit; others use one significant digit unless it is "1", or perhaps "2", and then express the uncertainty to two significant digits; and, still others always use two significant digits. In any case, the uncertainty and the value to which it applies are typically expressed to the same number of decimal places. Thus, in the present example, we would write either

$$x_{\text{exp}} = (4.23 \pm 0.13)\ \text{cm} \tag{12.23a}$$

or

$$x_{\text{exp}} = (4.2 \pm 0.2)\ \text{cm}. \tag{12.23b}$$

In particular, note that although the experimental value 4.23 cm was appropriately rounded to 4.2 cm, the uncertainty 0.1246 cm was *increased* to either 0.13 cm or 0.2 cm, rather than simply being rounded down in each case as one might expect. The reason is that these new uncertainties have been chosen to be as small as possible, with the desired number of significant digits, such that the uncertainty intervals in (12.23) fully encompass that in (12.22).

In contrast to making a measurement with an analog device (as in Figure 12.1), there is no uncertainty in *reading* the output of a digital device, at least if the display is stable.[43] Thus, if a digital voltmeter displays "32.49 volts" for an electric potential x_{exp}, then this is exactly what should be recorded—i.e.,

$$x_{\text{exp}} = 32.49\ \text{V}$$

---

43. If the output of either type of device contains a random variation (indicating a lack of equipment precision), then a visual estimate of the average value may be recorded, perhaps including a qualitative and/or quantitative description of the variation.

(unless the displayed resolution is greater than that needed for later calculations, in which case one might round to fewer digits). Afterward, the experimental uncertainty x_{unc} in x_{exp}, now due solely to other sources of error, can be found. For example, if the voltmeter is accurate to within $\pm 0.5\%$ of the reading plus ± 2 in the last displayed digit, then

$$x_{\text{unc}} = 0.5\% \times 32.49\ \text{V} + 2 \times 0.01\ \text{V} = 0.18245\ \text{V},$$

and so we could write

$$x_{\text{exp}} = (32.49 \pm 0.19)\ \text{V}.$$

12.2 Data Manipulation

The work of an experimenter rarely consists only of making measurements. One reason is that not all physical quantities are directly measurable. Another is that what is of primary interest might not be the values of these quantities themselves, but rather correlations among them. Additionally, an experimenter must often communicate their findings to others. Accordingly, calculations must often be performed on raw data to obtain desired information, which is then reported in an easily assimilable form.

12.2.1 Dimensional Analysis

Table 12.4 gives some commonly used symbols for the dimensions corresponding to the SI base units. Taking these dimensions as fundamental, the dimension of any nonzero physical quantity Q can be expressed in the form

$$\dim Q = \mathsf{L}^{\alpha}\mathsf{M}^{\beta}\mathsf{T}^{\gamma}\mathsf{I}^{\delta}\Theta^{\varepsilon}\mathsf{N}^{\zeta}\mathsf{J}^{\eta}, \tag{12.24}$$

for unique integers α, β, γ, δ, ε, ζ, and η.[44] In particular, from Tables 12.1 and 12.4 we conclude that

$$\dim 1\ \text{m} = \mathsf{L}, \quad \dim 1\ \text{kg} = \mathsf{M}, \quad \dim 1\ \text{s} = \mathsf{T},$$

$$\dim 1\ \text{A} = \mathsf{I}, \quad \dim 1\ \text{K} = \Theta, \quad \dim 1\ \text{mol} = \mathsf{N}, \quad \dim 1\ \text{cd} = \mathsf{J};$$

TABLE 12.4 Symbols for the SI Base Dimensions

Dimension	Symbol
Length, distance	L
Mass	M
Time	T
Electric current	I
Thermodynamic temperature	Θ
Amount of substance	N
Luminous intensity	J

44. The right side of (12.24) is merely symbolic, since dimension symbols such as those in Table 12.4 are not intended to take on numerical values (though the symbols are often manipulated *as if* they were numbers).

and, using these facts along with the rules

$$\dim Q_1Q_2 = (\dim Q_1)(\dim Q_2),$$
$$\dim Q_1/Q_2 = (\dim Q_1)/(\dim Q_2)$$

for any physical quantities Q_1 and Q_2, one can determine how any derived dimension is formed from the fundamental dimensions. For example, since $1\text{ N} = 1\text{ kg·m/s}^2 = (1\text{ kg})(1\text{ m})/(1\text{ s})^2$, we have

$$\dim 1\text{ N} = (\dim 1\text{ kg})(\dim 1\text{ m})/(\dim 1\text{ s})^2 = \mathsf{LM/T^2} = \mathsf{LMT^{-2}}$$

—i.e., $\alpha = \beta = 1$, $\gamma = -2$, and $\delta = \varepsilon = \zeta = \eta = 0$ in (12.24). Thus, symbolizing the dimension force by "F", we can write

$$\mathsf{F} = \mathsf{LMT^{-2}}. \tag{12.25}$$

Likewise, other dimension symbols can be introduced, such as one for each of the derived SI units. Furthermore, if desired, one can depart from the SI and choose dimensions other than those in Table 12.4 to be fundamental; the only requirement is that one's fundamental dimensions be independent, in that none is expressible in terms of the others. For example, in physics, the theory of mechanics can be fully developed in terms of the dimensions length, mass, and time alone; but, one can instead take length, force, and time as fundamental, since they are independent and it follows from (12.25) that $\mathsf{M} = \mathsf{L^{-1}FT^2}$.

By (12.24), the dimension of any dimensionless physical quantity (i.e., $\alpha = \beta = \cdots = \eta = 0$) is simply 1. Therefore, scaling a physical quantity Q by a pure number $\kappa \neq 0$ does not affect its dimension, since $\dim \kappa Q = (\dim \kappa)(\dim Q) = 1 \cdot \dim Q = \dim Q$ (e.g., $\mathsf{L} = \dim 1\text{ m} = \dim 5.2\text{ m} = \dim 10^{-3}\text{ m} = \dim -7\text{ m}$).

For a mathematical expression involving physical quantities to be physically meaningful, it must be **dimensionally homogeneous**; specifically: Both sides of an equation or inequality must have the same physical dimension. (For example, in Newton's second law of motion, $F = ma$—i.e., force equals mass times acceleration—the dimension of the left side is $\mathsf{F} = \mathsf{LMT^{-2}}$, and that of the right is $\mathsf{M} \cdot \mathsf{L/T^2} = \mathsf{LMT^{-2}}$.) Whenever two physical quantities are added or subtracted, their dimensions must be the same, and that becomes the dimension of the result. Arguments of all exponential, trigonometric, and hyperbolic functions, as well as their inverses (including logarithms) must be dimensionless.

Note that the above rules are stated in terms of dimensions, not units; because, although the dimension associated with a given physical unit is unique, a given dimension (e.g., length) can be quantified in terms of many different units (inch, foot, meter, ...). Thus, the equation

$$1\text{ in} = 2.54\text{ cm} \tag{12.26}$$

is dimensionally homogeneous because both sides have dimension L, even though different units are used. Of course, it is possible for a relation about physical quantities to be acceptable dimensionally without being correct (as would be the case in the above equation if the "1" were changed to "3"); that is, dimensional homogeneity is a necessary, but not sufficient, condition for overall correctness.

Nevertheless, being aware of the physical dimensions entering into a calculation is often helpful toward performing it correctly. For example, one might remember that the frequency f and wavelength λ of a beam of light are related to its speed c, in that one of these three quantities equals the product of the other two. Accordingly, since $\dim f = 1/\mathsf{T}$, $\dim \lambda = \mathsf{L}$, and $\dim c = \mathsf{L}/\mathsf{T} = (1/\mathsf{T})(\mathsf{L})$, it must be that $c = f\lambda$. (The same result also obtains by thinking in terms of units specifically, rather than dimensions generally, by taking f in Hz = 1/s, λ in m, and c in m/s.) This kind of reasoning is much less fatiguing than pondering, "How would f change if λ were increased with c held fixed?"

Turning from dimensions to units, recall that we can view (12.26) as

$$1 \times \text{in} = 2.54 \ \times \text{cm},$$

which in turn can be viewed as

$$1 \times \text{in} = 2.54 \ \times (10^{-2} \times \text{m}).$$

That is, physical units (e.g., inch, meter), as well as the scaled units (e.g., centimeter), can be treated algebraically as nonzero mathematical constants symbolized by abbreviations ("in", "m", "cm")—subject to the rules for dimensional homogeneity. Accordingly, various relationships among the SI units, beyond those defined in Table 12.2, are easily obtained; e.g., $\text{V} = \text{J}/\text{C}$, since $\text{V} \triangleq \text{W}/\text{A} = (\text{J}/\text{s})/\text{A} = \text{J}/\text{A}{\cdot}\text{s} = \text{J}/\text{C}$. It thus follows from (12.26) that

$$1 = 2.54\,\text{cm}/\text{in},$$

by which the quantity on the right can be used as a **conversion factor** to convert any length expressed in inches to the same length in centimeters; e.g.,

$$5\text{ in} = (5\text{ in})(1) = (5\text{ in})(2.54\,\text{cm}/\text{in}) = (5 \cdot 2.54)\text{ cm} = 7.7\text{ cm}.$$

In general, for any two nondimensionless units u_1 and u_2 having the same dimension, there exists a unique conversion factor from u_1 to u_2; and, its reciprocal is the unique conversion factor from u_2 to u_1 (e.g., the factor $\frac{1}{2.54}$ in/cm converts from centimeters to inches). Moreover, since every conversion factor is equivalent to the pure number 1, we may multiply or divide by it anywhere in a numerical calculation, as desired. Thus, using the above conversion factor from inches to centimeters, as well as those from miles (mi) to feet (ft), feet to inches, centimeters to meters, and kilometers to meters, we can calculate the conversion factor from miles to kilometers:

$$\begin{aligned} 1 &= \frac{(1)(1)(1)(1)}{1} = \frac{(5280\,\text{ft}/\text{mi})(12\,\text{in}/\text{ft})(2.54\,\text{cm}/\text{in})(10^{-2}\,\text{m}/\text{cm})}{10^3\,\text{m}/\text{km}} \\ &= \frac{(5280)(12)(2.54)(10^{-2})}{10^3}\,\text{km}/\text{mi} = 1.609\,344\,\text{km}/\text{mi}. \end{aligned}$$

In particular, observe how the conversion factors are placed in either the numerator or the denominator to obtain the desired final unit (km/mi), with all the other units canceling each other out.

Special attention is required, however, when performing calculations involving *dimensionless* physical quantities—for various reasons. One is that some dimensionless units are equivalent to the pure number 1, and therefore are optional; that is, they may pop up or disappear at any time. In particular, both the radian and the steradian are assigned the value 1 in the SI; thus, the equation 6 rad $= 6$ is correct, even though at first glance it might appear to be dimensionally inhomogeneous.

As a more applied example, suppose we have a **waveform** (or **signal**)—i.e., a function $x(t)$ of time t—that is periodic; and, let T be its period: $x(t) = x(t+T)$ (all t). Thus, since T can be added to t, its dimension must also be time. By definition, the **frequency** f of a periodic phenomenon is the rate at which it repeats; so, introducing the dimensionless unit **cycle** to indicate any contiguous portion of the waveform $x(t)$ that has duration T, the frequency of $x(t)$ is the number of cycles that occur per unit of time:

$$f = \frac{1 \text{ cycle}}{T} \tag{12.27}$$

—e.g., if $T = 25$ ms, then

$$f = \frac{1 \text{ cycle}}{25 \text{ ms}} = 40 \text{ cycle/s}.$$

However, neither the unit cycle nor the unit cycle/s—i.e., **cycle per second (cps)**—is recognized by the SI. But, since the cycle is independent of the SI units (i.e., it cannot be expressed in terms of them), it can be appended to the SI as a dimensionless unit that is equivalent to the number 1; accordingly, it can be dropped from (12.27) to give

$$f = \frac{1}{T} \tag{12.28}$$

(as is commonly done). Thus, we now write

$$f = \frac{1}{25 \text{ ms}} = 40 \text{ s}^{-1} = 40 \text{ Hz},$$

which, though fully SI-compliant, no longer contains the unit cycle to remind us of what this frequency is about. Such a hint is often helpful, because the unit hertz can be used to express various other periodic rates.[45]

A small adjustment is usually necessary, though, when one is using a non-SI dimensionless unit that is related to the SI units.[46] For example, besides the radian, another

45. As another example, a common operation prior to performing DSP is *sampling*, whereby a waveform $x(t)$ is converted into a sequence of "samples" $x_k = x(kT_s)$ ($k = 0, \pm1, \pm2, \ldots$). The time spacing $T_s > 0$ between adjacent samples is called the "sample period", and thus its reciprocal $1/T_s$ is the "sample rate", akin to (12.28). One can now introduce a dimensionless unit *sample* equivalent to the number 1, and express the sample rate as (1 sample)$/T_s$, akin to (12.27). Thus, e.g., the sample rate for compact disc (CD) audio is 44.1 kHz—i.e., 44 100 samples per second.

46. By contrast, as illustrated by (12.26) for the dimension length, there is no problem in mixing SI units with non-SI units that are *non*dimensionless, assuming none of the non-SI units has the same name as an SI unit.

popular unit for quantifying plane angles is the degree (°), which is not in the SI. As we know from (10.14) and (10.15), 1 radian equals $\frac{180}{\pi}$ degrees; hence, $\frac{180°}{\pi \text{ rad}} = 1$ and $\frac{\pi \text{ rad}}{180°} = 1$ can be used as conversion factors between degrees and radians. But, the radian, being an SI dimensionless unit, is equivalent to the pure number 1; therefore, for consistency, when the degree is used with the SI, it must be equivalent to the pure number $\pi/180 \approx 0.0175$, not 1.[47] Likewise, the percent (%) can be viewed as a dimensionless unit of value $1/100$ (since, e.g., $7\% = 7 \times 1/100 = 0.07$). Accordingly, unlike the radian, the dimensionless units degree and percent cannot be inserted or removed at will (e.g., "rad" can be omitted from the above two conversion factors, but "°" cannot).

As a final example, imagine a point moving around a circle at an angular speed of ω radians per second, for some constant $\omega > 0$ (e.g., let $\theta = \omega t$ in Figure 11.3, with t being time in seconds). Such motion is clearly periodic, with exactly 1 cycle (i.e., traversing the circle once) occurring for every 2π rad of movement; hence, we now have

$$1 \text{ cycle} = 2\pi \text{ rad}. \tag{12.29}$$

Therefore, since rad = 1 in the SI, we now conclude that cycle = 2π (not 1, as above). Accordingly, in the present context, the dimensionless unit cycle cannot be dropped whenever it appears; yet, this is precisely what is routinely done in practice! For $\omega = 6$, e.g., the frequency of the periodic motion would be typically calculated as follows:

$$\frac{6 \text{ rad/s}}{2\pi \text{ rad/cycle}} = (3/\pi) \text{ cycle/s} \stackrel{?}{=} (3/\pi) \text{ s}^{-1} = (3/\pi) \text{ Hz}. \tag{12.30}$$

That is, in general, if the frequency of the motion is f hertz, then

$$\omega/2\pi = f. \tag{12.31}$$

The source of the confusion is that the word "cycle" is now being used for two distinctly different purposes: first as a generic name for a one-period portion of a periodic phenomenon, and second as the measure of a particular plane angle. The first meaning makes no numerical connection with the SI units; however, the second does, via (12.29). Therefore, to clarify the calculation (12.31) of f from ω, let us retain the first meaning of "cycle", which is dimensionless unit of value 1; and, for the second meaning, let us instead use the dimensionless unit "rotation", thus replacing (12.29) with

$$1 \text{ rotation} = 2\pi \text{ rad}. \tag{12.32}$$

Now, since there is 1 cycle per rotation, the calculation in (12.30) can be unambiguously performed as

$$(6 \text{ rad/s})\frac{1 \text{ cycle/rotation}}{2\pi \text{ rad/rotation}} = (3/\pi) \text{ cycle/s} \stackrel{\checkmark}{=} (3/\pi) \text{ s}^{-1} = (3/\pi) \text{ Hz}. \tag{12.33}$$

---

47. An alternative approach would be to require *all* dimensionless units to evaluate to 1, the rationale being that in any system of units there can be only one "true" unit for each quantifiable physical feature, just as there is only one unit—viz., 1 (unity)—among the real numbers. But then, although dimensionless units such as the degree could still be used along with the SI, it would no longer be consistent to refer to them as "units". (Nor could, e.g., the inch be consistently called a "unit", because the one "true" unit of length is the meter.)

Likewise, the general formula (12.31) relating **angular frequency** ω in radians per second and **cyclic frequency** f in hertz (or cycles per second) is indeed valid.

A careful examination of (12.33), though, might still produce some puzzlement. For we have already stated that rad $=$ 1 and cycle $=$ 1; therefore, (12.32) implies that rotation $= 2\pi$. Hence, the quantity 2π rad/rotation $=$ 1 can be viewed as a conversion factor (from the unit rotation to the unit radian), which we are free to multiply or divide by anywhere. But, 1 cycle/rotation $= 1/2\pi \neq 1$; so, what is the justification for inserting this factor into (12.33)? First, though cycle and rotation are both dimensionless units (and, in nontechnical contexts are often synonymous), they do not (as defined above) quantify the same physical feature.[48] Again, a cycle is a one-period portion of some periodic phenomenon, whereas a rotation is the angular measure of a complete circle. Therefore, the factor 1 cycle/rotation cannot be viewed as a *general* conversion factor; rather, it is *specifically* applicable to the present situation (viz., a point moving around a circle, with each rotation being identified as a cycle).[49] Accordingly, equation (12.31) is not merely a unit conversion from frequency in radians per second to frequency in hertz; rather, angular frequency and cyclic frequency quantify different physical features, which happen to be related in the present situation.

Summarizing: Nondimensionless quantities having different dimensions can never be directly compared (e.g., asked if they are equal), because that would violate dimensional homogeneity; however, dimensionless quantities can always be compared, as can nondimensionless quantities having the same dimension (even if expressed in different units). The only units that are optional—i.e., can be included or not when expressing a physical quantity—are dimensionless units that are equivalent to the number 1 (e.g., the radian). Attempting to use more than one dimensionless unit for the same physical feature (e.g., appending to the SI another plane-angle unit) is potentially problematic, since it might not be consistent to interpret all of them as the number 1. Also, keep in mind that not all dimensionless units pertain to the same physical feature (whereas all nondimensionless units having the same dimension do); and, be aware that units taken from outside the SI may have multiple meanings.

Furthermore, although dimensionless units of value 1 are optional, they often provide helpful information, and therefore should not be dropped simply because it is allowed (e.g., recall the definition of "lumen" in Table 12.2); however, such a unit should be included only when appropriate. In particular, it is not appropriate to substitute the unit hertz for every occurrence of an inverse second; rather, this substitution should be done only to express the frequency of some periodic phenomenon. For example, although the equation 8 rad/s $=$ 8 Hz is both dimensionally homogeneous and numerically valid, it

---

48. By contrast, when two quantities have *different* dimensions, one can immediately see that they do not quantify the same physical feature.

49. In some other situation, we might stipulate (for whatever reason) that *five* rotations constitutes a cycle. Or, as a completely different example, multiplying the length of the side of a square by itself to obtain the square's area is not a unit conversion (obviously, because of the change in dimensions); rather, this calculation is dictated by the particular geometry of a square.

is doubtful to be applicable in practice.[50] Similarly, it is very unlikely that one would ever have an occasion to write "$1\ \mathrm{N} = 1\ \mathrm{kg{\cdot}m{\cdot}Hz^2}$".

Finally, it is important to observe that there are two different ways to represent a physical quantity by a variable. To wit, consider these statements:

(1a) The height of the picture in meters is 2.

(1b) The height of the picture is 2 meters.

Though nearly identical, (1a) and (1b) are not quite semantically equivalent; the precise meanings given to the word "height" differ. This subtlety becomes more apparent when a variable is introduced:

(2a) Let h be the height of the picture in meters; then, $h = 2$.

(2b) Let h be the height of the picture; then, $h = 2$ meters.

In (2a), the height of the picture is expressed by first specifying the physical unit to be used, then giving a pure number to multiply that unit; and, the number alone is represented by a *mathematical* variable "h". In (2b), a number and a unit are provided simultaneously, and these two entities together are represented by a *physical* variable "h". Thus, it would be incorrect after asserting (2a) to say that $h = 2$ meters, since 2 meters is not a pure number; but, it would be perfectly fine to say $h = \sqrt{4}$. Likewise, it would incorrect after asserting (2b) to say that $h = 2$, since 2 is not a physical quantity having the dimension length; but, it would be perfectly fine to say $h = \sqrt{4 \text{ square meters}}$ or $h \approx 78.74$ inches.

Both of these locution methods—(a) mathematical variables and (b) physical variables—have been used in this chapter, and each has its own advantages. In particular, consider using them to express a mathematical formula relating physical quantities:

(3a) Let h and w be, respectively, the height and width of the picture in meters, and let a be its area in meters squared; then, $a = hw$.

(3b) Let h and w be, respectively, the height and width of the picture, and let a be its area; then, $a = hw$.

Clearly, (3b) is less cumbersome; moreover, it allows total freedom in choosing, at a later time, the physical units to be used (as long as they have the appropriate dimensions)—e.g., h could be expressed in meters, w in feet, and a in square inches (with the numerical components of these variables adjusted accordingly). However, it is also possible to postpone the specification of units with method (a) as well:

50. It *is* conceivable, though: Suppose a point moves around a circle at an angular speed of 8 rad/s; and, let us count each time it travels 1 rad. Then, the count frequency is

$$(8\ \mathrm{rad/s})(1\ \mathrm{count/rad}) = 8\ \mathrm{count/s} = 8\ \mathrm{s}^{-1} = 8\ \mathrm{Hz},$$

since *count* is a dimensionless unit of value 1; and, for the same reason, the left side equals 8 rad/s.

(3a′) Using *consistent units* (to be specified later) for all physical quantities, let h and w be, respectively, the numerical height and width of the picture, and let a be its numerical area; then, $a = hw$.

Here, "consistent" means that the units are not only appropriate dimensionally, but also consistent with the given formula—e.g., if h is the height in meters, and w is the width in feet, then the area a must be in meter-feet.

12.2.2 Propagation of Error and Uncertainty

For the various reasons discussed in Subsection 12.1.3, experimental error is introduced whenever a measurement is made. The numerical value of the error is unknown; otherwise, it would be removed from the measurement by simple subtraction. Accordingly, when calculations are performed upon measurements, their errors are inextricably processed too, thereby causing the calculated values to be inaccurate as well; moreover, additional calculations performed using these results yield further inaccurate results.[51] In general, the passing of error from the inputs of a calculation to its output is called **error propagation**.

Definitions (12.6) and (12.7) for the absolute error and relative error in an experimental value x_{exp} apply just as well to processed data as to raw data. Although both of these expressions of experimental error are useful, the effect of a calculation on experimental error is sometimes more naturally expressed in terms of one rather than the other. Also, some calculations are more sensitive than others to input errors. For example, suppose two physical values $x_{\text{phy}} = \pi$ and $y_{\text{phy}} = 22/7$ are respectively measured as $x_{\text{exp}} = 3.142$ and $y_{\text{exp}} = 3.143$; thus, both measurements are accurate to four digits of resolution, the first being in error by about 0.013%, and the second by about 0.005%. Then, the relative error in the calculated value $x_{\text{exp}} + y_{\text{exp}} = 6.285$ is

$$\frac{x_{\text{exp}} + y_{\text{exp}}}{x_{\text{phy}} + y_{\text{phy}}} - 1 \approx 0.009\%; \tag{12.34}$$

whereas the relative error in $x_{\text{exp}} - y_{\text{exp}} = -0.001$ is

$$\frac{x_{\text{exp}} - y_{\text{exp}}}{x_{\text{phy}} - y_{\text{phy}}} - 1 \approx -20.9\% \tag{12.35}$$

—*much* worse. Indeed, a common source of relative-error magnification is the subtraction of two nearly equal, but slightly inaccurate, numbers; hence, this operation should be avoided. (It is *possible*, though, for input errors to balance each other, perhaps canceling out altogether.)

---

51. Of course, exceptions can be given to any of the statements just made: A measurement might be so trivial (e.g., counting a small number of objects) that no error is introduced. An error in a simple measurement (e.g., counting objects $\alpha_0, \alpha_1, \ldots, \alpha_n$ as being n in number) might be discovered later and then corrected (to $n + 1$). And, certain calculations performed on erroneous data (e.g., multiplying a measurement by 0) always yield error-free results. On the other hand, numerical calculations themselves are often imperfect (e.g., owing to rounding), thereby causing even *more* error to be introduced.

Consider, then, the error propagation that occurs in the basic arithmetic operations. In each case, let x be the result of processing real values a and b; also, let $x + \Delta x$ be the result of applying the same process to $a + \Delta a$ and $b + \Delta b$, for some real-valued errors Δa and Δb. Starting with the operation of addition, it is easily verified that

$$x = a + b \quad \wedge \quad x + \Delta x = (a + \Delta a) + (b + \Delta b) \qquad \Rightarrow \qquad \Delta x = \Delta a + \Delta b. \tag{12.36a}$$

Similarly, for subtraction,

$$x = a - b \quad \wedge \quad x + \Delta x = (a + \Delta a) - (b + \Delta b) \qquad \Rightarrow \qquad \Delta x = \Delta a - \Delta b. \tag{12.36b}$$

Thus, the error propagation properties of addition and subtraction are naturally expressed in terms of *absolute* error.

For multiplication, let $x = a \cdot b$, and thus

$$\begin{aligned} x + \Delta x = (a + \Delta a) \cdot (b + \Delta b) &= ab + (a\,\Delta b + b\,\Delta a + \Delta a\,\Delta b) \\ &\approx x + (b\,\Delta a + a\,\Delta b) \end{aligned}$$

—having assumed that $|\Delta a| \ll |a|$ and $|\Delta b| \ll |b|$, which will be done for the remainder of the paragraph. (Hence, it is implicit that $a, b, x \neq 0$.) It follows that $\Delta x / x \approx (b\,\Delta a + a\,\Delta b)/ab$, from which we obtain

$$x = a \cdot b \quad \wedge \quad x + \Delta x = (a + \Delta a) \cdot (b + \Delta b) \qquad \Rightarrow \qquad \frac{\Delta x}{x} \approx \frac{\Delta a}{a} + \frac{\Delta b}{b} \tag{12.36c}$$

—now using *relative* error to express the error propagation. In particular, if b is constant, thereby making $\Delta b = 0$, we have $(\Delta x)/x = (\Delta a)/a$; that is, mere scaling does not affect relative error. Also, before turning to division, consider the simpler operation of reciprocation; thus, letting $x = 1/b$ and

$$x + \Delta x = \frac{1}{b + \Delta b} = \frac{1}{b} + \left(\frac{1}{b + \Delta b} - \frac{1}{b} \right) = x - \frac{\Delta b}{b(b + \Delta b)} \approx x - \frac{\Delta b}{b^2},$$

we conclude that

$$x = 1/b \quad \wedge \quad x + \Delta x = 1/(b + \Delta b) \qquad \Rightarrow \qquad \frac{\Delta x}{x} \approx -\frac{\Delta b}{b}. \tag{12.36d}$$

So, for the division $a/b = a \cdot (1/b)$, it follows from (12.36c) and (12.36d) that

$$x = a/b \quad \wedge \quad x + \Delta x = (a + \Delta a)/(b + \Delta b) \qquad \Rightarrow \qquad \frac{\Delta x}{x} \approx \frac{\Delta a}{a} - \frac{\Delta b}{b}. \tag{12.36e}$$

As discussed in Subsection 12.1.4, although the exact value of a particular experimental error is typically unknown, it can usually be bounded. Specifically, an experimental value x_{exp} (raw or processed data) is often expressed as an interval (12.11) of possible values, the half-length of this interval being the absolute uncertainty x_{unc} in x_{exp}; moreover, when $x_{\text{exp}} \neq 0$, this value may be converted to the relative uncertainty $x_{\text{rel-unc}}$ in x_{exp}, by (12.14). In the notation of the preceding paragraph, the absolute uncertainty in x is $(\max \Delta x - \min \Delta x)/2$, where $\max \Delta x$ and $\min \Delta x$ are the upper and lower bounds on the absolute error Δx; and, when $x \neq 0$, we may also divide by $|x|$ to obtain the relative uncertainty in x. Accordingly, by analyzing the extremes in error propagation for a given operation, we can determine its **uncertainty propagation**—i.e., how the uncertainties in the inputs of a calculation transform into the uncertainty in the output. In particular, the following facts readily obtain from (12.36):

- For addition and subtraction, the absolute uncertainty of the output equals the sum of the absolute uncertainties of the inputs.
- For a reciprocal, the relative uncertainty of the output approximately equals the relative uncertainty of the input (when the latter is small).
- For multiplication and division, the relative uncertainty of the output approximately equals the sum of the relative uncertainties of the inputs (when the latter are small).

For example, for subtraction, (12.36b) ends with $\Delta x = \Delta a - \Delta b$; thus, $\max \Delta x = \max \Delta a - \min \Delta b$ and $\min \Delta x = \min \Delta a - \max \Delta b$, making the absolute uncertainty of the output $(\max \Delta x - \min \Delta x)/2 = (\max \Delta a - \min \Delta a)/2 + (\max \Delta b - \min \Delta b)/2$—i.e., the sum of the absolute uncertainties of the inputs.

When one is applying a *monotonic* function $f(\cdot)$ to an experimental value x_{exp}, the uncertainty interval of the output can be easily calculated from that of the input. Specifically, let x_{phy} be the true physical value of x_{exp}; and, suppose we are confident that

$$x_{\text{lb}} \leq x_{\text{phy}} \leq x_{\text{ub}},$$

per (12.11). Then, if $f(\cdot)$ is increasing, we have

$$f(x_{\text{lb}}) \leq f(x_{\text{phy}}) \leq f(x_{\text{ub}});$$

whereas if $f(\cdot)$ is decreasing,

$$f(x_{\text{ub}}) \leq f(x_{\text{phy}}) \leq f(x_{\text{lb}}).$$

It should be noted, however, that in neither case is $f(x_{\text{exp}})$ necessarily the exact midpoint of the interval. In other words, writing the uncertainty interval of the input as $x_{\text{phy}} = x_{\text{exp}} \pm x_{\text{unc}}$, per (12.12), we need not have $f(x_{\text{phy}}) = f(x_{\text{exp}}) \pm f(x_{\text{unc}})$; rather, in both cases the uncertainty interval of the output can be expressed as

$$f(x_{\text{phy}}) = [f(x_{\text{ub}}) + f(x_{\text{lb}})]/2 \pm |f(x_{\text{ub}}) - f(x_{\text{lb}})|/2.$$

For example, the function $f(x) = e^{-x}$ is decreasing (in fact, strictly); therefore, given a measurement $x_{\text{phy}} = 2.0 \pm 0.3$—i.e., its uncertainty interval is $1.7 \leq x_{\text{phy}} \leq 2.3$—the uncertainty interval for $f(\cdot)$ applied to this measurement is $e^{-2.3} \leq f(x_{phy}) \leq e^{-1.7}$. That is, the interval is $0.100 \lesssim f(x_{phy}) \lesssim 0.183$, of which $f(2.0) = e^{-2.0} \approx 0.135$ is not the midpoint; rather, $f(x_{phy}) \approx 0.141 \pm 0.041$.

When uncertainties in experimental quantities are expressed implicitly (and roughly) via the number of significant digits written, some widely used rules of thumb can be used to *estimate* the appropriate number of significant digits for the output of an arithmetic calculation:

- For addition and subtraction, first rewrite each input in exponential notation (without changing the number of significant digits) such that all input numerals have a decimal point and identical exponential parts; then, using the same exponential part for the output, the output should be rounded to the decimal place corresponding to the leftmost least significant digit of the inputs.
- For a reciprocal, the number of significant digits to which the output should be rounded is the same as the number of significant digits in the input.
- For multiplication and division, the number of significant digits to which the output should be rounded is the same as the smallest number of significant digits in the inputs.

These rules also apply to calculations involving mathematical constants (e.g., π) and other exact values, which can all be written with as many significant digits as desired.[52]

Some examples:

$$\begin{aligned} 9.9 \times 10^{-3} + 7.52 \times 10^{-4} &= 9.9 \times 10^{-3} + 0.752 \times 10^{-3} = (9.9 + 0.752) \times 10^{-3} \\ &= 10.652 \times 10^{-3} \approx 10.7 \times 10^{-3}, \end{aligned}$$

having rounded "10.652" to the first decimal place (which happens to result in three significant digits) since the least significant digits of "9.9" and "0.752" occur in the first and third decimal places, the former being more to the left. Another example,

$$9.9 \times 10^{-3} \cdot 7.52 \times 10^{-4} = 7.4448 \times 10^{-6} \approx 7.4 \times 10^{-6},$$

where we have rounded "7.4448" to two significant digits, since "9.9" and "7.52" have two and three significant digits, the former being smaller. Also, note the loss of significant digits that occurs upon calculating the difference of two nearly equal quantities.[53]

The above rules, which are fairly reliable, can be "derived" from (12.21) and (12.36); however, the rules do not always yield the desired results.[54] For example, suppose one's policy regarding significant digits is to always round to a single uncertain digit. Then, since

$$(7.00 \pm 0.05) - (4.00 \pm 0.05) = 3.00 \pm 0.10$$

52. Thus, e.g., the rule for reciprocation can be viewed as a special case of the rule for division, since the numerator (1) of a reciprocal is known exactly.

53. This explains the poor relative error obtained in (12.35) for the calculation $x_{exp} - y_{exp} = 3.142 - 3.143 = -0.001$, in which four significant digits in each input reduce to only one in the output.

54. Accordingly, alternative rules have been suggested, such as including an additional significant digit in the output (for "safety"). But, the fact remains that whenever uncertainties in experimental quantities are expressed and manipulated via significant digits alone, only fairly reliable results should be expected.

(because the two uncertainties of 0.05 add), the policy would be to write "7.00 – 4.00 = 3.0"; whereas the rule for subtraction says the result should be written "3.00" (one significant digit too many). Also, since

$$\frac{8.2 \pm 0.1}{6.9 \pm 0.1} \approx 1.1889 \pm 0.0317$$

(the uncertainty interval of this quotient being from $(8.2 - 0.1)/(6.9 + 0.1)$ to $(8.2 + 0.1)/(6.9 - 0.1)$), the policy would be to write "$8.2/6.9 \approx 1.19$"; whereas the rule for division says the result should be written "1.2" (one significant digit too few). Furthermore, the effect of such discrepancies usually worsens when several operations are performed at once.

Finally, before electronic calculators were prevalent, it was quite reasonable to keep track of experimental uncertainty via significant digits, since carrying along unneeded digits required significant extra work in both calculation and writing. Now, though, several successive calculations can be easily performed on a calculator, which usually has sufficient internal resolution to make negligible the accumulated roundoff error, as well as having internal memory to store intermediate results. Thus, rather than rounding the intermediate results of a long calculation, thereby introducing additional error, it is just as easy to have the calculator manage many more digits than are needed; then, only when the final result is recorded, do we select an appropriate number of significant digits.[55]

12.2.3 Data Plots

As discussed at the beginning of the chapter, the main purpose of experimentation is to either discover or verify correlations among physical observables. This is usually difficult to do, though, by simply perusing a table of numerical data (raw or processed). However, such correlations can often be recognized when the data are plotted, since relationships among numbers sometimes become more apparent when viewed geometrically.[56] Moreover, in addition to revealing interrelationships *within* a set of data, plots are often helpful for comparing one data set with another (e.g., it might be seen that the two data sets come into agreement when some variable is appropriately adjusted); or, for comparing experimental results with theoretical expectations (e.g., a proposed theory might be tested); or, for comparing the predictions of competing theories (e.g., they might agree over some range of measured values, but then diverge).

As any spreadsheet program will show, there are many kinds of plots—e.g., pie charts, bar graphs, contour plots, polar plots, and three-dimensional plots of various

---

55. This approach is common in DSP. For example, a digital audio workstation (a specially equipped computer for processing audio signals) might internally manipulate audio data with 24 bits of resolution, only to truncate the final result to 16 bits (as found on CDs).

56. Accordingly, even in this digital age, analog equipment is still widely used; because, an analog display, though not as convenient for making accurate numerical readings, includes a helpful spatial component not present in a digital display. For example, it is much easier for a sound engineer to monitor variations in the level of a signal by watching a VU meter ("VU" meaning "volume unit"), with its traditional analog display, than it would be by watching the changing numbers on a comparable digital meter.

kinds—by which data can be displayed visually. In this section, though, we restrict our attention to rectangular plots (as discussed in Section 5.1), which are by far the most frequently used. This type of plot is particularly useful for illustrating a correlation between a *pair* of variables; furthermore, correlations among more than two variables can sometimes be shown by rectangularly plotting a *family* of data sets.

Figure 12.2 shows a rectangular plot of data obtained from a fictitious experiment: one thermometer is placed inside a box, a second inside a ball, and both the box and the ball are put in an oven along with a third thermometer; then, the temperature reading of each thermometer is monitored versus time while the oven is rapidly heated, starting with all of the components at thermal equilibrium and room temperature. (Owing to experimental error, the initial temperature measurements do not agree exactly.) As expected, the temperatures inside the box and ball lag that of the oven; and, since the latter lags more, we conclude that the ball provides better thermal insulation for its contents.

The figure illustrates many features commonly employed to make the information on a data plot easier to assimilate: A brief but descriptive title (here, placed at the top) informs the reader what the plot is about. Each axis is labeled with either a name ("temp.") or a variable ("t") stating the corresponding physical quantity, along with the chosen unit of measurement (kelvin, seconds). (Here, a single variable is not appropriate for the vertical axis, since three different temperatures are being plotted.) The scale of each axis is indicated by small line segments called **tick marks**, which are labeled with respective numbers; additionally, horizontal and vertical **grid lines** further aid in reading numerical values from the plot. (Another graph style, which may be used either in lieu of or

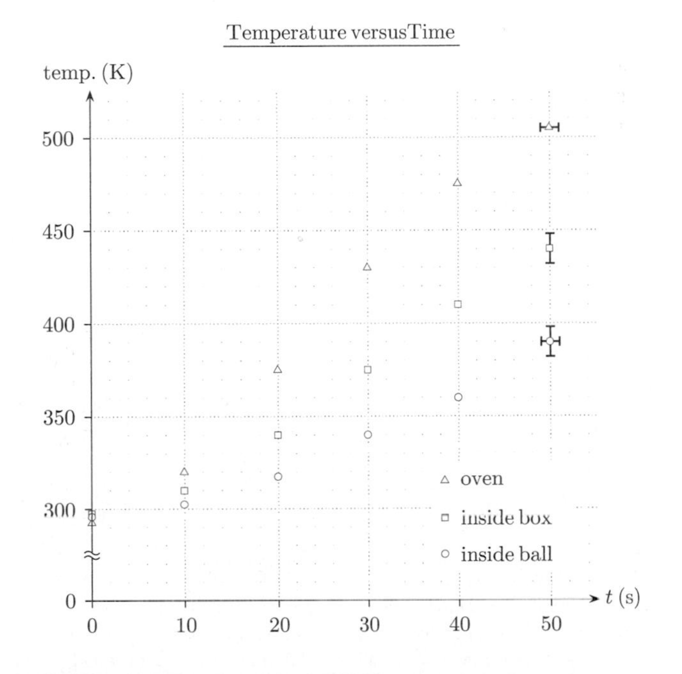

FIGURE 12.2 A Rectangular Plot of Experimental Data

along with a grid, is to plot the data within a rectangular frame having tick marks all around.) Also, rather than showing an entire region that includes both the data points and the origin where the axes intersect, the vertical axis is broken, thereby eliminating much blank space (from 0 K to about 300 K); this allows the plot to be made larger, which is beneficial because since the primary purpose of a data plot is to enable *visual* comparisons. An alternative would be to have the axes intersect at a point other than origin—e.g., (0 s, 300 K); but, for a linear scale, an explicit axis break alerts the reader that distances in that direction do not translate proportionally into numerical values. (Here, e.g., the distance from 0 K to 500 K is 5 times, not $\frac{5}{3}$ times, the distance from 0 K to 300 K.) It is unusual, though, to show a break in a logarithmic scale, since it has no point corresponding to zero; rather, any power of 10 will do for the position of the orthogonal axis.

Since Figure 12.2 shows multiple sets of data (viz., temperatures taken at three different locations), a separate symbol (triangle, square, circle) is used to plot the data points of each set; moreover, as a further aid to the reader, a square is used for the box, and a circle for the ball. These symbols are defined in a **legend** (here, placed in the lower right); but, when space allows, one may instead place the definition of each symbol next to one of the corresponding data points. Finally, the three rightmost data points in Figure 12.2 show how **error bars** (I-shaped symbols) can accompany the data points to indicate the corresponding experimental uncertainties.[57] Such horizontal and vertical bars may be used either alone or together, though most frequently only vertical bars are used. (Here, the horizontal and vertical uncertainties shown are 1 s and 8 K, respectively. Thus, e.g., the rightmost data circle indicates a temperature 390 ± 8 K taken at time $t = 50 \pm 1$ s.)

Based on the rectangular plot of some data, it might appear that the vertical axis value, say y, is some simple function $f(\cdot)$ of the horizontal-axis value, say x—i.e., $y = f(x)$. Or, this functional relationship might be posited based on other knowledge, such as the predictions of a physical theory. In any event, it is common to draw on the plot a curve that "fits" the data, thereby graphically expressing the function $f(\cdot)$; thus, the process of generating such a curve is called **curve fitting**. Likewise, for a plot of several data sets, as in Figure 12.2, a *family* of curves might be drawn, one curve for each set.

There are many methods by which a curve can be fitted to a data plot, the easiest being to simply connect neighboring data points with line segments. Alternatively, one or more curved segments can be drawn manually, either freehand or guided by a **spline** (a stiff but bendable length of some material) or a **French curve** (a flat rigid piece of material having several smoothly varying curves).[58] Analytically, a popular choice for the function $f(\cdot)$ is a polynomial, in which case the curve-fitting process amounts to choosing real coefficients $a_0, a_1, \ldots, a_m$ ($m \geq 0$) and plotting the equation

$$y = a_0 + a_1 x + \cdots + a_m x^m. \tag{12.37}$$

57. Hence, "uncertainty bars" would probably be a more appropriate name.

58. Admittedly, with the increasing quality of inexpensive plotting software, drawing curves manually is becoming much less common. Nevertheless, it is a basic skill that every scientist and engineer should have, if only to make quick sketches in a lab notebook.

Given $n \geq 1$ data points $(x_i, y_i) \in \mathbb{R}^2$ $(i = 1, \ldots, n)$ having distinct abscissas x_i, one way of choosing the coefficients is to use the Lagrange interpolation formula (6.25). Then, the resulting polynomial passes through all of the data points exactly; and, it is either the zero polynomial (when $y_i = 0$ for all i) or its degree is as small as possible (and therefore at most $n-1$). But, unless n is small, this method usually produces a polynomial of large degree, which might exhibit considerably more variation than desired.

Another way to choose the coefficients in (12.37) is to first fix, as desired, the maximum allowable degree m of the polynomial; then, a_0, a_1, ..., a_m are chosen to minimize the sum

$$\sum_{i=1}^{n}(a_0 + a_1 x_i + \cdots + a_m {x_i}^m - y_i)^2 \tag{12.38}$$

of squared "errors" (in points on the polynomial versus the data). That is, we perform a **least-squares fit** of a polynomial to the data. Specifically, upon introducing the matrices

$$\mathbf{a} = \begin{bmatrix} a_0 \\ a_1 \\ \vdots \\ a_m \end{bmatrix}, \quad \mathbf{X} = \begin{bmatrix} 1 & x_1 & {x_1}^2 & \cdots & {x_1}^m \\ 1 & x_2 & {x_2}^2 & \cdots & {x_2}^m \\ \vdots & \vdots & \vdots & & \vdots \\ 1 & x_n & {x_n}^2 & \cdots & {x_n}^m \end{bmatrix}, \quad \mathbf{y} = \begin{bmatrix} y_1 \\ y_2 \\ \vdots \\ y_n \end{bmatrix} \tag{12.39a}$$

(where $\mathbf{X}$ and $\mathbf{y}$ are determined by the data), it can be shown that if at least $m+1$ of the values $x_1, x_2, \ldots, x_n$ are distinct (hence, we must have $m \leq n-1$),[59] then the coefficients minimizing the sum (12.38) are unique and are provided by the matrix equation

$$\mathbf{a} = (\mathbf{X}^{\mathrm{T}}\mathbf{X})^{-1}\mathbf{X}^{\mathrm{T}}\mathbf{y}. \tag{12.39b}$$

In particular, for $m = 0$, the polynomial to be used in (12.37) is simply the constant

$$a_0 = \frac{1}{n}\sum_{i=1}^{n} y_i.$$

Whereas for $m = 1$, the polynomial is $a_0 + a_1 x$—i.e., (12.37) describes a line—with a_0 and a_1 being explicitly given in (8.18).

An example of such curve fitting is shown in Figure 12.3, for which the given data points (x_i, y_i) are $(1, 5)$, $(2, 17)$, $(3, 11)$, and $(4, 47)$. The Lagrange interpolation polynomial obtained from (6.25) is $10x^3 - 69x^2 + 149x - 85$, which has degree 3 (the maximum possible for $n = 4$ data points); and, as can be seen, this polynomial indeed interpolates the data points. By contrast, choosing $m = 2$ in (12.37)–(12.39) yields a least-squares-fit polynomial $6x^2 - 18x + 20$ (a parabola), for which the square-error sum in (12.38) evaluates to 180. But, even though the latter polynomial does not interpolate the data points, it might actually be preferred. For example, we might know

59. This condition on the abscissa values is necessary and sufficient for the $(m+1) \times (m+1)$ matrix $\mathbf{X}^{\mathrm{T}}\mathbf{X}$ appearing below to be nonsingular.

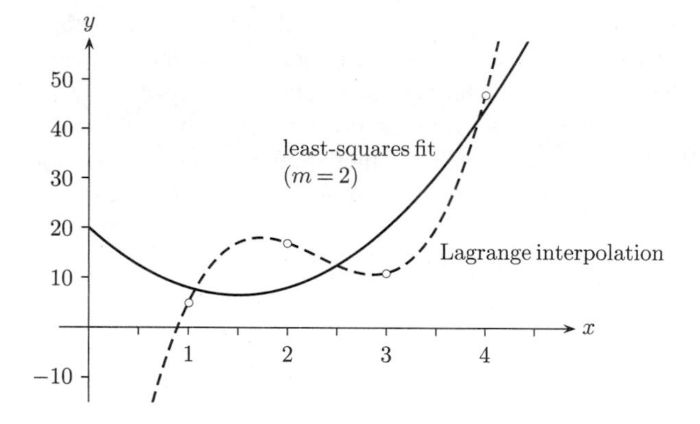

FIGURE 12.3 Fitting Curves to Data

from theory that the dependence of y on x is quadratic; or, there might be some reason to believe that $y \to \infty$ as $x \to -\infty$, which is not the case for the other polynomial. In these cases the least-squares-fit polynomial would enable us to more reliably **extrapolate** from the data—i.e., draw conclusions about the value of y outside the data range of x (here, from 1 to 4). For instance, it would be more likely that the value of y for $x = 0$ is near 20, as predicted by the least-squares-fit polynomial, rather than near -85 as predicted by the Lagrange interpolation polynomial.

Finally, as illustrated by Figure 12.3 and numerous other figures in this book, when several curves are plotted together, some visual cue—e.g., the use of solid, dashed, and dotted curves—is given to help the reader distinguish the curves. Also, whenever a curve is generated from a mathematical formula—be it a theoretical equation or a fit to given data points—no points on the curve itself (e.g., any calculated for purpose of drawing) should be highlighted, except when some exact points on the curve are particularly noteworthy (and the reason explicitly given). That is, normally, all that should be visible is the mathematically generated curve itself and any *data* points; because, highlighting any other points can either distract from the data or cause those points to be misread *as* data.

Epilogue: A Glimpse of Things to Come

As stated in the preface, a primary purpose of this book is to help science and engineering students review the mathematical knowledge and skills, sometimes referred to as "precalculus", generally expected to have been learned prior to attending college. Though this comprises much diverse material, many more mathematical tools must also be mastered to become a practicing scientist or engineer.

In this final section, then, we take a quick look at some college-level subjects, thereby giving the intended reader a glimpse of the mathematics yet encountered. Additionally, some ancillary topics and a few interesting philosophical issues are briefly discussed. For those who might like to perform some self-study in these areas, suggestions are made for further reading.

E.1 Calculus

By far, the portion of "higher mathematics" most used by scientists and engineers is **calculus**, the study of which occupies the typical undergraduate student for about two years. Originally called *the* calculus to distinguish it from other kinds of analysis, this area of mathematics comprises two complementary subjects:

- **Differential calculus**: the theory of **differentiation**, which is an operation to calculate the *instantaneous* rate of change in one variable versus another (e.g., speed is the rate of change in distance versus time);
- **Integral calculus**: the theory of **integration**, which is a *continuous* summation operation (enabling one to, e.g., calculate the area of a nonpolygonal region).

The fundamental concept common to these two subjects is the **differential**. Roughly put, for any real variable "x", the corresponding differential "dx" represents a "small" change Δx in the value x. Indeed, the change is considered so small as to be infinitesimal; that is, although dx is not the same as the number zero, we can view its size $|dx|$ as being less than any positive number. This apparent contradiction is resolved by treating dx not as a static number, but as the dynamic process of letting $\Delta x \to 0$. Moreover, differentials are not usually handled in isolation; rather, they typically appear within some limiting operation, such as differentiation or integration.

Given a real-valued function $f(\cdot)$ having domain $D \subseteq \mathbb{R}$, we define its **derivative** at a point $x \in D$ as

$$\frac{df(x)}{dx} \triangleq \lim_{\Delta x \to 0} \frac{f(x + \Delta x) - f(x)}{\Delta x} \tag{E.1}$$

—assuming that the limit exists, in which case we say that $f(\cdot)$ is **differentiable** *at the point* x. It follows easily from (E.1) that a necessary condition for $f(\cdot)$ to be differentiable at a point x is that $f(\cdot)$ be continuous at x. However, this condition is not sufficient; for example, the function $|x|$ of $x \in \mathbb{R}$ is continuous, but not differentiable at $x = 0$. We call $f(\cdot)$ a **differentiable function** if and only if it is differentiable at each point in D.

It is customary to mark a function symbol with a prime to indicate the derivative of that function, which is itself a function. Thus,

$$f'(x) \triangleq \frac{df(x)}{dx}$$

is a real-valued function defined at each $x \in D$ for which the above limit exists. Introducing another variable "y" and letting $y = f(x)$, one may also write (E.1) as

$$\frac{dy}{dx} \triangleq \lim_{\Delta x \to 0} \frac{\Delta y}{\Delta x}, \tag{E.2}$$

which is referred to as the derivative of y *with respect to* x. Comparing (E.2) with (5.5), illustrated in Figure 5.5, we conclude that the derivative of $f(\cdot)$ at a point $a \in D$—i.e., the value $f'(a)$—may be interpreted as the *slope* of the line drawn tangent to the plot of $f(x)$ versus x at the point $\big(a, f(a)\big)$; see Figure E.1.

The derivatives of some common functions are given in Table E.1 (where $p \in \mathbb{R}$ and $0^0 = 1$), with it understood that the derivative can be calculated only at those points $x \in \mathbb{R}$ for which the functions given for $f(x)$ and $f'(x)$ both exist (e.g., $d\sqrt{x}/dx = dx^{1/2}/dx = (1/2)x^{-1/2} = 1/2\sqrt{x}$ only for $x > 0$). For example, note how in Figure 11.8 the slope of $\sin\theta$ versus θ, at each value of θ, is indeed $\cos\theta$.

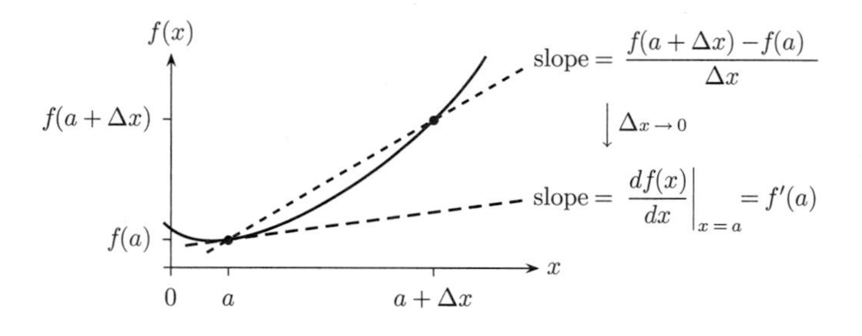

FIGURE E.1 A Geometric Interpretation of Differentiation

TABLE E.1 The Derivatives of Some Common Functions

$f(x)$	$f'(x)$	$f(x)$	$f'(x)$
1	0	x^p	px^{p-1}
e^x	e^x	$\ln x$	x^{-1}
$\sin x$	$\cos x$	$\sin^{-1} x$	$1/\sqrt{1-x^2}$
$\cos x$	$-\sin x$	$\cos^{-1} x$	$-1/\sqrt{1-x^2}$
$\tan x$	$\sec^2 x$	$\tan^{-1} x$	$1/(1+x^2)$

There are also various rules by which the derivatives of more complicated functions can be calculated. First, the **differentiation operator** "d/dx" (here, with respect to the variable "x") is "linear"; that is,

$$\frac{d}{dx}[af(x) + bg(x)] = a\frac{df(x)}{dx} + b\frac{dg(x)}{dx} \tag{E.3}$$

for any functions $f(\cdot)$ and $g(\cdot)$ having the same domain, and any real constants a and b. Furthermore, we have the following rules:

- **Product rule**:

$$\frac{d}{dx}f(x)g(x) = \frac{df(x)}{dx}g(x) + f(x)\frac{dg(x)}{dx}. \tag{E.4}$$

- **Quotient rule**:

$$\frac{d}{dx}\frac{f(x)}{g(x)} = \frac{g(x) \cdot df(x)/dx - f(x) \cdot dg(x)/dx}{g^2(x)}. \tag{E.5}$$

- **Chain rule** for a composite function $g[f(\cdot)]$, now assuming that the domain of $g(\cdot)$ contains the range of $f(\cdot)$: Letting $y = f(x)$,

$$\frac{d}{dx}g[f(x)] = \frac{dg(y)}{dy}\frac{df(x)}{dx} = g'(y)f'(x) = g'[f(x)]f'(x); \tag{E.6a}$$

therefore, also letting $z = g(y)$, we may more briefly write

$$\frac{dz}{dx} = \frac{dz}{dy} \cdot \frac{dy}{dx} \tag{E.6b}$$

(as might have been guessed intuitively).

Each of the equations in (E.3)–(E.6) holds whenever its right side exists.

Some examples: It follows from (E.3) with $b = 0$ and the first entry in Table E.1 that the derivative of any real constant a with respect to a real variable x is zero:

$$\frac{da}{dx} = \frac{d(a \cdot 1)}{dx} = a\frac{d1}{dx} = a \cdot 0 = 0$$

—as obtains directly from (E.2), since $\Delta a = 0$ for all values of Δx. By contrast, taking $p = 1$ in Table E.1 shows that $dx/dx = 1$, as one would expect. As an example of applying the product rule, we have

$$\frac{d}{dx}x^2 \ln x = \frac{dx^2}{dx}\ln x + x^2\frac{d \ln x}{dx} = 2x \cdot \ln x + x^2 \cdot x^{-1} = x(2\ln x + 1). \tag{E.7}$$

Also, we can use the quotient rule to obtain the derivative of the tangent function from those of the sine and cosine functions:

$$\frac{d \tan x}{dx} = \frac{d}{dx}\frac{\sin x}{\cos x} = \frac{(\cos x) \cdot d(\sin x)/dx - (\sin x) \cdot d(\cos x)/dx}{\cos^2 x}$$

$$= \frac{(\cos x)(\cos x) - (\sin x)(-\sin x)}{\cos^2 x} = \frac{\cos^2 x + \sin^2 x}{\cos^2 x} = \frac{1}{\cos^2 x} = \sec^2 x,$$

as stated in the table. Finally, for any real constant a, letting $y = ax$ and using the chain rule yields

$$\frac{d}{dx}e^{ax} = \frac{d}{dx}e^{y} = \frac{de^{y}}{dy} \cdot \frac{dy}{dx} = e^{y} \cdot a = ae^{ax};$$

therefore, for any constant $b > 0$, noting that $b^x = e^{(\ln b)x}$ yields

$$\frac{db^x}{dx} = b^x \ln b.$$ [1]

As an application, we now see that the surface area $4\pi r^2$ of a sphere having radius r is calculated in (10.26) by differentiating the sphere's volume $\frac{4}{3}\pi r^3$ with respect to r, which can be more quickly performed via Table E.1.

A frequent use of differentiation is to help determine the extrema of a real-valued function $f(\cdot)$ defined on an interval domain D; because, assuming that it is differentiable, a *necessary* condition for $f(\cdot)$ to have a *local* extremum at an *interior* point x of D is that $f'(x) = 0$ (which makes sense intuitively because we have interpreted derivatives as slopes). For example, for the function $f(x) = x^2 \ln x$ defined on $D = (0, \infty)$, of which every point is an interior point, it follows from (E.7) that local extrema can possibly occur only at those points $x \in D$ for which $x(2 \ln x + 1) = 0$; and, the one such point is $x = e^{-1/2} \approx 0.607$, which further analysis shows is indeed where a local minimum of $f(\cdot)$ occurs. It is important to observe, however, that the condition of having $f'(x) = 0$ at some interior point x of D is not generally sufficient for $f(\cdot)$ to have a local extremum there. For example, the function $f(x) = x^3$ $(x \in \mathbb{R})$ has derivative $f'(x) = 3x^2$, which equals 0 at $x = 0$; but, $f(x)$ has neither a local minimum nor a local maximum at this point (see Figure 6.1).

As another application, although the summation formula

$$\sum_{i=1}^{n} ic^i = c\frac{nc^{n+1} - (n+1)c^n + 1}{(c-1)^2} \qquad (c \in \mathbb{R} \text{ with } c \neq 1) \tag{E.8}$$

can be *verified* via mathematical induction, it is easily *derived* by differentiating the formula (4.37) for the sum of a geometric progression. Namely, we simply equate

$$\frac{d}{dc}\sum_{i=1}^{n} c^i = \sum_{i=1}^{n} \frac{d}{dc}c^i = \sum_{i=1}^{n} ic^{i-1}$$

(where bringing the differentiation operator inside the sum is justified by the linearity of the operator, since the sum is finite) with

$$\frac{d}{dc}\sum_{i=1}^{n} c^i = \frac{d}{dc}\frac{c^{n+1} - c}{c-1} = \frac{nc^{n+1} - (n+1)c^n + 1}{(c-1)^2}$$

1. This result, and the similar fact that $d(\log_b x)/dx = 1/x \ln b$ for any constant $b > 0$ other than 1, show why Napier's number e is the "preferred" base for exponentials and logarithms.

(the details of deriving the last equality being left to the reader); then, multiply the resulting equation by c.[2]

Also, it follows from (E.2) that

$$\Delta y \approx \frac{dy}{dx}\Delta x$$

when $\Delta x \approx 0$, and thus the derivative dy/dx tells us how "sensitive" y is to changes in x. Accordingly, error propagation rules such as those in (12.36) can be easily derived via differentiation. For example, if $y = \ln x$ ($x > 0$), then for $\Delta x \approx 0$ we have

$$\Delta y \approx \frac{d \ln x}{dx}\Delta x = \frac{\Delta x}{x},$$

thereby making $y + \Delta y \approx \ln x + (\Delta x)/x$. (Thus, e.g., taking $x = 2$ and $\Delta x = 0.1$, we find $\ln(2 + 0.1) \approx \ln 2 + 0.1/2 \approx 0.7431$, as compared with $\ln 2.1 \approx 0.7419$.)

Turning to integration, suppose a positive function $f(\cdot)$ on $\mathbb{R}$ is given, along with two constants $a, b \in \mathbb{R}$ such that $a < b$; and, as illustrated in Figure E.2, consider the problem of finding the area of the region in the xy-plane bounded by the x-axis, the curve described by the equation $y = f(x)$, and the vertical lines at $x = a$ and $x = b$. As shown on the left side of the figure (for $n = 5$), one approach is to sample $f(x)$ at n points $x = a + i\Delta x$ ($i = 0, 1, \ldots, n-1$), with $\Delta x = (b - a)/n$, then sum the areas of the corresponding rectangular strips of height $f(a + i\Delta x)$ and width Δx; because, one might expect that when n is chosen to be large, the sum

$$\sum_{i=0}^{n-1} f(a + i\Delta x)\Delta x$$

will nearly equal the desired area. Indeed, it can be shown that when $f(\cdot)$ is continuous, this sum converges to the desired area as $n \to \infty$ (equivalently, as $\Delta x \to 0$). That is,

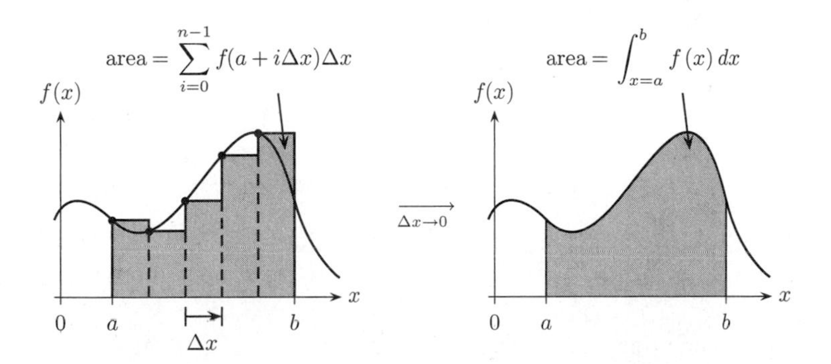

FIGURE E.2 A Geometric Interpretation of Integration

2. Alternatively, but not as easily, (E.8) can be derived from (4.37) via "summation by parts"; see Problem 4.29(b).

as shown on the right side of the figure, the desired area is given by the **integral** of $f(x)$ versus x from $x = a$ to $x = b$:

$$\int_{x=a}^{b} f(x)\,dx \triangleq \lim_{\Delta x \to 0} \sum_{i=0}^{n-1} f(a + i\Delta x)\Delta x. \tag{E.9}$$

The integral sign "$\int$" on the left looks like an elongated letter "S", standing for "sum", of which an integral can be viewed as a continuous version. For note that the variable of integration "x" takes on a continuum of values from a to b, whereas the index of summation "i" takes on only discrete values $0, 1, \ldots, n-1$. The values a and b are called the **limits** on the integral, and the function $f(\cdot)$ being integrated is the **integrand**.

The integration process just described can be generalized in many ways. First, although not every function can be integrated, the integrand need not be continuous; nor must it be positive. Nor must the limits a and b be finite. Nor must we have $a < b$; rather, it is consistent to define

$$\int_{x=a}^{a} f(x)\,dx \triangleq 0,$$

$$\int_{x=b}^{a} f(x)\,dx \triangleq -\int_{x=a}^{b} f(x)\,dx \quad (a < b).$$

It can then be shown that for any $a, b, c \in \overline{\mathbb{R}}$, we have

$$\int_{x=a}^{c} f(x)\,dx = \int_{x=a}^{b} f(x)\,dx + \int_{x=b}^{c} f(x)\,dx$$

(when the integrals on the right exist)—analogous to how the convention (4.43a) for performing a summation backward yields the similar result (4.42a).

Some basic integration facts, for any constant $c \in \mathbb{R}$ and functions $f\colon (a,b) \to \mathbb{R}$ and $g\colon (a,b) \to \mathbb{R}$ with $a, b \in \overline{\mathbb{R}}$, are:

$$\int_{x=a}^{b} 1\,dx = b - a, \tag{E.10a}$$

$$\int_{x=a}^{b} cf(x)\,dx = c\int_{x=a}^{b} f(x)\,dx, \tag{E.10b}$$

$$\int_{x=a}^{b} [f(x) + g(x)]\,dx = \int_{x=a}^{b} f(x)\,dx + \int_{x=a}^{b} g(x)\,dx \tag{E.10c}$$

—assuming for each line that the right side exists.

Returning to Figure E.2—again assuming, for simplicity, a positive continuous function $f(\cdot)$ and $a < b$—let us view the integration process dynamically; that is, think of the area under $f(\cdot)$ as being accumulated bit by bit as x moves from the lower limit a to the upper limit b. On the left side of the figure, each time x increases by Δx, the area increases by $f(x)\Delta x$. Similarly on the right side, but now moving in a continuous fashion rather than in steps, we may say that when x increases by a differential amount dx, the area increases by $f(x)\,dx$. The *rate*, then, at which area is accumulated during the integration process is this amount divided by the change dx in x; hence, this rate equals $f(x)$. Earlier, though, we associated rates with derivatives; thus, it is not surprising that there is an intimate connection between integration and differentiation; specifically, we now see that these two operations are essentially inverses of each other. This connection is made precise by the **fundamental theorem of calculus**, which has various forms. One form states that for any continuous function $f: [a, b] \to \mathbb{R}$ $(-\infty < a < b < \infty)$, if $F(\cdot)$ is an **antiderivative** of $f(\cdot)$—that is, $F: [a, b] \to \mathbb{R}$ is a differentiable function such that $F'(\cdot) = f(\cdot)$—then

$$\int_{x=a}^{b} f(x)\,dx = F(b) - F(a). \tag{E.11}$$

Another form states that for any continuous function $f: [a, b] \to \mathbb{R}$ $(-\infty < a < b < \infty)$, we have

$$\frac{d}{dx}\int_{t=a}^{x} f(t)\,dt = f(x) \quad (-\infty < a < b < \infty) \tag{E.12}$$

—which clearly shows that differentiation "undoes" integration.[3] The practical significance of the theorem is that since derivative tables such as Table E.1 can be used "in reverse" to find antiderivatives, they can also be used to perform integration; for example, since $d(\ln x)/dx = x^{-1}$, it follows from (E.11) that

$$\int_{x=2}^{3} x^{-1}\,dx = \ln 3 - \ln 2 = \ln \tfrac{3}{2} \approx 0.405.$$

Many mathematical tools, beyond differentiation and integration themselves, can be derived by using calculus. For one, a real function $f(\cdot)$ can often be expressed in the form of a **power series**

$$f(x) = \sum_{k=0}^{\infty} a_k x^k$$

(where $0^0 = 1$), for some real coefficients a_k $(k = 0, 1, \ldots)$, over some interval of real values x. For example, by the binomial expansion (8.4) we have

$$(1+x)^n = 1 + \sum_{k=1}^{n} \frac{n(n-1)(n-2)\cdots(n-k+1)}{k!} x^k \quad (-\infty < x < \infty),$$

3. In (E.12), a variable other than "x"—viz., "t"—is arbitrarily chosen to express the integration process, because one of the limits is "x", thereby making the integral itself a function of x. This change of variables is allowed because every variable of integration is a *dummy* variable in the same way that a summation index is (see Section 4.2).

for any *nonnegative integer* power n; but, further analysis using calculus shows that this finite sum can be generalized to any *real* power p as the infinite sum

$$(1+x)^p = 1 + \sum_{k=1}^{\infty} \frac{p(p-1)(p-2)\cdots(p-k+1)}{k!} x^k \quad (-1 < x < 1), \tag{E.13}$$

though the domain of the generalization is restricted.[4] In particular, upon taking $p = \pm 1/2$, it follows that $\sqrt{1+x} \approx 1 + x/2$ and $1/\sqrt{1+x} \approx 1 - x/2$ for $x \approx 0$; because, in the above sum, the terms for $k > 1$ become altogether negligible compared to the one term for $k = 1$ when $|x|$ is sufficiently small.

Another such tool is **L'Hospital's rule**, which provides an easy way to evaluate certain limits. It states that for any two differentiable functions $f\colon (a,b) \to \mathbb{R}$ and $g\colon (a,b) \to \mathbb{R}$ $(-\infty \le a < b \le \infty)$, if either

$$\lim_{x\to a} f(x) = \lim_{x\to a} g(x) = 0 \quad \text{or} \quad \lim_{x\to a} g(x) = \pm\infty,$$

then

$$\lim_{x\to a} \frac{f(x)}{g(x)} = \lim_{x\to a} \frac{f'(x)}{g'(x)}$$

when the limit on the right exists (infinite limits being allowed); and, likewise upon replacing each "$\to a$" with "$\to b$". For example,

$$\lim_{x\downarrow 0} x \ln x = \lim_{x\downarrow 0} \frac{\ln x}{x^{-1}} = \lim_{x\downarrow 0} \frac{d(\ln x)/dx}{d(x^{-1})/dx} = \lim_{x\downarrow 0} \frac{x^{-1}}{-x^{-2}} = \lim_{x\downarrow 0} -x = 0.$$

Finally, numerous books on basic calculus are in print, with more being written all the time. A good source for further information on this topic is the introductory calculus textbook used at any college or university.[5]

E.2 Differential Equations and Difference Equations

After differentiating a function $f(\cdot)$, the resulting function $f'(\cdot)$ might itself be differentiable, in which case another function $f''(\cdot)$ can be obtained by differentiating again:

$$f''(x) \triangleq \frac{df'(x)}{dx}$$

4. Examples of other power series—viz., for the exponential, cosine, and sine functions—are given in fn. 2 in Chapter 7.

5. The author still has his [55]! Indeed, it is a good idea to keep as many of your college textbooks as possible, for future reference.

(for each $x \in \mathbb{R}$ such that the right side exists). This *second* derivative of $f(\cdot)$ may also be written

$$\frac{d^2f(x)}{dx^2} \triangleq \frac{d}{dx}\frac{df(x)}{dx}.$$

In general, differentiating n times (as allowed), we obtain the ***n*th derivative** of $f(\cdot)$,

$$f^{(n)}(x) \triangleq \frac{d^nf(x)}{dx^n} \quad (n = 0, 1, 2, \ldots),$$

with the 0th derivative being the function $f(\cdot)$ itself. (The notation used on either side of this equation is preferable to the "prime notation" when one is differentiating a function more than two or three times.)

Often, a physical system is described by a **differential equation**, which is generally any equation relating one or more functions and their derivatives (including derivatives of derivatives, and so forth).[6] Usually, one is interested in how the system evolves with time; thus, a frequently occurring form for a differential equation describing a system—having input $x(t)$ and output $y(t)$, both real-valued functions of time t—is

$$b_ny^{(n)}(t) + \cdots + b_1y^{(1)}(t) + b_0y(t) = a_mx^{(m)}(t) + \cdots + a_1x^{(1)}(t) + a_0x(t), \quad \text{(E.14)}$$

where the coefficients $a_0, a_1, \ldots, a_m$ $(m \geq 0)$ and $b_0, b_1, \ldots, b_n$ $(n \geq 0)$ are real constants. Typically, the input $x(t)$ is known for some interval of time, and the objective mathematically is to solve the differential equation for the output $y(t)$ over the same interval.

As a simple example, suppose a mass $M > 0$ is falling to earth under the force $F > 0$ of gravity, which we assume to be constant (though it actually increases as the mass nears the ground); also, we will ignore all frictional forces such as the drag of air resistance. Then, letting $y(t)$ be the distance of the mass above the ground, it follows from Newton's second law of motion that $Ma(t) = -F$ (the minus sign is due to the force being downward whereas the distance measured is upward), where the acceleration $a(t)$ of the mass is the rate of change in its velocity $v(t)$, which in turn is the rate of change in the position $y(t)$. Therefore, $a(t) = dv(t)/dt = d[dy(t)/dt]/dt = d^2y(t)/dt^2$, and we obtain the differential equation

$$My^{(2)}(t) = -F. \quad \text{(E.15)}$$

(Hence, the input $x(t)$ to this system can be taken to be the constant force F, by letting $m = 0$ and $a_0 = -1$ in (E.14).) It can then be shown—e.g., by integrating (E.15) twice versus t—that all solutions to this equation have the form

$$y(t) = -(F/2M)t^2 + At + B \quad \text{(E.16)}$$

6. The kind of differential equation discussed in this brief introduction, in which all differentiation is performed with respect to a single variable, is called an **ordinary differential equation**. Alternatively, one can have a **partial differential equation**, in which the functions appearing may be (partially) differentiated with respect to each of several variables.

for some real coefficients A and B. The values of these constants are determined by imposing **boundary conditions** on the differential equation, which obtain from some additional information given about the system. Thus, we might know that the mass has initial position $y(0) = 3$ and initial velocity $y^{(1)}(0) = v(0) = 4$; then, since (E.16) yields

$$y(0) = B, \quad y^{(1)}(0) = A,$$

we conclude that $A = 4$ and $B = 3$, which substituted into (E.16) yield the unique solution to the differential equation and boundary conditions. (Another set of boundary conditions sufficient to determine the values A and B are the position $y(t)$ at two different times t. By contrast, knowing the velocity $y^{(1)}(t)$ at two different t is not sufficient.)

Whereas differential equations may describe *continuous*-time systems—i.e., systems having inputs and outputs that are functions of a time variable "t" that can take on a continuum of real values—it is also common to encounter systems having inputs and outputs that are functions of a time variable, say "n", restricted to integer values. Such a *discrete*-time system is often described by a **difference equation**. In particular, analogous to (E.14), a frequently occurring form for such an equation is

$$\begin{aligned} b_0y(n) + b_1y(n-1) + \cdots + b_Ny(n-N) \\ = a_0x(n) + a_1x(n-1) + \cdots + a_Mx(n-M), \end{aligned} \tag{E.17}$$

where the coefficients $a_0, a_1, \ldots, a_M$ $(M \geq 0)$ and $b_0, b_1, \ldots, b_N$ $(N \geq 0)$ are real constants. Again, the input $x(n)$ is typically known over some interval of time, and the objective mathematically is to solve the difference equation for the output $y(n)$ over the same interval.

As an example, consider for each length $n \geq 1$ all possible sequences $\{L_j\}_{j=1}^{n}$ of n letters L_j taken from the usual 26-letter alphabet $\{\text{A}, \text{B}, \ldots, \text{Z}\}$. Thus, for a particular length n, the number of such sequences is 26^n. Now, suppose we require that only the letter U is allowed to immediately follow an occurrence of the letter Q. Then, for each $n \geq 1$, letting $y(n)$ be the number of allowable sequences of length n, we must have

$$y(n) = 25y(n-1) + y(n-2) \quad (n \geq 3); \tag{E.18}$$

because, every allowable sequence of length $n \geq 3$ is either one of the 25 non-Q letters followed by any allowable sequence of length $n-1$, or the 2-letter sequence QU followed by any allowable sequence of length $n-2$. Also, initially we have $y(1) = 26$, since all 1-letter sequences are allowable; and, $y(2) = 25 \cdot 26 + 1 = 651$, since the only allowable 2-letter sequences are the 25 non-Q letters followed by any letter, plus the sequence QU. Using these initial values, we *could* use (E.18) recursively to find $y(3) = 16\,301$, $y(4) = 408\,176$, etc.; but, this iterative method is inconvenient for finding $y(n)$ when n is large (e.g., when studying the asymptotic behavior of $y(n)$ as $n \to \infty$). Instead, a formula for $y(n)$ in terms of n alone can be obtained by rearranging (E.18) into the form (E.17)—i.e.,

$$y(n) - 25y(n-1) - y(n-2) = 0 \quad (n \geq 3) \tag{E.19a}$$

(thus, one might say that the input $x(n) = 0$ for all n)—then solving this difference equation while imposing the boundary conditions

$$y(1) = 26, \quad y(2) = 651. \tag{E.19b}$$

The result is

$$y(n) = A_1 r_1{}^n + A_2 r_2{}^n \quad (n \geq 1)$$

for the constants

$$r_1 = \frac{25 - \sqrt{629}}{2}, \qquad r_2 = \frac{25 + \sqrt{629}}{2},$$
$$A_1 = \frac{\sqrt{629} - 27}{2\sqrt{629}}, \qquad A_2 = \frac{\sqrt{629} + 27}{2\sqrt{629}},$$

which the tenacious reader can verify indeed satisfies (E.19).[7] Thus, e.g., we can now quickly calculate that there are $y(50) \approx 8.87 \times 10^{69}$ sequences of length 50 in which no occurrence of Q is immediately followed by a letter other than U—as compared with the $26^{50} \approx 56.06 \times 10^{69}$ unconstrained sequences of length 50.

E.3 Transforms

Often, an analytical problem is made easier by converting the functions $x(\cdot)$ appearing in it into functions $X(\cdot)$ of another "kind", then directly manipulating the latter functions to essentially solve the problem. Such a conversion $\mathcal{T}$ of functions $x(\cdot)$ into corresponding functions $X(\cdot)$ is called a **transform**; and, the action of this mapping $\mathcal{T}$ may be expressed by writing either

$$x(\cdot) \xrightarrow{\mathcal{T}} X(\cdot) \tag{E.20a}$$

or

$$X(\cdot) = \mathcal{T}\{x(\cdot)\}. \tag{E.20b}$$

The general idea of using a transform to solve problems is illustrated in Figure E.3. All functions $x(\cdot)$ of the kind used to express the given problem are collected into an original domain of functions; in particular, this domain includes the unknown solution $y(\cdot)$ to the problem, which is to be found. (In the figure, each domain is indicated by a circle, with the points within representing individual functions.) Per (E.20), each function $x(\cdot)$ is mapped, via the chosen transform $\mathcal{T}$, to a function $X(\cdot)$ of another kind, the

7. Likewise, equation (P4.32.1) in Problem 4.32 for the Fibonacci numbers was obtained by solving the difference equation plus boundary conditions in (4.49).

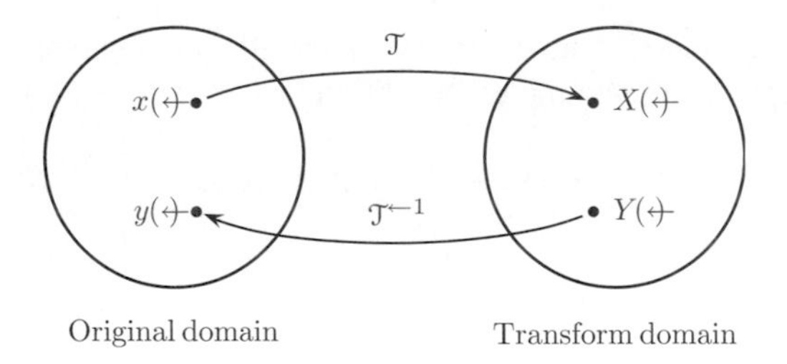

FIGURE E.3 The General Idea of a Transform $\mathcal{T}$

collection of all such functions being another domain—the transform domain.[8] Once the problem has been moved into the transform domain, the goal is to manipulate the functions there directly to obtain a "solution" $Y(\cdot)$; that is, $Y(\cdot) = \mathcal{T}\{y(\cdot)\}$. Finally, assuming (as usual) that the transform $\mathcal{T}$ is one-to-one—i.e., for each function $X(\cdot)$ in the transform domain there is exactly one function $x(\cdot)$ in the original domain such that $X(\cdot) = \mathcal{T}\{x(\cdot)\}$—we can convert $Y(\cdot)$ into the actual solution $y(\cdot)$ by applying the inverse transform $\mathcal{T}^{-1}$:

$$y(\cdot) = \mathcal{T}^{-1}\{Y(\cdot)\}.$$

One benefit of this approach is that if the work required to move between the two domains (i.e., the work of applying $\mathcal{T}$ and $\mathcal{T}^{-1}$) is relatively small, and if solving for $Y(\cdot)$ in the transform domain is significantly easier than solving for $y(\cdot)$ in the original domain, then the total work expended in the roundabout process just described will actually be less than what would be expended by solving the problem directly in the original domain. Additionally, viewing a physical situation from the perspective of a transform domain can often lead to conceptual insights that, though *possible* to obtain without utilizing a transform, would be more difficult to recognize directly in the original domain.

Among the many transforms that exist, six in particular receive the most use in science and engineering. For functions $x(\cdot)$ of a continuous variable—e.g., functions $x(t)$ of time $t \in \mathbb{R}$—there are the **Laplace transform**, the **Fourier transform**, and the **Fourier series**. Of these, the Laplace transform is the most powerful; the Fourier transform is a special case that is somewhat simpler. Likewise, the Fourier series, which applies only to periodic functions, can be viewed as a simplification of the Fourier transform. For functions $x(\cdot)$ of a discrete variable (i.e., sequences)—e.g., functions $x(n)$ of time $n \in \mathbb{Z}$—the analogous transforms are the z **transform**, the **discrete-time Fourier transform (DTFT)**, and the **discrete Fourier transform (DFT)**, respectively.

8. Regarding a function's "kind": The original domain and transform domain need not, in fact, differ with respect to what *functions* they contain; e.g., the function sine might very well belong to both domains. Rather, the two domains differ in that the *significance* of a function taken from one domain differs from that of a function (perhaps the same) taken from the other domain. For example, in signal processing problems, the original domain contains functions of time (and thus is referred to as the "time domain"), and the transform domain typically contains functions of frequency (and thus is referred to as the "frequency domain"). By contrast, in quantum mechanics, the original domain might contain functions of distance, with the transform domain containing functions of momentum.

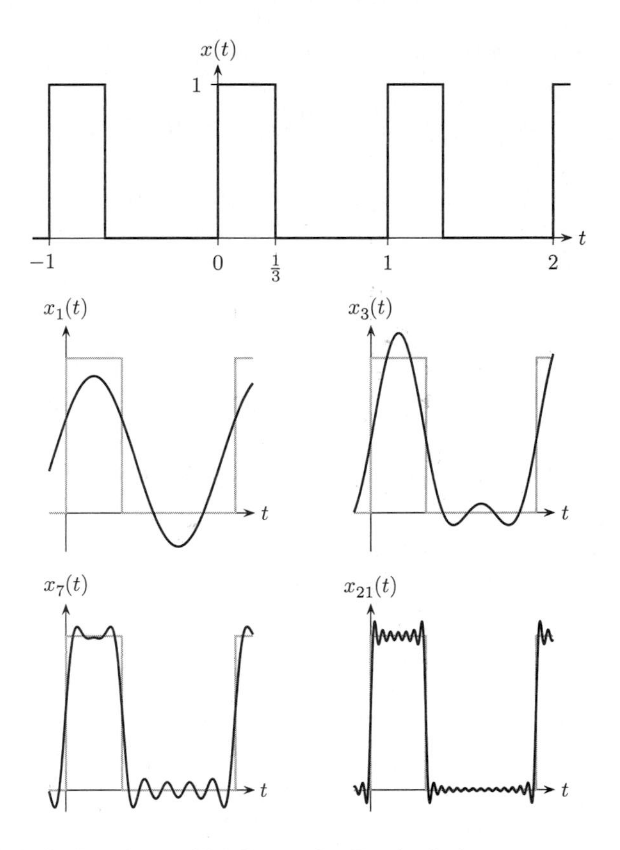

FIGURE E.4 A Periodic Function and Its Converging Fourier Series

Thus, the z transform is more powerful than the simpler DTFT, which is more powerful than the simpler DFT, which applies only to periodic functions.[9]

As an example, consider using the Fourier series to analyze the periodic function $x(\cdot)$ plotted at the top of Figure E.4. The period of this function is 1; and, over the one-period interval $0 \leq t < 1$ we have

$$x(t) = \begin{cases} 1 & 0 \leq t < \frac{1}{3} \\ 0 & \frac{1}{3} \leq t < 1 \end{cases},$$

thereby defining $x(t)$ for all t. In general, a T-periodic ($T > 0$) function $x(\cdot)$ can be expressed as a Fourier series

$$x(t) = \sum_{k=-\infty}^{\infty} X(k) e^{i2\pi kt/T} \quad (t \in \mathbb{R}) \tag{E.21a}$$

9. There is also the **fast Fourier transform (FFT)**; however, it is not really another transform, but an *algorithm* for calculating the DFT efficiently. For example, the DFT of a periodic sequence having period $2^{10} = 1024$ (a common value) can be calculated about 100 times faster by using an FFT.

(assuming the series converges) having coefficients

$$X(k) = \frac{1}{T}\int_{t=0}^{T} x(t)e^{-i2\pi kt/T}\,dt \quad (k = 0, \pm 1, \pm 2, \ldots) \tag{E.21b}$$

(assuming the integrals exist).[10] Hence, for the Fourier series, the original domain comprises periodic functions $x(\cdot)$ of a real variable, whereas the transform domain comprises functions $X(\cdot)$ of an integer variable.[11] Performing the integrals (E.21b) for our 1-periodic function $x(\cdot)$ yields

$$X(k) = \begin{cases} \frac{1}{3} & k = 0 \\ e^{-i\pi k/3}\dfrac{\sin \pi k/3}{\pi k} & k \neq 0 \end{cases};$$

and, substituting these results into (E.21a), then performing a bit of manipulation, we obtain

$$x(t) = c_0 + \sum_{k=1}^{\infty} c_k \cos(2\pi kt/T + \theta_k) \quad (t \in \mathbb{R})$$

with

$$c_0 = \tfrac{1}{3}, \quad c_k = \frac{2 \sin \pi k/3}{\pi k}, \quad \theta_k = -k\pi/3 \quad (k = 1, 2, \ldots).$$

Thus, by applying the Fourier series we have decomposed the periodic function $x(\cdot)$ into a constant plus an infinite sum of *sinusoids* (see Problem 11.21), which are fundamental waveforms in the theory and practice of signal processing. Finally, to see how this sum converges to $x(\cdot)$, Figure E.4 also shows plots of the partial sum

$$x_n(t) = c_0 + \sum_{k=1}^{n} c_k \cos(2\pi kt/T + \theta_k) \quad (t \in \mathbb{R})$$

for $n = 1, 3, 7$, and 21.

---

10. In contrast to our earlier discussion of integration, in which all integrands were real-valued, here that restriction is removed. This is done by simply defining, for any function $f : \mathbb{R} \to \mathbb{C}$ and limits $a, b \in \overline{\mathbb{R}}$, the integral

$$\int_{x=a}^{b} f(x)\,dx \triangleq \int_{x=a}^{b} \operatorname{Re} f(x)\,dx + i\int_{x=a}^{b} \operatorname{Im} f(x)\,dx$$

(when the integrals on the right exist)—as motivated by (E.10b), (E.10c), and the fact that $f(x) = \operatorname{Re} f(x) + i \operatorname{Im} f(x)$ (all x). Similarly, motivated by (E.3), the derivative of such a function $f(\cdot)$ is defined as

$$\frac{df(x)}{dx} \triangleq \frac{d \operatorname{Re} f(x)}{dx} + i\frac{d \operatorname{Im} f(x)}{dx}$$

(when the derivatives on the right exist).

11. The Fourier series (E.21a), which converts $X(\cdot)$ into $x(\cdot)$, actually performs the *inverse* transform; whereas the calculation in (E.21b) of the Fourier series coefficients, which converts $x(\cdot)$ into $X(\cdot)$, is the transform itself.

E.4 Probability and Random Processes

Many of the phenomena studied in science and engineering exhibit some form of randomness. As a practical example, turn on any ordinary radio and tune it to a place on its frequency band where no transmitted signal is present. What you will hear is a hissing sound caused by electrical **noise** originating from inside the radio itself, due mainly to the thermal jostling of electrons in the circuitry. You might also hear various buzzing, crackling, and clicking sounds due to electrical **interference** from outside the radio, often caused by electrical appliances, lightning, and automobile ignition systems. Although such noise and interference signals are random in that they cannot be exactly predicted before being encountered, they are not *wholly* unpredictable. For one thing, it is known from physics that the power of thermal noise is approximately proportional to absolute temperature. Thus, one way to improve the reception quality of a radio receiver (which is indeed done for deep-space communication) is to cool the portion of the circuitry that is most sensitive to thermal noise.

In physics theory, randomness is an essential aspect of quantum mechanics. Any particle (e.g., an electron) can be associated with a complex-valued function of both space and time called a **wave function**, which evolves according to a differential equation called the **Schrödinger wave equation**. The squared magnitude of this function tells us where the particle is likely to be, and the squared magnitude of its Fourier transform tells us what its momentum is likely to be. It then follows from the properties of the Fourier transform that the more certain we are of the particle's position (respectively, momentum), the less certain we must be of its momentum (position); this is the **Heisenberg uncertainty principle**. Philosophically, most scientists conclude from quantum mechanics that the real world is *inherently* random; others, though, still hold the classical belief in a deterministic world, viewing the randomness in quantum mechanics as reflecting our incomplete understanding of physical law.

To understand and analyze random phenomena, an undergraduate student in science or engineering might take a course in probability and random processes. Such a course typically starts with a study of random *events*, then progresses to random *variables*, and culminates with an introduction to random *processes*. For most students, the material is challenging; because, not only is it mathematically intense, but a special effort is usually required to reconcile one's intuitive conception of "likelihood" with the mathematical treatment of it.

A **(random) event** is simply something that could possibly happen, to which a probability can be assigned. For example, when flipping a coin, either the flip turns out to be a head (one event) or a tail (another event). In Section 8.2, a probability of $\frac{1}{2}$ was assigned to each of these events, as motivated by the principle of indifference; however, this principle applies only when the number of basic events is finite *and* there is no reason to suspect that any one of them is more likely to occur than any other. This, of course, is not always the case; e.g., for a coin that is known to be biased, we might instead have $\Pr(\text{flip} = \text{HEAD}) = \frac{1}{3}$ and $\Pr(\text{flip} = \text{TAIL}) = \frac{2}{3}$. The practical significance of the probability p of a random event is provided by the **law of large numbers**, which states what to expect on average if the same random experiment (e.g., flipping a coin) is performed repeatedly. Loosely put, if the experiment is performed n times for some large number n, and k is the number of times that the event occurs, then it is highly likely we will obtain a relative frequency $k/n \approx p$.[12]

In Section 8.2 we discussed conditions under which the probabilities of random events add or multiply when the principle of indifference is applicable; here, for the general case, these operations can be similarly justified. Specifically, let A and B be random events, let $A \vee B$ be the random event that *either A or B* occurs, and let $A \wedge B$ be the random event that *both A and B* occur; also, let their respective probabilities be $P(A)$, $P(B)$, $P(A \vee B)$, and $P(A \wedge B)$. Then, when A and B are **disjoint events**—i.e., it is impossible for both to occur simultaneously—we have $P(A \vee B) = P(A) + P(B)$ (cf. (8.7)). (Thus, e.g., if the probability of team X finishing first is 0.2, and the probability of team Y finishing first is 0.3, then the probability of team X or team Y finishing first is $0.2 + 0.3 = 0.5$.) Whereas when A and B are **(statistically) independent events**—i.e., the occurrence of one has no influence on the likelihood of the other—we have $P(A \wedge B) = P(A) \cdot P(B)$ (cf. (8.8)). (For example, if we flip our biased coin twice—with neither flip affecting the other—then the probability of the first flip being a head and the second flip being a tail is $\frac{1}{3} \cdot \frac{2}{3} = \frac{2}{9}$.)

A **random variable** is a mathematical variable "x" having a value x that varies randomly; in essence, it is a collection of random events that pertain to the value x. For example, for a **real random variable**, we must have $x \in \mathbb{R}$; thus, we might consider the probability of having $x = 0$, or of having $2 \leq x < 5$, or of x being prime—all of which are random events in terms of x. An example of a nonnumerical random variable is a coin flip, for which all events are expressed in terms of the nonnumerical outcomes HEAD and TAIL. For two **statistically independent random variables** x and y, every random event expressible in terms of x alone is statistically independent of every random event expressible in terms of y alone. (For example, two coin flips are statistically independent random variables, since the outcome of one has no influence on the outcome of the other.)

Usually, a real random variable "x" has an associated **probability density (function)** $p_x \colon \mathbb{R} \to [0, 1]$, from which all probabilities pertaining to the value x can be found. Specifically (see Figure E.5), for any constants a and b such that $-\infty \leq a \leq b \leq \infty$, the probability of having $a \leq x \leq b$ equals the area under the curve for $p_x(\cdot)$

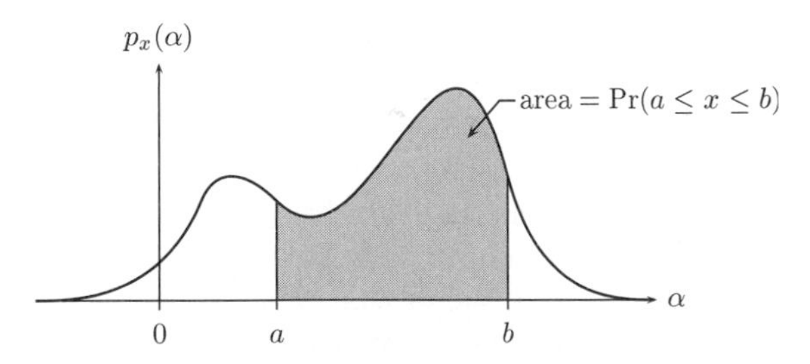

FIGURE E.5 Obtaining a Probability from a Probability Density

12. It would be mistaken, though, to conclude that it is also highly likely that $k \approx pn$ in the sense of having $k - pn \approx 0$; because, it is possible for $k/n \to p$ as $n \to \infty$ yet $k - pn \not\to 0$.

from a to b:

$$\Pr(a \leq x \leq b) = \int_{\alpha=a}^{b} p_x(\alpha)\, d\alpha$$

(the dummy variable "α" being arbitrarily chosen).[13] Hence, roughly speaking, the random value x is more likely to be found where $p_x(\cdot)$ is relatively large. Moreover, the probability density of a random variable can be used to calculate its **mean** (or **expected value**)

$$E\{x\} = \int_{\alpha=-\infty}^{\infty} \alpha p_x(\alpha)\, d\alpha$$

(cf. (8.9), where a sum is performed rather than an integral, and the factor $\frac{1}{n}$ plays the role of the function $p_x(\cdot)$ here), as well as its **variance**

$$E\left\{(x - E\{x\})^2\right\} = \int_{\alpha=-\infty}^{\infty} (\alpha - E\{x\})^2 p_x(\alpha)\, d\alpha$$

(cf. (8.13))—assuming in each case that the integral exists.

Two frequently used probability densities are plotted in Figure E.6. On the left is a **uniform probability density**,

$$p_x(\alpha) = \begin{cases} 1/(b-a) & a \leq \alpha \leq b \\ 0 & \text{otherwise} \end{cases},$$

for any constants a and b such that $-\infty < a < b < \infty$. (For example, one might use this density for the position of the second hand of a clock, with $a = 0$ s and $b = 60$ s.) A random variable with this probability density has mean $(b - a)/2$ and variance $(b - a)^2/12$.

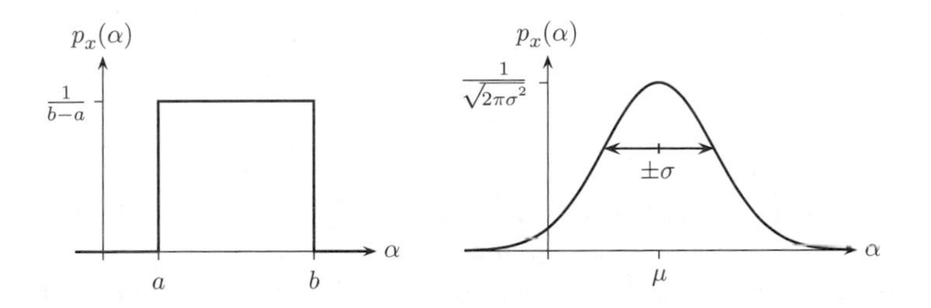

FIGURE E.6 Uniform and Gaussian Probability Densities

---

13. Note that since the area under any single point on the curve is zero, it follows that we now have $\Pr(x = a) = \Pr(a \leq x \leq a) = 0$ when a is finite; and, the same also holds when a is infinite, since the real random variable x cannot take on infinite values. Therefore, $\Pr(a < x < b) = \Pr(a \leq x < b) = \Pr(a < x \leq b) = \Pr(a \leq x \leq b)$ for all a and b.

On the right side of Figure E.6 is the bell-shaped **Gaussian probability density**,

$$p_x(\alpha) = \frac{1}{\sqrt{2\pi\sigma^2}} e^{-(\alpha-\mu)^2/2\sigma^2} \qquad (\alpha \in \mathbb{R}),$$

for any constants $\mu \in \mathbb{R}$ and $\sigma > 0$. In this case, "x" is a **Gaussian** (or **normal**) **random variable** having mean μ and variance σ^2. One reason the Gaussian probability density has special importance is the **central limit theorem**, which states that if many statistically independent, real random variables are added together, then the sum will be (approximately) a Gaussian random variable.[14] For example, the electric current flowing through a cross section of some wire obtains by summing the minute currents contributed by a very large number of electrons moving independently through the cross section; accordingly, the current due to thermal noise is, at each point in time, commonly modeled as a Gaussian random variable.

Finally, having just mentioned time, we have implicitly made the leap from random *variables* to random *processes*. Because, a **random process** "$x(t)$" is essentially a collection of random variables "x", with one at each time t.

E.5 Calculators and Computers

Many physical systems are so complex that it is impractical to completely model them exactly. Furthermore, even when an acceptable model is at hand, using it might be very difficult; that is, the mathematical problems the model produces might be "intractable". One way to proceed in such cases is to choose a simpler model, which is often done. For example, a simplified model is sometimes sufficient to show that even under the best conditions, a particular system design cannot possibly attain the desired behavior. Moreover, analyzing a simple model often gives some idea of what to expect from a better model, and thus can serve to check a more refined analysis of the system.

Fortunately, the availability of powerful, inexpensive, high-speed electronic calculators and computers gives scientists and engineers another avenue of attack on difficult mathematical problems. Namely, rather than (or, better, in addition to) attempting direct mathematical analyses, some problems can be "solved" by numerical calculation. Some examples:

- Weather can sometimes be forecast days in advance by performing simulations based on sophisticated climate models.
- The layout of an electrical circuit containing thousands of components can be optimized (nearly) by systematically examining numerous possibilities.
- The paths in space of several gravitationally interacting bodies can be predicted.[15]

---

14. Another reason is that the Fourier transform of any Gaussian probability density function is itself (essentially) a Gaussian probability density function. This fact has significance with regard to the Heisenberg uncertainty principle.

15. The general analytical problem of finding such paths is called the ***n*-body problem**, with $n \geq 2$ being the number of bodies. Although it involves only n simultaneous differential equations, each having the same fairly simple form, a solution is presently known only for $n = 2$.

Of course, calculators are also used for more mundane tasks, such as computing numerical values of mathematical functions (e.g., $\sin 13 \approx 0.420$).

The use of computing machines, however, should never become a substitute for *thinking*. For one thing, mindless "number crunching" is less likely than thoughtful analysis to provide conceptual insights helpful toward *discovering* a physical law or *designing* a physical system (in contrast to *applying* a given law or *simulating* a given system). Furthermore, many students are too quick to pull out a calculator, either to perform a calculation that can just as well be done mentally or to produce a result having unnecessary accuracy. For example, since π is slightly greater than 3, only a little thought is required to conclude that a cylinder of diameter 6 cm and height 7 cm has a volume of about 200 cm^2. Indeed, scientists and engineers are renowned for making "back-of-an-envelope" calculations, in which sufficiently accurate *estimates* of desired values are quickly obtained. For not only does this skill save time, it helps develop one's physical intuition, and is also good for checking more elaborate calculations.[16]

Some calculators are programmable; that is, the calculator can be given a set of instructions by which a desired series of calculations will be performed automatically. This capability is especially useful for repetitively performing lengthy computations, such as plotting a curve point by point from an equation or performing a recursion such as (E.18).

As for programming a computer, the typical undergraduate student in science or engineering takes at least one course in some **programming language**, of which there are many (BASIC, FORTRAN, Pascal, C, C++, Java, . . .). Happily, although each language has its strengths and weaknesses, much of what can be done with one can be done with any other; indeed, many key features such as variables, statements, and control structures (e.g., a "while loop") are pretty much universal, which makes learning an additional programming language much easier than learning one's first. Accordingly, whereas educators may disagree about *which* language is the best to start with,[17] it is important that the beginner learn programming from a source (e.g., a textbook) that assumes no previous experience.

In order to execute (i.e., to "run") a program, most programming languages require a **compiler**. This is *itself* a computer program, which converts ("compiles") a file provided by a programmer into another file that the computer actually runs. That is, the purpose of the compiler is to translate the "source code" written by the programmer, in a language convenient for *humans*, into a form that can be efficiently executed on a particular *machine*. For many programming languages, free compilers can be downloaded from the Internet.

There also exist various computer applications that provide programming environments in which higher-level operations (e.g., matrix arithmetic) and routines (e.g., numerical integration) are preprogrammed for the user. For example, the program Scilab, which was used to perform many of the calculations in this book (e.g., the matrix calculation (12.39) to prepare Figure 12.3), is freely available.[18] Such software, which ranges from

---

16. For practice, try to estimate how many jelly beans will fit into a typical candy jar.
17. See, e.g., [32] and [38].
18. See [56].

very general to very specialized, finds wide usage throughout science and engineering, though its added convenience usually comes at the cost of reduced computational speed.

Finally, speaking of computer programs, many mathematicians, scientists, and engineers choose to prepare their documents with either the typesetting program TEX or its easier-to-use offspring LATEX, rather than ordinary word processors.[19] These programs are especially suited for formatting mathematical expressions, and both are freely available.[20] Moreover, their use is encouraged by the American Mathematical Society (AMS),[21] which supplies its own related tools (some, along with LATEX, having been used to write this book).

E.6 Numerical Analysis

Whereas the real and complex variables in most mathematical analyses are allowed to vary continuously, the memory in a digital computer is discrete; therefore, the numbers that are expressible in a typical programming language have limited resolution. Accordingly, whenever a digital computer is used, one must be mindful of quantization error occurring, and accumulating, in the calculations performed.[22] A simple example obtains by evaluating $1 \div 3$ on a calculator, then multiplying the result by 3; for unless the calculator is unusual, the result will only approximate the correct value of 1.[23] As a less trivial example, calculating the derivative of a function on a digital computer cannot generally be done by numerically evaluating the limit in definition (E.1); because, both the numerator and denominator of the fraction there approach zero, whereas the limited resolution of the computer precludes having nonzero numbers of arbitrarily small magnitude. Nor is the obvious compromise of choosing $|\Delta x|$ as small as possible in (E.1) a good approach; for then we must calculate the difference $f(x + \Delta x) - f(x)$ of two nearly equal numbers, which (as discussed in Subsection 12.2.2) can greatly magnify any errors present in $f(x + \Delta x)$ and $f(x)$.

The study of techniques for solving computational problems both accurately and quickly is called **numerical analysis** (or **numerical methods**). Although this subject typically is not required coursework for science and engineering students, a basic knowledge is valuable nonetheless, especially for those who will regularly program numerical algorithms.[24] Moreover, it often enables one to make fast and reliable calculations conveniently on a calculator, rather than resorting to some software package or programming language on a computer. In this brief introduction, though, only a few topics in the area will be broached.

---

19. The main documentation for TEX (written "TeX" and pronounced *TECK*) is [34], and that of LATEX (written "LaTeX" and pronounced *LAY-teck* or *LA-teck*) is [36].

20. See [13].

21. See [2].

22. There do exist *analog* computers, which operate in a continuous fashion, and thus do not suffer from quantization error. However, their accuracy is limited by internal noise, a problem that is generally negligible in digital computers.

23. Should it appear that your calculator gives the correct result, subtract 1 from it to be sure.

24. An **algorithm** is any well-defined, step-by-step computational procedure.

As an example of a speed consideration, a power x^y ($x, y \in \mathbb{R}$ appropriately restricted) can be calculated by using the built-in power function of some programming language, but this is typically much slower than performing a single multiplication in that language. Therefore, when the power y is known at the outset, and is a small positive integer, it might be faster to explicitly program this calculation as a repeated multiplication—e.g., $x^3 = x \cdot x \cdot x$. Likewise, rather than evaluate a polynomial directly, it is generally more efficient to convert it into another form by using **Horner's rule**,

$$a_n x^n + a_{n-1} x^{n-1} + a_{n-2} x^{n-2} + \cdots + a_1 x + a_0 \\ = (\cdots((a_n x + a_{n-1})x + a_{n-2})x + \cdots + a_1)x + a_0 \qquad \text{(E.22)}$$

($n \geq 0$; $a_0, a_1, \ldots, a_n, x \in \mathbb{C}$), which avoids calculating powers altogether while leaving the number of multiplications and additions (n each) unchanged.

A major topic in numerical analysis is solving simultaneous linear equations. As discussed in Section 9.2, an equivalent problem is solving the matrix equation $\mathbf{Ax} = \mathbf{b}$ for an unknown $n \times 1$ matrix $\mathbf{x}$, given an $m \times n$ matrix $\mathbf{A}$ and an $m \times 1$ matrix $\mathbf{b}$ (m, $n \geq 1$). For the simple case of $m = n$, we know that if $\mathbf{A}$ is nonsingular, then $\mathbf{x} = \mathbf{A}^{-1}\mathbf{b}$. For example, since

$$\mathbf{A} = \begin{bmatrix} 9 & 8 \\ 8 & 7 \end{bmatrix} \quad \Rightarrow \quad \mathbf{A}^{-1} = \begin{bmatrix} -7 & 8 \\ 8 & -9 \end{bmatrix},$$

we obtain

$$\mathbf{A} = \begin{bmatrix} 9 & 8 \\ 8 & 7 \end{bmatrix} \quad \wedge \quad \mathbf{b} = \begin{bmatrix} 1 \\ 1 \end{bmatrix} \quad \Rightarrow \quad \mathbf{x} = \begin{bmatrix} 1 \\ -1 \end{bmatrix}.$$

But, note also that

$$\mathbf{A} = \begin{bmatrix} 9 & 8 \\ 8 & 7 \end{bmatrix} \quad \wedge \quad \mathbf{b} = \begin{bmatrix} 1 \\ 1.01 \end{bmatrix} \quad \Rightarrow \quad \mathbf{x} = \begin{bmatrix} 1.08 \\ -1.09 \end{bmatrix};$$

thus, a 1% variation in one of components components of $\mathbf{b}$ (e.g., due to quantization error or experimental error) has produced variations of 8% and 9% in the components of the solution $\mathbf{x}$. This is an example of an **ill-conditioned problem**, for which small numerical errors in the values given in the problem can lead to significantly larger errors in the solution. Thus, one of the concerns of numerical analysis is to recognize when an ill-conditioned problem is at hand.

Another major topic is how, given a function $f : D \to \mathbb{C}$ having domain $D \subseteq \mathbb{C}$, to find all values $x \in D$ for which $f(x) = 0$—each such value being called a **root** of $f(\cdot)$. As a simple example (with $D = \mathbb{R}$), suppose we wish to determine all of the values $x \in \mathbb{R}$ for which

$$3e^x = 5x + 6. \qquad \text{(E.23)}$$

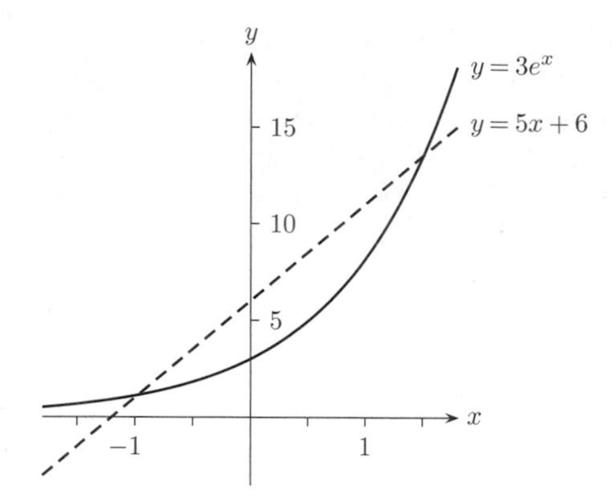

FIGURE E.7 Two Intersecting Functions

As is evident from Figure E.7, this condition is satisfied by some value $x \approx -1.0$ and another value $x \approx 1.5$. Letting

$$f(x) = 3e^x - 5x - 6 \quad (x \in \mathbb{R}), \tag{E.24}$$

the problem is equivalent to finding the roots of $f(\cdot)$.

One way to attack this problem is to perform a "brute force" search in the vicinities of $x = -1.0$ and $x = 1.5$, systematically hunting down each of the two roots by repeatedly guessing its value x and calculating $f(x)$; indeed, sometimes this is a reasonable approach. However, an easier tack that often works is to take the condition to be met—here (E.23), which is in terms of the variable "x"—and "solve" for one of the occurrences of the variable in terms of the others; then, the resulting equation is iterated recursively, starting with some initial value for the variable. Hence, solving for the "x" on the right side of (E.23), we obtain the recursion

$$x_{i+1} = \frac{3e^{x_i} - 6}{5} \quad (i = 0, 1, \ldots) \tag{E.25a}$$

—for some initial value x_0—whereas solving for the "x" on the left side of (E.23) yields the recursion

$$x_{i+1} = \ln \frac{5x_i + 6}{3} \quad (i = 0, 1, \ldots), \tag{E.25b}$$

in which it is required that $x_i > -6/5$ (all i).[25] The results of performing these recursions on a 10-digit calculator are shown in Table E.2 for $x_0 = 0$, with each column of values x_i being terminated upon the first occurrence of a previous value (which must happen eventually). Thus, for this initial value, (E.25a) converges to the root near −1 and (E.25b) converges to the root near 1.5.

25. As another example, the recursion (P4.8.1), which converges to $x = \sqrt{a}$ for any constant $a > 0$, obtains by this approach; because, since $a > 0$, we have $x = \sqrt{a} \Leftrightarrow x^2 = a \Leftrightarrow x = a/x \Leftrightarrow 2x = x + a/x \Leftrightarrow x = (x + a/x)/2$.

TABLE E.2 Results of the Recursions (E.25) for $x_0 = 0$

i	x_i from (E.25a)	x_i from (E.25a)
1	−0.6000000000	0.6931471806
2	−0.8707130184	1.149066243
3	−0.9488082386	1.364843530
4	−0.9676786792	1.452723099
5	−0.9720215798	1.486412321
6	−0.9730095206	1.499032218
7	−0.9732336630	1.503718895
8	−0.9732844854	1.505453818
9	−0.9732960072	1.506095292
10	−0.9732986192	1.506332368
11	−0.9732992114	1.506419972
12	−0.9732993456	1.506452342
13	−0.9732993760	1.506464302
14	−0.9732993830	1.506468721
15	−0.9732993846	1.506470354
16	−0.9732993850	1.506470957
17	−0.9732993850	1.506471180
18		1.506471263
19		1.506471294
20		1.506471305
21		1.506471309
22		1.506471311
23		1.506471311

A popular alternative approach is the **Newton-Raphson method**, which is known for its fast convergence—*when* it converges. Here, given a function $f : D \to \mathbb{R}$ having domain $D \subseteq \mathbb{R}$, one performs the recursion

$$x_{i+1} = x_i - \frac{f(x_i)}{f'(x_i)} \quad (i = 0, 1, \ldots)$$

(assuming division by 0 is not attempted), choosing the initial value x_0 so that x_i will approach the desired root of $f(\cdot)$. For $f(\cdot)$ given by (E.24), we obtain the derivative $f'(x) = 3e^x - 5$ $(x \in \mathbb{R})$, and thus

$$x_{i+1} = x_i - \frac{3e^{x_i} - 5x_i - 6}{3e^{x_i} - 5} \quad (i = 0, 1, \ldots). \tag{E.26}$$

The results of performing this recursion on a 10-digit calculator are shown in Table E.3 for $x_0 = 0$ and $x_0 = 1$. Comparing with Table E.2, we see that this method is indeed faster. Usually, when a Newton-Raphson recursion nears a root, the number of digits of accuracy roughly doubles for each iteration (as allowed by the resolution of the computing machine). Table E.3 also shows that there need not come a point at which x_i repeats the same value indefinitely; rather, for $x_0 = 1$, it happens that x_i eventually cycles between *two* values.

Finally, although a digital computer is a deterministic device—specifically, its next "state" is always uniquely determined by its external input (if any) and its present state

TABLE E.3 Results of the Newton-Raphson Recursion (E.26)

i	x_i for $x_0 = 0$	x_i for $x_0 = 1$
1	−1.500000000	1.901836407
2	−0.9990565485	1.598059956
3	−0.9733942696	1.512610039
4	−0.9732993863	1.506501027
5	−0.9732993850	1.506471312
6	−0.9732993850	1.506471311
7		1.506471310
8		1.506471311

(including the program in memory being run)—one can nevertheless *simulate* randomness on a digital computer, thereby providing a convenient alternative to flipping coins or throwing dice (as, e.g., in Table 8.1). This is typically done via a **pseudorandom number generator** (often simply called a **random number generator**), which is an algorithm that produces a sequence of numbers *appearing* to be random (perhaps satisfying some statistical tests for randomness). Such an algorithm usually requires an initial value called a **seed** to get started, after which it continues indefinitely with no further input by recycling the random numbers it generates as subsequent seed values. Being deterministic, an advantage of a *pseudo*random number generator is its repeatability, since the exact same number sequence results whenever the same seed value is used. Numerical methods that utilize pseudorandom numbers to solve computational problems (including problems such as integration that have no inherent randomness) are generally referred to as **Monte Carlo methods** (after the gambling resort).

One well-known pseudorandom number generator[26] is the simple recursion

$$seed_{i+1} = (16807 \cdot seed_i) \bmod 2147483647 \quad (i = 0, 1, \ldots), \tag{E.27}$$

where the initial seed, $seed_0$, is chosen to be any integer ranging from 1 to $2^{31} - 2 = 2\,147\,483\,646$. The result is a sequence of numbers $seed_i$ that take on every value in this range before repeating,[27] with no apparent correlation between successive values. Accordingly, to obtain a pseudorandom sequence of real numbers x_i that are uniformly distributed over a chosen interval $(a, b]$,[28] where $-\infty < a < b < \infty$, one can let

$$x_i = a + (b - a)(seed_i/2147483646) \quad (i = 1, 2, \ldots). \tag{E.28}$$

For example, the probability histogram shown on the left side of Figure E.8, which is plotted in gray over a desired uniform probability density (per the left side of Figure E.6)

---

26. See [44].

27. Therefore, the recursion must be run with the variable $seed_i$ having at least 46 bits of resolution, to be able to hold the number $16807 \cdot (2^{31} - 2) = 36\,092\,757\,638\,322 = 1\,000\,001\,101\,001\,101\,111\,111\,111\,111\,110\,111\,110\,010\,110\,010_2$. In particular, the recursion will eventually fail if run on a calculator having an internal resolution of less than 14 digits.

28. This interval from a to b excludes a but includes b because in the following equation, $x_i = a$ never occurs but $x_i = b$ does.

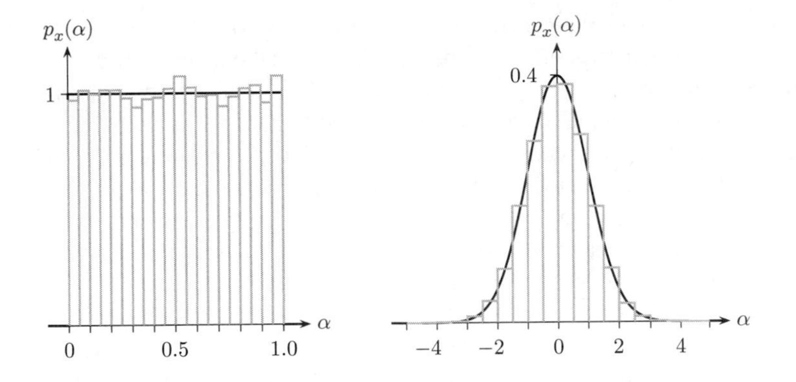

FIGURE E.8 Probability Histograms Generated Pseudorandomly

for $a = 0$ and $b = 1$, was obtained by starting (E.27) with $seed_0 = 1$ then taking the first 10 000 samples x_i from (E.28).

Likewise, a pseudorandom sequence of Gaussian samples x_i $(i = 1, 2, \ldots)$ having any mean $\mu \in \mathbb{R}$ and standard deviation $\sigma > 0$ can be generated via (E.27) by letting

$$w_i = \sqrt{-2\sigma^2 \ln(seed_i/2147483646)}, \quad w_{i+1} = 2\pi(seed_{i+1}/2147483646) \qquad (i = 1, 3, 5, \ldots) \quad \text{(E.29a)}$$

and then

$$x_i = w_i \cos w_{i+1} + \mu, \quad x_{i+1} = w_i \sin w_{i+1} + \mu \quad (i = 1, 3, 5, \ldots). \qquad \text{(E.29b)}$$

For example, the probability histogram shown on the right side of Figure E.8, which is plotted in gray over a desired Gaussian probability density (per the right side of Figure E.6) for $\mu = 0$ and $\sigma = 1$, was obtained by starting (E.27) with $seed_0 = 1$ then taking the first 10 000 samples x_i (two at a time) from (E.29).

E.7 The Foundations of Mathematics

As we have seen, many mathematical developments find significant use in science and engineering; moreover, many others (e.g., linear algebra, real analysis, complex analysis, the calculus of variations, graph theory) could be cited. However, mathematics can also be appreciated in its own right, irrespective of any potential application. In this final section, then, we broach a few thought-provoking fundamental issues, perhaps arousing the reader's curiosity enough for them to pursue additional mathematical knowledge for its own sake.[29]

29. A good place to start on the following topics is several readable articles that have appeared in *Scientific American* magazine: [10], [15], [17], [26], [61]. Also recommended are the books [7], [16], [33], [35], [50], and [53].

To begin, those who view mathematics as a towering edifice of certainty (as do most people) might be surprised to learn that not all of its foundational matters receive universal agreement, even among mathematicians. Philosophical controversies exist, including disagreements on which logical rules are acceptable, where the line separating logic and mathematics should be drawn, what mathematical objects exist, and how mathematical theorems should be interpreted. Though the working mathematician typically continues onward unperturbed, apparently convinced that however these matters are decided the vast majority of mathematics will be unaffected, it is still a bit unsettling to think that the ground on which the edifice rests might not be totally secure.

One school of thought in the philosophy of mathematics is **logicism**, the basic tenet of which is that mathematics is actually a part of logic. The first significant attempt toward reducing mathematics to logic was provided by Gottlob Frege, starting in 1879. Unfortunately (and giving an indication of the difficulty of this endeavor), a serious flaw in Frege's work was revealed in 1901 when Bertrand Russell discovered the paradox that now bears his name (see Chapter 2). Nevertheless, Russell himself worked around the problem, and along with Alfred North Whitehead published the three-volume opus *Principia Mathematica*. Though widely acclaimed, just how well the work achieves the goals of logicism is less than clear. In particular, logician Willard Van Orman Quine has argued that the set-membership predicate "$\in$" properly belongs to mathematics, not logic, and thus the latter cannot contain the former.[30]

Another school of thought is **formalism**, an extreme form of which asserts that mathematics is nothing more than a game of manipulating printed symbols on a page according to arbitrary rules; and, for whatever reasons, this game is deemed worthy of intense study. The main attraction of this position is that by attributing no meaning to mathematical objects and theorems, the mathematician leaves all responsibility of interpreting the theory to those who choose to use it.

Most mathematicians, though, seem to take the polar opposite position called **Platonism** (after Plato), which asserts that all possible mathematical objects and their interrelationships already exist in some ideal nonphysical realm, independent of human thought; therefore, all mathematical truth is objective (i.e., is the same for everyone) and has only to be discovered. Accordingly, when putting forth a mathematical axiom (e.g., that the union of two sets always exists), the Platonist cannot proceed arbitrarily, but must try to express what is *really* is the case. However, a problem with Platonism is that if all possible mathematical objects already exist, including all sets, then it would seem that we can conceive of collecting all these sets together into a single whole; but, as was seen in Chapter 2, assuming the existence of a set of all sets leads to a logical contradiction.

Yet another school of thought is **constructivism**, which particularly criticizes how the concept "infinity" is treated in modern mathematics. By this view, a mathematical object exists only upon being constructed (e.g., $3 = 1+1+1$), or upon giving an explicit procedure by which its construction can be performed, if only in principle (i.e., the procedure might require more steps than is humanly possible to carry out). Accordingly, a constructivist views infinity not as a completed object that *is* infinite, but rather as an

30. See, e.g., [48], p. 72.

unbounded process of *becoming* infinite (such as counting indefinitely). Thus, whereas the Platonist speaks of **actual infinity**, the constructivist speaks only of **potential infinity**.

The legitimacy of actual infinity was vigorously debated upon the remarkable discovery by Georg Cantor, in the 1870s and 1880s, that actual infinity (if accepted) comes in different sizes. First, two finite or infinite sets A and B have the same size, or cardinality—i.e., $|A| = |B|$—if and only if there exists a one-to-one correspondence between A and B. That is, there exists a set $R \subseteq A \times B$ such that for each $x \in A$ there is exactly one $y \in B$ for which $(x, y) \in R$, and for each $y \in B$ there is exactly one $x \in A$ for which $(x, y) \in R$. Notably, it is possible to have $|A| = |B|$ even when $A \subset B$; for example, although $\mathbb{N} \subset \mathbb{Z}^+$, we have $|\mathbb{N}| = |\mathbb{Z}^+|$, as seen by the one-to-one correspondence $R = \{(0, 1), (1, 2), (2, 3), \ldots\} = \{(x, x + 1) \mid x \in \mathbb{N}\} \subseteq \mathbb{N} \times \mathbb{Z}^+$. Indeed, one definition of an "infinite set" is a set that has a one-to-one correspondence with some proper subset.

Using a clever "diagonal argument", Cantor proved that $|\mathbb{Z}^+| < |\mathbb{R}|$, and thus $|\mathbb{Z}^+|$ and $|\mathbb{R}|$ represent *different* infinities. To wit: Let each nonzero number $y \in \mathbb{R}$ be expressed by its unique infinite decimal expansion,

$$y = \pm y_n \ldots y_1 y_0 . y_{-1} y_{-2} y_{-3} \ldots$$

($n \geq 0$), with $y_n = 0$ being allowed only when $n = 0$; and, let the number 0 be represented by the expansion

$$0 = 0.000 \ldots$$

(i.e., no minus sign and all digits zero). Now, *assuming* that a one-to-one correspondence exists between $\mathbb{Z}^+$ and $\mathbb{R}$, we can form an infinite list of the numbers $y \in \mathbb{R}$ versus the corresponding numbers $x \in \mathbb{Z}^+$, as illustrated in Table E.4. At the bottom of the table, another decimal expansion

$$z = 0.z_{-1} z_{-2} z_{-3} \ldots$$

is then constructed by simply changing the digits (shown underscored) along a diagonal in the list of y-expansions; namely, for each x we choose $z_{-x} \in \{1, 2\}$ to be other than y_{-x}. It follows that the resulting number z differs from every number y in the list. Because, the expansion for z is infinite, and thus $z \neq 0$; and, each nonzero y has only one infinite decimal expansion (viz., the expansion listed), with which the expansion for z differs by

TABLE E.4 Cantor's Diagonal Argument

x	y
1	$89.\underline{2}1132\ldots$
2	$-516.7\underline{5}604\ldots$
3	$1.00\underline{2}92\ldots$
4	$0.000\underline{0}0\ldots$
5	$-4039.6653\underline{8}\ldots$
⋮	⋮
—	$z = 0.12122\ldots$

at least one digit. But, z not being listed as value of y contradicts the assumed one-to-one correspondence between $\mathbb{Z}^+$ and $\mathbb{R}$; therefore, no such correspondence exists—i.e., $|\mathbb{Z}^+| \neq |\mathbb{R}|$. Hence, since $\mathbb{Z}^+ \subseteq \mathbb{R}$, we must have $|\mathbb{Z}^+| < |\mathbb{R}|$.

The existence of various actual infinities—i.e., **transfinite numbers**—is now widely accepted among mathematicians. Such numbers come in two kinds. The **cardinal numbers** (or **cardinals**)—viz., $0, 1, 2, \ldots, \aleph_0, \aleph_1, \aleph_2, \ldots$—are used to express the cardinality $|A|$ of any set A.[31] Those following the natural numbers (0, 1, 2, ...) are the **transfinite cardinals**, with $\aleph_0 \triangleq |\mathbb{N}|$ being the first transfinite cardinal.[32] By contrast, the **ordinal numbers** (or **ordinals**)—viz., $0, 1, 2, \ldots, \omega, \omega+1, \omega+2, \ldots, 2\omega, 2\omega+1, 2\omega+2, \ldots, 3\omega, \ldots, \omega^2, \omega^2+1, \omega^2+2, \ldots, \omega^2+\omega, \omega^2+\omega+1, \ldots$—express the idea of continuing the counting process beyond finite numbers. Those following the natural numbers are the **transfinite ordinals**, with ω being the first transfinite ordinal.[33]

A set A is said to be **countably infinite** when $|A| = \aleph_0$; in particular, $\mathbb{N}$, $\mathbb{Z}$, $\mathbb{Z}^+$, $\mathbb{Z}^-$, and $\mathbb{Q}$ are all countably infinite. A set A is **countable** if it is either finite—i.e., $|A| < \aleph_0$—or countably infinite; otherwise—i.e., $|A| > \aleph_0$—A is **uncountable**. In particular, every interval of length greater than zero—e.g., $[0, 1]$ or $\mathbb{R}$—is uncountable, as are the set of irrational numbers and $\mathbb{C}$; moreover, perhaps remarkably, all of these sets have the *same* cardinality.[34]

Besides its rejection of actual infinity, there are other reasons why constructivism is now out of the mathematics mainstream. One form, **intuitionism**, founded in 1907 by mathematician Luitzen E. J. Brouwer, rejects the law of excluded middle—i.e., the assertion that for every statement A either it or its negation $\neg A$ must be true—which almost all logicians use without question. Intuitionists also attack the validity of the **axiom of choice** used in set theory, one version of which states that if A is a set of disjoint nonempty sets (thus, each element of A is itself a set), then there exists another set C that contains exactly one element from each set in A; hence, C effectively *chooses* a single element from each of these sets. (The objection of the intuitionist is that the "choice set" C appears without having been constructed.) Although seemingly innocuous, as well as useful for proving many important mathematical theorems, the inclusion of this axiom in set theory has some rather bizarre consequences. For example, the **Banach-Tarski paradox** (which despite its name is actually a theorem, not a contradiction) states that any solid sphere can be partitioned into finitely many pieces, which can be rigidly repositioned in space to produce *two* solid spheres having the same radius as the original. (The "trick" is that not all of the pieces have *measurable* volumes.)

One of the most remarkable results in the history of logic, obtained in 1931 by logician Kurt Gödel,[35] was made possible by the emphasis of formalism on performing

---

31. In Chapter 2 we defined the cardinality $|A|$ of a set as the number of elements in A; this definition remains valid for infinite sets, once the transfinite cardinals are accepted.

32. The symbol "$\aleph$" is the first letter, aleph, of the Hebrew alphabet. Ironically, the symbol "ω" about to be used is the last letter, omega, of the Greek alphabet.

33. Thus, the natural numbers are both cardinals and ordinals. Also, the ordinary concept of infinity—viz., ∞—is akin to $\aleph_0$, not ω. For, just as $\infty + 1 = \infty$, we have $\aleph_0 + 1 = \aleph_0$; however, $\omega + 1 > \omega$.

34. Just what this cardinality is, though, is debatable (but, to a Platonist, it is nonetheless an objective fact). According to the **continuum hypothesis**, $|\mathbb{R}|$ equals the cardinal number $\aleph_1$ immediately following $\aleph_0$.

35. The name "Gödel" is pronounced *GIR-del*.

logical deductions *symbolically*. For the utmost precision, such deductions are carried out in a carefully defined language called a **formal language**. Each statement in a formal language is a finite string of symbols taken from a finite set of allowable symbols (e.g., "$\neg$", "$\wedge$", "$\forall$", comma, predicate symbols, left and right parentheses); also, which strings are recognized as statements in the language is determined by the strict rules of a fully specified grammar. Then, certain statements are designated as axioms, and the allowable rules of inference are given, thereby producing a **formal system** (such as that used in *Principia Mathematica*). Moreover, in keeping with the formalist view, the rules of grammar and inference are purely *syntactical*; that is, they refer only to how symbols are arranged within statements, not to what the symbols or statements formed from them may mean.[36] (Of course, how the axioms and rules are chosen typically *does* reflect intended meanings.) Now, what Gödel proved was that for any formal system in which basic arithmetic can be expressed, if the system is *consistent*—i.e., if it is impossible to prove, within the system, both some statement A and its negation $\neg A$—then the system must be *incomplete*; that is, there exists at least one statement A in the system such that, within the system, neither A nor $\neg A$ can be proved. This negative result, which shows that no formal system can capture all mathematical truth (even that of mere arithmetic) is known as **Gödel's (incompleteness) theorem.**

Gödel actually went a bit further, by constructing a particular statement in a formal system that cannot be proved within the system, but *can* be proved from *outside* the system, and therefore is true. Here is an analogy to Gödel's proof: Imagine that you, the reader, *are* the formal system, and you try to prove the following statement:

The reader cannot prove this statement.

Clearly, you cannot do so and be logically consistent; because, the statement says you cannot prove it, and proving it would imply that it is true. However, upon rereading the preceding sentence you will see that I (from outside the system) have just deduced that you cannot prove the statement, which is exactly what it says; hence, there exists a true statement that is impossible for you to prove.

Of course, the above statement is objectionable because it refers to itself, and self-referencing statements are generally not allowed in logic because they sometimes lead to paradoxes.[37] What Gödel did instead was construct a statement that is *effectively* self-referencing, as viewed from *outside* the system; whereas within the system, the statement does not refer to itself, nor is there any other such "vicious circle" of logic.[38] Specifically, by a simple procedure now called **Gödel numbering**, a unique natural

36. For example, note that in modus ponens—i.e., given that statements A and $A \Rightarrow B$ are true, conclude the truth of statement B—we need not know what any of these statements mean in order to apply the rule.

37. The most famous example is the **liar's paradox**, which obtains from the self-referencing statement "I am lying" (equivalently, "This statement is false"); because, assuming the statement is true, or false, leads one to logically draw the opposite conclusion. In ordinary exposition, though, self-referencing does occur occasionally; e.g., consider the instances (like now!) in which this book refers to itself.

38. Note that a vicious circle can be formed from more than one statement; e.g., though not self-referencing individually, the following two statements *together* yield a contradiction:

- The next statement is true.
- The preceding statement is false.

number was assigned to every possible statement in the formal system. A kind of loop was thereby formed, since arithmetical statements in the system can be interpreted as making assertions about numbers, and numbers could now be used (from outside the system) to refer to statements in the system. In particular, provability within the system could be viewed as a *mathematical* property; namely, the numbers assigned to statements provable within the system constitute some set P. The final step (and the difficult one) was to construct within the system a statement that says $n \notin P$, where n is the number assigned to this very statement. Such a statement can then be interpreted as saying

The statement having number n cannot be proved within the system.

But, since this *is* the statement having number n, it effectively say of itself that it is not provable within the system.[39]

As a practical application of Gödel numbering, consider the problem of determining whether a given algorithm (e.g., a computer program), with a given numerical input, will eventually halt (i.e., stop) rather than continue on indefinitely (e.g., enter an "infinite loop"). Just as with statements in a formal language, it can be shown that algorithms can also be numbered. Hence, one might consider designing an algorithm that takes the number x of the given algorithm, along with the number y of its input, and tells us whether algorithm x halts when given input y. This is known in computer science as the **halting problem**.[40]

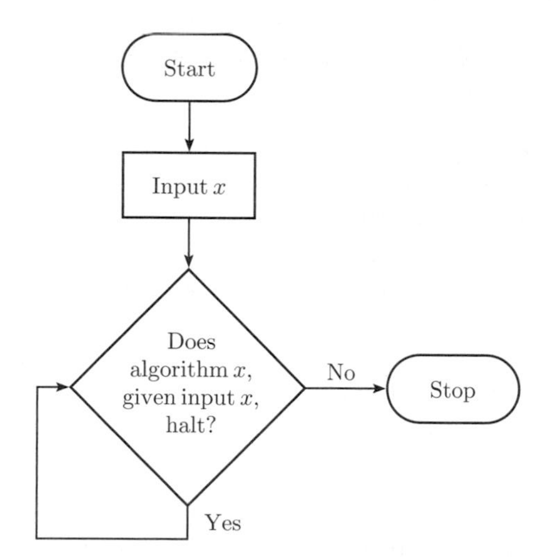

FIGURE E.9 An Algorithm Pertaining to the Halting Problem

39. A more complete overview of Gödel's proof can be found in [41]. For a fully detailed development, [8] is recommended.

40. The problem assumes that the computer on which algorithm x is run has infinite memory. For an algorithm run on a computer having finite memory, the halting problem is easily solved (at least in principle). Just execute the algorithm for a number of computer cycles equal to the number of internal states in the computer; if it hasn't halted by then, it never will.

Unfortunately, the solution to the halting problem is to conclude that no such algorithm exists; moreover, this is so even if we restrict our attention to the case of $y = x$, for which the algorithm is simply to determine whether a given algorithm x halts when its *own* number x is the input. To prove this fact, assume to the contrary that such an algorithm *does* exist for the case of $y = x$, and consider the related algorithm expressed by the **flow diagram** in Figure E.9. The algorithm in the figure utilizes the assumed algorithm, shown as a decision diamond. Now, let the number of the entire algorithm in the figure be n, and consider what happens when this algorithm is run with input $x = n$. If the answer to the question inside the decision diamond is "No"—i.e., algorithm n does *not* halt given input n—then, as the flow diagram shows, the algorithm *does* halt. Whereas if the answer to the question is "Yes"—i.e., algorithm n *does* halt given input n—then, the algorithm enters an infinite loop of asking this same question repeatedly, and thus it does *not* halt. Hence, in either case, a contradiction obtains; therefore, the assumption that the algorithm in the decision diamond exists must be false.

Appendix: Problems with Solutions

Problems

Chapter 1. Logic

In this section, A, B, and C are atomic statements in a sentential logic, unless stated otherwise.

Problem 1.1 Use truth tables to show that the statement $(A \Rightarrow B) \wedge (B \Rightarrow A)$ is logically equivalent to the statement $A \Leftrightarrow B$.

Problem 1.2 Clearly, the statements $A \wedge B$ and $B \wedge A$ are logically equivalent, as are the statements $A \vee B$ and $B \vee A$; thus, the logical connectives $\wedge$ and $\vee$ are *commutative*. Also, the statements $(A \wedge B) \wedge C$ and $A \wedge (B \wedge C)$ are logically equivalent, as are the statements $(A \vee B) \vee C$ and $A \vee (B \vee C)$; thus, $\wedge$ and $\vee$ are also *associative*.[1] Now, consider the *distributive* properties of these connectives with respect to each other:

(a) Are the statements $(A \wedge B) \vee C$ and $(A \vee C) \wedge (B \vee C)$ logically equivalent?

(b) Are the statements $(A \vee B) \wedge C$ and $(A \wedge C) \vee (B \wedge C)$ logically equivalent?

Problem 1.3 Use truth tables to show that the statement $A \Rightarrow (B \Rightarrow A)$ is a tautology.

Problem 1.4 The logical operation of disjunction, indicated by the connective "$\vee$" (read "or"), is sometimes called **inclusive disjunction**, to distinguish it from the operation of **exclusive disjunction**, which we will indicate by the connective "$\veebar$" (read "exclusive-or").[2] As shown in Table P1.4.1, the difference between the statements $A \vee B$ and $A \veebar B$ is that the truth of the latter statement *excludes* the possibility of A and B being *both* true.

(a) Explain why one of the statements $A \vee B$ and $A \veebar B$ logically implies the other, but not vice versa.

(b) Is the statement $A \veebar B$ logically equivalent to the statement $A \Leftrightarrow \neg B$?

Problem 1.5 How many truth tables are possible for a sentential-logic statement A constructed from $n \geq 1$ atomic statements $A_1, \ldots, A_n$? Of these, how many express tautologies?

TABLE P1.4.1 The Truth Table for Disjunction, Inclusive and Exclusive

A	B	$A \vee B$	$A \veebar B$
F	F	F	F
F	T	T	T
T	F	T	T
T	T	T	F

1. It follows that the parentheses in these expressions can be omitted—that is, "$A \wedge B \wedge C$" and "$A \vee B \vee C$" indicate statements having truth values that depend unambiguously on the truth values of A, B, and C—and likewise for any conjunction or disjunction of finitely many statements.

2. Other symbols—e.g., "+"—are also commonly used for this connective.

Problem 1.6 Prove via truth tables the following rule of inference, for which the statements A, B, and C need not be atomic: Given that the conditionals $A \Rightarrow B$ and $B \Rightarrow C$ are true, conclude that the conditional $A \Rightarrow C$ is true.

Problem 1.7 A collection of statements is said to be **logically consistent** if and only if it is *logically possible* for the statements to be simultaneously true, irrespective of their *actual* truth values; otherwise, the collection is **logically inconsistent**.

(a) Explain why $n \geq 1$ sentential-logic statements $A_1, \ldots, A_n$ are logically inconsistent if and only if the single statement $(\neg A_1) \vee \cdots \vee (\neg A_n)$ is a tautology.

(b) In light of DeMorgan's laws, one might suspect that $A_1, \ldots, A_n$ are logically consistent if and only if the statement $A_1 \wedge \cdots \wedge A_n$ is a tautology. Is this so?

Problem 1.8 In general, is the converse of the contrapositive of a conditional identical to the contrapositive of its converse?

Problem 1.9 Use the contrapositive form to obtain an equivalent, though less eloquent, expression of each of the following popular sayings.

(a) Where there's a will, there's a way.

(b) Those who can, do; those who can't, teach.

(c) Anyone who uses the phrase "easy as taking candy from a baby" has never tried taking candy from a baby.

Problem 1.10 Consider the following two statements about an infinite sequence $\{a_k\}_{k=1}^{\infty}$ of numbers a_k:

(i) $a_k = 0$ for all k.

(ii) $a_k \neq 0$ for all k.

(a) Give a sequence for which both statements are false, thereby proving that (ii) is not the negation of (i).

(b) One way to express the negation of (i) is:

We do not have $a_k = 0$ for all k.

Give another way, using "for some" rather than "for all".

Problem 1.11 Consider the following popular saying:

(i) If you have your health, then you have everything.

Letting "H" be a predicate such that "$H(x)$" means "you have x", and letting "h" be a name denoting your health, (i) can be translated into first-order logic as

$$H(h) \Rightarrow [\forall x, H(x)]. \qquad \text{(P1.11.1)}$$

Use this translation to solve what follows.

(a) Clearly, (i) is not true (e.g., one can have health without having money); therefore, it cannot be logically true either. Now, find the converse of (i). Is *it* logically true?

(b) Determine if (i) is logically equivalent to this statement:

(ii) If you have nothing, then you do not have your health.

Problem 1.12 If A is a logically true statement in a first-order logic, and B is any other statement in the same logic, then why must we have $B \models A$?

Problem 1.13 Suppose A, B, and C are statements in a first-order logic.

(a) If $A, B \models C$, then must we have $A \models (B \Rightarrow C)$?

(b) Conversely, if $A \models (B \Rightarrow C)$, then must we have $A, B \models C$?

Problem 1.14 Let the statement

$$\exists! x,\ P(x) \tag{P1.14.1}$$

mean "there exists a *unique* x such that $P(x)$"; that is, for one and only one thing x (in the universe under discussion) do we have $P(x)$. Show how to express this statement solely in terms of the logical connectives in Table 1.1, the universal and existential quantifiers, variables, and the identity predicate.

Problem 1.15 Consider the following observation:

Better is necessarily different, but different is not necessarily better.

(a) Is being different a *necessary* or *sufficient* condition (perhaps neither, or both) for being better?

(b) Conversely, is being better a necessary or sufficient condition for being different?

Problem 1.16 Prove the following useful fact about any two real numbers a and b:

$$\text{If } a < b + \varepsilon \text{ for all } \varepsilon > 0, \text{ then } a \leq b.$$

(The converse statement is obviously true.) Accomplish your proof by proving the contrapositive statement.

Problem 1.17 Use truth tables to prove the logical implication

$$(A \wedge \neg B) \Rightarrow C, \quad (A \wedge \neg B) \Rightarrow \neg C \quad \vDash \quad A \Rightarrow B, \tag{P1.17.1}$$

thereby justifying the method of proof by contradiction.

Problem 1.18 Consider the two statements $\forall x,\ \exists y,\ P(x, y)$ and $\exists y,\ \forall x,\ P(x, y)$ for the same binary predicate "P". Prove that one of these statements logically implies the other; then provide a counterexample showing that the reverse logical implication does not hold.

Problem 1.19 Suppose I make the following two assertions:

(i) All of my brothers are short.
(ii) All of my brothers are tall.

Now, even though being short and being tall are mutually exclusive, *both* of these statements are, in fact, true! How can that be? (*Suggestion*: Introduce predicates "B", "S", and "T"—with "$B(x)$", "$S(x)$", and "$T(x)$" respectively meaning "x is my brother", "x is short", and "x is tall"—and rewrite the two statements in the formal syntax of first-order logic.)

Problem 1.20 The logics discussed in Chapter 1 deal primarily with *truth*, which is an *objective* quality; for irrespective of what anyone knows (or believes), a given statement is either true (i.e., asserts a *fact*) or it is false (i.e., denies a fact). By contrast, **modal logic** allows for assertions containing the "modalities" of *possibility* and *necessity*, which are *subjective* qualities; for whether something in a given situation is possible or necessary depends upon what one knows about it. (For example, if a particular coin is flipped, and only you observe the result—say, HEAD—then to everyone but you it is possible that the result is TAIL.) Thus, besides truth, modal logic deals with *knowledge*.

Suppose, then, that for each statement A we let

$$\Diamond A$$

mean "it is *possible* that A is true"; and, we also let

$$\Box A$$

mean "it is *necessary* that A is true". Justify the logical equivalence

$$\Diamond \neg A \quad \vDash\!\!\!\dashv \quad \neg \Box A,$$

analogous to (1.5); that is, explain why, for any statement A and any particular state of knowledge, the truth of $\Diamond \neg A$ implies the truth of $\Box A$, and vice versa.

Chapter 2. Sets

In this section, A, B, and C are arbitrary sets, unless stated otherwise.

Problem 2.1 Here is a multiple-set situation akin to Russell's paradox: Show that it is impossible for there to exist three sets A, B, and C such that

$$A = \{\text{sets } x \mid x \notin B\}, \quad B = \{\text{sets } x \mid x \notin C\}, \quad C = \{\text{sets } x \mid x \notin A\}.$$

Problem 2.2 Suppose every element of A is an integer. How is it possible for every element of A to be even, yet every element of A to be odd?

Problem 2.3 Of what sets is $\emptyset$ a superset? A proper superset?

Problem 2.4 Recall that for any two real numbers a and b, if $a \not< b$ then $a \geq b$. Is it similarly true that if $A \not\subset B$ then $A \supseteq B$?

Problem 2.5[3] Clearly, the sets $A \cap B$ and $B \cap A$ are equal, as are the sets $A \cup B$ and $B \cup A$; thus, the set operations $\cap$ and $\cup$ are *commutative*. Also, the sets $(A \cap B) \cap C$ and $A \cap (B \cap C)$ are equal, as are the sets $(A \cup B) \cup C$ and $A \cup (B \cup C)$; thus, $\cap$ and $\cup$ are *associative*.[4] Now, consider the *distributive* properties of these operations with respect to each other:

(a) Are the sets $(A \cap B) \cup C$ and $(A \cup C) \cap (B \cup C)$ equal?

(b) Are the sets $(A \cup B) \cap C$ and $(A \cap C) \cup (B \cap C)$ equal?

Problem 2.6 A common set-operation is the **symmetric difference**:

$$A \triangle B \triangleq (A - B) \cup (B - A). \tag{P2.6.1}$$

(a) Draw a Venn diagram illustrating this operation.

(b) Utilizing the exclusive-or connective $\veebar$ defined in Problem 1.4, complete the following logic statement: $\forall x,\ x \in A \triangle B \Leftrightarrow [?]$.

(c) Let the set A have **indicator function**

$$\mathbf{1}_A(x) = \begin{cases} 1 & x \in A \\ 0 & x \notin A \end{cases};$$

and, likewise for the set B. Show that $x \in A \triangle B$ if and only if $\mathbf{1}_A(x) \oplus \mathbf{1}_B(x) = 1$, where "$\oplus$" indicates addition modulo 2 (defined in (3.24)).

(d) Observe that by the symmetry in (P2.6.1), the sets $A \triangle B$ and $B \triangle A$ must be equal (unlike, e.g., the sets $A - B$ and $B - A$); thus, the operation $\triangle$ is *commutative*. Show that $\triangle$ is also *associative*; that is, $(A \triangle B) \triangle C = A \triangle (B \triangle C)$.[5] (*Suggestion*: Show, by continuing part (c), that the indicator functions for these two sets are the same.)

Problem 2.7 **(a)** It is obvious from (P2.6.1) that $A - B \subseteq A \triangle B$. Give a simple necessary and sufficient condition on the sets A and B for equality to be attained.

(b) Show that $A \triangle B \subseteq (A \triangle C) \cup (B \triangle C)$.

Problem 2.8 Suppose $A = I_1 \times I_2$, where I_1 and I_2 are subintervals of the real line that have finite but nonzero lengths. What is the geometric appearance of A when viewed as subset of the plane $\mathbb{R}^2$?

Problem 2.9 If R and S are both ternary relations on a set A, then is $R \cup S$ also a ternary relation on A?

---

3. Compare Problem 1.2.

4. It follows that the parentheses in these expressions can be omitted—that is, "$A \cap B \cap C$" and "$A \cup B \cup C$" indicate unambiguous sets—and likewise for any intersection or union of finitely many sets.

5. It follows that the parentheses in these two expressions can be removed without ambiguity; that is, we may write "$A \triangle B \triangle C$" for the common set. Likewise, for any number $n \geq 1$ of sets $A_1, \ldots, A_n$, the set $A_1 \triangle \cdots \triangle A_n$ is unambiguously defined.

Problem 2.10 Suppose R and S are nonempty binary relations, with $R \subseteq S$.

- **(a)** Give an example showing that it is possible for R to be one-to-one yet S not be one-to-one.
- **(b)** Is it possible for S to be one-to-one yet R not be one-to-one?

Problem 2.11 In Section 2.1 it was observed that the less-than relation $<$ on the set $\{1, 2, 3\}$ is not reflexive or symmetric, but is transitive. This, though, is not the case for the less-than relation $<$ on the set $\{1\}$. Why not?

Problem 2.12 Give an example of a binary relation R on the set $\{1, 2, 3\}$ such that R is reflexive and symmetric, but not transitive.

Problem 2.13 Let R be the binary relation on $\mathbb{Z}$ such that for all $x, y \in \mathbb{Z}$ we have $x\,R\,y$ if and only if $x + y$ is even.

- **(a)** Show that R is an equivalence relation.
- **(b)** What are the corresponding equivalence classes?

Problem 2.14 Suppose that $\equiv_1$ and $\equiv_2$ are both equivalence relations on a set A, and that $\equiv_1$ implies $\equiv_2$; that is, for all $x, y \in A$, if $x \equiv_1 y$ then $x \equiv_2 y$. Show that every equivalence class under $\equiv_2$ can be expressed as a union of equivalence classes under $\equiv_1$.

Problem 2.15 As explained in Section 2.1, an equivalence relation on a nonempty set A induces a partitioning of A. Conversely, given a partition $\mathcal{P}$ of A, there exists an equivalence relation R on A that induces $\mathcal{P}$; namely, we can let R be the binary relation on A such that $x\,R\,y$ if and only if $x, y \in A$ belong to the same partition set:

$$R = \{(x, y) \in A^2 \mid \exists B \in \mathcal{P},\, x \in B \wedge y \in B\}. \tag{P2.15.1}$$

Show that this relation R is indeed an equivalence relation.

Chapter 3. Numbers

Problem 3.1 **(a)** Show that the sum of two rational numbers must be rational; that is, $\mathbb{Q}$ is closed under addition.

- **(b)** Give an example showing that the sum of two irrational numbers need not be irrational.
- **(c)** What, then, for the sum of a rational number and an irrational number?

Problem 3.2 Express the number $-98.76\overline{543}\ldots$ as a ratio of two integers, reducing the fraction as much as possible.

Problem 3.3 Give all answers below as *finite* expansions.

- **(a)** Express the number 110111001.10101_2 in decimal, octal, and hexadecimal forms.
- **(b)** Express the number 0.456_8 in decimal, binary, and hexadecimal forms.
- **(c)** Express the number -9D.C_{16} in decimal, binary, and octal forms.
- **(d)** Express the number 39.625_{10} in binary, octal, and hexadecimal forms.

Problem 3.4 **(a)** Prove that if a real number has a finite binary expansion, then it can be expressed as a finite decimal expansion.

(b) Give an example of a real number having a finite decimal expansion that does not have a finite binary expansion.

Problem 3.5 Suppose that a, b, and c are real numbers such that

$$a = bc. \tag{P3.5.1}$$

If this equation continues to hold upon changing the value of b to a different real number, what can we conclude about the value of c?

Problem 3.6 What does each of the intervals $[a, b]$, $(a, b]$, $[a, b)$, and (a, b) become when $a = b \in \overline{\mathbb{R}}$?

Problem 3.7 For each of the following number-sets A, find $\sup A$ and $\inf A$. Also, determine whether the supremum is a maximum, and whether the infimum is a minimum.

(a) $A = \{(-1)^n n \mid n = 1, 2, \dots\}$.

(b) $A = \{(-1)^n / n \mid n = 1, 2, \dots\}$.

(c) $A = [a, b) \cup [c, d)$, where $a, b, c, d \in \overline{\mathbb{R}}$ such that $a < b$ and $c < d$.

(d) $A = \emptyset$. (*Hint*: This is the only subset of $\overline{\mathbb{R}}$ for which $\inf A \nleq \sup A$.)

Problem 3.8 Suppose $\sup S \in \mathbb{R}$ for some set $S \subseteq \mathbb{R}$. Which (perhaps neither, or both) of the follow statements is necessarily true?

(i) $\sup S$ is the smallest real value x such that $y \notin S$ for all real values $y \geq x$.

(ii) $\sup S$ is the smallest real value x such that $y \notin S$ for all real values $y > x$.

Problem 3.9 Determine all pairs $(x, y) \in \mathbb{R}^2$ for which

$$xy = x^2. \tag{P3.9.1}$$

Problem 3.10 In contrast to (3.13c), give an example showing that addition is not distributive over multiplication.

Problem 3.11 Suppose $x \in \overline{\mathbb{R}}$.

(a) Is it necessarily true that $|x| \geq 0$?

(b) Do we have $|x| = 0$ if and only if $x = 0$?

(c) Is it necessarily true that $x \cdot \infty = (\operatorname{sgn} x) \cdot \infty$?

Problem 3.12 Express the absolute-value function (3.17) in terms of the signum function in (3.16).

Problem 3.13 Suppose $a, b, c \in \mathbb{R}$ are such that $b \neq c$ and $|a - b| = |a - c|$. Solve for a in terms of b and c.

Problem 3.14 Which pairs of the parentheses in (3.13c) can be omitted?

Problem 3.15 **(a)** Give a simple example of an arithmetic expression containing only additions and subtractions, and with no explicit groupings (via parentheses, etc.), for which the correct value is *not* obtained by first performing all additions left to right and then performing the remaining subtractions left to right.

(b) Using the fact stated in fn. 15 of Chapter 3, and the associative property of addition, show that $a + (b - c) = (a + b) - c$ for $a, b, c \in \mathbb{C}$.

(c) Despite the result of part (a), prove that for any such arithmetic expression, the correct value *is* obtained by first performing all subtractions left-to-right and then performing the remaining additions in any order.

Problem 3.16 For what integers x is $x - 1$ divisible by 2, and also $x - 2$ divisible by 3?

Problem 3.17 According to **Goldbach's conjecture** (stated in 1742, but yet to be proved or disproved), every even integer $n > 2$ can be expressed as the sum of two prime numbers. Verify the conjecture up to $n = 30$.

Problem 3.18 When one is verifying that an integer $x > 1$ is prime, why is it sufficient to only test x for divisibility by prime numbers no greater than $\sqrt{x}$?

Problem 3.19 Let x and y be two positive integers.

(a) Express $\operatorname{LCM}(x, y)$ and $\operatorname{GCD}(x, y)$ in terms of the prime factorizations of x and y.

(b) Note that $\operatorname{LCM}(9, 12) \cdot \operatorname{GCD}(9, 12) = 36 \cdot 3 = 108 = 9 \cdot 12$. Must we always have $\operatorname{LCM}(x, y) \cdot \operatorname{GCD}(x, y) = x \cdot y$?

Problem 3.20 Show that if two positive integers x and y are relatively prime, then $\{nx \bmod y \mid n = 0, 1, \dots, y - 1\} = \{0, 1, \dots, y - 1\}$. (*Aside*: An illustration of this fact is the **circle of fifths** for the standard 12-note musical scale: each of the $y = 12$ notes of the scale—viz., C, C♯, D, D♯, E, F, F♯, G, G♯, A, A♯, B—is visited exactly once by starting at any note and cyclically incrementing by $x = 7$ notes (called a "fifth") each step—e.g., C, G, D, A, E, B, F♯, C♯, G♯, D♯, A♯, F.)

Problem 3.21 Show that for any integers x, y, and $m, n \geq 1$, we have

$$x \bmod n = y \bmod n \quad \Leftrightarrow \quad (x - y) \bmod n = 0, \tag{P3.21.1a}$$

$$(xn + y) \bmod n = y \bmod n, \tag{P3.21.1b}$$

$$[x + (y \bmod n)] \bmod n = (x + y) \bmod n, \tag{P3.21.1c}$$

$$x(y \bmod n) \bmod n = xy \bmod n, \tag{P3.21.1d}$$

$$(x \bmod n)^m \bmod n = x^m \bmod n. \tag{P3.21.1e}$$

(*Suggestions*: First decompose x and y versus n as in (3.21). Also, for (P3.21.1e) use the binomial expansion (8.4).)

Problem 3.22 **(a)** Prove that for any fixed base $n \geq 2$, modular addition $\oplus$ (upon integers) is commutative and associative. (*Suggestion*: For the latter, use (P3.21.1c).)

(b) Analogous to $\oplus$, define **modular subtraction** (symbolized by "$\ominus$") with respect to a fixed integer base $n \geq 2$ as

$$x \ominus y \triangleq (x - y) \bmod n \quad (x, y \in \mathbb{Z}). \tag{P3.22.1}$$

Show, using the same base for $\oplus$, that $(x \oplus y) \ominus y = x$ $(x, y \in \mathbb{Z})$.

Problem 3.23 Per (3.4) and (3.5), let $\boldsymbol{x}_n \ldots \boldsymbol{x}_1\boldsymbol{x}_0$ $(n \geq 0)$ be a fractionless decimal expansion representing a given nonnegative integer

$$x = \sum_{i=0}^{n} x_i \cdot 10^i.$$

Likewise, when $n \geq 1$, let y be the number represented by the numeral $\boldsymbol{x}_n \ldots \boldsymbol{x}_1$. Prove the following facts about the divisibility of x.

(a) x is divisible by 2 if and only if x_0 is even.

(b) x is divisible by 3 if and only if $x_0 + \cdots + x_n$ is divisible by 3.[6] (*Suggestion*: Use (P3.21.1) to show $x \bmod 3 = (x_0 + \cdots + x_n) \bmod 3$.)

(c) x is divisible by 4 if and only if both x_0 and $x_0/2 + x_1$ are even (letting $x_1 = 0$ when $n < 1$).[7]

(d) x is divisible 5 if and only if $x_0 \in \{0, 5\}$.

(e) x is divisible by 6 if and only if x is divisible by both 2 and 3.

(f) x is divisible by 7 if and only if either $x \in \{0, 7\}$ or $y - 2x_0$ is divisible by 7.

(g) x is divisible by 8 if and only if $x_0, x_0/2+x_1$, and $(x_0/2+x_1)/2+x_2$ are all even (letting $x_1 = 0$ when $n < 1$, and $x_2 = 0$ when $n < 2$).

(h) x is divisible by 9 if and only if $x_0 + \cdots + x_n$ is divisible by 9.

(i) x is divisible by 10 if and only if $x_0 = 0$.

(j) x is divisible by 11 if and only if either $x = 0$ or $y - x_0$ is divisible by 11.

---

6. This fact, and others below, can be applied recursively. For example, 1234 is divisible by 3 if and only if $1+2+3+4 = 10$ is divisible by 3, which is the case if and only if $1 + 0 = 1$ is divisible by 3, which is not the case; therefore, 1234 is not divisible by 3.

7. An equivalent, briefer, condition is simply that $x_0/2+x_1$ be an even integer; however, the stated condition is more convenient to apply, since one would first check that x_0 is even, and only then go on to calculate $x_0/2 + x_1$. A similar comment applies to part (g).

Problem 3.24 If $x:y::x':y'$ for some nonzero numbers x, y, x', and y', then is it also true that $x:x'::y:y'$?

Problem 3.25 When a computer file is compressed from a size $x > 0$ to a smaller size $y > 0$ (both expressed in either bits or bytes), the amount of compression is commonly stated either as a percentage p $(0 \le p < 100)$ or as a ratio $r:1$ $(r > 1)$; namely, y is $p\%$ smaller than x, and $x:y::r:1$. Provide a formula for converting from p to r; and, vice versa.

Problem 3.26 Suppose three real variables x, y, and z are such that

$$x:y:z::a:b:c, \quad x+y+z=d,$$

for some real constants a, b, c, and $d \ne 0$. Solve for x, y, and z individually in terms of a, b, c, and d.

Problem 3.27 **(a)** Convert the complex number $-2\angle(\pi/3)$ into rectangular form.

(b) Convert the complex number $-2 + i3$ into polar form.

Problem 3.28 **(a)** With regard to its vector representation (i.e., Argand diagram), multiplying a complex number z by another complex number $Ae^{i\theta}$ $(A \ge 0, \theta \in \mathbb{R})$ has what geometric effect?

(b) In particular, what does multiplying z by i do?

Problem 3.29 Express the value $|1 + e^{i\theta}|^2$, where θ is real, without using i. (The trigonometric identities in Section 11.2 might be helpful.)

Problem 3.30 Let z be a complex number. Show that if $z = z^*$, then z is real (the converse of this statement being obvious). Also, state a similar necessary and sufficient condition for z to be imaginary.

Problem 3.31 Prove (3.36f) and (3.36g) for all $z \in \mathbb{C}$.

Problem 3.32 Suppose that for some constants $a, b \in \mathbb{C}$, the value $ae^{ix} + be^{-ix}$ is real for all $x \in \mathbb{R}$. Show that we must then have $a = b^*$.

Problem 3.33 **(a)** If $z_1, z_2 \in \mathbb{C}$ are such that $|z_1| = |z_2|$, then what is the relationship between z_1 and z_2? Specifically, given $|z_1|$, characterize the possible values for z_2.

(b) What if, instead, $\arg z_1 = \arg z_2$ (assuming $z_1, z_2 \ne 0$)?

Problem 3.34 Let z be a nonzero complex number.

(a) Show that $\operatorname{Re}(1/z) = (\operatorname{Re} z)/|z|^2$.

(b) Derive a similar expression for $\operatorname{Im}(1/z)$.

Problem 3.35 **(a)** Prove the triangle inequality (3.37).

(b) Show that upon restricting $\arg z_1$ and $\arg z_2$ to the same interval of length 2π—e.g., $[0, 2\pi)$ or $(-\pi, \pi]$—equality is attained in (3.37) if and only if $z_1 = 0$, $z_2 = 0$, or $\arg z_1 = \arg z_2$.

Problem 3.36 Use the triangle inequality (3.37) to prove the following assertions for $x, y, z \in \mathbb{C}$.

(a) $|x - y| \le |x - z| + |y - z|$.

(b) $\max(|x - z|, |y - z|) \ge |x - y|/2$. (*Suggestion*: Show, via the fact in part (a), that assuming the contrary yields a contradiction.)

Problem 3.37 **(a)** Use the triangle inequality (3.37) to show

$$|x - y| \ge \big||x| - |y|\big| \quad (x, y \in \mathbb{C}). \tag{P3.37.1}$$

(b) Conversely, derive (3.37) directly from (P3.37.1).

Chapter 4. Sequences

In this section, $\{x_i\}_{i=-\infty}^{\infty}$, $\{y_i\}_{i=\infty}^{\infty}$, and $\{z_i\}_{i=-\infty}^{\infty}$ are arbitrary complex sequences, unless stated otherwise.

Problem 4.1 Show that if a real sequence $\{x_i\}_{i=1}^{\infty}$ is increasing (respectively, strictly increasing), then the sequence $\{-x_i\}_{i=1}^{\infty}$ is decreasing (strictly decreasing).

Problem 4.2 Let $\{x_i\}_{i=1}^{\infty}$ be an increasing real sequence that is bounded. Prove $\lim_{i\to\infty} x_i = \sup_i x_i$.

Problem 4.3 For each part below, either find $\lim_{i\to\infty} x_i$, including any divergences to $\pm\infty$, or show that the limit does not exist.

(a) $x_i = i$ (all i).

(b) $x_i = \sin(i\pi/10)$ (all i).

(c) $x_i = [\sin(i\pi/10)]/i$ ($i \neq 0$).

(d) $x_i = i\sin(i\pi/10)$ (all i).

Problem 4.4 Show that if both $x_i \to x$ and $y_i \to y$ as $i \to \infty$, where $x, y \in \mathbb{C}$, then for each $\varepsilon > 0$ there exists an $n \in \mathbb{Z}^+$ such that *both* $|x_i - x| \leq \varepsilon$ and $|y_i - y| \leq \varepsilon$ for all $i \geq n$.

Problem 4.5 Suppose $x_i \leq y_i \leq z_i$ (all i) and $\lim_{i\to\infty} x_i = \lim_{i\to\infty} z_i$, with infinite limits being allowed. Show that $\lim_{i\to\infty} y_i$ must also exist and equal the other two limits. (*Hint*: The fact stated in Problem 4.4 might be helpful.)

Problem 4.6 Given a sequence $\{x_i\}_{i=1}^{\infty}$, suppose we form another sequence $\{y_i\}_{i=1}^{\infty}$ by letting $y_i = x_{i+l}$ ($i = 1, 2, \ldots$) for some $l \in \mathbb{N}$. Show that $\lim_{i\to\infty} y_i$ exists if and only if $\lim_{i\to\infty} x_i$ exists, in which case their values are the same. (Thus, shifting a sequence does not change the existence or value of its limit.)

Problem 4.7 Suppose that for each $i \in \mathbb{Z}^+$ we have $j_i \in \mathbb{Z}^+$, and that for each $j \in \mathbb{Z}^+$ there exists exactly one $i \in \mathbb{Z}^+$ such that $j = j_i$; that is, $(j_1, j_2, \ldots)$ is a reordering of the sequence $(1, 2, \ldots)$. Now, suppose $\{y_i\}_{i=1}^{\infty}$ is the corresponding reordering of $\{x_i\}_{i=1}^{\infty}$; namely, $y_i = x_{j_i}$ ($i = 1, 2, \ldots$). Show that $\lim_{i\to\infty} y_i$ exists if and only if $\lim_{i\to\infty} x_i$ exists, in which case their values are the same. (Thus, reordering a sequence does not change the existence or value of its limit.)

Problem 4.8 Suppose that for some constant $a > 0$ we have

$$x_{i+1} = (x_i + a/x_i)/2 \quad (i = 1, 2, \ldots), \tag{P4.8.1}$$

for some initial value $x_1 > 0$; hence, $x_i > 0$ for each successive value of i, so there are no divisions by 0. It can be shown that $X = \lim_{i\to\infty} x_i$ exists. Find X in terms of a and x_1.

Problem 4.9 A set $S \subseteq \mathbb{R}$ is a **closed set** if and only if it is impossible to escape it by a limiting process; that is, for every sequence $\{x_i\}_{i=1}^{\infty}$ of values $x_i \in S$, if $\lim_{i\to\infty} x_i = X$ for some $X \in \mathbb{R}$ then $X \in S$. Also, an **open set** is the complement (here, relative to $\mathbb{R}$) of a closed set.[8] It then follows from (2.3) that the complement of an open set must be closed.

(a) Is $\mathbb{R}$ closed or open (perhaps neither, or both)? What about, $\emptyset$?

(b) For each of the four types of an interval I given in (3.12), consider all values of the endpoints $a, b \in \overline{\mathbb{R}}$ for which $\emptyset \neq I \subseteq \mathbb{R}$, and state under what conditions on the endpoints I is closed, as well as under what conditions I is open.

Problem 4.10 **(a)** Prove that $\{z_i\}_{i=1}^{\infty}$ converges to a limit $z \in \mathbb{C}$ if and only if $\{\operatorname{Re} z_i\}_{i=1}^{\infty}$ converges to $\operatorname{Re} z$ and $\{\operatorname{Im} z_i\}_{i=1}^{\infty}$ converges to $\operatorname{Im} z$.

(b) Prove that every convergent real sequence must have a real limit.

Problem 4.11 Let S be the set of all complex sequences $\{x_i\}_{i=1}^{\infty}$; and, let $\simeq$ be the binary relation on S such that for any two such sequences $\{x_i\}_{i=1}^{\infty}$ and $\{y_i\}_{i=1}^{\infty}$ we have

$$\{x_i\}_{i=1}^{\infty} \simeq \{y_i\}_{i=1}^{\infty} \quad \text{if and only if} \quad \lim_{i\to\infty} (x_i - y_i) = 0.$$

Show that $\simeq$ is an equivalence relation.

---

8. These statements, though correct (and having heuristic benefit), do not reflect how the terms "closed set" and "open set" are typically defined (which is beyond the scope of this book); in particular, the latter term is usually defined first. Also, just as for the "complement" of a set, these terms are defined *relative* to some set—here, $\mathbb{R}$. For example, given $S \subseteq \mathbb{R}$, we also have $S \subseteq \mathbb{C}$; however, S might be open relative $\mathbb{R}$ without being open relative to $\mathbb{C}$.

Problem 4.12 Given a complex double sequence $\{x_{i,j}\}$ $(i = 1, 2, \ldots; j = 1, 2, \ldots, n)$, it follows from repeated application of (4.11a) that if $\lim_{i\to\infty} x_{i,j}$ exists for each j, then

$$\lim_{i\to\infty} \sum_{j=1}^{n} x_{i,j} = \sum_{j=1}^{n} \lim_{i\to\infty} x_{i,j}.$$

By contrast, give a double sequence $\{x_{i,j}\}$ $(i = 1, 2, \ldots; j = 1, 2, \ldots)$ for which all of the limits and sums below exist, yet we do *not* have

$$\lim_{i\to\infty} \sum_{j=1}^{\infty} x_{i,j} = \sum_{j=1}^{\infty} \lim_{i\to\infty} x_{i,j}.$$

Problem 4.13 In contrast to the example in (4.9), find a double sequence $\{x_{i,j}\}_{i,j=1}^{\infty}$ of complex numbers $x_{i,j}$ such that

$$\lim_{i\to\infty} \lim_{j\to\infty} x_{i,j} = \lim_{j\to\infty} \lim_{i\to\infty} x_{i,j}$$

(a finite value), yet $\lim_{i,j\to\infty} x_{i,j}$ does not exist. (*Suggestion*: Construct the sequence such that $x_{i,j} = c$ $(i = j)$ for some constant c that does not equal the above limits.)

Problem 4.14 **(a)** As done in (4.12), use a limiting argument to explain why it is reasonable to let $x/\infty = 0$ for any fixed $x \in \mathbb{R}$.

(b) Use another such argument to explain why ∞/∞ should be left undefined.

Problem 4.15 Express $\prod_{i=m}^{n} ax_i$ $(-\infty < m \le n < \infty,\ a \in \mathbb{C})$ in terms of $\prod_{i=m}^{n} x_i$.

Problem 4.16 **(a)** Show that if $|x_i| \le c$ (all i) for some constant $c \in [0, 1)$, then $\prod_{i=1}^{\infty} x_i = 0$.

(b) When does this product converge to 0, and when does it diverge to 0?

Problem 4.17 An interesting convergent infinite product is given by **Wallis's formula**:[9]

$$\frac{\pi}{2} = \frac{2}{1}\cdot\frac{2}{3}\cdot\frac{4}{3}\cdot\frac{4}{5}\cdot\frac{6}{5}\cdot\frac{6}{7}\cdot\cdots = \left(\frac{2}{1}\cdot\frac{2}{3}\right)\left(\frac{4}{3}\cdot\frac{4}{5}\right)\left(\frac{6}{5}\cdot\frac{6}{7}\right)\cdots$$

$$= \prod_{i=1}^{\infty} \frac{2i}{2i-1}\cdot\frac{2i}{2i+1} = \prod_{i=1}^{\infty} \frac{4i^2}{4i^2-1}.$$

Find the value of the similar product

$$\prod_{i=2}^{\infty} \frac{i^2}{i^2-1}.$$

Problem 4.18 **(a)** In the sum

$$\sum_{i=1}^{100} x_i,$$

change the index i to another integer index j, with $i = j/2 + 5$.

(b) By contrast, why would making this change with $i = 2j + 5$ be problematic?

---

9. The grouping of factors performed below is justified because the convergence and value of a convergent infinite sum or product are unaffected by any grouping of consecutive terms. (Reordering the terms, though, is another matter.) Contrast this fact with the observation that

$$[(1)(1)]\cdot[(-1)(-1)]\cdot[(1)(1)]\cdot[(-1)(-1)]\cdot\cdots = 1\cdot 1\cdot 1\cdot 1\cdot\cdots = 1,$$

whereas the *re*grouping

$$(1)\cdot[(1)(-1)]\cdot[(-1)(1)]\cdot[(1)(-1)]\cdot[(1)(-1)]\cdot\cdots = 1\cdot(-1)\cdot(-1)\cdot(-1)\cdot\cdots$$

does not converge.

Problem 4.19 Show that if a sum $\sum_{i=-\infty}^{\infty} x_i$ exists (infinite limits being allowed), then the sum $\sum_{i=-\infty}^{\infty} x_{i+l}$ has the same value for all integers l. (Thus, shifting a doubly infinite sequence does not change its existence or value.)

Problem 4.20 Perform the interchanging of sums

$$\sum_{i=2}^{5} \sum_{j=-\infty}^{9-i} x_{i,j} = \sum_{j=?}^{?} \sum_{i=?}^{?} x_{i,j}, \tag{P4.20.1}$$

assuming that the left side of the equation exists and that infinite limits are allowed.

Problem 4.21 Give four values $x_{i,j}$ $(i, j = 1, 2)$ for which

$$\sum_{i=1}^{2} \prod_{j=1}^{2} x_{i,j} \neq \prod_{j=1}^{2} \sum_{i=1}^{2} x_{i,j}.$$

Problem 4.22 It follows from (4.28) and (4.29) that if the sums $\sum_{j=1}^{\infty} \operatorname{Re} z_j$ and $\sum_{j=1}^{\infty} \operatorname{Im} z_j$ converge, then

$$\sum_{j=1}^{\infty} z_j = \sum_{j=1}^{\infty} \operatorname{Re} z_j + i \sum_{j=1}^{\infty} \operatorname{Im} z_j, \tag{P4.22.1}$$

$(i = \sqrt{-1})$. But, suppose all we know is that the sum on the left converges.

(a) Show directly that

$$\left(\sum_{j=1}^{\infty} z_j \right)^* = \sum_{j=1}^{\infty} z_j^*. \tag{P4.22.2}$$

(b) Using (P4.22.2), (3.36q), and (3.36r), show that

$$\operatorname{Re} \left(\sum_{j=1}^{\infty} z_j \right) = \sum_{j=1}^{\infty} \operatorname{Re} z_j, \quad \operatorname{Im} \left(\sum_{j=1}^{\infty} z_j \right) = \sum_{j=1}^{\infty} \operatorname{Im} z_j. \tag{P4.22.3}$$

(c) Using (P4.22.3), show that (P4.22.1) again holds.

Problem 4.23 **(a)** Let a real sequence $\{x_i\}_{i=1}^{\infty}$ be given. Show that the infinite sum $\sum_{i=1}^{\infty} x_i$ converges if and only if

$$\lim_{m \to \infty} \sum_{i=m}^{\infty} x_i = 0.$$

(b) Use (4.33a) to give a corresponding fact for infinite products.

Problem 4.24 Convert the harmonic series into a related infinite product, and thereby provide an example of a sequence $\{x_i\}_{i=1}^{\infty}$ of positive values x_i such that $x_i \to 1$ as $i \to \infty$, yet the product $\prod_{i=1}^{\infty} x_i$ diverges to ∞.

Problem 4.25 Calculate the values of each of the following series (infinite limits allowed):

$$x = \tfrac{1}{2} + \tfrac{1}{3} + \tfrac{1}{4} + \tfrac{1}{5} + \cdots,$$
$$y = \tfrac{1}{2} + \tfrac{1}{4} + \tfrac{1}{6} + \tfrac{1}{8} + \cdots,$$
$$z = \tfrac{1}{2} + \tfrac{1}{4} + \tfrac{1}{8} + \tfrac{1}{16} + \cdots.$$

Problem 4.26 A worker earning \$100/week is given weekly raises.

(a) Suppose that for the second week, the wage is increased by \$1/week to \$101/week; for the third week, it is increased \$1/week more; and, so forth. How much, then, will the worker earn after 50 weeks?

(b) In part (a), the worker's wage is \$1/week more for each successive week. Suppose instead that each week the wage is increased by 1%. Thus, the *first* raise is again \$1/week. Now how much (to the nearest dollar) will the worker earn after 50 weeks?

Problem 4.27 Use (4.38), the formula for a geometric series, to justify the particular conversion (3.9) of a repeating decimal expansion into a rational number.

Problem 4.28 **(a)** Show

$$\sum_{i=1}^{n} i \cdot i! = (n+1)! - 1 \quad (n = 1, 2, \dots).$$

(*Suggestion*: Rewrite the summand in terms of $(i+1)!$.)

(b) Why is it reasonable to say that this equation also holds for $n = 0$?

Problem 4.29 **(a)** Prove the following formula showing how to perform a **summation by parts**:

$$\sum_{i=m}^{n} x_i y_i = x_n \sum_{j=m}^{n} y_j - \sum_{i=m+1}^{n} (x_i - x_{i-1}) \sum_{j=m}^{i-1} y_j \quad (m, n \in \mathbb{Z} \text{ with } m \le n), \tag{P4.29.1}$$

where for $m = n$ the sum from $i = m+1$ to n is understood to equal 0, by convention (4.43a). (Also, notice that the last sum is within the scope of the immediately preceding sum.)

(b) Use (P4.29.1) to obtain a formula for $\sum_{i=m}^{n} ic^i$ ($m, n \in \mathbb{Z}$ with $m \le n$; $c \in \mathbb{C}$, with $c \neq 0$ for $m < 0$) in terms of m, n, and c.

Problem 4.30 **(a)** Use mathematical induction to prove

$$1 + 3 + 5 + \cdots + (2n-1) = n^2 \quad (n = 1, 2, \dots). \tag{P4.30.1}$$

(b) Explain this fact geometrically in terms of an n-by-n square.

Problem 4.31 Prove by mathematical induction that for each $n \in \mathbb{Z}^+$, the integer $6 + 8^n$ is divisible by 7.

Problem 4.32 Prove by mathematical induction that the solution to (4.49) is

$$f_n = \frac{1}{\sqrt{5}} \left[\left(\frac{1+\sqrt{5}}{2} \right)^n - \left(\frac{1-\sqrt{5}}{2} \right)^n \right] \quad (n = 1, 2, \dots). \tag{P4.32.1}$$

Chapter 5. Functions

Problem 5.1 What can be said about the range of

(a) a constant function?

(b) an identity function?

Problem 5.2 **(a)** Give an example of a function $f(\cdot)$ that is not a constant function for which the composite function $f[f(\cdot)]$ *is* a constant function.

(b) Give an example of a function $f(\cdot)$ that is not an identity function for which the composite function $f[f(\cdot)]$ *is* an identity function.

Problem 5.3 Let $h(\cdot)$ be the composition $g[f(\cdot)]$ of two functions $f(\cdot)$ and $g(\cdot)$. Express its inverse $h^{-1}(\cdot)$ in terms of the inverses $f^{-1}(\cdot)$ and $g^{-1}(\cdot)$.

Problem 5.4 Let $f(\cdot)$ be a function having domain D and range E. Determine which of the following statements must be true, and give counterexamples to the others.

(a) $f(A_1) \subseteq f(A_2)$ for all $A_1, A_2 \subseteq D$ such that $A_1 \subseteq A_2$.
(b) $f^{-1}(B_1) \subseteq f^{-1}(B_2)$ for all $B_1, B_2 \subseteq E$ such that $B_1 \subseteq B_2$.
(c) $f^{-1}[f(A)] = A$ for all $A \subseteq D$.
(d) $f[f^{-1}(B)] = B$ for all $B \subseteq E$.

Problem 5.5 For a one-to-one function $f: D \to S$, must we have $f^{-1}: S \to D$?

Problem 5.6 **(a)** Show that the function $f(x) = x$ $(x \in \mathbb{R})$ is strictly increasing.
(b) Show that the function $f(x) = 1/x$ $(x > 0)$ is strictly decreasing.

Problem 5.7 **(a)** Explain why it is impossible for a real sequence $\{x_i\}_{i=1}^{\infty}$ to be both strictly increasing and strictly decreasing.
(b) In contrast to part (a), show that there exists a real function $f(\cdot)$ that is both strictly increasing and strictly decreasing. (*Hint*: vacuous truth.)

Problem 5.8 Suppose $f: D \to \mathbb{R}$.

(a) Show that

$$\min_{x\in D} -f(x) = -\max_{x\in D} f(x),$$

when either the minimum or the maximum exists.

(b) When is it true that

$$\inf_{x\in D} -f(x) = -\sup_{x\in D} f(x)?$$

(c) Give an example of two functions $f: D \to \mathbb{R}$ and $g: D \to \mathbb{R}$ for which

$$\max_{x\in D}[f(x) + g(x)] \neq \max_{x\in D} f(x) + \max_{x\in D} g(x),$$

even though all three maxima exist.

Problem 5.9 The **inverse maximum** and **inverse minimum** operators, argmax and argmin, are defined as follows: Given a function $f: D \to \mathbb{R}$ and set $S \subseteq D$,

$$\operatorname*{argmax}_{x\in S} f(x) \triangleq \{x \in S \mid f(x) \geq f(x')\ (x' \in S)\},$$

$$\operatorname*{argmin}_{x\in S} f(x) \triangleq \{x \in S \mid f(x) \leq f(x')\ (x' \in S)\};$$

and, for the frequently occurring case of $S = D$, the left sides of these definitions may be more succinctly written

$$\operatorname*{argmax}_{x} f(x), \quad \operatorname*{argmin}_{x} f(x),$$

respectively. Thus, $\operatorname{argmax}_{x\in S} f(x)$ (respectively, $\operatorname{argmin}_{x\in S} f(x)$) is the set of points in S at which the function $f(\cdot)$, restricted to S, attains its global maximum (minimum), if any; accordingly, these sets always exist, though either might be empty.[10] Find $\operatorname{argmax}_{x\in\mathbb{R}} \cos x$, $\operatorname{argmin}_{x\in\mathbb{R}} \cos x$, $\operatorname{argmax}_{x\in\mathbb{R}} e^x$, and $\operatorname{argmin}_{x\in\mathbb{R}} e^x$.

---

10. When it is known that $f(\cdot)$ restricted to S attains its global maximum at a *single* point $X \in S$, it is common for $\operatorname{argmax}_{x\in S} f(x)$ to be defined instead as the *value* X, rather than the *set* $\{X\}$; and, likewise for argmin. One can then state that $\max_{x\in S} f(x) = f[\operatorname{argmax}_{x\in S} f(x)]$.

Problem 5.10 **(a)** Plot the following function $f : [1,5) \to \mathbb{R}$:

$$f(x) = \begin{cases} 4 & 1 \le x \le 2 \\ -x^2 + 6x - 4 & 2 < x < 5 \end{cases}. \tag{P5.10.1}$$

(b) Is this function continuous?

(c) What are its local and global extrema?

Problem 5.11 Two useful functions are: the **floor function** $\lfloor x \rfloor$ ($x \in \mathbb{R}$), which gives the largest integer $n \le x$; and, the **ceiling function** $\lceil x \rceil$ ($x \in \mathbb{R}$), which gives the smallest integer $n \ge x$. In other words, $\lfloor x \rfloor$ (respectively, $\lceil x \rceil$) is the best approximation to x from below (above) by an integer. These functions are often used to discretize real variables.

(a) Plot $\lfloor x \rfloor$ and $\lceil x \rceil$ for $-2.5 \le x \le 2.5$.

(b) Prove the following basic facts for $x \in \mathbb{R}$ and $m \in \mathbb{Z}$:

$$\lfloor -x \rfloor = -\lceil x \rceil, \tag{P5.11.1a}$$

$$x - 1 < \lfloor x \rfloor \le x \le \lceil x \rceil < x + 1, \tag{P5.11.1b}$$

$$\lfloor x \pm m \rfloor = \lfloor x \rfloor \pm m, \quad \lceil x \pm m \rceil = \lceil x \rceil \pm m \tag{P5.11.1c}$$

($\pm$ respectively in each equation).

(c) Regarding (P5.11.1b), show that if $x - 1 < m \le x$ (respectively, $x \le m < x + 1$) for some $x \in \mathbb{R}$ and $m \in \mathbb{Z}$, then we *must* have $m = \lfloor x \rfloor$ ($m = \lceil x \rceil$).

Problem 5.12 Here are some applications of Problem 5.11.

(a) Roughly, for any integer $b \ge 2$, the number of digits in the fractionless base-b expansion (3.5) of a positive integer x is $\log_b x$ (e.g., 123456789_{10} has $9 \approx \log_{10} 123456789 \approx 8.1$ digits). Give an exact formula for this number using either the floor or ceiling function.

(b) Which of the following two values is the integer quotient of x divided by y, for any positive integers x and y: $\lfloor x/y \rfloor$ or $\lceil x/y \rceil$?

(c) Note that for any numerical sequence $\{x_i\}_{i=-\infty}^{\infty}$, and any *real* numbers α and β such that $\alpha \le \beta$, we have

$$\sum_{\alpha \le i \le \beta} x_i = \sum_{i=\lceil \alpha \rceil}^{\lfloor \beta \rfloor} x_i,$$

where the sum on the right is empty if $\lceil \alpha \rceil > \lfloor \beta \rfloor$ (which does not require $\alpha > \beta$). Provide a similar expression for a sum over $\alpha < i < \beta$.

Problem 5.13 **(a)** Explain why an invertible bounded function $f : \mathbb{R} \to \mathbb{R}$ cannot have a bounded inverse.

(b) Give an example of an invertible bounded function $f : \mathbb{R} \to \mathbb{R}$ that has no local extrema.

Problem 5.14 **(a)** Clearly, any constant function is a periodic function. What are all possible periods of such a function? Why does "the" period of such a function not exist?

(b) Show that the function in (5.25) is periodic. What are all possible periods, and "the" period, of this function?

Problem 5.15 **(a)** Is it possible to have an odd function $f : \mathbb{R} \to \mathbb{R}$ for which $f(0) \ne 0$?

(b) What is the only function $f : \mathbb{R} \to \mathbb{R}$ that is both even and odd?

Problem 5.16 Consider an arbitrary function $f : \mathbb{R} \to \mathbb{R}$.

(a) An example of such a function that is both odd-symmetric and invertible is the identity function on $\mathbb{R}$. By contrast, prove that if $f(\cdot)$ is even-symmetric, then it cannot be invertible.

(b) Prove that if $f(\cdot)$ is odd-symmetric and invertible, then so is $f^{-1}(\cdot)$.

Problem 5.17 Given a function $f: \mathbb{R} \to \mathbb{R}$, we clearly have

$$f(x) = f_e(x) + f_o(x) \quad (x \in \mathbb{R}) \tag{P5.17.1a}$$

for

$$f_e(x) = \tfrac{1}{2}[f(x) + f(-x)], \quad f_o(x) = \tfrac{1}{2}[f(x) - f(-x)] \quad (x \in \mathbb{R}). \tag{P5.17.1b}$$

(a) Show that $f_e(\cdot)$ is an even function and $f_o(\cdot)$ is an odd function.

(b) Prove that the decomposition (P5.17.1a) of $f(\cdot)$ into even and odd components is unique. Specifically, show that if

$$f(x) = f_e{}'(x) + f_o{}'(x) \quad (x \in \mathbb{R}) \tag{P5.17.2}$$

for some even function $f_e{}': \mathbb{R} \to \mathbb{R}$ and odd function $f_o{}': \mathbb{R} \to \mathbb{R}$, then we must have $f_e{}'(\cdot) = f_e(\cdot)$ and $f_o{}'(\cdot) = f_o(\cdot)$.

(c) What are $f_e(\cdot)$ and $f_o(\cdot)$ for the exponential function $f(x) = e^x$ $(x \in \mathbb{R})$?

Problem 5.18 Suppose we wish to use a function $f(\cdot)$ to toggle back and forth between two distinct values $a, b \in \mathbb{R}$ that are given; that is, $f(a) = b$ and $f(b) = a$. Find such a function $f(\cdot)$ that plots to a line.

Problem 5.19 If for two variable quantities x and y we have $y \propto x$, then must we also have $x \propto y$?

Problem 5.20 Let x_1 and x_2 be any two numbers corresponding to distinct points on the real line; and, let γ be the number corresponding to the point midway between.

(a) Show that for a linear scale, $\gamma = (x_1 + x_2)/2$—i.e., the arithmetic mean of x_1 and x_2.

(b) Show that for a logarithmic scale (here, the numbers all being positive), $\gamma = \sqrt{x_1 x_2}$—i.e., the geometric mean of x_1 and x_2.

Problem 5.21 Equation (5.6a) characterizes all functions $f: \mathbb{R} \to \mathbb{R}$ for which the set

$$\{(x, y) \in \mathbb{R}^2 \mid y = f(x)\}$$

yields a line when its points (x, y) are plotted in the xy-plane; namely, this set plots to a line if and only if (5.6a) holds for some constants m, $Y \in \mathbb{R}$. However, this result assumes that the equation $y = f(x)$ is plotted using *linear* scales for both axes; that is, we have a **linear-linear plot**. In each of the alternative cases below, give an equation characterizing all functions $f: D \to \mathbb{R}$, having the largest domain $D \subseteq \mathbb{R}$ possible for that case, such that the above set plots to a line.

(a) A linear-scale versus log-scale plot—i.e., a **log-linear plot.**[11]

(b) A log-scale versus linear-scale plot—i.e., a **linear-log plot**.

(c) A log-scale versus log-scale plot—i.e., a **log-log plot**.

Problem 5.22 For a standard 12-note *musical* scale, each "octave"—which represents a factor-of-2 change in frequency (pitch)—is equally divided into 12 intervals (e.g., the interval from C to C♯) called "semitones". Psychologically, an octave anywhere in the human auditory range is judged to correspond to the same amount of change in frequency; accordingly, "equally divided" here means that the note frequencies are uniformly spaced on a *logarithmic* scale. Show that incrementing by one semitone (anywhere) corresponds to approximately a 6% increase in frequency.

Problem 5.23 Plot the family of functions $\frac{2}{\pi} \tan^{-1} \alpha x$ $(x \in \mathbb{R})$ versus the parameter $\alpha \in \{1, 3, 30\}$ over $-5 \le x \le 5$. Compare the results with a plot of $\operatorname{sgn} x$.

---

11. The order of the adjectives in this definition and the next follow from the fact that an xy-plot is a plot of y versus x.

Problem 5.24 Make a parametric plot in the xy-plane of the equations $x = \sin t$ and $y = \sin 2t$ for $0 \leq t < 2\pi$.

Problem 5.25 Does (5.17) hold for all $X \in D$ if $D = \mathbb{Q}$?

Problem 5.26 Suppose $f: D \to \mathbb{R}$, where $D \subseteq \mathbb{R}$ is such that (5.17) holds for $X = -\infty$. Give an expression akin to (5.15) and (5.16) stating the exact condition for having $f(x) \to -\infty$ as $x \to -\infty$.

Problem 5.27 For the function $f(x) = \tan^{-1}(\ln x) - 3x$ ($x > 0$), calculate each of the following limits.

(a) $\lim_{x\to 0} f(x)$.
(b) $\lim_{x\to 1} f(x)$.
(c) $\lim_{x\to \infty} f(x)$.

Problem 5.28 **(a)** Let $\{x_i\}_{i=1}^{\infty}$ be a real sequence such that $\lim_{i\to\infty} x_i$ exists. Prove that the sequence must be bounded.

(b) Let $f(\cdot)$ be a real function on $[1, \infty)$ such that $\lim_{x\to\infty} f(x)$ exists. Prove via an example that, in contrast to part (a), the function need not be bounded.

Problem 5.29 Give an expression akin to (5.19) stating the exact condition for having $Y = f(X-)$, given a function $f: D \to \mathbb{C}$, a point $X \in D \subseteq \mathbb{R}$, and a value $Y \in \mathbb{C}$.

Problem 5.30 Clearly, for any complex variable x and integer $n \geq 1$ we have $1 - (1 - x)^n \to 0$ as $x \to 0$. More precisely, use the binomial expansion (8.4) to prove that $1 - (1 - x)^n \sim nx$ as $x \to 0$.

Problem 5.31 **(a)** Suppose $f(x) = e^{1/x}$ and $g(x) = e^{1/x^2}$ ($x > 0$). Show that $f(x) \sim g(x)$ as $x \to \infty$, yet we do not have $\ln f(x) \sim \ln g(x)$ as $x \to \infty$.

(b) Likewise, give an example of two functions $f(\cdot)$ and $g(\cdot)$ such that $f(x) \sim g(x)$ as $x \to \infty$, yet we do not have $e^{f(x)} \sim e^{g(x)}$ as $x \to \infty$.

Problem 5.32 Suppose we have two functions $f: \mathbb{C} \to \mathbb{C}$ and $g: \mathbb{C} \to \mathbb{C}$, and a value $X \in \mathbb{C}$, such that $f(x) \sim g(x)$ as $x \to X$ *and* the limit $\lim_{x\to X} g(x)$ exists. Does it then follow that $\lim_{x\to X} f(x) = \lim_{x\to X} g(x)$?

Problem 5.33 Let $f(\cdot)$ and $g(\cdot)$ be complex-valued functions having a common domain $D \subseteq \mathbb{C}$; and, suppose that either $X \in \mathbb{C}$ or $X = \pm\infty$. Prove each of the following statements regarding the asymptotic behavior of these functions as $x \to X$.

(a) If $f(x) = o[g(x)]$, then $f(x) = O[g(x)]$.
(b) If $f(x) \sim g(x)$, then $f(x) = O[g(x)]$.
(c) If $f(x) = O[g(x)]$, then for any complex constant $a \neq 0$ we have $f(x) = O[ag(x)]$; and, likewise for $o(\cdot)$.
(d) If $f(x) = o[g(x)]$, and $g(x) \neq 0$ for all $x \in D - \{X\}$ sufficiently close to X, then $f(x) + g(x) \sim g(x)$.

Problem 5.34 At what points is the function

$$f(x) = \begin{cases} 0 & x \text{ rational} \\ x & x \text{ irrational} \end{cases}$$

continuous?

Problem 5.35 Give an example of a continuous invertible function $f(\cdot)$ for which the inverse function $f^{-1}(\cdot)$ is not continuous.

Problem 5.36 Suppose $f : \mathbb{C} \to \mathbb{C}$ and $g : \mathbb{C} \to \mathbb{C}$ are continuous functions. Prove that the composite function $f[g(\cdot)]$ is also continuous; specifically, show that for any $X \in \mathbb{C}$ and $\varepsilon > 0$, there exists a $\delta > 0$ such that $|f[g(x)] - f[g(X)]| \leq \varepsilon$ for all $x \in \mathbb{C}$ such that $|x - X| \leq \delta$.

Problem 5.37 **(a)** Give an example of a continuous increasing function $f : \mathbb{R} \to \mathbb{R}$ that is not strictly increasing, whereas its restriction to the interval $(0, 1)$ *is* strictly increasing.

(b) Show that no such example would exist in part (a) if the domain $\mathbb{R}$ of $f(\cdot)$ changed to $[0, 1]$; that is, show that if the restriction to $(0, 1)$ of a continuous function $f : [0, 1] \to \mathbb{R}$ is strictly increasing, then $f(\cdot)$ itself must be strictly increasing.

Problem 5.38 Suppose we restrict a function $f : D \to \mathbb{R}$ having domain $D \subseteq \mathbb{R}$ to obtain a function $g : D' \to \mathbb{R}$ having domain $D' \subseteq D$; and, let $X \in D'$.

(a) If $f(\cdot)$ is increasing or decreasing, then must $g(\cdot)$ also be increasing or decreasing, respectively?

(b) If $f(\cdot)$ is strictly increasing or strictly decreasing, then must $g(\cdot)$ also be strictly increasing or strictly decreasing, respectively?

(c) If $f(\cdot)$ has a global minimum or a global maximum at X, then must $g(\cdot)$ also have a global minimum or a global maximum at X, respectively?

(d) If $f(\cdot)$ has a local minimum or a local maximum at X, then must $g(\cdot)$ also have a local minimum or a local maximum at X, respectively?

(c) If $f(\cdot)$ is discontinuous at X, then must $g(\cdot)$ also be discontinuous at X?

Problem 5.39 Let the $f : \mathbb{R} \to \mathbb{R}$ and $g : \mathbb{R} \to \mathbb{R}$ be the similar functions

$$f(x) = \begin{cases} 0 & x \leq 0 \\ e^{-1/x} & \text{otherwise} \end{cases}, \qquad g(x) = \begin{cases} 0 & x = 0 \\ e^{-1/x} & \text{otherwise} \end{cases}.$$

(a) Find $f(0-)$ and $f(0+)$, and thereby show that $f(x)$ is continuous at $x = 0$.

(b) Find $g(0-)$ and $g(0+)$, and thereby show that $g(x)$ is discontinuous at $x = 0$.

Problem 5.40 Suppose a function $f : \mathbb{R} \to \mathbb{C}$ is discontinuous at a point $X \in \mathbb{R}$, yet $f(X-) = f(X+)$. If for each $x \in \mathbb{R}$ we let

$$g(x) = \begin{cases} f(X-) & x = X \\ f(x) & x \neq X \end{cases},$$

do we obtain a function $g : \mathbb{R} \to \mathbb{C}$ that is continuous at X?

Problem 5.41 Use the intermediate-value theorem to prove that, as stated in the discussion of (5.1), for each $x > 0$ there is exactly one $y \in \mathbb{R}$ such that $x = e^y + e^{6y}$.

Problem 5.42 Suppose $f : [a, b] \to \mathbb{C}$ $(-\infty < a < b < \infty)$ is a continuous—and, therefore, *uniformly* continuous—function.

(a) Show that $f(\cdot)$ can be uniformly approximated, as closely as desired, by a piecewise-constant function having finitely many pieces. Specifically, for each $\varepsilon > 0$ there exists a function

$$g(x) = \begin{cases} c_1 & x_0 \leq x < x_1 \\ c_2 & x_1 \leq x < x_2 \\ & \vdots \\ c_{n-1} & x_{n-2} \leq x < x_{n-1} \\ c_n & x_{n-1} \leq x \leq x_n \end{cases} \qquad \text{(P5.42.1)}$$

$(n \geq 1)$, with $c_i \in \mathbb{C}$ $(i = 1, 2, \ldots, n)$ and $a = x_0 < x_1 < \cdots < x_n = b$, such that $|g(x) - f(x)| \leq \varepsilon$ $(a \leq x \leq b)$.

(b) Show that $f(\cdot)$ must be bounded.[12]

Problem 5.43 Let $\{f_i(\cdot)\}_{i=1}^{\infty}$ be a sequence of functions $f_i: D \to \mathbb{C}$ having a common domain $D \subseteq \mathbb{C}$, such that the sequence converges uniformly to a function $f: D \to \mathbb{C}$.

(a) Show that if the functions $f_i(\cdot)$ are continuous, then so must be $f(\cdot)$.[13]

(b) Show that for $D = \mathbb{R}$,

$$\lim_{i\to\infty}\lim_{x\to\infty} f_i(x) = \lim_{x\to\infty}\lim_{i\to\infty} f_i(x) = \lim_{x\to\infty} f(x), \tag{P5.43.1}$$

when the limits $\lim_{x\to\infty} f_i(x)$ (all i) and $\lim_{x\to\infty} f(x)$ exist (and, likewise upon replacing every "$x \to \infty$" with "$x \to -\infty$").[14]

Chapter 6. Powers and Roots

Problem 6.1 **(a)** Show that each property in (6.2) would hold for all $x, y \in \mathbb{C}$ and $m, n \in \mathbb{N}$, were we to define 0^0 as 0; and, likewise, were we to define 0^0 as 1.

(b) Now, suppose 0^0 were defined to be a value other than 0 or 1. Which of the properties in (6.2) would still hold for all $x, y \in \mathbb{C}$ and $m, n \in \mathbb{N}$?

Problem 6.2 Prove that for any complex number $x \neq 0$, we have both $\operatorname{Re} x^n \neq 0$ and $\operatorname{Im} x^n \neq 0$ for all integers $n \neq 0$ if and only if $(2/\pi)\arg x$ is irrational.

Problem 6.3 Use (6.10), (6.11), and (6.14) to solve the following.

(a) Show that $x\sqrt[n]{y} = \sqrt[n]{x^n y}$ for all $n \in \mathbb{Z}^+$ and $x, y \geq 0$.

(b) Simplify $\sqrt[km]{x^{kn}}$, where $k, m, n \in \mathbb{Z}^+$ and $x \geq 0$.

(c) Given $m, n \in \mathbb{Z}^+$ and $x \geq 0$, find $p \in \mathbb{Q}$ and $k, l \in \mathbb{Z}^+$ such that

$$\sqrt[m]{x}\sqrt[n]{x} = x^p = \sqrt[k]{x^l}.$$

Problem 6.4 Prove that for any complex numbers a and b, and any positive number p (not necessarily an integer), $|a+b|^p \leq 2^p(|a|^p + |b|^p)$. (*Suggestion*: If you have difficulty, first consider the case of a and b positive, for which the absolute-value operations become superfluous.)

Problem 6.5 Since the only complex number that equals its negative is zero, it follows from (3.36a) that i does not equal $1/i$. What, then, is erroneous in the following "proof" that it does?

$$i = \sqrt{-1} = \sqrt{1/(-1)} = \sqrt{1}/\sqrt{-1} = 1/i. \tag{P6.5.1}$$

Problem 6.6 **(a)** Show that, in contrast to (6.15),

$$\sqrt{-x} = -i\sqrt{x} \quad (x \leq 0).$$

(b) Now, show that for all $x, y \in \mathbb{R}$, if x and y are not *both* negative, then $\sqrt{xy} = \sqrt{x}\sqrt{y}$. But, what if they are?

Problem 6.7 Suppose

$$f(x) = \frac{x + \sqrt[3]{x} + 4}{x^2 + \sqrt{x} + 1} \quad (x > 0).$$

---

12. A continuous function can be bounded, though, without being uniformly continuous: e.g., $\sin(1/x)$ for $x \in (0, 1]$.

13. By contrast, the example in Problem 5.23 shows that the limit of a sequence of continuous functions need not be continuous when the convergence is not uniform.

14. By contrast, for the functions in (5.27) we have $\lim_{x\to\infty} f_i(x) = 1$ (all i), thereby making $\lim_{i\to\infty}\lim_{x\to\infty} f_i(x) = 1$, whereas $\lim_{i\to\infty} f_i(x) = 0$ (all x), thereby making $\lim_{x\to\infty}\lim_{i\to\infty} f_i(x) = 0$.

(a) Find $\lim_{x\to 0} f(x)$.

(b) Find $\lim_{x\to\infty} f(x)$.

Problem 6.8 (a) Show that for any cubic polynomial $f(\cdot)$ having real coefficients, the plot of $y = f(x)$ versus $x \in \mathbb{R}$ displays odd symmetry about exactly one point (X, Y); that is, there exist unique constants $X, Y \in \mathbb{R}$ such that the function $g(x) = f(x+X) - Y$ of $x \in \mathbb{R}$ is odd.

(b) Can there also exist a point about which the plot displays even symmetry?

Problem 6.9 Prove that a polynomial $f(\cdot)$ has real coefficients if and only if $f^*(x) = f(x^*)$ $(x \in \mathbb{C})$.

Problem 6.10 Having radicals in the denominator of a fraction is generally discouraged, though at times it might be preferable to write, e.g., "$1/\sqrt{2}$" rather than "$\sqrt{2}/2$" or "$\frac{1}{2}\sqrt{2}$". Accordingly, rewrite the following fractions:

$$\frac{1}{\sqrt{2}+\sqrt{3}}, \quad \frac{1}{5-\sqrt{7}}.$$

Problem 6.11 For the polynomial $f(\cdot)$ in (6.16), suppose that $n \geq 1$ and $a_0, a_n \neq 0$; and, let another polynomial $g(\cdot)$ be formed by reversing the order of the coefficients:

$$g(x) = a_0 x^n + a_1 x^{n-1} + \cdots + a_{n-1}x + a_n = \sum_{j=0}^{n} a_{n-j} x^j \quad (x \in \mathbb{C}).$$

Show that if $f(\cdot)$ has distinct roots $\hat{x}_1, \ldots, \hat{x}_{\hat{n}}$ with respective multiplicities $m_1, \ldots, m_{\hat{n}}$, then $g(\cdot)$ has distinct roots $1/\hat{x}_1, \ldots, 1/\hat{x}_{\hat{n}}$ with the same respective multiplicities.

Problem 6.12 Suppose we plot $f(x)$ versus $x \in \mathbb{R}$ for an odd-degree polynomial $f(\cdot)$ having real coefficients. Why must the plot have at least one x-axis intercept?

Problem 6.13 The roots of the polynomial $f(x) = x^5 + x^4 - 15x^3 - x^2 + 38x - 24$ $(x \in \mathbb{C})$ are all integers. Find them.

Problem 6.14 (a) What are the second roots of unity?

(b) What are the fourth roots of unity?

Problem 6.15 Viewing complex numbers as vectors in the complex plane (per the Argand diagram in Figure 3.3), it would appear from Figure 6.5 that for $n \geq 2$, the nth roots of unity always sum to 0. Verify that this is indeed true by finding a general formula for the sum

$$\sum_{k=0}^{m-1} e^{i2\pi k/n} \quad (m, n \in \mathbb{Z}^+),$$

then evaluating it for $m = n \geq 2$. (*Hint*: geometric progression.)

Problem 6.16 As done on p. 95 for the nth roots of unity, describe the positioning in the complex plane of the values in (6.22b) for given complex number $c \neq 0$.

Problem 6.17 Use the quadratic formula (6.23c) to verify (6.21) in the case of degree $n = 2$ with coefficients $a_j \in \mathbb{C}$ $(j = 0, 1, 2)$.

Problem 6.18 Consider a quadratic polynomial (6.23a) with real coefficients $a \neq 0$, b, and c, having roots x_1 and x_2. Find necessary and sufficient conditions on the three coefficients for each of the following situations to hold.

(a) $|x_1| = |x_2|$.

(b) x_1 and x_2 are imaginary and nonzero.

Problem 6.19 Consider a monic quadratic polynomial

$$f(x) = x^2 + bx + c \quad (x \in \mathbb{C}) \tag{P6.19.1}$$

having real coefficients b and c. Solve the following *without* using the quadratic formula.

(a) Show how to **complete the square** of the polynomial $f(\cdot)$; that is, find real values β and γ, in terms of b and c, for which

$$f(x) = (x+\beta)^2 + \gamma \quad (x \in \mathbb{C}). \tag{P6.19.2}$$

(b) Now consider the following form similar to (P6.19.2):

$$f(x) = (x-Y)^2 + Z^2 \quad (x \in \mathbb{C}) \tag{P6.19.3}$$

for real values Y and $Z > 0$. Show that the two roots of $f(\cdot)$ are now $Y \pm iZ$, which are distinct and nonreal. Furthermore, show that if for some real coefficients b and c the polynomial in (P6.19.1) has nonreal roots, then it can be put into the form (P6.19.3) for some real values Y and $Z > 0$; and, find Y and Z in terms of b and c.

Problem 6.20 Find the roots of the quintic polynomial $f(x) = x^5 - 28x^3 + 36x$ $(x \in \mathbb{C})$.

Problem 6.21 Show that there exists no pair $(x, y) \in \mathbb{R}^2$ for which

$$\frac{1}{x+y} = \frac{1}{x} + \frac{1}{y}. \tag{P6.21.1}$$

Problem 6.22 Find all complex values x for which

$$\sqrt{2x+7} = x+2. \tag{P6.22.1}$$

Problem 6.23 Prove that $x + 1/x \geq 2$ $(x > 0)$. Also, determine a necessary and sufficient condition on x for equality to hold.

Problem 6.24 Based on the facts

$$\sum_{j=1}^{n} j^0 = n, \quad \sum_{j=1}^{n} j^1 = \tfrac{1}{2}n(n+1) \quad (n = 1, 2, \ldots),$$

it is reasonable to guess that

$$\sum_{j=1}^{n} j^2 = f(n) \quad (n = 1, 2, \ldots), \tag{P6.24.1}$$

for some third-degree polynomial $f(\cdot)$. Find such a polynomial by assuming (P6.24.1) and evaluating $f(n+1) - f(n)$ $(n \geq 1)$.

Problem 6.25 **(a)** The polynomial $x^3 - 8$ $(x \in \mathbb{C})$ clearly has a root at $x = \sqrt[3]{8} = 2$. After dividing by $x - 2$ to factor this root out, find the remaining two roots in rectangular form.

(b) Now, obtain the same result via (6.22).

Problem 6.26 Find the roots of the polynomial $f(x) = (1+i1)x^2 - i6x + (-44 + i28)$ $(x \in \mathbb{C})$.

Problem 6.27 Use the Lagrange interpolation formula (6.25) with $n = 2$ to obtain the function $f : \mathbb{R} \to \mathbb{R}$ for the line passing through two points (x_1, y_1) and (x_2, y_2), where $x_1 \neq x_2$. Then, use the result to express the function $f(\cdot)$ in the form (5.6a), for some constants m and Y.

Problem 6.28 Suppose $f(\cdot)$ and $g(\cdot)$ are both polynomials passing through the same $n \geq 1$ points $(x_j, y_j) \in \mathbb{C}^2$ $(j = 1, \ldots, n < \infty)$, where the values x_j are distinct. We know that if neither polynomial has a degree greater than $n - 1$, then $f(\cdot) = g(\cdot)$. Use this fact to prove that if $g(\cdot)$ is not the same as the polynomial $f(\cdot)$ given by the Lagrange interpolation formula (6.25), then either $f(\cdot)$ is the zero polynomial or $g(\cdot)$ has a degree greater than that of $f(\cdot)$.

Problem 6.29 Suppose $f(\cdot)$ and $g(\cdot)$ are both polynomials on $\mathbb{C}$, neither of which is the zero polynomial, having degrees m and n, respectively. For each of the following polynomials $h(\cdot)$, state the range of possible degrees.

(a) $h(x) = f(x) + g(x)$ (all x).

(b) $h(x) = f(x) \cdot g(x)$ (all x).

Problem 6.30 Let $f(\cdot)$ be a polynomial defined on $\mathbb{C}$; and, suppose that for some constant $c \in \mathbb{C}$ we divide $f(x)$ by $x - c$ $(x \in \mathbb{C})$. Use (6.27) to show that if $r(\cdot)$ is the resulting remainder polynomial, then $r(x) = f(c)$ $(x \in \mathbb{C})$.

Chapter 7. Exponentials and Logarithms

Problem 7.1 **(a)** Plot $y = b^x$ versus x for bases $b = \frac{1}{10}, \frac{1}{2}, 2$, and 10. Use the same pair of axes for all four plots, displaying the region for $-2.5 \le x \le 2.5$ and $0 < y \le 3$.

(b) Plot $y = \log_b x$ versus x for bases $b = 2$ and 10. Use the same pair of axes for both plots, displaying the region for $0 < x \le 4$ and $-2 \le y \le 2$.

Problem 7.2 **(a)** Suppose the equation $y = b^x$ $(x \in \mathbb{R})$ is plotted in the xy-plane for a fixed base $b > 0$. Show that shifting the plot to the left by an amount $\alpha \in \mathbb{R}$ is equivalent to vertically scaling the plot, and find the scale factor in terms of α.

(b) What is the corresponding fact for plotting the equation $y = \log_b x$ $(x > 0)$ in the xy-plane for a fixed base $b > 1$?

Problem 7.3 Suppose that for some $p > 0$ and $n \in \mathbb{Z}^+$, a bank account earns compound interest at an annual rate of $p\%$, compounded n times a year. That is, if we assume no deposits or withdrawals, then upon each accumulation period of one-nth year, the amount in the account is multiplied by $1 + p\%/n$; so, for a full year, the multiplier is $(1 + p\%/n)^n$. Of course, the larger the number n (i.e., the more frequently the interest is compounded), the faster the account grows.

(a) Taking this idea to the limit, use (7.2) to show that if the interest is compounded *continuously*—i.e., we let $n \to \infty$—then the annual multiplier becomes simply $e^{p\%}$. (For example, 5% interest compounded continuously produces an annual yield of $e^{5\%} - 1 = e^{0.05} - 1 \approx 0.0513 = 5.13\%$.)

(b) Therefore, earning interest at an annual rate of $p\%$, compounded n times a year, is equivalent to earning interest continuously at an annual rate of $q\%$ for what value of q?

Problem 7.4 Assuming (7.4h) and (7.4i) have already been proved, prove the remainder of (7.4).

Problem 7.5 Simplify $e^{z+in\pi/2}$ for $z \in \mathbb{C}$ and $n \in \mathbb{Z}$.

Problem 7.6 We know that if $x_1, x_2 \in \mathbb{R}$ are such that $e^{x_1} = e^{x_2}$, then we must have $x_1 = x_2$. Now, suppose $z_1, z_2 \in \mathbb{C}$ are such that $e^{z_1} = e^{z_2}$. What is the relationship between z_1 and z_2?

Problem 7.7 Show that there is no imaginary number z for which e^{e^z} is imaginary.[15]

Problem 7.8 **(a)** Explain how it is evident from Figure 7.1 that

$$\ln x \le x - 1 \quad (x > 0), \tag{P7.8.1a}$$

with equality obtaining if and only if $x = 1$. Then, use this upper bound on the natural logarithm function to obtain the lower bound

$$\ln x \ge 1 - 1/x \quad (x > 0); \tag{P7.8.1b}$$

and, determine which values of x attain equality.

(b) Again starting from Figure 7.1, obtain similar upper and lower bounds on the exponential function.

15. To be clear, "e^{e^z}" means $e^{(e^z)}$, not $(e^e)^z$.

Problem 7.9 **(a)** Akin to (7.1) and (7.7), obtain the limiting behavior of $a^x / \log_b x$ ($a > 0$ and $b > 1$ fixed, with $a \neq 1$) as $x \to \infty$.

(b) Now, find $\lim_{x\to\infty} f(x)$ for

$$f(x) = \frac{e^{2x} + 8e^{3x} + \log_7 x + 5x^9}{2e^{3x} + 4\ln x + 5\sqrt{6x}e^{-7x} + x^9} \quad (x \in \mathbb{R}).$$

Problem 7.10 **(a)** From (7.7) we know that $(\ln x)/x \to 0$ as $x \to \infty$. Use this fact to show that $x^{1/x} \to 1$ as $x \to \infty$.

(b) Use the result of part (a) to show that $(1/x)^{1/x} \to 1$ as $x \to \infty$.

(c) Continuing, show that for any real-valued sequence $\{x_n\}_{n=1}^{\infty}$ such that $\lim_{n\to\infty} x_n > 0$, we have both ${x_n}^{1/n} \to 1$ and $(1/x_n)^{1/n} \to 1$ as $n \to \infty$. (*Hint*: Note that we must have $1/n \leq x_n \leq n$ for n sufficiently large.)

(d) Apply the above results to Stirling's approximation (5.21) to show

$$e = \lim_{n\to\infty} \frac{n}{(n!)^{1/n}} \quad \text{(P7.10.1)}$$

—an alternative to (7.2)

Problem 7.11 **(a)** Given constants $a, b > 0$ and $c, d \in \mathbb{R}$, find constants $\alpha, \beta \in \mathbb{R}$ such that $ab^{cx+d} = e^{\alpha x+\beta}$ ($x \in \mathbb{R}$).

(b) Given constants $a \in \mathbb{R}, b > 1, c > 0$, and $d \in \mathbb{R}$, find constants $\alpha, \beta \in \mathbb{R}$ such that $a \log_b cx+d = \alpha \ln x + \beta$ ($x > 0$).

Problem 7.12 **(a)** Show that for any $x, y > 0$ with $y \neq 1$, the ratio

$$\frac{\log_b x}{\log_b y}$$

has the same value no matter what positive base $b \neq 1$ is chosen for both logarithms.

(b) Show that for any $x, y > 0$ and positive base $b \neq 1$, we have $x^{\log_b y} = y^{\log_b x}$.

Problem 7.13 **(a)** It can be shown that $f(x) = (\ln x)/x$ is a strictly decreasing function of $x \geq e$. Use this fact to show

$$a^b > b^a \quad (e \leq a < b).$$

(b) By contrast, find two real numbers a and b with $2 \leq a < b$ for which $a^b < b^a$.

Problem 7.14 Let $f(\cdot)$ be a positive function on $\mathbb{R}$ such that

$$f(x)f(y) = f(x+y) \quad (x, y \in \mathbb{R}). \quad \text{(P7.14.1)}$$

(a) Prove that if $f(\cdot)$ is continuous at some point $X \in \mathbb{R}$, then

$$f(x) = b^x \quad (x \in \mathbb{R}) \quad \text{(P7.14.2)}$$

for some constant $b > 0$ (thereby making $f(\cdot)$ continuous at *every* point). Proceed as follows:

- Show that $f(0) = 1$.
- Applying (P7.14.1) successively m times gives

$$f(x)f^m(y) = f(x+my) \quad (x \in \mathbb{R}, y \in \mathbb{R}, m \in \mathbb{Z}^+). \quad \text{(P7.14.3)}$$

Show that this equality continues to hold if m is allowed to be *any* integer.

- Use the above facts to show

$$f^n(1/n) = f(1) \quad (n \in \mathbb{Z}^+), \tag{P7.14.4}$$

and therefore

$$f(x)f^{m/n}(1) = f(x + m/n) \quad (x \in \mathbb{R}, m \in \mathbb{Z}, n \in \mathbb{Z}^+). \tag{P7.14.5}$$

- Let $m/n \to X - x$ in (P7.14.5), and simplify.

(b) Explain why it follows from part (a) that for any positive function $f(\cdot)$ on $\mathbb{R}$, if (P7.14.1) holds and $f(\cdot)$ is *discontinuous* at some point, then it is discontinuous at *every* point.

Problem 7.15 Prove (7.11) from (7.4).

Chapter 8. Possibility and Probability

Problem 8.1 Let

$$(2n+1)!! = 1 \cdot 3 \cdot 5 \cdot \cdots \cdot (2n+1) \quad (n = 0, 1, \ldots);$$

that is, $(2n+1)!!$ is the product of the *odd* positive integers through $2n+1$.

(a) Express $(2n+1)!!$ $(n = 0, 1, \ldots)$ in terms of the factorials.

(b) Use the result of part (a) and Stirling's approximation (5.21) to find an asymptotic expression for $(2n+1)!!$ as $n \to \infty$.

Problem 8.2 Suppose that for two finite sets A and B, the number of permutations of A is 20 times that of B. What does this tell us about the cardinalities of these sets?

Problem 8.3 **(a)** Use the binomial expansion (8.4) to prove

$$\sum_{k=0}^{n} \binom{n}{k} = 2^n \quad (n = 0, 1, \ldots). \tag{P8.3.1}$$

(b) Note that the left side of (P8.3.1) is the *total* number of combinations of elements taken from an n-element set A. Accordingly, provide an alternative justification of (P8.3.1) based on forming the combinations of A one element at a time.

Problem 8.4 Shown in Figure P8.4.1 is **Pascal's triangle**, which is an infinite triangular array of all possible binomial coefficients; specifically, for all integers n and k such that $0 \le k \le n$, the $(n+1)$th row of the array contains $n+1$ elements, with the $(k+1)$th element being $\binom{n}{k}$.

(a) That the left and right sides of the triangle are composed of all 1's is a reflection of the fact that

$$\binom{n}{0} = \binom{n}{n} = 1 \quad (n = 0, 1, \ldots). \tag{P8.4.1a}$$

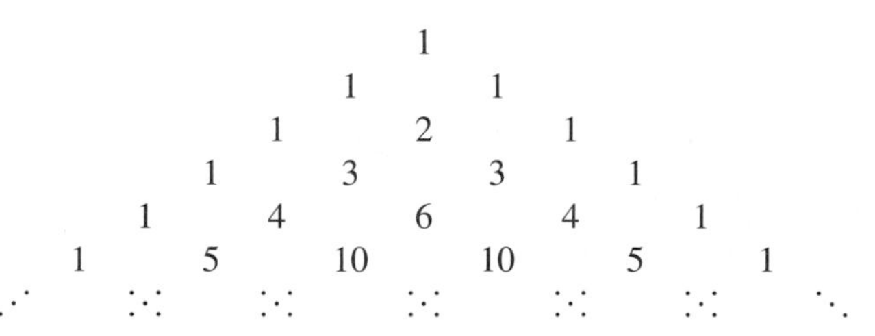

FIGURE P8.4.1 Pascal's Triangle

It is also evident from the figure that once these sides are in place, the remainder of the triangle can be generated row by row, starting with the third row, by simply summing adjacent elements in the previous row (e.g., the sixth row obtains from the fifth by the calculations $1+4=5$, $4+6=10$, etc.); that is,

$$\binom{n}{k} = \binom{n-1}{k} + \binom{n-1}{k-1} \quad (n = 2, 3, \ldots; k = 1, 2, \ldots, n-1). \tag{P8.4.1b}$$

Prove (P8.4.1b) from (8.2).

(b) Alternatively, explain (P8.4.1b) in terms of selecting k objects from a set A of n objects. (*Suggestion*: Arbitrarily specify a particular object $a \in A$, and consider the two cases of whether or not a ends up in the selection.)

Problem 8.5 Use (P8.4.1) to prove

$$\sum_{k=0}^{m}(-1)^k\binom{n}{k} = (-1)^m\binom{n-1}{m} \quad (n = 1, 2, \ldots; m = 0, 1, \ldots, n-1). \tag{P8.5.1}$$

Also, what does the left side equal when $m = n \geq 0$?

Problem 8.6 Generalizing the binomial expansion (8.4), use mathematical induction (versus m) to prove the **multinomial expansion**:

$$(x_1 + \cdots + x_m)^n = \sum_{(n_1,\ldots,n_m)\in I_{m,n}} \frac{n!}{n_1!\cdots n_m!} x_1{}^{n_1}\cdots x_m{}^{n_m}, \tag{P8.6.1a}$$

for the index set

$$I_{m,n} = \{(n_1, \ldots, n_m) \in \mathbb{N}^m \mid n_1 + \cdots + n_m = n\} \tag{P8.6.1b}$$

($m \in \mathbb{Z}^+$; $x_1, \ldots, x_m \in \mathbb{C}$; $n \in \mathbb{N}$).

Problem 8.7 Consider finding all permutations of a collection of objects where some of the objects are identical (i.e., indistinguishable). Specifically, calculate how many permutations there are of the six objects 1, 2, 2, 3, 3, and 3. (*Suggestion*: First suppose that the objects *were* distinguishable.)

Problem 8.8 In poker, how many different hands constitute a full house (i.e., five cards consisting of three of a kind in one rank along with a pair in another)?

Problem 8.9 Consider forming "words" of length $n \geq 1$, each being a sequence of "letters" taken from an "alphabet" (set) A of size $m \geq 1$. (Here, *every* finite sequence of letters is viewed as a valid word; thus, there are a total of m^n words of length n.)

(a) For each length n, how many words are such that no two adjacent letters are the same?

(b) For each length n, how many words are palindromes (i.e., each word reads the same backward and forward)?

Problem 8.10 Just as a permutation of a set A can be visualized as an ordering of the elements of A along a line, so a **circular permutation** of A is a particular ordering of the elements of A around a circle (in which case every element has a unique "next" element—e.g., in the clockwise direction—but none of the elements is designated "first"). Given a finite set A having $n \geq 1$ elements, how may circular permutations does it have?

Problem 8.11 Which has greater probability, an event with odds of 3 in 7, or an event with odds of 8 to 5 against?

Problem 8.12 Suppose a fair coin is flipped $n > 0$ times, where n is even; and, let p_n be the probability that *exactly* the same number of heads occur as tails.

TABLE P8.13.1

Flip 1	Flip 2	Flip 3	Choice
TAIL	TAIL	TAIL	A
TAIL	TAIL	HEAD	B
TAIL	HEAD	TAIL	C
TAIL	HEAD	HEAD	D
HEAD	TAIL	TAIL	E

(a) Explain why

$$p_n = \frac{n!}{2^n (n/2)!^2} \quad (n = 2, 4, 6, \ldots).$$

(b) Use Stirling's approximation (5.21) to show

$$p_n \sim \sqrt{2/\pi n} \quad \text{as} \quad n \to \infty.$$

Hence, $p_n \to 0$ as $n \to \infty$ (contrary to what many lay people would think).

(c) Calculate and tabulate both p_n and $\sqrt{2/\pi n}$ to three decimal places for $n = 2, 4, 10, 20$, and 40.

Problem 8.13 A person who wishes to randomly choose one of five possibilities—say, A, B, C, D, and E—with equal probability for each, decides to flip a fair coin three times, then use Table P8.13.1 to convert the results into one of the choices. Suppose, then, that the first flip happens to be HEAD, thereby making A, B, C, and D impossible. Why, though, would it be erroneous at this point to select choice E? What should the person do instead?

Problem 8.14 Let a finite set V of $n \geq 1$ objects be given, along with an integer $m \in \{0, 1, \ldots, n\}$; and, suppose we randomly choose from V a *subset* $U \subseteq V$ of m objects, in such a way that every m-element subset is equally likely to be chosen. Thus, for each particular m-element set $B \subseteq V$, we have

$$\Pr(U = B) = 1 \bigg/ \binom{n}{m},$$

by (8.5). Prove the "obvious" fact that for any particular *element* $b \in V$,

$$\Pr(b \in U) = \frac{m}{n}.$$

Do so by using (8.6) with A being the set of all possible subsets U; that is,

$$\Pr(b \in U) = \frac{\text{number of possible } U \text{ for which } b \in U}{\text{number of possible } U}. \quad \text{(P8.14.1)}$$

Problem 8.15 Suppose a finite set $\{1, 2, \ldots, n\}$ $(n \geq 1)$ is randomly sampled with replacement $m \geq 1$ times. What is the probability that all of the samples differ?

Problem 8.16 Find the arithmetic mean, geometric mean, harmonic mean, median, modes, second moment, variance, standard deviation, and root mean square for the following samples of a random variable x: 37, 12, 34, 46, 26, 14, 34, 15.

Problem 8.17 Show that if M is the harmonic mean of $n \geq 2$ numbers $x_i > 0$ $(i = 1, 2, \ldots, n)$, then

$$M < n \min_i x_i \quad \text{(P8.17.1)}$$

—which, if $(\max_i x_i)/(\min_i x_i) \geq n$, is tighter than the more obvious bound that $M \leq \max_i x_i$.

Problem 8.18 Suppose a train makes a trip in $n \geq 1$ parts, with $d_i > 0$ being the distance traveled and $t_i > 0$ being the time expended during the ith part ($i = 1, 2, \ldots, n$). Thus, the train travels a total distance $D = d_1 + d_2 + \cdots + d_n$, in a total time $T = t_1 + t_2 + \cdots + t_n$, at an average speed $S = D/T$; whereas the average speed for the ith part is $s_i = d_i/t_i$ (all i).

(a) Express S as a weighted arithmetic mean of the speeds s_i, with the weights being solely in terms of the times t_i (all i) and T.

(b) Now, express S as a "weighted harmonic mean" of the speeds s_i, with the weights being solely in terms of the distances d_i (all i) and D.

Problem 8.19 **(a)** Show that sample averaging, indicated by an overbar in (8.9), is a linear operation; that is, for any real-valued random variables x and y, and any real constant a, we have $\overline{ax} = a\,\overline{x}$ and $\overline{x+y} = \overline{x} + \overline{y}$.

(b) Use part (a) to derive the result in (8.15), without rewriting the averages indicated by overbars in terms of sums.

Problem 8.20 **(a)** Plot a frequency histogram of the samples in Table 8.1 for the random variable y; use the same bin intervals as on the left side of Figure 8.1.

(b) Plot a relative-frequency histogram of the same data, but with bin intervals four times as wide.

Problem 8.21 **(a)** Plot the probability histogram corresponding to the frequency histogram on the left side of Figure 8.1.

(b) Repeat part (a) using these nonuniform bin intervals instead: $[2.5, 6.5)$, $[6.5, 9.5)$, $[9.5, 11.5)$, $[11.5, 14.5)$, $[14.5, 18.5)$.

Problem 8.22 Table P8.22.1 shows 20 blood pressure measurements on a particular patient, where s is the systolic pressure and d is the diastolic pressure, both measured in millimeters of mercury (mmHg).

(a) Plot a scattergram of d versus s.

(b) Calculate the corresponding regression line, and plot it on the scattergram. Also, calculate the centroid and the correlation coefficient of the data.

(c) As can be seen from the scattergram, one of the plotted points is not closely grouped with the others; such an anomalous data point (of which there can be more that one) is called an **outlier**.[16] Remove the outlier from the data and repeat part (b).

Problem 8.23 Sometimes, given $n \geq 1$ data points $(x_i, y_i) \in \mathbb{R}^2$ ($i = 1, 2, \ldots, n$), one wishes to perform a linear regression in such a way that the plotted line will pass through a particular point;

TABLE P8.22.1 Blood Pressure Data

s (mmHg)	d (mmHg)	s (mmHg)	d (mmHg)
126	72	122	64
125	74	116	70
124	69	110	61
116	62	105	65
116	65	100	61
117	70	119	68
121	63	96	62
125	74	92	67
92	56	95	72
105	66	93	66

16. In experimental work, just what constitutes an outlier is a subjective matter. Generally, such an anomaly is attributed to some sporadic error in measurement, thereby justifying its being discarded.

specifically, suppose that this point is the origin, (0, 0). Thus, we set $a_0 = 0$ in (8.16) to get the regression function $f(x) = a_1 x$ ($x \in \mathbb{R}$). Show that the mean square error $\overline{[f(x) - y]^2}$ in (8.17) is now minimized if and only if

$$a_1 = \frac{\overline{xy}}{\overline{x^2}} = \frac{\sum_{i=1}^n x_i y_i}{\sum_{i=1}^n x_i{}^2}$$

(which, not surprisingly, is what obtains from (8.18) for $\overline{x} = \overline{y} = 0$)—assuming $x_i \neq 0$ for some i, in order to avoid division by 0. (*Suggestion*: Express $n\overline{[f(x) - y]^2}$ as a quadratic polynomial in the variable a_1; then, after making the polynomial monic, complete the square—see Problem 6.19.)

Problem 8.24 **(a)** Derive these alternative expressions for the correlation coefficient defined in (8.19):

$$r = \frac{\overline{xy} - \overline{x} \cdot \overline{y}}{\sqrt{\left(\overline{x^2} - \overline{x}^2\right)\left(\overline{y^2} - \overline{y}^2\right)}}$$

$$= \frac{n\sum_{i=1}^n x_i y_i - \sum_{i=1}^n x_i \sum_{i=1}^n y_i}{\sqrt{\left[n\sum_{i=1}^n x_i{}^2 - \left(\sum_{i=1}^n x_i\right)^2\right]\left[n\sum_{i=1}^n y_i{}^2 - \left(\sum_{i=1}^n y_i\right)^2\right]}}. \tag{P8.24.1}$$

(b) Express the linear-regression coefficient a_1 in (8.18b) in terms of r, assuming both exist.

Chapter 9. Matrices

In this section, the elements of all matrices are complex numbers.

Problem 9.1 For the 2×3 matrices

$$\mathbf{A} = \begin{bmatrix} 1 & -9 & 7 \\ 0 & 5 & -3 \end{bmatrix}, \quad \mathbf{B} = \begin{bmatrix} -2 & 8 & 0 \\ 0 & 4 & -6 \end{bmatrix},$$

find $-5\mathbf{A}$, $\mathbf{B}/2$, $\mathbf{A} + \mathbf{B}$, $\mathbf{A} - \mathbf{B}$, $\mathbf{AB}^\mathrm{T}$, $\mathbf{A}^\mathrm{T}\mathbf{B}$, $\mathbf{BA}^\mathrm{T}$, and $\mathbf{B}^\mathrm{T}\mathbf{A}$.

Problem 9.2 When is a zero matrix $\mathbf{0}$ a diagonal matrix?

Problem 9.3 A matrix $\mathbf{A}$ is an **antisymmetric matrix** if and only if $\mathbf{A} = -\mathbf{A}^\mathrm{T}$; whereas $\mathbf{A}$ is an **anti-Hermitian matrix** if and only if $\mathbf{A} = -\mathbf{A}^*$. Thus, antisymmetric and anti-Hermitian matrices are necessarily square.

(a) What can be said about the main diagonals of such matrices? And, of symmetric and Hermitian matrices?

(b) For any square matrix $\mathbf{A}$, letting

$$\mathbf{A}_s = \tfrac{1}{2}(\mathbf{A} + \mathbf{A}^\mathrm{T}), \quad \mathbf{A}_{as} = \tfrac{1}{2}(\mathbf{A} - \mathbf{A}^\mathrm{T}) \tag{P9.3.1a}$$

clearly yields

$$\mathbf{A} = \mathbf{A}_s + \mathbf{A}_{as} \tag{P9.3.1b}$$

(cf. (P5.17.1)); also, it is easily shown that $\mathbf{A}_s$ is symmetric and $\mathbf{A}_{as}$ is antisymmetric. Prove that this decomposition of $\mathbf{A}$ is unique; specifically, show that if

$$\mathbf{A} = \mathbf{A}_s{}' + \mathbf{A}_{as}{}' \tag{P9.3.2}$$

for some symmetric matrix $\mathbf{A}_s{}'$ and antisymmetric matrix $\mathbf{A}_{as}{}'$, then we must have $\mathbf{A}_s{}' = \mathbf{A}_s$ and $\mathbf{A}_{as}{}' = \mathbf{A}_{as}$.

(c) State and prove the corresponding facts regarding the expression of a square matrix $\mathbf{A}$ as a sum

$$\mathbf{A} = \mathbf{A}_H + \mathbf{A}_{aH} \tag{P9.3.3}$$

of a Hermitian matrix $\mathbf{A}_H$ and an anti-Hermitian matrix $\mathbf{A}_{aH}$.

Problem 9.4 Prove that if $\mathbf{A}$ is an $m \times n$ matrix ($m, n \geq 1$) such that $\mathbf{Ax} = \mathbf{0}$ for all $n \times 1$ matrices $\mathbf{x}$, then $\mathbf{A} = \mathbf{0}$.[17] Then, use this result to show that if $\mathbf{B}$ and $\mathbf{C}$ are both $m \times n$ matrices such that $\mathbf{Bx} = \mathbf{Cx}$ for all $\mathbf{x}$, then $\mathbf{B} = \mathbf{C}$.

Problem 9.5 **(a)** Prove that for any column vector $\mathbf{y}$ we have $\mathbf{y}^*\mathbf{y} \geq 0$, with equality obtaining if and only if $\mathbf{y} = \mathbf{0}$.

(b) Use this fact to prove that for any $m \times n$ matrix $\mathbf{A}$ and $n \times 1$ matrix $\mathbf{x}$, if $\mathbf{A}^*\mathbf{Ax} = \mathbf{0}$ then $\mathbf{Ax} = \mathbf{0}$ (the converse being obvious).

Problem 9.6 For any real value α, we have $\alpha^2 \geq 0$; and, equality is attained if and only if $\alpha = 0$. By contrast:

(a) Give a real matrix $\mathbf{A}$ for which $\mathbf{A}^2$ has at least one negative element.

(b) Give a real matrix $\mathbf{A} \neq \mathbf{0}$ for which $\mathbf{A}^2 = \mathbf{0}$.

Problem 9.7 Prove that for any $m \times n$ matrix $\mathbf{A}$ with $m > n$, its rows must be linearly dependent. (*Suggestion*: Show that assuming the contrary enables one to express row $n + 1$ of $\mathbf{A}$ as a linear combination of rows 1 through n.) Likewise, the columns of $\mathbf{A}$ must be linearly dependent when $m < n$.

Problem 9.8 Find $\mathbf{A}^{-1}$ for the 2×2 matrix

$$\mathbf{A} = \begin{bmatrix} 1 & 2 \\ 3 & 4 \end{bmatrix};$$

then, verify (9.13).

Problem 9.9 **(a)** Show that the 3×3 matrix

$$\mathbf{A} = \begin{bmatrix} 1 & 2 & 3 \\ 4 & 5 & 6 \\ 7 & 8 & 9 \end{bmatrix}$$

is singular by calculating its determinant.

(b) Accordingly, express one of the columns of $\mathbf{A}$ as a linear combination of the other columns.

(c) Also, express one of the rows of $\mathbf{A}$ as a linear combination of the other rows.

Problem 9.10 **(a)** As done in (9.9) for 1×1 and 2×2 matrices, find a general expression for the determinant of a 3×3 matrix $\mathbf{A}$ specified by (9.1a). Do so by expanding the determinant about the third row of $\mathbf{A}$.

(b) Repeat part (a) by expanding the determinant about the second column of $\mathbf{A}$.

Problem 9.11 Given two square matrices $\mathbf{A}$ and $\mathbf{B}$ of the same order, if $\mathbf{A}$ is singular, then must $\mathbf{AB}$ be singular?

Problem 9.12 An **upper triangular matrix** is a square matrix $\mathbf{A}$ for which every element below the main diagonal is 0—i.e., in (9.1a), $m = n$ and $a_{i,j} = 0$ $(i > j)$:

$$\mathbf{A} = \begin{bmatrix} a_{1,1} & a_{1,2} & \cdots & a_{1,n-1} & a_{1,n} \\ 0 & a_{2,2} & \cdots & a_{2,n-1} & a_{2,n} \\ \vdots & \vdots & & \vdots & \vdots \\ 0 & 0 & \cdots & a_{n-1,n-1} & a_{n-1,n} \\ 0 & 0 & \cdots & 0 & a_{n,n} \end{bmatrix}.$$

17. Note that the two instances of "$\mathbf{0}$" here do not stand for the same zero matrix; rather, it is understood from the dimensions of $\mathbf{A}$ and $\mathbf{x}$ that the first $\mathbf{0}$ is $m \times 1$ and the second is $m \times n$.

Similarly, a **lower triangular matrix** is a square matrix $\mathbf{A}$ for which every element above the main diagonal is 0—i.e., in (9.1a), $m = n$ and $a_{i,j} = 0$ $(i < j)$:

$$\mathbf{A} = \begin{bmatrix} a_{1,1} & 0 & \cdots & 0 & 0 \\ a_{2,1} & a_{2,2} & \cdots & 0 & 0 \\ \vdots & \vdots & & \vdots & \vdots \\ a_{n-1,1} & a_{n-1,2} & \cdots & a_{n-1,n-1} & 0 \\ a_{n,1} & a_{n,2} & \cdots & a_{n,n-1} & a_{n,n} \end{bmatrix}.$$

Thus, a matrix is diagonal if and only if it is *both* upper triangular and lower triangular; whereas a **triangular matrix** is a matrix that is *either* upper triangular or lower triangular. Show that the determinant of a triangular matrix equals the product of the elements on the main diagonal.

Problem 9.13 Given

$$\mathbf{A} = \begin{bmatrix} 1 & 2 & 3 \\ 8 & 9 & 4 \\ 7 & 6 & 5 \end{bmatrix}, \quad \mathbf{b} = \begin{bmatrix} 10 \\ 11 \\ 12 \end{bmatrix},$$

solve the matrix equation (9.17) for $\mathbf{x}$ by first finding $\mathbf{A}^{-1}$; then, verify that your solution $\mathbf{x}$ satisfies (9.17).

Problem 9.14 For the matrices $\mathbf{A}$ and $\mathbf{b}$ given in Problem 9.13, solve the matrix equation (9.17) for $\mathbf{x}$ by finding its components via Cramer's rule.

Problem 9.15 For the matrices $\mathbf{A}$ and $\mathbf{b}$ given in Problem 9.13, write the $n = 3$ simultaneous equations (9.16) corresponding to matrix equation (9.17); then, find the value of the variable x_3 by manipulating the equations directly.

Problem 9.16 Regarding the simultaneous linear equations (9.16):

(a) Describe all the ways that a *single* linear equation (i.e., $m = 1$ and $n \geq 1$) can be inconsistent.

(b) Show that for each pair (m, n) of positive integers, it is possible for the equations (9.16) to be inconsistent.

Problem 9.17 Show that if the simultaneous linear equations (9.16) are consistent, then they have either a *unique* solution or *infinitely many* solutions. (*Suggestion*: Consider the difference of two distinct solution vectors.)

Problem 9.18 Find for each $\phi \in [0, 2\pi)$ the eigenvalues of the matrix

$$\mathbf{A} = \begin{bmatrix} \cos\phi & -\sin\phi \\ \sin\phi & \cos\phi \end{bmatrix}$$

(the significance of which is discussed in Section 11.4 with regard to (11.78)). In particular, for what values of ϕ are the eigenvalues degenerate?

Problem 9.19 Find the column eigenvectors and associated eigenvalues of the matrix

$$\mathbf{A} = \begin{bmatrix} 3 & 1 & -1 \\ 2 & 2 & -2 \\ 1 & -1 & 1 \end{bmatrix}.$$

Problem 9.20 Prove that if λ is an eigenvalue of a Hermitian matrix $\mathbf{A}$, then λ must be real. (*Suggestion*: Consider $\mathbf{x}^*\mathbf{A}\mathbf{x}$ for an associated eigenvector $\mathbf{x}$.)

Problem 9.21 Show how to express the eigenvalues λ_1 and λ_2 of a 2×2 matrix $\mathbf{A} = [a_{i,j}]_{i,j=1}^2$ in terms of $\det \mathbf{A}$ and $\operatorname{tr} \mathbf{A}$.

Chapter 10. Euclidean Geometry

In this section, each problem is to be solved without the use of analytic geometry (i.e., the methods of coordinate geometry discussed in Chapter 11).

Problem 10.1 **(a)** Prove that, in either plane or solid geometry, $n \geq 3$ distinct points $P_1, P_2, \ldots, P_n$ on a plane, or in space, are collinear if every subcollection of three of these points is collinear.

(b) What is the corresponding statement for coplanar points?

Problem 10.2 **(a)** Let l and m be two skew lines, with P_1 and P_2 being two distinct points on l, and Q_1 and Q_2 being two distinct points on m. Explain why these four points cannot be coplanar.

(b) Let l, m, and n be lines in space, with n crossing l and m at distinct points. Prove that if l and m and coplanar, then all three lines are coplanar.

Problem 10.3 Suppose a line l and a plane p intersect at a single point P. Suppose also that there exist $n \geq 1$ *other* points $Q_i \in p$ $(i = 1, 2, \ldots, n)$ such that no two of these points and P are collinear, and $l \perp \overleftrightarrow{PQ_i}$ (all i). What is the smallest value of n for which we can thereby conclude that $l \perp p$?

Problem 10.4 **(a)** As stated in Section 10.2, both similarity and congruence are equivalence relations. Determine which of the following geometric relations are also equivalence relations: betweenness of points, coplanarity, parallelness of lines, perpendicularity of lines, perpendicularity of a line and a plane.

(b) In plane geometry, is the set of all circles an equivalence class under the similarity relation $\sim$? Under the congruence relation $\cong$?.

Problem 10.5 **(a)** Show that it is possible to have two distinct but similar geometric figures f and g, and a fixed scale factor $\kappa > 0$, such that (10.2) holds for more than one allowable function $\varphi(\cdot)$.

(b) If f and g are congruent, is it possible for (10.2) to hold for some allowable function $\varphi(\cdot)$ and a scale factor $\kappa > 0$ other than 1?

Problem 10.6 **(a)** The prefix "poly-" means "many". Look up in a dictionary the meaning of the suffix "-gon", and thereby determine whether "polygon" fundamentally means a figure having many *sides* or many *angles*.

(b) Now, look up the meaning of the prefix "iso-", and determine what kind of polygon an "isogon" is.

Problem 10.7 **(a)** For the general definition of a "polygon" given in (10.3) and what follows, one additional constraint that might be placed on the points (10.4) is that each pair of the segments (10.5) have at most one point in common; that is, the intersection of any two of these segments cannot itself be a segment. (Then, although the points (10.4) need not be distinct, the segments (10.5) would be, thereby making $n \geq 3$.) Show that if this constraint is imposed, then the figure on the left of Figure P10.7.1 is still a polygon, but the figure on the right is not.

(b) Alternatively, one might require that the points (10.4) be such that for each $i \in \{1, 2, \ldots, n\}$, the three consecutive points A_i, A_{i+1}, and A_{i+2} (letting $A_{n+1} = A_1$ and $A_{n+2} = A_2$) be noncollinear. The idea here is that if A_i, A_{i+1}, and A_{i+2} were collinear, then $\overline{A_iA_{i+1}} \cup \overline{A_{i+1}A_{i+2}}$ would equal $\overline{A_iA_{i+2}}$, and thus the point A_{i+1} could be eliminated. Give an example of a polygon that has (at least

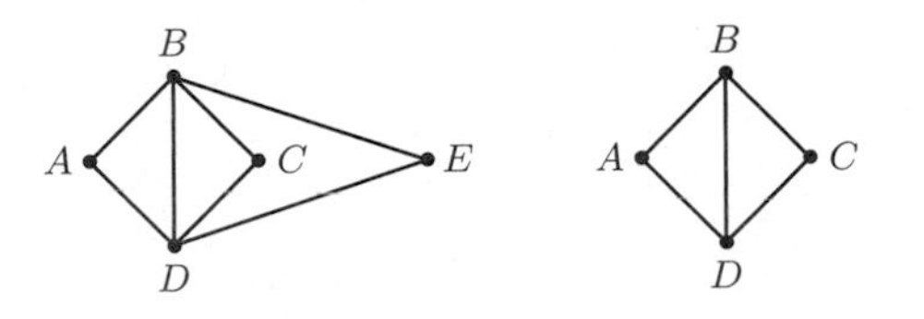

FIGURE P10.7.1

two different expressions (10.4) satisfying this constraint such that:

- The value of n for each expression is the smallest possible.
- When the segments (10.5) are taken together as a *set*, these sets for the two expressions are *different*.

Hence, it could now be argued that this polygon does not have a *unique* set of sides. (*Hint*: If this constraint were applied to the polygon on the right of Figure 10.9, then the intersection of the segments $\overline{A_1A_2}$ and $\overline{A_3A_4}$ could not be designated as another vertex.)

Problem 10.8 **(a)** By the definition given in Section 10.1, the empty set $\emptyset$ is a geometric figure; but, is it a convex set?

(b) If $f = \{P\}$ for some point P, then is f a convex set?

Problem 10.9 **(a)** State which of the following geometric figures is always a convex set: a point, a line, a half-line, a ray, a segment, a plane, a half-plane, a closed half-plane, space, a half-space, a closed half-space, a proper angle, the interior of a proper angle, the exterior of a proper angle, a circle, the interior of a circle, the exterior of a circle, a closed disk, an arc.

(b) The **convex hull** of a figure f is the intersection of all convex subsets g of the space under consideration for which $f \subseteq g$. Prove that a convex hull is itself a convex set. (Hence, the "convex hull" of a figure f is often defined as the *smallest* convex set containing f.)

Problem 10.10 Find a formula for the number of diagonals of a simple n-gon in terms of $n \geq 3$.

Problem 10.11 Prove that every triangle has at least two acute angles.

Problem 10.12 Prove that two simple polygons that are similar must be congruent if their perimeters are equal.

Problem 10.13 Suppose that for some $n \geq 3$, we have two simple polygons $A_1A_2 \cdots A_n$ and $B_1B_2 \cdots B_n$. For what values of n is (10.8a) a sufficient condition to have $A_1A_2 \cdots A_n \sim B_1B_2 \cdots B_n$? What about, (10.8b)?

Problem 10.14 Use the Pythagorean theorem to prove that if d is the distance from a given point P to a given line l, then for every point P' on l we have $d(P, P') \geq d$.

Problem 10.15 Given a triangle $\triangle ABC$, let D be an interior point of the side $\overline{AB}$. Prove the "obvious" fact that $d(C, D) < \max[d(C, A), d(C, B)]$. (*Suggestion*: Let P be the unique point on $\overleftrightarrow{AB}$ such that $\overline{CP} \perp \overleftrightarrow{AB}$; then, apply the Pythagorean theorem.)

Problem 10.16 Given a segment $\overline{AB}$ lying in a plane p, the **perpendicular bisector** of $\overline{AB}$ in p is the unique line l in p that is perpendicular to $\overline{AB}$ and crosses $\overline{AB}$ at its midpoint. Prove that l is also the set of points on p that are equidistant from A and B.

Problem 10.17 Find all Pythagorean triples (x, y, z) such that $0 < x \leq y \leq 13$.

Problem 10.18 Prove that the segment joining the midpoints of two sides of a triangle has half the length of the remaining side.

Problem 10.19 **(a)** Identify three similar triangles in Figure P10.19.1. Then, use them to perform the following proofs.

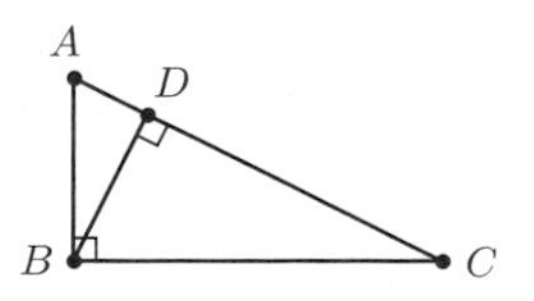

FIGURE P10.19.1

(b) Prove that $|\overline{BC}|$ equals the geometric mean of $|\overline{AC}|$ and $|\overline{DC}|$.

(c) Prove the Pythagorean theorem (10.12).

Problem 10.20 In Figure 10.10, vertex A_4 of the equilateral pentagon in the middle can be moved to produce the regular pentagon on the right. Use this fact to perform the following tasks regarding the middle pentagon.

(a) Determine, in degrees, the measures of the interior angles.

(b) Prove that the vertices A_1, A_3, and A_4 are collinear.

Problem 10.21 Prove these "obvious" facts about a parallelogram:

(a) Opposite sides have the same length.

(b) The two diagonals bisect each other.

Problem 10.22 Prove that if a quadrilateral is inscribed in a circle, then each pair of opposite angles is supplementary.

Problem 10.23 Let the regular hexagon in Figure 10.18, which has radius r and apothem a, have sides of length s.

(a) Prove that $r = s$.

(b) Express a in terms of s.

Problem 10.24 A **geometric construction** is typically understood to mean a drawing on a plane that is constructed in finitely many steps using only a straightedge and a **compass** (a tool for drawing arcs and circles of a chosen radius about a chosen point). The ability to construct a variety of geometric figures as the need arises is a useful drafting skill by which much analytical work (as in Chapter 11) can be avoided. As an example, Figure P10.24.1 shows a simple procedure by which, given a segment $\overline{AB}$, one can construct its perpendicular bisector l (defined in Problem 10.16): First the compass is set to any radius larger than $d(A, B)/2$. Then, it is used to draw circles of this radius centered at A and B; or, as in the figure, enough of the circles are drawn to see where they intersect. Finally, it follows from the fact stated in Problem 10.16 that by using the straightedge to draw a line through the two points of intersection, we obtain l.

Likewise, explain how to perform each of the following tasks using just a straightedge and a compass.

(a) Given a circle, locate its center.

(b) Given a triangle, circumscribe it with a circle. (*Suggestion*: Apply the fact stated in Problem 10.16.)

(c) Given a circle, its center, and point Q on it, inscribe a regular hexagon that also contains Q. (*Suggestion*: Apply the fact stated in part (a) of Problem 10.23.)

(d) Given a proper angle, construct a ray that bisects its interior. (*Aside*: To *tri*sect an angle, using only a straightedge and a compass, has been proved to be impossible.)

Problem 10.25 What kind of figure is the section of a solid sphere?

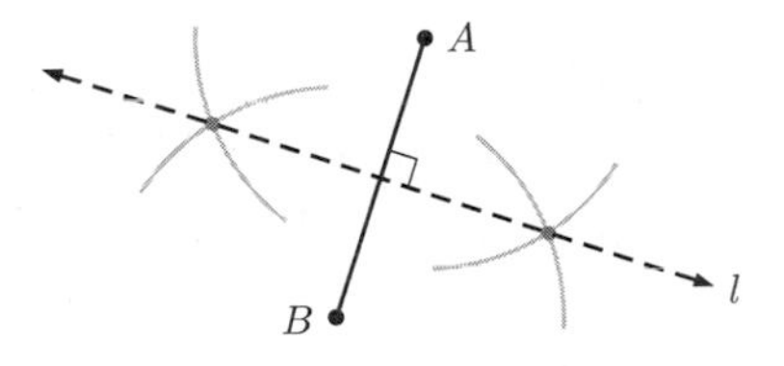

FIGURE P10.24.1 Constructing the Perpendicular Bisector of a Segment

Problem 10.26 For every conic section other than a circle, a **latus rectum** of the section is defined as a segment that has its endpoints on the section, passes through a focus, and is parallel to the corresponding directrix.[18] Thus, a parabola has only one latus rectum, whereas an ellipse and a hyperbola each have two, which have the same length. Let l be the half-length of each latus rectum of a given conic section.

(a) Find a general formula for l in terms of the eccentricity e and focal length p of the section.

(b) For an ellipse, express l in terms of the semimajor axis a and the semiminor axis b.

(c) For a hyperbola, express l in terms of the semitransverse axis a and the semiconjugate axis b.

Problem 10.27 It is sometimes said that the eccentricity e of a circle is 0; however, although setting $e = 1$ in (10.18) yields the unique parabola f in a plane q generated by a given focus F and directrix d, simply setting $e = 0$ in (10.18) produces the trivial figure $f = \{P \in q \mid d(P, F) = 0\} = \{F\}$. On the other hand, letting $F' = F$ in (10.19), which for $F' \neq F$ provides an alternative to (10.18) for describing an ellipse of a given semimajor axis $a > 0$, we obtain $f = \{P \in q \mid d(P, F) = a\}$—i.e., a circle in p of radius a centered at F. Additionally, show that if we let $d(F, F') \to 0$ while holding F and a fixed, then the eccentricity e of the ellipse approaches 0. Also, what happens to its focal parameter p and semimajor axis b?

Problem 10.28 Verify Euler's formula (10.22) for:

(a) the pyramid on the right side of Figure 10.26.

(b) the prism on the right side of Figure 10.27.

Problem 10.29 It might be thought that more work than necessary was performed in the beginning of Subsection 10.4.1 to define a "simple polyhedron", which is often described simply as a closed polygonal surface that has no "holes" like a donut (as in Figure 10.25). However, to see that this description fails as a precise definition, prove that the polyhedron on the right side of Figure P10.29.1—obtained as shown by "pinching together" the two bases of a triangular prism to produce a new vertex—is not a simple polyhedron. Do this implicitly by showing that Euler's formula (10.22) does not apply.

Problem 10.30 How many diagonals does each of the regular polyhedra have?

Problem 10.31 This problem shows that we cannot define a "regular polyhedron" as simply a convex simple polyhedron for which the face boundaries are congruent regular polygons.

(a) It is clear from the fact stated in part (a) of Problem 10.23 that if a regular pentagon having sides of length s has radius r, then $s > r$. Accordingly, show that there exists a regular pyramid having five congruent lateral faces, each bounded by an equilateral triangle.

(b) Use two such pyramids to show that there exists a nonregular convex polyhedron, the faces of which are bounded by congruent regular polygons.

Problem 10.32 What is another name for a "right rectangular prism"?

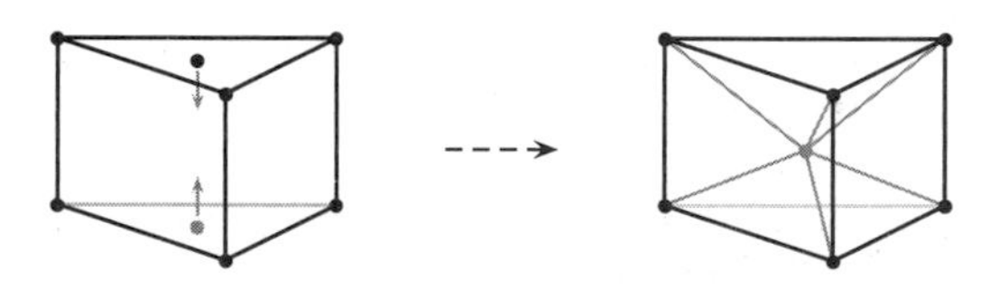

FIGURE P10.29.1 A Nonsimple Polyhedron with No Holes

18. The latus recta are not shown in Figures 10.21–10.23, but can be easily added as vertical segments through the points F and F'.

Problem 10.33 Answer each of the following questions regarding plane and solid geometry with either "yes" or "no"; and, explain your "no" answers.

(a) Let A and B be distinct points, with neither lying on a particular line l; and, suppose all of these figures are contained in a plane p. Regarding the two half-planes of p having edge l, is it true that A and B lie on the same half-plane if and only if $\overline{AB}$ does not intersect l?

(b) Recall that the intersection of two lines that are not parallel must be a single point. Is it also true that the intersection of *three* planes that are not *pairwise* parallel is a single point?

(c) Does the interior of a proper angle $\angle ABC$ always equal the set of points D such that D lies between some point on $\overleftarrow{AB}$ and another point on $\overrightarrow{BC}$?

(d) Suppose p and q are planes crossing at a line l. Must there exist lines $l_p \subseteq p$ and $l_q \subseteq q$ such that $l_p \neq l$, $l_q \neq l$, and $l_p \perp l_q$? Also, if $p \perp q$, must there exist crossing lines $l_p \subseteq p$ and $l_q \subseteq q$ such that $l_p \neq l$, $l_q \neq l$, and l_p and l_q are not perpendicular (i.e., $l_p \not\perp l_q$)?

(e) Must two similar simple polygons be congruent if the members of at least one pair of corresponding diagonals are equal?

(f) If a chord of a circle is not a diameter, then must the segment from the midpoint of the chord to the center of the circle be perpendicular to the chord?

(g) If a simple polygon is inscribed in a circle, then must the polygon be regular if it is equilateral? What if the polygon is circumscribed about the circle instead?

(h) Repeat part (f) with "equilateral" replaced by "equiangular".

(i) Can every solid polyhedron be expressed as the intersection of finitely many closed half-planes?

(j) In solid geometry, is every nonempty intersection of finitely many closed half-planes a solid polyhedron?

(k) Are two right circular cones similar if and only if the ratio of the base radius to the height has the same value for both cones?

(l) Are two right circular cylinders congruent if and only if their base radii are equal and their heights are equal?

(m) If f and g are two similar figures, with f being a convex set, then must g be a convex set?

Problem 10.34 Prove that the following two conditions for a nonempty figure f in space to be **bounded** are equivalent:

(i) $\sup\{d(P, Q) \mid P, Q \in f\} < \infty$.

(ii) $f \subseteq g$ for some solid sphere g.

Problem 10.35 Let f and g be coplanar figures, and let $\alpha(\cdot)$ be an area measure. Give an equation relating $\alpha(f)$, $\alpha(g)$, $\alpha(f \cup g)$, and $\alpha(f \cap g)$, assuming that these areas exist.

Problem 10.36 Given a triangle $\triangle ABC$, let D be the midpoint of the side $\overline{AB}$; that is, as in Figure 10.13, $\overline{CD}$ is a median of the triangle. Prove that $\triangle ADC$ and $\triangle BDC$ have equal areas.

Problem 10.37 **(a)** As is easily verified, the area of a circle equals the product of its semiperimeter and its radius. Prove that, similarly, if a simple polygon has an inscribed circle, then the area of the polygon equals the product of its semiperimeter and the radius of the circle.

(b) Accordingly, given a triangle having sides of lengths a, b, and c, what is the radius of the inscribed circle?

Problem 10.38 Show that for a given perimeter, a rectangular has the largest possible area if and only if it is a square. (*Suggestion*: Starting with an arbitrary rectangle, subtract its area from the area of a square having the same perimeter.)

Problem 10.39 Suppose two simple polygons f and g are similar, and have respective perimeters p_f and p_g, and areas α_f and α_g. How are the ratios p_g/p_f and α_g/α_f related?

Problem 10.40 **(a)** Find the area α of an equilateral triangle having sides of length l by first finding its height; then use Heron's formula to verify your result.

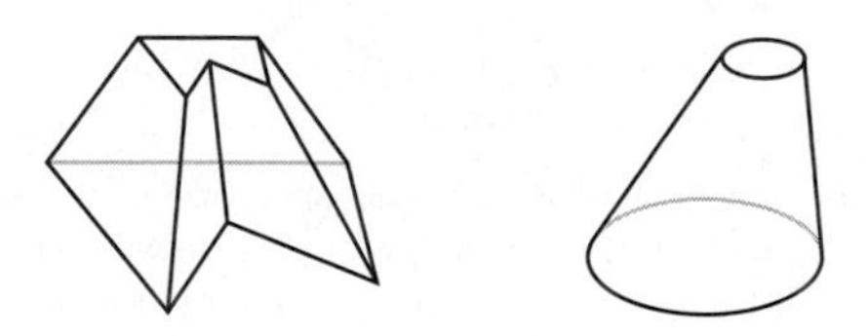

FIGURE P10.43.1 Frustums of a Pyramid and a Cone

(b) Find the volume v of a regular tetrahedron having edges of length l by first using the result of part (a) and Heron's formula to express the radius r of the base in terms of l; then use the result to determine the height of the tetrahedron.

Problem 10.41 Why is there no entry in Table 10.6 under "surface area" for a pyramid, or for a circular cone?

Problem 10.42 Suppose f is a right circular cylinder and g is a sphere. If the base radius of f equals the radius of g, and the lateral surface area of f equals the surface area of g, then how does the height of f compare to g?

Problem 10.43 In general, a **frustum** f of a solid figure g is the portion of g lying between two parallel planes p_1 and p_2, specified as desired. Included in f are the intersection of p_1 with g and its interior, and the intersection of p_2 with g and its interior, each of the two intersections (which we assume are nonempty) being a **base** of the frustum. Often, g is either a pyramid or some kind of cone, in which case *its* base typically lies in one of the planes, thereby becoming a base of f; see Figure P10.43.1 (cf. the right side of Figure 10.26 and the right side of 10.28). Excluding from the surface of f the interiors of its bases yields the **lateral surface** of the frustum; and, the **height** of f is the distance between the planes.

(a) Suppose f is a frustum of a right circular cone. Let its bases have radii r_1 and r_2; and, let h be its height. After defining the "slant height" h_s of f in a reasonable way, express h_s in terms of r_1, r_2, and h. Then, show that f has volume $v = \frac{1}{3}\pi({r_1}^2 + r_1 r_2 + {r_2}^2)h$ and lateral surface area $\alpha_l = \pi(r_1 + r_2)h_s$.

(b) Using the results of part (a), determine the limits of h_s, v, and α_l that obtain by letting $r_2 \to r_1$ with r_1 and h fixed; and, state their significance.

Problem 10.44 As on the left side of Figure 10.32, consider the solid angle subtended by the interior of a (finite) right circular cone from its apex. Use (10.28) to express the measure Ω of this angle in terms of the height h and slant height h_s of the cone.

Chapter 11. Analytic Geometry

Problem 11.1 Refer to the three coordinate axes for xyz-space on the right of Figure 11.2: if the axis pointing to the right were instead chosen to be the z-axis, then which axis would be the x-axis and which would be the y-axis?

Problem 11.2 In the polar coordinate system of points (r, θ) shown in Figure 11.3, suppose that we allow the distance r to take on negative values. How can the angle θ be restricted so that every point in the xy-plane has exactly one polar representation?

Problem 11.3 Prove (11.3) for all real values R, R', Θ, and Θ'.

Problem 11.4 The **spherical coordinate system** is shown in Figure P11.4.1. Here, each point P in xyz-space is represented by a triple $(\rho, \theta, \phi) \in [0, \infty) \times \mathbb{R} \times \mathbb{R}$. Specifically, letting P' be the point in this space having the same x and y coordinates as P but a z-coordinate of 0, then $\rho \geq 0$ is the distance from the origin O to P, $\theta \in \mathbb{R}$ is the counterclockwise angle of the segment $\overline{OP'}$ relative to the positive portion of the x-axis, and $\phi \in \mathbb{R}$ is the angle of the segment $\overline{OP}$ relative to the positive portion of the

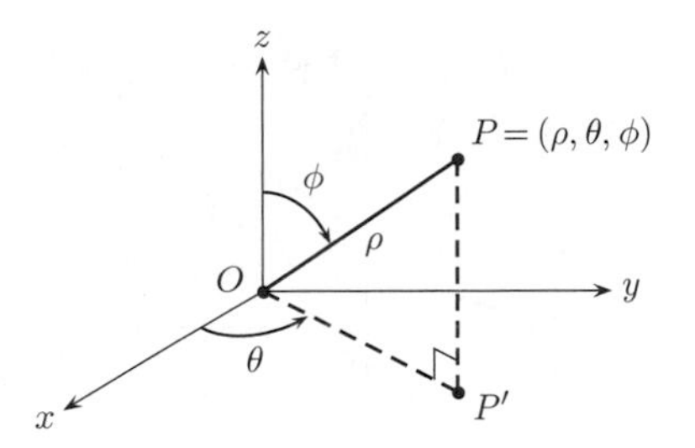

FIGURE P11.4.1 The Spherical Coordinate System

TABLE P11.10.1 Additions to Table 11.1

θ	$\sin\theta$	$\cos\theta$	$\tan\theta$
$\pi/12$ (15°)	$\frac{\sqrt{6}-\sqrt{2}}{4}$	$\frac{\sqrt{6}+\sqrt{2}}{4}$	$2-\sqrt{3}$
$\pi/8$ (22.5°)	$\frac{\sqrt{2-\sqrt{2}}}{2}$	$\frac{\sqrt{2+\sqrt{2}}}{2}$	$\sqrt{2}-1$

z-axis; accordingly, ϕ is undefined when $\rho = 0$ (since then P is O), and θ is undefined when either $\rho = 0$ or ϕ is a multiple of π (since then P lies on the z-axis).[19] Show how to convert from spherical coordinates ρ, θ, and ϕ to cylindrical coordinates $r \geq 0$, θ, and z; and, vice versa.

Problem 11.5 Simplify $\sin(19\pi/2 - \theta)$ ($\theta \in \mathbb{R}$).

Problem 11.6 Prove that for any real number $x \neq -1$, we have $\sin \pi x/(x+1) = \sin \pi/(x+1)$.

Problem 11.7 A useful approximation for the function sine is that $\sin\theta \approx \theta$ for $\theta \approx 0$; more precisely,

$$\lim_{\theta \to 0} \frac{\sin\theta}{\theta} = 1, \tag{P11.7.1}$$

which is to say $\sin\theta \sim \theta$ as $\theta \to 0$. Use this fact to show that, similarly, $\tan\theta \sim \theta$ as $\theta \to 0$, and thus $\tan\theta \approx \theta$ for $\theta \approx 0$.

Problem 11.8 Find all $t \in \mathbb{R}$ for which

$$2\sin t = 3 + 4\cos t. \tag{P11.8.1}$$

Start by squaring the equation.

Problem 11.9 Both of the double-angle identities (11.19a) and (11.19b) are valid for all $\theta \in \mathbb{R}$. What about (11.19c)?

Problem 11.10 Prove the additions to Table 11.1 shown in Table P11.10.1.

Problem 11.11 Prove that the general triangle in Figure 11.12 has area

$$\tfrac{1}{2}ab\sin C = \tfrac{1}{2}c^2 \frac{\sin A \sin B}{\sin C}. \tag{P11.11.1}$$

---

19. If desired, one may restrict (ρ, θ, ϕ) to, e.g., $[0, \infty) \times [0, 2\pi) \times [0, \pi]$, in which case each point in the space has exactly such one representation. Or, one may allow ρ to be negative.

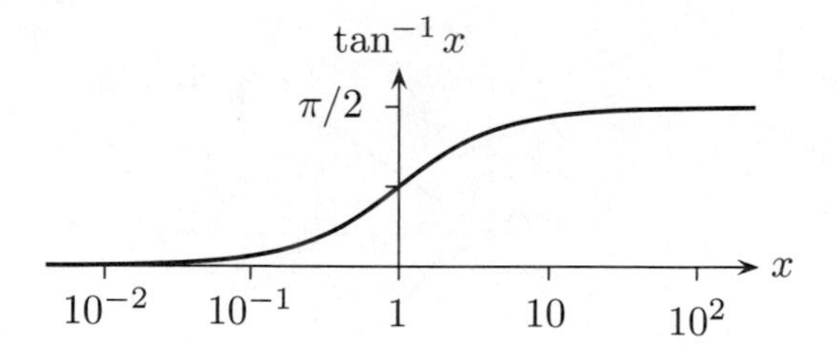

FIGURE P11.15.1 Arctangent Plotted versus a Logarithmic Scale

(*Aside*: It follows from the left side of this equation that the area of a parallelogram equals the product of any two adjacent sides times the sine of the included angle.)

Problem 11.12 Regarding the general triangle in Figure 11.12, prove

$$a = b \cos C + c \cos B. \tag{P11.12.1}$$

Problem 11.13 Use the law of cosines to prove the following converse to the Pythagorean theorem: If (10.12) holds for distinct coplanar points A, B, and C, then $\angle ABC$ is a right angle.

Problem 11.14 Derive, for the general triangle in Figure 11.12 the **law of tangents**:

$$\frac{a-b}{a+b} = \frac{\tan(A-B)/2}{\tan(A+B)/2}. \tag{P11.14.1}$$

Problem 11.15 As shown in Figure P11.15.1 (cf. bottom of Figure 11.11), a plot of $\tan^{-1} x$ versus $x > 0$ on a *logarithmic* scale appears to show odd symmetry about the point $(1, \tan^{-1} 1) = (1, \pi/4)$. That is, letting $x' = \ln x$ (choosing the natural logarithm for convenience), and thus $x = e^{x'}$, it appears that the function $f(x') = \tan^{-1} e^{x'} - \pi/4$ of $x' \in \mathbb{R}$ is odd symmetric. Prove that this is indeed so.

Problem 11.16 **(a)** To better understand (11.14), and the restrictions on θ therein, plot the three functions $\sin^{-1}(\sin\theta)$, $\cos^{-1}(\cos\theta)$, and $\tan^{-1}(\tan\theta)$ versus θ for $-2.2\pi < \theta < 2.2\pi$.

(b) Now, *without* using trigonometric functions, provide an alternative expression for each of these three functions of θ, with each expression being valid for all $\theta \in \mathbb{R}$ (as allowed); that is, θ is now unrestricted (except as necessary for each function to be calculable). In other words, provide for each function a rule that states precisely how to convert a given value of $\theta \in \mathbb{R}$ into the corresponding value of the function; but, do not use any trigonometric functions to express the rule.

(c) Use the results of part (b) to find $\sin^{-1}(\sin 9)$, $\cos^{-1}(\cos 9)$, and $\tan^{-1}(\tan 9)$.

Problem 11.17 Verify the following alternative to (11.4b) for performing a rectangular-to-polar conversion from (x, y) to (r, θ), again with $r \geq 0$: Obtain r as in (11.4b). If $r = 0$, then declare θ to be undefined; otherwise, let

$$\theta = \begin{cases} \cos^{-1} x/r & y \geq 0 \\ -\cos^{-1} x/r & y < 0 \end{cases}. \tag{P11.17.1}$$

(Besides breaking the calculation of θ into fewer cases, this method makes $\theta \in (-\pi, \pi]$ a continuous function of (x, y), except on the negative portion of the x-axis; whereas (11.4b) yields $\theta \in [-\pi/2, 3\pi/2)$ with continuity, except on the negative portion of the y-axis.)

Problem 11.18 **(a)** Show that these identities hold for all $\theta \in \mathbb{R}$:

$$\sin(\theta + \sin^{-1} x) = \cos(\theta - \cos^{-1} x) \quad (-1 \le x \le 1), \tag{P11.18.1a}$$

$$\sin(\theta + \cos^{-1} x) = \cos(\theta - \sin^{-1} x) \quad (-1 \le x \le 1), \tag{P11.18.1b}$$

$$\sin(\theta + \tan^{-1} x) = \cos(\theta - \tan^{-1} 1/x) \quad (x > 0). \tag{P11.18.1c}$$

(b) State and prove similar identities for $\sin(\theta - \sin^{-1} x)$, $\sin(\theta - \cos^{-1} x)$, and $\sin(\theta - \tan^{-1} x)$.

Problem 11.19 Many trigonometric identities have analogs for the hyperbolic functions; for example, it follows immediately from their definitions (11.33) that

$$\tanh x = \frac{\sinh x}{\cosh x},$$

$$(\cosh\theta + \sinh\theta)^n = \cosh n\theta + \sinh n\theta$$

($x \in \mathbb{R}, n \in \mathbb{Z}$), the latter identity being analogous to DeMoivre's theorem (11.24b). Additionally, prove

$$\cosh^2 x - \sinh^2 x = 1,$$

$$\sinh(x \pm y) = \sinh x \cosh y \pm \cosh x \sinh y,$$

$$\cosh(x \pm y) = \cosh x \cosh y \pm \sinh x \sinh y$$

($x, y \in \mathbb{R}$).

Problem 11.20 As noted in Section 3.3, it is generally easier to add complex numbers when they are in rectangular form rather than polar form. Indeed, show that given $A_1, A_2 \in \mathbb{R}$ and $\theta_1, \theta_2 \in \mathbb{R}$ such that $A_1 \ge A_2 \ge 0$, the equation

$$A_1 e^{i\theta_1} + A_2 e^{i\theta_2} = A e^{i\theta} \tag{P11.20.1a}$$

is satisfied for

$$A = \sqrt{{A_1}^2 + {A_2}^2 + 2A_1A_2\cos(\theta_2 - \theta_1)}, \tag{P11.20.1b}$$

$$\theta = \begin{cases} \theta_1 + \tan^{-1}\frac{A_2\sin(\theta_2-\theta_1)}{A_1+A_2\cos(\theta_2-\theta_1)} & A \ne 0 \\ \text{arbitrary} & A = 0 \end{cases}. \tag{P11.20.1c}$$

Problem 11.21 In general, a **sinusoid** (or **sine wave**), is any vertically scaled and horizontally shifted version of the sine function; see Figure P11.21.1. Hence, not only is sine a sinusoid, so is cosine, by (11.10b). Indeed, it is often more convenient to express a sinusoid in terms of cosine than sine; specifically, for any sinusoid $x(t)$ that is a function of time $t \in \mathbb{R}$ (say, in seconds, as usual), we can write

$$x(t) = A\cos(\omega t + \theta) \quad (t \in \mathbb{R}) \tag{P11.21.1}$$

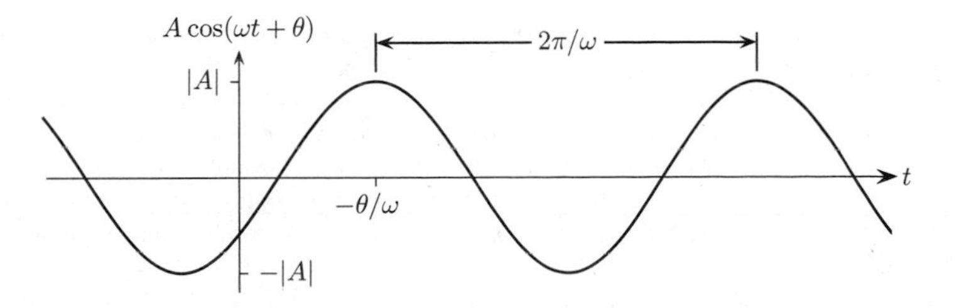

FIGURE P11.21.1 A General Sinusoid

for some constants $A \in \mathbb{R}$, $\omega > 0$, and $\theta \in \mathbb{R}$. So written, A is the **amplitude** of the sinusoid, $|A|$ is its **magnitude**, ω is its **angular frequency** (in radians per second), and θ is its **phase** (in radians).[20] The function $x(t)$ is clearly periodic, its period being $T = 2\pi/\omega$ (in seconds).[21] A **cycle** of the sinusoid is any contiguous portion having a duration of one period; thus, the sinusoid has **cyclic frequency** $f = 1/T = \omega/2\pi$ (in cycles per second, or hertz), thereby making $T = 1/f$ as well.

(a) The **phasor** associated with the sinusoid $x(t)$ in (P11.21.1) is the complex number

$$\mathbf{x} = Ae^{i\theta}; \tag{P11.21.2}$$

thus, the *function* $x(t)$ is now being represented by a *unique* complex-valued *constant* $\mathbf{x}$.[22] In fact, for any fixed frequency ω, there is a *one-to-one correspondence* between sinusoids and phasors; because, for each complex constant $\mathbf{x}$, there is exactly one sinusoid $x(t)$ of frequency ω for which $\mathbf{x}$ is the associated phasor. Specifically, show that

$$x(t) = \mathrm{Re}\left(\mathbf{x}\, e^{i\omega t}\right) \quad (t \in \mathbb{R}). \tag{P11.21.3}$$

(b) For any frequency ω, the phasor associated with cosine is simply 1; because, letting $x(t) = \cos \omega t$ we can satisfy (P11.21.1) by taking $A = 1$ and $\theta = 0$, thereby obtaining $\mathbf{x} = 1$ from (P11.21.2). What is the phasor associated with sine?

(c) In (P11.21.2), the complex number $\mathbf{x}$ is expressed in polar form; accordingly, we may refer to the right side of (P11.21.1) as the **polar form** *of a sinusoid* $x(t)$. We can just as well, though, express $\mathbf{x}$ in the rectangular form

$$\mathbf{x} = a + ib \tag{P11.21.4}$$

for some $a, b \in \mathbb{R}$. Show that we then have

$$x(t) = a \cos \omega t - b \sin \omega t \quad (t \in \mathbb{R}), \tag{P11.21.5}$$

which expresses the **rectangular form** *of a sinusoid.*

(d) In many physical situations, all of the sinusoids under consideration have the *same* frequency ω. The main benefit of using phasors is that linear combinations of such sinusoids can be more easily calculated via complex arithmetic than directly via trigonometric identities. Specifically, show that if

$$ax(t) + by(t) = z(t) \quad (t \in \mathbb{R}), \tag{P11.21.6a}$$

where $x(t)$ and $y(t)$ are sinusoids of some frequency ω, and $a, b \in \mathbb{R}$, then $z(t)$ is *also* a sinusoid of frequency ω, with its phasor $\mathbf{z}$ being given by the corresponding equation

$$a\mathbf{x} + b\mathbf{y} = \mathbf{z}, \tag{P11.21.6b}$$

where $\mathbf{x}$ and $\mathbf{y}$ are the phasors of $x(t)$ and $y(t)$, respectively.

---

20. The magnitude of each sinusoid is unique, as is its angular frequency when its magnitude is nonzero. However, since amplitudes may be negative, a sinusoid that is not identically zero (i.e., its magnitude is nonzero) has *two* possible amplitudes; because, given (P11.21.1) we also have $x(t) = -A\cos(\omega t + \theta + \pi)$ $(t \in \mathbb{R})$. Hence, the phase of a sinusoid is not unique either, being determined only up to a multiple of π when a nonzero magnitude is given, and up to a multiple of 2π when a nonzero amplitude is given.

21. The trivial case of $A = 0$, which is ignored both here and in the following definitions of "cycle" and "cycle frequency", requires special consideration since $x(t)$ is then a constant function. If interested, see Problem 5.14(a).

22. Regarding the uniqueness of $\mathbf{x}$, suppose we also have $x(t) = A' \cos(\omega' t + \theta')$ $(t \in \mathbb{R})$ for some $A' \in \mathbb{R}$, $\omega' > 0$, and $\theta' \in \mathbb{R}$; then, although it is possible that $A' \neq A$ or $\theta' \neq \theta$ (see fn. 20 above), or even $\omega' \neq \omega$ if $A = 0$, we must nevertheless have $A' e^{i\theta'} = Ae^{i\theta}$.

(e) As an example of using (P11.21.6b) to find $z(t)$ in (P11.21.6a), let a frequency $\omega > 0$, amplitudes $A_1, A_2 \in \mathbb{R}$, and phases $\theta_1, \theta_2 \in \mathbb{R}$ be given; then

$$A_1 \cos(\omega t + \theta_1) + A_2 \cos(\omega t + \theta_2) = A \cos(\omega t + \theta) \quad (t \in \mathbb{R})$$

for any real constants A and θ satisfying (P11.20.1a). In particular, it follows from Problem 11.20 that if $A_1 \geq A_2 \geq 0$, then these constants can be taken from (P11.20.1b) and (P11.20.1c). Similarly, use (P11.21.6) to obtain a simpler solution to Problem 11.8.

Problem 11.22 Let V be a vector space with associated scalar field F. Prove directly from the basic vector-space properties (11.36) that (11.37) holds for all $\vec{x} \in V$ and $a \in F$.

Problem 11.23 For each of the following pairs (V, F), state "yes" or "no" as to whether the pair constitutes a vector space V with scalar field F. Also, for each "yes" answer, provide the corresponding zero vector $\vec{0}$; and, explain each "no" answer.

(a) $V = \mathbb{R}, F = \mathbb{C}$.

(b) V is the set of all complex functions on $[0, 1]$; and, $F = \mathbb{C}$.

(c) V is the set of all functions $f : (0, \infty) \to \mathbb{R}$ such that $f(x) = \alpha \log_b x$ $(x > 0)$ for some $\alpha \in \mathbb{R}$ and $b > 1$; and, $F = \mathbb{R}$.

(d) V is the set of all functions $f : \mathbb{R} \to \mathbb{R}$ such that $f(x) = \alpha b^x$ $(x \in \mathbb{R})$ for some $\alpha \in \mathbb{R}$ and $b > 1$; and, $F = \mathbb{R}$.

Problem 11.24 Let $\vec{x}$ and $\vec{y}$ be vectors in an inner product space. Prove the **parallelogram law**,

$$\|\vec{x} + \vec{y}\|^2 + \|\vec{x} - \vec{y}\|^2 = 2\|\vec{x}\|^2 + 2\|\vec{y}\|^2, \tag{P11.24.1}$$

and state its significance regarding parallelograms.

Problem 11.25 Consider the vector space $V = \mathbb{C}$ with scalar field $F = \mathbb{R}$.

(a) Prove that the function $\|\vec{x}\| = \max(|\operatorname{Re}\vec{x}|, |\operatorname{Im}\vec{x}|)$ $(\vec{x} \in V)$ is a valid norm on V.

(b) Does applying the appropriate polarization law in (11.44) to this norm yield a valid inner product on V?

Problem 11.26 **(a)** State the Schwarz inequality (11.47), and its condition for equality, specifically for the vector space $V = \mathbb{C}$ with scalar field $F = \mathbb{R}$ and inner product given by (11.45).

(b) Verify the result of part (a) directly via complex arithmetic.

Problem 11.27 Let $\vec{x}$, $\vec{y}$, and $\vec{z}$ be vectors in an inner product space V with scalar field F. Also, let $\vec{z}_x$ and $\vec{z}_y$ be the projections of $\vec{z}$ onto $\vec{x}$ and $\vec{y}$, respectively. Show that if $\vec{x} \perp \vec{y}$, then $\vec{z} - \vec{z}_x - \vec{z}_y \perp b_x\vec{x} + b_y\vec{y}$ for all $b_x, b_y \in F$.

Problem 11.28 Regarding Argand diagrams of complex numbers, as in Figure 3.3:

(a) Suppose a vector $\vec{z}$ in the complex plane is drawn from the point corresponding to a complex number a to the point corresponding to a complex number b; thus, the tail of the vector is at a and its head is at b. What complex number z does $\vec{z}$ represent?

(b) Explain via Figure 11.16 why, if $\vec{z}_1$ and $\vec{z}_2$ are vectors in the complex plane representing complex numbers z_1 and z_2, respectively, then the vector $\vec{z}_1 + \vec{z}_2$ represents the complex number $z_1 + z_2$.

Problem 11.29 Use the distance measure in (11.56) to determine the common length of the diagonals of a rectangular parallelepiped for which the three edges that meet at any vertex have lengths a, b, and c. Start by representing the parallelepiped simply in $\mathbb{E}^3$.

Problem 11.30 Apply the Schwarz inequality (11.64) to the statistical correlation coefficient r defined in (8.19) to show that $|r| \leq 1$. Then, determine under what conditions we have $r = 1$, in terms of the sample pairs (x_i, y_i) $(i = 1, 2, \ldots, n)$; and, likewise for having $r = -1$.

Problem 11.31 **(a)** Show that if $\mathbf{x}, \mathbf{y} \in \mathbb{V}^2$ with $\mathbf{x} = x_1\hat{\mathbf{\imath}} + x_2\hat{\mathbf{\jmath}}$ $(x_1, x_2 \in \mathbb{R})$ and $\mathbf{y} = y_1\hat{\mathbf{\imath}} + y_2\hat{\mathbf{\jmath}}$ $(y_1, y_2 \in \mathbb{R})$, then $\|\mathbf{x}\| = \sqrt{x_1{}^2 + x_2{}^2}$ and $\mathbf{x} \bullet \mathbf{y} = x_1y_1 + x_2y_2$.

(b) What is the corresponding fact for $\mathbb{V}^3$?

Problem 11.32 Let $\mathbf{x} = \begin{bmatrix} x_1 & x_2 & \cdots & x_n \end{bmatrix}^{\mathrm{T}}$ be a nonzero vector in $\mathbb{V}^n$ for some $n \geq 1$. By (11.66), we have

$$\cos \measuredangle(\mathbf{x}, \mathbf{y}) = \frac{\mathbf{x} \bullet \mathbf{y}}{\|\mathbf{x}\|}$$

for any unit vector $\mathbf{y} \in \mathbb{V}^n$. In particular, letting $\mathbf{e}_j$ $(j = 1, 2, \ldots, n)$ be the standard-basis vectors of $\mathbb{V}^n$, per (11.67), the n numbers

$$v_j = \cos \measuredangle(\mathbf{x}, \mathbf{e}_j) = \frac{\mathbf{x} \bullet \mathbf{e}_j}{\|\mathbf{x}\|} = \frac{x_j}{\|\mathbf{x}\|} \quad (j = 1, 2, \ldots, n) \tag{P11.32.1a}$$

are called the **direction cosines** of $\mathbf{x}$. Indeed, show that

$$\mathbf{v} = \sum_{j=1}^{n} v_j \mathbf{e}_j \tag{P11.32.1b}$$

is a unit vector and that it points in the same direction as $\mathbf{x}$ (i.e., it equals a positive scalar times $\mathbf{x}$). (Hence, the direction cosines of a nonzero vector $\mathbf{x}$ provide enough information to determine its direction; but, they say nothing about the magnitude of $\mathbf{x}$, because they are unchanged when $\mathbf{x}$ is multiplied by a positive scalar.)

Problem 11.33 For each $n \geq 1$, the n-dimensional Euclidean vector space $\mathbb{V}^n$ (a real space) has an analogous complex n-dimensional space $\overline{\mathbb{V}}^n$, the vectors of which are $n \times 1$ complex matrices. The inner product on this space is

$$[\mathbf{x}, \mathbf{y}] = \sum_{j=1}^{n} x_j y_j{}^* = \mathbf{y}^* \mathbf{x} \quad (\mathbf{x}, \mathbf{y} \in \overline{\mathbb{V}}^n) \tag{11.2}$$

(cf. (11.60a)), where $\mathbf{x}$ and $\mathbf{y}$ are given by (11.58) and (11.57).

(a) Show that the function $[\cdot\,, \cdot]$ on $\overline{\mathbb{V}}^n \times \overline{\mathbb{V}}^n$ specified in (P11.33.1) is indeed a valid inner product.

(b) What is the induced norm?

Problem 11.34 Prove the following facts for any vectors $\mathbf{x}, \mathbf{y}, \mathbf{z} \in \mathbb{V}^3$.

(a) $(\mathbf{x} \times \mathbf{y}) \bullet \mathbf{x} = (\mathbf{x} \times \mathbf{y}) \bullet \mathbf{y} = 0$.

(b) $(\mathbf{x} \times \mathbf{y}) \bullet \mathbf{z} = (\mathbf{y} \times \mathbf{z}) \bullet \mathbf{x} = (\mathbf{z} \times \mathbf{x}) \bullet \mathbf{y}$.

Problem 11.35 Prove that two vectors $\mathbf{x}, \mathbf{y} \in \mathbb{V}^n$ $(n \geq 1)$ are collinear if and only if $|\mathbf{x} \bullet \mathbf{y}| = \|\mathbf{x}\| \cdot \|\mathbf{y}\|$.

Problem 11.36 Suppose $\varphi_1(\cdot)$ and $\varphi_2(\cdot)$ are transformations of $\mathbb{E}^n$ $(n \geq 1)$; thus, so is their composition $\varphi(\cdot) = \varphi_2[\varphi_1(\cdot)]$.

(a) Use (11.75) to prove that, for $n = 2$, if $\varphi_1(\cdot)$ and $\varphi_2(\cdot)$ are both translations, or both rotations, or both dilations, then so is $\varphi(\cdot)$.

(b) Prove that, in general, if $\varphi_1(\cdot)$ and $\varphi_2(\cdot)$ are both isometries, then so is $\varphi(\cdot)$.

Problem 11.37 Transforming $\mathbb{E}^2$ by first translating it by a vector $\mathbf{v} \in \mathbb{V}^2$ then rotating the result counterclockwise by an angle $\phi \in \mathbb{R}$ is equivalent to first rotating $\mathbb{E}^2$ counterclockwise by ϕ then translating the result by what vector $\mathbf{v}' \in \mathbb{V}^2$? Also, under what conditions on $\mathbf{v}$ and ϕ does $\mathbf{v}' = \mathbf{v}$?

Problem 11.38 Another transformation of the plane $\mathbb{E}^2$ is **point reflection** with respect to a given point $Q \in \mathbb{E}^2$. Specifically, as shown on the left side of Figure P11.38.1, the reflection of a point $P \in \mathbb{E}^2$ *about* (or *through*) Q is, if $P \neq Q$, the unique point $P' \in \mathbb{E}^2$ for which Q is the midpoint of $\overline{PP'}$; otherwise, $P' = P$. Also shown in the figure is **line reflection** with respect to a given line $l \subseteq \mathbb{E}^2$.

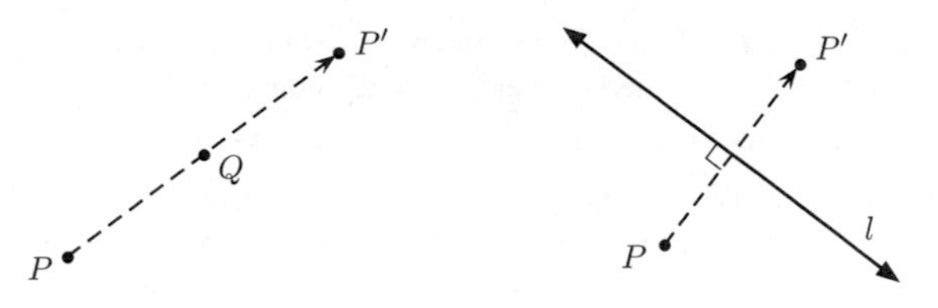

FIGURE P11.38.1 Reflection about a Point, and about a Line

Here, the reflection of a point $P \in \mathbb{E}^2$ *about* l is, if $P \notin l$, the unique point $P' \in \mathbb{E}^2$ for which l is the perpendicular bisector (see Problem 10.16) of $\overline{PP'}$; otherwise, $P' = P$.

(a) Reflection about the origin $(0, 0)$ corresponds to mapping each point $P = (x, y)$ to the point $P' = (x', y')$ such that

$$\begin{bmatrix} x' \\ y' \end{bmatrix} = \begin{bmatrix} -x \\ -y \end{bmatrix},$$

which can be expressed via (11.75) by taking $\mathbf{A} = -\mathbf{I}$ and $\mathbf{b} = \mathbf{0}$. (Thus, this transformation is equivalent to inverting both the x- and y-axes). What are the matrices $\mathbf{A}$ and $\mathbf{b}$ in (11.75) for expressing reflection about an arbitrary point $(X, Y) \in \mathbb{E}^2$?

(b) What are the matrices $\mathbf{A}$ and $\mathbf{b}$ in (11.75) for expressing reflection about a line described by equation (11.88)?

Problem 11.39 Show that a triangle $\triangle ABC$ in the xy-plane having vertices $A = (a_1, a_2)$, $B = (b_1, b_2)$, and $C = (c_1, c_2)$ has area

$$\alpha = \tfrac{1}{2} \left| \det \begin{bmatrix} a_1 & a_2 & 1 \\ b_1 & b_2 & 1 \\ c_1 & c_2 & 1 \end{bmatrix} \right|. \tag{P11.39.1}$$

(*Suggestion*: Compare (P11.11.1) and (11.72).)

Problem 11.40 Suppose two distinct points in the xy-plane are expressed in polar coordinates $r \geq 0$ and $\theta \in \mathbb{R}$ as (r_1, θ_1) and (r_2, θ_2). Use (11.96) to show that the line l passing through the points is described by the equation

$$r[r_1 \sin(\theta - \theta_1) - r_2 \sin(\theta - \theta_2)] + r_1 r_2 \sin(\theta_1 - \theta_2) = 0 \tag{P11.40.1}$$

—which is not of the form (11.98).

Problem 11.41 Per (5.6a), suppose two lines l_1 and l_2 in the xy-plane are described by the equations $y = m_1 x + Y_1$ and $y = m_2 x + Y_2$, for some constants $m_1, Y_1, m_2, Y_2 \in \mathbb{R}$. Use the arctangent expression in (11.100) to express the angle measure between the lines in terms of m_1 and m_2.

Problem 11.42 In terms of their x-, y-, and z-intercepts, which we assume to exist and to be nonzero, give necessary and sufficient conditions for two planes in xyz-space to be parallel, or perpendicular.

Problem 11.43 State a procedure for finding the equation of a plane determined by two crossing lines in xyz-space, where each line is described by a pair of equations of the form (11.111).

Problem 11.44 With the generic point (x, y, z) in xyz-space being specified by the vector $\mathbf{x} = \begin{bmatrix} x & y & z \end{bmatrix}^{\mathrm{T}}$, let l be the line in the space that is described by the parametric equation

$$\mathbf{x} = \mathbf{p} + r\mathbf{u}, \quad (r \in \mathbb{R}) \tag{P11.44.1}$$

for some fixed vectors $\mathbf{p}, \mathbf{u} \in \mathbb{V}^3$; and, let p be the plane in the space that is described by the parametric equation

$$\mathbf{x} = \mathbf{q} + s\mathbf{v} + t\mathbf{w} \quad (s, t \in \mathbb{R}), \tag{P11.44.2}$$

for some fixed vectors $\mathbf{q}$, $\mathbf{v}$, $\mathbf{w} \in \mathbb{V}^3$. Thus, $\mathbf{v}$ and $\mathbf{w}$ are assumed to be noncoplanar; but, furthermore, suppose $\mathbf{u}$, $\mathbf{v}$, and $\mathbf{w}$ are noncoplanar, thereby making l and p nonparallel (i.e., $l \nparallel p$). Show how to find the vector $\mathbf{x}' = \begin{bmatrix} x' & y' & z' \end{bmatrix}^{\mathrm{T}}$ of the point (x', y', z') at which l and p cross.

Problem 11.45 **(a)** Let $P = (x, y)$ be the generic point in xy-plane; and, let $P_1 = (0, 0)$ and $P_2 = (1, 0)$. Show that for any constant $\alpha > 0$ other than 1, the equation

$$d(P, P_1) = \alpha d(P, P_2) \qquad \text{(P11.45.1)}$$

describes a circle. Also, find its center and radius.

(b) For $\alpha = 0$, equation (P11.45.1) clearly describes (the set comprising) the single point P_1. What does it describe for $\alpha = 1$? And, for $\alpha < 0$?

Problem 11.46 Show that, given three values $a, b, c > 0$ such that the sum of any two is greater than the third, there exists a triangle having sides of lengths a, b, and c. (*Suggestion*: Use intersecting circles.)

Problem 11.47 Suppose one of the directrices d of an ellipse in the xy-plane is described by the equation $Ax + By + C = 0$, for some coefficients A, B, and C; and, the corresponding focus F is (X, Y). If the ellipse has eccentricity e, then what are the other directrix d' and focus F'?

Problem 11.48 Use Table 11.3 to prove that for any real constants $a \neq 0$ and $\theta \in (0, \pi)$, the parametric equations

$$x = a\cos(t + \theta), \quad y = \cos t \quad (0 \le t < 2\pi) \qquad \text{(P11.48.1)}$$

describe either a circle or an ellipse in the xy-plane; and, show how to easily determine which from a and θ.

Problem 11.49 Prove via Table 11.3 that the equation $xy = 1$ describes a hyperbola in the xy-plane. Then, find its semitransverse axis, semiconjugate axis, foci, center, directrices, asymptotes, eccentricity, and focal parameter.

Chapter 12. Mathematics in Practice

Problem 12.1 When a long suspension bridge is built, its towers might be so far separated that, the curvature of the earth prevents them from being made parallel; for otherwise each tower would not be sufficiently close to vertical. Accordingly, if an architect designing such a bridge were to assume the surface of the earth in the vicinity of the bridge to be perfectly flat, what kind of error would they be making?

Problem 12.2 **(a)** Express each of the following physical quantities in both scientific notation and engineering notation:

$$0.5940\ \mathrm{Sv}, \quad 1.00\ \mathrm{S{\cdot}s}, \quad -71\,030\,856.2\ \mathrm{Wb/H}.$$

(b) Using numeric prefixes, express the same quantities in decimal notation with no exponential part, such that the value of the integer part lies somewhere from 1 to 999. Use abbreviations for both prefixes and units.

(c) Express the results in part (b) with no abbreviations.

Problem 12.3 Show that if a number $x_{\mathrm{exp}} \neq 0$ is expressed in scientific notation as in (12.3), then its order of magnitude $p = \lfloor \log_{10} |x_{\mathrm{exp}}| \rfloor$. (Thus, scientific notation implicitly provides a quick estimate of the common logarithm of a number.)

Problem 12.4 Express each of the following SI units solely in terms of the SI base units: pascal, joule, watt, volt, ohm, siemens, farad, weber, tesla, henry, lux, gray, sievert.

Problem 12.5 Correct the expression of units in each of the following sentences to be SI-compliant.

(a) The mathematician Blaise Pascal measured the pressure to be 6.2 Pascals.

(b) An illuminance of 4.5 Lx., previously acceptable, now is not.

(c) The noise floor was at 3 watts per hertz = 3 J.

(d) It's easy to lift a mass of 19 gs.

(e) Water boils at about 373 degrees Kelvin.

Problem 12.6 The **Mach number** of a speed is that speed divided by the speed of sound (for the medium under consideration). What is the physical dimension of this unit of measurement?

Problem 12.7 Regarding the accuracy versus resolution of numerical data, which pertains more to numbers and which more to numerals?

Problem 12.8 Suppose an electric potential of -6.1 V is measured as -5.90 V.

(a) What is the absolute error of the measurement?

(b) What is the relative error of the measurement expressed as a percentage?

Problem 12.9 **(a)** Express the power $x = 7.3$ W in decibels relative to 1 mW.

(b) Calculate directly the percentage by which x exceeds $y = 5$ W; then, obtain the same result by first expressing x relative to y in decibels.

Problem 12.10 Recall the various ways—(12.12), (12.15), and (12.17a)—to commonly express the uncertainty interval (12.11) of an experimental value. Convert each of the following expressions into the other two forms, rounding (where necessary) to completely include the given interval.

(a) (65.43 ± 0.08) mT.

(b) -7.65 Tm $\pm\, 4\%$.

(c) $98.7(21) \times 10^6$ T.

Problem 12.11 One might suspect that we always have $|x_{\text{q-err}}| \le |x_{\text{err}}|$; that is, the magnitude of the quantization error incurred by recording a measurement cannot exceed the magnitude of the error in the measurement. After all, the former error contributes to the latter. To the contrary, explain why this inequality need not hold, and provide a numerical example.

Problem 12.12 **(a)** Determine which digits are significant in each of the following numerals:

$$275\,300, \quad -0.001\,70, \quad 5.286\,491 \times 10^{-3}.$$

(b) How many digits of resolution does each numeral have?

(c) Successively round each numeral to one fewer significant digit until only one digit remains.

Problem 12.13 In this problem we compare the use of percentages and decibels to express relative changes.

(a) First, note that for all $p \in \mathbb{R}$ other than $p = 0$, increasing a number $x > 0$ by $p\%$ to obtain $y = (1 + p\%)x$, then decreasing y by $p\%$ to produce $z = (1 - p\%)y$, we do not obtain $z = x$; and, likewise if a decrease is followed by the corresponding percentage increase. (Thus, e.g., increasing x by 10%, then decreasing the result by 10%, we get $0.99x$, which is 1% smaller than x.) By contrast, show that for any $\rho \in \mathbb{R}$, if $x > 0$ is increased by ρ dB to produce y, then y is decreased by ρ dB to produce z, we *do* obtain $z = x$ (and, likewise if a decrease is followed by the corresponding decibel increase).

(b) What change in decibels corresponds to a 100% increase? And, a 100% decrease?

(c) From Figure 7.1 it is evident that $\ln x \approx x - 1$ for $x \approx 1$. Accordingly, an increase of $p\%$ for $p \approx 0$ corresponds to approximately what change in decibels?

Problem 12.14 Explain why force, mass, and velocity (length divided by time) can be taken as the fundamental dimensions for the theory of mechanics, but not force, mass, and acceleration (length divided by time squared). What about force, velocity, and acceleration?

Problem 12.15 A simple way to design an electronic oscillator—i.e., an electrical circuit for which the output is a current or voltage waveform that is a sinusoid (see Problem 11.21)—is to connect together an inductor and a capacitor. The frequency of the oscillation f (in cycles per second, or hertz) is then determined by the inductor value L (in henrys) and the capacitor value C (in farads).

(a) What are the physical dimensions of L and C in terms of the SI base dimensions?

(b) It can be shown that $f = L^a C^b / 2\pi$ for some real powers a and b. What are they?

Problem 12.16 The CGS unit for force is the dyne (abbreviated "dyn"), which is defined as one gram-centimeter per second squared. What is the conversion factor from dynes to newtons?

Problem 12.17 In the Fahrenheit and Celsius temperature scales, water freezes at 32°F and 0°C, and boils at 212°F and 100°C. Accordingly, derive a formula for converting any temperature t°F into its corresponding value t'°C. Specifically, find constants a and b such that $t' = at + b$.

Problem 12.18 In air, the speed of light is about 3.0×10^8 m/s, whereas the speed of sound is only about 1100 ft/s.

(a) How many times faster is light than sound?

(b) If a lightning strike is seen x seconds before its thunder is heard, then how many miles away was the strike?

Problem 12.19 Since each byte (abbreviated "B") of computer memory comprises 8 bits (abbreviated "b"), the conversion factor from bytes to bits is 8 b/B. Also, byte and bit are dimensionless units, so each can be interpreted as the number 1; therefore, this conversion factor is simply 8. But, every conversion factor is just another name for the number 1. So, what is wrong here?

Problem 12.20 Suppose the experimental data $a = 1.23$ m and $b = 4.56$ m are in error by -0.07 m and 0.08 m, respectively. Use (12.36) to calculate the absolute errors in $a + b$ and $a - b$, as well as to estimate the relative errors in $a \cdot b$ and a/b.

Problem 12.21 In this problem we append to (12.36) the error propagation properties of the nth-power function ($n \in \mathbb{Z}$) and the nth-root function ($n \in \mathbb{Z}^+$). Thus, assume a and Δa are real numbers such that $|\Delta a| \ll |a|$.

(a) Use (12.36c) to show

$$x = a^n \quad \wedge \quad x + \Delta x = (a + \Delta a)^n \quad \Rightarrow \quad \frac{\Delta x}{x} \approx n \frac{\Delta a}{a} \tag{P12.21.1}$$

($n \geq 1$).

(b) Obtain the same fact by performing a binomial expansion of $(a + \Delta a)^n$.

(c) Note that (P12.21.1) holds for $n = 0$ as well, since then x is constant (1), which makes $\Delta x = 0$. Use (12.36d) to show (P12.21.1) also holds for $n \leq -1$, and therefore for all $n \in \mathbb{Z}$.

(d) Derive from (P12.21.1) the similar fact

$$x = \sqrt[n]{a} \quad \wedge \quad x + \Delta x = \sqrt[n]{a + \Delta a} \quad \Rightarrow \quad \frac{\Delta x}{x} \approx (1/n) \frac{\Delta a}{a} \tag{P12.21.2}$$

($a > 0, n \geq 1$).

(e) As an application of (P12.21.1) and (P12.21.2), suppose $x = \sqrt{a^2 + b^2}$ where a, b, Δa, and Δb are real numbers such that $|\Delta a| \ll |a|$ and $|\Delta b| \ll |b|$. Express Δx approximately in terms of x, a, b, Δa, and Δb.

Problem 12.22 In (12.34) the relative error in $x_{\text{exp}} + y_{\text{exp}}$ was found to be roughly equal to the average of the relative errors in x_{exp} and y_{exp}. Show that this was no accident but, rather, the result of the values x_{exp}, y_{exp}, x_{phy}, and y_{phy} being nearly all the same.

Problem 12.23 Suppose a man weighs himself on a scale that displays accurately to the nearest 0.5 pound. According to the scale, his weight one day is 191.0 pounds, and 188.5 pounds the next. Based on this information, what is the range of his possible weight change?

Problem 12.24 A common method used in construction to obtain a right angle is to form a triangle having sides with lengths a, b, and c such that $a : b : c = 3 : 4 : 5$, thereby getting a right triangle (by Problem 11.13, since $3^2 + 4^2 = 5^2$). Suppose the sides of the triangle are all measured to an uncertainty of 0.1%. What is the uncertainty interval, in degrees, of the purported right angle?

Problem 12.25 Suppose $x = 123$ MBq and $y = 76.54$ GBq. Apply the "rules" stated on p. 257 to express $x+y$, $x-y$, xy, and x/y in scientific notation with the "appropriate" number of significant digits.

Problem 12.26 When one is taking data that are to be plotted, measurements are typically made so that the data points will be uniformly distributed, at least approximately, along the horizontal axis of the plot. In particular, when plotting in the xy-plane from $x = a$ to $x = b$ $(-\infty < a < b < \infty)$ using a linear scale on the x-axis, the ideal is to obtain $n + 1 \geq 2$ data points $(x_0, y_0), (x_1, y_1), \ldots, (x_n, y_n)$ for which

$$x_i = a + i(b - a)/n \quad (i = 0, 1, \ldots, n). \tag{P12.26.1}$$

As an example of uniformly distributing points along a *logarithmic* scale, it is common for the values on an equipment dial to form a "1-3 sequence" (e.g., 0.1 V, 0.3 V, 1 V, 3 V, 10 V, 30 V, 100 V) or a "1-2-5 sequence" (e.g., 0.1 V, 0.2 V, 0.5 V, 1 V, 2 V, 5 V, 10 V, 20 V, 50 V, 100 V), thereby dividing each decade into two or three subintervals of nearly equally length. Similarly, find to two significant digits of resolution a sequence of six values ranging from 100 to 1000 that are nearly uniformly distributed on a logarithmic scale.

Problem 12.27 Suppose $n \geq 1$ data points $(x_1, y_1), (x_2, y_2), \ldots, (x_n, y_n)$ have distinct abscissa values $x_1, x_2, \ldots, x_n$. Fixing an integer m such that $0 \leq m \leq n-1$, let d_m be the degree of the unique polynomial in (12.37) that satisfies the least-squares-fit criterion; hence, assuming that this polynomial is not the zero polynomial (the degree of which does not exist), we have $0 \leq d_m \leq m$. Also, let d_L be the degree of the polynomial obtained from the Lagrange interpolation (6.25); hence, assuming that this polynomial is not the zero polynomial, we have $0 \leq d_L \leq n - 1$. Given only the values of m and d_L, how can one determine whether the least-squares-fit criterion will yield $d_m = d_L$?

Problem 12.28 According to **Moore's law**, the number of transistors that can be put onto a single integrated circuit roughly doubles every two years. This **empirical law** (i.e., a law induced from experience rather than deduced from theory) is supported by the data in Table P12.28.1. Verify Moore's law by plotting the data and fitting a line to the resulting points. Specifically, since the law states that the number of transistors varies exponentially versus time, make a linear-log plot (see Problem 5.21) of number of transistors versus year. Then, perform a least-squares fit of a line to the *plot* rather than directly to the data; that is, letting x be the year and y be the corresponding number of transistors, and thus the data comprise a collection of points (x_i, y_i) $(i = 1, 2, \ldots, n)$, find the values of real coefficients

TABLE P12.28.1 Data Supporting Moore's Law

Intel Microprocessor	Year Introduced	Transistors
4004	1971	2 300
8008	1972	2 500
8080	1974	4 500
8086	1978	29 000
286	1982	134 000
386	1985	275 000
486	1989	1 200 000
Pentium	1993	3 100 000
Pentium II	1997	7 500 000
Pentium III	1999	9 500 000
Pentium 4	2000	42 000 000
Itanium	2001	25 000 000
Itanium 2	2002	220 000 000
Itanium 2 (9 MB cache)	2004	592 000 000

TABLE S1.1.1

A	B	$A \Rightarrow B$	$B \Rightarrow A$	$(A \Rightarrow B) \wedge (B \Rightarrow A)$
F	F	T	T	T
F	T	T	F	F
T	F	F	T	F
T	T	T	T	T

a_0 and a_1 that minimize

$$\sum_{i=1}^{n}(a_0 + a_1 x_i - \log_{10} y_i)^2.$$

Finally, use this line to determine, to two significant digits of resolution, roughly how many years are required for the number of transistors to double.

Problem 12.29 Show that for the coefficients $a_0, a_1, \ldots, a_m$ obtained from (12.39), the squared error in (12.38) equals $\mathbf{y}^{\mathrm{T}}(\mathbf{y} - \mathbf{Xa})$.

Problem 12.30 Simplify the matrix $(\mathbf{X}^{\mathrm{T}}\mathbf{X})^{-1}\mathbf{X}^{\mathrm{T}}$ in (12.39b) for the extreme case of $m = n - 1$.

Solutions

Chapter 1. Logic

Solution 1.1 Using the truth tables in Table 1.2 for the logical connectives $\Rightarrow$ and $\wedge$, we obtain the truth tables in Table S1.1.1. The truth values in last column are, respectively, the same as those given in Table 1.2 for the statement $A \Leftrightarrow B$; therefore, the statements $(A \Rightarrow B) \wedge (B \Rightarrow A)$ and $A \Leftrightarrow B$ are logically equivalent.

Solution 1.2 **(a)** Yes, the statements $(A \wedge B) \vee C$ and $(A \vee C) \wedge (B \vee C)$ are logically equivalent, since their truth tables are identical, as shown in the upper half of Table S1.2.1.

(b) Likewise, the statements $(A \vee B) \wedge C$ and $(A \wedge C) \vee (B \wedge C)$ are logically equivalent, as shown in the lower half of Table S1.2.1. This conclusion also obtains by applying DeMorgan's laws to the result of part (a). Namely, by DeMorgan's laws, $(A \vee B) \wedge C$ is logically equivalent to $\neg\{[\neg(A \vee B)] \vee (\neg C)\}$, and then to $\neg\{[(\neg A) \wedge (\neg B)] \vee (\neg C)\}$, which by the result of part (a) is logically equivalent to $\neg\{[(\neg A) \vee (\neg C)] \wedge [(\neg B) \vee (\neg C)]\}$, which by DeMorgan's laws is logically equivalent to $\{\neg[(\neg A) \vee (\neg C)]\} \vee \{\neg[(\neg B) \vee (\neg C)]\}$, and then to $(A \wedge C) \vee (B \wedge C)]$—having used the fact that for any statement D, the statement $\neg(\neg D)$ is logically equivalent.

Solution 1.3 As shown in Table S1.3.1, the statement $A \Rightarrow (B \Rightarrow A)$ is a tautology because it is always true.

Solution 1.4 **(a)** Comparing the truth tables for the connectives $\vee$ and $\veebar$, we see that $A \vee B$ is true in all cases for which $A \veebar B$ is true; therefore, $A \veebar B \models A \vee B$. However, $A \vee B \not\models A \veebar B$; because, when A and B are both true, $A \vee B$ is true but $A \veebar B$ is false.

(b) Yes, the statements $A \veebar B$ and $A \Leftrightarrow \neg B$ are logically equivalent, since their truth tables are identical.

Solution 1.5 Since each of the constituent statements A_i $(i = 1, \ldots, n)$ has 2 possible truth values, the truth table for A has 2^n rows; therefore, since in each row the truth value assigned to A has 2 possibilities, there are 2^{2^n} possible truth tables for A. Of these, only the one table for which A is assigned a truth value of T throughout expresses a tautology.

Solution 1.6 Table S1.6.1 shows the truth tables for the conditionals $A \Rightarrow B$, $B \Rightarrow C$, and $A \Rightarrow C$ versus their constituent statements A, B, and C. If $A \Rightarrow B$ and $B \Rightarrow C$ are true, then the third, fifth,

TABLE S1.2.1

A	B	C	$(A \wedge B) \vee C$	$(A \vee C) \wedge (B \vee C)$
F	F	F	F	F
F	F	T	T	T
F	T	F	F	F
F	T	T	T	T
T	F	F	F	F
T	F	T	T	T
T	T	F	T	T
T	T	T	T	T

A	B	C	$(A \vee B) \wedge C$	$(A \wedge C) \vee (B \wedge C)$
F	F	F	F	F
F	F	T	F	F
F	T	F	F	F
F	T	T	T	T
T	F	F	F	F
T	F	T	T	T
T	T	F	F	F
T	T	T	T	T

TABLE S1.3.1

A	B	$B \Rightarrow A$	$A \Rightarrow (B \Rightarrow A)$
F	F	T	T
F	T	F	T
T	F	T	T
T	T	T	T

TABLE S1.6.1

A	B	C	$A \Rightarrow B$	$B \Rightarrow C$	$A \Rightarrow C$
F	F	F	T	T	T
F	F	T	T	T	T
F	T	F	T	F	T
F	T	T	T	T	T
T	F	F	F	T	F
T	F	T	F	T	T
T	T	F	T	F	F
T	T	T	T	T	T

sixth, and seventh rows can be eliminated as impossible; therefore, since the truth value of $A \Rightarrow C$ for each of the remaining rows is T, it must also be true.

Solution 1.7 **(a)** The statements A_i $(i = 1, \ldots, n)$ are logically inconsistent if and only if it is logically impossible for them to be simultaneously true; that is, for each assignment of truth values to their constituent atomic statements, at least one of the statements A_i is false. But, this is the case if and only if for each such assignment, at least one of the statements $\neg A_1, \ldots, \neg A_n$ is true, which is the case if and only if the disjunction $(\neg A_1) \vee \cdots \vee (\neg A_n)$ is true. Therefore, $A_1, \ldots, A_n$ are logically inconsistent if and only if the statement $(\neg A_1) \vee \cdots \vee (\neg A_n)$ is a tautology.

(b) No, it is possible for $A_1, \ldots, A_n$ to be logically consistent without $A_1 \wedge \cdots \wedge A_n$ being a tautology; because, the latter is the case if and only if $A_1, \ldots, A_n$ are all tautologies.

Solution 1.8 Yes, the two statements are identical. Because, for the contrapositive of a conditional $A \Rightarrow B$ is the conditional $(\neg B) \Rightarrow (\neg A)$, for which the converse is $(\neg A) \Rightarrow (\neg B)$; whereas the converse of the conditional $A \Rightarrow B$ is the conditional $B \Rightarrow A$, for which the contrapositive is $(\neg A) \Rightarrow (\neg B)$.

Solution 1.9 **(a)** Where there's no way, there's no will.

(b) Those who don't, can't; those who don't teach, can do.

(c) Anyone who has ever tried taking candy from a baby does not use the phrase "easy as taking candy from a baby".

Solution 1.10 **(a)** Let, e.g., $a_k = k - 1$ $(k = 1, 2, \ldots)$; then, statement (i) is false because $a_2 \neq 0$, whereas statement (ii) is false because $a_1 = 0$.

(b) Since (i) states that $\forall k,\ a_k = 0$, its negation is $\neg\forall k,\ a_k = 0$, which by (1.5) is logically equivalent to $\exists k,\ a_k \neq 0$; that is: For some k we have $a_k \neq 0$.

Solution 1.11 **(a)** The converse of (P1.11.1) is

$$[\forall x,\ H(x)] \Rightarrow H(h); \tag{S1.11.1}$$

therefore, the converse of (i) is:

If you have everything, then you have your health.

This statement *is* logically true, since (S1.11.1) is true for any meanings of the name "h" and predicate "H".

(b) No, (i) is not logically equivalent to (ii); because, (ii) translates as

$$[\neg\exists x,\ H(x)] \Rightarrow \neg H(h),$$

which is logically true, as is clear from its contrapositive:

$$H(h) \Rightarrow [\exists x,\ H(x)].$$

Solution 1.12 Being logically true, the statement A is true by virtue of its logical structure alone; and, this structure is independent of the statement B. Therefore, we have $B \vDash A$ because no more than the logical structure of both A and B is required to be able to conclude that A is true whenever B is true.

Solution 1.13 **(a)** Yes. For suppose $A, B \vDash C$; that is, solely by virtue of the logical structure of the statements A, B, and C, if A and B are both true, then so is C. Therefore, owing solely to the logical structure of these statements, whenever A is true, it is impossible to have B true and C false, which is the only case under which the conditional $B \Rightarrow C$ is false. Hence, $A \vDash (B \Rightarrow C)$.

(b) Yes. For suppose $A \vDash (B \Rightarrow C)$; that is, solely by virtue of the logical structure of the statements A, B, and C, if A is true, then so is $B \Rightarrow C$—i.e., it is impossible to have B true and C false. Therefore, owing solely to the logical structure of these statements, whenever A and B are both true, C must be true as well. Hence, $A, B \vDash C$.

Solution 1.14 One way to so express (P1.14.1) is

$$\exists x,\ \forall y,\ P(y) \Leftrightarrow y = x. \tag{S1.14.1}$$

For if (P1.14.1) is true—i.e., there exists a unique x such that $P(x)$ holds—then there exists something X such that $P(x)$ holds if and only if $x = X$; equivalently, in terms of the variable "y", there exists something X such that $P(y)$ holds if and only if $y = X$—i.e.,

$$\forall y,\ P(y) \Leftrightarrow y = X \tag{S1.14.2}$$

TABLE S1.17.1

A	B	C	$A \wedge \neg B$	$(A \wedge \neg B) \Rightarrow C$	$(A \wedge \neg B) \Rightarrow \neg C$	$A \Rightarrow B$
F	F	F	F	T	T	T
F	F	T	F	T	T	T
F	T	F	F	T	T	T
F	T	T	F	T	T	T
T	F	F	T	F	T	F
T	F	T	T	T	F	F
T	T	F	F	T	T	T
T	T	T	F	T	T	T

—and thus (S1.14.1) obtains. Conversely, if (S1.14.1) is true, then (S1.14.2) holds for something X—i.e., $P(y)$ holds if and only if $y = X$—and thus there exists a unique y such that $P(y)$ holds; and, this is what (P1.14.1) states, in terms of the variable "x".

Solution 1.15 **(a)** Since "better is necessarily different", being different is a necessary condition for being better. However, being different is not a sufficient condition for being better, since "different is not necessarily better".

(b) Based on the result to part (a): Being better is not a necessary condition for being different, because being different is not a sufficient condition for being better. However, being better is a sufficient condition for being different, because being different is a necessary condition for being better.

Solution 1.16 Suppose $a \nleq b$; that is, $a > b$. Then, letting $\varepsilon = a - b$, we have $\varepsilon > 0$; and, $a = b + \varepsilon$, so $a \nless b + \varepsilon$. Thus, we have proved that if $a \nleq b$, then $a \nless b + \varepsilon$ for some $\varepsilon > 0$; and, this assertion is equivalent to the contrapositive of the statement given in the problem: If $a \nleq b$, then we do not have $a < b + \varepsilon$ for all $\varepsilon > 0$. Therefore, the given statement is proved.

Solution 1.17 For the truth tables in Table S1.17.1, we see that if the statements $(A \wedge \neg B) \Rightarrow C$ and $(A \wedge \neg B) \Rightarrow \neg C$ are both true, then the fifth and sixth rows of this table can be eliminated from being possible; therefore, since the statement $A \Rightarrow B$ is true in all of the remaining rows, the logical implication (P1.17.1) holds.

Solution 1.18 Suppose that the statement $\exists y,\ \forall x,\ P(x, y)$ is true. Then, for something Y, the statement $\forall x,\ P(x, Y)$ is true; that is, $P(x, Y)$ holds for all x. Therefore, for each x there exists a y (in particular, Y) for which we have $P(x, y)$; so, the statement $\forall x,\ \exists y,\ P(x, y)$ is also true. Hence, since the argument just given is valid regardless of what the predicate "P" means, we obtain the logical implication $\exists y,\ \forall x,\ P(x, y) \models \forall x,\ \exists y,\ P(x, y)$.

The reverse of this implication, though, does not hold. As a counterexample, suppose that the universe under discussion is the set of integers, and the predicate "P" is such that "$P(x, y)$" means "$x < y$". Then, since for each integer x the value $y = x + 1$ is an integer for which $x < y$, the statement $\forall x,\ \exists y,\ P(x, y)$ is true; however, the statement $\exists y,\ \forall x,\ P(x, y)$ is not true, because there is no integer y that is greater than *every* integer x. Hence, $\forall x,\ \exists y,\ P(x, y) \not\models \exists y,\ \forall x,\ P(x, y)$.

Aside: The statement $\forall x,\ \exists y,\ P(x, y)$ asserts that for *each* value of x, a value of y can be found such that $P(x, y)$ is true. Thus, the value of y can change to suit the value x, since y is chosen *after* x. (Note that the quantification over y lies within the "scope" of the quantification over x.) By contrast, the statement $\exists y,\ \forall x,\ P(x, y)$, in which y is chosen *before* x, asserts that a *single* value of y can be found to make $P(x, y)$ true for *every* value of x. (Here, the quantification over x lies inside the scope of the quantification over y.) Accordingly, the latter statement makes a stronger assertion, since for it a single value of y must "work" for all x; hence the above results.

Solution 1.19 Using the suggested predicates, the two statements can be rewritten as

(i) $\forall x,\ B(x) \Rightarrow S(x)$.

(ii) $\forall x,\ B(x) \Rightarrow T(x)$.

Now, *suppose* there exists a person X that is my brother; that is, assume that the statement $B(X)$ is true. Then, it follows from (i) and (ii) that both $S(X)$ and $T(X)$ are true; that is, person X is both short and tall, which is impossible. Therefore, the assumption that such a person X exists has led to a contradiction; so, we must conclude that this assumption is false—i.e., I have no brothers! That is, $\forall x,\ \neg B(x)$, which makes the statements (i) and (ii) *vacuously* true.

Solution 1.20 If $\neg A$ is possibly true—i.e., $\Diamond \neg A$ is true—then A is possibly false; because, the truth of $\neg A$ implies the falsity of A. But, if A is possibly false, then A is not necessarily true—i.e., $\neg \Box A$ is true. Hence, whenever $\Diamond \neg A$ is true, $\neg \Box A$ is true; so, $\Diamond \neg A \vDash \neg \Box A$. Conversely, proceeding similarly in reverse yields $\neg \Box A \vDash \Diamond \neg A$; because, $\neg \Box A$ states that A is not necessarily true, which implies the possible falsity of A, which in turn implies $\Diamond \neg A$.

Chapter 2. Sets

Solution 2.1 Suppose that three such sets A, B, and C all exist; and, let x be some set. If $x \in A$, then by the description of A, we have $x \notin B$; then, by the description of B, we have $x \in C$; then, by the description of C, we have $x \notin A$. Hence, assuming $x \in A$ leads to a contradiction; thus, we must have $x \notin A$. But, by the description of A, this implies that $x \in B$, which by the description of B implies that $x \notin C$, which by the description of C implies that $x \in A$, contradicting our conclusion that $x \notin A$. Therefore, the assumption that three such sets A, B, and C exist itself leads to a contradiction; so, this assumption must be false.

Solution 2.2 Since no integer can be both even and odd, the one and only way is to have $A = \emptyset$; then, each of the stated conditions about A is vacuously true.

Solution 2.3 The only set A for which we have $\emptyset \supseteq A$ is $A = \emptyset$. On the other hand, there is no set A for which $\emptyset \supset A$; because, if there were, then $\emptyset$ would have an element that is not an element of A, whereas $\emptyset$ has no elements at all.

Solution 2.4 No. For example, if $A = \{0\}$ and $B = \{1\}$, then $A \not\subset B$; but we do not have $A \supseteq B$.

Solution 2.5 **(a)** Yes, the sets $(A \cap B) \cup C$ and $(A \cup C) \cap (B \cup C)$ must be equal. Because, for any object x, if $x \in C$, then x is a member of both of these sets; whereas if $x \notin C$, then for each of the above two sets, x is a member if and only if both $x \in A$ and $x \in B$.

(b) Likewise, the sets $(A \cup B) \cap C$ and $(A \cap C) \cup (B \cap C)$ must be equal. Because, for any object x, if $x \notin C$, then x is a member of neither of these sets; whereas if $x \in C$, then for each of the above two sets, x is a member if and only if either $x \in A$ or $x \in B$.

The same conclusion also obtains from the result of part (a). Namely, replacing the sets A, B, and C there with their complements (relative to any set D that is a superset of all three sets—e.g., $A \cup B \cup C$), it follows from part (a) that

$$(A^c \cap B^c) \cup C^c = (A^c \cup C^c) \cap (B^c \cup C^c).$$

Then, complementing both sides of this equality, and applying DeMorgan's laws several times, we obtain $(A \cup B) \cap C = (A \cap C) \cup (B \cap C)$.

Solution 2.6 **(a)** For any object x, we have $x \in A \triangle B$ if and only if either $x \in A$ or $x \in B$, but not both; thus, we obtain the Venn diagram in Figures S2.6.1.

(b) $\forall x,\ x \in A \triangle B \Leftrightarrow (x \in A \veebar x \in B)$.

(c) Considering all possibilities, we obtain the results in Table S2.6.1. Thus, $x \in A \triangle B$ if and only if $\mathbf{1}_A(x) \oplus \mathbf{1}_B(x) = 1$.

(d) In part (c) we found

$$\mathbf{1}_A(x) \oplus \mathbf{1}_B(x) = \begin{cases} 1 & x \in A \triangle B \\ 0 & x \notin A \triangle B \end{cases};$$

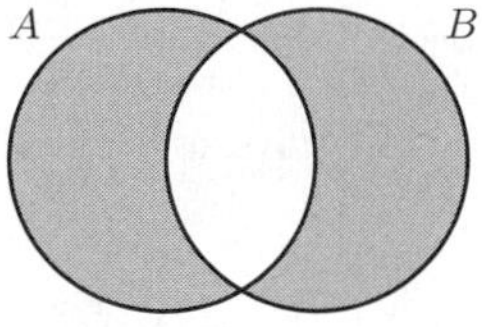

FIGURE S2.6.1 Venn Diagram for $A \triangle B$

TABLE S2.6.1

$x \in A$?	$x \in B$?	$x \in A \triangle B$?	$\mathbf{1}_A(x)$	$\mathbf{1}_B(x)$	$\mathbf{1}_A(x) \oplus \mathbf{1}_B(x)$
No	No	No	0	0	0
No	Yes	Yes	0	1	1
Yes	No	Yes	1	0	1
Yes	Yes	No	1	1	0

hence,

$$\mathbf{1}_{A\triangle B}(x) = \mathbf{1}_A(x) \oplus \mathbf{1}_B(x).$$

Since modulo-2 addition is associative, it follows that for all x we have

$$\begin{aligned}\mathbf{1}_{(A\triangle B)\triangle C}(x) &= \mathbf{1}_{A\triangle B}(x) \oplus \mathbf{1}_C(x) = [\mathbf{1}_A(x) \oplus \mathbf{1}_B(x)] \oplus \mathbf{1}_C(x)\\ &= \mathbf{1}_A(x) \oplus [\mathbf{1}_B(x) \oplus \mathbf{1}_C(x)] = \mathbf{1}_A(x) \oplus \mathbf{1}_{B\triangle C}(x)\\ &= \mathbf{1}_{A\triangle(B\triangle C)}(x).\end{aligned}$$

Thus, the indicator functions for the sets $(A \triangle B) \triangle C$ and $A \triangle (B \triangle C)$ are identical; therefore, the sets themselves are equal. (*Aside*: Continuing this reasoning, it can be shown that for each $n \geq 1$ we have $x \in A_1 \triangle \cdots \triangle A_n$ if and only if x is an element of an *odd* number of the sets A_i $(i = 1, \ldots, n)$.)

Solution 2.7 **(a)** If for some object x we have $x \in B$ but not $x \in A$, then $x \in B - A$ but $x \notin A - B$; therefore, by (P2.6.1), we have $x \in A \triangle B$, and so $A - B \neq A \triangle B$. Therefore, a necessary condition for having $A - B = A \triangle B$ is that for all x, if $x \in B$ then $x \in A$; that is, we need $B \subseteq A$. Now, suppose that this condition holds; then, $B - A = \emptyset$, and it follows from (P2.6.1) that $A - B = A \triangle B$. Hence, a necessary and sufficient condition for having $A - B = A \triangle B$ is that $B \subseteq A$. (This result also obtains by comparing Figure S2.6.1 with the right side of Figure 2.1.)

(b) Assuming $x \in A \triangle B$, we have either $x \in A$ or $x \in B$, but not both. Therefore, in the former case, we have $x \in A \triangle C$ if $x \notin C$, and $x \in B \triangle C$ if $x \in C$; whereas in the latter case, we have $x \in A \triangle C$ if $x \in C$, and $x \in B \triangle C$ if $x \notin C$. Thus, in each case, either $x \in A \triangle C$ or $x \in B \triangle C$, making $x \in (A \triangle C) \cup (B \triangle C)$. So $A \triangle B \subseteq (A \triangle C) \cup (B \triangle C)$.

Solution 2.8 Roughly put, the set A has the appearance of a rectangular region, the sides of which are parallel (and perpendicular) to the coordinate axes of $\mathbb{R}^2$. More precisely, A includes the interior of the rectangle, but just how much of the boundary of this region (i.e., the rectangle itself) is included depends upon which (if any) endpoints of the intervals I_1 and I_2 belong to the intervals themselves.

Solution 2.9 Yes, because $R, S \subseteq A^3$, making $R \cup S \subseteq A^3$.

Solution 2.10 **(a)** Let, e.g., $R = \{1\} \times \{1\} = \{(1, 1)\}$ and $S = \{1\} \times \{1, 2\} = \{(1, 1), (1, 2)\}$.

(b) No. Suppose S is one-to-one. If $x\,R\,y$ and $x'\,R\,y'$—i.e., $(x, y), (x', y') \in R$—then $x\,S\,y$ and $x'\,S\,y'$ as well, since $R \subseteq S$; therefore, $x = x'$ if and only if $y = y'$, since S is one-to-one. Hence, R is also one-to-one.

Solution 2.11 Since $1 \not< 1$, the less-than relation $<$ on $A = \{1\}$ is not reflexive, as before. However, this relation *is* symmetric; because, for no $x, y \in A$ do we have $x < y$, and so it is vacuously true that $\forall x, y \in A,\ x < y \Rightarrow y < x$. Likewise, $<$ is transitive, as before.

Solution 2.12 The binary relation

$$R = \{(1,1), (1,2), (2,1), (2,2), (2,3), (3,2), (3,3)\}$$

is clearly reflexive and symmetric; however, it is not transitive, since $(1,2), (2,3) \in R$ but $(1,3) \notin R$.

Solution 2.13 Let $x, y, z \in \mathbb{N}$ be arbitrarily chosen.

(a) The relation R is reflexive because $x + x = 2x$ is even; it is symmetric since $x + y = y + x$; and, it is transitive, since if $x + y$ and $y + z$ are even, then so are $(x + y) + (y + z) = x + 2y + z$ and $2y$, and thus so is $(x + 2y + z) - 2y = x + z$. Therefore, R is an equivalence relation.

(b) For this relation, there are two equivalence classes: the set of even integers and the set of odd integers. Because, if x and y are both even, then so is $x + y$, thereby putting x and y in the same equivalence class; and, likewise if x and y are both odd. Whereas if x is even and y is odd, then so is $x + y$, and therefore x and y are not in the same equivalence class.

Solution 2.14 For each $x \in A$, the set $\{y \in A \mid y \equiv_1 x\}$ is an equivalence class under $\equiv_1$; and, if C is an equivalence class under $\equiv_2$, then

$$C = \bigcup_{x \in C} \{y \in A \mid y \equiv_1 x\}. \tag{S2.14.1}$$

Because, for all $x, y \in A$, if $y \equiv_1 x$ then $y \equiv_2 x$; therefore, for each $x \in C$ we have $\{y \in A \mid y \equiv_1 x\} \subseteq \{y \in A \mid y \equiv_2 x\} = C$, making the right side of (S2.14.1) a subset of C. But, for each $x \in C$ we have $x \in \{y \in A \mid y \equiv_1 x\}$; so, the right side of (S2.14.1) is also a superset of C.

Solution 2.15 The relation R possesses the defining properties of an equivalence relation:

- R reflexive: Given $x \in A$, there exists a $B \in \mathcal{P}$ for which $x \in B$, since $\mathcal{P}$ is a partition of A; therefore, $x\,R\,x$, by (P2.15.1).
- R symmetric: Given $x, y \in A$ such that $x\,R\,y$, there exists a $B \in \mathcal{P}$ for which $x \in B$ and $y \in B$, by (P2.15.1); therefore, $y\,R\,x$, again by (P2.15.1).
- R transitive: Given $x, y, z \in A$ such that $x\,R\,y$ and $y\,R\,z$, there exist $B, C \in \mathcal{P}$ for which $x, y \in B$ and $y, z \in C$, by (P2.15.1). Thus, $y \in B \cap C$, making $B \cap C \neq \emptyset$, from which it follows that $B = C$ because $\mathcal{P}$ is a partition. Therefore, $x, z \in B$ where $B \in \mathcal{P}$; so, by (P2.15.1), we have $x\,R\,z$.

Chapter 3. Numbers

Solution 3.1 **(a)** Suppose, $x, y \in \mathbb{Q}$; that is, $x = m_x/n_x$ and $y = m_y/n_y$ for some $m_x, n_x, m_y, n_y \in \mathbb{Z}$ with $n_x, n_y \neq 0$. Then,

$$x + y = \frac{m_x}{n_x} + \frac{m_y}{n_y} = \frac{m_x n_y + m_y n_x}{n_x n_y},$$

where the numerator and denominator of the last fraction are both integers; therefore, $x + y \in \mathbb{Q}$.

(b) Clearly, a number is rational if and only if its negative is rational; therefore, since $\sqrt{2}$ is irrational, so is $-\sqrt{2}$. However, $\sqrt{2} + (-\sqrt{2}) = 0$, which is rational.

(c) The sum of a rational number and an irrational number must be *irrational*. For suppose that for some rational numbers x and y, and irrational number z, we have $x + z = y$; then, $z = x - y$, which (as explained above) is rational, thereby contradicting the assumption that z is irrational.

Solution 3.2 Evaluating the given repeating decimal expansion, we obtain

$$\begin{aligned} -98.76\overline{543}\ldots &= -(98.76 + 0.00543/0.999) = -9866667/99900 \\ &= -3288889/33300, \end{aligned}$$

where the next-to-last fraction was reduced to lowest terms by test-dividing its numerator by the prime factors of its denominator, $99900 = 2^2 3^3 5^2 37^1$.

Solution 3.3 **(a)** The number 110111001.10101_2 equals $110\,111\,001.101\,010_2 = 671.52_8$ and $1\,1011\,1001.1010\,1000_2 = 1B9.A8_{16}$; thus, it also equals $1 \cdot 16^2 + 11 \cdot 16^1 + 9 \cdot 16^0 + 10 \cdot 16^{-1} + 8 \cdot 16^{-2} = 441.65625_{10}$.[23]

(b) The number 0.456_8 equals $4 \cdot 8^{-1} + 5 \cdot 8^{-2} + 6 \cdot 8^{-3} = 0.58984375_{10}$, as well as $0.100\,101\,110_2 = 0.10010111_2$ and $0.1001\,0111_2 = 0.97_{16}$.

(c) The number $-9D.C_{16}$ equals $-(9 \cdot 16^1 + 13 \cdot 16^0 + 12 \cdot 16^{-1}) = -157.75_{10}$ and $-1001\,1101.1100_2 = -10011101.11_2$; thus, it also equals $-10\,011\,101.110_2 = -235.6_8$.

(d) Conversion *from* a decimal form to these other forms, however, generally requires more work.

- One way is to first divide (if necessary) the given number $x \neq 0$ by a power of 2 large enough to produce a number y having a magnitude less than 1; hence, since here $x = 39.625_{10}$, and thus $2^5 = 32 \le x < 64 = 2^6$, let $y = x/2^6 = 0.619140625$. We now convert y to binary form $0.\mathbf{y}_{-1}\mathbf{y}_{-2}\mathbf{y}_{-3}\cdots$ by performing the following simple algorithm—stated here assuming that $y > 0$ (otherwise, apply the algorithm to $-y$):

  1. Let $z_1 = 2y$ and $i = 1$.
  2. If $z_i \ge 1$, then $y_{-i} = 1$ and $\mathbf{y}_{-i}$ is "1"; otherwise, $y_{-i} = 0$ and $\mathbf{y}_{-i}$ is "0".
  3. If $z_i = 1$, then stop; otherwise, let $z_{i+1} = 2(z_i - y_{-i})$, add 1 to i, and go to step 2.

  Applying this algorithm to our value of y yields the results in Table S3.3.1, which shows that $y = 0.100111101_2$. Finally, since multiplying a number y by 2^m (integer m) effects the binary representation of y by simply shifting the binary point m places to the right, we have $x = 2^6 \cdot y = 100111.101_2$.

- Another way to convert a decimal numeral to binary form is to separately convert the integer and fractional parts. As for the fractional part of 39.625_{10}, the above algorithm yields $0.625_{10} = 0.101_2$; and, here is a similar algorithm to convert an integer $y > 0$ to binary form $\mathbf{y}_n \ldots \mathbf{y}_1\mathbf{y}_0$:

  1. Let $z_0 = y$ and $i = 0$.
  2. If z_i is odd, then $y_i = 1$ and $\mathbf{y}_i$ is "1"; otherwise, $y_i = 0$ and $\mathbf{y}_i$ is "0".
  3. If $z_i = 1$, then stop (with $n = i$); otherwise, let $z_{i+1} = (z_i - y_{-i})/2$, add 1 to i, and go to step 2.

---

23. Note that this computation can be efficiently performed, without directly calculating powers, by rewriting it as follows:

$$
\begin{aligned}
1 \cdot 16^2 &+ 11 \cdot 16^1 + 9 \cdot 16^0 + 10 \cdot 16^{-1} + 8 \cdot 16^{-2} \\
&= (1 \cdot 16^2 + 11 \cdot 16^1 + 9 \cdot 16^0) + (8 \cdot 16^{-2} + 10 \cdot 16^{-1}) \\
&= (1 \cdot 16 + 11) \cdot 16 + 9 + (8/16 + 10)/16.
\end{aligned}
$$

In general, applying Horner's rule (E.22) to the integer and fractional parts of a finite base-b expansion yields

$$
\begin{aligned}
(\mathbf{x}_n \ldots \mathbf{x}_1\mathbf{x}_0.\mathbf{x}_{-1}\mathbf{x}_{-2} \ldots \mathbf{x}_{-m})_b &= \sum_{i=-m}^{n} x_i \cdot b^i = \sum_{i=0}^{n} x_i \cdot b^i + \sum_{i=-m}^{-1} x_i \cdot b^i \\
&= (\cdots((x_n b + x_{n-1})b + x_{n-2})b + \cdots + x_1)b + x_0 \\
&\quad + (\cdots((x_{-m}/b + x_{-m+1})/b + x_{-m+2})/b + \cdots + x_{-1})/b,
\end{aligned}
$$

where the sum from $i = -m$ to -1 and its expression via Horner's rule are omitted when the numeral has no fractional part (i.e., $m = 0$). For example, try applying this method to the numerals "110111001.10101_2" and "671.52_8", in each case obtaining the same base-10 result.

TABLE S3.3.1

i	z_i	y_{-i}
1	1.23828125	1
2	0.4765625	0
3	0.953125	0
4	1.90625	1
5	1.8125	1
6	1.625	1
7	1.25	1
8	0.5	0
9	1	1

TABLE S3.3.2

i	z_i	y_{-i}
0	39	1
1	19	1
2	9	1
3	4	0
4	2	0
5	1	1

Applying this algorithm to $y = 39$ yields the results in Table S3.3.2,which shows that $y = 100111_2$. Hence, $39.625_{10} = 100111_2 + 0.101_2 = 100111.101_2$.

Having found $39.625_{10} = 100111.101_2$, we also have $39.625_{10} = 100\,111.101_2 = 47.5_8$ and $39.625_{10} = 0010\,0111.1010_2 = 27.\mathrm{A}_{16}$.

Solution 3.4 **(a)** If a real number x has a finite binary expansion, then for some integers $m, n \geq 0$ and values $x_i \in \{0, 1\}$ $(i = -m, -m+1, \ldots, n)$ we have

$$x = \pm \sum_{i=-m}^{n} x_i \cdot 2^i.$$

Therefore, letting $y = 10^m \cdot x = 2^m 5^m \cdot x$, we obtain

$$y = \pm 2^m 5^m \sum_{i=-m}^{n} x_i \cdot 2^i = \pm 5^m \sum_{i=-m}^{n} x_i \cdot 2^{i+m} = \pm 5^m \sum_{j=0}^{n+m} x_{j-m} \cdot 2^j$$

($\pm$ respectively), which is an clearly integer, and thus has a finite decimal expansion. Hence, $x = y/10^m$ also has a finite decimal expansion, since a decimal expansion for $y/10^m$ obtains from that for y by shifting the decimal point m places to the left.

(b) One such number is $\frac{1}{5} = 0.2_{10} = 0.\overline{0011}\ldots_2$, where the binary expansion obtains by applying the first algorithm given in Solution 3.3(d).

Solution 3.5 Suppose $a = b'c$, where $b' \neq b$. Then, subtracting this equation from (P3.5.1) yields

$$0 = (b - b')c,$$

where $b - b' \neq 0$; therefore, $c = 0$.

Solution 3.6 By (3.12), $[a,a]$ is simply the set $\{a\}$ consisting of the one element a, whereas $(a,a] = [a,a) = (a,a) = \emptyset$.

Solution 3.7 **(a)** For all $x > 0$ (no matter how large), we have $(-1)^n n \geq x$ for even $n \geq x$, as well as $(-1)^n n \leq -x$ for odd $n \geq x$; therefore, $\sup A = \infty$ and $\inf A = -\infty$. Moreover, since A contains neither of these values, the supremum is not a maximum and the infimum is not a minimum.

(b) For all $n \in \{1,2,\ldots\}$, we have $-1 = (-1)^1/1 \leq (-1)^n/n \leq (-1)^2/2 = \frac{1}{2}$; therefore, $\sup A = \frac{1}{2}$ and $\inf A = -1$, with A containing both of these values. Hence, the supremum is a maximum and the infimum is a minimum.

(c) Here, $\sup A = \max(b,d)$, which does not belong to A, and thus the maximum of A does not exist; whereas $\inf A = \min(a,c)$, which does belong to A, and thus is the minimum of A.

(d) Since $-\infty$ is the smallest value $x \in \overline{\mathbb{R}}$ such that $y \leq x$ for all $y \in A$ (vacuously), we have $\sup A = -\infty$; similarly, $\inf A = \infty$. Also, since A contains neither of these values, the supremum is not a maximum and the infimum is not a minimum.

Solution 3.8 Although statement (i) *can* be true—e.g., for $S = [0,1)$—it is not *necessarily* true. For example, if $S = [0,1]$, then $\sup S = 1$, but it is not the case that $y \notin S$ for all real values $y \geq 1$; namely, take $y = 1$. In fact, for this set S, the value x described in statement (i) does not even exist.

Statement (ii), though, *is* necessarily true. Because, if $S \subseteq \mathbb{R}$ with $\sup S \in \mathbb{R}$, then $\sup S$ is indeed a real value x such that $y \notin S$ for all real values $y > x$; otherwise, there would exist a real value $y > \sup S$ for which $y \in S$, thereby contradicting the fact that $\sup S$ is an upper bound on S. Also, $\sup S$ must be the *smallest* such value x; otherwise, there would exist a value $x' < \sup S$ such that $y \notin S$ for all real values $y > x'$, and thus $y \leq x'$ for all $y \in S$, thereby making x' an upper bound on S smaller than its *least* upper bound, $\sup S$.

Solution 3.9 If $x = 0$, then (P3.9.1) is satisfied for all $y \in \mathbb{R}$; whereas for $x \neq 0$ we can divide (P3.9.1) by x to conclude that, in this case, the equality holds if and only if $y = x$. Hence, (P3.9.1) is satisfied for only those pairs $(x,y) \in \mathbb{R}^2$ such that either $x = 0$ or $x = y$.

Aside: As this problem shows, care must be taken not to inadvertently divide by zero. Had we immediately divided both sides of (P3.9.1) by x, and thereby concluded that we *must* have $y = x$, we would have missed all of the valid pairs $(0,y)$ with $y \neq 0$.

Solution 3.10 One example: $2 + (1 \cdot 0) = 2 \neq 6 = (2+1) \cdot (2+0)$.

Solution 3.11 **(a)** Yes, by (3.17) and the fact that if $x < 0$ then $-x > 0$.

(b) Yes, by (3.17) and the fact that if $x \neq 0$ then $-x \neq 0$.

(c) Yes; because, from (3.15) and (3.17) we obtain

$$x \cdot \infty = \begin{cases} \infty & x > 0 \\ 0 & x = 0\,, \\ -\infty & x < 0 \end{cases}$$

including when x is infinite.

Solution 3.12 $|x| = x \operatorname{sgn} x$ $(x \in \overline{\mathbb{R}})$.

Solution 3.13 Squaring both sides of the given equality, we obtain $|a-b|^2 = |a-c|^2$, which since a, b, and c are real is equivalent to having $(a-b)^2 = (a-c)^2$. Thus, $a^2 - 2ab + b^2 = a^2 - 2ac + c^2$, which rearranged yields $2a(b-c) = b^2 - c^2 = (b+c)(b-c)$. Since $b \neq c$, we can divide by $b - c \neq 0$, thereby obtaining $a = (b+c)/2$. (*Aside*: This result could have been predicted by plotting on the real line two points b and c, then another point a equidistant from both.)

Solution 3.14 Since multiplication takes precedence over addition, the two pairs of the parentheses on the right side of (3.13c) can be omitted; however, for the same reason, the pair on the left side cannot (in general).

Solution 3.15 **(a)** One example: whereas $1 - 2 + 3 = 2$, performing the addition first yields $1 - (2 + 3) = 1 - 5 = -4$.

(b) $a + (b - c) = a + [b + (-c)] = (a + b) + (-c) = (a + b) - c$.

(c) After the subtractions have been performed, the remaining additions may be performed in any order because given an expression containing only additions, these operations can be reordered left to right as desired by using the commutative and associative properties of addition. As for the subtractions, it is sufficient to show that for any expression containing only additions and subtractions, we may first perform the leftmost subtraction; for then, performing this operation repeatedly, we eventually perform left to right all the subtractions in the original expression without performing any of the additions. Thus, consider performing the leftmost subtraction in the following expression for some integer $n \geq 1$ and numbers x_i $(i = 1, 2, \ldots, n+1)$:

$$x_1 + \cdots + x_n - x_{n+1} \cdots \text{(remaining terms).} \tag{S3.15.1}$$

In fact, we may assume $n > 1$, since for $n = 1$ the leftmost subtraction is the leftmost operation overall, which would be performed first anyway by the usual evaluation rules stated in Subsection 3.2.1. Therefore, performing the leftmost subtraction in (S3.15.1) yields

$$\begin{aligned} &x_1 + \cdots + (x_n - x_{n+1}) \cdots \text{(remaining terms)} \\ &\qquad = [(x_1 + \cdots + x_{n-1}) + (x_n - x_{n+1})] \cdots \text{(remaining terms),} \end{aligned}$$

with the equality holding by the usual rule of evaluating all additions and subtractions left-to-right. This is to be compared with applying the usual rules directly to (S3.15.1), which gives

$$\{[(x_1 + \cdots + x_{n-1}) + x_n] - x_{n+1}\} \cdots \text{(remaining terms).}$$

Hence, we must show (via the usual rules) that

$$(x_1 + \cdots + x_{n-1}) + (x_n - x_{n+1}) = [(x_1 + \cdots + x_{n-1}) + x_n] - x_{n+1};$$

but, this follows from the result of part (b) by taking $a = x_1 + \cdots + x_{n-1}$, $b = x_n$, and $c = x_{n+1}$.

Solution 3.16 It is given that $x - 1 = 2m$ and $x - 2 = 3n$ for some integers m and n; so, $2m = 3n + 1$. Hence, since $2m$ must be even, n must be odd; that is, $n = 2k - 1$ for some integer k. Therefore, a necessary condition for x to possess the desired properties is that $x - 2 = 3(2k - 1)$ for some integer k; that is, $x = 6k - 1$ for some integer k. Furthermore, this condition is also sufficient; for then $x - 1 = 6k - 2 = 2(3k - 1)$ is divisible by 2, and $x - 2 = 3(2k - 1)$ is divisible by 3.

Solution 3.17 Here are all such sums for $n \leq 30$:

$$\begin{aligned} 4 &= 2 + 2, & 18 &= 5 + 13 = 7 + 11, \\ 6 &= 3 + 3, & 20 &= 3 + 17 = 7 + 13, \\ 8 &= 3 + 5, & 22 &= 3 + 19 = 5 + 17 = 11 + 11, \\ 10 &= 3 + 7 = 5 + 5, & 24 &= 5 + 19 = 7 + 17 = 11 + 13, \\ 12 &= 5 + 7, & 26 &= 3 + 23 = 7 + 19 = 13 + 13, \\ 14 &= 3 + 11 = 7 + 7, & 28 &= 5 + 23 = 11 + 17, \\ 16 &= 3 + 13 = 5 + 11, & 30 &= 7 + 23 = 11 + 19 = 13 + 17. \end{aligned}$$

Solution 3.18 Given an integer $x > 1$ (thus $\sqrt{x} < x$), suppose there is no prime number $p \leq \sqrt{x}$ that divides x. Then, there can be no integer n that divides x for which $1 < n \leq \sqrt{x}$; for otherwise n would have a prime factor $p \leq n \leq \sqrt{x}$ that divides x (viz., let p be any prime divisor of n). It follows that there can be no integer n that divides x for which $\sqrt{x} < n < x$; for otherwise x/n would be an integer that divides x (since $x/(x/n) = n$) for which $1 < x/n < \sqrt{x}$. Hence, if there is no prime number $p \leq \sqrt{x}$ that divides x, then there is no integer n that divides x for which $1 < n < x$; therefore, x must be prime, because clearly no integer greater than x can divide it.

Solution 3.19 **(a)** Let

$$x = p_1{}^{m_1} p_2{}^{m_2} p_3{}^{m_3} \cdots, \quad y = p_1{}^{n_1} p_2{}^{n_2} p_3{}^{n_3} \cdots,$$

where p_i $(i = 1, 2, \dots)$ is the ith prime number and $m_i, n_1 \in \mathbb{N}$ (all i). Then,

$$\mathrm{LCM}(x, y) = p_1{}^{\max(m_1,n_1)} p_2{}^{\max(m_2,n_2)} p_3{}^{\max(m_3,n_3)} \cdots,$$

$$\mathrm{GCD}(x, y) = p_1{}^{\min(m_1,n_1)} p_2{}^{\min(m_2,n_2)} p_3{}^{\min(m_3,n_3)} \cdots.$$

(b) Yes. For any $m, n \in \mathbb{N}$ we have $\max(m, n) + \min(m, n) = m + n$; therefore, it follows from the result of part (a) that

$$\begin{aligned}\mathrm{LCM}(x, y) \cdot \mathrm{GCD}(x, y) &= p_1{}^{\max(m_1,n_1)+\min(m_1,n_1)} p_2{}^{\max(m_2,n_2)+\min(m_2,n_2)} \cdots \\ &= p_1{}^{m_1+n_1} p_2{}^{m_2+n_2} \cdots = xy.\end{aligned}$$

Solution 3.20 Since each of the y numbers $nx \bmod y$ $(n = 0, 1, \dots, y-1)$ is an element of the set $\{0, 1, \dots, y-1\}$, we need only show that these numbers are distinct. Therefore, assume the contrary; that is, suppose $n_1, n_2 \in \{0, 1, \dots, y-1\}$ with $n_1 \neq n_2$, yet $n_1 x \bmod y = n_2 x \bmod y$. Then, by (3.21), there exist integers q_1, r_1, q_2, and r_1 such that

$$n_1 x = q_1 y + r_1, \quad n_2 x = q_2 y + r_2;$$

and, $r_1 = r_2$ by (3.22), so $q_1 \neq q_2$. Now, subtracting these two equations yields $(n_1 - n_2)x = (q_1 - q_2)y$; hence, $|n_1 - n_2|x/y$ equals a positive integer $|q_1 - q_2|$. But, this is impossible; because, the relatively prime integers x and y have no common factors, and y cannot divide a positive integer $|n_1 - n_2| < y$. Thus, having been led to a contradiction, we conclude that no such numbers n_1 and n_2 exist.

Solution 3.21 Let q_x and q_y be, respectively, the integer quotients obtained by dividing x and y by n; thus, per (3.21),

$$x = q_x n + r_x, \quad y = q_y n + r_y, \quad \text{where} \quad r_x, r_y \in \{0, 1, \dots, n-1\}.$$

Therefore, $x - y = (q_x - q_y)n + (r_x - r_y)$ is a multiple of n if and only if $r_x - r_y$ is a multiple of n; and, this is the case if and only if $r_x = r_y$, since $-(n-1) \leq r_x - r_y \leq n - 1$. Also, $r_x = x \bmod n$ and $r_y = y \bmod n$; so, $x - y$ is a multiple of n if and only if $x \bmod n = y \bmod n$, from which (P3.21.1a) follows. Now, (P3.21.1b) obtains, since $[(xn + y) - y] \bmod n = xn \bmod n = 0$. As for (P3.21.1c),

$$\begin{aligned}[x + (y \bmod n)] \bmod n &= [(q_x n + r_x) + r_y] \bmod n \\ &= \{q_y n + [(q_x n + r_x) + r_y]\} \bmod n \\ &= [(q_x + q_y)n + (r_x + r_y)] \bmod n = (x + y) \bmod n,\end{aligned}$$

having used (P3.21.1b) for the second equality. (*Aside*: Taking $x = 0$ in (P3.21.1c) shows that (y mod n) mod $n = y$ mod n.) Again using (P3.21.1b), we obtain

$$\begin{aligned}x(y \bmod n) \bmod n &= (q_x n + r_x) r_y \bmod n \\ &= [(q_x n + r_x) q_y n + (q_x n + r_x) r_y] \bmod n \\ &= (q_x n + r_x)(q_y n + r_y) \bmod n = xy \bmod n,\end{aligned}$$

proving (P3.21.1d). Finally, using (8.4) and then (P3.21.1b) gives

$$\begin{aligned}x^m \bmod n = (q_x n + r_x)^m \bmod n &= \left[\sum_{k=0}^{m} \binom{m}{k} (q_x n)^k r_x{}^{m-k}\right] \bmod n \\ &= \left[n \sum_{k=1}^{m} \binom{m}{k} q_x{}^k n^{k-1} r_x{}^{m-k} + r_x{}^m\right] \bmod n = r_x{}^m \bmod n \\ &= (x \bmod n)^m \bmod n,\end{aligned}$$

with the second equality holding since the numbers $\binom{m}{k} q_x{}^k n^{k-1} r_x{}^{m-k}$ $(1 \le k \le m)$ are all integers, and thus so is their sum; this proves (P3.21.1e).

Solution 3.22 **(a)** By the commutative property of ordinary addition,

$$x \oplus y = (x + y) \bmod n = (y + x) \bmod n = y \oplus x \quad (x, y \in \mathbb{Z});$$

therefore, $\oplus$ is commutative. Whereas by (P3.21.1c), along with the commutative and associative properties of ordinary addition,

$$\begin{aligned}(x \oplus y) \oplus z &= \{[(x + y) \bmod n] + z\} \bmod n = (x + y + z) \bmod n \\ &= \{x + [(y + z) \bmod n]\} \bmod n = x \oplus (y \oplus z) \quad (x, y, z \in \mathbb{Z});\end{aligned}$$

therefore, $\oplus$ is associative.

(b) Comparing (P3.22.1) with (3.23), and using the associative property of $\oplus$, we obtain

$$(x \oplus y) \ominus y = (x \oplus y) \oplus (-y) = x \oplus [y \oplus (-y)] = x \oplus (y \ominus y) = x \oplus 0 = x$$

$(x, y \in \mathbb{Z})$.

Solution 3.23 **(a)** We have

$$x/2 = \frac{1}{2}\left(\sum_{i=0}^{n} x_i \cdot 10^i\right) = \frac{1}{2}\left(x_0 + \sum_{i=1}^{n} x_i \cdot 2^i 5^i\right) = x_0/2 + \sum_{i=1}^{n} x_i \cdot 2^{i-1} 5^i, \qquad \text{(S3.23.1)}$$

where the sums from $i = 1$ to n equal 0 when $n = 0$, because they are then empty (likewise for similar sums below). The numbers $x_i \cdot 2^{i-1} 5^i$ $(1 \le i \le n)$ are all integers, and thus so is their sum; therefore, $x/2$ is an integer (i.e., x is divisible by 2) if and only if $x_0/2$ is an integer (i.e., x_0 is even).

(b) An integer m is divisible by 3 if and only if $m \bmod 3 = 0$; therefore, it is sufficient to show $x \bmod 3 = (x_0 + \cdots + x_n) \bmod 3$. Indeed, using (P3.21.1c)–(P3.21.1e) we obtain

$$\begin{aligned} x \bmod 3 &= \left(\sum_{i=0}^{n} x_i \cdot 10^i\right) \bmod 3 = \left(\sum_{i=0}^{n} \{x_i \cdot 10^i\} \bmod 3\right) \bmod 3 \\ &= \left(\sum_{i=0}^{n} x_i \{10^i \bmod 3\} \bmod 3\right) \bmod 3 \\ &= \left(\sum_{i=0}^{n} x_i \{[10 \bmod 3]^i \bmod 3\} \bmod 3\right) \bmod 3 \\ &= \left(\sum_{i=0}^{n} x_i \{1^i \bmod 3\} \bmod 3\right) \bmod 3 = \left(\sum_{i=0}^{n} x_i \bmod 3\right) \bmod 3 \\ &= \left(\sum_{i=0}^{n} x_i\right) \bmod 3. \end{aligned} \qquad \text{(S3.23.2)}$$

(c) Proceeding similarly to (S3.23.1) yields

$$x/4 = (x_0 + 10x_1)/4 + \sum_{i=2}^{n} x_i \cdot 2^{i-2} 5^i,$$

where the numbers $x_i \cdot 2^{i-2} 5^i$ $(2 \le i \le n)$ are all integers, and thus so is their sum. Therefore, $x/4$ is an integer (i.e., x is divisible by 4) if and only if $(x_0 + 10x_1)/4 = (x_0/2 + x_1)/2 + 2x_1$ is an integer. But, since $2x_1$ must be an integer, this is the case if and only if $(x_0/2 + x_1)/2$ is an integer, which is the case if and only if $x_0/2+x_1$ is an even integer; and, this is the case if and only if x_0 and $x_0/2+x_1$ are both even.

(d) Proceeding similarly to (S3.23.1) yields

$$x/5 = x_0/5 + \sum_{i=1}^{n} x_i \cdot 2^i 5^{i-1},$$

where the numbers $x_i \cdot 2^i 5^{i-1}$ $(1 \leq i \leq n)$ are all integers, and thus so is their sum. Therefore, $x/5$ is an integer (i.e., x is divisible by 5) if and only if $x_0/5$ is an integer, which is the case if and only if $x_0 \in \{0, 5\}$.

(e) If x is divisible by 6, then $x = 6k$ for some integer k; hence, we then have $x/2 = 3k$ and $x/3 = 2k$, where $3k$ and $2k$ are both integers, so x is divisible by both 2 and 3. Conversely, if x is divisible by both 2 and 3, which are prime numbers, then the prime factorization of x must contain at least one factor each; thus, for some integer l we have $x = 2 \cdot 3 \cdot l = 6l$, making x divisible by 6.

(f) If $n = 0$, then x is divisible by 7 if and only if $x \in \{0, 7\}$; therefore, assume $n \geq 1$. We now have

$$y = \sum_{i=1}^{n} x_i \cdot 10^{i-1} = \frac{1}{10} \sum_{i=1}^{n} x_i \cdot 10^i = (x - x_0)/10, \tag{S3.23.3}$$

and thus $10(y - 2x_0) = x - 21x_0$. Therefore, since 10 and 7 are relatively prime, $y - 2x_0$ is divisible by 7 if and only if $x - 21x_0$ is divisible by 7; and, since $x - 21x_0 = x - 7(3x_0)$, where $3x_0$ must be an integer, this is the case if and only if x is divisible by 7.

(g) Proceeding similarly to (S3.23.1) yields

$$x/8 = (x_0 + 10x_1 + 100x_2)/8 + \sum_{i=3}^{n} x_i \cdot 2^{i-3} 5^i,$$

where the numbers $x_i \cdot 2^{i-3}5^i$ $(3 \leq i \leq n)$ are all integers, and thus so is their sum. Therefore, $x/8$ is an integer (i.e., x is divisible by 8) if and only if $(x_0 + 10x_1 + 100x_2)/8 = (x_0/4 + x_1/2 + x_2)/2 + (x_1 + 12x_2)$ is an integer. But, since $x_1 + 12x_2$ must be an integer, this is the case if and only if $(x_0/4 + x_1/2 + x_2)/2 = [(x_0/2 + x_1)/2 + x_2]/2$ is an integer, which is the case if and only if $(x_0/2 + x_1)/2 + x_2$ is an even integer; and, this is the case if and only if x_0, $x_0/2 + x_1$, and $(x_0/2 + x_1)/2 + x_2$ are all even.

(h) Replacing each "3" in (S3.23.2) with "9" yields an argument showing that $x \bmod 9 = (x_0 + \cdots + x_n) \bmod 9$; thus, x is divisible by 9 (i.e., $x \bmod 9 = 0$) if and only if $x_0 + \cdots + x_n$ is divisible by 9.

(i) Proceeding similarly to (S3.23.1) yields

$$x/10 = x_0/10 + \sum_{i=1}^{n} x_i \cdot 10^{i-1},$$

where the numbers $x_i \cdot 10^{i-1}$ $(1 \leq i \leq n)$ are all integers, and thus so is their sum. Therefore, $x/10$ is an integer (i.e., x is divisible by 10) if and only if $x_0/10$ is an integer, which is the case if and only if $x_0 = 0$.

(j) If $n = 0$, then x is divisible by 11 if and only if $x = 0$; therefore, assume $n \geq 1$. We now have (S3.23.3), and thus $10(y - x_0) = x - 11x_0$. Therefore, since 10 and 11 are relatively prime, $y - x_0$ is divisible by 11 if and only if $x - 11x_0$ is divisible by 11, which is the case if and only if x is divisible by 11.

Solution 3.24 Yes. For if $x:y::x':y'$ with x, y, x', $y' \neq 0$, then $x/y = x'/y'$, thereby making $x/x' = y/y'$, and thus $x:x'::y:y'$.

Solution 3.25 It is given that $y = (1 - p/100)x$ and $x/y = r/1 = r$; therefore, $r = 1/(1 - p/100) = 100/(100 - p)$, making $p = 100 - 100/r = 100(r - 1)/r$.

Solution 3.26 We know that $a = \alpha x$, $b = \alpha y$, and $c = \alpha z$, for some constant $\alpha \neq 0$; hence,

$$a + b + c = \alpha(x + y + z) = \alpha d,$$

making $\alpha = (a + b + c)/d$ (where division by d is allowed because it is nonzero). Therefore, since $x = a/\alpha$, $y = b/\alpha$, and $z = c/\alpha$, we have

$$x = \frac{ad}{a+b+c}, \quad y = \frac{bd}{a+b+c}, \quad z = \frac{cd}{a+b+c}.$$

Solution 3.27 **(a)** By (3.31), $-2\angle(\pi/3) = -2\cos(\pi/3) + i[-2\sin(\pi/3)] = -2(1/2) + i\big[-2(\sqrt{3}/2)\big] = -1 - i\sqrt{3} \approx -1 - i1.73$.

(b) By (3.30), $-2 + i3 = Ae^{i\theta} = A\angle\theta$, where $A = \sqrt{2^2 + 3^2} = \sqrt{13} \approx 3.61$ and (observing that the real part, -2, is negative) $\theta = \tan^{-1}(-3)/2 + \pi = \pi - \tan^{-1}\frac{3}{2} \approx 2.16$.

Solution 3.28 **(a)** Using the polar form for z, we obtain

$$z \cdot Ae^{i\theta} = |z|e^{i\arg z} \cdot Ae^{i\theta} = A|z| \cdot e^{i(\arg z + \theta)};$$

therefore, the length of the vector z is scaled by A (which, if negative, causes the resulting vector to point in the opposite direction to z) and then the result is rotated counterclockwise by θ.

(b) Since $i = 1e^{i\pi/2}$, the vector z is merely rotated counterclockwise by $\pi/2$ (90°).

Solution 3.29 One way:

$$\begin{aligned}|1 + e^{i\theta}|^2 &= (1 + e^{i\theta})(1 + e^{i\theta})^* = (1 + e^{i\theta})(1 + e^{-i\theta}) \\ &= 1 \cdot 1 + 1 \cdot e^{-i\theta} + e^{i\theta} \cdot 1 + e^{i\theta} \cdot e^{-i\theta} = 1 + e^{-i\theta} + e^{i\theta} + 1 \\ &= 2 + (e^{i\theta} + e^{-i\theta}) = 2 + 2\cos\theta = 2(1 + \cos\theta).\end{aligned}$$

Another way (a bit more tedious): Using Euler's identity,

$$\begin{aligned}|1 + e^{i\theta}|^2 &= |1 + (\cos\theta + i\sin\theta)|^2 = |(1 + \cos\theta) + i\sin\theta|^2 \\ &= (1 + \cos\theta)^2 + \sin^2\theta = (1 + 2\cos\theta + \cos^2\theta) + \sin^2\theta \\ &= 1 + 2\cos\theta + (\cos^2\theta + \sin^2\theta) = 1 + 2\cos\theta + 1 = 2(1 + \cos\theta).\end{aligned}$$

Yet another way (a little tricky):

$$\begin{aligned}|1 + e^{i\theta}|^2 &= \left|e^{i\theta/2}\left(e^{-i\theta/2} + e^{i\theta/2}\right)\right|^2 = \left|e^{i\theta/2} \cdot 2\cos\theta/2\right|^2 = \left[\left|e^{i\theta/2}\right| \cdot |2\cos\theta/2|\right]^2 \\ &= 1^2 \cdot |2\cos\theta/2|^2 = 4\cos^2\theta/2 = 4[(1 + \cos\theta)/2] = 2(1 + \cos\theta).\end{aligned}$$

Solution 3.30 If $z = z^*$, then $0 = z - z^* = (\operatorname{Re} z + i\operatorname{Im} z) - (\operatorname{Re} z - i\operatorname{Im} z) = 2i\operatorname{Im} z$; therefore, $\operatorname{Im} z = 0$, so z is real. Similarly, z is imaginary if and only if $z = -z^*$.

Solution 3.31 Since $iz = i(\operatorname{Re} z + i\operatorname{Im} z) = i\operatorname{Re} z - \operatorname{Im} z$, we have $\operatorname{Re} iz = -\operatorname{Im} z$ and $\operatorname{Im} iz = \operatorname{Re} z$.

Solution 3.32 For $x = 0$, it follows from the stated condition that $a+b$ is real; hence, $0 = \operatorname{Im}(a+b) = \operatorname{Im} a + \operatorname{Im} b$, and so $\operatorname{Im} a = -\operatorname{Im} b$. Similarly, taking $x = \pi/2$ shows that $ai - bi = i(a - b)$ is real, making $a - b$ imaginary; hence, $0 = \operatorname{Re}(a - b) = \operatorname{Re} a - \operatorname{Re} b$, and so $\operatorname{Re} a = \operatorname{Re} b$. Therefore, $\operatorname{Re} a + i\operatorname{Im} a = \operatorname{Re} b - i\operatorname{Im} b$; that is, $a = b^*$.

Solution 3.33 **(a)** $z_2 = |z_1|e^{i\theta}$ for some $\theta \in \mathbb{R}$.

(b) $z_2 = Ae^{i\arg z_1}$ for some $A > 0$.

Solution 3.34 **(a)** Multiplying the numerator and denominator of $1/z$ by z^*, we obtain

$$\operatorname{Re}\frac{1}{z} = \operatorname{Re}\frac{z^*}{zz^*} = \operatorname{Re}\frac{z^*}{|z|^2} = \frac{\operatorname{Re} z^*}{|z|^2} = \frac{\operatorname{Re} z}{|z|^2},$$

where the third equality is valid since $|z|^2$ is real.
(b) Similarly, $\operatorname{Im}(1/z) = (-\operatorname{Im} z)/|z|^2$, since

$$\operatorname{Im}\frac{1}{z} = \operatorname{Im}\frac{z^*}{zz^*} = \operatorname{Im}\frac{z^*}{|z|^2} = \frac{\operatorname{Im} z^*}{|z|^2} = \frac{-\operatorname{Im} z}{|z|^2}.$$

Solution 3.35 **(a)** For all $z_1, z_2 \in \mathbb{C}$,

$$\begin{aligned}
|z_1 + z_2|^2 &= \left||z_1|e^{i\arg z_1} + |z_2|e^{i\arg z_2}\right|^2 \\
&= [|z_1|e^{i\arg z_1} + |z_2|e^{i\arg z_2}] \cdot [|z_1|e^{i\arg z_1} + |z_2|e^{i\arg z_2}]^* \\
&= [|z_1|e^{i\arg z_1} + |z_2|e^{i\arg z_2}] \cdot [|z_1|e^{-i\arg z_1} + |z_2|e^{-i\arg z_2}] \\
&= |z_1|^2 + |z_1||z_2|e^{i(\arg z_1 - \arg z_2)} + |z_1||z_2|e^{i(\arg z_2 - \arg z_1)} + |z_2|^2 \\
&= |z_1|^2 + |z_1||z_2|[e^{i(\arg z_1 - \arg z_2)} + e^{-i(\arg z_1 - \arg z_2)}] + |z_2|^2 \\
&= |z_1|^2 + |z_1||z_2| \cdot 2\cos(\arg z_1 - \arg z_2) + |z_2|^2,
\end{aligned}$$

where

$$|z_1||z_2| \cdot 2\cos(\arg z_1 - \arg z_2) \le 2|z_1||z_2| \tag{S3.35.1}$$

since $|z_1||z_2| \ge 0$ and $\cos\alpha \le 1$ ($\alpha \in \mathbb{R}$); hence,

$$|z_1 + z_2|^2 \le |z_1|^2 + 2|z_1||z_2| + |z_2|^2 = (|z_1| + |z_2|)^2,$$

giving (3.37).
(b) Assume (without loss of generality) that for some $\phi \in \mathbb{R}$, either $\arg z_1, \arg z_2 \in [\phi, \phi + 2\pi)$ or $\arg z_1, \arg z_2 \in (\phi, \phi + 2\pi]$. It follows from earlier statements above that equality is attained in (3.37) if and only if it is attained in (S3.35.1); that is, $|z_1| = 0$, $|z_2| = 0$, or $\cos(\arg z_1 - \arg z_2) = 1$. But, having $|z_j| = 0$ ($j = 1$ or 2) is equivalent to having $z_j = 0$. Furthermore, our initial restrictions on $\arg z_1$ and $\arg z_2$, constrain us to have $-2\pi < \arg z_1 - \arg z_2 < 2\pi$; so (see Figure 11.8) $\cos(\arg z_1 - \arg z_2) = 1$ if and only if $\arg z_1 = \arg z_2$. Therefore, equality attains in (3.37) if and only if $z_1 = 0$, $z_2 = 0$, or $\arg z_1 = \arg z_2$.

Solution 3.36 **(a)** Using (3.37) yields

$$\begin{aligned}
|x - y| = |(x - z) - (y - z)| &= |(x - z) + [-(y - z)]| \le |x - z| + |-(y - z)| \\
&= |x - z| + |y - z|,
\end{aligned}$$

and thus $|x - y| \le |x - z| + |y - z|$.
(b) Suppose $\max(|x-z|, |y-z|) \not\ge |x-y|/2$; that is, both $|x-z| < |x-y|/2$ and $|y-z| < |x-y|/2$. Then,

$$|x - z| + |y - z| < |x - y|/2 + |x - y|/2 = |x - y|,$$

contradicting the result of part (a); therefore, we must have $\max(|x - z|, |y - z|) \ge |x - y|/2$.

Solution 3.37 **(a)** Let $x, y \in \mathbb{C}$ be given. Using (3.37) we obtain

$$|x| = |y + (x - y)| \le |y| + |x - y|;$$

thus, $|x-y| \geq |x| - |y|$. Hence, $|x-y| = |y-x| \geq |y| - |x|$. Therefore,

$$|x-y| \geq \max(|x|-|y|, |y|-|x|) = \big||x|-|y|\big|,$$

proving (P3.37.1).

(b) Given $z_1, z_2 \in \mathbb{C}$, it follows from (P3.37.1) that

$$|z_1| = |(z_1+z_2) - z_2| \geq \big||z_1+z_2| - |z_2|\big| \geq |z_1+z_2| - |z_2|,$$

from which (3.37) follows.

Chapter 4. Sequences

Solution 4.1 If the sequence $\{x_i\}_{i=1}^{\infty}$ is increasing (respectively, strictly increasing), then $x_i \leq x_{i+1}$ ($x_i < x_{i+1}$) for all i, and thus $-x_i \geq -x_{i+1}$ ($-x_i > -x_{i+1}$) for all i, thereby making the sequence $\{-x_i\}_{i=1}^{\infty}$ decreasing (strictly decreasing).

Solution 4.2 Let $X = \sup_i x_i$, which is a finite value since the sequence $\{x_i\}_{i=1}^{\infty}$ is bounded. By the definition of a supremum, X is the smallest value $x \in \overline{\mathbb{R}}$ such that $x_i \leq x$ (all i). Accordingly, $x_i \leq X$ (all i), and for each $\varepsilon > 0$ there exists an $n \in \mathbb{Z}^+$ such that $x_n \geq X - \varepsilon$. (Otherwise, for some $\varepsilon > 0$ we would have $x_i < X - \varepsilon$ for all i, with $X - \varepsilon < \sup_i x_i$, which is impossible.) Furthermore, since the sequence is increasing, for these same values of ε and n we have $x_i \geq x_n$ ($i \geq n$), thereby making $x_i \geq X - \varepsilon$ ($i \geq n$). Hence, it has been shown that for each $\varepsilon > 0$, there exists an $n \in \mathbb{Z}^+$ such that $X - \varepsilon \leq x_i \leq X$ ($i \geq n$), and thus $|x_i - X| \leq \varepsilon$ ($i \geq n$); therefore, $\lim_{i\to\infty} x_i = X$.

Solution 4.3 **(a)** Here, $\lim_{i\to\infty} x_i = \infty$; because, for each $\xi > 0$ we have $x_i \geq \xi$ when i is sufficiently large (specifically, when $i \geq \xi$).

(b) Here, $\lim_{i\to\infty} x_i$ does not exist. First, since the sequence is bounded, it cannot diverge to $\pm\infty$; so if the limit exists, then it must be finite. Also, (noting that the given sinusoid has period 20), for $i = 5, 25, 45, 65, \ldots$, we have $x_i = \sin \pi/2 = 1$; whereas for $i = 15, 35, 55, 75, \ldots$, we have $x_i = \sin 3\pi/2 = -1$. Therefore, fixing $\varepsilon < 1$, it is impossible to choose $x \in \mathbb{R}$ and $n \in \mathbb{Z}^+$ such that $|x_i - x| \leq \varepsilon$ ($i \geq n$).[24]

(c) Here, $\lim_{i\to\infty} x_i = 0$; because, $|x_i| \leq 1/i$ ($i \neq 0$), and clearly $1/i \to 0$ as $i \to \infty$.

(d) Here, $\lim_{i\to\infty} x_i$ does not exist. Because, for $i = 5, 25, 45, 65, \ldots$, we have $x_i = i$ (which becomes arbitrarily large in the positive direction); whereas for $i = 15, 35, 55, 75, \ldots$, we have $x_i = -i$ (which becomes arbitrarily large in the negative direction). Therefore, the sequence neither converges to a finite value nor diverges to $\pm\infty$.

Solution 4.4 Given $\varepsilon > 0$, there exist $n_x, n_y \in \mathbb{Z}^+$ such that $|x_i - x| \leq \varepsilon$ for $i \geq n_x$, and $|y_i - y| \leq \varepsilon$ for $i \geq n_y$; therefore, taking $n = \max(n_x, n_y)$, we have both $|x_i - x| \leq \varepsilon$ and $|y_i - y| \leq \varepsilon$ for $i \geq n$.

Solution 4.5 Let $Y = \lim_{i\to\infty} x_i = \lim_{i\to\infty} z_i$. If $Y = \infty$, then for each $\xi > 0$ we have $x_i \geq \xi$ when i is sufficiently large; so, since $y_i \geq x_i$ (all i), we also have $y_i \geq \xi$ when i is sufficiently large. Therefore, $\lim_{i\to\infty} y_i = \infty = Y$. Similarly, if $Y = -\infty$ instead, then for each $\xi > 0$ we have $z_i \leq -\xi$ when i is sufficiently large; so, since $y_i \leq z_i$ (all i), we also have $y_i \leq -\xi$ when i is sufficiently large. Therefore, $\lim_{i\to\infty} y_i = -\infty = Y$.

Suppose, then, that Y is finite. By Problem 4.4 we know that for each $\varepsilon > 0$ there exists an $n \in \mathbb{Z}^+$ such that $|x_i - Y| \leq \varepsilon$ and $|z_i - Y| \leq \varepsilon$ for all $i \geq n$. Hence, for $i \geq n$ we have $Y - \varepsilon \leq x_i \leq y_i \leq z_i \leq Y + \varepsilon$, making $|y_i - Y| \leq \varepsilon$. Therefore, $\lim_{i\to\infty} y_i = Y$.

Solution 4.6 Suppose $\lim_{i\to\infty} x_i = x$ for some $x \in \mathbb{C}$; then, given $\varepsilon > 0$, there exists an $n \in \mathbb{Z}^+$ such that $|x_i - x| \leq \varepsilon$ ($i \geq n$). Therefore, since $l \geq 0$, we also have $|y_i - x| = |x_{i+l} - x| \leq \varepsilon$ ($i \geq n$). Thus, given $\varepsilon > 0$, there exists an $n \in \mathbb{Z}^+$ such that $|y_i - x| \leq \varepsilon$ ($i \geq n$); so, $\lim_{i\to\infty} y_i = x$.

Conversely, suppose $\lim_{i\to\infty} y_i = y$ for some $y \in \mathbb{C}$; then, given $\varepsilon > 0$, there exists an $n \in \mathbb{Z}^+$ such that $|y_i - y| \leq \varepsilon$ ($i \geq n$). Therefore, for $n' = n + l$, we also have $|x_i - y| = |y_{i-l} - y| \leq \varepsilon$

---

24. If this is not obvious, then note that by part (b) of Problem 3.36 we have $\max[|1-x|, |(-1)-x|] \geq |1-(-1)|/2 = 1$ for all $x \in \mathbb{R}$.

$(i \geq n')$, since having $i \geq n'$ implies that $i - l \geq n$. Thus, given $\varepsilon > 0$, there exists an $n' \in \mathbb{Z}^+$ such that $|x_i - y| \leq \varepsilon$ $(i \geq n')$; so, $\lim_{i\to\infty} x_i = y$.

Solution 4.7 Suppose $\lim_{i\to\infty} x_i = x$ for some $x \in \mathbb{C}$. Given $\varepsilon > 0$, there then exists an $n \in \mathbb{Z}^+$ such that $|x_i - x| \leq \varepsilon$ $(i \geq n)$. Now, since each of the values $1, 2, \ldots, n$ appears only once in the sequence $(j_1, j_2, \ldots)$, there must exist an $n' \in \mathbb{Z}^+$ such that $j_i \geq n$ for all $i \geq n'$; hence, if $i \geq n'$, then $|y_i - x| = |x_{j_i} - x| \leq \varepsilon$. Thus, given $\varepsilon > 0$, there exists an $n' \in \mathbb{Z}^+$ such that $|y_i - x| \leq \varepsilon$; therefore, $\lim_{i\to\infty} y_i = x$. It remains to show that if we suppose instead that $\lim_{i\to\infty} y_i = y$ for some $y \in \mathbb{C}$, then we must have $\lim_{i\to\infty} x_i = y$; but, this fact follows immediately from the result just obtained, because $\{x_i\}_{i=1}^{\infty}$ can be viewed as a reordering of $\{y_i\}_{i=1}^{\infty}$.

Solution 4.8 Taking the limit of each side of (P4.8.1), we obtain

$$X = \lim_{i\to\infty} x_{i+1} = \lim_{i\to\infty} (x_i + a/x_i)/2 = \left[\left(\lim_{i\to\infty} x_i\right) + \left(\lim_{i\to\infty} a/x_i\right)\right]/2$$
$$= (X + a/X)/2,$$

having applied (4.5a) and (4.6). Multiplying the resulting equation by $2X$ and combining terms yields $X^2 = a$. Therefore, $X = \sqrt{a}$ (independent of x_1); because, we cannot have $X < 0$, since $x_i > 0$ (all i).

Aside: Having shown that $\lim_{i\to\infty} x_i = \sqrt{a}$ for any initial value $x_1 > 0$ *if* the limit exists, here is a proof that it does: Letting $z_i = x_i/\sqrt{a}$ $(i = 1, 2, \ldots)$, it is sufficient to show $\lim_{i\to\infty} z_i = 1$ for any $z_1 > 0$; and, dividing (P4.8.1) by $\sqrt{a}$ yields

$$z_{i+1} = (z_i + 1/z_i)/2 \quad (i \geq 1).$$

Since $z_i > 0$ $(i \geq 1)$, it follows from Problem 6.23 that $z_i + 1/z_i \geq 2$ $(i \geq 1)$, and thus $z_i \geq 1$ $(i \geq 2)$; hence, $1/z_i \leq 1$ $(i \geq 2)$. Therefore,

$$1 \leq z_{i+1} \leq (z_i + 1)/2 \quad (i \geq 2),$$

which gives

$$0 \leq z_{i+1} - 1 \leq (z_i - 1)/2 \quad (i \geq 2).$$

By repeated application of the last bound we obtain

$$0 \leq z_i - 1 \leq (z_2 - 1)/2^{i-2} \quad (i \geq 2),$$

where the right side approaches 0 as $i \to \infty$; so, by Problem 4.5, $z_i - 1 \to 0$, which shows that $z_i \to 1$.

Solution 4.9 **(a)** Since in the given description of a closed set we have $X \in \mathbb{R}$, we must have $X \in S$ when $S = \mathbb{R}$; therefore, $\mathbb{R}$ is closed. So, its complement $\emptyset$ is open. On the other hand, there exists no sequence $\{x_i\}_{i=1}^{\infty}$ of values $x_i \in \emptyset$; therefore (vacuously), $\emptyset$ is closed. So, its complement $\mathbb{R}$ is open. Thus, $\mathbb{R}$ and $\emptyset$ are each both closed and open.

(b) See Table S4.9.1.

TABLE S4.9.1 Nonempty Intervals $I \subseteq \mathbb{R}$

Interval I	Possible Endpoints	I Closed	I Open
(a, b)	$-\infty \leq a < b \leq \infty$	$a = -\infty$ and $b = \infty$	Always
$(a, b]$	$-\infty \leq a < b < \infty$	$a = -\infty$	Never
$[a, b)$	$-\infty < a < b \leq \infty$	$b = \infty$	Never
$[a, b]$	$-\infty < a \leq b < \infty$	Always	Never

Solution 4.10 **(a)** First, observe that

$$\begin{aligned}|z_i - z|^2 &= |(\operatorname{Re} z_i + i \operatorname{Im} z_i) - (\operatorname{Re} z + i \operatorname{Im} z)|^2 \\ &= |(\operatorname{Re} z_i - \operatorname{Re} z) + i(\operatorname{Im} z_i - \operatorname{Im} z)|^2 \\ &= (\operatorname{Re} z_i - \operatorname{Re} z)^2 + (\operatorname{Im} z_i - \operatorname{Im} z)^2 \quad (\text{all } i).\end{aligned} \tag{S4.10.1}$$

Therefore, if $\lim_{i\to\infty} \operatorname{Re} z_i = \operatorname{Re} z$ and $\lim_{i\to\infty} \operatorname{Im} z_i = \operatorname{Im} z$, then by (4.5a) we have

$$\lim_{i\to\infty} |z_i - z|^2 = \lim_{i\to\infty} (\operatorname{Re} z_i - \operatorname{Re} z)^2 + \lim_{i\to\infty} (\operatorname{Im} z_i - \operatorname{Im} z)^2 = 0 + 0 = 0,$$

and thus $\lim_{i\to\infty} z_i = z$. Conversely, (S4.10.1) implies

$$|z_i - z| \geq |\operatorname{Re} z_i - \operatorname{Re} z| \quad (\text{all } i);$$

therefore, if we assume instead that $\lim_{i\to\infty} z_i = z$, then as $i \to \infty$ we obtain $|z_i - z| \to 0$, and thus $|\operatorname{Re} z_i - \operatorname{Re} z| \to 0$, from which it follows that $\lim_{i\to\infty} \operatorname{Re} z_i = \operatorname{Re} z$. Similarly, if $\lim_{i\to\infty} z_i = z$, then $\lim_{i\to\infty} \operatorname{Im} z_i = \operatorname{Im} z$.

(b) If $\lim_{i\to\infty} z_i = z$ with $z_i \in \mathbb{R}$ (all i), then it follows from part (a) that $\operatorname{Im} z = \lim_{i\to\infty} \operatorname{Im} z_i = \lim_{i\to\infty} 0 = 0$, and thus z must be real.

Solution 4.11 For any sequence $\{x_i\}_{i=1}^{\infty}$ in S, we have

$$\lim_{i\to\infty} (x_i - x_i) = \lim_{i\to\infty} 0 = 0;$$

therefore, $\{x_i\}_{i=1}^{\infty} \simeq \{x_i\}_{i=1}^{\infty}$, so $\simeq$ is reflexive. Also, if $\{y_i\}_{i=1}^{\infty}$ is a sequence in S with $\{x_i\}_{i=1}^{\infty} \simeq \{y_i\}_{i=1}^{\infty}$, then

$$\lim_{i\to\infty} (y_i - x_i) = -\lim_{i\to\infty} (x_i - y_i) = -0 = 0;$$

therefore, $\{y_i\}_{i=1}^{\infty} \simeq \{x_i\}_{i=1}^{\infty}$, so $\simeq$ is symmetric. Finally, if $\{z_i\}_{i=1}^{\infty}$ is also sequence in S with $\{x_i\}_{i=1}^{\infty} \simeq \{y_i\}_{i=1}^{\infty}$ and $\{y_i\}_{i=1}^{\infty} \simeq \{z_i\}_{i=1}^{\infty}$, then

$$\begin{aligned}\lim_{i\to\infty} (x_i - z_i) &= \lim_{i\to\infty} [(x_i - y_i) + (y_i - z_i)] = \lim_{i\to\infty} (x_i - y_i) + \lim_{i\to\infty} (y_i - z_i) \\ &= 0 + 0 = 0;\end{aligned}$$

therefore, $\{x_i\}_{i=1}^{\infty} \simeq \{z_i\}_{i=1}^{\infty}$, so $\simeq$ is transitive. Hence, since the relation $\simeq$ is reflexive, symmetric, and transitive, it is an equivalence relation.

Solution 4.12 Let, e.g.,

$$x_{i,j} = \begin{cases} 1 & i = j \\ 0 & \text{otherwise} \end{cases} \qquad (i, j = 1, 2, \ldots).$$

Then, $\sum_{j=1}^{\infty} x_{i,j} = 1$ (all i), and thus

$$\lim_{i\to\infty} \sum_{j=1}^{\infty} x_{i,j} = 1;$$

but, $\lim_{i\to\infty} x_{i,j} = 0$ (all j), and thus

$$\sum_{j=1}^{\infty} \lim_{i\to\infty} x_{i,j} = 0.$$

Solution 4.13 Let, e.g., $x_{i,j} = 1/(i - j + \frac{1}{2})^2$ $(i, j = 1, 2, \ldots)$, where the addition of $\frac{1}{2}$ precludes division by 0; then,

$$\lim_{i\to\infty}\lim_{j\to\infty} x_{i,j} = 0 = \lim_{j\to\infty}\lim_{i\to\infty} x_{i,j}.$$

These limits show that for any choice of $\varepsilon > 0$ and $n \in \mathbb{Z}^+$, there exist integers $i, j \geq n$ such that $|x_{i,j} - 0| < \varepsilon$; however, since $x_{i,j} = 4$ $(i = j)$, there also exist integers $i, j \geq n$ such that $|x_{i,j} - 4| < \varepsilon$. Therefore, $\lim_{i,j\to\infty} x_{i,j}$ cannot exist, because it cannot equal both 0 and 4.

Solution 4.14 **(a)** Given $x \in \mathbb{R}$, it follows from the left side of (4.6) that

$$\lim_{i\to\infty} x/y_i = x \lim_{i\to\infty} 1/y_i = 0,$$

for any real sequence $\{y_i\}_{i=1}^{\infty}$ such that $\lim_{i\to\infty} y_i = \infty$; therefore, it is reasonable say that $x/\infty = 0$.

(b) We do not define ∞/∞ because the multiple limit $\lim_{i,j\to\infty} x_i/y_j$ need not have the same value for any two real sequences $\{x_i\}_{i=1}^{\infty}$ and $\{y_i\}_{i=1}^{\infty}$ diverging to ∞. Indeed, the limit might not even exist; for example, take $x_i = y_i = i$ (all i); then, $x_i/y_j = i/j$, which for i fixed approaches 0 as $j \to \infty$, whereas for j fixed diverges to ∞ as $i \to \infty$.

Solution 4.15 By (4.30),

$$\prod_{i=m}^{n} ax_i = \left(\prod_{i=m}^{n} a\right)\left(\prod_{i=m}^{n} x_i\right) = a^{n-m+1}\prod_{i=m}^{n} x_i,$$

since the product of the constant a has $n - m + 1$ terms. (Thus, in contrast to (4.28), we do *not* generally have $\prod_{i=m}^{n} ax_i = a\prod_{i=m}^{n} x_i$.)

Solution 4.16 **(a)** Using the continuity of absolute-value function, and (4.7b), we obtain

$$\left|\prod_{i=1}^{\infty} x_i\right| = \left|\lim_{n\to\infty}\prod_{i=1}^{n} x_i\right| = \lim_{n\to\infty}\left|\prod_{i=1}^{n} x_i\right| = \lim_{n\to\infty}\prod_{i=1}^{n}|x_i| \leq \lim_{n\to\infty}\prod_{i=1}^{n} c = \lim_{n\to\infty} c^n$$
$$= 0,$$

since $|c| < 1$; therefore, $\prod_{i=1}^{\infty} x_i = 0$. Alternatively, by (4.33b),

$$\prod_{i=1}^{\infty}|x_i| = \exp\left(\sum_{i=1}^{\infty}\ln|x_i|\right)$$

if the sum exists—here, allowing infinite limits. (The absolute value is included here because $\ln x_i$ is not defined for $x_i < 0$.) Since the natural logarithm function is increasing, we have $\ln|x_i| \leq \ln c < 0$ (all i); therefore,

$$\sum_{i=1}^{\infty}\ln|x_i| = \lim_{n\to\infty}\sum_{i=1}^{n}\ln|x_i| \leq \lim_{n\to\infty}\sum_{i=1}^{n}\ln c = \lim_{n\to\infty} n\ln c = -\infty,$$

making $\prod_{i=1}^{\infty}|x_i| = e^{-\infty} = 0$. So, from (4.32b) we have $\left|\prod_{i=1}^{\infty} x_i\right| = 0$, and again we obtain $\prod_{i=1}^{\infty} x_i = 0$.

(b) The divergence of the related sum above is a clue that the product $\prod_{i=1}^{\infty} x_i$ must diverge as well. Indeed, suppose that only finitely many of the terms x_i equal 0 (otherwise, the product necessarily diverges); then, changing these x_i to 1, we still have $\ln|x_i| \leq \ln c$ for i sufficiently large, and thus still obtain $\prod_{i=1}^{\infty} x_i = 0$. Therefore, the product $\prod_{i=1}^{\infty} x_i$ must diverge to 0.

Solution 4.17 Factoring the multiplicand yields

$$\prod_{i=2}^{\infty}\frac{i^2}{i^2 - 1} = \lim_{n\to\infty}\prod_{i=2}^{n}\frac{i}{i-1}\cdot\frac{i}{i+1},$$

where for each $n \geq 3$ we have

$$\begin{aligned}\prod_{i=2}^{n} \frac{i}{i-1} \cdot \frac{i}{i+1} &= \left(\frac{2}{1} \cdot \frac{2}{3}\right) \cdot \left(\frac{3}{2} \cdot \frac{3}{4}\right) \cdot \left(\frac{4}{3} \cdot \frac{4}{5}\right) \cdot \cdots \cdot \left(\frac{n}{n-1} \cdot \frac{n}{n+1}\right) \\ &= \frac{2}{1} \cdot \left[\left(\frac{2}{3} \cdot \frac{3}{2}\right)\left(\frac{3}{4} \cdot \frac{4}{3}\right) \cdots \left(\frac{n-1}{n} \cdot \frac{n}{n-1}\right)\right] \cdot \frac{n}{n+1} \\ &= 2 \cdot [(1)(1) \cdots (1)] \cdot \frac{n}{n+1} = 2\frac{n}{n+1}.\end{aligned}$$

(Here the regrouping of factors is allowed because the product is finite.) Hence,

$$\prod_{i=2}^{\infty} \frac{i^2}{i^2-1} = \lim_{n \to \infty} 2\frac{n}{n+1} = 2.$$

Solution 4.18 **(a)** Proceeding in steps, starting with the substitution of $i = j/2 + 5$, we obtain

$$\sum_{i=1}^{100} x_i = \sum_{j/2+5=1}^{100} x_{j/2+5} = \sum_{j/2=-4}^{95} x_{j/2+5} = \sum_{\substack{j=-8 \\ j \text{ even}}}^{190} x_{j/2+5}.$$

(b) Changing from i to j with $i = 2j + 5$ instead would be problematic because it is then impossible to obtain all $i \in \{1, 2, \ldots, 100\}$ with *integer* values j (e.g, $i = 6$ corresponds to $j = \frac{1}{2}$).

Solution 4.19 If $\sum_{i=-\infty}^{\infty} x_i$ exists, then for each integer l we have

$$\begin{aligned}\sum_{i=-\infty}^{\infty} x_{i+l} &= \lim_{m,n\to\infty} \sum_{i=-m}^{n} x_{i+l} = \lim_{m,n\to\infty} \sum_{j=-(m-l)}^{n+l} x_j = \lim_{m,n\to\infty} \sum_{j=-m}^{n} x_j \\ &= \sum_{j=-\infty}^{\infty} x_j,\end{aligned}$$

where the next-to-last equality holds because if m and n simultaneously approach ∞, then so do $m - l$ and $n + l$.

Solution 4.20 After introducing the indicator function

$$\begin{aligned}\mathbf{1}(i,j) &= \begin{cases} 1 & 2 \leq i \leq 5, -\infty < j \leq 9 - i \\ 0 & \text{otherwise} \end{cases} \\ &= \begin{cases} 1 & -\infty < j \leq 7, 2 \leq i \leq \min(5, 9-j) \\ 0 & \text{otherwise} \end{cases},\end{aligned}$$

we have

$$\begin{aligned}\sum_{i=2}^{5} \sum_{j=-\infty}^{9-i} x_{i,j} &= \sum_{i=-\infty}^{\infty} \sum_{j=-\infty}^{\infty} x_{i,j}\mathbf{1}(i,j) = \sum_{j=-\infty}^{\infty} \sum_{i=-\infty}^{\infty} x_{i,j}\mathbf{1}(i,j) \\ &= \sum_{j=-\infty}^{7} \sum_{i=2}^{\min(5,9-j)} x_{i,j}\end{aligned}$$

if the interchanging of infinite sums to obtain the second equality can be justified. Indeed, it can, because we could just as well perform this calculation with the sums over i being finite throughout:

$$\sum_{i=2}^{5}\sum_{j=-\infty}^{9-i} x_{i,j} = \sum_{i=2}^{5}\sum_{j=-\infty}^{\infty} x_{i,j}\mathbf{1}(i,j) = \sum_{j=-\infty}^{\infty}\sum_{i=2}^{5} x_{i,j}\mathbf{1}(i,j)$$
$$= \sum_{j=-\infty}^{7}\sum_{i=2}^{\min(5,9-j)} x_{i,j}$$

—the second equality being justified because the left side of (P4.20.1) is assumed to exist.

Alternatively, after introducing the indicator functions

$$\mathbf{1}_n(i,j) = \begin{cases} 1 & 2 \le i \le 5, n \le j \le 9-i \\ 0 & \text{otherwise} \end{cases}$$
$$= \begin{cases} 1 & n \le j \le 7, 2 \le i \le \min(5, 9-j) \\ 0 & \text{otherwise} \end{cases}$$

($n = 7, 6, 5, \ldots$), we have

$$\sum_{i=2}^{5}\sum_{j=-\infty}^{9-i} x_{i,j} = \sum_{i=2}^{5}\left(\lim_{n\to-\infty}\sum_{j=n}^{9-i} x_{i,j}\right) = \lim_{n\to-\infty}\sum_{i=2}^{5}\sum_{j=n}^{9-i} x_{i,j}$$
$$= \lim_{n\to-\infty}\sum_{i=-\infty}^{\infty}\sum_{j=-\infty}^{\infty} x_{i,j}\mathbf{1}_n(i,j) = \lim_{n\to-\infty}\sum_{j=-\infty}^{\infty}\sum_{i=-\infty}^{\infty} x_{i,j}\mathbf{1}_n(i,j)$$
$$= \lim_{n\to-\infty}\sum_{j=n}^{7}\sum_{i=2}^{\min(5,9-j)} x_{i,j} = \sum_{j=-\infty}^{7}\sum_{i=2}^{\min(5,9-j)} x_{i,j},$$

where interchanging the two infinite sums is allowed since for each n the summand $x_{i,j}\mathbf{1}_n(i,j)$ is nonzero for only finitely many pairs (i,j). Here, it is the interchanging of the sum over i and the limit versus n that is justified because the left side of (P4.20.1) is assumed to exist.

Solution 4.21 Taking, e.g., $x_{1,1} = x_{2,2} = 0$ and $x_{1,2} = x_{2,1} = 1$, we have

$$\sum_{i=1}^{2}\prod_{j=1}^{2} x_{i,j} = x_{1,1}x_{1,2} + x_{2,1}x_{2,2} = 0 + 0 = 0,$$

which does not equal

$$\prod_{j=1}^{2}\sum_{i=1}^{2} x_{i,j} = (x_{1,1} + x_{2,1})(x_{1,2} + x_{2,2}) = 1 \cdot 1 = 1.$$

Solution 4.22 **(a)** Having assumed that the sum on the left of (P4.22.1) converges, there exists a $Z \in \mathbb{C}$ such that

$$Z = \sum_{j=1}^{\infty} z_j = \lim_{n\to\infty}\sum_{j=1}^{n} z_j.$$

That is, for each $\varepsilon > 0$ there exists an $N \in \mathbb{Z}^+$ such that $\left|\sum_{j=1}^{n} z_j - Z\right| \le \varepsilon$ for all $n \ge N$. But,

$$\left|\sum_{j=1}^{n} z_j - Z\right| = \left|\left(\sum_{j=1}^{n} z_j - Z\right)^*\right| = \left|\sum_{j=1}^{n} z_j{}^* - Z^*\right|;$$

therefore, for each $\varepsilon > 0$, there exists an $N \in \mathbb{Z}^+$ such that $\left|\sum_{j=1}^{n} z_j{}^* - Z^*\right| \leq \varepsilon$ for all $n \geq N$. Hence,

$$\sum_{j=1}^{\infty} z_j{}^* = \lim_{n\to\infty} \sum_{j=1}^{n} z_j{}^* = Z^* = \left(\sum_{j=1}^{\infty} z_j\right)^*.$$

(b) By (3.36q), (P4.22.2), (4.29), and (4.28),

$$\begin{aligned}\operatorname{Re}\left(\sum_{j=1}^{\infty} z_j\right) &= \frac{1}{2}\left[\sum_{j=1}^{\infty} z_j + \left(\sum_{j=1}^{\infty} z_j\right)^*\right] = \frac{1}{2}\left(\sum_{j=1}^{\infty} z_j + \sum_{j=1}^{\infty} z_j{}^*\right) \\ &= \frac{1}{2}\sum_{j=1}^{\infty}(z_j + z_j{}^*) = \sum_{j=1}^{\infty}(z_j + z_j{}^*)/2 = \sum_{j=1}^{\infty} \operatorname{Re} z_j.\end{aligned}$$

Similarly, by (3.36r), (P4.22.2), (4.29), and (4.28),

$$\begin{aligned}\operatorname{Im}\left(\sum_{j=1}^{\infty} z_j\right) &= \frac{1}{i2}\left[\sum_{j=1}^{\infty} z_j - \left(\sum_{j=1}^{\infty} z_j\right)^*\right] = \frac{1}{i2}\left(\sum_{j=1}^{\infty} z_j - \sum_{j=1}^{\infty} z_j{}^*\right) \\ &= \frac{1}{i2}\sum_{j=1}^{\infty}(z_j - z_j{}^*) = \sum_{j=1}^{\infty}(z_j - z_j{}^*)/i2 = \sum_{j=1}^{\infty} \operatorname{Im} z_j.\end{aligned}$$

(c) By (P4.22.3),

$$\sum_{j=1}^{\infty} z_j = \operatorname{Re}\left(\sum_{j=1}^{\infty} z_j\right) + i\operatorname{Im}\left(\sum_{j=1}^{\infty} z_j\right) = \sum_{j=1}^{\infty} \operatorname{Re} z_j + i\sum_{j=1}^{\infty} \operatorname{Im} z_j,$$

giving (P4.22.1).

Solution 4.23 **(a)** If the sum $\sum_{i=1}^{\infty} x_i$ converges, then

$$\sum_{i=1}^{\infty} x_i = \sum_{i=1}^{m-1} x_i + \sum_{i=m}^{\infty} x_i \quad (m = 2, 3, \ldots), \tag{S4.23.1}$$

and thus

$$\begin{aligned}\lim_{m\to\infty} \sum_{i=m}^{\infty} x_i &= \lim_{m\to\infty}\left(\sum_{i=1}^{\infty} x_i - \sum_{i=1}^{m-1} x_i\right) = \lim_{m\to\infty} \sum_{i=1}^{\infty} x_i - \lim_{m\to\infty} \sum_{i=1}^{m-1} x_i \\ &= \sum_{i=1}^{\infty} x_i - \sum_{i=1}^{\infty} x_i = 0.\end{aligned}$$

Conversely, if $\lim_{m\to\infty} \sum_{i=m}^{\infty} x_i = 0$, then there must be an integer $m \geq 2$ for which the value of the sum $\sum_{i=m}^{\infty} x_i$ is finite; therefore, since the value of the finite sum $\sum_{i=1}^{m-1} x_i$ is certainly finite, using this integer m in (S4.23.1) shows that the value of the sum on the left side is also finite (i.e., the sum converges).

(b) By (4.33a),

$$\exp\left(\sum_{i=1}^{\infty} x_i\right) = \prod_{i=1}^{\infty} e^{x_i};$$

hence, the sum converges if and only if the product converges to a positive value. Also, by part (a), and again using (4.33a), the sum converges if and only if

$$1 = e^0 = \exp\left(\lim_{m\to\infty}\sum_{i=m}^{\infty} x_i\right) = \lim_{m\to\infty}\exp\left(\sum_{i=m}^{\infty} x_i\right) = \lim_{m\to\infty}\prod_{i=m}^{\infty} e^{x_i}$$

—having used the continuity of the exponential function to interchange it and a limit. Therefore—as seen by letting $x_i = \ln y_i$, and thus $y_i = e^{x_i}$, for each i—we obtain the corresponding fact for infinite products: Given a positive sequence $\{y_i\}_{i=1}^{\infty}$, the infinite product $\prod_{i=1}^{\infty} y_i$ converges to a positive value if and only if $\lim_{m\to\infty}\prod_{i=m}^{\infty} y_i = 1$.

Solution 4.24 Let $x_i = e^{1/i}$ ($i = 1, 2, \dots$). Then, $x_i > 0$ (all i); therefore, by the right side of (4.33b) and the divergence of the harmonic series,

$$\prod_{i=1}^{\infty} x_i = \exp\left(\sum_{i=1}^{\infty} \ln x_i\right) = \exp\left(\sum_{i=1}^{\infty} 1/i\right) = e^{\infty} = \infty.$$

Yet, $x_i \to 1$ as $i \to \infty$, since $1/i \to 0$, $e^0 = 1$, and the exponential function is continuous.

Solution 4.25 The first series is the harmonic series ($1 + \frac{1}{2} + \frac{1}{3} + \cdots = \infty$) minus its first term; therefore, $x = \infty - 1 = \infty$. The second series is the harmonic series divided by 2; therefore, $y = \infty/2 = \infty$. The third series, though, is a geometric series

$$z = \left(\tfrac{1}{2}\right)^1 + \left(\tfrac{1}{2}\right)^2 + \left(\tfrac{1}{2}\right)^3 + \cdots = \tfrac{1}{2}\left[1 + \tfrac{1}{2} + \left(\tfrac{1}{2}\right)^2 + \cdots\right] = \tfrac{1}{2}\frac{1}{1 - \frac{1}{2}} = 1,$$

where we used (4.38) with $c = \frac{1}{2}$.

Solution 4.26 **(a)** After 50 weeks, the worker will earn

$$\begin{aligned} 100 + (100 + 1) + (100 + 2) + \cdots + (100 + 49) &= \sum_{i=100}^{149} i \\ &= \tfrac{1}{2}[(149)(150) - (99)(100)] = 6225 \text{ dollars,} \end{aligned}$$

where we have used (4.36) with $m = 100$ and $n = 149$.

(b) Here, after 50 weeks, the worker will earn

$$\begin{aligned} 100 + 100 \cdot 1.01 + 100 \cdot 1.01^2 + \cdots + 100 \cdot 1.01^{49} &= 100\sum_{i=0}^{49} 1.01^i \\ &= 100\frac{1.01^{50} - 1}{1.01 - 1} \approx 6446 \text{ dollars,} \end{aligned}$$

where we have used (4.37) with $c = 1.01$, $m = 0$, and $n = 49$.

Solution 4.27 From (4.38) with $c = 10^{-4}$ we obtain

$$\begin{aligned} 12345\overline{6.789}\ldots &= 123450 + (6.789 + 0.000\,678\,9 + 0.000\,000\,067\,89 + \cdots) \\ &= 123450 + 6.789(1 + 10^{-4} + 10^{-8} + \cdots) \\ &= 123450 + 6.789\frac{1}{1 - 10^{-4}} = 123450 + \frac{6.789}{0.9999}, \end{aligned}$$

as stated in (3.9).

Solution 4.28 **(a)** For each $n \geq 1$,

$$\sum_{i=1}^{n} i \cdot i! = \sum_{i=1}^{n} [(i+1) - 1]i! = \sum_{i=1}^{n} [(i+1)! - i!],$$

which by (4.34a) telescopes to $(n+1)! - 1! = (n+1)! - 1$.

(b) When $n = 0$, the lower limit of the given sum exceeds its upper limit by 1; therefore, upon invoking either convention (4.41) or convention (4.43a), we find that the sum evaluates to 0, which equals $(n+1)! - 1$.

Solution 4.29 **(a)** Starting with the second term on right side of (P4.29.1), we obtain

$$\begin{aligned}
\sum_{i=m+1}^{n} (x_i - x_{i-1}) \sum_{j=m}^{i-1} y_j &= \sum_{i=m+1}^{n} x_i \sum_{j=m}^{i-1} y_j - \sum_{i=m+1}^{n} x_{i-1} \sum_{j=m}^{i-1} y_j \\
&= \sum_{i=m+1}^{n} x_i \left(\sum_{j=m}^{i} y_j - y_i \right) - \sum_{i=m}^{n-1} x_i \sum_{j=m}^{i} y_j \\
&= \left(\sum_{i=m+1}^{n} x_i \sum_{j=m}^{i} y_j - \sum_{i=m}^{n-1} x_i \sum_{j=m}^{i} y_j \right) - \sum_{i=m+1}^{n} x_i y_i \\
&= \left(x_n \sum_{j=m}^{n} y_j - x_m \sum_{j=m}^{m} y_j \right) - \sum_{i=m+1}^{n} x_i y_i \\
&= x_n \sum_{j=m}^{n} y_j - x_m y_m - \sum_{i=m+1}^{n} x_i y_i \\
&= x_n \sum_{j=m}^{n} y_j - \sum_{i=m}^{n} x_i y_i \quad (m, n \in \mathbb{Z} \text{ with } m \leq n),
\end{aligned}$$

which yields (P4.29.1). It should be noted that the cases above in which the lower limit of a sum exceeds its upper limit are handled automatically by invoking convention (4.43a). (As it happens, for each sum the lower limit at most equals the upper limit plus 1, in which case the sum equals 0.)

(b) Let m, n, and c be as stated; and, let $x_i = i$ and $y_i = c^i$ (all i). Then, using (P4.29.1) and (4.37) yields

$$\begin{aligned}
\sum_{i=m}^{n} ic^i &= n \sum_{j=m}^{n} c^j - \sum_{i=m+1}^{n} [i - (i-1)] \sum_{j=m}^{i-1} c^j \\
&= n \frac{c^{n+1} - c^m}{c-1} - \sum_{i=m+1}^{n} 1 \cdot \frac{c^i - c^m}{c-1} \\
&= n \frac{c^{n+1} - c^m}{c-1} - \frac{1}{c-1} \sum_{i=m+1}^{n} c^i + \sum_{i=m+1}^{n} \frac{c^m}{c-1} \\
&= n \frac{c^{n+1} - c^m}{c-1} - \frac{c^{n+1} - c^{m+1}}{(c-1)^2} + (n-m) \frac{c^m}{c-1} \\
&= \frac{nc^{n+1} - mc^m}{c-1} + \frac{c^{m+1} - c^{n+1}}{(c-1)^2}
\end{aligned}$$

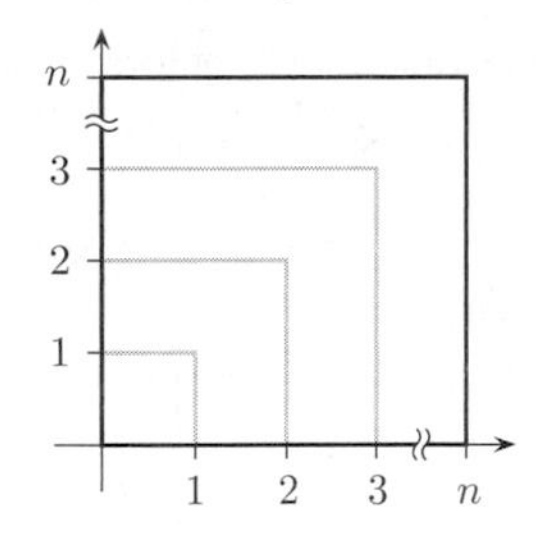

FIGURE S4.30.1 A Geometric Interpretation of (P4.30.1)

$$= \frac{nc^{n+1}(c-1) - mc^m(c-1)}{(c-1)^2} + \frac{c^{m+1} - c^{n+1}}{(c-1)^2}$$

$$= \frac{c^{n+1}[n(c-1)-1] - c^m[m(c-1)-c]}{(c-1)^2} \quad (c \neq 1) \tag{S4.29.1a}$$

(with $c \neq 0$ for $m < 0$); whereas (4.36) directly yields

$$\sum_{i=m}^{n} ic^i = \tfrac{1}{2}[n(n+1) - (m-1)m] \quad (c = 1). \tag{S4.29.1b}$$

The same comment about empty sums applies to this derivation as well. (Indeed, note that both parts of (S4.29.1) also hold when $m = n + 1$.)

Solution 4.30 **(a)** For $n = 1$, we have $2n - 1 = 1$, and both sides of (P4.30.1) equal 1. Now, assume (induction hypothesis) that the equation in (P4.30.1) holds for some particular integer $n \geq 1$; then,

$$1 + 3 + 5 + \cdots + [2(n+1) - 1] = [1 + 3 + 5 + \cdots + (2n-1)] + [2(n+1) - 1]$$
$$= n^2 + 2n + 1 = (n+1)^2,$$

and thus the equation in (P4.30.1) holds upon replacing each "n" with "$n + 1$". Therefore, by mathematical induction, the equation in (P4.30.1) holds for all $n \geq 1$.

(b) As shown in Figure S4.30.1, an $n \times n$ square can be partitioned into n square and L-shaped regions having areas 1, 3, 5, ..., $(2n - 1)$; therefore, since the area of the square is n^2, we have (P4.30.1).

Solution 4.31 For $n = 1$, we have $6 + 8^n = 14$, which is divisible by 7. Now, assume (induction hypothesis) that $6 + 8^n$ is divisible by 7 for some particular integer $n \geq 1$; that is, $6 + 8^n = 7k$ for some integer k. Then,

$$6 + 8^{n+1} = 6 + 8^n + (8^{n+1} - 8^n) = 7k + (8-1)8^n = 7(k + 8^n),$$

where $k + 8^n$ must be an integer; so, $6 + 8^{n+1}$ is also divisible by 7. Therefore, by mathematical induction, $6 + 8^n$ is divisible by 7 for all integers $n \geq 1$.

Solution 4.32 First, (P4.32.1) yields

$$f_1 = \frac{1}{\sqrt{5}}\left[\left(\frac{1+\sqrt{5}}{2}\right) - \left(\frac{1-\sqrt{5}}{2}\right)\right] = 1,$$

$$f_2 = \frac{1}{\sqrt{5}}\left[\left(\frac{1+\sqrt{5}}{2}\right)^2 - \left(\frac{1-\sqrt{5}}{2}\right)^2\right] = \frac{1}{\sqrt{5}}\left(\frac{6+2\sqrt{5}}{4} - \frac{6-2\sqrt{5}}{4}\right) = 1,$$

in agreement with (4.49a). Now, assume (induction hypothesis) that for a particular integer $N \geq 2$, the equation in (P4.32.1) holds for all $n \in \{1, 2, \ldots, N\}$ (as we now know is the case for $N = 2$); then, since $N + 1 \geq 3$, we can use (4.49b) to obtain

$$
\begin{aligned}
f_{N+1} &= f_N + f_{N-1} \\
&= \frac{1}{\sqrt{5}}\left[\left(\frac{1+\sqrt{5}}{2}\right)^N - \left(\frac{1-\sqrt{5}}{2}\right)^N\right] \\
&\quad + \frac{1}{\sqrt{5}}\left[\left(\frac{1+\sqrt{5}}{2}\right)^{N-1} - \left(\frac{1-\sqrt{5}}{2}\right)^{N-1}\right] \\
&= \frac{1}{\sqrt{5}}\left[\frac{2}{1+\sqrt{5}}\left(\frac{1+\sqrt{5}}{2}\right)^{N+1} - \frac{2}{1-\sqrt{5}}\left(\frac{1-\sqrt{5}}{2}\right)^{N+1}\right] \\
&\quad + \frac{1}{\sqrt{5}}\left[\left(\frac{2}{1+\sqrt{5}}\right)^2\left(\frac{1+\sqrt{5}}{2}\right)^{N+1} - \left(\frac{2}{1-\sqrt{5}}\right)^2\left(\frac{1-\sqrt{5}}{2}\right)^{N+1}\right] \\
&= \frac{1}{\sqrt{5}}\left[\frac{2+2\sqrt{5}}{(1+\sqrt{5})^2}\left(\frac{1+\sqrt{5}}{2}\right)^{N+1} - \frac{2-2\sqrt{5}}{(1-\sqrt{5})^2}\left(\frac{1-\sqrt{5}}{2}\right)^{N+1}\right] \\
&\quad + \frac{1}{\sqrt{5}}\left[\frac{4}{(1+\sqrt{5})^2}\left(\frac{1+\sqrt{5}}{2}\right)^{N+1} - \frac{4}{(1-\sqrt{5})^2}\left(\frac{1-\sqrt{5}}{2}\right)^{N+1}\right] \\
&= \frac{1}{\sqrt{5}}\left[\frac{6+2\sqrt{5}}{(1+\sqrt{5})^2}\left(\frac{1+\sqrt{5}}{2}\right)^{N+1} - \frac{6-2\sqrt{5}}{(1-\sqrt{5})^2}\left(\frac{1-\sqrt{5}}{2}\right)^{N+1}\right] \\
&= \frac{1}{\sqrt{5}}\left[\left(\frac{1+\sqrt{5}}{2}\right)^{N+1} - \left(\frac{1-\sqrt{5}}{2}\right)^{N+1}\right],
\end{aligned}
$$

and thus the equation in (P4.32.1) also holds for $n = N + 1$. Therefore, by mathematical induction (using the strong induction principle), the inequality in (P4.32.1) holds for all integers $n \geq 1$.

Chapter 5. Functions

Solution 5.1 **(a)** The range of a constant function consists of a single element (and, a function having such a range must be constant).

(b) The range of an identity function is the same as its domain (which is also the case for some other functions).

Solution 5.2 **(a)** Let, e.g., $f(\cdot)$ be the function with domain $D \in \{1, 2, 3\}$ given by

$$
f(x) = \begin{cases} 2 & x = 1 \\ 3 & x = 2, 3 \end{cases};
$$

then, $f[f(x)] = 3$ $(x \in D)$.

(b) Let, e.g., $f(\cdot)$ be the function with domain $D \in \{1, 2\}$ given by

$$f(x) = \begin{cases} 2 & x = 1 \\ 1 & x = 2 \end{cases};$$

then, $f[f(x)] = x$ $(x \in D)$.

Solution 5.3 $h^{-1}(\cdot)$ is the composite function $f^{-1}[g^{-1}(\cdot)]$ (notice the order reversal of "f" and "g").

Solution 5.4 **(a)** This statement must be true; because: if $y \in f(A_1)$, then there exists an $x \in A_1$ such that $f(x) = y$; but then, since $A_1 \subseteq A_2$, we also have $x \in A_2$ with $f(x) = y$, and thus $y \in f(A_2)$. Therefore, $f(A_1) \subseteq f(A_2)$.

(b) This statement must be true; because: if $x \in f^{-1}(B_1)$, then $f(x) \in B_1$; but then, since $B_1 \subseteq B_2$, we also have $f(x) \in B_2$, and thus $x \in f^{-1}(B_2)$. Therefore, $f^{-1}(B_1) \subseteq f^{-1}(B_2)$.

(c) This statement need not be true; for example: if $D = \{1, 2\}, f(x) = 0$ $(x \in D)$, and $A = \{1\} \subseteq D$, then $f^{-1}[f(A)] = f^{-1}[\{0\}] = D \neq A$. (*Aside*: It must be true, though, that $A \subseteq f^{-1}[f(A)]$; because, if $x \in A \subseteq D$, then $f(x) \in f(A)$, and thus $x \in f^{-1}[f(A)]$. Moreover, if $f(\cdot)$ is invertible, then $f^{-1}[f(A)] \subseteq A$ as well. For if $x \in f^{-1}[f(A)]$, then $f(x) \in f(A)$, from which we can now conclude that $x \in A$; because, $A \subseteq D$ and, since $f(\cdot)$ is invertible, there is only one $x' \in D$ for which $f(x') = f(x)$. Therefore, when $f(\cdot)$ is invertible, we *do* have $f^{-1}[f(A)] = A$ for all $A \subseteq D$.)

(d) This statement must be true; because: If $y \in B \subseteq E$, then there exists an $x \in f^{-1}(B)$ such that $f(x) = y$, and thus $y \in f[f^{-1}(B)]$; so $B \subseteq f[f^{-1}(B)]$. Conversely, if $y \in f[f^{-1}(B)]$, then there exists an $x \in f^{-1}(B)$ such that $f(x) = y$, and thus $y \in B$; so $f[f^{-1}(B)] \subseteq B$. Therefore, $f[f^{-1}(B)] = B$.

Solution 5.5 No, because S might contain points outside the range E of $f(\cdot)$. (However, it *is* true that $f^{-1}: E \to D$.)

Solution 5.6 **(a)** For all $x_1, x_2 \in \mathbb{R}$ such that $x_1 < x_2$, we have $f(x_1) = x_1 < x_2 = f(x_2)$, and thus $f(x_1) < f(x_2)$; therefore, $f(\cdot)$ is strictly increasing.

(b) For all $x_1, x_2 > 0$ such that $x_1 < x_2$, dividing both sides of this inequality by $x_1x_2 > 0$ yields $1/x_2 < 1/x_1$; hence, $f(x_1) = 1/x_1 > 1/x_2 = f(x_2)$, and thus $f(x_1) > f(x_2)$. Therefore, $f(\cdot)$ is strictly decreasing.

Solution 5.7 **(a)** A real sequence $\{x_i\}_{i=1}^{\infty}$ is both strictly increasing and strictly decreasing if and only if we have both $x_i < x_{i+1}$ and $x_i > x_{i+1}$ for all i, which is impossible.

(b) A real function $f(\cdot)$ with domain D is both strictly increasing and strictly decreasing if and only if we have both $f(x_1) > f(x_2)$ and $f(x_1) < f(x_2)$ for all $x_1, x_2 \in D$ such that $x_1 < x_2$. This is the case if and only if D has exactly one element; for then it is impossible to have $x_1, x_2 \in D$ such that $x_1 < x_2$, and thus the conditions for being strictly increasing and strictly decreasing each hold vacuously.

Solution 5.8 **(a)** If $\max_{x \in D} f(x)$ exists, then there is an $X \in D$ such that $f(x) \leq f(X)$ $(x \in D)$, and thus $\max_{x \in D} f(x) = f(X)$. Therefore, $-f(x) \geq -f(X)$ $(x \in D)$, making $\min_{x \in D} -f(x) = -f(X) = -\max_{x \in D} f(x)$. A similar argument shows that the same result obtains if we assume instead that $\min_{x \in D} f(x)$ exists.

(b) Always. For let $Y = \sup_{x \in D} f(x)$; then, $f(x) \leq Y$ $(x \in D)$, and for no $y \in \overline{\mathbb{R}}$ do we have $f(x) \leq y < Y$ $(x \in D)$. Hence, $-f(x) \geq -Y$ $(x \in D)$, and for no $y \in \overline{\mathbb{R}}$ do we have $-f(x) \geq -y > -Y$ $(x \in D)$; moreover, since $\{-y \mid y \in \overline{\mathbb{R}}\} = \overline{\mathbb{R}}$, the latter condition is equivalent to saying that for no $y \in \overline{\mathbb{R}}$ do we have $-f(x) \geq y > -Y$ $(x \in D)$. Therefore, $\inf_{x \in D} -f(x) = -Y = -\sup_{x \in D} f(x)$.

(c) Let, e.g., $D = \{-1, 1\}, f(x) = x$ $(x \in D)$, and $g(x) = -x$ $(x \in D)$. Then, $f(x) + g(x) = 0$ $(x \in D)$, and therefore

$$\max_{x \in D}[f(x) + g(x)] = 0,$$

whereas

$$\max_{x \in D} f(x) + \max_{x \in D} g(x) = 1 + 1 = 2.$$

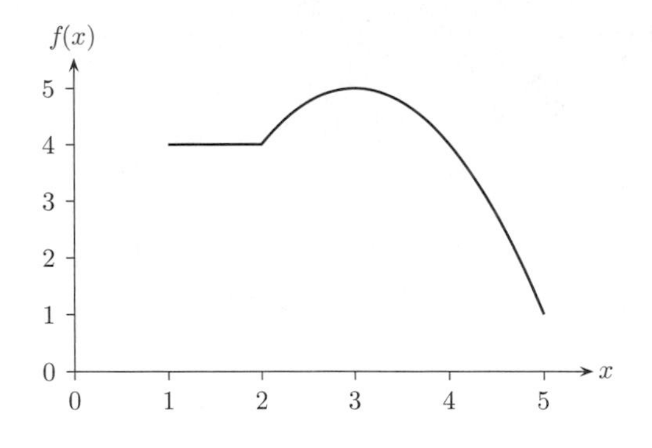

FIGURE S5.10.1 A Plot of (P5.10.1)

Solution 5.9 Since $\cos x$ ($x \in \mathbb{R}$) attains its maximum (respectively, minimum) value of 1 (-1) if and only if $x = n\pi$ for some even (odd) integer n, we have

$$\operatorname*{argmax}_{x\in\mathbb{R}} \cos x = \{0, \pm 2\pi, \pm 4\pi, \pm 6\pi, \dots\},$$

$$\operatorname*{argmin}_{x\in\mathbb{R}} \cos x = \{\pm\pi, \pm 3\pi, \pm 5\pi, \dots\};$$

whereas

$$\operatorname*{argmax}_{x\in\mathbb{R}} e^x = \operatorname*{argmin}_{x\in\mathbb{R}} e^x = \emptyset,$$

since neither $\max_{x\in\mathbb{R}} e^x$ nor $\min_{x\in\mathbb{R}} e^x$ exists.

Solution 5.10 **(a)** See Figure S5.10.1.

(b) It is clear from (P5.10.1) that $f(\cdot)$ is continuous at each point $x \neq 2$ in its domain $[1, 5)$; and, since $4 = -x^2 + 6x - 4$ for $x = 2$, it is also continuous at this point. Therefore, $f(\cdot)$ is continuous.

(c) From Figure S5.10.1 we see that $f(\cdot)$ has *both* a local maximum and a local minimum of 4 at each point in $[1, 2)$; in addition, it has a local minimum (but not a local maximum) of 4 at 2. It also appears that $f(\cdot)$ has a local maximum of 5 at 3; indeed, this is so since for all $x \in [2, 5)$ we have $-x^2 + 6x - 4 = 5 - (x - 3)^2 \leq 5$, with equality obtaining only for $x = 3$. Moreover, since $f(x) = 4 < 5$ for $x \in [1, 2)$, this local maximum is a global maximum. There are no other extrema of $f(\cdot)$; in particular, it has no global minimum (for note that 5 is not a point in the domain $[1, 5)$ of the function).

Solution 5.11 **(a)** See Figure S5.11.1. (Observe how solid and open dots are used to clearly indicate the function values at jump discontinuities.)

(b) Given $x \in \mathbb{R}$, there can be no integer k such that $-\lceil x\rceil < k \leq -x$; for then we would have an integer $-k$ such that $\lceil x\rceil > -k \geq x$, contradicting the fact that $\lceil x\rceil$ is the smallest integer $n \geq x$. Also, the same fact implies that $-\lceil x\rceil$ is an integer and $-\lceil x\rceil \leq -x$. Therefore, $-\lceil x\rceil$ is the largest integer $n \leq -x$, which by definition is $\lfloor -x\rfloor$; so, we have (P5.11.1a).

Since $\lfloor x\rfloor$ is the largest integer $n \leq x$, we must have $\lfloor x\rfloor \leq x$; but, we cannot have $\lfloor x\rfloor \leq x - 1$, for then $\lfloor x\rfloor + 1$ would be an integer $n > \lfloor x\rfloor$ such that $n \leq x$. Hence, $x - 1 < \lfloor x\rfloor \leq x$ for all $x \in \mathbb{R}$. It follows that $-x - 1 < \lfloor -x\rfloor \leq -x$ for all $x \in \mathbb{R}$, which upon changing signs throughout yields $x + 1 > \lceil x\rceil \geq x$, by (P5.11.1a). Thus, we have (P5.11.1b).

Now, let $m \in \mathbb{Z}$ be given too. Since $\lfloor x+m\rfloor$ is the largest integer $n \leq x+m$, and (P5.11.1b) yields $\lfloor x\rfloor + m \leq x + m$ where $\lfloor x\rfloor + m$ is also an integer, we must have $\lfloor x + m\rfloor \geq \lfloor x\rfloor + m$. Likewise, since $\lfloor x\rfloor$ is the largest integer $n \leq x$, and (P5.11.1b) yields $\lfloor x+m\rfloor - m \leq (x+m) - m = x$ where $\lfloor x + m\rfloor - m$ is also an integer, we must have $\lfloor x\rfloor \geq \lfloor x + m\rfloor - m$, and thus $\lfloor x + m\rfloor \leq \lfloor x\rfloor + m$. Hence, $\lfloor x + m\rfloor = \lfloor x\rfloor + m$. Therefore, since $-m \in \mathbb{Z}$, we have $\lfloor x - m\rfloor = \lfloor x + (-m)\rfloor =$

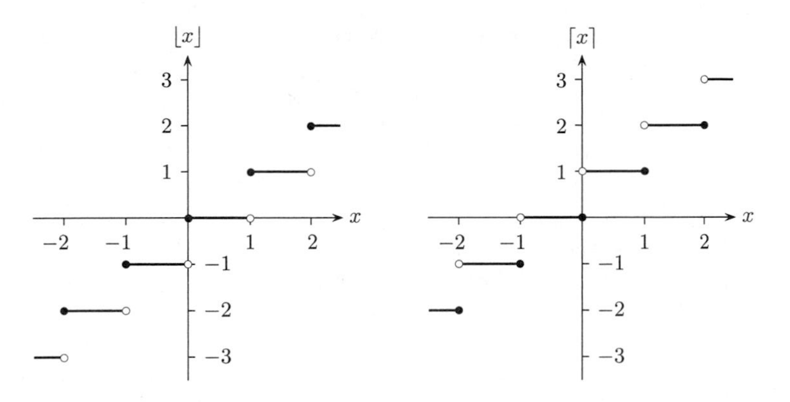

FIGURE S5.11.1 The Floor and Ceiling Functions

$\lfloor x \rfloor + (-m) = \lfloor x \rfloor - m$ as well, and the left side of (P5.11.1c) is proved. Finally, using (P5.11.1a) we obtain $\lceil x \pm m \rceil = -\lfloor -(x \pm m) \rfloor = -\lfloor -x \mp m \rfloor = -(\lfloor -x \rfloor \mp m) = -\lfloor -x \rfloor \pm m = \lceil x \rceil \pm m$ ($\pm$ and $\mp$ respectively), completing the proof of (P5.11.1c).

(c) Let $x \in \mathbb{R}$ be given. Since $x-(x-1) = 1$, there is exactly one integer m for which $x-1 < m \leq x$, which by the lower two inequalities in (P5.11.1b) is $\lfloor x \rfloor$. Similarly, since $(x + 1) - x = 1$, there is exactly one integer m for which $x \leq m < x + 1$, which by the upper two inequalities in (P5.11.1b) is $\lceil x \rceil$.

Solution 5.12 **(a)** It follows from (3.10) that for the fractionless base-b expansion (3.5) of a positive integer x, we have

$$x = \sum_{i=0}^{n} x_i \cdot b^i,$$

where $x_i \in \{0, 1, \ldots, b-1\}$ is the value of digit $\boldsymbol{x}_i$ in the expansion ($i = 0, 1, \ldots, n \geq 0$), and $x_n \neq 0$. The expansion has $n + 1$ digits; and, x can be as small as b^n (when $x_n = 1$ and all other $x_i = 0$) or as large as $b^{n+1} - 1$ (when all $x_i = b - 1$). Hence, if m is the number of digits in the fractionless base-b expansion of a positive integer x, then

$$b^{m-1} \leq x \leq b^m - 1. \tag{S5.12.1}$$

Hence, $x < b^m = b^{m-1}b \leq xb$, making

$$\log_b x < m \leq \log_b x + 1;$$

therefore, by part (c) of Problem 5.11 we have $m = \lfloor \log_b x + 1 \rfloor$, which by (P5.11.1c) equals $\lfloor \log_b x \rfloor + 1$. Alternatively, from (S5.12.1) we obtain $x + 1 \leq b^m \leq xb < (x + 1)b$, making

$$\log_b(x + 1) \leq m < \log_b(x + 1) + 1;$$

so, again by part (c) of Problem 5.11, we also have $m = \lceil \log_b(x + 1) \rceil$.

(b) Let q be the integer quotient of x divided by y. Then, by (3.21), we have $x/y = q + r/y$, where $0 \leq r/y < 1$ (since here $y > 0$); therefore, $q = \lfloor x/y \rfloor$. However, we need not have $q = \lceil x/y \rceil$, as seen by taking $x = 3$ and $y = 2$, for which $q = 1$ and $\lceil x/y \rceil = 2$. (*Aside*: It now follows from (3.22) that $x \bmod y = x - \lfloor x/y \rfloor y$ ($x, y \in \mathbb{Z}^+$).)

(c) By (P5.11.1b),

$$\lfloor \alpha + 1 \rfloor - 1 \leq \alpha < \lfloor \alpha + 1 \rfloor, \quad \lceil \beta - 1 \rceil < \beta \leq \lceil \beta - 1 \rceil + 1$$

($\alpha, \beta \in \mathbb{R}$); therefore, also using (P5.11.1c), we obtain

$$\sum_{\alpha<i<\beta} x_i = \sum_{i=\lfloor\alpha+1\rfloor}^{\lceil\beta-1\rceil} x_i = \sum_{i=\lfloor\alpha\rfloor+1}^{\lceil\beta\rceil-1} x_i,$$

where the two rightmost sums are empty if $\lfloor\alpha+1\rfloor > \lceil\beta-1\rceil$ (equivalently, $\lfloor\alpha\rfloor+1 > \lceil\beta\rceil - 1$).

Solution 5.13 **(a)** The range of $f^{-1}(\cdot)$ is the domain $\mathbb{R}$ of $f(\cdot)$; therefore, the function $f^{-1}(\cdot)$ cannot be bounded, since it takes on every possible real value.

(b) Let, e.g., $f(x) = \tan^{-1} x$ ($x \in \mathbb{R}$), which is plotted in Figure 11.11. This function is bounded since $|f(x)| < \pi/2$ (all x); however, being strictly increasing, it is invertible but has no local extrema.

Solution 5.14 **(a)** If $f(\cdot)$ is a constant function, then every positive number X satisfies (5.3), so every positive number is a period. But, none of these periods is smaller than all the others; therefore, "the" period of this function $f(\cdot)$ does not exist.

(b) Let $f(\cdot)$ be the function in (5.25), which has domain $D = \mathbb{R}$. If $X > 0$ is rational, then (per Problem 3.1) $x+X$ is rational (respectively, irrational) if and only if $x \in \mathbb{R}$ is rational (irrational); so, (5.3) is satisfied. However, if $X > 0$ is irrational, then $x + X$ is is irrational when $x \in \mathbb{R}$ is rational; so, (5.3) is not satisfied. Therefore, $f(\cdot)$ is periodic, with every rational number being a possible period. However, "the" period of this function does not exist, since none of its periods is smaller than all the others.

Solution 5.15 **(a)** No; because, if $f(\cdot)$ is odd, then by (5.4b) we have $f(0) = -f(-0) = -f(0)$, which implies that $f(0) = 0$.

(b) The one and only such function is that which is identically 0; because: If $f(\cdot)$ is both even and odd, then by (5.4) we have $f(x) = f(-x) = -f(x)$ (all x), making $f(x) = 0$ (all x); conversely, this function is clearly both even and odd.

Solution 5.16 **(a)** Suppose $f(\cdot)$ is even. Then, $f(1) = f(-1)$; that is, $f(\cdot)$ maps both 1 and -1 to the same value. So, the function $f(\cdot)$ is not one-to-one, and therefore not invertible.

(b) Suppose $f(\cdot)$ is odd and invertible; and, let this function have range E, which is also the domain of $f^{-1}(\cdot)$. Given $y \in E$, there exists an $x \in \mathbb{R}$ such that $y = f(x)$; so, since $f(\cdot)$ is odd, we also have $-y = f(-x)$, and thus $-y \in E$ as well. Furthermore,

$$-y = -f[f^{-1}(y)] = f[-f^{-1}(y)],$$

where the latter equality obtains from the odd symmetry of $f(\cdot)$; therefore,

$$f^{-1}(-y) = f^{-1}\{f[-f^{-1}(y)]\} = -f^{-1}(y),$$

which shows that $f^{-1}(\cdot)$ is odd.

Solution 5.17 **(a)** Since for all $x \in \mathbb{R}$ we have

$$f_e(-x) = \tfrac{1}{2}[f(-x) + f(x)] = \tfrac{1}{2}[f(x) + f(-x)] = f_e(x),$$
$$f_o(-x) = \tfrac{1}{2}[f(-x) - f(x)] = -\tfrac{1}{2}[f(x) - f(-x)] = -f_o(-x),$$

the function $f_e(\cdot)$ is even and the function $f_o(\cdot)$ is odd.

(b) Substituting (P5.17.2) into (P5.17.1b), and using the assumed symmetries of $f_e'(\cdot)$ and $f_o'(\cdot)$, we obtain

$$\begin{aligned} f_e(x) &= \tfrac{1}{2}\{[f_e'(x) + f_o'(x)] + [f_e'(-x) + f_o'(-x)]\} \\ &= \tfrac{1}{2}\{[f_e'(x) + f_o'(x)] + [f_e'(x) - f_o'(x)]\} = f_e'(x), \\ f_o(x) &= \tfrac{1}{2}\{[f_e'(x) + f_o'(x)] - [f_e'(-x) + f_o'(-x)]\} \\ &= \tfrac{1}{2}\{[f_e'(x) + f_o'(x)] - [f_e'(x) - f_o'(x)]\} = f_o'(x) \end{aligned}$$

($x \in \mathbb{R}$); therefore, the decomposition (P5.17.1a) of $f(\cdot)$ into even and odd functions is unique.

(c) By (P5.17.1b) and (11.33), the even component $f_e(\cdot)$ of $f(x) = e^x$ ($x \in \mathbb{R}$) is the hyperbolic cosine function $\cosh x$ ($x \in \mathbb{R}$), and its odd component $f_o(\cdot)$ is the hyperbolic sine function $\sinh x$ ($x \in \mathbb{R}$), both of which are plotted in Figure 11.13.

Solution 5.18 Based on equation (5.6a) for a function $f(\cdot)$ that plots to a line, we wish to find values m and Y such that $b = f(a) = ma + Y$ and $a = f(b) = mb + Y$. It readily follows that these constraints are satisfied if and only if $m = -1$ and $Y = a + b$. Hence, the desired function is $f(x) = a + b - x$ (all x).

Solution 5.19 Yes. Because, if $y \propto x$, then there exists a constant $m > 0$ such that $y = mx$ for all possible pairs (x, y); therefore, taking $m' = 1/m > 0$, we also have $x = m'y$ for all possible pairs (x, y).

Solution 5.20 **(a)** Using (5.10) with $\kappa > 0$ fixed, we now have $\kappa(\gamma - x_1) = d(x_1, \gamma) = d(\gamma, x_2) = \kappa(x_2 - \gamma)$, making $\gamma - x_1 = x_2 - \gamma$, from which it follows that $\gamma = (x_1 + x_2)/2$.

(b) Using (5.11) with $\kappa > 0$ and $b > 1$ fixed, we now have $\kappa \log_b(\gamma/x_1) = d(x_1, \gamma) = d(\gamma, x_2) = \kappa \log_b(x_2/\gamma)$, making $\gamma/x_1 = x_2/\gamma$, from which it follows that $\gamma = \sqrt{x_1 x_2}$.

Solution 5.21 In each part below, the solution follows from (5.6a). Also, although natural logarithms are shown, any logarithm base greater than 1 could be used.

(a) Here, $f(x) = a \ln x + b$ ($x > 0$) for some constants $a, b \in \mathbb{R}$.

(b) Here, $\ln f(x) = ax + b$ ($x \in \mathbb{R}$) for some constants $a, b \in \mathbb{R}$ (thus, $f(\cdot)$ must be a positive function). For each x, we then have $f(x) = e^{ax+b} = e^b(e^a)^x$; so, an equivalent condition is that $f(x) = AB^x$ ($x \in \mathbb{R}$) for some constants $A, B > 0$.

(c) Here, $\ln f(x) = a \ln x + b$ ($x > 0$) for some constants $a, b \in \mathbb{R}$ (thus, $f(\cdot)$ must be a positive function). For each x, we then have $f(x) = e^{a \ln x + b} = e^b(e^{\ln x})^a = e^b x^a$; so, an equivalent condition is that $f(x) = Ax^B$ ($x > 0$) for some constants $A > 0$ and $B \in \mathbb{R}$.

Solution 5.22 Let f_1 and f_2 be two note frequencies, the latter being a semitone higher. Then, since a semitone on a logarithmic scale is one-twelfth of an octave, it follows from (5.11) that

$$\kappa \log_b(f_2/f_1) = d(f_1, f_2) = \tfrac{1}{12} d(f_1, 2f_1) = \tfrac{1}{12} \kappa \log_b[(2f_1)/f_1],$$

where $\kappa > 0$ and $b > 1$ determine how frequencies are logarithmically mapped into positions on the real line. Therefore, $\log_b(f_2/f_1) = \log_b 2^{1/12}$, making $f_2/f_1 = 2^{1/12} \approx 1.0595$. So, expressed as a percentage, incrementation by one semitone corresponds to a frequency increase of $2^{1/12} - 1 \approx 0.0595 = 5.95\%$.

Solution 5.23 Figure S5.23.1 shows the plotted family of curves, along with dashed lines indicating $\operatorname{sgn} x$. The figure displays visually the analytically verifiable fact that $\lim_{\alpha \to \infty} \frac{2}{\pi} \tan^{-1} \alpha x = \operatorname{sgn} x$ ($x \in \mathbb{R}$).

Solution 5.24 See Figure S5.24.1.

Solution 5.25 Yes, because for each $X \in D = \mathbb{Q}$ and $\varepsilon > 0$, there exists an $x \in D - \{X\}$ such that $|x - X| < \varepsilon$.

Solution 5.26 $\forall \xi > 0, \exists \eta > 0, \forall x \in D, x \leq -\eta \Rightarrow f(x) \leq -\xi$.

Solution 5.27 **(a)** Since $\ln x \to -\infty$ as $x \downarrow 0$, and $\tan^{-1} y \to -\pi/2$ as $y \to -\infty$, we have $\lim_{x \to 0} f(x) = -\pi/2 - 3(0) = -\pi/2$.

(b) Since $\ln x \to 0$ as $x \to 1$, and $\tan^{-1} y \to 0$ as $y \to 0$, we have $\lim_{x \to 1} f(x) = 0 - 3(1) = -3$.

(c) Since $\ln x \to \infty$ as $x \to \infty$, and $\tan^{-1} y \to \pi/2$ as $y \to \infty$, we *would* have $\lim_{x \to \infty} f(x) = \pi/2 - 3(\infty) = -\infty$ if infinite limits were allowed; however, no such allowance has been explicitly made. Therefore, this limit does not exist.

Solution 5.28 **(a)** Since $\lim_{i \to \infty} x_i = x$ for some $x \in \mathbb{R}$, then for each $\varepsilon > 0$ there exists an $n \in \mathbb{Z}^+$ such that $|x_i - x| \leq \varepsilon$ for $i \geq n$; in particular, fixing n suitable for $\varepsilon = 1$, we have $|x_i - x| \leq 1$,

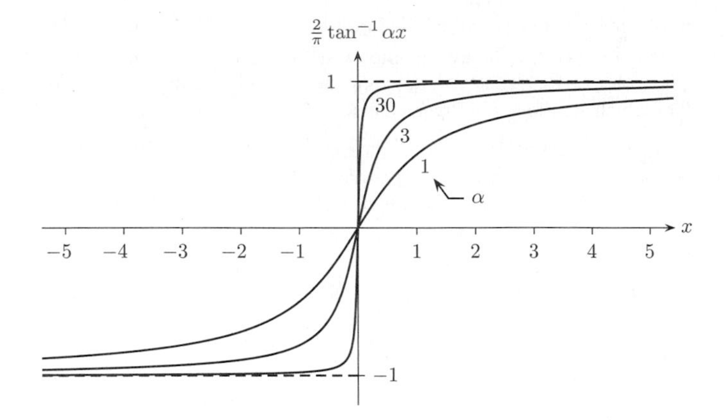

FIGURE S5.23.1 A Family of Curves Approximating the Function Signum

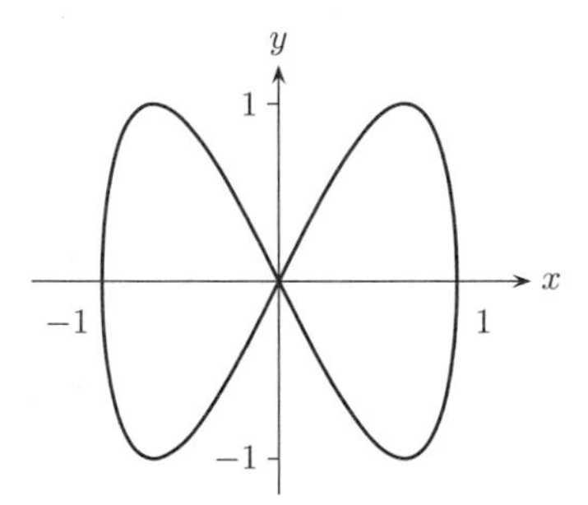

FIGURE S5.24.1 A Parametric Plot

and thus $|x_i| \leq |x| + 1$,[25] for $i \geq n$. Hence, including consideration of the finitely many values $|x_1|, |x_2|, \ldots, |x_{n-1}|$, it follows that

$$|x_i| \leq \max(|x_1|, |x_2|, \ldots, |x_{n-1}|, |x| + 1) \quad \text{(all } i\text{)};$$

therefore, the sequence $\{x_i\}_{i=1}^{\infty}$ is bounded.

(b) Let, e.g.,

$$f(x) = \begin{cases} 0 & x = 1 \\ 1/(x-1) & x > 1 \end{cases}.$$

This function $f\colon [1, \infty) \to \mathbb{R}$ is unbounded, since $f(x)$ diverges to ∞ as $x \to 1$; yet, $\lim_{x\to\infty} f(x) = 0$.

Solution 5.29 $\forall \varepsilon > 0, \exists \delta > 0, \forall x \in D, X - \delta \leq x < X \Rightarrow |f(x) - Y| \leq \varepsilon$.

Solution 5.30 By (8.4),

$$(1-x)^n = \sum_{k=0}^{n} \binom{n}{k} 1^k (-x)^{n-k} = \sum_{k=0}^{n} \frac{n!}{k!(n-k)!}(-x)^{n-k}$$

25. If this is not obvious, then note by (P3.37.1) that $|x_i - x| \geq \big||x_i| - |x|\big| \geq |x_i| - |x|$.

($x \in \mathbb{C}, n \in \mathbb{Z}^+$), which upon changing the summation index to $j = n - k$ gives

$$(1-x)^n = \sum_{j=0}^{n} \frac{n!}{(n-j)!j!}(-x)^j = \sum_{j=0}^{n} \binom{n}{j}(-x)^j = 1 - nx + \sum_{j=2}^{n} \binom{n}{j}(-x)^j$$

—the last sum being empty (and therefore equal to 0) for $n = 1$. Hence, fixing $n \geq 1$, we obtain

$$\lim_{x\to 0} \frac{1-(1-x)^n}{nx} = \lim_{x\to 0} \frac{nx - \sum_{j=2}^{n}\binom{n}{j}(-x)^j}{nx} = 1 - \lim_{x\to 0} \frac{1}{n}\sum_{j=2}^{n}\binom{n}{j}(-x)^{j-1}$$
$$= 1 - 0 = 1,$$

and thus $1-(1-x)^n \sim nx$ as $x \to 0$.

Solution 5.31 **(a)** Since the exponential function is continuous, and both $1/x$ and $1/x^2$ approach 0 as $x \to \infty$, we have

$$\frac{f(x)}{g(x)} = \frac{e^{1/x}}{e^{1/x^2}} \xrightarrow[x\to\infty]{} \frac{e^0}{e^0} = 1;$$

thus, $f(x) \sim g(x)$ as $x \to \infty$. However, we do not have $\ln f(x) \sim \ln g(x)$ as $x \to \infty$, since

$$\frac{\ln f(x)}{\ln g(x)} = \frac{1/x}{1/x^2} = x \quad (x > 0),$$

which does not approach 1 as $x \to \infty$.

(b) Let, e.g., $f(x) = x + \ln x$ and $g(x) = x$ $(x > 0)$. Then,

$$\frac{f(x)}{g(x)} = \frac{x + \ln x}{x} = 1 + \frac{\ln x}{x} \xrightarrow[x\to\infty]{} 1 + 0 = 1,$$

and thus $f(x) \sim g(x)$ as $x \to \infty$. However, we do not have $e^{f(x)} \sim e^{g(x)}$ as $x \to \infty$, since

$$\frac{e^{f(x)}}{e^{g(x)}} = \frac{e^x x}{e^x} = x \quad (x > 0),$$

which does not approach 1 as $x \to \infty$.

Solution 5.32 Yes, because

$$\lim_{x\to X} f(x) = \lim_{x\to X} \frac{f(x)}{g(x)} g(x) = \left[\lim_{x\to X} \frac{f(x)}{g(x)}\right] \cdot \left[\lim_{x\to X} g(x)\right] = 1 \cdot \lim_{x\to X} g(x)$$
$$= \lim_{x\to X} g(x).$$

Solution 5.33 **(a)** If $f(x) = o[g(x)]$, then for any $\varepsilon > 0$ we have $|f(x)| \leq \varepsilon|g(x)|$ for all $x \in D - \{X\}$ sufficiently close to X; in particular, taking $\varepsilon = 1$ shows that $|f(x)| \leq |g(x)|$ for all $x \in D - \{X\}$ sufficiently close to X, and thus $f(x) = O[g(x)]$.

(b) If $f(x) \sim g(x)$, then $\lim_{x\to X} f(x)/g(x) = 1$; that is, for any $\varepsilon > 0$ we have $|f(x)/g(x) - 1| \leq \varepsilon$ for all $x \in D - \{X\}$ sufficiently close to X. Hence, taking $\varepsilon = 1$, we have $0 \leq f(x)/g(x) \leq 2$, and therefore $|f(x)| \leq 2|g(x)|$, for all $x \in D - \{X\}$ sufficiently close to X; thus, $f(x) = O[g(x)]$.

(c) Let a complex constant $a \neq 0$ be given. If $f(x) = O[g(x)]$, then there exists an $M > 0$ such that $|f(x)| \leq M|g(x)|$ for all $x \in D - \{X\}$ sufficiently close to X; therefore, for $M' = M/|a| > 0$ we have $|f(x)| \leq M'|ag(x)|$ for all $x \in D - \{X\}$ sufficiently close to X, thereby showing that $f(x) = O[ag(x)]$. Similarly, if $f(x) = o[g(x)]$, then for any $\varepsilon > 0$ we have $|f(x)| \leq \varepsilon|g(x)|$ for all $x \in D - \{X\}$ sufficiently close to X; therefore, given $\varepsilon > 0$ we can let $\varepsilon' = \varepsilon|a| > 0$, and thus obtain $|f(x)| \leq \varepsilon'|g(x)| = \varepsilon|ag(x)|$ for all $x \in D - \{X\}$ sufficiently close to X, thereby showing that $f(x) = o[ag(x)]$.

(d) If $f(x) = o[g(x)]$, and $g(x) \neq 0$ for $x \in D - \{X\}$ sufficiently close to X, then $f(x)/g(x) \to 0$ as $x \to X$; therefore,

$$\lim_{x\to X} \frac{f(x)+g(x)}{g(x)} = \lim_{x\to X}\left(\frac{f(x)}{g(x)}+1\right) = \lim_{x\to X}\frac{f(x)}{g(x)} + \lim_{x\to X} 1 = 0+1 = 1,$$

and thus $f(x)+g(x) \sim g(x)$.

Solution 5.34 Only at 0. First, suppose $X = 0$, and thus $f(X) = 0$. Now, given $\varepsilon > 0$, take $\delta = \varepsilon$; then, for all $x \in \mathbb{R}$ such that $|x - X| \leq \delta$, we have $|f(x) - f(X)| = |f(x)| \leq |x| \leq \delta = \varepsilon$, making $|f(x) - f(X)| \leq \varepsilon$. Hence, given $\varepsilon > 0$, there exists a $\delta > 0$ such that $|f(x) - f(X)| \leq \varepsilon$ for all $x \in \mathbb{R}$ such that $|x - X| \leq \delta$; therefore, $f(\cdot)$ is continuous at $X = 0$.

On the other hand, suppose $X \in \mathbb{R}$ but $X \neq 0$; and, let $\varepsilon = |X|/2 > 0$. Then, no matter how small we fix $\delta > 0$, there exists a rational number x_r such that $|x_r - X| \leq \delta$, as well as an irrational number x_i such that $|x_i - X| \leq \delta$ and $|x_i| > |X|/2$. So, if X is rational, then taking $x = x_i$ yields both $|x - X| \leq \delta$ and $|f(x) - f(X)| = |x_i| > \varepsilon$; whereas if X is irrational, then taking $x = x_r$ yields both $|x - X| \leq \delta$ and $|f(x) - f(X)| = |X| > \varepsilon$. Hence, for any $X \neq 0$, there exists an $\varepsilon > 0$ such that for each $\delta > 0$, we do not have $|f(x) - f(X)| \leq \varepsilon$ for all $x \in \mathbb{R}$ such that $|x - X| \leq \delta$; therefore, $f(\cdot)$ is not continuous at any point other than 0.

Solution 5.35 Let, e.g., $f(\cdot)$ be the continuous function on $[0,1) \cup [2,3]$ given by

$$f(x) = \begin{cases} x & 0 \leq x < 1 \\ x-1 & 2 \leq x \leq 3 \end{cases}.$$

The range of this function is $[0,2]$, on which $f^{-1}(\cdot)$ is given by

$$f^{-1}(x) = \begin{cases} x & 0 \leq x < 1 \\ x+1 & 1 \leq x \leq 2 \end{cases}.$$

Thus, $f^{-1}(\cdot)$ is not continuous at 1.

Solution 5.36 Let $X \in \mathbb{C}$ and $\varepsilon > 0$ be given. Since $f(\cdot)$ is continuous, there exists an $\eta > 0$ for which $|f(y) - f[g(X)]| \leq \varepsilon$ for all $y \in \mathbb{C}$ such that $|y - g(X)| \leq \eta$. But then for this η, since $g(\cdot)$ is continuous, there exists an $\delta > 0$ for which $|g(x) - g(X)| \leq \eta$ for all $x \in \mathbb{C}$ such that $|x - X| \leq \delta$. Therefore, letting $y = g(x)$ shows that for this δ we have $|f[g(x)] - f[g(X)]| \leq \varepsilon$ for all $x \in \mathbb{C}$ such that $|x - X| \leq \delta$.

Solution 5.37 **(a)** Let, e.g.,

$$f(x) = \begin{cases} 0 & x \leq 0 \\ x & 0 < x < 1 \\ 1 & x \geq 1 \end{cases}.$$

(b) Suppose $f(\cdot)$ is a continuous real-valued function on $[0,1]$ such that its restriction to $(0,1)$ is strictly increasing. Then, to prove that $f(\cdot)$ is itself strictly increasing, we need only show that $f(0) < f(x)$ for all $x \in (0,1]$, and that $f(x) < f(1)$ for all $x \in [0,1)$. Thus, let $x \in (0,1]$ be given. Arbitrarily choosing $x' \in (0,x)$, we have $f(x) - f(x') > 0$, since the restriction of $f(\cdot)$ to $(0,1)$ is strictly increasing; so, since $f(\cdot)$ is continuous at 0, there exists an $X \in (0,x')$ such that $|f(X) - f(0)| \leq f(x) - f(x')$. Therefore, now having $f(0) - f(X) \leq |f(0) - f(X)| \leq f(x) - f(x')$, along with $f(X) < f(x')$ since the restriction of $f(\cdot)$ to $(0,1)$ is strictly increasing, we obtain

$$f(0) = [f(0) - f(X)] + f(X) < [f(x) - f(x')] + f(x') = f(x).$$

Hence, $f(0) < f(x)$ for all $x \in (0,1]$; and, a similar argument shows that $f(x) < f(1)$ for all $x \in [0,1)$.

Solution 5.38 **(a)** Yes. If $f(\cdot)$ is increasing, then $f(x_1) \leq f(x_2)$ for all $x_1, x_2 \in D$ such that $x_1 < x_2$; therefore, $g(x_1) \leq g(x_2)$ for all $x_1, x_2 \in D'$ such that $x_1 < x_2$, and thus $g(\cdot)$ is increasing. Similarly, if $f(\cdot)$ is decreasing, then so is $g(\cdot)$.

(b) Yes. If $f(\cdot)$ is strictly increasing, then $f(x_1) < f(x_2)$ for all $x_1, x_2 \in D$ such that $x_1 < x_2$; therefore, $g(x_1) < g(x_2)$ for all $x_1, x_2 \in D'$ such that $x_1 < x_2$, and thus $g(\cdot)$ is strictly increasing. Similarly, if $f(\cdot)$ is strictly decreasing, then so is $g(\cdot)$.

(c) Yes. If $f(\cdot)$ has a global minimum at X, then $f(x) \geq f(X)$ for all $x \in D$; therefore, $g(x) \geq g(X)$ for all $x \in D'$, and thus $g(\cdot)$ has a global minimum at X. Similarly, if $f(\cdot)$ has a global maximum at X, then so does $g(\cdot)$.

(d) Yes. If $f(\cdot)$ has a local minimum at X, then for some $\varepsilon > 0$ we have $f(x) \geq f(X)$ for all $x \in D$ such that $|x - X| \leq \varepsilon$; therefore, for the same ε we have $g(x) \geq g(X)$ for all $x \in D'$ such that $|x - X| \leq \varepsilon$, and thus $g(\cdot)$ has a local minimum at X. Similarly, if $f(\cdot)$ has a local maximum at X, then so does $g(\cdot)$.

(e) No. For example, the function $f(\cdot)$ on $\mathbb{R}$ given by

$$f(x) = \begin{cases} 0 & x < 0 \\ 1 & x \geq 0 \end{cases}$$

is discontinuous at 0, whereas its restriction to $[0, \infty)$ is a continuous function.

Solution 5.39 **(a)** Since $f(x) = 0$ for $x < 0$, we clearly have $f(0-) = 0$. Also, since $-1/x \to -\infty$ as $x \to 0$ from the right, and $e^y \to 0$ as $y \to -\infty$, we have $f(0+) = 0$. Hence, $f(0-) = f(0) = f(0+)$, so $f(x)$ is continuous at $x = 0$.

(b) Since $-1/x \to \infty$ as $x \to 0$ from the left, and $e^y \to \infty$ as $y \to \infty$, we have $g(x)$ diverging to ∞ as $x \to 0$ from the left; therefore, since infinite limits are not allowed in this context, $g(0-)$ does not exist. Hence, we cannot have $g(0-) = g(0+)$, so $g(x)$ is not continuous at $x = 0$.

Solution 5.40 Yes. Since $g(x) = f(x)$ $(x < X)$, we have $g(X-) = f(X-)$; and, since $g(x) = f(x)$ $(x > X)$, we have $g(X+) = f(X+)$. Therefore, since $f(X-) = f(X+)$ and $g(X) = f(X-)$, we obtain $g(X-) = g(X) = g(X+)$; so, $g(\cdot)$ is continuous at X. (*Aside*: Because the discontinuity in $f(\cdot)$ at X can be removed by changing the one value $f(X)$, this function is said to have a **removable discontinuity** at X.)

Solution 5.41 Since e^y and e^{6y} are both strictly increasing and continuous functions of $y \in \mathbb{R}$, so is $g(y) = e^y + e^{6y}$. Also,

$$\lim_{y\to-\infty} g(y) = \lim_{y\to-\infty} e^y + \lim_{y\to-\infty} e^{6y} = 0 + 0 = 0,$$

$$\lim_{y\to\infty} g(y) = \lim_{y\to\infty} e^y + \lim_{y\to\infty} e^{6y} = \infty + \infty = \infty$$

(allowing infinite limits for the present argument only); therefore, given $x > 0$, there exist $y_1, y_2 \in (0, \infty)$ with $y_1 < y_2$ such that $g(y_1) \leq x \leq g(y_2)$. So, by the continuity of $g(\cdot)$ and the intermediate-value theorem, there must exist a value $y \in [y_1, y_2]$ such that $g(y) = x$. Moreover, being strictly increasing, $g(\cdot)$ is invertible, and thus one-to-one; hence, there can be at most one value y for which $g(y) = x$.

Solution 5.42 **(a)** Given $\varepsilon > 0$, it follows from the uniform continuity of $f(\cdot)$ that we can choose an integer $n \geq 1$ sufficiently large that $|f(x) - f(X)| \leq \varepsilon$ for all $x, X \in [a, b]$ such that $|x - X| \leq (b - a)/n$. With n so fixed, let $x_i = a + i \cdot (b - a)/n$ $(i = 0, 1, \ldots, n)$ and $c_i = f(x_i)$ $(i = 1, \ldots, n)$ in (P5.42.1). Then, for each $x \in [a, b]$ we have $x_{i-1} \leq x \leq x_i$ for some $i > 1$, making $|x - x_i| \leq x_i - x_{i-1} = (b - a)/n$, and thus $|g(x) - f(x)| = |f(x_i) - f(x)| = |f(x) - f(x_i)| \leq \varepsilon$.

(b) By part (a) and the triangle inequality,

$$|f(x)| = |[f(x) - g(x)] + g(x)| \leq |f(x) - g(x)| + |g(x)| \leq \varepsilon + \max_i |c_i| < \infty$$

$(a \leq x \leq b)$; therefore, $f(\cdot)$ is bounded.

Solution 5.43 **(a)** Let $\varepsilon > 0$ and $X \in D$ be given. Since $f_i(\cdot)$ converges uniformly to $f(\cdot)$, we can choose an integer i such that $|f_i(x) - f(x)| \leq \varepsilon/3$ for all $x \in D$. Also, since $f_i(\cdot)$ is continuous, there exists a $\delta > 0$ such that $|f_i(x) - f_i(X)| \leq \varepsilon/3$ for all $x \in D$ such that $|x - X| \leq \delta$; therefore, it follows from the triangle inequality that, for all such x,

$$\begin{aligned} |f(x) - f(X)| &= |[f(x) - f_i(x)] + [f_i(x) - f_i(X)] + [f_i(X) - f(X)]| \\ &\leq |f(x) - f_i(x)| + |f_i(x) - f_i(X)| + |f_i(X) - f(X)| \\ &\leq \varepsilon/3 + \varepsilon/3 + \varepsilon/3 = \varepsilon. \end{aligned}$$

Hence, for all $\varepsilon > 0$ and $X \in D$, there exists a $\delta > 0$ such that $|f(x) - f(X)| \leq \varepsilon$ for all $x \in D$ such that $|x - X| \leq \delta$; so, $f(\cdot)$ is continuous.

(b) Suppose $\lim_{x\to\infty} f_i(x)$ (all i) and $\lim_{x\to\infty} f(x)$ exist; also, let $\varepsilon > 0$ be given. Since $f_i(\cdot)$ converges to $f(\cdot)$, the second equality in (P5.43.1) holds; moreover, since the convergence is uniform, there exists an $n \in \mathbb{Z}^+$ such that $-\varepsilon \leq f_i(x) - f(x) \leq \varepsilon$ for all $i \geq n$ and $x \in \mathbb{R}$. Therefore, since $\lim_{x\to\infty}[f_i(x) - f(x)] = \lim_{x\to\infty} f_i(x) - \lim_{x\to\infty} f(x)$ (all i), we have $-\varepsilon \leq \lim_{x\to\infty} f_i(x) - \lim_{x\to\infty} f(x) \leq \varepsilon$ for $i \geq n$. Hence, for each $\varepsilon > 0$ we have $|\lim_{x\to\infty} f_i(x) - \lim_{x\to\infty} f(x)| \leq \varepsilon$ when i is sufficiently large; so, $\lim_{x\to\infty} f_i(x)$ converges to $\lim_{x\to\infty} f(x)$ as $i \to \infty$, and the first equality in (P5.43.1) also obtains.

Chapter 6. Powers and Roots

Solution 6.1 **(a)** Since (6.2) is known to hold for any $x, y \in \mathbb{C}$ and $m, n \in \mathbb{Z}^+$, we need consider only the cases for which $m = 0$ or $n = 0$. Also, we know that when $x, y \neq 0$, (6.2) holds for $m, n \in \mathbb{Z}$; therefore, we can further restrict our attention to having $x = 0$ or $y = 0$. Hence:

- For $m = 0$, (6.2a) with $x = 0$ becomes $0^0 0^n = 0^n$, which holds for all $n \in \mathbb{N}$ if either $0^0 = 0$ or $0^0 = 1$; and, likewise if we assume $n = 0$ instead.
- For $m = 0$, (6.2b) with $x = 0$ becomes $(0^0)^n = 0^0$, which holds for all $n \in \mathbb{N}$ if either $0^0 = 0$ or $0^0 = 1$. Whereas for $n = 0$, (6.2b) with $x = 0$ becomes $(0^m)^0 = 0^0$, which holds for all $m \in \mathbb{N}$ if either $0^0 = 0$ or $0^0 = 1$.
- For $n = 0$, (6.2c) with $x = 0$ becomes $0^0 = 0^0 y^0$, which holds for all $y \in \mathbb{C}$ if either $0^0 = 0$ or $0^0 = 1$; and, likewise with $y = 0$ instead.
- For $n = 0$, (6.2d) with $x = 0$ and $y \neq 0$ becomes $0^0 = 0^0/1$, which holds regardless of how 0^0 is defined.

(b) Only (6.2d) would still hold, for the following reasons:

- For $x = 0$ and $m = n = 0$, (6.2a) becomes $0^0 0^0 = 0^0$, which implies that either $0^0 = 0$ or $0^0 = 1$.
- For $x = 0$ and $m = n = 0$, (6.2b) becomes $(0^0)^0 = 0^0$, which implies that either $0^0 = 0$ or $0^0 = 1$; because, if $0^0 \neq 0$, then $(0^0)^0 = 1$.
- For $x = y = 0$ and $n = 0$, (6.2c) becomes $0^0 = 0^0 0^0$, which implies that either $0^0 = 0$ or $0^0 = 1$.

Solution 6.2 Let a complex number $x \neq 0$ be given. By (6.8a) and (6.8b), we have both $\operatorname{Re} x^n \neq 0$ and $\operatorname{Im} x^n \neq 0$ for all integers $n \neq 0$ if and only if $\cos(n \arg x) \neq 0$ and $\sin(n \arg x) \neq 0$ for all such n. But, $\cos(n \arg x) = 0$ if and only if $n \arg x$ is an odd multiple of $\pi/2$, and $\sin(n \arg x) = 0$ if and only if $n \arg x$ is an even multiple of $\pi/2$; therefore, we have both $\cos(n \arg x) \neq 0$ and $\sin(n \arg x) \neq 0$ if and only if $n \arg x$ is not a multiple of $\pi/2$. Combining these facts, we conclude that $\operatorname{Re} x^n \neq 0$ and $\operatorname{Im} x^n \neq 0$ for all integers $n \neq 0$ if and only if for all such n we have $n \arg x \neq m(\pi/2)$ for all integers m; that is, for no integers m and $n \neq 0$ do we have $(2/\pi) \arg x = m/n$, which is to say that $(2/\pi) \arg x$ is irrational.

Solution 6.3 **(a)** By (6.10a) and (6.10c), $x\sqrt[n]{y} = \sqrt[n]{x^n}\sqrt[n]{y} = \sqrt[n]{x^n y}$.

(b) By (6.11), $\sqrt[km]{x^{kn}} = x^{kn/km} = x^{n/m} = \sqrt[m]{x^n}$.

(c) By (6.11) and (6.14a), $\sqrt[m]{x}\sqrt[n]{x} = x^{1/m}x^{1/n} = x^{1/m+1/n}$; therefore, since $1/m + 1/n = (m + n)/mn \in \mathbb{Q}$, we must take $p = 1/m + 1/n$. It now follows from (6.11) that we can take any $k, l \in \mathbb{Z}^+$ such that $p = l/k$—e.g., $k = mn$, $l = m + n$.

Solution 6.4 By the triangle equality, $|a + b| \le |a| + |b|$; therefore, since x^p is an increasing function of $x \ge 0$, we have

$$|a + b|^p \le (|a| + |b|)^p. \tag{S6.4.1}$$

If $|a| \ge |b|$, then $(|a| + |b|)^p \le (|a| + |a|)^p = (2|a|)^p = 2^p|a|^p \le 2^p(|a|^p + |b|^p)$, and thus

$$(|a| + |b|)^p \le 2^p(|a|^p + |b|^p); \tag{S6.4.2}$$

likewise, the same result obtains if $|b| \ge |a|$. Finally, combining (S6.4.1) and (S6.4.2) yields $|a + b|^p \le 2^p(|a|^p + |b|^p)$. (*Moral*: Don't expect to be able to solve every problem by merely writing down formulas and manipulating them mechanically. Often, some *thought* is required to determine what path to take.)

Solution 6.5 All of the equalities in (P6.5.1) are valid except the third, which *assumes* that (6.10d) holds with $x = 1$, $y = -1$, and $n = 2$.

Solution 6.6 **(a)** If $x \le 0$, then $-x \ge 0$, and thus by (6.15) we have $\sqrt{x} = \sqrt{-(-x)} = i\sqrt{-x}$; therefore, multiplying through by $-i$, we obtain $-i\sqrt{x} = -i \cdot i\sqrt{-x} = \sqrt{-x}$.

(b) If $x, y \ge 0$, then it follows from (6.10c) with $n = 2$ that $\sqrt{xy} = \sqrt{x}\sqrt{y}$. If $x < 0$ and $y \ge 0$, then $-x > 0$ and $-xy \ge 0$, and thus it follows from (6.15), (6.10c), and part (a) that

$$\sqrt{xy} = \sqrt{-(-xy)} = i\sqrt{-xy} = i\sqrt{(-x)y} = i\sqrt{-x}\sqrt{y} = i\big(-i\sqrt{x}\big)\sqrt{y} = \sqrt{x}\sqrt{y};$$

and, by symmetry, the same result obtains when $x \ge 0$ and $y < 0$. But, if $x, y < 0$, then $-x, -y > 0$, by which it follows from (6.10c) and (6.15) that

$$\sqrt{xy} = \sqrt{(-x)(-y)} = \sqrt{-x}\sqrt{-y} = \big(-i\sqrt{x}\big)\big(-i\sqrt{y}\big) = -\sqrt{x}\sqrt{y}.$$

Solution 6.7 **(a)** Using (4.5d), as generalized for function limits, we have

$$\lim_{x\to 0} f(x) = \frac{\lim_{x\to 0}(x + \sqrt[3]{x} + 4)}{\lim_{x\to 0}(x^2 + \sqrt{x} + 1)} = \frac{0 + 0 + 4}{0 + 0 + 1} = 4.$$

Specifically, the first equality holds because the numerator and denominator limits exist, with the latter being nonzero.

(b) Since the numerator and denominator of $f(x)$ become infinite as $x \to \infty$, we cannot apply (4.5d) directly here; however, if we first divide both by x^2, then their limits become finite, with the latter being nonzero, so (4.5d) applies:

$$\begin{aligned}\lim_{x\to\infty} f(x) &= \frac{\lim_{x\to\infty}(x + \sqrt[3]{x} + 4)/x^2}{\lim_{x\to\infty}(x^2 + \sqrt{x} + 1)/x^2} = \frac{\lim_{x\to\infty}(x^{-1} + x^{-5/3} + 4x^{-2})}{\lim_{x\to\infty}(1 + x^{-3/2} + x^{-2})}\\ &= \frac{0 + 0 + 0}{1 + 0 + 0} = 0.\end{aligned}$$

Alternatively, one can first determine the asymptotic behavior of the numerator and denominator of $f(x)$ as $x \to \infty$—viz.,

$$x + \sqrt[3]{x} + 4 \sim x, \quad x^2 + \sqrt{x} + 1 \sim x^2$$

—and thereby obtain

$$\lim_{x\to\infty} f(x) = \lim_{x\to\infty} \frac{x}{x^2} = \lim_{x\to\infty} x^{-1} = 0.$$

Solution 6.8 Let $f(x) = a_3x^3 + a_2x^2 + a_1x + a_0$ ($x \in \mathbb{R}$) for some real coefficients a_0, a_1, a_2, and $a_3 \neq 0$; and, suppose $X, Y \in \mathbb{R}$.

(a) Now,

$$\begin{aligned} g(x) &= f(x+X) - Y = a_3(x+X)^3 + a_2(x+X)^2 + a_1(x+X) + a_0 - Y \\ &= a_3x^3 + (3a_3X + a_2)x^2 + (3a_3X^2 + 2a_2X + a_1)x \\ &\quad + (a_3X^3 + a_2X^2 + a_1X + a_0 - Y) \quad (x \in \mathbb{R}), \end{aligned}$$

and thus

$$\begin{aligned} g(-x) = -a_3x^3 &+ (3a_3X + a_2)x^2 - (3a_3X^2 + 2a_2X + a_1)x \\ &+ (a_3X^3 + a_2X^2 + a_1X + a_0 - Y) \quad (x \in \mathbb{R}). \end{aligned}$$

Therefore, $g(\cdot)$ has odd symmetry—i.e., $g(x) = -g(-x)$ (all x)—if and only if

$$\begin{aligned} 0 &= g(x) + g(-x) \\ &= 2(3a_3X + a_2)x^2 + 2(a_3X^3 + a_2X^2 + a_1X + a_0 - Y) \quad (x \in \mathbb{R}), \end{aligned}$$

which is the case if and only if the coefficients of the polynomial both equal 0:

$$2(3a_3X + a_2) = 0, \quad 2(a_3X^3 + a_2X^2 + a_1X + a_0 - Y) = 0.$$

Solving these equations, we conclude that $g(\cdot)$ is odd if and only if

$$X = -a_2/3a_3, \quad Y = a_3X^3 + a_2X^2 + a_1X + a_0.$$

(b) For any X and Y, the function $g(\cdot)$ cannot have even symmetry—i.e., $g(x) = g(-x)$ (all x)—simply because $g(x) \sim a_3x^3$ and $g(-x) \sim -a_3x^3$ as $x \to \infty$, where $a_3 \neq 0$. Alternatively, here we need

$$0 = g(x) - g(-x) = 2a_3x^3 + 2(3a_3X^2 + 2a_2X + a_1)x \quad (x \in \mathbb{R}),$$

which cannot be since the coefficient $2a_3 \neq 0$.

Solution 6.9 Letting

$$f(x) = \sum_{j=0}^{n} a_jx^j \quad (x \in \mathbb{C}) \tag{S6.9.1}$$

for some coefficients $a_j \in \mathbb{C}$ ($j = 0, 1, \ldots, n$), we have

$$f^*(x) = \left(\sum_{j=0}^{n} a_jx^j\right)^* = \sum_{j=0}^{n} (a_jx^j)^* = \sum_{j=0}^{n} a_j{}^*(x^*)^j \quad (x \in \mathbb{C}). \tag{S6.9.2}$$

Therefore, if the coefficients of $f(\cdot)$ are real, then $a_j = a_j{}^*$ (all j), thereby making $f^*(x) = f(x^*)$ ($x \in \mathbb{C}$). Conversely, if $f^*(x) = f(x^*)$ ($x \in \mathbb{C}$), then conjugating this equation and using (S6.9.2) yields

$$f(x) = f^*(x^*) = \sum_{j=0}^{n} a_j{}^*[(x^*)^*]^j = \sum_{j=0}^{n} a_j{}^*x^j \quad (x \in \mathbb{C});$$

hence, upon comparing with (S6.9.1) and using the fact that two polynomials that are equal on all of $\mathbb{C}$ must be the same, we conclude that $a_j = a_j{}^*$ (all j), and thus the coefficients of $f(\cdot)$ are real.

Solution 6.10 Using the factorization (6.18) of a difference of two squares, we obtain

$$\frac{1}{\sqrt{2}+\sqrt{3}} = \frac{1}{\sqrt{2}+\sqrt{3}} \cdot \frac{\sqrt{2}-\sqrt{3}}{\sqrt{2}-\sqrt{3}} = \frac{\sqrt{2}-\sqrt{3}}{2-3} = \sqrt{3}-\sqrt{2},$$

$$\frac{1}{5-\sqrt{7}} = \frac{1}{5-\sqrt{7}} \cdot \frac{5+\sqrt{7}}{5+\sqrt{7}} = \frac{5+\sqrt{7}}{5^2-7} = \frac{5+\sqrt{7}}{18}.$$

Solution 6.11 For $x \neq 0$ we have

$$g(x) = x^n[g(x)/x^n] = x^n(a_0 + a_1x + \cdots + a_{n-1}x^{n-1} + a_nx^{-n}) = x^nf(x^{-1}).$$

Also, since $a_0 \neq 0$, all of the roots $\hat{x}_1, \ldots, \hat{x}_{\hat{n}}$ of $f(\cdot)$ are nonzero. So, it follows from the factorization (6.19) of $f(\cdot)$, along with (6.20), that

$$\begin{aligned} g(x) &= x^n \cdot a_n \prod_{k=1}^{\hat{n}} (x^{-1} - \hat{x}_k)^{m_k} = \left(\prod_{k=1}^{\hat{n}} x^{m_k}\right) a_n \prod_{k=1}^{\hat{n}} (x^{-1} - \hat{x}_k)^{m_k} \\ &= a_n \prod_{k=1}^{\hat{n}} (1 - \hat{x}_k x)^{m_k} = a_n \left[\prod_{k=1}^{\hat{n}} (-\hat{x}_k)^{m_k}\right] \prod_{k=1}^{\hat{n}} (x - 1/\hat{x}_k)^{m_k} \quad (x \neq 0), \end{aligned}$$

where by (6.21b) the factor in brackets equals a_0/a_n; thus,

$$g(x) = a_0 \prod_{k=1}^{\hat{n}} (x - 1/\hat{x}_k)^{m_k} \quad (x \neq 0).$$

Moreover, since $a_n \neq 0$, all of the roots of $g(\cdot)$ are nonzero; hence, $g(x) = 0$ for some $x \in \mathbb{C}$ if and only if $x = 1/\hat{x}_{\hat{k}}$ for some k. Finally, since the values $\hat{x}_1, \ldots, \hat{x}_{\hat{n}}$ are distinct, so are their reciprocals; therefore, $g(\cdot)$ has distinct roots $1/\hat{x}_1, \ldots, 1/\hat{x}_{\hat{n}}$ with respective multiplicities $m_1, \ldots, m_{\hat{n}}$.

Solution 6.12 Since the coefficients of $f(\cdot)$ are real, its nonreal roots come in conjugate pairs; therefore, the number of nonreal roots is even. But, the degree of $f(\cdot)$, which equals the total number of roots, is odd; hence, there must exist at least one real root. So, since the x-axis intercepts are precisely the real roots of $f(\cdot)$, there must be at least one such intercept.

Alternatively, $f(\cdot)$ can be expressed as in (6.16) with n odd and $a_n \neq 0$. If $a_n > 0$, then $\lim_{x\to-\infty} f(x) = -\infty$, and $\lim_{x\to\infty} f(x) = \infty$; whereas these limits reverse if $a_n < 0$. Hence, for $a_n > 0$ we can choose a value $X > 0$ such that $f(-X) < 0 < f(X)$, whereas for $a_n < 0$ we can choose $X > 0$ such that $f(-X) > 0 > f(X)$. Therefore, in either case, since the polynomial $f(\cdot)$ is a continuous function, it follows from the intermediate-value theorem that exists an $x \in [-X, X]$ such that $f(x) = 0$.

Solution 6.13 The fifth-degree polynomial $f(\cdot)$ is monic and has integer coefficients; so each of its five roots must be a factor—viz., $\pm1, \pm2, \pm3, \pm4, \pm6, \pm8, \pm12$, or ±24—of its last coefficient, -24. To get these roots by trial and error, there are many paths to take. Starting at the low end, we find $f(1) = 0$, making 1 a root; so, we divide $f(x)$ by $x-1$, as in Figure S6.13.1, to obtain the fourth-degree polynomial $x^4 + 2x^3 - 13x^2 - 14x + 24$. It also evaluates to 0 at $x = 1$; therefore, we can again divide by $x - 1$, as in Figure S6.13.2, obtaining the third-degree polynomial $g(x) = x^3 + 3x^2 - 10x - 24$. We then find $g(1) = -30 \neq 0$, $g(-1) = -12 \neq 0$, and $g(2) = -24 \neq 0$; but, $g(-2) = 0$. So, we divide $g(x)$ by $x + 2$, as in Figure S6.13.2, obtaining a second-degree polynomial that can be factored by inspection: $x^2 + x - 12 = (x - 3)(x + 4)$. Therefore, $f(\cdot)$ has roots (including repetitions) at 1, 1, -2, 3, and -4.

Solution 6.14 **(a)** ±1, since the square of each equals 1.

(b) ±1 and $\pm i$, since the fourth-power of each equals 1.

$$
\begin{array}{r}
x^4 + 2x^3 - 13x^2 - 14x + 24 \\
x - 1\overline{)x^5 + x^4 - 15x^3 - x^2 + 38x - 24} \\
\underline{x^5 - x^4} \qquad\qquad\qquad\qquad\quad \\
2x^4 - 15x^3 \qquad\qquad\qquad \\
\underline{2x^4 - 2x^3} \qquad\qquad\qquad \\
-13x^3 - x^2 \qquad\qquad \\
\underline{-13x^3 + 13x^2} \qquad\qquad \\
-14x^2 + 38x \qquad \\
\underline{-14x^2 + 14x} \qquad \\
24x - 24 \\
\underline{24x - 24} \\
0
\end{array}
$$

FIGURE S6.13.1 First Division in Solution 6.13

$$
\begin{array}{r}
x^3 + 3x^2 - 10x - 24 \\
x - 1\overline{)x^4 + 2x^3 - 13x^2 - 14x + 24} \\
\underline{x^4 - x^3} \qquad\qquad\qquad\qquad \\
3x^3 - 13x^2 \qquad\qquad\qquad \\
\underline{3x^3 - 3x^2} \qquad\qquad\qquad \\
-10x^2 - 14x \qquad\quad \\
\underline{-10x^2 + 10x} \qquad\quad \\
-24x + 24 \\
\underline{-24x + 24} \\
0
\end{array}
\qquad
\begin{array}{r}
x^2 + x - 12 \\
x + 2\overline{)x^3 + 3x^2 - 10x - 24} \\
\underline{x^3 + 2x^2} \qquad\qquad\quad \\
x^2 - 10x \qquad\quad \\
\underline{x^2 + 2x} \qquad\quad \\
-12x - 24 \\
\underline{-12x - 24} \\
0
\end{array}
$$

FIGURE S6.13.2 Other Divisions in Solution 6.13

Solution 6.15 Rewriting the given sum as

$$\sum_{k=0}^{m-1} e^{i2\pi k/n} = \sum_{k=0}^{m-1} \left(e^{i2\pi/n}\right)^k,$$

we have a sum of a geometric progression. Hence, since $e^{i2\pi/n} = 1$ only for $n = 1$, it follows from (4.37) that for $n > 1$ we have

$$
\begin{aligned}
\sum_{k=0}^{m-1} e^{i2\pi k/n} &= \frac{e^{i2\pi m/n} - 1}{e^{i2\pi/n} - 1} = \frac{e^{i\pi m/n}(e^{i\pi m/n} - e^{-i\pi m/n})}{e^{i\pi/n}(e^{i\pi/n} - e^{-i\pi/n})} \\
&= \frac{e^{i\pi m/n}}{e^{i\pi/n}} \cdot \frac{i2 \sin \pi m/n}{i2 \sin \pi/n} = e^{i\pi(m-1)/n} \frac{\sin \pi m/n}{\sin \pi/n} \qquad (m \geq 1).
\end{aligned}
$$

Therefore, including the trivial case of $n = 1$ yields

$$\sum_{k=0}^{m-1} e^{i2\pi k/n} = \begin{cases} m & n = 1 \\ e^{i\pi(m-1)/n} \dfrac{\sin \pi m/n}{\sin \pi/n} & n \geq 2 \end{cases} \qquad (m \geq 1);$$

and, evaluating for $m = n \geq 2$, we obtain

$$\sum_{k=0}^{m-1} e^{i2\pi k/n} = e^{i\pi(n-1)/n} \frac{\sin \pi}{\sin \pi/n} = e^{i\pi(n-1)/n} \frac{0}{\sin \pi/n} = 0,$$

as expected.

Solution 6.16 When viewed in the complex plane, the n distinct values in (6.22b) are uniformly spaced on the circle of radius $|c|^{1/n}$ that is centered at the origin, with one of the values (for $k = 0$) being $|c|^{1/n} e^{i(\arg c)/n}$, which is on the circle at an angle of $(\arg c)/n$.

Solution 6.17 Comparing (6.23a) with (6.16), we now have $a = a_2 \neq 0, b = a_1, c = a_0$. Therefore, (6.23c) gives

$$x_1 + x_2 = \frac{-b + \sqrt{b^2 - 4ac}}{2a} + \frac{-b - \sqrt{b^2 - 4ac}}{2a} = -\frac{b}{a} = -\frac{a_1}{a_2},$$

$$x_1 \cdot x_2 = \frac{-b + \sqrt{b^2 - 4ac}}{2a} \cdot \frac{-b - \sqrt{b^2 - 4ac}}{2a} = \frac{b^2 - (b^2 - 4ac)}{4a^2} = \frac{c}{a} = \frac{a_0}{a_2},$$

as would be obtained from (6.21) with $n = 2$.

Solution 6.18 **(a)** If the discriminant $b^2 - 4ac > 0$, then x_1 and x_2 are real and distinct; therefore, in this case we have $|x_1| = |x_2|$ if and only if $x_1 = -x_2$, which by (6.23c) is so if and only if $b = 0$. Whereas if $b^2 - 4ac \leq 0$, then $x_1 = x_2{}^*$, making $|x_1| = |x_2|$. Thus, a necessary and sufficient condition to have $|x_1| = |x_2|$ is that $b = 0$ or $b^2 \leq 4ac$.

(b) If the discriminant $b^2 - 4ac \geq 0$, then the roots x_1 and x_2 are real; therefore, for the roots to be imaginary and nonzero we must have $b^2 - 4ac < 0$, in which case they are conjugate with real parts $-b/2a$. Thus, a necessary and sufficient condition for x_1 and x_2 to be imaginary and nonzero is that $b = 0$ and $ac > 0$.

Solution 6.19 **(a)** Since

$$(x + \beta)^2 + \gamma = x^2 + 2\beta x + (\beta^2 + \gamma) \quad (x \in \mathbb{C}),$$

we can convert $f(\cdot)$ from form (P6.19.1) to form (P6.19.2) by taking $\beta = b/2$ and $\gamma = c - \beta^2 = c - (b/2)^2$, which are real.

(b) By (P6.19.3),

$$f(Y \pm iZ) = [(Y \pm iZ) - Y]^2 + Z^2 = (\pm iZ)^2 + Z^2 = -Z^2 + Z^2 = 0$$

($\pm$ respectively); therefore, since Y is real and $Z > 0$, the values $Y + iZ$ and $Y - iZ$ are distinct nonreal roots of $f(\cdot)$. Also, if for some real coefficients b and c the polynomial in (P6.19.1) has nonreal roots, then the roots must form a conjugate pair—say, $Y \pm iZ$, for real values Y and $Z > 0$—and thus, by (6.17),

$$\begin{aligned} f(x) &= [x - (Y + iZ)][x - (Y - iZ)] = [(x - Y) - iZ][(x - Y) + iZ] \\ &= (x - Y)^2 - (iZ)^2 \quad (x \in \mathbb{C}), \end{aligned}$$

which gives (P6.19.3). And, by the result of part (a) with $\beta = -Y$ and $\gamma = Z^2$, we have $Y = -b/2$ and $Z = \sqrt{c - Y^2} = \sqrt{c - (b/2)^2}$ (cf. (6.23c) with $a = 1$).

Aside: In general, the roots of a polynomial $f(\cdot)$ having real coefficients can be separated into $n' \geq 0$ distinct real roots $\hat{x}_k$ having respective multiplicities m_k $(k = 1, \ldots, n')$, along with $N' \geq 0$ distinct nonreal roots $\hat{y}_l \pm i\hat{z}_l$ (with $\hat{y}_l$ real and $\hat{z}_l > 0$) having respective multiplicities M_l $(l = 1, \ldots, N')$. Hence, if $\deg f(\cdot) = n$ then

$$\sum_{k=1}^{n'} m_k + 2\sum_{l=1}^{N'} M_l = n,$$

per (6.20). (If $n' = 0$, then the first sum is empty and is understood to equal 0; and, likewise for the second sum when $N' = 0$.) It then follows from above that we can convert $f(\cdot)$ from the expanded form (6.16) into the factored form

$$f(x) = a_n \prod_{k=1}^{n'} (x - \hat{x}_k)^{m_k} \prod_{l=1}^{N'} [(x - \hat{y}_l)^2 + \hat{z}_l{}^2]^{M_l} \quad (x \in \mathbb{C}). \tag{S6.9.3}$$

(If $n' = 0$, then the first product is empty and understood to equal 1; and, likewise for the second product when $N' = 0$.) This is an alternative to the factorization (6.19), which is the same if all of the roots are real (i.e., $N' = 0$). However, if some roots are nonreal, then only (S6.9.1)—which decomposes the polynomial $f(x)$ $(x \in \mathbb{C})$ into its **first-order factors** $(x - \hat{x}_k)$ and its **second-order factors** $[(x - \hat{y}_l)^2 + \hat{z}_l{}^2]$—is expressed solely in terms of *real*-valued constants (viz., a_n and the $\hat{x}_k$, $\hat{y}_l$, and $\hat{z}_l$), which is often preferred in physical applications.

Solution 6.20 Since $f(x) = x^5 - 28x^3 + 36x = x(x^4 - 28x^2 + 36)$, the polynomial $f(x)$ has exactly one root at 0. Noticing that the remaining fourth-degree polynomial $x^4 - 28x^2 + 36$ has no odd-power terms, we can rewrite it as $(x^2)^2 - 28(x^2) + 36$—i.e., a quadratic polynomial in "x^2". Thus, by using the quadratic formula to solve for x^2, the remaining four roots x of $f(x)$ satisfy the equation

$$x^2 = \frac{28 \pm \sqrt{28^2 - 4 \cdot 36}}{2} = 14 \pm 4\sqrt{10}. \tag{S6.20.1}$$

Therefore, the five roots of $f(x)$ are 0 and $\pm\sqrt{14 \pm 4\sqrt{10}}$.

Alternatively, we can manipulate the quadratic polynomial in x^2 to obtain

$$\begin{aligned}(x^2)^2 - 28(x^2) + 36 &= [(x^2)^2 + 36] - 28(x^2) = \{[(x^2) + 6]^2 - 12(x^2)\} - 28(x^2) \\ &= (x^2 + 6)^2 - 40x^2 = (x^2 + 6)^2 - \left(\sqrt{40}\,x\right)^2 \\ &= \left(x^2 + \sqrt{40}\,x + 6\right)\left(x^2 - \sqrt{40}\,x + 6\right),\end{aligned}$$

having used (6.18) to factor a difference of two squares.[26] Now, applying the quadratic formula to each of the last two factors yields the four roots

$$x = \frac{\pm\sqrt{40} \pm \sqrt{40 - 4 \cdot 6}}{2} = \pm\sqrt{10} \pm 2.$$

Therefore, the five roots of $f(x)$ are also 0 and $\pm 2 \pm \sqrt{10}$.

---

26. Contrast this computation with completing the square (see Problem 6.19) to obtain

$$\begin{aligned}(x^2)^2 - 28(x^2) + 36 &= [(x^2)^2 - 28(x^2)] + 36 = \{[(x^2) - 14]^2 - 196\} + 36 = [(x^2) - 14]^2 - 160 \\ &= [(x^2) - 14]^2 - \left(4\sqrt{10}\right)^2,\end{aligned}$$

which leads to (S6.20.1).

Solution 6.21 Suppose (P6.21.1) does hold for some pair $(x, y) \in \mathbb{R}^2$. Then, since it is implicit that $1/x$, $1/y$, and $1/(x+y)$ exist, we must have $x \neq 0$, $y \neq 0$, and $x+y \neq 0$. Hence, multiplying (P6.21.1) by $xy(x+y)$ shows that we must have

$$xy = y(x+y) + x(x+y),$$

yielding

$$x^2 + xy + y^2 = 0.$$

Now, solving for x by the quadratic formula, we obtain

$$x = \frac{-y \pm \sqrt{y^2 - 4y^2}}{2} = \frac{-1 \pm i\sqrt{3}}{2}y;$$

but, from this it follows that $x \notin \mathbb{R}$ since $y \neq 0$. Therefore, having obtained a contradiction, we conclude that (P6.21.1) cannot hold for any pair $(x, y) \in \mathbb{R}^2$.

Solution 6.22 Squaring (P6.22.1) gives $2x + 7 = (x+2)^2 = x^2 + 4x + 4$, and thus $x^2 + 2x - 3 = 0$. Since the left side of this equation factors to $(x-1)(x+3)$, its solutions x are 1 and -3. Therefore, we conclude that *if* (P6.22.1) has any solutions, then each must be either 1 or -3; however, at this point we do not know whether either of these values is actually a solution. Because, thus far we have only shown that a *necessary* condition for x to be a solution to (P6.22.1) is that $x \in \{1, -3\}$; but, we have not shown that this is a sufficient condition. In fact, by direct substitution we find that $x = -1$ is *not* a solution to (P6.22.1), whereas $x = 3$ is; hence, 3 is the one and only value $x \in \mathbb{C}$ satisfying (P6.22.1).

Aside: Here we see the occurrence of an **extraneous solution**,[27] which is a bogus solution that slips into an analysis as a result of actions performed. Above, the first step of squaring (P6.22.1) produced another equation that contained not only the actual solution $x = 3$ to the original equation, but also an extraneous solution $x = -1$. Logically, extraneous solutions arise because steps of reasoning often have the form "every solution satisfying conditions A, B, and C must also satisfy conditions X, Y, and Z", which is effectively a conditional: A, B, and C imply X, Y, and Z. But, it might not be that X, Y, and Z imply A, B, and C; above, being a solution to the equation $x^2+2x-3 = 0$ does not imply being a solution to (P6.22.1). Extraneous solutions may be avoided by making each analytical step a biconditional—i.e., X, Y, and Z hold if *and only if* A, B, and C hold. Otherwise, any "solutions" obtained must be checked for validity, since all that has been shown is the necessity that any actual solution must come from these "solutions"; their sufficiency for being actual solutions is yet to be established.

Solution 6.23 For $x > 0$ we have

$$x + 1/x - 2 = (x^2 + 1 - 2x)/x = (x-1)^2/x \geq 0,$$

which proves that $x + 1/x \geq 2$ $(x > 0)$, and also shows that equality is attained if and only if $x = 1$.

Solution 6.24 Letting $f(x) = a_3x^3 + a_2x^2 + a_1x + a_0$ $(x \in \mathbb{C})$ for some complex coefficients a_i $(i = 0, 1, 2, 3)$, it follows from (P6.24.1) that

$$\begin{aligned}(n+1)^2 &= f(n+1) - f(n)\\ &= [a_3(n+1)^3 + a_2(n+1)^2 + a_1(n+1) + a_0]\\ &\quad - (a_3n^3 + a_2n^2 + a_1n + a_0) \quad (n \geq 1), \end{aligned} \tag{S6.24.1}$$

and thus

$$n^2 + 2n + 1 = (3a_3)n^2 + (3a_3 + 2a_2)n + (a_3 + a_2 + a_1) \quad (n \geq 1).$$

---

27. This is often referred to as an **extraneous root**. Strictly speaking, though, a "root" is a special kind of solution—namely, a value for which a given function (e.g., a polynomial) evaluates to 0.

The polynomials on the two sides of this equation are equal at the infinitely many integers $n \geq 1$ if and only if the corresponding coefficients are equal:

$$1 = 3a_3, \quad 2 = 3a_3 + 2a_2, \quad 1 = a_3 + a_2 + a_1;$$

and, these equations hold if and only if

$$a_3 = \tfrac{1}{3}, \quad a_2 = (2 - 3a_3)/2 = \tfrac{1}{2}, \quad a_1 = 1 - a_3 - a_2 = \tfrac{1}{6}.$$

Also, evaluating (P6.24.1) at $n = 1$ yields $1 = a_3 + a_2 + a_1 + a_0$, and thus

$$a_0 = 1 - a_3 + a_2 + a_1 = 0;$$

therefore, *if* there is a polynomial $f(\cdot)$ satisfying (P6.24.1), then

$$f(x) = \tfrac{1}{3}x^3 + \tfrac{1}{2}x^2 + \tfrac{1}{6}x + 0 = \tfrac{1}{6}x(2x^2 + 3x + 1) = \tfrac{1}{6}x(x+1)(2x+1)$$

(all x). To prove that this polynomial indeed satisfies (P6.24.1), we could proceed by mathematical induction (as done in Section 4.3 for a similar problem). Instead, we shall note that since $f(1) = 1$, the equation in (P6.24.1) now holds for $n = 1$; and, for $n \geq 2$, it follows from (S6.24.1) that

$$\sum_{j=1}^{n} j^2 = 1 + \sum_{j=2}^{n} j^2 = 1 + \sum_{k=1}^{n-1} (k+1)^2 = 1 + \sum_{k=1}^{n-1} [f(k+1) - f(k)]$$
$$= 1 + [f(n) - f(1)] = 1 + f(n) - 1 = f(n).$$

Hence, we can now conclude that

$$\sum_{j=1}^{n} j^2 = f(n) = \tfrac{1}{6}n(n+1)(2n+1) \quad (n = 1, 2, \ldots).$$

Solution 6.25 **(a)** The stated division is performed in Figure S6.25.1. Now, completing the square (see Problem 6.19) of the resulting quadratic polynomial yields $x^2 + 2x + 4 = (x+1)^2 + (\sqrt{3})^2$; thus, the remaining two roots are $x = -1 \pm i\sqrt{3}$. (Alternatively, the quadratic formula could have been used.)

(b) With $n = 3$ and $c = 8$, we have $|c|^{1/n} = 2$ and $\arg c = 0$; therefore, (6.22b) yields the three roots $2e^0 = 2$, $2e^{i2\pi/3}$, and $2e^{i4\pi/3} = 2e^{-i2\pi/3}$. So, other than 2, we have roots at

$$2e^{\pm i2\pi/3} = 2(\cos 2\pi/3 \pm i \sin 2\pi/3) = 2(-1/2 \pm i\sqrt{3}/2) = -1 \pm i\sqrt{3}.$$

$$\begin{array}{r}
x^2 + 2x + 4 \\
x - 2 \,\overline{)\, x^3 + 0x^2 + 0x - 8} \\
\underline{x^3 - 2x^2 \qquad\qquad} \\
2x^2 + 0x \qquad \\
\underline{2x^2 - 4x \qquad} \\
4x - 8 \\
\underline{4x - 8} \\
0
\end{array}$$

FIGURE S6.25.1

Solution 6.26 Applying the quadratic formula (6.23c), we find that the given polynomial has roots

$$x_1, x_2 = \frac{i6 \pm \sqrt{(-i6)^2 - 4(1+i)(-44+i28)}}{2(1+i)} = \frac{i6 \pm \sqrt{-36 - 4(-72 - i16)}}{2(1+i)}$$
$$= \frac{i3 \pm \sqrt{-9 - (-72 - i16)}}{1+i} = \frac{i3 \pm \sqrt{63 + i16}}{1+i},$$

where (6.24) gives

$$\sqrt{63 + i16} = \sqrt{\frac{\sqrt{63^2 + 16^2} + 63}{2}} + i\sqrt{\frac{\sqrt{63^2 + 16^2} - 63}{2}} = 8 + i.$$

Hence,

$$x_1, x_2 = \frac{i3 \pm (8+i)}{1+i} = \frac{(8 + i4, -8 + i2)}{1+i} \cdot \frac{1-i}{1-i} = (4 + i2, -4 + i) \cdot (1 - i)$$
$$= 6 - i2, -3 + i5.$$

Solution 6.27 For $n = 2$, (6.25) becomes

$$f(x) = \frac{x - x_2}{x_1 - x_2} y_1 + \frac{x - x_1}{x_2 - x_1} y_2 \quad (x \in \mathbb{R}).$$

(*Aside*: Letting $y = f(x)$, this equation is essentially equivalent to the determinant equation (11.96), which is also valid when $x_1 = x_2$.) Then, taking

$$m = \frac{y_1 - y_2}{x_1 - x_2}, \quad Y = \frac{x_1 y_2 - x_2 y_1}{x_1 - x_2}$$

yields (5.6a).

Solution 6.28 First, $g(\cdot)$ cannot be the zero polynomial, since we would then have $y_j = g(x_j) = 0$ (all j), and thus $f(\cdot)$ from (6.25) would be the same as $g(\cdot)$. Also, since each of the products in (6.25) has $n - 1$ terms, $f(\cdot)$ either is the zero polynomial or has a degree no greater than $n - 1$. In the latter case, it follows from the fact stated in the problem that $g(\cdot)$ cannot have a degree less than n, since we would then have $f(\cdot) = g(\cdot)$.

Solution 6.29 **(a)** If $m = n$, and $h(\cdot)$ is not the zero polynomial (which could happen), then its possible degrees are $0, 1, \ldots, n$, since any like-power terms in $f(\cdot)$ and $g(\cdot)$ might cancel each other when they are summed; whereas if $m \neq n$, then $h(\cdot)$ cannot be the zero polynomial, and the only possible degree is $\max(m, n)$, since the highest-power term of $f(\cdot)$ or $g(\cdot)$ (whichever is higher) cannot be canceled.

(b) Here, the only possible degree is $m + n$, since the highest-power term of $h(\cdot)$ equals the product of the highest-power terms of $f(\cdot)$ and $g(\cdot)$.

Solution 6.30 Using (6.27) with $g(x) = x - c$ (all x), we have

$$f(x) = q(x)(x - c) + r(x) \quad (x \in \mathbb{C}),$$

where $r(\cdot)$ either is identically 0 or has a degree less than that of $g(\cdot)$. Hence, since $\deg g(\cdot) = 1$, we have $\deg r(\cdot) = 0$ in the latter case. In either case, then, $r(\cdot)$ is a constant function, and the above equation evaluated at $x = c$ yields $f(c) = r(c)$; therefore, $r(x) = f(c)$ $(x \in \mathbb{C})$.

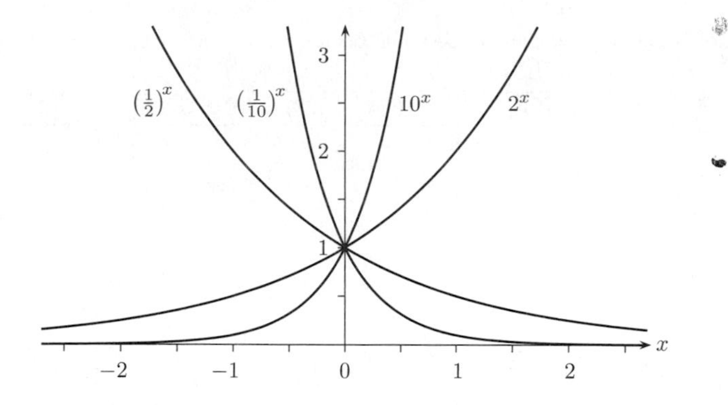

FIGURE S7.1.1 Four Exponential Functions

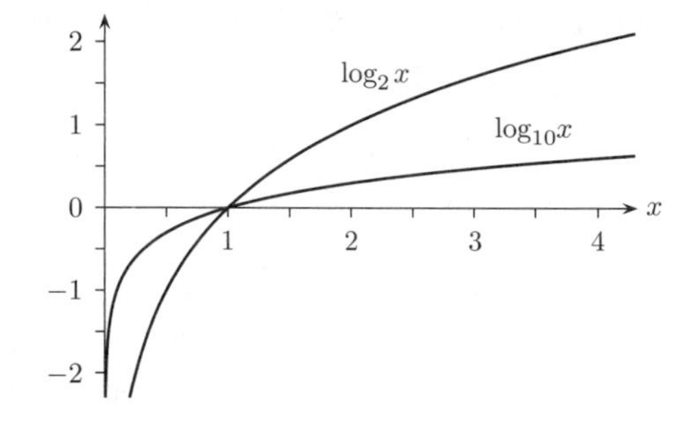

FIGURE S7.1.2 Two Logarithm Functions

Chapter 7. Exponentials and Logarithms

Solution 7.1 **(a)** See Figure S7.1.1.

(b) See Figure S7.1.2.

Solution 7.2 From Section 5.1 (regarding Figure 5.4) we know that the plot of a function scales vertically by a certain factor when the function itself is multiplied by that factor, whereas the plot scales horizontally by a certain factor when the function argument is divided by that factor.

(a) Since $b^{x+\alpha} = b^\alpha b^x$ ($x \in \mathbb{R}$), left-shifting the plot by α is equivalent to vertically scaling it by a factor of b^α.

(b) For $b > 1$, the base-b logarithm is the inverse function of the base-b exponential; so (per Figure 5.6), the corresponding operations in the horizontal and vertical directions are now the reverse of those in part (a). Thus, down-shifting the plot by α is equivalent to horizontally scaling it by a factor of b^α; in other words (letting $\beta = b^\alpha$), horizontally scaling the plot by a factor of $\beta > 0$ is equivalent to down-shifting it by $\log_b \beta$. This conclusion also obtains directly by observing that $\log_b x - \alpha = \log_b(x/b^\alpha)$ and $\log_b(x/\beta) = \log_b x - \log_b \beta$ ($x > 0$).

Solution 7.3 **(a)** Using the continuity of the $(p/100)$th-power function, and (7.2), we obtain

$$\lim_{n\to\infty}(1 + p\%/n)^n = \lim_{n\to\infty}(1 + p/100n)^n = \lim_{n\to\infty}\left[(1 + p/100n)^{100n/p}\right]^{p/100}$$

$$= \left[\lim_{n\to\infty}(1 + p/100n)^{100n/p}\right]^{p/100} = \left[\lim_{\alpha\to\infty}(1 + 1/\alpha)^\alpha\right]^{p/100}$$

$$= e^{p/100} = e^{p\%}.$$

(b) Letting $e^{q\%} = (1 + p\%/n)^n$, we have $q\% = \ln(1 + p\%/n)^n$, and thus $q = 100n \ln(1 + p\%/n)$.

Solution 7.4 Suppose $z, z_1, z_2 \in \mathbb{C}$ and $n \in \mathbb{Z}$. By (7.3),

$$e^z = e^{\operatorname{Re} z}[\cos(\operatorname{Im} z) + i \sin(\operatorname{Im} z)] = e^{\operatorname{Re} z} \cos(\operatorname{Im} z) + ie^{\operatorname{Re} z} \sin(\operatorname{Im} z), \tag{S7.4.1}$$

where the exponential and sinusoidal factors on the right are all real-valued; therefore, (7.4a) and (7.4b) follow. We *could* substitute these results into (3.30a) to obtain (7.4c); alternatively, proceeding via (7.4h) and (3.33) gives

$$|e^z| = \left|e^{\operatorname{Re} z + i \operatorname{Im} z}\right| = \left|e^{\operatorname{Re} z} e^{i \operatorname{Im} z}\right| = e^{\operatorname{Re} z},$$

having used the fact the $e^x > 0$ $(x \in \mathbb{R})$. Similarly,

$$\arg e^z = \arg e^{\operatorname{Re} z + i \operatorname{Im} z} = \arg e^{\operatorname{Re} z} e^{i \operatorname{Im} z} = \arg e^{i \operatorname{Im} z} = \operatorname{Im} z,$$

proving (7.4d). Conjugating (S7.4.1) yields (7.4e):

$$\begin{aligned}(e^z)^* &= e^{\operatorname{Re} z} \cos(\operatorname{Im} z) - ie^{\operatorname{Re} z} \sin(\operatorname{Im} z) = e^{\operatorname{Re} z} \cos(-\operatorname{Im} z) + ie^{\operatorname{Re} z} \sin(-\operatorname{Im} z) \\ &= e^{\operatorname{Re} z^*} \cos(\operatorname{Im} z^*) + ie^{\operatorname{Re} z^*} \sin(\operatorname{Im} z^*) = e^{z^*},\end{aligned}$$

where the last equality obtains by replacing "z" with "z^*" throughout (S7.4.1). Finally, since $e^z \neq 0$, it follows from (6.7), (7.4c), (7.4d), and (7.4h) that

$$\begin{aligned}(e^z)^n &= |e^z|^n \exp(in \arg e^z) = \left(e^{\operatorname{Re} z}\right)^n \exp(in \operatorname{Im} z) = e^{n \operatorname{Re} z} e^{in \operatorname{Im} z} \\ &= e^{n \operatorname{Re} z + in \operatorname{Im} z} = e^{\operatorname{Re} nz + i \operatorname{Im} nz} = e^{nz},\end{aligned}$$

proving (7.4f); then, taking $n = -1$ gives (7.4g).

Solution 7.5 $e^{z + in\pi/2} = e^z e^{in\pi/2} = e^z \left(e^{i\pi/2}\right)^n = e^z i^n$ $(z \in \mathbb{C}, n \in \mathbb{Z})$.

Solution 7.6 A necessary and sufficient condition to have $e^{z_1} = e^{z_2}$ for some $z_1, z_2 \in \mathbb{C}$ is that $|e^{z_1}| = |e^{z_2}|$ and $\arg e^{z_1} = \arg e^{z_2}$; therefore, by (7.4c) and (7.4d), a simpler equivalent condition is: $\operatorname{Re} z_1 = \operatorname{Re} z_2$, and $\operatorname{Im} z_1 = \operatorname{Im} z_1 + 2\pi n$ for some $n \in \mathbb{Z}$ (the addition of $2\pi n$ being due to the ambiguity of the function arg).

Solution 7.7 Suppose z is an imaginary number; that is, $z = i\theta$ for some $\theta \in \mathbb{R}$. Then, $e^{e^z} = e^{e^{i\theta}} = e^{\cos\theta + i \sin\theta}$, and thus $\operatorname{Re} e^{e^z} = e^{\cos\theta} \cos(\sin\theta)$, by (7.4a). Since $e^{\cos\theta} \neq 0$, the number e^{e^z} is imaginary—i.e., $\operatorname{Re} e^{e^z} = 0$—if and only if $\cos(\sin\theta) = 0$, which is the case if and only if $\sin\theta = n\pi/2$ for some odd integer n. But, since $|\sin\theta| \leq 1$ and $\pi/2 > 1$, this condition cannot be satisfied. Therefore, e^{e^z} cannot be imaginary.

Solution 7.8 **(a)** Since the dashed line on the right side Figure 7.1 has slope 1 and x-intercept 1, it follows from (5.6b) that it is a plot of $x - 1$ versus x. We see from the figure that the plot of $\ln x$ versus x is always below this line, except at $x = 1$ where the plot and the line intersect; therefore, we have (P7.8.1a), with equality obtaining if and only if $x = 1$.

Now, letting $y = 1/x$ makes $y > 0$ when $x > 0$. Hence, we can apply (P7.8.1a) to obtain

$$-\ln x = \ln(1/x) = \ln y \leq y - 1 = 1/x - 1,$$

which upon changing signs yields (P7.8.1b); and, since $y = 1$ if and only if $x = 1$, it follows from the condition for equality in (P7.8.1a) that equality attains in (P7.8.1b) if and only if $x = 1$.

(b) Similarly, the left side of Figure 7.1 yields the lower bound

$$e^x \geq x + 1 \quad (x \in \mathbb{R}),$$

with equality being attained if and only if $x = 0$. Now, restricting our attention to $x > -1$, for which both sides of the above inequality are positive, we can take reciprocals and change the sign of x to obtain the (restricted) upper bound

$$e^x \leq 1/(1 - x) \quad (x < 1),$$

with equality being attained if and only if $x = 0$. (These bounds also obtain by replacing "x" throughout (P7.8.1) with "e^x".)

Solution 7.9 **(a)** Since for each $p \in \mathbb{R}$ we have

$$\frac{a^x}{\log_b x} = \frac{a^x}{x^p} \Big/ \frac{\log_b x}{x^p} \quad (x > 0),$$

it follows from (7.1) and (7.7) that

$$\lim_{x\to\infty} \frac{a^x}{\log_b x} = \begin{cases} 0 & 0 < a < 1 \\ \infty & a > 1 \end{cases}$$

($b > 1$); specifically, we can take $p = 0$ for the first case (giving $a^x/x^p \to 0$ and $(\log_b x)/x^p \to \infty$), and take $p = 1$ for the second case (giving $a^x/x^p \to \infty$ and $(\log_b x)/x^p \to 0$). Thus, for all $b > 1$, and every $a > 0$ other than the uninteresting case of $a = 1$, the limiting behavior as $x \to \infty$ of $a^x/\log_b x$ is exactly the same as that of a^x alone.

(b) First observing the asymptotic behavior of the numerator and denominator of $f(x)$ as $x \to \infty$ (as done in the Solution 6.7(b))—viz.,

$$e^{2x} + 8e^{3x} + \log_7 x + 5x^9 \sim 8e^{3x}, \quad 2e^{3x} + 4\ln x + 5\sqrt{6x}e^{-7x} + x^9 \sim 2e^{3x},$$

—we quickly obtain

$$\lim_{x\to\infty} f(x) = \lim_{x\to\infty} \frac{8e^{3x}}{2e^{3x}} = \lim_{x\to\infty} 4 = 4.$$

Solution 7.10 **(a)** Letting $y = (\ln x)/x = \ln x^{1/x}$, we know that $y \to 0$ as $x \to \infty$; therefore, since the exponential function is continuous, we have $x^{1/x} = e^y \to e^0 = 1$ as $y \to 0$. Hence, $x^{1/x} \to 1$ as $x \to \infty$.

(b) With $x > 0$, it follows from the above result that

$$\left(\frac{1}{x}\right)^{1/x} = \frac{1^{1/x}}{x^{1/x}} = \frac{1}{x^{1/x}} \xrightarrow[x\to\infty]{} \frac{1}{1} = 1. \tag{S7.10.1}$$

(c) Since $\lim_{n\to\infty} x_n > 0$, there must exist an integer N such that $1/n \le x_n \le n$ for all $n \ge N$; hence, $(1/n)^{1/n} \le {x_n}^{1/n} \le n^{1/n}$ ($n \ge N$), because every nth-root function is increasing. Therefore, letting $n \to \infty$, these inequalities yield $1 \le \lim_{n\to\infty} {x_n}^{1/n} \le 1$, making $\lim_{n\to\infty} {x_n}^{1/n} = 1$. Then, as in (S7.10.1), since $(1/x_n)^{1/n} = 1\big/{x_n}^{1/n}$ when $x_n > 0$, we also have $\lim_{n\to\infty}(1/x_n)^{1/n} = 1$.

(d) By (5.21), we have

$$\lim_{n\to\infty} \frac{n!}{\sqrt{2\pi n}(n/e)^n} = 1. \tag{S7.10.2}$$

Thus, it follows from the second result of part (c) that

$$\begin{aligned} 1 = \lim_{n\to\infty} \left[\frac{\sqrt{2\pi n}(n/e)^n}{n!}\right]^{1/n} &= \lim_{n\to\infty} \frac{(2\pi n)^{1/2n}(n/e)}{(n!)^{1/n}} \\ &= \frac{1}{e}\left[\lim_{n\to\infty}(2\pi n)^{1/2n}\right]\left[\lim_{n\to\infty}\frac{n}{(n!)^{1/n}}\right], \end{aligned} \tag{S7.10.3}$$

where

$$\lim_{n\to\infty}(2\pi n)^{1/2n} = \lim_{n\to\infty}\left[(2\pi n)^{1/2\pi n}\right]^{\pi} = \left[\lim_{n\to\infty}(2\pi n)^{1/2\pi n}\right]^{\pi} = 1^{\pi} = 1,$$

since x^π is a continuous function of $x > 0$, and by part (a). Hence, multiplying (S7.10.3) by e, we obtain (P7.10.1).

Aside: An alternative (perhaps simpler) derivation of (P7.10.1) starts by taking the natural logarithm of (S7.10.2). To see what to do next, take the natural logarithm of (P7.10.1) itself.

Solution 7.11 **(a)** Taking the natural logarithm of both sides of the equation shows that we have $ab^{cx+d} = e^{\alpha x+\beta}$ ($x \in \mathbb{R}$) if and only if $\ln a + (cx + d)\ln b = \alpha x + \beta$ ($x \in \mathbb{R}$); and, this is the case if and only if $\alpha = c \ln b$ and $\beta = \ln a + d \ln b$.

(b) By (7.8), $a \log_b cx + d = a(\ln cx)/\ln b + d = a(\ln c + \ln x)/\ln b + d$ ($x > 0$); therefore, we have $a \log_b cx + d = \alpha \ln x + \beta$ ($x > 0$) if and only if $\alpha = a/\ln b$ and $\beta = a(\ln c)/\ln b + d$ (i.e., $\beta = a \log_b c + d$).

Solution 7.12 **(a)** For all positive constants $a, b \neq 1$, and values $x, y > 0$ with $y \neq 1$ (thereby making the denominators below nonzero), (7.8) yields

$$\frac{\log_a x}{\log_a y} = \frac{(\log_a b)\log_b x}{(\log_a b)\log_b y} = \frac{\log_b x}{\log_b y}.$$

(b) For $x \neq 1$ it follows from (7.8) that $x^{\log_b y} = x^{(\log_b x)\log_x y} = (x^{\log_x y})^{\log_b x} = y^{\log_b x}$, making $x^{\log_b y} = y^{\log_b x}$; and, this result also obtains for $x = 1$, since then $x^{\log_b y} = 1^{\log_b y} = 1$ and $y^{\log_b x} = y^0 = 1$.

Solution 7.13 **(a)** Given a and b such that $e \le a < b$, it follows from the given fact that $(\ln a)/a > (\ln b)/b$. Hence, $b \ln a > a \ln b$, and thus $\ln a^b > \ln b^a$. Therefore, since the natural logarithm function is strictly increasing, we have $a^b > b^a$.

(b) Taking, e.g., $a = 2$ and $b = 3$, we have $a^b = 8$ and $b^a = 9$, making $a^b < b^a$.

Solution 7.14 **(a)** Taking $x = y = 0$ in (P7.14.1) shows that $f^2(0) = f(0)$; therefore, since $f(\cdot)$ is positive, we must have $f(0) = 1$. Also, since the 0th power of any positive number equals 1, $f^0(y) = 1$ ($y \in \mathbb{R}$); so, (P7.14.3) holds for $m = 0$. Furthermore, (P7.14.1) yields

$$f(x-y)f(y) = f[(x-y)+y] = f(x) \quad (x, y \in \mathbb{R}),$$

and thus

$$f(x)f^{-1}(y) = f(x-y) \quad (x, y \in \mathbb{R})$$

(where the superscript "−1" indicates a reciprocal, not a function inverse), which applied successively m times gives

$$f(x)f^{-m}(y) = f(x-my) \quad (x \in \mathbb{R}, y \in \mathbb{R}, m \in \mathbb{Z}^+);$$

therefore, (P7.14.3) holds for any $m \in \mathbb{Z}$. In particular, for $y = 1/n$ we have

$$f(x)f^m(1/n) = f(x+m/n) \quad (x \in \mathbb{R}, m \in \mathbb{Z}, n \in \mathbb{Z}^+), \tag{S7.14.1}$$

which for $x = 0$ and $m = n$ yields (P7.14.4); so, we can substitute $f(1/n) = f^{1/n}(1)$ into (S7.14.1), thereby obtaining (P7.14.5).

For any rational number q, there exist $m \in \mathbb{Z}$ and $n \in \mathbb{Z}^+$ such that $q = m/n$. Furthermore, for any $x \in \mathbb{R}$ there exists a sequence of rational numbers converging to the real number $X - x$; thus, there exists a sequence of pairs $(m, n) \in \mathbb{Z} \times \mathbb{Z}^+$ such that $m/n \to X - x$. Then, $x + m/n \to X$, making $f(x + m/n) \to f(X)$, since $f(\cdot)$ is continuous at X; so, in the limit, (P7.14.5) becomes

$$f(x)f^{X-x}(1) = f(X) \quad (x \in \mathbb{R}). \tag{S7.14.2}$$

In particular, taking $x = 0$ shows that $f^X(1) = f(X)$; therefore, letting $b = f(1) > 0$, we can manipulate (S7.14.2) to obtain

$$f(x) = \frac{f(X)}{f^X(1)}f^x(1) = b^x \quad (x \in \mathbb{R}).$$

(b) Given a positive function $f(\cdot)$ on $\mathbb{R}$ such that (P7.14.1) holds, and that is discontinuous at some point, suppose that it is *not* discontinuous at every point. Then, $f(\cdot)$ is continuous at some point, and it follows from the result of part (a) that $f(\cdot)$ must be continuous at every point—which it is not. Thus, the assumption that $f(\cdot)$ is not discontinuous at every point leads to a contradiction; therefore, $f(\cdot)$ must be discontinuous at every point.

Solution 7.15 Throughout the derivations to follow, $a > 0$ and $z, z_1, z_2 \in \mathbb{C}$; and, definition (7.10) is invoked as needed. By (7.4a),

$$\begin{aligned}\operatorname{Re} a^z &= \operatorname{Re} e^{(\ln a)z} = e^{\operatorname{Re}[(\ln a)z]} \cos\{\operatorname{Im}[(\ln a)z]\} = e^{(\ln a)\operatorname{Re} z} \cos[(\ln a)\operatorname{Im} z] \\ &= a^{\operatorname{Re} z} \cos[(\ln a)\operatorname{Im} z],\end{aligned}$$

proving (7.11a); similarly, by (7.4b),

$$\begin{aligned}\operatorname{Im} a^z &= \operatorname{Im} e^{(\ln a)z} = e^{\operatorname{Re}[(\ln a)z]} \sin\{\operatorname{Im}[(\ln a)z]\} = e^{(\ln a)\operatorname{Re} z} \sin[(\ln a)\operatorname{Im} z] \\ &= a^{\operatorname{Re} z} \sin[(\ln a)\operatorname{Im} z],\end{aligned}$$

proving (7.11b). By (7.4c),

$$|a^z| = \left|e^{(\ln a)z}\right| = e^{\operatorname{Re}[(\ln a)z]} = e^{(\ln a)\operatorname{Re} z} = a^{\operatorname{Re} z},$$

proving (7.11c); and, by (7.4d),

$$\arg a^z = \arg e^{(\ln a)z} = \operatorname{Im}[(\ln a)z] = (\ln a)\operatorname{Im} z,$$

proving (7.11d). By (7.4e),

$$\left(a^z\right)^* = \left[e^{(\ln a)z}\right]^* = e^{[(\ln a)z]^*} = e^{(\ln a)z^*} = a^{z^*},$$

proving (7.11e). By (7.4f),

$$(a^z)^n = \left[e^{(\ln a)z}\right]^n = e^{n(\ln a)z} = e^{(\ln a)(nz)} = a^{nz} \quad (n \in \mathbb{Z}),$$

proving (7.11f); and, taking $n = -1$ yields (7.11g). Finally, by (7.4h),

$$a^{z_1} a^{z_2} = e^{(\ln a)z_1} e^{(\ln a)z_2} = e^{(\ln a)z_1 + (\ln a)z_2} = e^{(\ln a)(z_1+z_2)} = a^{z_1+z_2},$$

proving (7.11h); similarly, by (7.4i),

$$a^{z_1}/a^{z_2} = e^{(\ln a)z_1} \big/ e^{(\ln a)z_2} = e^{(\ln a)z_1 - (\ln a)z_2} = e^{(\ln a)(z_1-z_2)} = a^{z_1-z_2},$$

proving (7.11i).

Chapter 8. Possibility and Probability

Solution 8.1 **(a)** Inserting the missing even positive integers, we obtain

$$\begin{aligned}(2n+1)!! &= [1\cdot 3\cdot 5\cdot \cdots \cdot (2n+1)]\frac{2\cdot 4\cdot 6\cdot \cdots \cdot (2n)}{2\cdot 4\cdot 6\cdot \cdots \cdot (2n)} = \frac{(2n+1)!}{2^n(1\cdot 2\cdot 3\cdot \cdots \cdot n)} \\ &= \frac{(2n+1)!}{2^n n!} \quad (n = 1, 2, \dots);\end{aligned}$$

and, this result is also valid for $n = 0$, as can be directly verified.

(b) It now follows from (5.21) that, as $n \to \infty$,

$$\begin{aligned}(2n+1)!! &\sim \frac{\sqrt{2\pi(2n+1)}\,[(2n+1)/e]^{(2n+1)}}{2^n\cdot\sqrt{2\pi n}\,(n/e)^n} = \frac{\sqrt{2n+1}\,[(2n+1)/e]^{(2n+1)}}{\sqrt{n}(2n/e)^n} \\ &= \sqrt{2+1/n}\,\frac{(2n+1)^{(2n+1)}}{(2n)^n e^{n+1}},\end{aligned}$$

where $\sqrt{2+1/n} \to \sqrt{2}$; therefore,

$$(2n+1)!! \sim \sqrt{2}\,\frac{(2n+1)^{(2n+1)}}{(2n)^n e^{n+1}} \quad \text{as} \quad n \to \infty.$$

Solution 8.2 Letting $m = |A|$ and $n = |B|$, we have $m! = 20 \cdot n!$. Dividing by $n!$—which must be less than $m!$, and thus $m > n \geq 0$—we have $m(m-1)(m-2)\cdots(n+1) = 20$. Thus, considering all m, $n \in \{0, 1, \ldots, 20\}$ such that $m > n$, we find that either $|A| = 5$ and $|B| = 3$, or $|A| = 20$ and $|B| = 19$.

Solution 8.3 **(a)** Taking $x = y = 1$ in (8.4) yields (P8.3.1).

(b) When forming a combination of an n-element set A ($n \in \mathbb{N}$), there are two possibilities for each element $a \in A$; namely, either a is in the combination, or it is not. Thus, for $n \geq 1$, combining these possibilities over the n elements yields $2 \cdot 2 \cdot \cdots \cdot 2 = 2^n$ total combinations; whereas for $n = 0$, there is only the $1 = 2^n$ possible combination $\emptyset$.

Solution 8.4 **(a)** For all $n \in \{2, 3, \ldots\}$ and $k \in \{1, 2, \ldots, n-1\}$, we have

$$\begin{aligned}\binom{n-1}{k} + \binom{n-1}{k-1} &= \frac{(n-1)!}{k!(n-1-k)!} + \frac{(n-1)!}{(k-1)!(n-k)!} \\ &= \frac{n-k}{n} \cdot \frac{n!}{k!(n-k)!} + \frac{k}{n} \cdot \frac{n!}{k!(n-k)!} \\ &= \left(\frac{n-k}{n} + \frac{k}{n}\right)\frac{n!}{k!(n-k)!} = \frac{n!}{k!(n-k)!} = \binom{n}{k}.\end{aligned}$$

(b) The quantity $\binom{n}{k}$ is the number of ways of selecting k objects from a set A of n objects. Now, suppose $a \in A$; then, a either is, or is not, in the selected k objects. The number of ways of selecting the k objects such that a is *not* in the selection is $\binom{n-1}{k}$, since we can first discard a and then choose k objects from the remaining $n-1$ objects in A. Whereas, the number of ways of selecting the k objects such that a *is* in the selection is $\binom{n-1}{k-1}$, since we can first select a and then choose $k-1$ more objects from the remaining $n-1$ objects in A. Hence, (P8.4.1b) holds.

Solution 8.5 For $m = 0$, both sides of (P8.5.1) equal 1 for all $n \geq 1$, by (P8.4.1a). Whereas for $m \geq 1$ we must have $n \geq 2$, thereby enabling the use of (P8.4.1b) to obtain

$$\begin{aligned}\sum_{k=0}^{m}(-1)^k\binom{n}{k} &= \binom{n}{0} + \sum_{k=1}^{m}(-1)^k\binom{n}{k} = 1 + \sum_{k=1}^{m}(-1)^k\left[\binom{n-1}{k} + \binom{n-1}{k-1}\right] \\ &= 1 + \sum_{k=1}^{m}(-1)^k\binom{n-1}{k} + \sum_{k=1}^{m}(-1)^k\binom{n-1}{k-1} \\ &= 1 + \sum_{k=1}^{m}(-1)^k\binom{n-1}{k} - \sum_{l=0}^{m-1}(-1)^l\binom{n-1}{l} \\ &= 1 + (-1)^m\binom{n-1}{m} - \binom{n-1}{0} = (-1)^m\binom{n-1}{m}\end{aligned}$$

($n = 2, 3, \ldots; m = 1, 2, \ldots, n-1$), proving (P8.5.1). Finally, for $m = n = 0$, the left side of (P8.5.1) equals $(-1)^0\binom{0}{0} = 1$; whereas for $m = n \geq 1$, it follows from (P8.5.1) that

$$\begin{aligned}\sum_{k=0}^{n}(-1)^k\binom{n}{k} &= \sum_{k=0}^{n-1}(-1)^k\binom{n}{k} + (-1)^n\binom{n}{n} = (-1)^{n-1}\binom{n-1}{n-1} + (-1)^n\binom{n}{n} \\ &= (-1)^{n-1} + (-1)^n = 0.\end{aligned}$$

(These last results also obtain from the binomial expansion (8.4) by taking $x = -1$ and $y = 1$.)

Solution 8.6 Since

$$I_{1,n} = \{n_1 \in \mathbb{N} \mid n_1 = n\} = \{n\} \quad (n \in \mathbb{N}),$$

(P8.6.1) for $m = 1$ becomes

$$x_1{}^n = \sum_{n_1 = n} \frac{n!}{n_1!} x_1{}^{n_1} \quad (x_1 \in \mathbb{C}),$$

which is true. Now, assume (induction hypothesis) that (P8.6.1) holds for a particular integer $m \geq 1$ (with any $x_1, \ldots, x_m \in \mathbb{C}$ and $n \in \mathbb{N}$); then, we can use the binomial expansion (8.4) to obtain

$$\begin{aligned}
(x_1 + \cdots + x_{m+1})^n &= [(x_1 + \cdots + x_m) + x_{m+1}]^n \\
&= \sum_{k=0}^{n} \frac{n!}{k!(n-k)!} (x_1 + \cdots + x_m)^k x_{m+1}{}^{n-k} \\
&= \sum_{k=0}^{n} \frac{n!}{k!(n-k)!} \left[\sum_{(n_1,\ldots,n_m) \in I_{m,k}} \frac{k!}{n_1! \cdots n_m!} x_1{}^{n_1} \cdots x_m{}^{n_m} \right] x_{m+1}{}^{n-k} \\
&= \sum_{k=0}^{n} \sum_{(n_1,\ldots,n_m) \in I_{m,k}} \frac{n!}{n_1! \cdots n_m!(n-k)!} x_1{}^{n_1} \cdots x_m{}^{n_m} x_{m+1}{}^{n-k} \\
&= \sum_{(n_1,\ldots,n_m,n_{m+1}) \in I_{m+1,n}} \frac{n!}{n_1! \cdots n_m! n_{m+1}!} x_1{}^{n_1} \cdots x_m{}^{n_m} x_{m+1}{}^{n_{m+1}}
\end{aligned}$$

$(x_1, \ldots, x_{m+1} \in \mathbb{C};\ n \in \mathbb{N})$, because

$$\begin{aligned}
I_{m+1,n} &= \{(n_1, \ldots, n_m, n_{m+1}) \in \mathbb{N}^{m+1} \mid n_1 + \cdots + n_m + n_{m+1} = n\} \\
&= \bigcup_{k=0}^{n} \{(n_1, \ldots, n_m, n-k) \in \mathbb{N}^{m+1} \mid n_1 + \cdots + n_m = k\} \\
&= \bigcup_{k=0}^{n} \bigcup_{(n_1,\ldots,n_m) \in I_{m,k}} \{(n_1, \ldots, n_m, n-k)\}
\end{aligned}$$

$(n \in \mathbb{N})$. Thus, in addition to showing that (P8.6.1) holds for $m = 1$, we have shown that if (P8.6.1) holds for a particular integer $m \geq 1$, then it also holds upon replacing each "m" with "$m+1$"; therefore, by mathematical induction, (P8.6.1) holds for all $m \geq 1$ (with any $x_1, \ldots, x_m \in \mathbb{C}$ and $n \in \mathbb{N}$).

Solution 8.7 Suppose that the six objects to be permuted were instead 1, 2, 2′, 3, 3′, and 3″, which are all distinct. Then, there will be 6! total permutations. Within each of these permutations, there are 2! ways to rearrange the objects 2 and 2′ while holding the other objects fixed; likewise, there are 3! ways to rearrange the objects 3, 3′, and 3″. Therefore, now treating 2 and 2′ as indistinguishable, and likewise for 3, 3′, and 3″, we conclude that the number of permutations of 1, 2, 2, 3, 3, and 3 is $6!/2!3! = 60$.

Solution 8.8 There are 13 ways of choosing the rank from which the three-of-a-kind is taken, after which there are 12 ways of choosing the rank from which the pair is taken; therefore, the total number of ways of specifying these ranks is $13 \cdot 12 = 156$—not $\binom{13}{2}$, because here order *is* significant. Each rank consists of four cards (one of each suit); so, within a given rank, there are $\binom{4}{3} = 4$ ways of having three of a kind, and $\binom{4}{2} = 6$ ways of having a pair. Hence, the number of hands that constitute a full house is $156 \cdot 4 \cdot 6 = 3744$.

Solution 8.9 Let the generic word of length n be the n-tuple $(a_1, a_2, \ldots, a_n)$, where $a_i \in A$ $(i = 1, 2, \ldots, n)$.

(a) For the stated condition, there are exactly m ways to select a_1; but then there are exactly $m-1$ ways to select each of $a_2, \dots, a_n$. Therefore, the number of allowable words is $m(m-1)^{n-1}$.

(b) Here, for n even, we can freely choose $a_1, a_2, \dots, a_{n/2}$, after which we must take $a_n = a_1$, $a_{n-1} = a_2$, $\dots$, $a_{n/2+1} = a_{n/2}$; therefore, the number of palindromes is $m^{n/2}$ when n is even. Whereas for n odd, we can freely choose $a_1, a_2, \dots, a_{(n+1)/2}$, after which (if $n \neq 1$) we must take $a_n = a_1, a_{n-1} = a_2, \dots, a_{(n+3)/2} = a_{(n-1)/2}$; therefore, the number of palindromes is $m^{(n+1)/2}$ when n is odd.

Solution 8.10 A circular permutation of A can be formed by first arbitrarily choosing an element from A and placing it on the circle; then, continuing to place objects from A along the circle in one direction, there are $n-1$ possibilities for the next object, after which there are $n-2$ possibilities for the next object, and so forth, until finally there is a single possibility for the last object. Therefore, there are $(n-1)\cdot(n-2)\cdots 1 = (n-1)!$ circular permutations of A. Alternatively, this result obtains by noting that each circular permutation essentially comprises n ordinary permutations of A, each obtainable by a circular shift of any other, but each having a different element that is designated as "first"; therefore, since A has $n!$ ordinary permutations, it has $n!/n = (n-1)!$ circular permutations.

Solution 8.11 Odds of 3 in 7 correspond to a probability of $3/7$, whereas odds of 8 to 5 against correspond to a probability of $5/(8+5) = 5/13$; therefore, since $3/7 = 39/91 > 35/91 = 5/13$, an event with the former odds is more probable.

Solution 8.12 **(a)** Since each coin flip has two possible outcomes, the total number coin-flip sequences of length $n > 0$ is 2^n; and, for n even, $\binom{n}{n/2}$ of these sequences have exactly as many heads as tails. Therefore, by (8.6) we have

$$p_n = \binom{n}{n/2} \Big/ 2^n = \frac{n!}{2^n[(n/2)!]^2} \quad (n = 2, 4, 6, \dots).$$

(b) Applying (5.21) to this result, we obtain

$$p_n \sim \frac{\sqrt{2\pi n}(n/e)^n}{2^n\{\sqrt{2\pi(n/2)}\,[(n/2)/e]^{n/2}\}^2} = \frac{\sqrt{2\pi n}(n/e)^n}{2^n[\pi n(n/2e)^n]} = \frac{\sqrt{2\pi n}}{\pi n} = \sqrt{2/\pi n}$$

as $n \to \infty$.

(c) See Table S8.12.1.

Solution 8.13 If the person were to immediately select choice E after observing the first flip to be HEAD, then the probability of choosing E would be $\frac{1}{2}$ rather than the desired $\frac{1}{5}$. Instead, the person should flip the coin again to see whether a second HEAD occurs; and, if so, flip it yet again to see whether a third HEAD occurs. If three HEADs indeed occur, then choice E should be selected; otherwise, the person should conduct another three-flip experiment, and repeatedly perform this experiment until the three flips correspond to one of the five choices. Note, though, that if both flips 1 and 2 of a particular experiment yield HEAD, then no harm is done in aborting that experiment—i.e., not actually performing its flip 3—thereby making the next flip the first of the succeeding three-flip experiment.

Solution 8.14 If $m = 0$—i.e., $\emptyset$ is the only possible U—then the numerator and denominator in (P8.14.1) equal 0 and 1 respectively, making $\Pr(b \in U) = 0 = m/n$; therefore, assume $m \geq 1$, thereby

TABLE S8.12.1

n	p_n	$\sqrt{2/\pi n}$
2	0.500	0.564...
4	0.375	0.398...
10	0.246...	0.252...
20	0.176...	0.178...
40	0.125...	0.126...

making $m-1 \geq 0$ and $n-1 \geq 0$ in what follows. The denominator in (P8.14.1) equals $\binom{n}{m}$, as implicitly observed in the problem statement. Similarly, the numerator now equals $\binom{n-1}{m-1}$; because, once we have set aside the particular element b of V that is required to be in U, we can freely choose the other $m-1$ elements of U from the remaining $n-1$ elements of V. Hence, (P8.14.1) yields

$$\Pr(b \in U) = \binom{n-1}{m-1} \Big/ \binom{n}{m} = \frac{(n-1)!}{(m-1)!(n-m)!} \Big/ \frac{n!}{m!(n-m)!}$$
$$= \frac{m!/(m-1)!}{n!/(n-1)!} = \frac{m}{n}.$$

Solution 8.15 Given $m, n \geq 1$, there are n^m sequences $\{x_i\}_{i=1}^{m}$ composed of m samples $x_i \in \{1, 2, \ldots, n\}$. If $m \leq n$, then per the derivation of (8.3), $n!/(n-m)!$ of these sequences have all different values for the x_i, and thus the probability of choosing any such sequence is $n!/(n-m)!\,n^m$; whereas for $m > n$, no such sequence exists, so the probability of choosing one is 0. (Note that for $m \leq n$, the correct answer is not $\binom{n}{m}/n^m$; because, each sequence—as *ordered*—has the same probability of being selected as any other.)

Solution 8.16 The given eight samples have arithmetic mean

$$\overline{x} = (37 + 12 + 34 + 46 + 26 + 14 + 34 + 15)/8 = 27.25,$$

geometric mean

$$\sqrt[8]{37 \cdot 12 \cdot 34 \cdot 46 \cdot 26 \cdot 14 \cdot 34 \cdot 15} \approx 24.48,$$

and harmonic mean

$$8/(1/37 + 1/12 + 1/34 + 1/46 + 1/26 + 1/14 + 1/34 + 1/15) \approx 21.77.$$

Also, upon ordering the samples by value—viz., 12, 14, 15, 26, 34, 34, 37, 46—we find that they have median $(26 + 34)/2 = 30$; and, there is only one mode, of value 34. Finally, the samples have second moment

$$\overline{x^2} = (37^2 + 12^2 + 34^2 + 46^2 + 26^2 + 14^2 + 34^2 + 15^2)/8 = 879.75,$$

variance ${\sigma_x}^2 = \overline{x^2} - \overline{x}^2 \approx 137.19$, standard deviation $\sigma_x \approx 11.71$, and root mean square $\sqrt{\overline{x^2}} \approx 29.66$.

Solution 8.17 Since $1/x_i > 0$ (all i) and $n \geq 2$, we have

$$\sum_{i=1}^{n} \frac{1}{x_i} > \max_i \frac{1}{x_i} = 1 \Big/ \min_i x_i,$$

because the reciprocal function is strictly decreasing. Hence,

$$1 \Big/ \sum_{i=1}^{n} \frac{1}{x_i} < \min_i x_i,$$

which multiplied by n yields (P8.17.1).

Solution 8.18 **(a)** Since $d_i = s_i t_i$ (all i), we have

$$S = \frac{D}{T} = \frac{d_1 + d_2 + \cdots + d_n}{T} = \frac{s_1 t_1 + s_2 t_2 + \cdots + s_n t_n}{T} = \sum_{i=1}^{n} w_i s_i$$

for weights $w_i = t_i/T$ ($i = 1, 2, \ldots, n$)—thereby making $w_i \geq 0$ (all i) and $\sum_{i=1}^{n} w_i = 1$, as required.

(b) Since $t_i = d_i/s_i$ (all i), we also have

$$S = \frac{D}{T} = \frac{D}{t_1 + t_2 + \cdots + t_n} = \frac{D}{d_1/s_1 + d_2/s_2 + \cdots + d_n/s_n} = 1 \Big/ \sum_{i=1}^{n} w_i \frac{1}{s_i}$$

for weights $w_i = d_i/D$ ($i = 1, 2, \ldots, n$)—again making $w_i \geq 0$ (all i) and $\sum_{i=1}^{n} w_i = 1$.

Solution 8.19 **(a)** Per (8.9),

$$\overline{ax} = \frac{1}{n}\sum_{i=1}^{n} ax_i = a \cdot \frac{1}{n}\sum_{i=1}^{n} x_i = a\,\overline{x},$$

$$\overline{x+y} = \frac{1}{n}\sum_{i=1}^{n}(x_i + y_i) = \frac{1}{n}\sum_{i=1}^{n} x_i + \frac{1}{n}\sum_{i=1}^{n} y_i = \overline{x} + \overline{y}.$$

(b) First, observe that it follows from part (a) that we also have

$$\overline{x-y} = \overline{x + (-1)y} = \overline{x} + \overline{(-1)y} = \overline{x} + (-1)\overline{y} = \overline{x} - \overline{y}.$$

Therefore, using the fact that $\overline{a} = a$ for any real constant a, we obtain

$$\sigma_x{}^2 \triangleq \overline{(x-\overline{x})^2} = \overline{x^2 - 2\overline{x}x + \overline{x}^2} = \overline{x^2} - \overline{2\overline{x}x} + \overline{\overline{x}^2} = \overline{x^2} - 2\overline{x}\cdot\overline{x} + \overline{x}^2 = \overline{x^2} - \overline{x}^2,$$

as compared with (8.15).

Solution 8.20 **(a)** See the left side of Figure S8.20.1.

(b) See the right side of Figure S8.20.1, where each bin interval comprises four of those in the histogram on the left.

Solution 8.21 **(a)** Since the frequency histogram on the left side of Figure 8.1 has a uniform bin width of 1, we simply divide vertically by the number of samples (40) to obtain the corresponding probability histogram, shown on the left side of Figure S8.21.1. (Regarding why the vertical axis is labeled "probability density", see the discussion of this topic in Section E.4, along with Figure E.8.)

(b) Here we divide the frequency of each bin by both its width and its number of samples, thereby obtaining the right side of Figure S8.21.1. (Note that the total area under both probability histograms is 1, and that their shapes are roughly the same.)

Solution 8.22 **(a)** See Figure S8.22.1.

(b) The coefficients of the regression line described by (8.16) are calculated from (8.18) to be $a_0 \approx 45.8$ and $a_1 \approx 0.186$; these produce the solid line in Figure S8.22.1. Also, the centroid

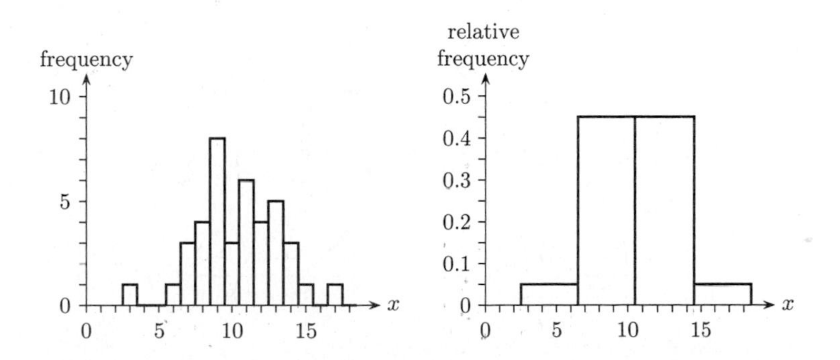

FIGURE S8.20.1 A Frequency Histogram and a Relative-Frequency Histogram

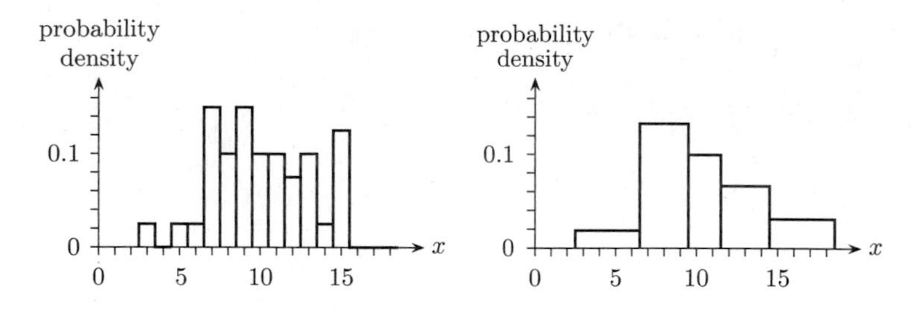

FIGURE S8.21.1 Two Probability Histograms

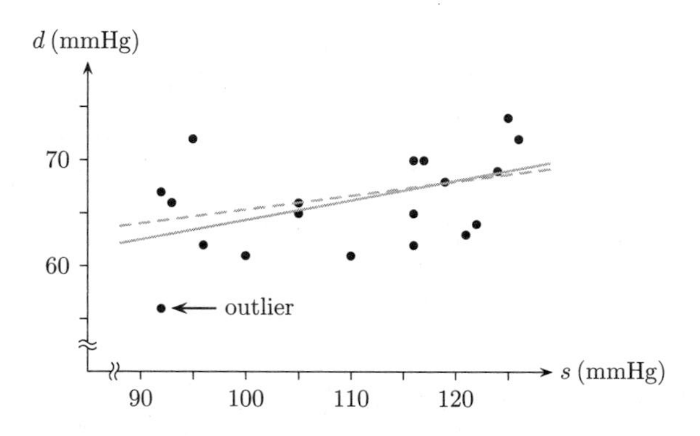

FIGURE S8.22.1 A Scattergram and Two Regression Lines

$\overline{(s,d)} \approx (110.8, 66.4)$, and the correlation coefficient is calculated from (8.19) to be $r \approx 0.476$ (a positive correlation between the two blood pressure measurements, as one might expect).

(c) The outlier point (s,d) is $(92, 56)$. After it is removed, we find $a_0 \approx 52.1$ and $a_1 \approx 0.132$, which produce the dashed line in Figure S8.22.1; also, $\overline{(s,d)} \approx (111.7, 66.9)$ and $r \approx 0.367$.

Solution 8.23 Following the suggestion—and, noting that with n fixed, minimizing $n\overline{[f(x)-y]^2}$ is equivalent to minimizing $\overline{[f(x)-y]^2}$—we obtain

$$\begin{aligned}
n\overline{[f(x)-y]^2} &= \sum_{i=1}^{n}(a_1x_i - y_i)^2 = \sum_{i=1}^{n}({a_1}^2{x_i}^2 - 2a_1x_iy_i + {y_i}^2) \\
&= {a_1}^2\sum_{i=1}^{n}{x_i}^2 - 2a_1\sum_{i=1}^{n}x_iy_i + \sum_{i=1}^{n}{y_i}^2 \\
&= \left(\sum_{i=1}^{n}{x_i}^2\right)\left({a_1}^2 - 2a_1\sum_{i=1}^{n}x_iy_i \Big/ \sum_{i=1}^{n}{x_i}^2\right) + \sum_{i=1}^{n}{y_i}^2 \\
&= \left(\sum_{i=1}^{n}{x_i}^2\right)\left(a_1 - \sum_{i=1}^{n}x_iy_i \Big/ \sum_{i=1}^{n}{x_i}^2\right)^2 - \left(\sum_{i=1}^{n}x_iy_i\right)^2 \Big/ \sum_{i=1}^{n}{x_i}^2 \\
&\quad + \sum_{i=1}^{n}{y_i}^2,
\end{aligned}$$

which, since $\sum_{i=1}^{n} x_i^2 > 0$, is minimized versus a_1 if and only if

$$a_1 = \sum_{i=1}^{n} x_i y_i \bigg/ \sum_{i=1}^{n} x_i^2 = \frac{1}{n}\sum_{i=1}^{n} x_i y_i \bigg/ \frac{1}{n}\sum_{i=1}^{n} x_i^2 = \overline{xy} \big/ \overline{x^2}.$$

Solution 8.24 **(a)** In (8.15) it was found that $\overline{(x-\overline{x})^2} = \overline{x^2} - \overline{x}^2$; likewise, $\overline{(y-\overline{y})^2} = \overline{y^2} - \overline{y}^2$. Also, a similar analysis yields

$$\begin{aligned}\overline{(x-\overline{x})(y-\overline{y})} &= \frac{1}{n}\sum_{i=1}^{n}(x_i-\overline{x})(y_i-\overline{y}) = \frac{1}{n}\sum_{i=1}^{n}(x_i y_i - x_i\overline{y} - \overline{x}y_i + \overline{x}\cdot\overline{y}) \\ &= \frac{1}{n}\sum_{i=1}^{n} x_i y_i - \overline{y}\cdot\frac{1}{n}\sum_{i=1}^{n} x_i - \overline{x}\cdot\frac{1}{n}\sum_{i=1}^{n} y_i + \frac{1}{n}\sum_{i=1}^{n}\overline{x}\cdot\overline{y} \\ &= \overline{xy} - \overline{y}\cdot\overline{x} - \overline{x}\cdot\overline{y} + \overline{x}\cdot\overline{y} = \overline{xy} - \overline{x}\cdot\overline{y};\end{aligned}$$

more simply, it follows from Problem 8.19(a) that

$$\begin{aligned}\overline{(x-\overline{x})(y-\overline{y})} &= \overline{xy - \overline{y}x - \overline{x}y + \overline{x}\cdot\overline{y}} = \overline{xy} - \overline{\overline{y}x} - \overline{\overline{x}y} + \overline{\overline{x}\cdot\overline{y}} \\ &= \overline{xy} - \overline{y}\cdot\overline{x} - \overline{x}\cdot\overline{y} + \overline{x}\cdot\overline{y} = \overline{xy} - \overline{x}\cdot\overline{y}.\end{aligned}$$

Substituting these results into the first part of (8.19) gives the first part of (P8.24.1); and, expressing each average indicated by an overbar as a corresponding sum divided by n, then multiplying numerator and denominator by n^2, we obtain the remainder of (P8.24.1).

(b) By (8.18b) and the above results,

$$\begin{aligned}a_1 &= \frac{\overline{xy} - \overline{x}\cdot\overline{y}}{\overline{x^2} - \overline{x}^2} = \frac{\overline{(x-\overline{x})(y-\overline{y})}}{\overline{(x-\overline{x})^2}} \\ &= \frac{\overline{(x-\overline{x})(y-\overline{y})}}{\sqrt{\overline{(x-\overline{x})^2}\cdot\overline{(y-\overline{y})^2}}}\cdot\frac{1}{\sqrt{\overline{(x-\overline{x})^2}\big/\overline{(y-\overline{y})^2}}} = r\sqrt{\overline{(y-\overline{y})^2}\big/\overline{(x-\overline{x})^2}},\end{aligned}$$

by (8.19).

Chapter 9. Matrices

Solution 9.1 For the given matrices **A** and **B**,

$$-5\mathbf{A} = \begin{bmatrix} -5(1) & -5(-9) & -5(7) \\ -5(0) & -5(5) & -5(-3) \end{bmatrix} = \begin{bmatrix} -5 & 45 & -35 \\ 0 & -25 & 15 \end{bmatrix},$$

$$\mathbf{B}/2 = \begin{bmatrix} (-2)/2 & (8)/2 & (0)/2 \\ (0)/2 & (4)/2 & (-6)/2 \end{bmatrix} = \begin{bmatrix} -1 & 4 & 0 \\ 0 & 2 & -3 \end{bmatrix},$$

$$\mathbf{A}+\mathbf{B} = \begin{bmatrix} 1+(-2) & -9+8 & 7+0 \\ 0+0 & 5+4 & -3+(-6) \end{bmatrix} = \begin{bmatrix} -1 & -1 & 7 \\ 0 & 9 & -9 \end{bmatrix},$$

$$\mathbf{A}-\mathbf{B} = \begin{bmatrix} 1-(-2) & -9-8 & 7-0 \\ 0-0 & 5-4 & -3-(-6) \end{bmatrix} = \begin{bmatrix} 3 & -17 & 7 \\ 0 & 1 & 3 \end{bmatrix},$$

$$\begin{aligned}\mathbf{A}\mathbf{B}^{\mathrm{T}} &= \begin{bmatrix} 1 & -9 & 7 \\ 0 & 5 & -3 \end{bmatrix}\begin{bmatrix} -2 & 0 \\ 8 & 4 \\ 0 & -6 \end{bmatrix} \\ &= \begin{bmatrix} (1)(-2)+(-9)(8)+(7)(0) & (1)(0)+(-9)(4)+(7)(-6) \\ (0)(-2)+(5)(8)+(-3)(0) & (0)(0)+(5)(4)+(-3)(-6) \end{bmatrix} \\ &= \begin{bmatrix} -74 & -78 \\ 40 & 38 \end{bmatrix},\end{aligned}$$

$$\mathbf{A}^{\mathrm{T}}\mathbf{B} = \begin{bmatrix} 1 & 0 \\ -9 & 5 \\ 7 & -3 \end{bmatrix} \begin{bmatrix} -2 & 8 & 0 \\ 0 & 4 & -6 \end{bmatrix}$$

$$= \begin{bmatrix} (1)(-2)+(0)(0) & (1)(8)+(0)(4) & (1)(0)+(0)(-6) \\ (-9)(-2)+(5)(0) & (-9)(8)+(5)(4) & (-9)(0)+(5)(-6) \\ (7)(-2)+(-3)(0) & (7)(8)+(-3)(4) & (7)(0)+(-3)(-6) \end{bmatrix}$$

$$= \begin{bmatrix} -2 & 8 & 0 \\ 18 & -52 & -30 \\ -14 & 44 & 18 \end{bmatrix},$$

$$\mathbf{B}\mathbf{A}^{\mathrm{T}} = \left(\mathbf{A}\mathbf{B}^{\mathrm{T}}\right)^{\mathrm{T}} = \begin{bmatrix} -74 & 40 \\ -78 & 38 \end{bmatrix},$$

$$\mathbf{B}^{\mathrm{T}}\mathbf{A} = \left(\mathbf{A}^{\mathrm{T}}\mathbf{B}\right)^{\mathrm{T}} = \begin{bmatrix} -2 & 18 & -14 \\ 8 & -52 & 44 \\ 0 & -30 & 18 \end{bmatrix}.$$

Solution 9.2 A zero matrix $\mathbf{0}$ is a diagonal matrix if and only if it is square (as a diagonal matrix must be), since then all of the elements off the main diagonal are 0 (as happen to be the other elements).

Solution 9.3 **(a)** Let $\mathbf{A} = [a_{i,j}]$. If $\mathbf{A}$ is antisymmetric, then $a_{i,j} = -a_{j,i}$ (all i, j); thus, on the main diagonal we have $a_{i,i} = -a_{i,i}$ (all i), which implies that all of the elements there are 0. Whereas if $\mathbf{A}$ is anti-Hermitian, then, $a_{i,j} = -a_{j,i}{}^*$ (all i, j), and thus $a_{i,i} = -a_{i,i}{}^*$ (all i), which implies that the elements on the main diagonal are all imaginary. By contrast, $\mathbf{A}$ is symmetric if and only if $a_{i,j} = a_{j,i}$ (all i, j), which does not constrain the elements on the main diagonal. However, if $\mathbf{A}$ is Hermitian, then, $a_{i,j} = a_{j,i}{}^*$ (all i, j), and thus $a_{i,i} = a_{i,i}{}^*$ (all i), which implies that the elements on the main diagonal are all real.

(b) Substituting (P9.3.2) into (P9.3.1a), and using the assumed symmetries of $\mathbf{A}_s{}'$ and $\mathbf{A}_{as}{}'$, we obtain

$$\begin{aligned} \mathbf{A}_s &= \tfrac{1}{2}[(\mathbf{A}_s{}' + \mathbf{A}_{as}{}') + (\mathbf{A}_s{}' + \mathbf{A}_{as}{}')^{\mathrm{T}}] = \tfrac{1}{2}(\mathbf{A}_s{}' + \mathbf{A}_{as}{}' + \mathbf{A}_s{}'^{\mathrm{T}} + \mathbf{A}_{as}{}'^{\mathrm{T}}) \\ &= \tfrac{1}{2}(\mathbf{A}_s{}' + \mathbf{A}_{as}{}' + \mathbf{A}_s{}' - \mathbf{A}_{as}{}') = \mathbf{A}_s{}', \qquad \text{(S9.3.1a)} \\ \mathbf{A}_{as} &= \tfrac{1}{2}[(\mathbf{A}_s{}' + \mathbf{A}_{as}{}') - (\mathbf{A}_s{}' + \mathbf{A}_{as}{}')^{\mathrm{T}}] = \tfrac{1}{2}(\mathbf{A}_s{}' + \mathbf{A}_{as}{}' - \mathbf{A}_s{}'^{\mathrm{T}} - \mathbf{A}_{as}{}'^{\mathrm{T}}) \\ &= \tfrac{1}{2}(\mathbf{A}_s{}' + \mathbf{A}_{as}{}' - \mathbf{A}_s{}' + \mathbf{A}_{as}{}') = \mathbf{A}_{as}{}', \qquad \text{(S9.3.1b)} \end{aligned}$$

which shows that the decomposition (P9.3.1b) of $\mathbf{A}$ into symmetric and antisymmetric matrices is unique.

(c) For any square matrix $\mathbf{A}$, letting

$$\mathbf{A}_H = \tfrac{1}{2}(\mathbf{A} + \mathbf{A}^*), \quad \mathbf{A}_{aH} = \tfrac{1}{2}(\mathbf{A} - \mathbf{A}^*)$$

clearly yields (P9.3.3); and, it is easily shown that $\mathbf{A}_H$ is Hermitian and $\mathbf{A}_{aH}$ is anti-Hermitian. Also, this decomposition is unique; that is, if $\mathbf{A} = \mathbf{A}_H{}' + \mathbf{A}_{aH}{}'$ for some Hermitian matrix $\mathbf{A}_H{}'$ and anti-Hermitian matrix $\mathbf{A}_{aH}{}'$, then we must have $\mathbf{A}_H{}' = \mathbf{A}_H$ and $\mathbf{A}_{aH}{}' = \mathbf{A}_{aH}$. A proof of this fact obtains from (S9.3.1) upon replacing each subscripted "s" with "H", and each transposition operation with a transpose-conjugation.

Solution 9.4 Let $\mathbf{A} = [a_{i,j}]$ and $\mathbf{x} = [x_i]$; then, if $\mathbf{Ax} = \mathbf{0}$ for all $\mathbf{x}$, it follows from (9.6b) that

$$\sum_{k=1}^{n} \begin{bmatrix} a_{1,k} \\ a_{2,k} \\ \vdots \\ a_{m,k} \end{bmatrix} x_k = \begin{bmatrix} 0 \\ 0 \\ \vdots \\ 0 \end{bmatrix} \quad \text{(all } \mathbf{x}\text{)}.$$

Hence, for each $k \in \{1, 2, \ldots, n\}$, choosing $x_k = 1$ and $x_i = 0$ $(i \neq k)$ shows that the kth column of $\mathbf{A}$ is composed of all zeros; therefore, $\mathbf{A} = \mathbf{0}$. It follows that if $\mathbf{Bx} = \mathbf{Cx}$ for all $\mathbf{x}$, and thus $(\mathbf{B} - \mathbf{C})\mathbf{x} = \mathbf{Bx} - \mathbf{Cx} = \mathbf{0}$ for all $\mathbf{x}$, then $\mathbf{B} - \mathbf{C} = \mathbf{0}$; so, $\mathbf{B} = \mathbf{C}$.

Solution 9.5 **(a)** Letting

$$\mathbf{y} = \begin{bmatrix} y_1 \\ y_2 \\ \vdots \\ y_m \end{bmatrix}$$

for some $m \geq 1$, the 1×1 matrix $\mathbf{y}^*\mathbf{y}$ has the single component

$$\sum_{i=1}^{m} |y_i|^2.$$

Hence, treating this 1×1 matrix as simply a number, we must have $\mathbf{y}^*\mathbf{y} \geq 0$; and, equality is attained if and only if $y_i = 0$ (all i)—i.e., $\mathbf{y} = \mathbf{0}$.

(b) If $\mathbf{A}^*\mathbf{A}\mathbf{x} = \mathbf{0}$ (an $n \times 1$ matrix), then

$$\mathbf{0} = \mathbf{x}^*\mathbf{0} = \mathbf{x}^*\left(\mathbf{A}^*\mathbf{A}\mathbf{x}\right) = \left(\mathbf{x}^*\mathbf{A}^*\right)\mathbf{A}\mathbf{x} = \left(\mathbf{A}\mathbf{x}\right)^*\mathbf{A}\mathbf{x};$$

therefore, letting $\mathbf{y} = \mathbf{Ax}$, it follows from part (a) that $\mathbf{Ax} = \mathbf{0}$.

Solution 9.6 **(a)** Let, e.g.,

$$\mathbf{A} = \begin{bmatrix} 1 & -1 \\ 0 & 0 \end{bmatrix};$$

then, $\mathbf{A}^2 = \mathbf{A}$.

(b) Let, e.g.,

$$\mathbf{A} = \begin{bmatrix} 0 & 1 \\ 0 & 0 \end{bmatrix}.$$

Solution 9.7 Suppose that the m rows of $\mathbf{A}$ in (9.1a) are linearly *in*dependent; then, since $m > n$, rows 1 through n must be linearly independent as well. Hence, the $n \times n$ matrix $\mathbf{B} = [a_{i,j}]_{i,j=1}^{n}$ comprising these n rows is nonsingular; so, its inverse $\mathbf{B}^{-1}$ exists, and we can right-multiply row $n+1$ of $\mathbf{A}$ by $\mathbf{B}^{-1}$ to get

$$\begin{bmatrix} c_1 & c_2 & \cdots & c_n \end{bmatrix} = \begin{bmatrix} a_{n+1,1} & a_{n+1,2} & \cdots & a_{n+1,n} \end{bmatrix} \mathbf{B}^{-1}.$$

It then follows that

$$\begin{aligned} \begin{bmatrix} a_{n+1,1} & a_{n+1,2} & \cdots & a_{n+1,n} \end{bmatrix} &= \begin{bmatrix} c_1 & c_2 & \cdots & c_n \end{bmatrix} \mathbf{B} \\ &= \sum_{i=1}^{m} c_i \begin{bmatrix} a_{i,1} & a_{i,2} & \cdots & a_{i,n} \end{bmatrix}, \end{aligned}$$

akin to (9.6a). But, this shows that row $n+1$ of $\mathbf{A}$ can be expressed as a linear combination of rows 1 through n, which contradicts the assumption that the rows of $\mathbf{A}$ are linearly independent; therefore, the rows of $\mathbf{A}$ must be linearly dependent.

Solution 9.8 By (9.15),

$$\mathbf{A}^{-1} = \frac{1}{1 \cdot 4 - 2 \cdot 3} \begin{bmatrix} 4 & -2 \\ -3 & 1 \end{bmatrix} = \begin{bmatrix} -2 & 1 \\ \frac{3}{2} & -\frac{1}{2} \end{bmatrix}.$$

Then,

$$\mathbf{AA}^{-1} = \begin{bmatrix} 1 & 2 \\ 3 & 4 \end{bmatrix} \begin{bmatrix} -2 & 1 \\ \frac{3}{2} & -\frac{1}{2} \end{bmatrix} = \begin{bmatrix} (1)(-2) + (2)(\frac{3}{2}) & (1)(1) + (2)(-\frac{1}{2}) \\ (3)(-2) + (4)(\frac{3}{2}) & (3)(1) + (4)(-\frac{1}{2}) \end{bmatrix} = \begin{bmatrix} 1 & 0 \\ 0 & 1 \end{bmatrix}$$

and

$$\mathbf{A}^{-1}\mathbf{A} = \begin{bmatrix} -2 & 1 \\ \frac{3}{2} & -\frac{1}{2} \end{bmatrix} \begin{bmatrix} 1 & 2 \\ 3 & 4 \end{bmatrix} = \begin{bmatrix} (-2)(1) + (1)(3) & (-2)(2) + (1)(4) \\ (\frac{3}{2})(1) + (-\frac{1}{2})(3) & (\frac{3}{2})(2) + (-\frac{1}{2})(4) \end{bmatrix}$$

$$= \begin{bmatrix} 1 & 0 \\ 0 & 1 \end{bmatrix},$$

per (9.13).

Solution 9.9 **(a)** Expanding the determinant about the first row yields

$$\det \mathbf{A} = 1(5 \cdot 9 - 6 \cdot 8) - 2(4 \cdot 9 - 6 \cdot 7) + 3(4 \cdot 8 - 5 \cdot 7) = 0,$$

so $\mathbf{A}$ is singular.

(b) Let us try to express the third column of $\mathbf{A}$ as a linear combination of its first two column that is, let us try to find scalars x_1 and x_2 such that

$$x_1 \begin{bmatrix} 1 \\ 4 \\ 7 \end{bmatrix} + x_2 \begin{bmatrix} 2 \\ 5 \\ 8 \end{bmatrix} = \begin{bmatrix} 3 \\ 6 \\ 9 \end{bmatrix}. \tag{S9.9.1}$$

(If such scalars do not exist, then either the first or the second column of $\mathbf{A}$ can be expressed as a linear combination of the other two columns, and we would likewise proceed as follows.) Accordingly, x_1 and x_2 must solve the matrix equation

$$\begin{bmatrix} 1 & 2 \\ 4 & 5 \end{bmatrix} \begin{bmatrix} x_1 \\ x_2 \end{bmatrix} = \begin{bmatrix} 3 \\ 6 \end{bmatrix},$$

obtained by arbitrarily ignoring the third row of $\mathbf{A}$. As it happens, the above 2×2 matrix is nonsingular, and therefore this equation has a unique solution; namely, using (9.15) we obtain

$$\begin{bmatrix} x_1 \\ x_2 \end{bmatrix} = \begin{bmatrix} 1 & 2 \\ 4 & 5 \end{bmatrix}^{-1} \begin{bmatrix} 3 \\ 6 \end{bmatrix} = \frac{1}{-3} \begin{bmatrix} 5 & -2 \\ -4 & 1 \end{bmatrix} \begin{bmatrix} 3 \\ 6 \end{bmatrix} = \begin{bmatrix} -1 \\ 2 \end{bmatrix}.$$

As can be verified, this solution indeed satisfies (S9.9.1).

(c) Similarly, to express (if possible) the third row of $\mathbf{A}$ in terms of its first two row—i.e.,

$$x_1 \begin{bmatrix} 1 & 2 & 3 \end{bmatrix} + x_2 \begin{bmatrix} 4 & 5 & 6 \end{bmatrix} = \begin{bmatrix} 7 & 8 & 9 \end{bmatrix} \tag{S9.9.2}$$

—we might solve the matrix equation

$$\begin{bmatrix} x_1 & x_2 \end{bmatrix} \begin{bmatrix} 1 & 2 \\ 4 & 5 \end{bmatrix} = \begin{bmatrix} 7 & 8 \end{bmatrix},$$

obtaining

$$\begin{bmatrix} x_1 & x_2 \end{bmatrix} = \begin{bmatrix} 7 & 8 \end{bmatrix} \begin{bmatrix} 1 & 2 \\ 4 & 5 \end{bmatrix}^{-1} = \begin{bmatrix} 7 & 8 \end{bmatrix} \cdot \frac{1}{-3} \begin{bmatrix} 5 & -2 \\ -4 & 1 \end{bmatrix} = \begin{bmatrix} -1 & 2 \end{bmatrix}.$$

(Coincidentally, x_1 and x_2 have the same values as before.) As can be verified, this solution indeed satisfies (S9.9.2).

Solution 9.10 **(a)** Per (9.10) and (9.11a) with $j = 3$, expanding the determinant of

$$\mathbf{A} = \begin{bmatrix} a_{1,1} & a_{1,2} & a_{1,3} \\ a_{2,1} & a_{2,2} & a_{2,3} \\ a_{3,1} & a_{3,2} & a_{3,3} \end{bmatrix}$$

about the third row yields

$$\begin{aligned} \det \mathbf{A} &= a_{3,1} \det \begin{bmatrix} a_{1,2} & a_{1,3} \\ a_{2,2} & a_{2,3} \end{bmatrix} - a_{3,2} \det \begin{bmatrix} a_{1,1} & a_{1,3} \\ a_{2,1} & a_{2,3} \end{bmatrix} + a_{3,3} \det \begin{bmatrix} a_{1,1} & a_{1,2} \\ a_{2,1} & a_{2,2} \end{bmatrix} \\ &= a_{3,1}(a_{1,2}a_{2,3} - a_{1,3}a_{2,2}) - a_{3,2}(a_{1,1}a_{2,3} - a_{1,3}a_{2,1}) \\ &\quad + a_{3,3}(a_{1,1}a_{2,2} - a_{1,2}a_{2,1}) \\ &= a_{1,1}a_{2,2}a_{3,3} - a_{1,1}a_{2,3}a_{3,2} - a_{1,2}a_{2,1}a_{3,3} + a_{1,2}a_{2,3}a_{3,1} + a_{1,3}a_{2,1}a_{3,2} \\ &\quad - a_{1,3}a_{2,2}a_{3,1}. \end{aligned}$$

(b) Per (9.10) and (9.11b) with $i = 2$, expanding instead about the second column yields

$$\begin{aligned} \det \mathbf{A} &= -a_{1,2} \det \begin{bmatrix} a_{2,1} & a_{2,3} \\ a_{3,1} & a_{3,3} \end{bmatrix} + a_{2,2} \det \begin{bmatrix} a_{1,1} & a_{1,3} \\ a_{3,1} & a_{3,3} \end{bmatrix} - a_{3,2} \det \begin{bmatrix} a_{1,1} & a_{1,3} \\ a_{2,1} & a_{2,3} \end{bmatrix} \\ &= -a_{1,2}(a_{2,1}a_{3,3} - a_{2,3}a_{3,1}) + a_{2,2}(a_{1,1}a_{3,3} - a_{1,3}a_{3,1}) \\ &\quad - a_{3,2}(a_{1,1}a_{2,3} - a_{1,3}a_{2,1}) \\ &= a_{1,1}a_{2,2}a_{3,3} - a_{1,1}a_{2,3}a_{3,2} - a_{1,2}a_{2,1}a_{3,3} + a_{1,2}a_{2,3}a_{3,1} + a_{1,3}a_{2,1}a_{3,2} \\ &\quad - a_{1,3}a_{2,2}a_{3,1}, \end{aligned}$$

as before.

Solution 9.11 Yes, because $\det \mathbf{AB} = \det \mathbf{A} \cdot \det \mathbf{B} = 0 \cdot \det \mathbf{B} = 0$.

Solution 9.12 If $\mathbf{A} = [a_{i,j}]_{i,j=1}^n$ is an upper triangular matrix of order $n \geq 1$, then the only nonzero element in the first column is $a_{1,1}$. If $n = 1$, then $\det \mathbf{A} = a_{1,1}$; otherwise, evaluating $\det \mathbf{A}$ by expanding about the first column yields $\det \mathbf{A} = a_{1,1} \det \mathbf{A}'$, where $\mathbf{A}'$ is the $(n-1) \times (n-1)$ the matrix obtained by deleting the first row and first column of $\mathbf{A}$. But then, $\mathbf{A}' = [a_{i,j}]_{i,j=2}^n$ is also upper triangular, and the only nonzero element in the first column is $a_{2,2}$. Therefore, repeatedly applying this same step, we eventually obtain $\det \mathbf{A} = a_{1,1}a_{2,2} \cdots a_{n-1,n-1} \det \left[a_{n,n}\right] = a_{1,1}a_{2,2} \cdots a_{n,n}$. Similarly, the same result obtains for a lower triangular matrix $\mathbf{A}$ by repeatedly expanding about the first row; or, we may simply observe that $\mathbf{A}^{\mathrm{T}}$ is now an upper triangular matrix having the same main diagonal as $\mathbf{A}$, and $\det \mathbf{A} = \det \mathbf{A}^{\mathrm{T}}$.

Solution 9.13 Expanding the determinant about the first row yields

$$\det \mathbf{A} = 1(9 \cdot 5 - 4 \cdot 6) - 2(8 \cdot 5 - 4 \cdot 7) + 3(8 \cdot 6 - 9 \cdot 7) = -48.$$

Also, per (9.10), the cofactor matrix $\mathbf{C} = [c_{i,j}]$ for $\mathbf{A}$ is

$$\mathbf{C} = \begin{bmatrix} \det \begin{bmatrix} 9 & 4 \\ 6 & 5 \end{bmatrix} & -\det \begin{bmatrix} 8 & 4 \\ 7 & 5 \end{bmatrix} & \det \begin{bmatrix} 8 & 9 \\ 7 & 6 \end{bmatrix} \\ -\det \begin{bmatrix} 2 & 3 \\ 6 & 5 \end{bmatrix} & \det \begin{bmatrix} 1 & 3 \\ 7 & 5 \end{bmatrix} & -\det \begin{bmatrix} 1 & 2 \\ 7 & 6 \end{bmatrix} \\ \det \begin{bmatrix} 2 & 3 \\ 9 & 4 \end{bmatrix} & -\det \begin{bmatrix} 1 & 3 \\ 8 & 4 \end{bmatrix} & \det \begin{bmatrix} 1 & 2 \\ 8 & 9 \end{bmatrix} \end{bmatrix} = \begin{bmatrix} 21 & -12 & -15 \\ 8 & -16 & 8 \\ -19 & 20 & -7 \end{bmatrix}.$$

Therefore, by (9.14),

$$\mathbf{A}^{-1} = \frac{1}{-48} \begin{bmatrix} 21 & -12 & -15 \\ 8 & -16 & 8 \\ -19 & 20 & -7 \end{bmatrix}^{\mathrm{T}} = \frac{1}{48} \begin{bmatrix} -21 & -8 & 19 \\ 12 & 16 & -20 \\ 15 & -8 & 7 \end{bmatrix}.$$

Hence,

$$\mathbf{x} = \mathbf{A}^{-1}\mathbf{b} = \frac{1}{48}\begin{bmatrix} -21 & -8 & 19 \\ 12 & 16 & -20 \\ 15 & -8 & 7 \end{bmatrix}\begin{bmatrix} 10 \\ 11 \\ 12 \end{bmatrix}$$

$$= \frac{1}{48}\begin{bmatrix} (-21)(10) + (-8)(11) + (19)(12) \\ (12)(10) + (16)(11) + (-20)(12) \\ (15)(10) + (-8)(11) + (7)(12) \end{bmatrix} = \frac{1}{24}\begin{bmatrix} -35 \\ 28 \\ 73 \end{bmatrix} = \begin{bmatrix} -35/24 \\ 7/6 \\ 73/24 \end{bmatrix}.$$

Indeed, for this vector $\mathbf{x}$ we obtain

$$\mathbf{Ax} = \begin{bmatrix} 1 & 2 & 3 \\ 8 & 9 & 4 \\ 7 & 6 & 5 \end{bmatrix} \cdot \frac{1}{24}\begin{bmatrix} -35 \\ 28 \\ 73 \end{bmatrix} = \frac{1}{24}\begin{bmatrix} (1)(-35) + (2)(28) + (3)(73) \\ (8)(-35) + (9)(28) + (4)(73) \\ (7)(-35) + (6)(28) + (5)(73) \end{bmatrix} = \begin{bmatrix} 10 \\ 11 \\ 12 \end{bmatrix}$$

$$= \mathbf{b},$$

thus satisfying (9.17).

Solution 9.14 As found in Solution 9.13, $\det \mathbf{A} = -48$. Therefore, letting

$$\mathbf{x} = \begin{bmatrix} x_1 \\ x_2 \\ x_3 \end{bmatrix},$$

it follows from Cramer's rule (9.18) that

$$x_1 = \frac{1}{\det \mathbf{A}} \det \begin{bmatrix} 10 & 2 & 3 \\ 11 & 9 & 4 \\ 12 & 6 & 5 \end{bmatrix}$$

$$= \frac{10(9 \cdot 5 - 4 \cdot 6) - 2(11 \cdot 5 - 4 \cdot 12) + 3(11 \cdot 6 - 9 \cdot 12)}{-48} = -35/24,$$

$$x_2 = \frac{1}{\det \mathbf{A}} \det \begin{bmatrix} 1 & 10 & 3 \\ 8 & 11 & 4 \\ 7 & 12 & 5 \end{bmatrix}$$

$$= \frac{1(11 \cdot 5 - 4 \cdot 12) - 10(8 \cdot 5 - 4 \cdot 7) + 3(8 \cdot 12 - 11 \cdot 7)}{-48} = 7/6,$$

$$x_3 = \frac{1}{\det \mathbf{A}} \det \begin{bmatrix} 1 & 2 & 10 \\ 8 & 9 & 11 \\ 7 & 6 & 12 \end{bmatrix}$$

$$= \frac{1(9 \cdot 12 - 11 \cdot 6) - 2(8 \cdot 12 - 11 \cdot 7) + 10(8 \cdot 6 - 9 \cdot 7)}{-48} = 73/24,$$

where we have expanded each determinant about the first row.

Solution 9.15 The equations (9.16) corresponding to (9.17) are:

$$x_1 + 2x_2 + 3x_3 = 10, \tag{S9.15.1a}$$

$$8x_1 + 9x_2 + 4x_3 = 11, \tag{S9.15.1b}$$

$$7x_1 + 6x_2 + 5x_3 = 12. \tag{S9.15.1c}$$

Multiplying (S9.15.1a) by 8 and subtracting (S9.15.1b) to eliminate x_1 yields

$$(8 \cdot 2 - 9)x_2 + (8 \cdot 3 - 4)x_3 = (8 \cdot 10 - 11),$$

and thus

$$7x_2 + 20x_3 = 69; \tag{S9.15.2a}$$

whereas multiplying (S9.15.1a) by 7 and subtracting (S9.15.1c) to eliminate x_1 yields

$$(7 \cdot 2 - 6)x_2 + (7 \cdot 3 - 5)x_3 = (7 \cdot 10 - 12),$$

and thus

$$8x_2 + 16x_3 = 58. \tag{S9.15.2b}$$

Now, taking 8 times (S9.15.2a) minus 7 times (S9.15.2b) to eliminate x_2, we obtain

$$(8 \cdot 20 - 7 \cdot 16)x_3 = (8 \cdot 69 - 7 \cdot 58),$$

and thus $48x_3 = 146$; therefore, $x_3 = 73/24$.

Solution 9.16 **(a)** For $m = 1$, (9.16) becomes

$$a_{1,1}x_1 + a_{1,2}x_2 + \cdots + a_{1,n}x_n = b_1,$$

for some $n \geq 1$. This equation is inconsistent—i.e., has no solutions—if and only if $a_{1,j} = 0$ (all j) and $b_1 \neq 0$. For suppose this condition holds; then, the left side equals 0 for all choices of the values $x_1, \ldots, x_n$, and thus the equality must fail. On the other hand, if $a_{1,j} \neq 0$ for some j, then a solution obtains by letting $x_k = 0$ $(k \neq j)$ and $x_j = b_1/a_{1,j}$; whereas if $b_1 = 0$, then $x_1 = \cdots = x_n = 0$ is a solution.

(b) Let $\mathbf{A}$ be the $m \times n$ zero matrix, and take any $m \times 1$ vector $\mathbf{b} \neq \mathbf{0}$; then, for all $n \times 1$ vectors $\mathbf{x}$ we have $\mathbf{Ax} = \mathbf{0} \neq \mathbf{b}$. Hence, in this case there are no solutions to (9.17), and thus no solutions to (9.16).

Solution 9.17 Assuming that the equations (9.16) are consistent (i.e., there exists *at least* one solution), suppose there is no *unique* solution; that is, (9.16) has more than one solution. Equivalently, there exist two distinct vectors $\mathbf{x}_0$ and $\mathbf{x}_1$ that are solutions $\mathbf{x}$ to the matrix equation (9.17). Now, for each $\alpha \in \mathbb{R}$, let

$$\mathbf{x}_\alpha = \mathbf{x}_0 + \alpha(\mathbf{x}_1 - \mathbf{x}_0);$$

then, since $\mathbf{x}_1 - \mathbf{x}_0 \neq \mathbf{0}$, there are infinitely such many vectors $\mathbf{x}_\alpha$. Also, each is a solution to (9.17), since

$$\mathbf{Ax}_\alpha = \mathbf{A}[\mathbf{x}_0 + \alpha(\mathbf{x}_1 - \mathbf{x}_0)] = \mathbf{Ax}_0 + \alpha(\mathbf{Ax}_1 - \mathbf{Ax}_0) = \mathbf{b} + \alpha(\mathbf{b} - \mathbf{b}) = \mathbf{b}.$$

Therefore, converting back from vectors $\mathbf{x}_\alpha$ to variables $x_1, \ldots, x_n$, there are infinitely many solutions to (9.16). (*Aside*: It can also be shown that a unique solution is impossible if $m < n$—i.e., there are fewer equations that variables.)

Solution 9.18 For each $\phi \in [0, 2\pi)$, the given matrix $\mathbf{A}$ has characteristic polynomial

$$\det(\lambda\mathbf{I} - \mathbf{A}) = \det\begin{bmatrix} \lambda - \cos\phi & \sin\phi \\ -\sin\phi & \lambda - \cos\phi \end{bmatrix} = (\lambda - \cos\phi)^2 + (\sin\phi)^2,$$

which equals 0 if and only if $\lambda - \cos\phi = \pm i \sin\phi$. Thus, the eigenvalues of $\mathbf{A}$ are $\lambda = \cos\phi \pm i\sin\phi$—i.e., $\lambda = e^{\pm i\phi}$—which are degenerate (i.e., not distinct) if and only if $\phi = 0$ or π.

Solution 9.19 Expanding the determinant about the first row, we obtain the characteristic polynomial

$$\begin{aligned}
\det(\lambda\mathbf{I} - \mathbf{A}) &= \det\begin{bmatrix} \lambda - 3 & -1 & 1 \\ -2 & \lambda - 2 & 2 \\ -1 & 1 & \lambda - 1 \end{bmatrix} \\
&= (\lambda - 3)[(\lambda - 2)(\lambda - 1) - (2)(1)] - (-1)[(-2)(\lambda - 1) - (2)(-1)] \\
&\quad + (1)[(-2)(1) - (\lambda - 2)(-1)] \\
&= (\lambda - 3)(\lambda^2 - 3\lambda) + (-2\lambda + 4) + (\lambda - 4) = \lambda^3 - 6\lambda^2 + 8\lambda \\
&= \lambda(\lambda - 2)(\lambda - 4);
\end{aligned}$$

thus, the eigenvalues of $\mathbf{A}$ are $\lambda = 0, 2, 4$.

For $\lambda = 0$, the column vector

$$\mathbf{x} = \begin{bmatrix} x_1 \\ x_2 \\ x_3 \end{bmatrix} \tag{S9.19.1}$$

is an eigenvector of $\mathbf{A}$ if and only if

$$\mathbf{0} = (0\mathbf{I} - \mathbf{A})\mathbf{x} = \begin{bmatrix} -3 & -1 & 1 \\ -2 & -2 & 2 \\ -1 & 1 & -1 \end{bmatrix} \begin{bmatrix} x_1 \\ x_2 \\ x_3 \end{bmatrix} = x_1 \begin{bmatrix} -3 \\ -2 \\ -1 \end{bmatrix} + (x_2 - x_3) \begin{bmatrix} -1 \\ -2 \\ 1 \end{bmatrix}.$$

This condition is satisfied for $x_1 = 0$ if and only if $x_2 = x_3$; whereas for $x_1 \neq 0$, no values of x_2 and x_3 will satisfy it. Therefore, $\mathbf{x}$ is an eigenvector of $\mathbf{A}$ associated with the eigenvalue 0 if and only if

$$\mathbf{x} = \begin{bmatrix} 0 \\ \beta \\ \beta \end{bmatrix} \quad (\text{nonzero } \beta \in \mathbb{C}).$$

For $\lambda = 2$, the vector $\mathbf{x}$ in (S9.19.1) is an eigenvector of $\mathbf{A}$ if and only if

$$\mathbf{0} = (2\mathbf{I} - \mathbf{A})\mathbf{x} = \begin{bmatrix} -1 & -1 & 1 \\ -2 & 0 & 2 \\ -1 & 1 & 1 \end{bmatrix} \begin{bmatrix} x_1 \\ x_2 \\ x_3 \end{bmatrix} = x_2 \begin{bmatrix} -1 \\ 0 \\ 1 \end{bmatrix} + (x_1 - x_3) \begin{bmatrix} -1 \\ -2 \\ -1 \end{bmatrix}.$$

This condition is satisfied for $x_2 = 0$ if and only if $x_1 = x_3$; whereas for $x_2 \neq 0$, no values of x_1 and x_3 will satisfy it. Therefore, $\mathbf{x}$ is an eigenvector of $\mathbf{A}$ associated with the eigenvalue 2 if and only if

$$\mathbf{x} = \begin{bmatrix} \beta \\ 0 \\ \beta \end{bmatrix} \quad (\text{nonzero } \beta \in \mathbb{C}).$$

Finally, for $\lambda = 4$, the vector $\mathbf{x}$ in (S9.19.1) is an eigenvector of $\mathbf{A}$ if and only if

$$\mathbf{0} = (4\mathbf{I} - \mathbf{A})\mathbf{x} = \begin{bmatrix} 1 & -1 & 1 \\ -2 & 2 & 2 \\ -1 & 1 & 3 \end{bmatrix} \begin{bmatrix} x_1 \\ x_2 \\ x_3 \end{bmatrix} = x_3 \begin{bmatrix} 1 \\ 2 \\ 3 \end{bmatrix} + (x_1 - x_2) \begin{bmatrix} 1 \\ -2 \\ -1 \end{bmatrix}.$$

This condition is satisfied for $x_3 = 0$ if and only if $x_1 = x_2$; whereas for $x_3 \neq 0$, no values of x_1 and x_2 will satisfy it. Therefore, $\mathbf{x}$ is an eigenvector of $\mathbf{A}$ associated with the eigenvalue 4 if and only if

$$\mathbf{x} = \begin{bmatrix} \beta \\ \beta \\ 0 \end{bmatrix} \quad (\text{nonzero } \beta \in \mathbb{C}).$$

Solution 9.20 Suppose $\mathbf{Ax} = \lambda\mathbf{x}$ for some $\mathbf{x} \neq \mathbf{0}$. Then, $\mathbf{x}^*\mathbf{Ax} = \mathbf{x}^*(\lambda\mathbf{x}) = \lambda(\mathbf{x}^*\mathbf{x})$, where $\mathbf{x}^*\mathbf{x} \neq 0$ (as shown in part (a) of Solution 9.5); hence,

$$\lambda = \frac{\mathbf{x}^*\mathbf{Ax}}{\mathbf{x}^*\mathbf{x}}.$$

Therefore, since it is assumed that $\mathbf{A} = \mathbf{A}^*$, we have

$$\lambda^* = \left(\frac{\mathbf{x}^*\mathbf{Ax}}{\mathbf{x}^*\mathbf{x}}\right)^* = \frac{(\mathbf{x}^*\mathbf{Ax})^*}{(\mathbf{x}^*\mathbf{x})^*} = \frac{\mathbf{x}^*\mathbf{A}^*\mathbf{x}}{\mathbf{x}^*\mathbf{x}} = \frac{\mathbf{x}^*\mathbf{Ax}}{\mathbf{x}^*\mathbf{x}} = \lambda;$$

so, λ is real.

Solution 9.21 We know that $\operatorname{tr}\mathbf{A} = \lambda_1 + \lambda_2$ and $\det\mathbf{A} = \lambda_1\lambda_2$. Hence,

$$\lambda_1 \operatorname{tr}\mathbf{A} = \lambda_1(\lambda_1 + \lambda_2) = \lambda_1{}^2 + \det\mathbf{A};$$

and, likewise with "λ_1" and "λ_2" interchanged. Thus, λ_1 and λ_2 are the two roots of the quadratic polynomial

$$\lambda^2 - (\operatorname{tr}\mathbf{A})\lambda + \det\mathbf{A}$$

—as could also have been deduced from (6.21). (Actually, all that has been explicitly shown is that *if* λ is an eigenvalue of $\mathbf{A}$, then it is also a root of this polynomial; it has not yet been shown that every root of the polynomial must be an eigenvalue of $\mathbf{A}$. Clearly, this must be the case if $\lambda_1 \neq \lambda_2$, since the polynomial only has two roots; but, it is also the case if $\lambda_1 = \lambda_2$, since then $\operatorname{tr}\mathbf{A} = 2\lambda_1$ and $\det\mathbf{A} = \lambda_1{}^2$, thereby making the polynomial equal $(\lambda - \lambda_1)^2$.) Applying the quadratic formula (6.23c), we now obtain

$$\lambda_1, \lambda_2 = (\operatorname{tr}\mathbf{A})/2 \pm \sqrt{[(\operatorname{tr}\mathbf{A})/2]^2 - \det\mathbf{A}}.$$

Chapter 10. Euclidean Geometry

Solution 10.1 **(a)** Proceeding by mathematical induction, suppose (induction hypothesis) that the assertion in the problem is true for some particular $n \geq 3$. Letting $P_1, P_2, \ldots, P_n, P_{n+1}$ be a collection of $n + 1$ distinct points for which every subcollection of three is collinear, we can then conclude that the n points $P_1, P_2, \ldots, P_n$ all lie on some line l_1; also, the three points P_1, P_2, and P_{n+1} all lie on some line l_2. But, the two points P_1 and P_2 cannot lie on more than one line; hence, $l_1 = l_2$, making all $n + 1$ points collinear. Thus, we have shown that if the assertion in the problem is true for a particular $n \geq 3$, then it is also true for $n + 1$; moreover, the assertion is clearly true for $n = 3$. Therefore, by mathematical induction, the assertion is true for all $n \geq 3$.

(b) In solid geometry, $n \geq 4$ distinct points $P_1, P_2, \ldots, P_n$ in space are coplanar if every subcollection of four of these points is coplanar.

Solution 10.2 **(a)** Suppose points P_1, P_2, Q_1, and Q_2 all lie on a plane p. Then, since for any two distinct points lying on a plane the line passing through those points must lie in that plane, the lines $l = \overleftrightarrow{P_1P_2}$ and $m = \overleftrightarrow{Q_1Q_2}$ lie in p, and thus are coplanar; but then, m and l are not skew lines. Therefore, since the assumption that the points P_1, P_2, Q_1, and Q_2 are coplanar leads to a contradiction, they cannot be coplanar.

(b) Suppose lines l and m both lie in some plane p. Let n cross l at point P_1, and let n cross m at point P_2; thus, $P_1 \in p$ and $P_2 \in p$. By assumption, $P_1 \neq P_2$; therefore, as explained in part (a), the line $n = \overleftrightarrow{P_1P_2}$ also lies in p. Hence, l, m, and n are coplanar.

Solution 10.3 Two. One point Q_1 is not sufficient because if $l \not\perp p$, then there exists a line in p that is perpendicular to l, and Q_1 might lie on that line. Whereas for two points Q_1 and Q_2, the lines $\overleftrightarrow{PQ_1}$ and $\overleftrightarrow{PQ_2}$ must cross at P (since P, Q_1, and Q_2 are noncollinear), and thus there exists exactly one line l perpendicular to both.

Solution 10.4 **(a)** Of the five stated relations, betweenness of points cannot be an equivalence relation because it is ternary. Nor can perpendicularity of a line and a plane be an equivalence relation; because, although this relation is binary, the set of all lines and the set of all planes are not the same.

Coplanarity is trivially an equivalence relation in plane geometry, where all figures are coplanar; however, it is not an equivalence relation in solid geometry. As a counterexample, let p and q be planes crossing at line l; then, p and l are coplanar, as are l and q, but p and q are not. Therefore, in solid geometry, the coplanarity relation is not a transitive.

In either plane or solid geometry, parallelness of lines *is* an equivalence relation; but, perpendicularity of lines is not. For let l, m, and n be lines in the space under consideration. Then, $l \parallel l$; and, if $l \parallel m$, then $m \parallel l$; and, if $l \parallel m$ and $m \parallel n$, then $l \parallel n$. Therefore, the parallelness-of-lines relation is reflexive, symmetric, and transitive. However, in either space, the perpendicularity-of-lines relation is not reflexive, since no line is perpendicular to itself.

(b) For any two figures f and g in the space under consideration, if f is a circle, then we have $g \sim f$ if and only if g is also a circle; therefore, yes, the set of all circles is an equivalence class under the similarity relation $\sim$. However, no, this set is not an equivalence class under the congruence relation $\cong$, since not all circles are congruent.

Solution 10.5 **(a)** Let f and g be two distinct lines; thus, $f \sim g$. Clearly, a one-to-one mapping $\varphi(\cdot)$ of f onto g obtains by translating and rotating f to become g; then, (10.2) holds for $\kappa = 1$. Moreover, appending to this transformation a shift along the line yields a different such function $\varphi(\cdot)$.

(b) Yes. Let f and g be two distinct lines; thus, $f \cong g$. Then, let $\varphi(\cdot)$ be a transformation of f into g obtained by translating, rotating, *and* scaling along the line by a scale factor $\kappa > 0$ other than 1.

Solution 10.6 **(a)** The suffix "-gon" means "angled figure". Therefore, a "polygon" is fundamentally a figure having many angles.

(b) The prefix "iso-" means "equal". Therefore, an isogon is a polygon for which the angles are equal (in measure)—i.e., an equiangular polygon.

Solution 10.7 **(a)** Let

$$A_1 = A, \quad A_2 = B, \quad A_3 = C, \quad A_4 = D, \quad A_5 = B, \quad A_6 = E, \quad A_7 = D.$$

Then, $A_1 \neq A_2, A_2 \neq A_3, \ldots, A_6 \neq A_7$, and $A_7 \neq A_1$; and, no pair of the $n = 7$ segments (10.5) have more than one point in common. Therefore, even with the additional constraint on the points (10.4), the figure f expressed by (10.3)—viz., that on the left of Figure P10.7.1—is a polygon. However, not so for the figure on the right side of Figure P10.7.1. Because for it, the path in (10.3) of segments (10.5) is now required to pass through the interior of each of the segments $\overline{BA}$, $\overline{BC}$, and $\overline{BD}$ exactly once; but, this is impossible since the number of these segments is odd, and there are no other ways for the path to get into or out of point B. (Without the additional constraint, the figure on the right of Figure P10.7.1 could be expressed as $\overline{AB} \cup \overline{BC} \cup \overline{CD} \cup \overline{DB} \cup \overline{BC} \cup \overline{CD} \cup \overline{DA}$.)

(b) One example is shown in Figure S10.7.1, where the points A, E, and D are collinear, as are B, E, and C. Here, in order for (10.3) to hold, the points (10.4) must include A through G, with $A_i = E$ for more than one $i \in \{1, 2, \ldots, n\}$; hence, n can be no smaller than 8. Furthermore, this figure can be expressed by (10.3) with either

$$\begin{aligned} &A_1 = A, \quad A_2 = B, \quad A_3 = C, \quad A_4 = D \\ &A_5 = E, \quad A_6 = F, \quad A_7 = G, \quad A_8 = E, \end{aligned}$$

or

$$\begin{aligned} &A_1 = A, \quad A_2 = B, \quad A_3 = E, \quad A_4 = F \\ &A_5 = G, \quad A_6 = E, \quad A_7 = C, \quad A_8 = D. \end{aligned}$$

In each case, $n = 8$; also, $A_1 \neq A_2, A_2 \neq A_3, \ldots, A_7 \neq A_8$, and $A_8 \neq A_1$, with no three consecutive points being collinear. However, in the first case $\overline{BC}$ belongs to the segments (10.5), but not so for the second case.

Solution 10.8 **(a)** Yes. Because, there are *no* points in $\emptyset$; so, it is vacuously true that $\overline{AB} \in \emptyset$ for any two distinct points $A, B \in \emptyset$.

(b) Yes, similarly, but now because it is impossible to find two *distinct* points $A, B \in f$.

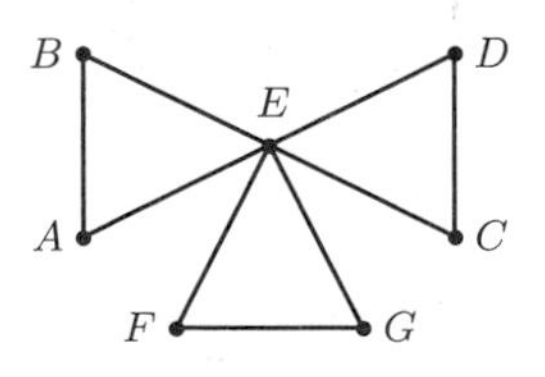

FIGURE S10.7.1

Solution 10.9 **(a)** All of the stated figures are convex except for a proper angle, the exterior of a proper angle, a circle, the exterior of a circle, and an arc.

(b) Letting $S = \{\text{convex sets } g \mid f \subseteq g\}$, the convex hull of f is

$$h = \bigcap_{g \in S} g.$$

Hence, given distinct points $A, B \in h$, we have $A, B \in g$ for each figure $g \in S$; and, since each such g is convex, we also have $\overline{AB} \subseteq g$ for all $g \in S$. Therefore, $\overline{AB} \subseteq h$, so h is itself convex.

Solution 10.10 For each vertex A of the n-gon, there are two vertices B for which $\overline{AB}$ is a side of the n-gon. Thus, for each of the n vertices A, there are $n-3$ vertices B for which $\overline{AB}$ is a diagonal (because, we also cannot have $B = A$). Therefore, since each diagonal has two vertices as endpoints, we divide by 2 and conclude that the number of diagonals is $n(n-3)/2$. This result also obtains by noting that there are $\binom{n}{2}$ ways to choose a pair of vertices from the n-gon, though n of these pairs correspond to sides; therefore, the number of diagonals is

$$\binom{n}{2} - n = \frac{n!}{2!(n-2)!} - n = \frac{n(n-1)}{2} - n = \frac{n(n-3)}{2}.$$

Solution 10.11 Let the three angles of a given triangle have measures α, β, and γ; thus, each of these values lies in the interval $(0, \pi)$, and $\alpha + \beta + \gamma = \pi$. Now, suppose that no pair of these angles are acute; hence, for two of the angles, say α and β, we have $\alpha, \beta \geq \pi/2$. Then, $\gamma = \pi - \alpha - \beta \leq \pi - \pi/2 - \pi/2 = 0$, implying that $\gamma \leq 0$, which is impossible. Therefore, the triangle must have at least two acute angles, since assuming the contrary leads to a contradiction.

Solution 10.12 Suppose that for some points A_i and B_i ($i = 1, 2, \ldots, n$) we have (polygon $A_1A_2 \cdots A_n$) $\sim$ (polygon $B_1B_2 \cdots B_n$); then, (10.8b) implies that $|\overline{B_1B_2}|/|\overline{A_1A_2}| = |\overline{B_2B_3}|/|\overline{A_2A_3}| = \cdots = |\overline{B_nB_1}|/|\overline{A_nA_1}| = \alpha$ for some constant α. Hence, the perimeter of polygon $B_1B_2 \cdots B_n$ is

$$\begin{aligned} |\overline{B_1B_2}| + |\overline{B_2B_3}| + \cdots + |\overline{B_nB_1}| &= \alpha|\overline{A_1A_2}| + \alpha|\overline{A_2A_3}| + \cdots + \alpha|\overline{A_nA_1}| \\ &= \alpha\big(|\overline{A_1A_2}| + |\overline{A_2A_3}| + \cdots + |\overline{A_nA_1}|\big), \end{aligned}$$

which is α times the perimeter of polygon $A_1A_2 \cdots A_n$. Therefore, if the two perimeters are equal, then $\alpha = 1$, which makes (10.11b) hold, and thus the polygons are congruent.

Solution 10.13 Only for $n = 3$, in either case. For $n = 3$, (10.8a) is sufficient by the AA condition for triangle similarity, and (10.8b) is sufficient by the SSS condition for triangle similarity. Whereas (briefly) for each $n > 3$, (10.8a) is not sufficient because it is possible for an n-gon to be such that the lengths of some (but not all) of its sides can be changed while leaving the measures of its angles fixed (e.g., compare the left and right sides polygons in Figure 10.10); and, (10.8b) is not sufficient because it is possible for an n-gon to be such that the measures of some of its angles can be changed while leaving the lengths of its sides fixed (e.g., compare the middle and right polygons in Figure 10.10).

Solution 10.14 If P is on l, then $d = 0$, and thus for any point P' we have $d(P, P') \geq d$; so, suppose P is not on l. As shown in Figure 10.8, the distance d from P to l is now $d(P, Q)$, where Q is a point on l for which $\overline{PQ} \perp l$. Hence, if $P' = Q$, then $d(P, P') = d$; so, let P' be any point on l other than Q. Now, the three points P, Q, and P' are distinct, and $\overline{PQ} \perp \overline{QP'}$. Therefore, it follows from the Pythagorean theorem (10.12) that $d^2(P, P') = d^2(P, Q) + d^2(Q, P')$, where $d(Q, P') > 0$, thereby making $d^2(P, P') > d^2(P, Q) = d^2$, and thus $d(P, P') > d$.

Solution 10.15 Let P be the suggested point. If $P = D$, then $\angle CDA$ is a right angle; hence, applying the Pythagorean theorem to $\triangle CDA$ yields $d^2(C, D) + d^2(D, A) = d^2(C, A)$, making $d^2(C, D) < d^2(C, A)$, so $d(C, D) < d(C, A) \leq \max[d(C, A), d(C, B)]$. Thus, suppose instead that $P \neq D$. Now, since D lies between A and B, either D lies between A and P, or D lies between B and P. In the former case, we have $d(P, A) > d(P, D)$; and, applying the Pythagorean theorem to the right triangles $\triangle CPA$ and

$\triangle CPD$ yields

$$d^2(C,P) + d^2(P,A) = d^2(C,A),$$
$$d^2(C,P) + d^2(P,D) = d^2(C,D).$$

Therefore, $d^2(C,A) - d^2(C,D) = d^2(P,A) - d^2(P,D) > 0$, making $d^2(C,D) < d^2(C,A)$, and thus $d(C,D) < d(C,A)$. Likewise, $d(C,D) < d(C,B)$ when D lies between B and P. So, in either case, $d(C,D) < \max[d(C,A), d(C,B)]$.

Solution 10.16 Letting M be the midpoint of $\overline{AB}$, and P be an arbitrary point on p, our task is to show that $P \in l$ if and only if $d(P,A) = d(P,B)$. First, note that for $P = M$ we have both $P \in l$ and $d(P,A) = d(P,B)$, whereas neither of these conditions holds for any other point $P \in \overleftrightarrow{AB}$. So, assume $P \notin \overleftrightarrow{AB}$; and, let Q be the unique point on $\overleftrightarrow{AB}$ such that $\overleftrightarrow{PQ} \perp \overleftrightarrow{AB}$. It then follows from the Pythagorean theorem that

$$d^2(P,Q) + d^2(Q,A) = d^2(P,A), \tag{S10.16.1a}$$
$$d^2(P,Q) + d^2(Q,B) = d^2(P,B) \tag{S10.16.1b}$$

(the cases of $Q = A$ and $Q = B$ being treated separately). If $P \in l$, then $Q = M$; because, $\overleftrightarrow{PM}$ is now l, $l \perp \overleftrightarrow{AB}$, and Q is unique. Conversely, since l is the unique perpendicular bisector of $\overline{AB}$, if $Q = M$, then $\overleftrightarrow{PQ} = l$, so $P \in l$. Therefore, $P \in l$ if and only if $Q = M$; but, $Q = M$ if and only if $d(Q,A) = d(Q,B)$, which by (S10.16.1) is the case if and only if $d(P,A) = d(P,B)$.

Solution 10.17 See Table S10.17.1 (obtained by trial and error), where boldface type indicates a triple that is "new" in that it is not a multiple of a previous triple.

Solution 10.18 Given a triangle $\triangle ABC$, let D and E be the midpoints of the segments $\overline{AB}$ and $\overline{BC}$, respectively. Then, since $|\overline{DB}|/|\overline{AB}| = \frac{1}{2} = |\overline{BE}|/|\overline{BC}|$ and $\measuredangle DBE = \measuredangle ABC$, it follows from the SAS condition for triangle similarity that $\triangle DBE \sim \triangle ABC$. Therefore, by (10.8b), $|\overline{DE}|/|\overline{AC}| = |\overline{DB}|/|\overline{AB}| = \frac{1}{2}$.

Solution 10.19 **(a)** Noting that neither of the right triangles $\triangle ABC$ and $\triangle BDC$ has its right angle at the point C, but $\angle ACB = \angle BCD$, it follows from the AA condition for triangle similarity that $\triangle ABC \sim \triangle BDC$; and, similar reasoning (now vis-à-vis the point A) shows that $\triangle ABC \sim \triangle ADB$.

TABLE S10.17.1 The First Several Pythagorean Triples (x, y, z)

x	y	z
3	**4**	**5**
5	**12**	**13**
6	8	10
7	**24**	**25**
8	**15**	**17**
9	12	15
9	**40**	**41**
10	24	26
11	**60**	**61**
12	16	20
12	**35**	**37**
13	**84**	**85**

(b) Since $\triangle ABC \sim \triangle BDC$, it follows from (10.8b) that

$$|\overline{BC}|/|\overline{AC}| = |\overline{DC}|/|\overline{BC}|; \tag{S10.19.1}$$

therefore, $|\overline{BC}|^2 = |\overline{AC}| \cdot |\overline{DC}|$, yielding $|\overline{BC}| = \sqrt{|\overline{AC}| \cdot |\overline{DC}|}$.

(c) Since $\triangle ABC \sim \triangle ADB$, we likewise have

$$|\overline{AB}|/|\overline{AC}| = |\overline{AD}|/|\overline{AB}|,$$

which together with (S10.19.1) and the fact that $|\overline{AD}| + |\overline{DC}| = |\overline{AC}|$ yields

$$|\overline{AB}|^2 + |\overline{BC}|^2 = |\overline{AC}| \cdot |\overline{AD}| + |\overline{AC}| \cdot |\overline{DC}| = |\overline{AC}|(|\overline{AD}| + |\overline{DC}|) = |\overline{AC}|^2,$$

proving (10.12).

Solution 10.20 **(a)** Since $\angle A_1$ and $\angle A_2$ of the middle pentagon are congruent to the corresponding angles of the regular pentagon on the right, and the measures of the angles of any pentagon sum to $(5-2)\cdot 180° = 540°$, we have $\measuredangle A_1 = \measuredangle A_2 = 540°/5 = 108°$. Also, since $|\overline{A_3A_4}|$, $|\overline{A_3A_5}|$, and $|\overline{A_4A_5}|$ have, respectively, the same values for both pentagons, it follows from the SSS condition for triangle congruence that the corresponding triangles $\triangle A_3A_4A_5$ are congruent. Hence, since $\measuredangle A_4 = 108°$ for the pentagon on the right, for the middle pentagon we have $\measuredangle A_4 = 360° - 108° = 252°$, because $\angle A_4$ is measured as an interior angle of the pentagon. Finally, by symmetry, $\angle A_3$ and $\angle A_5$ must have the same measure; therefore, $\measuredangle A_3 = \measuredangle A_5 = (540° - \measuredangle A_1 - \measuredangle A_2 - \measuredangle A_4)/2 = 36°$.

(b) Since the angles of $\triangle A_1A_2A_3$ have measures summing to 180°, we have $\measuredangle A_1A_2A_3 + \measuredangle A_2A_3A_1 + \measuredangle A_3A_1A_2 = 180°$. Also, $\measuredangle A_1A_2A_3 = \measuredangle A_2 = 108°$, and $\measuredangle A_2A_3A_1 = \measuredangle A_3A_1A_2$ by symmetry; therefore, $\measuredangle A_2A_3A_1 = (180° - \measuredangle A_1A_2A_3)/2 = 36° = \measuredangle A_3 = \measuredangle A_2A_3A_4$. Hence, $\measuredangle A_2A_3A_1 = \measuredangle A_2A_3A_4$, so A_1, A_3, and A_4 must be collinear.

Solution 10.21 Let a parallelogram $\square ABCD$ be given.

(a) As shown in Figure S10.21.1, let E be a point on $\overleftrightarrow{AB}$ lying on the side of A that is opposite from B, let F be a point on $\overleftrightarrow{DC}$ lying on the side of D that is opposite from C, and let G be a point on $\overleftrightarrow{AD}$ lying on the side of D that is opposite from A. Viewing $\overleftrightarrow{AB}$ as a transversal across the parallel lines $\overleftrightarrow{AD}$ and $\overleftrightarrow{BC}$, we conclude that the corresponding angles $\angle ABC$ and $\angle EAD$ are congruent. Similarly, viewing $\overleftrightarrow{AD}$ as a transversal across the parallel lines $\overleftrightarrow{AB}$ and $\overleftrightarrow{DC}$, we conclude that the corresponding angles $\angle EAD$ and $\angle FDG$ are congruent; furthermore, the vertical angles $\angle FDG$ and $\angle ADC$ are congruent. Hence, $\angle ABC \cong \angle EAD \cong \angle FDG \cong \angle ADC$, making

$$\measuredangle ABC = \measuredangle ADC. \tag{S10.21.1}$$

Also, viewing $\overleftrightarrow{BD}$ as a transversal across the parallel lines $\overleftrightarrow{AD}$ and $\overleftrightarrow{BC}$, we conclude that

$$\measuredangle ADB = \measuredangle CBD; \tag{S10.21.2}$$

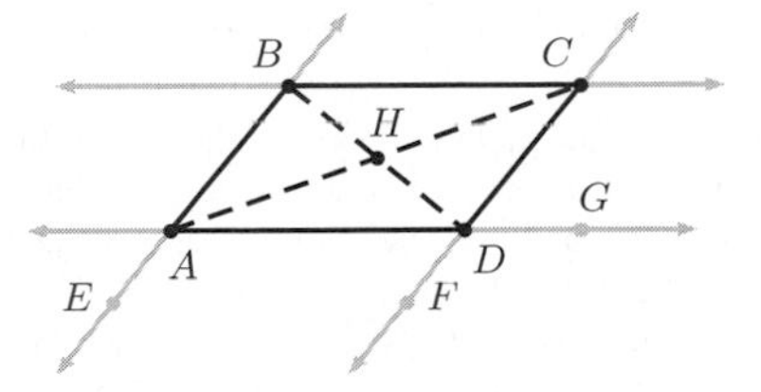

FIGURE S10.21.1

so, since (by the additivity of the angle measure) $\measuredangle ABC = \measuredangle ABD + \measuredangle CBD$ and $\measuredangle ADC = \measuredangle ADB + \measuredangle CDB$, adding (S10.21.1) and (S10.21.2) yields

$$\measuredangle ABD = \measuredangle CDB. \tag{S10.21.3}$$

By the ASA condition for triangle congruence, it follows from (S10.21.2) and (S10.21.3) that $\triangle ABD \cong CDB$; therefore, $|\overline{AB}| = |\overline{CD}|$ and $|\overline{AD}| = |\overline{CB}|$.

(b) Viewing $\overleftrightarrow{AC}$ as a transversal across the parallel lines $\overleftrightarrow{AD}$ and $\overleftrightarrow{BC}$, we conclude that

$$\measuredangle DAC = \measuredangle BCA. \tag{S10.21.4}$$

Also, letting H be the point where the diagonals of $\square ABCD$ intersect, we have $\angle ADB = \angle ADH$, $\angle CBD = \angle CBH$, $\angle DAC = \angle DAH$, and $\angle BCA = \angle BCH$; and, applying these facts to (S10.21.2) and (S10.21.4) yields $\measuredangle ADH = \measuredangle CBH$ and $\measuredangle DAH = \measuredangle BCH$. Therefore, since $|\overline{AD}| = |\overline{CB}|$, it follows from the ASA condition for triangle congruence that $\triangle AHD \cong CHB$. Accordingly, by (10.11b) we have $|\overline{AH}| = |\overline{CH}|$ and $|\overline{DH}| = |\overline{BH}|$; that is, the point H where the diagonals intersect is the midpoint of each.

Solution 10.22 Suppose a quadrilateral $\square ABCD$ is inscribed in a circle. Since $\angle ABC$ is an interior angle of the circle, $\measuredangle ABC$ equals one-half the angle measure of the intercepted arc $\widehat{CDA}$. Likewise, the opposite angle $\angle CDA$ has a measure $\measuredangle CDA$ that is one-half the angle measure of $\widehat{ABC}$. So, since the angle measures of $\widehat{CDA}$ and $\widehat{ABC}$ sum to 2π, we have $2\measuredangle ABC + 2\measuredangle CDA = 2\pi$, and thus $\measuredangle ABC + \measuredangle CDA = \pi$. Therefore, $\angle ABC$ and $\angle CDA$ are supplementary angles; and, likewise, so are the other opposite angles, $\angle BCD$ and $\angle DAB$.

Solution 10.23 **(a)** Adding to the figure the three diagonals that pass through the center P, we obtain six triangles. For each triangle, two sides have length r; the remaining side has length s; and, by symmetry, the two angles opposite the sides of length r are congruent. (The fact that the angles opposite the congruent sides of an isosceles triangle must be congruent also obtains from the SSS condition for triangle congruence.) Thus, since $\measuredangle P = 360°/6 = 60°$, the other two angles have measure $(180° - 60°)/2 = 60°$, and the triangle is equiangular. Therefore, the triangle is also equilateral, making $r = s$.

(b) For each of the preceding six equilateral triangles, the median having endpoint P is, by symmetry, perpendicular to the opposite side, which it bisects. We thus obtain, as shown by the dashed segments in Figure 10.18, a right triangle having legs of lengths a and $s/2$, and a hypotenuse of length $r = s$. Therefore, by the Pythagorean theorem, $a^2 + (s/2)^2 = r^2 = s^2$, yielding $a = (\sqrt{3}/2)s$.

Solution 10.24 **(a)** Use the straightedge to draw two nonparallel chords of the circle; then, use the procedure stated in the problem to draw their perpendicular bisectors. Since the chords are not parallel, neither are their perpendicular bisectors; therefore, they cross, and the point of intersection is the center of the circle.

(b) Since the circumscribed circle passes through the vertices of the triangle, its center is equidistant from them; moreover, it follows from Problem 10.16 that for each pair of vertices, the set all points equidistant from them is the perpendicular bisector of the corresponding side. Therefore, since no two of these perpendicular bisectors are parallel (because the same is true of the sides of the triangle), the center P of the circumscribed circle can located by drawing the perpendicular bisectors of any two sides, and finding where they cross. (We thus see that all *three* perpendicular bisectors must cross at a single point.) Now, use the compass to draw a circle centered at P that passes through any vertex of the triangle.

(c) From Problem 10.23(a) we know that the length of the side of the hexagon equals the radius of the circle. Thus, we can use the circle and its center to set the compass to the radius of the circle; then, we can use it to find a point Q' on the circle Q for which $d(Q, Q')$ equals the radius of the circle. Now, not only is Q a vertex of the desired hexagon, so also is Q'. Upon repeating this step four more times (the next step being applied to Q'), the remaining vertices of the hexagon can be found; and, by using the straightedge to draw segments between adjacent vertices, the desired hexagon is obtained.

(d) Given a proper angle $\angle ABC$, use the compass to draw a circle centered at B; this circle will intersect $\angle ABC$ at two points P and Q. Then, draw two circles centered at P and Q having the same radii (not necessarily the same as before), and sufficiently large for there to be at least one point D of intersection in the interior of $\angle ABC$. Finally, use the straightedge to draw the ray $\overrightarrow{AD}$, thereby (by symmetry) bisecting the interior of $\angle ABC$.

Solution 10.25 The section is either a single point (when the intersecting plane is a tangent plane of the solid sphere) or a closed disk (for a secant plane).

Solution 10.26 **(a)** Using (10.18) and Figure 10.21 (regardless of the value of e)—taking P to be an endpoint of the latus rectum corresponding to focus F, thereby making $d(P, F) = l$ and $d(P, Q) = p$—we obtain $l = ep$.
(b) With $c = \sqrt{a^2 - b^2}$, we obtain $l = ep = (c/a)(b^2/c) = b^2/a$.
(c) With $c = \sqrt{a^2 + b^2}$, we obtain $l = ep = (c/a)(b^2/c) = b^2/a$.

Solution 10.27 We know that for the ellipse, $d(F, F') = 2c$ where $e = c/a$; therefore, we now have $c \to 0$ with $a > 0$ fixed, making $e \to 0$. Also, since $a = ep/(1 - e^2)$, we have $p = a(1 - e^2)/e \to \infty$ (i.e., the directrix d moves infinitely far away from the focus F); and, since $b = ep/\sqrt{1 - e^2}$, and thus $b/a = \sqrt{1 - e^2} \to 1$, we have $b \to a$. (*Aside*: Looking ahead to Chapter 11, note that upon setting $b = a$, equation (11.134) describing an ellipse in the xy-plane becomes equation (11.133) describing a circle. Also, for any of the equations (11.141) describing an ellipse in polar coordinates r and θ, letting $e \to 0$ and $p \to \infty$ in such a way that $ep \to a > 0$, we obtain an equation describing a circle of radius a.)

Solution 10.28 Counting the V vertices, E edges, and F faces in each case, we obtain:

(a) $V - E + F = 6 - 10 + 6 = 2$, as expected.
(b) $V - E + F = 10 - 15 + 7 = 2$, as expected.

Solution 10.29 Counting the V vertices, E edges, and F faces, we obtain $V - E + F = 7 - 15 + 9 = 1 \neq 2$; therefore, Euler's formula does not apply to the given polyhedron, so it cannot be simple.

Solution 10.30 The first two columns of Table 10.4 are essentially repeated in Table S10.30.1. Also, for each of the five polyhedra, let N be the number of vertices that belong to each face, and again let M be the number of faces to which each vertex belongs (which is also the number of edges to which each vertex belongs). The next two columns of Table S10.30.1, then, obtain from the last two columns of Table 10.4.

Now, let A be a vertex of one of these polyhedra. Each of the M faces to which A belongs contains $N - 3$ vertices that are neither A nor adjacent to A; and, none of these vertices belong to more than one such face. Therefore, since the polyhedron has M vertices adjacent to A, each of which belongs to a face to which A belongs, there are $M(N - 3) + M = M(N - 2)$ vertices other than A that belong to faces to which A belongs. So, A is an endpoint of $V - 1 - M(N - 2)$ diagonals. Hence, since each diagonal has two vertices as endpoints, the number of diagonals is $V[V - 1 - M(N - 2)]/2$, which is evaluated in Table S10.30.1.

Solution 10.31 **(a)** Let P be the center of a regular pentagon $A_1A_2A_3A_4A_5$ having radius r, with each side having length s. Since $s > r$, and thus $s^2 - r^2 > 0$, there exists a point A such that $\overline{AP}$ is perpendicular to the plane containing the pentagon, and $|\overline{AP}| = \sqrt{s^2 - r^2}$. Then, since $|\overline{PA_i}| = r$

TABLE S10.30.1

Polyhedron	V	N	M	$V[V - 1 - M(N - 2)]/2$
Regular tetrahedron	4	3	3	0
Regular hexahedron	8	4	3	4
Regular octahedron	6	3	4	3
Regular dodecahedron	20	5	3	100
Regular icosahedron	12	3	5	36

and $\overline{AP} \perp \overline{PA_i}$ ($i = 1, 2, 3, 4, 5$), it follows from the Pythagorean theorem that

$$|\overline{AA_i}| = \sqrt{|\overline{AP}|^2 + |\overline{PA_i}|^2} = \sqrt{(s^2 - r^2) + r^2} = s \quad (\text{all } i).$$

Thus, the pyramid having the pentagon as its base and A as its apex is a regular pyramid; and, letting $A_6 = A_1$, its lateral faces are bounded by the five equilateral triangles $\triangle AA_iA_{i+1}$ ($i = 1, 2, 3, 4, 5$), all having sides of length s.

(b) Let A' be the point on $\overleftrightarrow{AP}$ that lies on the opposite side of P than A, such that $|\overline{A'P}| = |\overline{AP}|$. By symmetry (or, by arguing again as above), $\triangle A'A_iA_{i+1}$ ($i = 1, 2, 3, 4, 5$) are five more equilateral triangles having sides of length s. Hence, considering the simple polyhedron having vertices A_1, A_2, A_3, A_4, A_5, A, and A', its faces are bounded by the 10 congruent regular polygons $\triangle AA_iA_{i+1}$ and $\triangle A'A_iA_{i+1}$ ($i = 1, 2, 3, 4, 5$). This polyhedron is convex because for any pair of vertices, the segment having those vertices as endpoints does not pass through the exterior of the polyhedron. However, it is not regular, since five edges meet at each of the vertices A and A', whereas only four edges meet at each of the vertices A_i.

Solution 10.32 "Rectangular parallelepiped".

Solution 10.33 **(a)** Yes.

(b) No, we could have three distinct planes intersecting at a line.

(c) Yes.

(d) Yes and yes.

(e) Yes.

(f) Yes.

(g) Yes for inscribed, no for circumscribed. In the latter case, the polygon could be a rhombus that is not a square.

(h) No for inscribed, yes for circumscribed. In the former case, the polygon could be a rectangle that is not a square.

(i) No, this is the case only for a solid polyhedron that is convex.

(j) No. Let p be a plane in space, and let f and g be the two closed half-planes having p as their face; then, $f \cap g = p$, which is not a solid polyhedron.

(k) Yes.

(l) Yes.

(m) Yes.

Solution 10.34 Let $a = \sup\{d(P,Q) \mid P, Q \in f\}$. If $a < \infty$, then for any point $P \in f$ (of which there is at least one, since $f \neq \emptyset$) we have $d(P,Q) \leq a < a + 1 \neq 0$, for all points $Q \in f$; therefore, f is contained in the solid sphere of radius $a + 1$ centered at P. Hence, condition (i) implies condition (ii).

Conversely, if f is contained in the solid sphere of some radius r centered at a point C, then $d(C,P) \leq r$ for all $P \in f$; therefore, it follows from the triangle inequality that $d(P,Q) \leq d(C,P) + d(C,Q) = r + r = 2r < \infty$ ($P, Q \in f$). Hence, condition (ii) implies condition (i).

Solution 10.35 Since $f \cup g$ is the union of the disjoint sets $f \cap g$, $f^c \cap g$, and $f \cap g^c$ (taking complementation to be relative to a plane containing f and g), it follows from the additivity of $\alpha(\cdot)$ that

$$\alpha(f \cup g) = \alpha(f \cap g) + \alpha(f^c \cap g) + \alpha(f \cap g^c). \tag{S10.35.1}$$

Likewise, since $f = (f \cap g) \cup (f \cap g^c)$ and $g = (f \cap g) \cup (f^c \cap g)$, we have

$$\alpha(f) = \alpha(f \cap g) + \alpha(f^c \cap g), \quad \alpha(g) = \alpha(f \cap g) + \alpha(f \cap g^c).$$

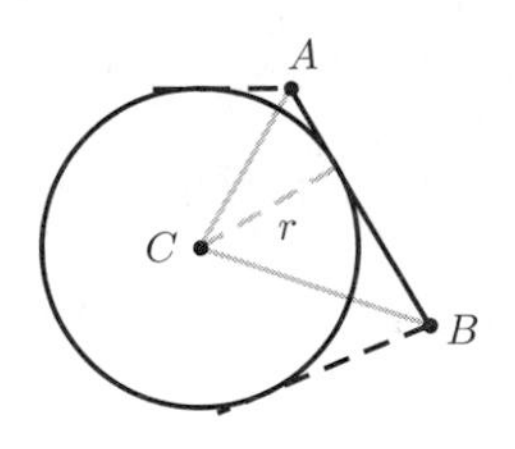

FIGURE S10.37.1

Therefore, adding $\alpha(f \cap g)$ to both sides of (S10.35.1) yields

$$\alpha(f \cup g) + \alpha(f \cap g) = \alpha(f) + \alpha(g).$$

(Note though that, having allowed infinite areas—e.g., all of the above areas equal ∞ if f and g are the same plane—we do not necessarily have $\alpha(f \cup g) = \alpha(f) + \alpha(g) - \alpha(f \cap g)$, since we cannot evaluate "$\infty - \infty$".)

Solution 10.36 The heights of the triangles $\triangle ADC$ and $\triangle BDC$, relative to their respective bases $\overline{AD}$ and $\overline{BD}$, both equal the distance from C to $\overleftrightarrow{AB}$; and, since D is the midpoint of $\overline{AB}$, the bases have equal lengths. Therefore, since the area of a triangle is uniquely determined by its height and the length of the corresponding base, $\triangle ADC$ and $\triangle BDC$ must have equal areas.

Solution 10.37 **(a)** Figure S10.37.1 shows a single side $\overline{AB}$ of a simple polygon that has been inscribed with a circle having center C and radius r. Since $\triangle ABC$ has height r relative to $\overline{AB}$ as the base, its area equals $\frac{1}{2}|\overline{AB}|r$. Summing the areas of all such triangles $\triangle APQ$ for which $\overline{PQ}$ is a side of the polygon yields the area α of the polygon; and, this sum equals $\frac{1}{2}pr$, where p the perimeter. Thus, $\alpha = sr$, where $s = \frac{1}{2}p$ is the semiperimeter.

(b) Using Heron's formula (10.25) for the area α of the triangle, it follows from part (a) that the inscribed circle has radius $r = \alpha/s = \sqrt{(s-a)(s-b)(s-c)/s}$, where $s = (a+b+c)/2$.

Solution 10.38 A rectangle having adjacent sides of length a and b has perimeter $p = 2(a+b)$ and area $\alpha = ab$. In particular, for a square ($a = b$) we have $p = 4a^2$ and $\alpha = a^2$, and thus a square of perimeter p has area $p^2/16$. Also, for the arbitrary rectangle we have

$$\begin{aligned} p^2/16 - \alpha &= (a+b)^2/4 - ab = (a^2 + 2ab + b^2)/4 - ab = (a^2 - 2ab + b^2)/4 \\ &= (a-b)^2/4 \geq 0, \end{aligned}$$

with equality being attained if and only if $a = b$; hence, $\alpha \leq p^2/16$, with equality if and only if the rectangle is a square.

Solution 10.39 Per (10.2), let $\kappa > 0$ be the scale factor from f to g. For each segment s in the polygon f, the corresponding segment in the polygon g has length κ times the length of s; hence, $p_g = \kappa p_f$. Also, we know that $\alpha_g = \kappa^2 \alpha_f$. Therefore, $(p_g/p_f)^2 = \kappa^2 = \alpha_g/\alpha_f$.

Solution 10.40 **(a)** Let the triangle be $\triangle ABC$, with side $\overline{AB}$ as the base; thus, the length of the base is $b = l$. Also, let the base have midpoint D; then, $\angle ADC \cong BDC$ by symmetry, so these supplementary angles are both right angles. Therefore, applying the Pythagorean theorem to the triangle $\triangle ADC$, the height of $\triangle ABC$ is

$$h = |\overline{DC}| = \sqrt{|\overline{AC}|^2 - |\overline{AD}|^2} = \sqrt{l^2 - (l/2)^2} = \sqrt{3l^2/4} = (\sqrt{3}/2)l;$$

hence, $\alpha = bh/2 = (\sqrt{3}/4)l^2$. Indeed, using Heron's formula (10.25) with $a = b = c = l$, and thus $s = (a+b+c)/2 = 3l/2$, we obtain

$$\alpha = \sqrt{s(s-a)(s-b)(s-c)} = \sqrt{(3l/2)(3l/2-l)(3l/2-l)(3l/2-l)}$$
$$= \sqrt{(3l/2)(l/2)^3} = \sqrt{3}(l/2)^2 = (\sqrt{3}/4)l^2.$$

(b) The base of the tetrahedron is an equilateral triangle having sides of length l; and, one *could* apply trigonometry (see Section 11.2) to easily express r in terms of l: $l/2 = r\cos\pi/6 = r\sqrt{3}/2$, making $r = l/\sqrt{3} = (\sqrt{3}/3)l$. Instead, note that one-third of the area of the base is consumed by a triangle having sides of length r, r, and l; hence, using Heron's formula (10.25) with $a = b = r$ and $c = l$, and thus $s = (a+b+c)/2 = r + l/2$, it follows from part (a) that

$$\tfrac{1}{3}(\sqrt{3}/4)l^2 = \sqrt{s(s-a)(s-b)(s-c)} = \sqrt{(r+l/2)(l/2)(l/2)(r-l/2)}$$
$$= \sqrt{r^2 - l^2/4}\cdot l/2,$$

making $r^2 - l^2/4 = l^2/12$, so $r = \sqrt{l^2/3} = (\sqrt{3}/3)l$. Now, by the Pythagorean theorem, if h is the height of the tetrahedron, then $h^2 + r^2 = l^2$, making $h = \sqrt{l^2 - r^2} = \sqrt{2/3}\,l$. Therefore, again using the result of part (a) for the area α of the base, we have $v = \frac{1}{3}\alpha h = \frac{1}{3}(\sqrt{3}/4)l^2 \cdot \sqrt{2/3}\,l = (\sqrt{2}/12)l^3$.

Solution 10.41 Because, for each of these two figures, the surface area is not generally determined by its height h and the area b of its base(s); nor has its slant height h_s been generally defined.

Solution 10.42 Let r be the common radius of f and g; and, let h be the height of f. Then, since the lateral surface area $2\pi r(r+h) - 2\cdot\pi r^2 = 2\pi rh$ of f equals the surface area $4\pi r^2$ of g, we have $h = 2r$; that is, the height of f equals the diameter of g. (Thus, we might say that the cylinder "circumscribes" the sphere.)

Solution 10.43 **(a)** The stated frustum f is shown in Figure S10.43.1, where it has been assumed without loss of generality that $r_1 < r_2$. Also shown is the original right circular cone g, having apex A and height h'. As indicated, let the bases have centers C_1 and C_2; thus, A, C_1, and C_2 are collinear. Additionally, taking any half-plane q having $\overleftrightarrow{C_1C_2}$ as its edge, let P_1 and P_2 be the unique points on the base boundaries that also lie on q; then, since $\angle P_1C_1A$ and $\angle P_2C_2A$ are right angles,

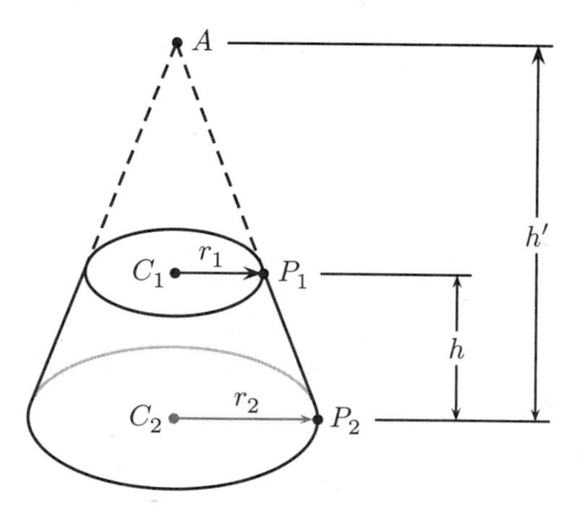

FIGURE S10.43.1 A Frustum of a Right Circular Cone

and $\angle C_1AP_1 = \angle C_2AP_2$, it follows from the AA condition for triangle similarity that $\triangle C_1AP_1 \sim \triangle C_2AP_2$. Hence, by (10.8b), $r_1/r_2 = |\overline{C_1P_1}|/|\overline{C_2P_2}| = |\overline{C_1A}|/|\overline{C_2A}| = (h'-h)/h' = 1 - h/h'$, making

$$h' = \frac{r_2}{r_2 - r_1} h. \tag{S10.43.1}$$

Therefore, defining the slant height h_s of the frustum f as $d(P_1, P_2)$, which equals the difference of the slant heights $h_s{}' = d(A, P_2)$ and $d(A, P_1)$ of the right circular cones g and $g - f$ (essentially), we obtain

$$h_s \triangleq d(P_1,P_2) = d(A,P_2) - d(A,P_1) = \sqrt{h'^2 + r_2{}^2} - \sqrt{(h'-h)^2 + r_1{}^2}$$

$$= \sqrt{\left(\frac{r_2}{r_2 - r_1}h\right)^2 + r_2{}^2} - \sqrt{\left(\frac{r_1}{r_2 - r_1}h\right)^2 + r_1{}^2}$$

$$= \frac{r_2}{r_2 - r_1}\sqrt{h^2 + (r_2 - r_1)^2} - \frac{r_1}{r_2 - r_1}\sqrt{h^2 + (r_2 - r_1)^2} = \sqrt{h^2 + (r_2 - r_1)^2}.$$

(This result can also be obtained, more easily, by applying the Pythagorean theorem to $\triangle P_1QP_2$, where Q is the point on $\overline{C_2P_2}$ for which $\overline{P_1Q} \perp \overline{QP_2}$; for then $|\overline{P_1Q}| = h$ and $|\overline{QP_2}| = r_2 - r_1$.) Also, using Table 10.6, the difference of the volumes of g and $g - f$ is

$$v = \tfrac{1}{3}\pi r_2{}^2 h' - \tfrac{1}{3}\pi r_1{}^2 (h' - h) = \tfrac{1}{3}\pi[(r_2{}^2 - r_1{}^2)h' + r_1{}^2 h]$$

$$= \tfrac{1}{3}\pi[(r_2 + r_1)(r_2 - r_1) \cdot r_2/(r_2 - r_1) + r_1{}^2]h = \tfrac{1}{3}\pi(r_1{}^2 + r_1 r_2 + r_2{}^2)h.$$

Finally, since (by proportionality) $h_s{}'/h_s = h'/h$, it follows from (S10.43.1) that $(r_2 - r_1)h_s{}' = r_2 h_s$, and thus the difference of lateral surface areas of g and $g - f$ is

$$\alpha_l = \pi r_2 h_s{}' - \pi r_1(h_s{}' - h_s) = \pi[(r_2 - r_1)h_s{}' + r_1 h_s] = \pi(r_1 + r_2)h_s$$

(to which we would add the total base area $\pi(r_1{}^2 + r_2{}^2)$ to get the total surface area of the frustum f).

(b) Letting $r_2 \to r_1$ makes $h_s \to h$, $v \to \pi r_1{}^2 h$, and $\alpha_l \to 2\pi r_1 h$, which are as expected; because, these limits are the height, volume, and lateral surface area of a right circular cylinder having base radius r_1 and height h.

Solution 10.44 In Figure 10.32, let C be the center of the circle c, which we take to be the boundary of the base of the given cone, having apex P. As indicated, let θ be the interior measure of the plane angle formed by the cone at its apex. For any point Q on c we have $\measuredangle CPQ = \theta/2$, with $\measuredangle PCQ$ being a right angle; hence, $\cos\theta/2 = d(P,C)/d(P,Q) = h/h_s$. Therefore, using (10.28) we obtain $\Omega = 2\pi(1 - h/h_s)$.

Chapter 11. Analytic Geometry

Solution 11.1 According to the right-hand rule, the vertical axis would be the x-axis and the axis pointing out of the page would be the y-axis.

Solution 11.2 Require θ to take its value from an interval I that is either $(a, a+\pi]$ or $[a, a+\pi)$ for some $a \in \mathbb{R}$. Because then, for each pair $(x,y) \in \mathbb{R} \times \mathbb{R}$ other than $(0,0)$, there is exactly one pair $(r,\theta) \in \mathbb{R} \times I$ for which (11.4a) holds, whereas $(x,y) = (0,0)$ again corresponds to having $r = 0$ with θ undefined.

Solution 11.3 Let $P = (X, Y)$ and $P' = (X', Y')$ be two points in the xy-plane having respective polar coordinate representations (R, Θ) and (R', Θ'), where R, R', Θ, $\Theta' \in \mathbb{R}$. To begin, suppose R, $R' > 0$ and $0 < |\Theta - \Theta'| < \pi$. Then P, P', and the origin O of the plane are distinct and noncollinear; so, since $\measuredangle POP' = |\Theta - \Theta'|$, applying the law of cosines (11.32) to $\triangle POP'$ yields

$$d(P,P') = \sqrt{d^2(P,O) + d^2(P',O) - 2d(P,O)d(P',O)\cos\measuredangle POP'}$$

$$= \sqrt{R^2 + R'^2 - 2RR'\cos|\Theta - \Theta'|} = \sqrt{R^2 + R'^2 - 2RR'\cos(\Theta - \Theta')},$$

per (11.3). Furthermore, (11.3) holds if R, $R' > 0$ and $\Theta - \Theta' = 0$, for then

$$\begin{aligned}\sqrt{R^2 + R'^2 - 2RR'\cos(\Theta - \Theta')} &= \sqrt{R^2 + R'^2 - 2RR'} = \sqrt{(R - R')^2} \\ &= |R - R'| = d(P, P');\end{aligned}$$

and, it also holds if R, $R' > 0$ and $\Theta - \Theta' = \pi$, for then

$$\begin{aligned}\sqrt{R^2 + R'^2 - 2RR'\cos(\Theta - \Theta')} &= \sqrt{R^2 + R'^2 + 2RR'} = \sqrt{(R + R')^2} \\ &= R + R' = d(P, P').\end{aligned}$$

Hence, we have thus far shown that (11.3) holds if R, $R' > 0$ and $-\pi < \Theta - \Theta' \le \pi$.

Now, suppose R, $R' \ne 0$ and Θ, $\Theta' \in \mathbb{R}$. Letting

$$\overline{R} = \begin{cases} R & R > 0 \\ -R & R < 0 \end{cases}, \qquad \overline{\Theta} = \begin{cases} \Theta & R > 0 \\ \Theta + \pi & R < 0 \end{cases},$$

the pair $(\overline{R}, \overline{\Theta})$ is also a polar coordinate representation of P, with $\overline{R} > 0$. Additionally, let

$$\overline{R}' = \begin{cases} R' & R' > 0 \\ -R' & R' < 0 \end{cases}, \qquad \overline{\Theta}' = \begin{cases} \Theta + 2n\pi & R' > 0 \\ \Theta + (2n+1)\pi & R' < 0 \end{cases},$$

where the integer n is chosen to make $-\pi < \overline{\Theta} - \overline{\Theta}' \le \pi$; then, the pair $(\overline{R}', \overline{\Theta}')$ is also a polar coordinate representation of P', with $\overline{R}' > 0$. Since we have $\overline{R}^2 = R^2$ and $\overline{R}'^2 = R'^2$, as well as $\overline{R}\,\overline{R}' \cos\left(\overline{\Theta} - \overline{\Theta}'\right) = RR'\cos(\Theta - \Theta')$ by the left equality in (11.10b) and the 2π-periodicity of cosine, it follows from the preceding paragraph that

$$d(P, P') = \sqrt{\overline{R}^2 + \overline{R}'^2 - 2\overline{R}\,\overline{R}'\cos\left(\overline{\Theta} - \overline{\Theta}'\right)} = \sqrt{R^2 + R'^2 - 2RR'\cos(\Theta - \Theta')};$$

so, (11.3) continues to hold. Finally, (11.3) holds whenever $R = 0$, for then $d(P, P') = d(O, P') = |R'|$; and, likewise whenever $R' = 0$. Therefore, (11.3) holds for all R, R', Θ, $\Theta' \in \mathbb{R}$.

Solution 11.4 Spherical-to-cylindrical conversion: First, suppose $\rho > 0$ and $0 < \phi < \pi$, thereby making the points P, P', and O distinct. Then, noting from Figure 11.6 that $r = d(O, P')$ and $z = d(P, P')$, it follows from the right triangle $\triangle OPP'$—for which $\angle OPP' = \phi$, since $\overline{PP'}$ is parallel to the z-axis—that

$$r = \rho\sin\phi, \quad \theta \text{ is as given}, \quad z = \rho\cos\phi. \tag{S11.4.1a}$$

A similar argument again yields (S11.4.1a) when $\rho > 0$ and $\pi < \phi < 2\pi$, though now $z = -d(P, P')$ and $\angle OPP' = \pi - \phi$; in particular, by (11.11) we have $\sin(\pi - \phi) = \sin\phi$ and $\cos(\pi - \phi) = -\cos\phi$. If, instead, $\rho > 0$ and $\phi = 0$, then (S11.4.1a) holds because $r = 0$ and $z = \rho$; whereas for $\rho > 0$ and $\phi = \pi$, (S11.4.1a) holds because $r = 0$ and $z = -\rho$. Furthermore, having considered $0 \le \phi \le \pi$, all other angles $\phi \in \mathbb{R}$ are handled similarly, and again yield (S11.4.1a). Finally, for $\rho = 0$ we have $r = z = 0$, and thus (S11.4.1a) can still be used even though ϕ is now undefined, since zero times anything equals zero.

Cylindrical-to-spherical conversion: We now show that

$$\rho = \sqrt{r^2 + z^2}, \quad \theta \text{ is as given}, \quad \phi = \begin{cases} \cos^{-1}(z/\rho) & \rho > 0 \\ \text{undefined} & \rho = 0 \end{cases}, \tag{S11.4.1b}$$

assuming $0 \le \phi \le \pi$. First, we see from Figure 11.6 that if $r = 0$, then P lies on the z-axis, making $\rho = |z|$ and

$$\phi = \begin{cases} 0 & z > 0 \\ \text{undefined} & z = 0 \\ \pi & z < 0 \end{cases}.$$

These results obtain from (S11.4.1b) as well; therefore, suppose $r > 0$. Moreover, if $z = 0$, then P lies on the xy-plane, making $\rho = r$ and $\phi = \pi/2$, which also obtain from (S11.4.1b); so, suppose $z \neq 0$ as well. The points P, P', and O and now distinct, and the expression for ρ in (S11.4.1b) obtains by applying the Pythagorean theorem to the right triangle $\Delta OPP'$. Also, for $z > 0$ we have $\measuredangle OPP' = \phi$ and thus $\cos\phi = \cos\measuredangle OPP' = d(P,P')/d(O,P) = z/\rho$; whereas for $z < 0$ we have $\measuredangle OPP' = \pi - \phi$ and thus $-\cos\phi = \cos(\pi - \phi) = \cos\measuredangle OPP' = d(P,P')/d(O,P) = -z/\rho$. Therefore, in either case, $\cos\phi = z/\rho$, thereby yielding the expression for ϕ in (S11.4.1b). (*Aside*: Note that we cannot state, based on (S11.4.1a), that $\phi = \tan^{-1}(r/z)$ when $z \neq 0$; e.g., consider converting $(r,\theta,z) = (1,1,-1)$ to (ρ,θ,ϕ).)

Solution 11.5 By the odd symmetry and 2π-periodicity of sine, along with (11.10b), we have $\sin(19\pi/2 - \theta) = -\sin(\theta - 19\pi/2) = -\sin(\theta - 5\cdot 2\pi + \pi/2) = -\sin(\theta + \pi/2) = -\cos\theta$.

Solution 11.6 For all such x,

$$\sin\pi x/(x+1) = \sin\pi\left(\frac{x+1}{x+1} - \frac{1}{x+1}\right) = \sin[\pi - \pi/(x+1)] = \sin\pi/(x+1),$$

by (11.11a).

Solution 11.7 By the continuity of the cosine function, $\lim_{\theta\to 0}\cos\theta = \cos 0 = 1$; therefore,

$$\lim_{\theta\to 0}\frac{\tan\theta}{\theta} = \lim_{\theta\to 0}\frac{\sin\theta/\cos\theta}{\theta} = \lim_{\theta\to 0}\frac{(\sin\theta)/\theta}{\cos\theta} = \frac{\lim_{\theta\to 0}(\sin\theta)/\theta}{\lim_{\theta\to 0}\cos\theta} = \frac{1}{1} = 1.$$

Hence, $\tan\theta \sim \theta$ as $\theta \to 0$.

Solution 11.8 Squaring equation (P11.8.1) and using (11.15a) shows that *if* a solution $t \in \mathbb{R}$ to (P11.8.1) exists, then we must have

$$4(1 - \cos^2 t) = 4\sin^2 t = 9 + 24\cos t + 16\cos^2 t,$$

which simplifies to

$$20\cos^2 t + 24\cos t + 5 = 0.$$

It then follows from the quadratic formula (6.23c) that

$$\cos t = \frac{-24 \pm \sqrt{24^2 - 4\cdot 20\cdot 5}}{2\cdot 20} = \alpha_1, \alpha_2,$$

where

$$\alpha_1 = \frac{-6 + \sqrt{11}}{10}, \quad \alpha_2 = \frac{-6 - \sqrt{11}}{10}. \tag{S11.8.1}$$

Since $3 = \sqrt{9} < \sqrt{11} < \sqrt{16} = 4$, both α_1 and α_2 lie between ± 1; so, neither can be immediately dismissed as a possible value for $\cos t$. But, we *can* say that $t \neq 0, \pi$. Moreover, it is clear from Figure 11.8 that there is exactly one $t \in (0,\pi)$ such that $\cos t = \alpha_1$, which by (11.14b) is $\cos^{-1}\alpha_1$; thus, by the even symmetry of cosine, $-\cos^{-1}\alpha_1$ is the one value $t \in (-\pi, 0)$ such that $\cos t = \alpha_1$. Likewise, the only values $t \in (-\pi,\pi]$ such that $\cos t = \alpha_2$ are $\pm\cos^{-1}\alpha_2$. Therefore, by the 2π-periodicity of cosine, the only values $t \in \mathbb{R}$ such that either $\cos t = \alpha_1$ or $\cos t = \alpha_2$ are

$$n\cdot 2\pi \pm \cos^{-1}\alpha_1, \quad n\cdot 2\pi \pm \cos^{-1}\alpha_2 \quad (n \in \mathbb{Z}).$$

Substituting $t = n\cdot 2\pi \pm \cos^{-1}\alpha_1$ into (P11.8.1) and using the 2π-periodicity of sine and cosine, we find that such a value t is a solution to (P11.8.1) for some $n \in \mathbb{Z}$ if and only if

$$2\sin(\pm\cos^{-1}\alpha_1) = 3 + 4\cos(\pm\cos^{-1}\alpha_1)$$

($\pm$ respectively); equivalently, by the odd symmetry of sine, (11.30c), the even symmetry of cosine, and (11.13b),

$$\pm 2\sqrt{1-\alpha_1{}^2} = 3 + 4\alpha_1$$

($\pm$ respectively), which is satisfied by (S11.8.1) if and only if "$\pm$" is taken to be "+". Likewise, $t = n \cdot 2\pi \pm \cos^{-1} \alpha_2$ is a solution to (P11.8.1) for some $n \in \mathbb{Z}$ if and only if $\pm 2\sqrt{1-\alpha_2{}^2} = 3 + 4\alpha_2$ ($\pm$ respectively), which is satisfied by (S11.8.1) if and only if "$\pm$" is taken to be "$-$". Therefore, the solutions $t \in \mathbb{R}$ of (P11.8.1) are

$$n \cdot 2\pi + \cos^{-1} \frac{-6+\sqrt{11}}{10}, \quad n \cdot 2\pi - \cos^{-1} \frac{-6-\sqrt{11}}{10} \quad (n \in \mathbb{Z}). \tag{S11.8.2}$$

Solution 11.9 As stated on p. 179, each of the identities in Section 11.2 that followed are valid for those $\theta, \phi \in \mathbb{R}$ for which all of the operations performed in the expression are defined. Thus, since "$\tan 2\theta$" appears in (11.19c), 2θ cannot be an odd multiple of $\pi/2$, and thus θ cannot be an odd multiple of $\pi/4$. Likewise, since "$\tan \theta$" also appears, θ cannot be an odd multiple of $\pi/2$. Furthermore, division by 0 is disallowed, which in (11.19c) precludes having $\tan \theta = \pm 1$; but, from Figure 11.9 and Table 11.1, along with the even symmetry and π-periodicity of tangent, we see that this occurs only if θ is an odd multiple of $\pi/4$. Therefore, (11.19c) is valid for all $\theta \in \mathbb{R}$ other than odd multiples of either $\pi/4$ or $\pi/2$; that is, we cannot have $\theta = (k/4 + n)\pi$ $(k = 1, 2, 3;\ n \in \mathbb{Z})$.

Solution 11.10 First, note that $\left(\sqrt{6} \pm \sqrt{2}\right)^2 = 6 + 2 \pm 2\sqrt{6}\sqrt{2} = 4\left(2 \pm \sqrt{3}\right)$ ($\pm$ respectively), and thus $\sqrt{2 \pm \sqrt{3}} = \left(\sqrt{6} \pm \sqrt{2}\right)/2$ ($\pm$ respectively). Therefore, using the half-angle identities (11.20), while noting from Table 11.2 that $\sin \pi/12$, $\cos \pi/12$, $\sin \pi/8$, and $\cos \pi/8$ are all positive, it follows from Table 11.1 that

$$\sin \pi/12 = \sin \tfrac{1}{2}(\pi/6) = \sqrt{\frac{1-\cos \pi/6}{2}} = \sqrt{\frac{1-\sqrt{3}/2}{2}} = \frac{\sqrt{2-\sqrt{3}}}{2} = \frac{\sqrt{6}-\sqrt{2}}{4},$$

$$\cos \pi/12 = \cos \tfrac{1}{2}(\pi/6) = \sqrt{\frac{1+\cos \pi/6}{2}} = \sqrt{\frac{1+\sqrt{3}/2}{2}} = \frac{\sqrt{2+\sqrt{3}}}{2} = \frac{\sqrt{6}+\sqrt{2}}{4},$$

$$\tan \pi/12 = \tan \tfrac{1}{2}(\pi/6) = \frac{1-\cos \pi/6}{\sin \pi/6} = \frac{1-\sqrt{3}/2}{1/2} = 2 - \sqrt{3},$$

and

$$\sin \pi/8 = \sin \tfrac{1}{2}(\pi/4) = \sqrt{\frac{1-\cos \pi/4}{2}} = \sqrt{\frac{1-\sqrt{2}/2}{2}} = \frac{\sqrt{2-\sqrt{2}}}{2},$$

$$\cos \pi/8 = \cos \tfrac{1}{2}(\pi/4) = \sqrt{\frac{1+\cos \pi/4}{2}} = \sqrt{\frac{1+\sqrt{2}/2}{2}} = \frac{\sqrt{2+\sqrt{2}}}{2},$$

$$\tan \pi/8 = \tan \tfrac{1}{2}(\pi/4) = \frac{1-\cos \pi/4}{\sin \pi/4} = \frac{1-\sqrt{2}/2}{\sqrt{2}/2} = \sqrt{2} - 1.$$

Solution 11.11 Taking the base of the triangle to be the side of length a, its height is then $b \sin C$; therefore, its area equals $\frac{1}{2}ab \sin C$. Also, multiplying this result by

$$\frac{(c/\sin C)^2}{(a/\sin A)(b/\sin B)},$$

which by the law of sines (11.31) equals 1, we obtain (P11.11.1).

Solution 11.12 Using the fact that $A+B+C=\pi$, along with (11.11a) and (11.16a), we obtain

$$\sin A = \sin[\pi - (B+C)] = \sin(B+C) = \sin B\cos C + \cos B\sin C;$$

therefore, multiplying through by $a/\sin A$ and applying by the law of sines (11.31), we get

$$\begin{aligned} a &= \left(\frac{a}{\sin A}\sin B\right)\cos C + \left(\frac{a}{\sin A}\sin C\right)\cos B \\ &= \left(\frac{b}{\sin B}\sin B\right)\cos C + \left(\frac{c}{\sin C}\sin C\right)\cos B = b\cos C + c\cos B. \end{aligned}$$

Solution 11.13 Let A, B, and C be distinct coplanar points. If these points are collinear, then either $|\overline{BC}| = |\overline{AB}| + |\overline{AC}|$ (when A lies between B and C), or $|\overline{AC}| = |\overline{AB}| + |\overline{BC}|$ (when B lies between A and C), or $|\overline{AB}| = |\overline{AC}| + |\overline{BC}|$ (when C lies between A and B); but, as can be verified by substitution, in none of these cases can (10.12) hold. Therefore, suppose that A, B, and C are noncollinear, and thus $\triangle ABC$ exists. Then, the law of cosines (11.32) yields

$$|\overline{AC}|^2 = |\overline{AB}|^2 + |\overline{BC}|^2 - 2|\overline{AB}|\cdot|\overline{BC}|\cos\angle ABC;$$

so, since $|\overline{AB}|$ and $|\overline{BC}|$ are nonzero, (10.12) holds only if $\cos\angle ABC = 0$. But, the only value $\theta \in (0,\pi)$ for which $\cos\theta = 0$ is $\pi/2$; therefore, if (10.12) holds, then $\angle ABC$ is a right angle.

Solution 11.14 By the law of sines (11.31), we can multiply the numerator and denominator on the left of (P11.14.1) by $(\sin A)/a = (\sin B)/b$; then, using (11.18) and (11.9), we obtain

$$\begin{aligned} \frac{a-b}{a+b} &= \frac{\sin A - \sin B}{\sin A + \sin B} = \frac{2\sin\frac{1}{2}(A-B)\cos\frac{1}{2}(A+B)}{2\cos\frac{1}{2}(A-B)\sin\frac{1}{2}(A+B)} \\ &= \frac{\sin\frac{1}{2}(A-B)}{\cos\frac{1}{2}(A-B)}\Big/\frac{\sin\frac{1}{2}(A+B)}{\cos\frac{1}{2}(A+B)} = \frac{\tan\frac{1}{2}(A-B)}{\tan\frac{1}{2}(A+B)}. \end{aligned}$$

(Note that since $A, B < \pi$ and $A+B = \pi - C < \pi$, thereby making $-\pi/2 < \frac{1}{2}(A-B) < \pi/2$ and $0 < \frac{1}{2}(A+B) < \pi/2$, there are no divisions by 0.)

Solution 11.15 Using (11.28c) yields

$$\begin{aligned} f(-x') &= \tan^{-1}\left(e^{-x'}\right) - \pi/4 = \tan^{-1}\left(1/e^{x'}\right) - \pi/4 \\ &= \left[\pi/2 - \tan^{-1}\left(e^{x'}\right)\right] - \pi/4 = -\tan^{-1}\left(e^{x'}\right) + \pi/4 = -f(x') \end{aligned}$$

($x' \in \mathbb{R}$); thus, $f(\cdot)$ is an odd function.

Solution 11.16 **(a)** See Figure S11.16.1.

(b) Given $\theta \in \mathbb{R}$, let

$$\begin{aligned} m &= (\text{unique integer for which } -\pi/2 < \theta + m\pi \le \pi/2), \\ n &= (\text{unique integer for which } 0 \le \theta + n\pi < \pi); \end{aligned}$$

then, it is evident from Figure S11.16.1 that

$$\sin^{-1}(\sin\theta) = \begin{cases} \theta + m\pi & m \text{ even} \\ -(\theta + m\pi) & m \text{ odd} \end{cases},$$

$$\cos^{-1}(\cos\theta) = \begin{cases} \theta + n\pi & n \text{ even} \\ \pi - (\theta + n\pi) & n \text{ odd} \end{cases},$$

$$\tan^{-1}(\tan\theta) = \begin{cases} \theta + m\pi & \theta \text{ not an odd multiple of } \pi/2 \\ \text{undefined} & \text{otherwise} \end{cases}.$$

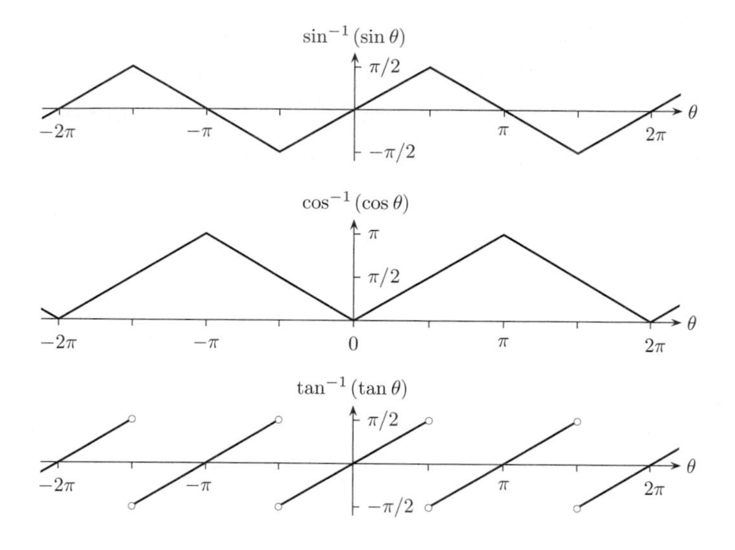

FIGURE S11.16.1

(c) For $\theta = 9$ we have $m = -3$, making $\sin^{-1}(\sin 9) = -(9 - 3\pi) \approx 0.425$ and $\tan^{-1}(\tan 9) = 9 - 3\pi \approx -0.425$; whereas $n = -2$, making $\cos^{-1}(\cos 9) = 9 - 2\pi \approx 2.717$.

Solution 11.17 By the left side of (11.4b), we have $r = 0$ if and only if $x = y = 0$; therefore, the proposed method declares θ to be undefined precisely when the right side of (11.4b) does. Now, suppose that a given rectangular pair (x, y) is converted as proposed to a polar pair (r, θ) with $r \neq 0$; then, it is sufficient for us to show that the polar-to-rectangular conversion (11.4a) returns (r, θ) to (x, y). To wit, let $s = 1$ if $y \geq 0$, and $s = -1$ if $y < 0$; thus, (P11.17.1) states that $\theta = s\cos^{-1} x/r$. Then, noting that the left side of (11.4b) implies that $|x| \leq r$, thereby making $-1 \leq x/r \leq 1$, it follows from the even symmetry of cosine, along with (11.13b), that

$$r\cos\theta = r\cos(s\cos^{-1} x/r) = r\cos(\cos^{-1} x/r) = r(x/r) = x,$$

per (11.4a). Also, the odd symmetry of sine along with (11.30c) yield

$$\begin{aligned} r\sin\theta &= r\sin(s\cos^{-1} x/r) = sr\sin(\cos^{-1} x/r) = sr\sqrt{1 - (x/r)^2} = s\sqrt{r^2 - x^2} \\ &= s|y| = y, \end{aligned}$$

per (11.4a).

Solution 11.18 **(a)** Let $\theta \in \mathbb{R}$ be given. This problem *could* be solved by using (11.16) to expand both sides of each equation in (P11.18.1), then using (11.13) and (11.30) to show that the two sides of each equation are equal. Instead, note that it follows from (11.10a) and (11.28a) that

$$\sin(\theta + \sin^{-1} x) = \cos(\theta + \sin^{-1} x - \pi/2) = \cos(\theta - \cos^{-1} x) \quad (-1 \leq x \leq 1),$$

and thus (P11.18.1a) holds. Likewise,

$$\sin(\theta + \cos^{-1} x) = \cos(\theta + \cos^{-1} x - \pi/2) = \cos(\theta - \sin^{-1} x) \quad (-1 \leq x \leq 1),$$

giving (P11.18.1b); and, also using (11.28c), we obtain

$$\sin(\theta + \tan^{-1} x) = \cos(\theta + \tan^{-1} x - \pi/2) = \cos(\theta - \tan^{-1} 1/x) \quad (x > 0),$$

giving (P11.18.1c).

(b) For all $\theta \in \mathbb{R}$:

$$\sin(\theta - \sin^{-1} x) = -\cos(\theta + \cos^{-1} x) \quad (-1 \le x \le 1),$$

$$\sin(\theta - \cos^{-1} x) = -\cos(\theta + \sin^{-1} x) \quad (-1 \le x \le 1),$$

$$\sin(\theta - \tan^{-1} x) = -\cos(\theta + \tan^{-1} 1/x) \quad (x > 0).$$

These identities obtain by replacing each "θ" in (P11.18.1) with "$\theta - \pi/2$", then using (11.10) to simplify.

Solution 11.19 For $x, y \in \mathbb{R}$, it follows from (11.33) that

$$\cosh^2 x - \sinh^2 x = \left(\frac{e^x + e^{-x}}{2}\right)^2 - \left(\frac{e^x - e^{-x}}{2}\right)^2$$
$$= \frac{e^{2x} + 2 + e^{-2x}}{4} - \frac{e^{2x} - 2 + e^{-2x}}{4} = \frac{4}{4} = 1;$$

and, by the even symmetry of hyperbolic cosine and the odd symmetry of hyperbolic sine,

$$\sinh x \cosh y \pm \cosh x \sinh y = \sinh x \cosh(\pm y) + \cosh x \sinh(\pm y)$$
$$= \frac{e^x - e^{-x}}{2} \cdot \frac{e^{\pm y} + e^{-(\pm y)}}{2} + \frac{e^x + e^{-x}}{2} \cdot \frac{e^{\pm y} - e^{-(\pm y)}}{2}$$
$$= \frac{2e^x e^{\pm y} - 2e^{-x} e^{-(\pm y)}}{4} = \frac{e^{x \pm y} - e^{-(x \pm y)}}{2} = \sinh(x \pm y)$$

($\pm$ respectively); and, likewise,

$$\cosh x \cosh y \pm \sinh x \sinh y = \cosh x \cosh(\pm y) + \sinh x \sinh(\pm y)$$
$$= \frac{e^x + e^{-x}}{2} \cdot \frac{e^{\pm y} + e^{-(\pm y)}}{2} + \frac{e^x - e^{-x}}{2} \cdot \frac{e^{\pm y} - e^{-(\pm y)}}{2}$$
$$= \frac{2e^x e^{\pm y} + 2e^{-x} e^{-(\pm y)}}{4} = \frac{e^{x \pm y} + e^{-(x \pm y)}}{2} = \cosh(x \pm y)$$

($\pm$ respectively).

Solution 11.20 Taking the real and imaginary parts of (P11.20.1a) yields

$$A_1 \cos\theta_1 + A_2 \cos\theta_2 = A \cos\theta,$$
$$A_1 \sin\theta_1 + A_2 \sin\theta_2 = A \sin\theta.$$

Squaring each of these equations, adding the results, and using (11.15a), we then obtain

$$A_1{}^2 + A_2{}^2 + 2A_1A_2(\cos\theta_1 \cos\theta_2 + \sin\theta_1 \sin\theta_2) = A^2,$$

where $\cos\theta_1 \cos\theta_2 + \sin\theta_1 \sin\theta_2 = \cos(\theta_2 - \theta_1)$ by (11.16b). Therefore, assuming $A \ge 0$, we must have (P11.20.1b).

If $A = 0$, then the right side of (P11.20.1a) does not depend on θ, which thus can be chosen arbitrarily, as stated in (P11.20.1c); therefore, suppose $A > 0$, thereby making $A_1 > 0$, since $A_1 \ge A_2 \ge 0$. Furthermore, suppose $A_2 > 0$ as well; for otherwise (P11.20.1b) gives $A = A_1$ and (P11.20.1c) gives $\theta = \theta_1$, thereby satisfying (P11.20.1a). We now have the situation illustrated by the Argand diagram in Figure S11.20.1, which makes clear that since $A_1 \ge A_2$, the contribution of θ_2 to the resultant angle θ is limited. Accordingly, in calculating θ, it behooves us to use θ_1 as a "reference"; thus, dividing (P11.20.1a) by $e^{i\theta_1}$, we obtain

$$A_1 + A_2 e^{i(\theta_2 - \theta_1)} = A e^{i(\theta - \theta_1)}. \tag{S11.20.1}$$

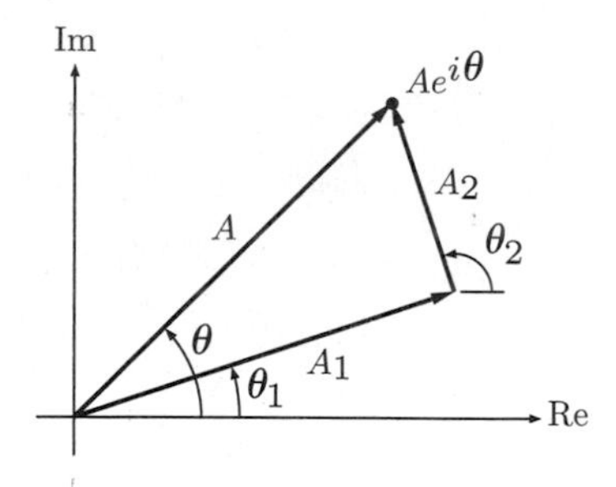

FIGURE S11.20.1

Since $\cos(\theta_2 - \theta_1) \geq -1$, we have

$$\text{Re}[A_1 + A_2 e^{i(\theta_2-\theta_1)}] = A_1 + A_2 \cos(\theta_2 - \theta_1) \geq 0;$$

moreover, this inequality cannot attain equality, since that would require $A_1 = A_2$ and $\cos(\theta_2 - \theta_1) = -1$, thereby making $A = 0$ (as can be seen from Figure S11.20.1, now with the vectors for $A_1 e^{i\theta_1}$ and $A_2 e^{i\theta_2}$ pointing in opposite directions). Therefore, applying (3.30b) to (S11.20.1) yields

$$\theta - \theta_1 = \arg[Ae^{i(\theta-\theta_1)}] = \tan^{-1} \frac{\text{Im}\left[A_1 + A_2 e^{i(\theta_2-\theta_1)}\right]}{\text{Re}\left[A_1 + A_2 e^{i(\theta_2-\theta_1)}\right]},$$

from which the first part of (P11.20.1c) obtains. (*Aside*: Note that although (P11.20.1b) is always valid (assuming $A \geq 0$), (P11.20.1c) need not hold when $A_1 < A_2$; e.g., for $A_1 = 1, A_2 = 2, \theta_1 = 0$, and $\theta_2 = \pi$, we have $A_1 e^{i\theta_1} + A_2 e^{i\theta_2} = -1 = 1\angle\pi$, whereas (P11.20.1c) states that θ equals 0.)

Solution 11.21 **(a)** Substituting (P11.21.2) into the right side of (P11.21.3), and using Euler's identity (3.32), we obtain

$$\begin{aligned}
\text{Re}\left(\mathbf{x}\, e^{i\omega t}\right) &= \text{Re}\left(Ae^{i\theta} e^{i\omega t}\right) = \text{Re}\left(Ae^{i(\omega t+\theta)}\right) \\
&= \text{Re}[A\cos(\omega t + \theta) + iA\sin(\omega t + \theta)] = A\cos(\omega t + \theta) \\
&= x(t) \quad (t \in \mathbb{R}).
\end{aligned}$$

(b) Letting $x(t) = \sin \omega t$, it follows from (11.10a) that (P11.21.1) is satisfied by taking $A = 1$ and $\theta = -\pi/2$, which substituted into (P11.21.2) gives $\mathbf{x} = e^{-i\pi/2} = -i$; therefore, the phasor associated with sine is $-i$, as can be verified by substitution for $\mathbf{x}$ in (P11.21.3).

(c) Given ω, substituting (P11.21.4) into (P11.21.3) yields

$$\begin{aligned}
x(t) &= \text{Re}\left[(a + ib)e^{i\omega t}\right] = \text{Re}[(a + ib)(\cos \omega t + i \sin \omega t)] \\
&= \text{Re}[(a\cos \omega t - b \sin \omega t) + i(a \sin \omega t + b \cos \omega t)] \\
&= a\cos \omega t - b \sin \omega t \quad (t \in \mathbb{R}).
\end{aligned}$$

(d) Using (P11.21.3) and (P11.21.6b), we obtain

$$\begin{aligned}
ax(t) + by(t) &= a\,\text{Re}\left(\mathbf{x}\, e^{i\omega t}\right) + b\,\text{Re}\left(\mathbf{y}\, e^{i\omega t}\right) = \text{Re}\left(a\mathbf{x}\, e^{i\omega t}\right) + \text{Re}\left(b\mathbf{y}\, e^{i\omega t}\right) \\
&= \text{Re}\left(a\mathbf{x}\, e^{i\omega t} + b\mathbf{y}\, e^{i\omega t}\right) = \text{Re}\left(\mathbf{z}\, e^{i\omega t}\right) \quad (t \in \mathbb{R}),
\end{aligned}$$

and thus $z(t) = \text{Re}\left(\mathbf{z}\, e^{i\omega t}\right)$ in (P11.21.6a); therefore, $z(t)$ is a sinusoid of frequency ω, with its phasor $\mathbf{z}$ being given by (P11.21.6b).

(e) Letting $\omega = 1, a = 4, x(t) = \cos \omega t, b = -2$, and $y(t) = \sin \omega t$ $(t \in \mathbb{R})$ in (P11.21.6a), it follows from parts (b) and (d) that $z(t) = 4\cos \omega t - 2 \sin \omega t$ is a sinusoid of frequency ω having phasor

$\mathbf{z} = 4(1) - 2(-i) = 4 + i2$. More simply, this result obtains immediately by converting the sinusoid $z(t)$ in rectangular form (P11.21.5) into the corresponding phasor (P11.21.4). Thus, by (3.30), $|\mathbf{z}| = \sqrt{4^2 + 2^2} = 2\sqrt{5}$ and $\arg \mathbf{z} = \tan^{-1} 2/4 = \tan^{-1} 1/2$; so, converting the phasor $\mathbf{z}$ in polar form (P11.21.2) into the corresponding sinusoid (P11.21.1), we have $z(t) = 2\sqrt{5}\cos(t + \tan^{-1} 1/2)$. Therefore, since (P11.8.1) is equivalent to the equation $z(t) = -3$, it is satisfied for some $t \in \mathbb{R}$ if and only if $2\sqrt{5}\cos(t + \tan^{-1} 1/2) = -3$; that is,

$$\cos(t + \tan^{-1} 1/2) = -3/2\sqrt{5} = -3\sqrt{5}/10.$$

Since $\sqrt{5} < \sqrt{9} = 3$, we have $\left|-3\sqrt{5}/10\right| < 1$, and thus there exist exactly two values $\alpha \in (-\pi, \pi]$ such that $\cos\alpha = -3\sqrt{5}/10$; namely, by the even symmetry of cosine, $\alpha = \pm\cos^{-1}\left(-3\sqrt{5}/10\right)$. Hence, letting $\alpha = t + \tan^{-1} 1/2$, it follows from the 2π-periodicity of cosine that (P11.8.1) is satisfied if and only if

$$t = n \cdot 2\pi - \tan^{-1} 1/2 \pm \cos^{-1}\left(-3\sqrt{5}/10\right) \quad (n \in \mathbb{Z})$$

(cf. (S11.8.2)).

Solution 11.22 We first show that for all $\vec{x}, \vec{y} \in V$,

$$\vec{x} + \vec{y} = \vec{x} \quad \Rightarrow \quad \vec{y} = \vec{0}. \tag{S11.22.1}$$

This can seen by adding $-\vec{x}$ to both sides of the left equation.[28] To wit, using (11.36a), the first part of (11.36b), (11.36f), and (11.36e), we obtain

$$(\vec{x} + \vec{y}) + (-\vec{x}) = (\vec{y} + \vec{x}) + (-\vec{x}) = \vec{y} + [\vec{x} + (-\vec{x})] = \vec{y} + \vec{0} = \vec{y};$$

therefore, if $\vec{x} + \vec{y} = \vec{x}$, then this result must equal $\vec{x} + (-\vec{x}) = \vec{0}$, thereby making $\vec{y} = \vec{0}$ by uniqueness of the vector $\vec{0}$ satisfying (11.36e).

Now, let $\vec{x} \in V$ and $a \in F$ be given. By (11.36f), $\vec{0} + (-\vec{0}) = \vec{0}$, and thus it follows from (S11.22.1) that $-\vec{0} = \vec{0}$, proving (11.37a). The second part of (11.36c) gives $(1+0)\vec{x} = 1\vec{x} + 0\vec{x}$, and thus $\vec{x} = \vec{x} + 0\vec{x}$ by (11.36d); so, by (S11.22.1) we have $0\vec{x} = \vec{0}$, proving (11.37b). The second part of (11.36b) gives $a(0\vec{x}) = (a \cdot 0)\vec{x}$, which by (11.37b) yields $a\vec{0} = \vec{0}$, proving (11.37c). The second part of (11.36c) gives $[1 + (-1)]\vec{x} = 1\vec{x} + (-1)\vec{x}$, which by (11.37b) and (11.36d) yields $\vec{0} = \vec{x} + (-1)\vec{x}$; therefore, since the vector $-\vec{x}$ satisfying (11.36f) is unique, we must have $(-1)\vec{x} = -\vec{x}$, proving (11.37d). The second part of (11.36b) gives $(-1)[(-1)\vec{x}] = [(-1) \cdot (-1)]\vec{x}$, which by (11.37d) and (11.36d) yields $-(-\vec{x}) = \vec{x}$, proving (11.37e). Finally, the second part of (11.36b) gives both $a[(-1)\vec{x}] = [a(-1)]\vec{x} = (-a)\vec{x}$ and $(-1)(a\vec{x}) = [(-1)a]\vec{x} = (-a)\vec{x}$; therefore, using (11.37d), $(-a)\vec{x}$ equals both $a[(-1)\vec{x}] = a(-\vec{x})$ and $(-1)(a\vec{x}) = -(a\vec{x})$, proving (11.37f).

Solution 11.23 **(a)** No; because, e.g., for $\vec{x} = 1 \in V$ and $a = i \in F$, we have $a\vec{x} = i \notin V$.

(b) Yes, with $\vec{0}$ being the function $f(\cdot)$ on $[0, 1]$ for which $f(x) = 0$ $(0 \le x \le 1)$.

(c) Yes, with $\vec{0}$ being the function $f(\cdot)$ on $(0, \infty)$ for which $f(x) = 0$ $(x > 0)$. (In particular, note that for any $\alpha, \beta \in F$ and $a, b > 1$, we have $\alpha \log_a x + \beta \log_b x = \alpha(\log_a b)\log_b x + \beta \log_b x = [\alpha(\log_a b) + \beta]\log_b x$ $(x > 0)$, where $\alpha(\log_a b) + \beta \in F$.)

(d) No; because, e.g., no $\alpha \in \mathbb{R}$ and $b > 1$ exist for which $2^x + 4^x = \alpha b^x$ $(x \in \mathbb{R})$. For suppose such α and b do exist. Then, since $(2^x + 4^x)/4^x = 2^{-x} + 1 \to 1$ as $x \to \infty$, we must have $\alpha b^x/4^x \to 1$ as $x \to \infty$; but, $\alpha b^x/4^x = \alpha b^x / b^{(\log_b 4)x} = \alpha b^{(1-\log_b 4)x}$, which approaches 1 as $x \to \infty$ only if $\alpha = 1$ and $1 - \log_b 4 = 0$, the latter condition implying $b = 4$. Hence, we must now have $2^x + 4^x = 4^x$ $(x \in \mathbb{R})$, which is impossible; therefore, no such α and b exist.

---

28. Observe that the antecedent in (S11.22.1) does not state that $\vec{y} \in V$ is such that $\vec{x} + \vec{y} = \vec{x}$ for *all* $\vec{x} \in V$. Had that been the case, then the validity of the implication would follow immediately from the uniqueness of the vector $\vec{0}$ satisfying (11.36e).

Solution 11.24 Using (11.43), (11.41c), and (11.42), we obtain

$$\|\vec{x} \pm \vec{y}\,\|^2 = [\vec{x} \pm \vec{y}, \vec{x} \pm \vec{y}\,] = [\vec{x}, \vec{x}\,] \pm [\vec{x}, \vec{y}\,] \pm [\vec{y}, \vec{x}\,] + [\vec{y}, \vec{y}\,]$$
$$= \|\vec{x}\,\|^2 \pm [\vec{x}, \vec{y}\,] \pm [\vec{y}, \vec{x}\,] + \|\vec{y}\,\|^2$$

($\pm$ respectively), which upon summation yields (P11.24.1). This result implies that for any parallelogram, the sum of the squares of the diagonals equals the sum of the squares of the sides.

Solution 11.25 **(a)** Let $\vec{x}, \vec{y} \in V$ and $a \in F$ be given. It is clear that the stated function $\|\cdot\|$ is nonnegative, with $\|\vec{x}\,\| = 0$ if and only if $\operatorname{Re}\vec{x} = \operatorname{Im}\vec{x} = 0$, which is the case if and only if $\vec{x} = 0 = \vec{0}$; hence, $\|\cdot\|$ satisfies the first condition (11.38a) for being a norm. Also, since $a \in \mathbb{R}$ we have

$$\|a\vec{x}\,\| = \max(|\operatorname{Re} a\vec{x}\,|, |\operatorname{Im} a\vec{x}\,|) = \max(|a \operatorname{Re}\vec{x}\,|, |a \operatorname{Im}\vec{x}\,|)$$
$$= \max(|a| \cdot |\operatorname{Re}\vec{x}\,|, |a| \cdot |\operatorname{Im}\vec{x}\,|) = |a| \max(|\operatorname{Re}\vec{x}\,|, |\operatorname{Im}\vec{x}\,|) = |a| \cdot \|\vec{x}\,\|,$$

thereby satisfying (11.38b). Finally, using the triangle inequality (3.37) for complex numbers, we obtain

$$\|\vec{x} + \vec{y}\,\| = \max(|\operatorname{Re}(\vec{x} + \vec{y})\,|, |\operatorname{Im}(\vec{x} + \vec{y})\,|) = \max(|\operatorname{Re}\vec{x} + \operatorname{Re}\vec{y}\,|, |\operatorname{Im}\vec{x} + \operatorname{Im}\vec{y}\,|)$$
$$\leq \max(|\operatorname{Re}\vec{x}\,| + |\operatorname{Re}\vec{y}\,|, |\operatorname{Im}\vec{x}\,| + |\operatorname{Im}\vec{y}\,|)$$
$$\leq \max(|\operatorname{Re}\vec{x}\,|, |\operatorname{Im}\vec{x}\,|) + \max(|\operatorname{Re}\vec{y}\,|, |\operatorname{Im}\vec{y}\,|) = \|\vec{x}\,\| + \|\vec{y}\,\|,$$

thereby satisfying (11.38c). Therefore, the given function $\|\cdot\|$ is indeed a norm on V.

(b) Applying (11.44a) to $\|\cdot\|$ yields

$$[\vec{x}, \vec{y}\,] = \max{}^2\left(\left|\operatorname{Re}\frac{\vec{x} + \vec{y}}{2}\right|, \left|\operatorname{Im}\frac{\vec{x} + \vec{y}}{2}\right|\right) - \max{}^2\left(\left|\operatorname{Re}\frac{\vec{x} - \vec{y}}{2}\right|, \left|\operatorname{Im}\frac{\vec{x} - \vec{y}}{2}\right|\right)$$

($\vec{x}, \vec{y} \in V$). It follows that, e.g., $[1 + i2, 1] = 0$, which does not equal $[1, 1] + [i2, 1] = 1 + 0 = 1$; therefore, (11.41c) is not satisfied, so this function $[\cdot, \cdot]$ is not an inner product on V.

Solution 11.26 **(a)** For $V = \mathbb{C}$, $F = \mathbb{R}$, and inner product in (11.45), the Schwarz inequality (11.47) is

$$|\operatorname{Re}(\vec{x}\vec{y}^{\,*})| \leq |\vec{x}| \cdot |\vec{y}| \quad (\vec{x}, \vec{y} \in \mathbb{C}), \tag{S11.26.1}$$

with equality being attained if and only if either $\vec{y} = 0$ or $\vec{x} = a\vec{y}$ for some $a \in \mathbb{R}$.

(b) Given $\vec{x}, \vec{y} \in \mathbb{C}$, we have $\vec{x} = Ae^{i\alpha}$ and $\vec{y} = Be^{i\beta}$ for some $A, B \geq 0$ and $\alpha, \beta \in \mathbb{R}$. Hence, $|\vec{x}| = A$, $|\vec{y}| = B$, and $|\operatorname{Re}(\vec{x}\vec{y}^{\,*})| = AB|\cos(\alpha - \beta)|$, by (11.49b); so (S11.26.1) becomes

$$AB|\cos(\alpha - \beta)| \leq AB. \tag{S11.26.2}$$

Since $AB \geq 0$ and the range of cosine is $[-1, 1]$, this inequality indeed holds. Moreover, equality attains if either $\vec{x} = 0$ or $\vec{y} = 0$, since then either $A = 0$ or $B = 0$; therefore, assume $\vec{x}, \vec{y} \neq 0$, and thus $A, B > 0$. We must show that (S11.26.2) attains equality if and only if $\vec{x} = a\vec{y}$ for some nonzero $a \in \mathbb{R}$; and, indeed, since now $AB > 0$, (S11.26.2) attains equality if and only if $\alpha - \beta = n\pi$ for some integer n, which is the case if and only if $e^{i\alpha} = \pm e^{i\beta}$, which is the case if and only if $\vec{x} = a\vec{y}$ for some nonzero $a \in \mathbb{R}$.

Solution 11.27 From (11.50) we know that there exist $a_x, a_y \in F$ for which $\vec{z}_x = a_x\vec{x}$ and $\vec{z}_y = a_y\vec{y}$, with $\vec{z} - \vec{z}_x \perp \vec{x}$ and $\vec{z} - \vec{z}_y \perp \vec{y}$. Therefore, given $b_x, b_y \in F$, we have

$$[\vec{z} - \vec{z}_x - \vec{z}_y, b_x\vec{x} + b_y\vec{y}\,] = [\vec{z} - \vec{z}_x, b_x\vec{x}\,] - [\vec{z}_y, b_x\vec{x}\,] + [\vec{z} - \vec{z}_y, b_y\vec{y}\,] - [\vec{z}_x, b_y\vec{y}\,]$$
$$= b_x{}^*[\vec{z} - \vec{z}_x, \vec{x}\,] - b_x{}^*[a_y\vec{y}, \vec{x}\,] + b_y{}^*[\vec{z} - \vec{z}_y, \vec{y}\,] - b_y{}^*[a_x\vec{x}, \vec{y}\,]$$
$$= b_x{}^*[\vec{z} - \vec{z}_x, \vec{x}\,] - a_yb_x{}^*[\vec{y}, \vec{x}\,] + b_y{}^*[\vec{z} - \vec{z}_y, \vec{y}\,] - a_xb_y{}^*[\vec{x}, \vec{y}\,]$$
$$= b_x{}^* \cdot 0 - a_yb_x{}^* \cdot 0 + b_y{}^* \cdot 0 - a_xb_y{}^* \cdot 0 = 0,$$

where $[\vec{y}, \vec{x}\,] = [\vec{x}, \vec{y}\,] = 0$ since $\vec{x} \perp \vec{y}$; thus $\vec{z} - \vec{z}_x - \vec{z}_y \perp b_x\vec{x} + b_y\vec{y}$.

Solution 11.28 **(a)** It is given that a complex number z is represented by a vector $\vec{z}$ in the complex plane, which is expressed as a pair (a, b) of points a and b. Subtracting the initial value a of this pair from both itself and the terminal value b yields another representation of $\vec{z}$—viz., the corresponding position vector $(0, b - a)$—the terminal value of which gives the value of z (per Figure 3.3). Hence, $z = b - a$.

(b) The vectors $\vec{z}_1$ and $\vec{z}_2$ are represented by position vectors $(0, z_1)$ and $(0, z_2)$. As shown on the left side of Figure 11.16, the position vector for $\vec{z}_1 + \vec{z}_2$ obtains by shifting either $(0, z_1)$ or $(0, z_2)$ so that the tail of one of these vectors coincides with the head of the other. Thus, shifting $(0, z_2)$ by adding z_1, we obtain $(z_1, z_1 + z_2)$, the terminal point of which coincides with the head of the position vector for $\vec{z}_1 + \vec{z}_2$. Hence, the vector $\vec{z}_1 + \vec{z}_2$ represents the complex number $z_1 + z_2$.

Solution 11.29 One such parallelepiped has vertices in $\mathbb{E}^3$ at $(0, 0, 0)$, $(a, 0, 0)$, $(0, b, 0)$, $(0, 0, c)$, $(a, b, 0)$, $(a, 0, c)$, $(0, b, c)$, and (a, b, c), with each face being parallel to either the xy-, xz-, or yz-plane. Each of the four diagonals is a segment having endpoints $P = (x_1, x_2, x_3)$ and $P' = (x_1', x_2', x_3')$ for which $|x_1 - x_1'| = a$, $|x_2 - x_2'| = b$, and $|x_3 - x_3'| = c$; for example, we could take $P = (0, 0, 0)$ and $P' = (a, b, c)$. It then follows from (11.56) that the length of each diagonal is $\sqrt{a^2 + b^2 + c^2}$.

Solution 11.30 Given the sample pairs $(x_i, y_i) \in \mathbb{R}^2$ $(i = 1, 2, \ldots, n)$, let

$$\mathbf{x} = \begin{bmatrix} x_1 - \overline{x} \\ x_2 - \overline{x} \\ \vdots \\ x_n - \overline{x} \end{bmatrix}, \quad \mathbf{y} = \begin{bmatrix} y_1 - \overline{y} \\ y_2 - \overline{y} \\ \vdots \\ y_n - \overline{y} \end{bmatrix},$$

where $\overline{x}$ and $\overline{y}$ are the corresponding sample means. Then, using the norm (11.59) and dot product (11.60) of n-dimensional Euclidean space, we have

$$\|\mathbf{x}\| = \sqrt{\sum_{j=1}^{n} (x_j - \overline{x})^2}, \quad \|\mathbf{y}\| = \sqrt{\sum_{j=1}^{n} (y_j - \overline{y})^2}, \quad \mathbf{x} \bullet \mathbf{y} = \sum_{j=1}^{n} (x_j - \overline{x})(y_j - \overline{y});$$

therefore, (8.19) states

$$r = \frac{\mathbf{x} \bullet \mathbf{y}}{\|\mathbf{x}\| \cdot \|\mathbf{y}\|},$$

and so it follows from the Schwarz inequality (11.64) that $|r| \leq 1$. Now, to have $r = 1$ or $r = -1$, we must have $|r| = 1$; hence, the Schwarz inequality must attain equality, which occurs if and only if either $\mathbf{y} = \mathbf{0}$ or $\mathbf{x} = a\mathbf{y}$ for some $a \in \mathbb{R}$. The former condition, though, cannot hold, because it would require division by 0 in (8.19); whereas the latter condition makes $\|\mathbf{x}\| = a^2 \|\mathbf{y}\|$ and $\mathbf{x} \bullet \mathbf{y} = a\|\mathbf{y}\|^2$, and thus $r = 1/a$. Therefore, we have $r = 1$ if and only if r exists and $\mathbf{x} = \mathbf{y}$; that is, $x_i - \overline{x} = y_i - \overline{y}$ (all i). Whereas we have $r = -1$ if and only if r exists and $\mathbf{x} = -\mathbf{y}$; that is, $x_i - \overline{x} = -(y_i - \overline{y})$ (all i).

Solution 11.31 **(a)** By (11.60) and (11.69), $\hat{\mathbf{i}} \bullet \hat{\mathbf{i}} = \hat{\mathbf{j}} \bullet \hat{\mathbf{j}} = 1$ and $\hat{\mathbf{i}} \bullet \hat{\mathbf{j}} = \hat{\mathbf{j}} \bullet \hat{\mathbf{i}} = 0$. Therefore, for $\mathbf{x}, \mathbf{y} \in \mathbb{V}^2$ as stated, (11.61c) and (11.62) yield

$$\begin{aligned} \mathbf{x} \bullet \mathbf{y} &= (x_1\hat{\mathbf{i}} + x_2\hat{\mathbf{j}}) \bullet (y_1\hat{\mathbf{i}} + y_2\hat{\mathbf{j}}) \\ &= x_1y_1(\hat{\mathbf{i}} \bullet \hat{\mathbf{i}}) + x_1y_2(\hat{\mathbf{i}} \bullet \hat{\mathbf{j}}) + x_2y_1(\hat{\mathbf{j}} \bullet \hat{\mathbf{i}}) + x_2y_2(\hat{\mathbf{j}} \bullet \hat{\mathbf{j}}) = x_1y_1 + x_2y_2, \end{aligned}$$

as contended. Hence, letting $\mathbf{y} = \mathbf{x}$, we have $\mathbf{x} \bullet \mathbf{x} = {x_1}^2 + {x_2}^2$, and thus $\|\mathbf{x}\| = \sqrt{{x_1}^2 + {x_2}^2}$ by (11.63). More quickly, these results also obtain by using (11.68) to express the vectors $\mathbf{x}$ and $\mathbf{y}$ in matrix form, as in (11.57) and (11.58), then calculating the norm and dot product by (11.59) and (11.60).

(b) If $\mathbf{x}, \mathbf{y} \in \mathbb{V}^3$ with $\mathbf{x} = x_1\hat{\mathbf{i}} + x_2\hat{\mathbf{j}} + x_3\hat{\mathbf{k}}$ $(x_1, x_2, x_3 \in \mathbb{R})$ and $\mathbf{y} = y_1\hat{\mathbf{i}} + y_2\hat{\mathbf{j}} + y_3\hat{\mathbf{k}}$ $(y_1, y_2, y_3 \in \mathbb{R})$, then $\|\mathbf{x}\| = \sqrt{{x_1}^2 + {x_2}^2 + {x_3}^2}$ and $\mathbf{x} \bullet \mathbf{y} = x_1y_1 + x_2y_2 + x_3y_3$.

Solution 11.32 Substituting (P11.32.1a) into (P11.32.1b) yields

$$\mathbf{v} = \sum_{j=1}^{n} \frac{x_j}{\|\mathbf{x}\|}\mathbf{e}_j = \frac{1}{\|\mathbf{x}\|}\sum_{j=1}^{n} x_j\mathbf{e}_j = \frac{1}{\|\mathbf{x}\|}\mathbf{x},$$

by (11.68). Thus, $\mathbf{v}$ is a unit vector; and, since $1/\|\mathbf{x}\| > 0$, it points in the same direction as $\mathbf{x}$.

Solution 11.33 **(a)** Let $\mathbf{x}, \mathbf{y}, \mathbf{z} \in \overline{\mathbb{V}}^n$ and $a, b \in \mathbb{C}$ be given. Then,

$$[\mathbf{x}, \mathbf{x}] = \sum_{j=1}^{n} x_j x_j^* = \sum_{j=1}^{n} |x_j|^2, \qquad \text{(S11.33.1)}$$

making $[\mathbf{x}, \mathbf{x}] \geq 0$, with equality being attained if and only if $\mathbf{x} = 0$; hence, (11.2) satisfies the first defining property (11.41a) of an inner product. Also,

$$[\mathbf{x}, \mathbf{y}] = \mathbf{y}^*\mathbf{x} = (\mathbf{x}^*\mathbf{y})^* = [\mathbf{x}, \mathbf{y}]^*,$$

thereby satisfying (11.41b). Finally,

$$[a\mathbf{x} + b\mathbf{y}, \mathbf{z}] = \mathbf{z}^*(a\mathbf{x} + b\mathbf{y}) = a\,\mathbf{z}^*\mathbf{x} + b\,\mathbf{z}^*\mathbf{y} = a[\mathbf{x}, \mathbf{z}] + b[\mathbf{y}, \mathbf{z}],$$

thereby satisfying (11.41c). So, the function $[\cdot, \cdot]$ specified in (11.2) is a valid inner product.

(b) By (S11.33.1) and (11.43),

$$\|\mathbf{x}\| = \sqrt{\sum_{j=1}^{n} |x_j|^2} = \sqrt{\mathbf{x}^*\mathbf{x}} \quad (\mathbf{x} \in \overline{\mathbb{V}}^n)$$

(cf. 11.59).

Solution 11.34 **(a)** These two equalities follow immediately from the fact that $\mathbf{x} \times \mathbf{y}$ is orthogonal to both $\mathbf{x}$ and $\mathbf{y}$.

(b) Let

$$\mathbf{x} = \begin{bmatrix} x_1 \\ x_2 \\ x_3 \end{bmatrix}, \quad \mathbf{y} = \begin{bmatrix} y_1 \\ y_2 \\ y_3 \end{bmatrix}, \quad \mathbf{z} = \begin{bmatrix} z_1 \\ z_2 \\ z_3 \end{bmatrix}.$$

Then (see Problem 11.31), for any vector $\mathbf{v} = v_1\hat{\mathbf{i}} + v_2\hat{\mathbf{j}} + v_3\hat{\mathbf{k}}$ ($v_1, v_2, v_3 \in \mathbb{R}$), we have $\mathbf{v} \bullet \mathbf{z} = v_1 z_1 + v_2 z_2 + v_3 z_3$; and thus it follows from (11.71) that

$$(\mathbf{x} \times \mathbf{y}) \bullet \mathbf{z} = \det \begin{bmatrix} z_1 & z_2 & z_3 \\ x_1 & x_2 & x_3 \\ y_1 & y_2 & y_3 \end{bmatrix}, \quad (\mathbf{y} \times \mathbf{z}) \bullet \mathbf{x} = \det \begin{bmatrix} x_1 & x_2 & x_3 \\ y_1 & y_2 & y_3 \\ z_1 & z_2 & z_3 \end{bmatrix}.$$

The 3×3 matrix on the left can be converted into the one on the right by first interchanging rows 1 and 2 (which changes the sign of the determinant), then interchanging rows 2 and 3 (which changes the sign again); therefore, the two determinants are equal. Hence, $(\mathbf{x} \times \mathbf{y}) \bullet \mathbf{z} = (\mathbf{y} \times \mathbf{z}) \bullet \mathbf{x}$; and, performing upon this expression the circular permutation "$\mathbf{x}$" $\to$ "$\mathbf{y}$" $\to$ "$\mathbf{z}$" $\to$ "$\mathbf{x}$", we obtain $(\mathbf{y} \times \mathbf{z}) \bullet \mathbf{x} = (\mathbf{z} \times \mathbf{x}) \bullet \mathbf{y}$.

Solution 11.35 We know that $\mathbf{x}$ and $\mathbf{y}$ are collinear if and only if one of these vectors is a scalar multiple of the other; that is, either $\mathbf{x} = a\mathbf{y}$ for some $a \in \mathbb{R}$ or $\mathbf{y} = b\mathbf{x}$ for some $b \in \mathbb{R}$. But, if we do not have $\mathbf{x} = a\mathbf{y}$ for some $a \in \mathbb{R}$, then the only way we can have $\mathbf{y} = b\mathbf{x}$ for some $b \in \mathbb{R}$ is with $b = 0$—i.e., $\mathbf{y} = \mathbf{0}$. Hence, $\mathbf{x}$ and $\mathbf{y}$ are collinear if and only if $\mathbf{y} = \mathbf{0}$ or $\mathbf{x} = a\mathbf{y}$ for some $a \in \mathbb{R}$, which is precisely the necessary and sufficient condition to attain equality in the Schwarz inequality (11.64). Therefore, $\mathbf{x}, \mathbf{y} \in \mathbb{V}^n$ ($n \geq 1$) are collinear if and only if $|\mathbf{x} \bullet \mathbf{y}| = \|\mathbf{x}\| \cdot \|\mathbf{y}\|$.

Solution 11.36 **(a)** Per (11.75), each $\varphi_i(\cdot)$ $(i = 1, 2)$ may be represented by the equation $\mathbf{x}' = \mathbf{A}_i\mathbf{x} + \mathbf{b}_i$ $(\mathbf{x}', \mathbf{x} \in \mathbb{V}^2)$, for some real 2×2 matrix $\mathbf{A}_i$ and real 2×1 matrix $\mathbf{b}_i$. It follows that $\varphi(\cdot)$ is represented by the matrix equation

$$\mathbf{x}' = \mathbf{A}_2(\mathbf{A}_1\mathbf{x} + \mathbf{b}_1) + \mathbf{b}_2$$

$(\mathbf{x}', \mathbf{x} \in \mathbb{V}^2)$, where for each $\mathbf{x}$ the right side equals $(\mathbf{A}_2\mathbf{A}_1)\mathbf{x} + (\mathbf{A}_2\mathbf{b}_1 + \mathbf{b}_2)$. Hence, $\varphi(\cdot)$ is represented by the matrix equation $\mathbf{x}' = \mathbf{A}\mathbf{x} + \mathbf{b}$ $(\mathbf{x}', \mathbf{x} \in \mathbb{V}^2)$ with

$$\mathbf{A} = \mathbf{A}_2\mathbf{A}_1, \quad \mathbf{b} = \mathbf{A}_2\mathbf{b}_1 + \mathbf{b}_2.$$

Now, suppose $\varphi_1(\cdot)$ is a translation by $\mathbf{v}_1 \in \mathbb{V}^2$, and $\varphi_2(\cdot)$ is a translation by $\mathbf{v}_2 \in \mathbb{V}^2$; then, by (11.77) we have $\mathbf{A}_1 = \mathbf{A}_2 = \mathbf{I}$, $\mathbf{b}_1 = \mathbf{v}_1$, and $\mathbf{b}_2 = \mathbf{v}_2$, thereby making $\mathbf{A} = \mathbf{I}$ and $\mathbf{b} = \mathbf{v}_1 + \mathbf{v}_2$, and thus $\varphi_1(\cdot)$ is a translation by $\mathbf{v}_1 + \mathbf{v}_2$. Or, suppose $\varphi_1(\cdot)$ is a counterclockwise rotation by $\theta_1 \in \mathbb{R}$, and $\varphi_2(\cdot)$ is a counterclockwise rotation by $\theta_2 \in \mathbb{R}$; then, by (11.78) we have

$$\mathbf{A}_1 = \begin{bmatrix} \cos\theta_1 & -\sin\theta_1 \\ \sin\theta_1 & \cos\theta_1 \end{bmatrix}, \quad \mathbf{A}_2 = \begin{bmatrix} \cos\theta_2 & -\sin\theta_2 \\ \sin\theta_2 & \cos\theta_2 \end{bmatrix}, \quad \mathbf{b}_1 = \mathbf{b}_2 = \mathbf{0},$$

thereby making

$$\begin{aligned} \mathbf{A} &= \begin{bmatrix} \cos\theta_1\cos\theta_2 - \sin\theta_1\sin\theta_2 & -\cos\theta_1\sin\theta_2 - \sin\theta_1\cos\theta_2 \\ \sin\theta_1\cos\theta_2 + \cos\theta_1\sin\theta_2 & -\sin\theta_1\sin\theta_2 + \cos\theta_1\cos\theta_2 \end{bmatrix} \\ &= \begin{bmatrix} \cos(\theta_1 + \theta_2) & -\sin(\theta_1 + \theta_2) \\ \sin(\theta_1 + \theta_2) & \cos(\theta_1 + \theta_2) \end{bmatrix} \end{aligned}$$

by (11.16), and $\mathbf{b} = \mathbf{0}$, and thus $\varphi_1(\cdot)$ is a counterclockwise rotation by $\theta_1 + \theta_2$. Or, suppose $\varphi_1(\cdot)$ is a dilation by $\kappa_1 > 0$, and $\varphi_2(\cdot)$ is a dilation by $\kappa_2 > 0$; then, by (11.81) we have $\mathbf{A}_1 = \kappa_1\mathbf{I}$, $\mathbf{A}_2 = \kappa_2\mathbf{I}$, and $\mathbf{b}_1 = \mathbf{b}_2 = \mathbf{0}$, thereby making $\mathbf{A} = \kappa_1\kappa_2\mathbf{I}$ and $\mathbf{b} = \mathbf{0}$, and thus $\varphi_1(\cdot)$ is a dilation by $\kappa_1\kappa_2$.

(b) Suppose $\varphi_1(\cdot)$ and $\varphi_2(\cdot)$ are isometries; hence, for all points P, $Q \in \mathbb{E}^2$ we have $d[\varphi_i(P), \varphi_i(Q)] = d(P, Q)$ $(i = 1, 2)$. Then,

$$d[\varphi(P), \varphi(Q)] = d\{\varphi_2[\varphi_1(P)], \varphi_2[\varphi_1(Q)]\} = d[\varphi_1(P), \varphi_1(Q)] = d(P, Q)$$

$(P, Q \in \mathbb{E}^2)$, and thus $\varphi(\cdot)$ is an isometry.

Solution 11.37 By (11.75), (11.77), and (11.78), translating $\mathbb{E}^2$ by $\mathbf{v} \in \mathbb{V}^2$ then rotating counterclockwise by $\phi \in \mathbb{R}$ maps each point $\mathbf{x} \in \mathbb{V}^2$ to the point

$$\mathbf{x}' = \begin{bmatrix} \cos\phi & -\sin\phi \\ \sin\phi & \cos\phi \end{bmatrix}\left(\begin{bmatrix} 1 & 0 \\ 0 & 1 \end{bmatrix}\mathbf{x} + \mathbf{v}\right) + \begin{bmatrix} 0 \\ 0 \end{bmatrix} = \begin{bmatrix} \cos\phi & -\sin\phi \\ \sin\phi & \cos\phi \end{bmatrix}(\mathbf{x} + \mathbf{v}),$$

whereas first rotating by ϕ then translating $\mathbf{v}' \in \mathbb{V}^2$ yields

$$\mathbf{x}' = \begin{bmatrix} 1 & 0 \\ 0 & 1 \end{bmatrix}\left(\begin{bmatrix} \cos\phi & -\sin\phi \\ \sin\phi & \cos\phi \end{bmatrix}\mathbf{x} + \begin{bmatrix} 0 \\ 0 \end{bmatrix}\right) + \mathbf{v}' = \begin{bmatrix} \cos\phi & -\sin\phi \\ \sin\phi & \cos\phi \end{bmatrix}\mathbf{x} + \mathbf{v}'.$$

Thus, the same overall transformation obtains if and only if

$$\mathbf{v}' = \begin{bmatrix} \cos\phi & -\sin\phi \\ \sin\phi & \cos\phi \end{bmatrix}\mathbf{v};$$

that is, $\mathbf{v}'$ is $\mathbf{v}$ rotated counterclockwise by ϕ. Hence, $\mathbf{v}' = \mathbf{v}$ if and only if either $\mathbf{v} = \mathbf{0}$ or ϕ is a multiple of 2π; that is, either the translation by $\mathbf{v}$ or the rotation by ϕ is the identity transformation.

Solution 11.38 **(a)** Letting $\mathbf{v} = \begin{bmatrix} X & Y \end{bmatrix}^{\mathrm{T}}$, this point reflection can be performed by first translating $\mathbb{E}^2$ by $-\mathbf{v}$, performing a reflection about the origin, then translating the result back by $\mathbf{v}$. Hence,

$$\begin{bmatrix} x' \\ y' \end{bmatrix} = -\left(\begin{bmatrix} x \\ y \end{bmatrix} - \begin{bmatrix} X \\ Y \end{bmatrix}\right) + \begin{bmatrix} X \\ Y \end{bmatrix} = \begin{bmatrix} -x \\ -y \end{bmatrix} + \begin{bmatrix} 2X \\ 2Y \end{bmatrix},$$

which is equivalent to taking

$$\mathbf{A} = -\mathbf{I} = \begin{bmatrix} -1 & 0 \\ 0 & -1 \end{bmatrix}, \quad \mathbf{b} = 2\mathbf{v} = \begin{bmatrix} 2X \\ 2Y \end{bmatrix}$$

in (11.75).

(b) From the discussion of (11.88) we know that the vector $\mathbf{n} = A\hat{\imath} + B\hat{\jmath}$ is normal to l, and the distance from $P = (x, y)$ to l is $|Ax+By+C|/\sqrt{A^2+B^2}$; also, when $Ax+By+C < 0$ (respectively, > 0), the vector $\mathbf{n}$ $(-\mathbf{n})$ points from P to l. Therefore, when $P \notin l$ (i.e., $Ax + By + C \neq 0$), the reflection point $P' = (x', y')$ obtains by translating P by the vector

$$\begin{aligned}\mathbf{v} &= -2\frac{Ax+By+C}{\sqrt{A^2+B^2}} \cdot \frac{\mathbf{n}}{\|\mathbf{n}\|} = -2\frac{Ax+By+C}{\sqrt{A^2+B^2}} \cdot \frac{1}{\sqrt{A^2+B^2}} \begin{bmatrix} A \\ B \end{bmatrix} \\ &= -2\frac{Ax+By+C}{A^2+B^2} \begin{bmatrix} A \\ B \end{bmatrix};\end{aligned}$$

moreover, this result is also valid when $P \in l$, for then $P' = P$, which corresponds to translating P by $\mathbf{0}$. Hence,

$$\begin{aligned}\begin{bmatrix} x' \\ y' \end{bmatrix} &= \begin{bmatrix} x \\ y \end{bmatrix} + \mathbf{v} = \begin{bmatrix} x \\ y \end{bmatrix} - 2\frac{Ax+By}{A^2+B^2}\begin{bmatrix} A \\ B \end{bmatrix} - 2\frac{C}{A^2+B^2}\begin{bmatrix} A \\ B \end{bmatrix} \\ &= \frac{A^2+B^2}{A^2+B^2}\begin{bmatrix} 1 & 0 \\ 0 & 1 \end{bmatrix}\begin{bmatrix} x \\ y \end{bmatrix} - \frac{2}{A^2+B^2}\begin{bmatrix} A^2 & AB \\ AB & B^2 \end{bmatrix}\begin{bmatrix} x \\ y \end{bmatrix} - \frac{2C}{A^2+B^2}\begin{bmatrix} A \\ B \end{bmatrix} \\ &= \frac{1}{A^2+B^2}\begin{bmatrix} B^2-A^2 & -2AB \\ -2AB & A^2-B^2 \end{bmatrix}\begin{bmatrix} x \\ y \end{bmatrix} - \frac{2C}{A^2+B^2}\begin{bmatrix} A \\ B \end{bmatrix},\end{aligned}$$

which is equivalent to taking

$$\mathbf{A} = \frac{1}{A^2+B^2}\begin{bmatrix} B^2-A^2 & -2AB \\ -2AB & A^2-B^2 \end{bmatrix}, \quad \mathbf{b} = -\frac{2C}{A^2+B^2}\begin{bmatrix} A \\ B \end{bmatrix}$$

in (11.75b).

Solution 11.39 Upon introducing the vectors

$$\mathbf{x} = \begin{bmatrix} a_1 - c_1 \\ a_2 - c_2 \\ 0 \end{bmatrix}, \quad \mathbf{y} = \begin{bmatrix} b_1 - c_1 \\ b_2 - c_2 \\ 0 \end{bmatrix},$$

we have $|\overline{AC}| = \|\mathbf{x}\|$, $|\overline{BC}| = \|\mathbf{y}\|$, and $\measuredangle ACB = \measuredangle(\mathbf{x}, \mathbf{y})$. Therefore, by Problem 11.11 (specifically, the left side of (P11.11.1)), (11.72), and (11.71), we conclude that $\triangle ABC$ has area

$$\begin{aligned}\alpha &= \tfrac{1}{2}|\overline{AC}| \cdot |\overline{BC}| \cdot \sin\measuredangle ACB = \tfrac{1}{2}\|\mathbf{x}\| \cdot \|\mathbf{y}\| \cdot \sin\measuredangle(\mathbf{x}, \mathbf{y}) = \tfrac{1}{2}\|\mathbf{x} \times \mathbf{y}\| \\ &= \tfrac{1}{2}\left\|\det\begin{bmatrix} \hat{\imath} & \hat{\jmath} & \hat{\mathbf{k}} \\ a_1-c_1 & a_2-c_2 & 0 \\ b_1-c_1 & b_2-c_2 & 0 \end{bmatrix}\right\| = \tfrac{1}{2}\left\|\det\begin{bmatrix} a_1-c_1 & a_2-c_2 \\ b_1-c_1 & b_2-c_2 \end{bmatrix}\hat{\mathbf{k}}\right\| \\ &= \tfrac{1}{2}\left|\det\begin{bmatrix} a_1-c_1 & a_2-c_2 \\ b_1-c_1 & b_2-c_2 \end{bmatrix}\right|.\end{aligned}$$

Also, since subtracting one row of a matrix from the other rows does not affect its determinant,

$$\det\begin{bmatrix} a_1 & a_2 & 1 \\ b_1 & b_2 & 1 \\ c_1 & c_2 & 1 \end{bmatrix} = \det\begin{bmatrix} a_1-c_1 & a_2-c_2 & 0 \\ b_1-c_1 & b_2-c_2 & 0 \\ c_1 & c_2 & 1 \end{bmatrix} = \det\begin{bmatrix} a_1-c_1 & a_2-c_2 \\ b_1-c_1 & b_2-c_2 \end{bmatrix};$$

hence, we obtain (P11.39.1).

Solution 11.40 For each $i \in \{1,2\}$, let P_i be the point expressed in polar coordinates by the pair (r_i, θ_i); thus, P_i is expressed in rectangular coordinates by the pair $(x_i, y_i) = (r_i \cos\theta_i, r_i \sin\theta_i)$. Hence, by (11.96) and (11.16a), the line l is described by the equation

$$\begin{aligned}
0 &= \det\begin{bmatrix} x & y & 1 \\ x_1 & y_1 & 1 \\ x_2 & y_2 & 1 \end{bmatrix} = \det\begin{bmatrix} r\cos\theta & r\sin\theta & 1 \\ r_1\cos\theta_1 & r_1\sin\theta_1 & 1 \\ r_2\cos\theta_2 & r_2\sin\theta_2 & 1 \end{bmatrix} \\
&= \det\begin{bmatrix} r\cos\theta & r\sin\theta \\ r_1\cos\theta_1 & r_1\sin\theta_1 \end{bmatrix} - \det\begin{bmatrix} r\cos\theta & r\sin\theta \\ r_2\cos\theta_2 & r_2\sin\theta_2 \end{bmatrix} + \det\begin{bmatrix} r_1\cos\theta_1 & r_1\sin\theta_1 \\ r_2\cos\theta_2 & r_2\sin\theta_2 \end{bmatrix} \\
&= rr_1(\cos\theta\sin\theta_1 - \sin\theta\cos\theta_1) - rr_2(\cos\theta\sin\theta_2 - \sin\theta\cos\theta_2) \\
&\quad + r_1r_2(\cos\theta_1\sin\theta_2 - \sin\theta_1\cos\theta_2) \\
&= -rr_1\sin(\theta - \theta_1) + rr_2\sin(\theta - \theta_2) - r_1r_2\sin(\theta_1 - \theta_2),
\end{aligned}$$

from which (P11.40.1) obtains.

Solution 11.41 Each line l_i $(i = 1, 2)$ can be described by the equation $A_ix + B_iy + C_i = 0$ with $A_i = m_i$, $B_i = -1$, and $C_i = Y_i$. Therefore, by (11.100), the angle between l_1 and l_2 has measure

$$\tan^{-1}\left|\frac{A_1B_2 - B_1A_2}{A_1A_2 + B_1B_2}\right| = \tan^{-1}\left|\frac{m_1 - m_2}{m_1m_2 + 1}\right|$$

—except when $m_1m_2 = -1$, in which case the lines are perpendicular.

Solution 11.42 For each $i \in \{1,2\}$, suppose a plane p_i has x-intercept X_i, y-intercept Y_i, and z-intercept Z_i; then, comparing (11.104) with (11.101), we conclude that p_i can be described by the equation $A_ix + B_iy + C_iz + D_i = 0$ with $A_i = 1/X_i$, $B_i = 1/Y_i$, $C_i = 1/Z_i$, and $D_i = -1$. We know that p_1 and p_2 are parallel if and only if $A_2 = \alpha A_1$, $B_2 = \alpha B_1$, and $C_2 = \alpha C_1$—i.e., $X_1 = \alpha X_2$, $Y_1 = \alpha Y_2$, and $Z_1 = \alpha Z_2$—for some constant α. Also, p_1 and p_2 are perpendicular if and only if $A_1A_2 + B_1B_2 + C_1C_2 = 0$—i.e., $1/X_1X_2 + 1/Y_1Y_2 + 1/Z_1Z_2 = 0$.

Solution 11.43 Here is one way to proceed: Let one line be described by the two equations

$$A_1x + B_1y + C_1z + D_1 = 0, \tag{S11.43.1a}$$

$$A_2x + B_2y + C_2z + D_2 = 0; \tag{S11.43.1b}$$

and, let the other line be described by the two equations

$$A_1'x + B_1'y + C_1'z + D_1' = 0, \tag{S11.43.2a}$$

$$A_2'x + B_2'y + C_2'z + D_2' = 0. \tag{S11.43.2b}$$

Since the lines cross, they have a unique point $P_1 = (x_1, y_1, z_1)$ of intersection, which can be found by solving the equations. Specifically, since P_1 is unique, three of the four equations must be such that none of them can be derived from the other two; thus, e.g., if these equations were found to be (S11.43.1a), (S11.43.1b), and (S11.43.2a), then we would have

$$\begin{bmatrix} x_1 \\ y_1 \\ z_1 \end{bmatrix} = -\begin{bmatrix} A_1 & B_1 & C_1 \\ A_2 & B_2 & C_2 \\ A_1' & B_1' & C_1' \end{bmatrix}^{-1} \begin{bmatrix} D_1 \\ D_2 \\ D_1' \end{bmatrix}.$$

Next, find two points $P_2 = (x_2, y_2, z_2)$ and $P_3 = (x_3, y_3, z_3)$, both different than P_1, with P_2 on the line given by (S11.43.1) and P_3 on the line given by (S11.43.2). For example, there must exist a point $(X, Y, Z) \neq (0, 0, 0)$ such that

$$A_1X + B_1Y + C_1Z = 0,$$

$$A_2X + B_2Y + C_2Z = 0,$$

because it follows from Problem 9.7 with

$$\mathbf{A} = \begin{bmatrix} A_1 & B_1 & C_1 \\ A_2 & B_2 & C_2 \end{bmatrix}$$

that the three vectors

$$\begin{bmatrix} A_1 \\ A_2 \end{bmatrix}, \quad \begin{bmatrix} B_1 \\ B_2 \end{bmatrix}, \quad \begin{bmatrix} C_1 \\ C_2 \end{bmatrix}$$

are linearly dependent; therefore, we can let $P_2 = (x_1 + X, y_1 + Y, z_1 + Z)$. We now have three noncollinear points P_1, P_2, and P_3, each lying on at least one of the two given lines; therefore, the equation for the desired plane is given by (11.110).

Solution 11.44 Since the point at which l and p cross is unique, $\mathbf{x}'$ is the unique vector $\mathbf{x}$ that satisfies both (P11.44.1) and (P11.44.2) for some choice of r, s, and t; hence, r, s, and t are the unique values that satisfy the equation

$$r\mathbf{u} - s\mathbf{v} - t\mathbf{w} = \mathbf{q} - \mathbf{p}. \tag{S11.44.1}$$

Letting

$$\mathbf{p} = \begin{bmatrix} x_p \\ y_p \\ z_p \end{bmatrix}, \quad \mathbf{u} = \begin{bmatrix} x_u \\ y_u \\ z_u \end{bmatrix}, \quad \mathbf{q} = \begin{bmatrix} x_q \\ y_q \\ z_q \end{bmatrix}, \quad \mathbf{v} = \begin{bmatrix} x_v \\ y_v \\ z_v \end{bmatrix}, \quad \mathbf{w} = \begin{bmatrix} x_w \\ y_w \\ z_w \end{bmatrix},$$

we can rewrite (S11.44.1) in matrix form as

$$\begin{bmatrix} x_u & y_u & z_u \\ x_v & y_v & z_v \\ x_w & y_w & z_w \end{bmatrix} \begin{bmatrix} r \\ -s \\ -t \end{bmatrix} = \begin{bmatrix} x_q - x_p \\ y_q - z_p \\ z_q - z_p \end{bmatrix},$$

where the 3×3 matrix is nonsingular since $\mathbf{u}$, $\mathbf{v}$, and $\mathbf{w}$ are noncoplanar (and thus linearly independent). Hence,

$$\begin{bmatrix} r \\ -s \\ -t \end{bmatrix} = \begin{bmatrix} x_u & y_u & z_u \\ x_v & y_v & z_v \\ x_w & y_w & z_w \end{bmatrix}^{-1} \begin{bmatrix} x_q - x_p \\ y_q - z_p \\ z_q - z_p \end{bmatrix},$$

from which we get $\mathbf{x}'$ by substituting the value of r obtained into (P11.44.1).

Solution 11.45 **(a)** By (11.1), (P11.45.1) states that

$$\sqrt{x^2 + y^2} = \alpha\sqrt{[(x-1)^2 + y^2]}, \tag{S11.45.1}$$

which squared becomes

$$x^2 + y^2 = \alpha^2[(x-1)^2 + y^2].$$

(Since it would be impossible for (S11.45.1) to hold if either side were multiplied by -1, this squaring operation cannot introduce any extraneous solutions—e.g., as happened in Solution 6.22.) Equivalently,

$$(\alpha^2 - 1)x^2 - 2\alpha^2 x + (\alpha^2 - 1)y^2 = -\alpha^2, \tag{S11.45.2}$$

which divided by $\alpha^2 - 1 \neq 0$ gives

$$x^2 - \frac{2\alpha^2}{1-\alpha^2}x + y^2 = -\frac{\alpha^2}{\alpha^2 - 1};$$

therefore, we complete the square with respect to "x" to obtain

$$\left(x - \frac{\alpha^2}{\alpha^2-1}\right)^2 + y^2 = -\frac{\alpha^2}{\alpha^2-1} + \left(\frac{\alpha^2}{\alpha^2-1}\right)^2 = \left(\frac{\alpha}{\alpha^2-1}\right)^2.$$

Upon comparing this result with (11.133), we conclude that equation (P11.45.1) describes a circle having center $[\alpha^2/(\alpha^2-1), 0]$ and radius $\alpha/|\alpha^2-1|$.

(b) For $\alpha = 1$, (S11.45.2) becomes the equation $2x = 1$—i.e., $x = 1/2$—and thus (P11.45.1) describes a vertical line having x-intercept 1/2. Also, for any $\alpha < 0$, (P11.45.1) describes the empty set. Because, having $d(P, P_2) > 0$ would imply $d(P, P_1) < 0$, which is impossible; whereas having $d(P, P_2) = 0$ would imply $d(P, P_1) = 0$, which is impossible since $P_1 \neq P_2$.

Solution 11.46 In the xy-plane, let A be the point $(0, 0)$, and let B be the point $(c, 0)$; thus, $d(A, B) = c$. We now will show that there exists a point $C = (x, y)$ such that $d(A, C) = b$ and $d(B, C) = a$, with $y > 0$ to make the points A, B, and C noncollinear. This will imply the existence of a triangle $\triangle ABC$ with sides of lengths a, b, and c.

Using the equation (11.133) for a circle, the points (x, y) in the xy-plane that are a distance b from A are those for which

$$(x-0)^2 + (y-0)^2 = b^2;$$

likewise, translating (11.133), the points that are a distance a from B are those for which

$$(x-c)^2 + (y-0)^2 = a^2.$$

Thus, the intersection of these two circles is the set of all points (x, y) for which

$$x^2 + y^2 = b^2, \quad (x^2 - 2cx + c^2) + y^2 = a^2. \tag{S11.46.1}$$

Taking the difference of these two equations yields $2cx - c^2 = b^2 - a^2$, and thus we must have

$$x = (b^2 + c^2 - a^2)/2c; \tag{S11.46.2a}$$

then, (S11.46.1) can be satisfied by also taking

$$y = \sqrt{b^2 - x^2}. \tag{S11.46.2b}$$

To show that $y > 0$, and thereby obtain the desired point $C = (x, y)$, it is sufficient to show that $-b < x < b$; and, by (S11.46.2a), this is the case if

$$-2bc < b^2 + c^2 - a^2 < 2bc. \tag{S11.46.3}$$

Indeed, rearranging and combining terms, we see that the lower bound in (S11.46.3) is equivalent to $a^2 < (b+c)^2$, which holds since $a < b + c$; likewise, the upper bound in (S11.46.3) is equivalent to $(b-c)^2 < a^2$, which holds since $c < a + b$ and $b < a + c$, thereby making $-a < b - c < a$.

Aside: Substituting (S11.46.2a) into (S11.46.2b) and manipulating the result yields

$$\begin{aligned} y &= \sqrt{(a+b+c)(-a+b+c)(a-b+c)(a+b-c)}\big/2c \\ &= 2\sqrt{s(s-a)(s-b)(s-c)}\big/c, \end{aligned}$$

where $s = (a+b+c)/2$. Therefore, since $\triangle ABC$ has height y when its base is taken to be $\overline{AB}$, thereby making its area $\frac{1}{2}d(A, B)y = cy/2$, we thus obtain a proof of Heron's formula (10.25) for the area of an arbitrary triangle.

Solution 11.47 The focal parameter p of the ellipse is the distance from F to d; hence, from the discussion of (11.88), we know $p = |AX + BY + C|\big/\sqrt{A^2 + B^2}$. Furthermore, we know that the vector $\mathbf{n} = A\hat{\imath} + B\hat{\jmath}$ is orthogonal to d; and, if $AX + BY + C < 0$ (respectively, > 0), then $\mathbf{n}$ $(-\mathbf{n})$ points from F to d. Additionally, we know from Subsection 10.3.2 that F and F' are separated by a distance of $2e^2p/(1-e^2)$. Therefore—in light of Figure 10.22, which shows the direction from F to F' to be opposite that from F to d—F' obtains by translating $F = (X, Y)$ by the vector

$$\frac{2e^2}{1-e^2} \cdot \frac{AX+BY+C}{\sqrt{A^2+B^2}} \cdot \frac{\mathbf{n}}{\|\mathbf{n}\|} = \frac{2e^2}{1-e^2} \cdot \frac{AX+BY+C}{\sqrt{A^2+B^2}} \cdot \frac{\mathbf{n}}{\sqrt{A^2+B^2}}$$
$$= \frac{2e^2}{1-e^2} \cdot \frac{AX+BY+C}{A^2+B^2}(A\hat{\mathbf{\imath}} + B\hat{\mathbf{\jmath}});$$

so,

$$F' = \left(X + A\frac{2e^2}{1-e^2} \cdot \frac{AX+BY+C}{A^2+B^2},\ Y + B\frac{2e^2}{1-e^2} \cdot \frac{AX+BY+C}{A^2+B^2}\right).$$

Finally, d' obtains by translating d in the same direction by a distance of $2e^2p/(1-e^2)+2p = 2p/(1-e^2)$; hence, d' is described by the equation

$$A\left(x - A\frac{2}{1-e^2} \cdot \frac{AX+BY+C}{A^2+B^2}\right) + B\left(y - B\frac{2}{1-e^2} \cdot \frac{AX+BY+C}{A^2+B^2}\right) + C = 0,$$

which simplifies to

$$Ax + By + \left[C - \frac{2}{1-e^2}(AX+BY+C)\right] = 0.$$

Solution 11.48 Using (11.16b) we obtain $a\cos(t+\theta) = a\cos t\cos\theta - a\sin t\sin\theta$ (all t), and therefore (P11.48.1) implies that

$$x = ay\cos\theta - a\sin t\sin\theta.$$

Thus,

$$(x - ay\cos\theta)^2 = a^2\sin^2 t\sin^2\theta = a^2(1-\cos^2 t)\sin^2\theta, \tag{S11.48.1}$$

by (11.15a), yielding

$$x^2 - (2a\cos\theta)xy + (a\cos\theta)^2y^2 = a^2(1-y^2)\sin^2\theta, \tag{S11.48.2}$$

which simplifies to

$$x^2 - (2a\cos\theta)xy + a^2y^2 - (a\sin\theta)^2 = 0. \tag{S11.48.3}$$

Comparing this equation with (11.128), we have $A = 1$, $B = -2a\cos\theta$, $C = a^2$, $D = E = 0$, and $F = -(a\sin\theta)^2$, by which (11.130) gives $\Delta = -(a\sin\theta)^2(a^2 - a^2\cos^2\theta) = -(a\sin\theta)^4 < 0$, $J = a^2 - a^2\cos^2\theta = (a\sin\theta)^2 > 0$, and $I = 1 + a^2 > 0$, making $I/\Delta < 0$. Accordingly, Table 11.3 states that equation (S11.48.3) describes either a circle or an ellipse. Also, since $\sin^2\theta \le 1$, we have $I^2 - 4J = (1 + 2a^2 + a^4) - 4a^2\sin^2\theta \ge (1 + 2a^2 + a^4) - 4a^2 = 1 - 2a^2 + a^4 = (1-a^2)^2 \ge 0$, making $I^2 \ge 4J$; and equality is attained only if $|a| = 1$ and $\theta = \pi/2$, in order that $a^2 = 1$ and $\sin^2\theta = 1$. Hence, Table 11.3 states that (S11.48.3) describes a circle if $|a| = 1$ and $\theta = \pi/2$, and an ellipse otherwise.

It remains to show that any point $(x, y) \in \mathbb{E}^2$ satisfying (S11.48.3) also satisfies (P11.48.1); because, only the converse was implied by deriving (S11.48.3). First, no point (x, y) with $|y| > 1$ can satisfy (S11.48.3); because, (S11.48.3) is equivalent to (S11.48.2), the right side of which is negative for $|y| > 1$, with the left side equaling $(x - ay\cos\theta)^2 \ge 0$. Hence, as in (P11.48.1), we may restrict our attention to $y = \cos t$ for values $t \in [0, 2\pi)$. Making this substitution in (S11.48.2) returns us the first equality in (S11.48.1), which holds if and only if

$$x - ay\cos\theta = \pm a\sin t\sin\theta;$$

equivalently,

$$x = ay\cos\theta \pm a\sin t\sin\theta = a\cos t\cos\theta \pm a\sin t\sin\theta = a\cos(t \mp \theta).$$

It might therefore appear that there are more points (x, y) satisfying (S11.48.3) than (P11.48.1). But, suppose $t \in (0, 2\pi)$; then, for $t' = 2\pi - t$ we have $t' \in (0, 2\pi)$, and $[a\cos(t-\theta), \cos t] = [a\cos(2\pi - t' - \theta), \cos(2\pi - t')] = [a\cos(-t' - \theta), \cos(-t')] = [a\cos(t' + \theta), \cos t']$, by the 2π-periodicity and even symmetry of cosine. Furthermore, $[a\cos(t-\theta), \cos t] = [a\cos(t+\theta), \cos t]$ for $t = 0$. So, for any $t \in [0, 2\pi)$, there exists a $t' \in [0, 2\pi)$ such that $[a\cos(t-\theta), \cos t] = [a\cos(t'+\theta), \cos t']$; accordingly, no points (x, y) are lost by neglecting the minus sign in "$\mp$" above, and thus we obtain (P11.48.1).

Solution 11.49 Comparing the given equation with (11.128), we have $A = C = D = E = 0$, $B = 1$, and $F = -1$, and thus by (11.130) we have $\Delta = 1/4 \neq 0$ and $J = -1/4 < 0$; therefore, Table 11.3 states the given equation describes a hyperbola in the xy-plane. Since $A = C$, we can choose $\phi = \pi/4$ to satisfy (11.138); then, performing a counterclockwise rotation by ϕ to convert to an $x'y'$-coordinate system, it follows from (11.137) that

$$A' = 1/2, \quad B' = 0, \quad C' = -1/2, \quad D' = 0, \quad E' = 0, \quad F' = -1$$

for the coefficients satisfying (11.136). We now have

$$\frac{x'^2}{2} - \frac{y'^2}{2} = 1,$$

which in comparison to (11.135) shows that the new hyperbola has semitransverse axis $a = \sqrt{2}$, and semiconjugate axis $b = \sqrt{2}$, with the x'-axis being the transverse axis and the y'-axis being the conjugate axis. Therefore, letting $c = \sqrt{a^2 + b^2} = 2$, we find that the new hyperbola has foci at $(x', y') = (\pm c, 0) = (\pm 2, 0)$, center at $(x', y') = (0, 0)$, directrices being the lines described by the equations $x' = \pm a^2/c = \pm 1$, asymptotes being the lines described by the equations $x'/\sqrt{2} \pm y'/\sqrt{2} = x'/a \pm y'/b = 0$, eccentricity $e = c/a = \sqrt{2}$, and focal parameter $p = b^2/c = 1$.

The original hyperbola, then, has semitransverse axis $\sqrt{2}$, semiconjugate axis $\sqrt{2}$, eccentricity $\sqrt{2}$, and focal parameter 1, since these parameters are unaffected by rotation. Also, by (11.85), the rotation performed corresponds to having

$$x = (x' - y')\big/\sqrt{2}, \quad y = (x' + y')\big/\sqrt{2}; \tag{S11.49.1}$$

therefore, the original hyperbola has foci at $(x, y) = \left(\sqrt{2}, \sqrt{2}\right), \left(-\sqrt{2}, -\sqrt{2}\right)$ and center at $(x, y) = (0, 0)$, and its asymptotes are the lines described by the equations $x = 0$ and $y = 0$ (i.e., the x- and y-axes). Additionally, by either solving (S11.49.1) or using (11.75) with (11.84) we obtain $x' = (x + y)\big/\sqrt{2}$; therefore, the directrices of the original hyperbola are the lines described by the equations $x + y = \pm\sqrt{2}$.

Chapter 12. Mathematics in Practice

Solution 12.1 A modeling error.

Solution 12.2 **(a)** Scientific notation:

$$5.940 \times 10^{-1}\ \text{Sv}, \quad 1.00 \times 10^{0}\ \text{S·s}, \quad -7.103\,085\,62 \times 10^{7}\ \text{Wb/H}.$$

Engineering notation:

$$594.0 \times 10^{-3}\ \text{Sv}, \quad 1.00 \times 10^{0}\ \text{S·s}, \quad -71.030\,856\,2 \times 10^{6}\ \text{Wb/H}.$$

(b) From the expressions in engineering notation we readily obtain 594.0 mSv, 1.00 S·s, −71.030 856 2 MWb/H.

(c) 594.0 millisieverts, 1.00 siemens-second, −71.030 856 2 megawebers per henry.

Solution 12.3 By (12.2) with $n = 0$, (12.3) asserts

$$x_{\text{exp}} = \pm\left(\sum_{i=-m}^{0} x_i \cdot 10^i\right) \times 10^p,$$

and thus

$$\log_{10}|x_{\text{exp}}| = \log_{10}\left(\sum_{i=-m}^{0} x_i \cdot 10^i\right) + p.$$

The quantity in parentheses is at least 1 (when $x_0 = 1$ with $x_i = 0$ otherwise), and at most $10 - 10^{-m}$ (when $x_i = 9$ for all i); hence, its common logarithm is at least 0, but less than 1. Therefore, since p is an integer, the closest approximation to $\log_{10}|x_{\text{exp}}|$ from below by an integer—i.e., $\lfloor \log_{10}|x_{\text{exp}}| \rfloor$—is $0 + p = p$.

Solution 12.4 Using Tables 12.1 and 12.2, we obtain:

$$\begin{aligned}
\text{Pa} &\triangleq \text{N/m}^2 = (\text{kg·m/s}^2)/\text{m}^2 = \text{kg/m·s}^2,\\
\text{J} &\triangleq \text{N·m} = (\text{kg·m/s}^2)\text{m} = \text{kg·m}^2/\text{s}^2,\\
\text{W} &\triangleq \text{J/s} = (\text{kg·m}^2/\text{s}^2)/\text{s} = \text{kg·m}^2/\text{s}^3,\\
\text{V} &\triangleq \text{W/A} = (\text{kg·m}^2/\text{s}^3)/\text{A} = \text{kg·m}^2/\text{A·s}^3,\\
\Omega &\triangleq \text{V/A} = (\text{kg·m}^2/\text{A·s}^3)/\text{A} = \text{kg·m}^2/\text{A}^2\text{·s}^3,\\
\text{S} &\triangleq 1/\Omega = \text{A}^2\text{·s}^3/\text{kg·m}^2,\\
\text{F} &\triangleq \text{C/V} = (\text{A·s})/(\text{kg·m}^2/\text{A·s}^3) = \text{A}^2\text{·s}^4/\text{kg·m}^2,\\
\text{Wb} &\triangleq \text{V·s} = (\text{kg·m}^2/\text{A·s}^3)\text{s} = \text{kg·m}^2/\text{A·s}^2,\\
\text{T} &\triangleq \text{Wb/m}^2 = (\text{kg·m}^2/\text{A·s}^2)/\text{m}^2 = \text{kg/A·s}^2,\\
\text{H} &\triangleq \text{Wb/A} = (\text{kg·m}^2/\text{A·s}^2)/\text{A} = \text{kg·m}^2/\text{A}^2\text{·s}^2,\\
\text{lx} &\triangleq \text{lm/m}^2 = (\text{cd·sr})/\text{m}^2 = \text{cd/m}^2,\\
\text{Gy} &\triangleq \text{J/kg} = (\text{kg·m}^2/\text{s}^2)/\text{kg} = \text{m}^2/\text{s}^2,\\
\text{Sv} &\triangleq \text{J/kg} = \text{m}^2/\text{s}^2.
\end{aligned}$$

Solution 12.5 **(a)** Replace “Pascals” with either “pascals” or “Pa”.

(b) Replace “Lx.” with “lx”.

(c) Replace the equation with either “3 watts per hertz = 3 joules” or “3 W/Hz = 3 J”.

(d) Replace “19 gs” with, e.g., “1.9×10^{-2} kg”.

(e) Replace “degrees Kelvin” with “Kelvin” or “K”.

Solution 12.6 Since it is calculated as the ratio of two speeds, Mach number is a dimensionless unit.

Solution 12.7 For numerical data, accuracy pertains more to numbers, and resolution pertains more to numerals; because, the accuracy of a datum is how close its *value* is to the true value, whereas its resolution depends how well that value is represented in *notation*.

Solution 12.8 **(b)** Here a physical quantity $x_{\text{phy}} = -6.1$ V is measured as $x_{\text{exp}} = -5.90$ V; therefore, by (12.6), the absolute error in the measurement is $x_{\text{err}} = (-5.90\text{ V}) - (-6.1\text{ V}) = 0.20$ V.

(b) By (12.7), the relative error of the measurement is $(0.20\text{ V})/(-6.10\text{ V}) \approx -0.033 = -3.3\%$.

Solution 12.9 **(a)** Using (12.8) with a reference power $x_{\text{ref}} = 1$ mW yields $x|_{\text{dB}} = 10\log_{10}[(7.3\text{ W})/(1\text{ mW})]\text{ dB} \approx 38.63$ dB.

(b) Calculated directly, x exceeds y by $(x - y)/y = (2.3\text{ W})/(5\text{ W}) = 0.46 = 46\%$. Alternatively, x is $10\log_{10}(x/y)$ dB ≈ 1.64 dB greater than y; therefore, by essentially solving (12.10) for $x_{\text{rel-err}}$, we find that x exceeds y by about $10^{(1.64\text{ dB})/10} - 1 \approx 0.459 = 45.9\%$ (the discrepancy versus the previous calculation being due to roundoff error).

Solution 12.10 **(a)** Since $0.0012 < 0.08/65.43 \le 0.0013$, the given uncertainty interval is approximately the same as 65.43 mT $\pm$0.13% (here expressing the uncertainty to the nearest 0.01% because 1 part in 6543 is about 0.015%). Also, the given uncertainty is the same as 65.43(8) mT.

(b) Since $7.65 \times 4\% = 0.306$, the given uncertainty interval is approximately the same as (-7.65 ± 0.31) Tm, which is the same as $-7.65(31)$ Tm.

(c) The given uncertainty interval is the same as $(98.7 \pm 2.1) \times 10^6$ T; therefore, since $0.021 < 2.1/98.7 \le 0.022$, it is also approximately the same as 98.7×10^6 T $\pm$ 2.2% (here expressing the uncertainty to the nearest 0.1% because 1 part in 987 is about 0.10%).

Solution 12.11 It is possible to have $|x_{\text{q-err}}| > |x_{\text{err}}|$ for a particular measurement because the quantization error $x_{\text{q-err}}$ from recording the measurement might offset the error in the equipment output x_{out}, thereby making the error x_{err} in the recorded value x_{exp} smaller in magnitude. (On the other hand, as stated in (12.19), the resolution x_{res} of the measurement—i.e., its worst-case quantization error—cannot exceed its uncertainty x_{unc} of the measurement—i.e., its worst-case error.) For example, suppose a physical length $x_{\text{phy}} = 0.294$ m is read from a ruler as $x_{\text{out}} = 0.29$ m, which is recorded to the nearest 0.1 m as $x_{\text{exp}} = 0.3$ m; then, $x_{\text{q-err}} = x_{\text{exp}} - x_{\text{out}} = 0.01$ m and $x_{\text{err}} = x_{\text{exp}} - x_{\text{phy}} = 0.006$ m, making $|x_{\text{q-err}}| > |x_{\text{err}}|$.

Solution 12.12 **(a)** For the first numeral, the first four digits—i.e., the portion "2753"—must be significant; however, since there is no decimal point, the trailing "0" digits may or may not be significant, though both must be significant if the rightmost "0" is. Since the second numeral has a decimal point, its significant digits are those from the first nonzero digit forward—i.e., the portion "170". Finally, all of the digits in the decimal portion of the third numeral are significant, since it is in scientific notation.

(b) Each numeral has d digits of resolution where d is the number of significant digits; thus: the first numeral has either 4, 5, or 6 digits of resolution; the second numeral has 3 digits of resolution; and, the third numeral has 7 digits of resolution.

(c) Successively rounding to one fewer significant digit, we convert "275 300" to "275 000" to "280 000" to "300 000"; and, we convert "$-0.001\,70$" to "$-0.001\,7$" to "-0.002"; and, we convert "$5.286\,491 \times 10^{-3}$" to "$5.286\,49 \times 10^{-3}$" to "$5.286\,5 \times 10^{-3}$" to "$5.286 \times 10^{-3}$" to "$5.29 \times 10^{-3}$" to "$5.3 \times 10^{-3}$" to "$5 \times 10^{-3}$".

Solution 12.13 **(b)** For ρ, x, y, and z as stated, it follows from (12.8) that $\rho = 10\log_{10}(y/x)$ and $-\rho = 10\log_{10}(z/y)$, thereby making

$$0 = 10\log_{10}(y/x) + 10\log_{10}(z/y) = 10\log_{10}[(y/x)(z/y)] = 10\log_{10}(z/x);$$

thus, $z/x = 1$, making $z = x$.

(b) If y obtains from increasing x by 100%, then this corresponds to an increase of

$$\left(10\log_{10}\frac{y}{x}\right)\text{ dB} = \left[10\log_{10}\frac{(1+100\%)x}{x}\right]\text{ dB} = \left(10\log_{10}\frac{2x}{x}\right)\text{ dB}$$
$$= (10\log_{10}2)\text{ dB} \approx 3.01\text{ dB}.$$

But, if y obtains from decreasing x by 100%, then $y = (1 - 100\%)x = 0$, making $\log_{10}(y/x)$ undefined; so, the change in decibels is also undefined. However, letting $y = (1 + p\%)x$ yields $10\log_{10}(y/x) = 10\log_{10}(1 + p\%)$, which approaches $-\infty$ as $p \downarrow -100$; hence, one might instead say that a decrease of 100% corresponds to an infinite decrease in decibels.

(c) For $p \approx 0$ we have $1 + p\% \approx 0$, and thus

$$10\log_{10}\frac{(1+p\%)x}{x} = 10\frac{\ln(1+p\%)}{\ln 10} \approx 10\frac{p\%}{\ln 10} = \frac{p}{10\ln 10};$$

therefore, an increase of $p\%$ for $p \approx 0$ corresponds to an increase of about $(p/10\ln 10)$ dB $\approx 0.043p$ dB. (Likewise, an increase of ρ dB for $\rho \approx 0$ corresponds to an increase of about $(10\ln 10)\rho\% \approx 23\rho\%$.)

Solution 12.14 Let V and A be the physical dimensions velocity and acceleration, respectively; thus, in terms of the SI base dimensions in Table 12.4, $\mathsf{V} = \mathsf{LT}^{-1}$ and $\mathsf{A} = \mathsf{LT}^{-2}$. As stated in Subsection 12.2.1, the theory of mechanics can be fully developed in terms of the dimensions length, mass, and time alone. Furthermore, using (12.25) for the dimension F of force, we obtain

$$\begin{aligned} \mathsf{L} &= (\mathsf{M})(\mathsf{LT}^{-1})^2/(\mathsf{LMT}^{-2}) = \mathsf{MV}^2/\mathsf{F} = \mathsf{MV}^2\mathsf{F}^{-1}, \\ \mathsf{M} &= \mathsf{M}, \\ \mathsf{T} &= (\mathsf{M})(\mathsf{LT}^{-1})/(\mathsf{LMT}^{-2}) = \mathsf{MV}/\mathsf{F} = \mathsf{MVF}^{-1}; \end{aligned}$$

therefore, we can instead take force, mass, and velocity as fundamental. However, this not the case for force, mass, and acceleration, since these dimensions are not independent: $\mathsf{F} = \mathsf{M} \cdot \mathsf{LT}^{-2} = \mathsf{MA}$ (which also follows from Newton's second law of motion, though not vice versa). On the other hand, force, velocity, and acceleration *can* be taken as fundamental, because

$$\begin{aligned} \mathsf{L} &= (\mathsf{LT}^{-1})^2/(\mathsf{LT}^{-2}) = \mathsf{V}^2/\mathsf{A} = \mathsf{V}^2\mathsf{A}^{-1}, \\ \mathsf{M} &= \mathsf{FA}^{-1}, \\ \mathsf{T} &= (\mathsf{LT}^{-1})/(\mathsf{LT}^{-2}) = \mathsf{V}/\mathsf{A} = \mathsf{VA}^{-1}. \end{aligned}$$

Solution 12.15 **(a)** From Solution 12.4 we know that the SI units for inductance and capacitance are $\mathrm{H} = \mathrm{kg{\cdot}m^2/A^2{\cdot}s^2}$ and $\mathrm{F} = \mathrm{A^2{\cdot}s^4/kg{\cdot}m^2}$, respectively; therefore, by Tables 12.1 and 12.4, the physical dimension of the inductance L is $\mathsf{L}^2\mathsf{MT}^{-2}\mathsf{I}^{-2}$, and that of the capacitance C is $\mathsf{L}^{-2}\mathsf{M}^{-1}\mathsf{T}^4\mathsf{I}^2$.

(b) Since the SI unit for the frequency f is $\mathrm{Hz} = 1/\mathrm{s}$, its dimension is T^{-1}; therefore, in order for the given equation relating f, L, and C to be dimensionally homogeneous, we must have

$$\mathsf{T}^{-1} = (\mathsf{L}^2\mathsf{MT}^{-2}\mathsf{I}^{-2})^a \cdot (\mathsf{L}^{-2}\mathsf{M}^{-1}\mathsf{T}^4\mathsf{I}^2)^b.$$

In particular, the dimensions of length, mass, and electric current on the right side must all cancel out, which occurs if and only if $a = b$, thereby leaving $\mathsf{T}^{-1} = \mathsf{T}^{2a}$; therefore, $a = b = -1/2$, making $f = 1/2\pi\sqrt{LC}$.

Solution 12.16 Since

$$\begin{aligned} 1 = \frac{1}{1} = \frac{1\ \mathrm{N/(kg{\cdot}m/s^2)}}{1\ \mathrm{dyn/(g{\cdot}cm/s^2)}} &= (\mathrm{g/kg})(\mathrm{cm/m})\ \mathrm{N/dyn} = (10^{-3})(10^{-2})\ \mathrm{N/dyn} \\ &= 10^{-5}\ \mathrm{N/dyn}, \end{aligned}$$

the conversion factor from dynes to newtons is 10^{-5} N/dyn.

Solution 12.17 Since a change of $(212 - 32)$°F = 180°F corresponds to a change of $(100 - 0)$°C = 100°C, an increase in temperature by 1°F corresponds to an increase of $\frac{100}{180}$°C $= \frac{5}{9}$°C. Therefore, if $t' = at + b$ for any temperature that is both t°F and t'°C, then we must have $a = 5/9$; and, since 32°F corresponds to 0°C, we also have $0 = a \cdot 32 + b$, making $b = -32a = -160/9$. Hence, $t' = \frac{5}{9}t - \frac{160}{9} = \frac{5}{9}(t - 32)$.

Solution 12.18 **(a)** The ratio of the speed of light to the speed of sound is

$$\frac{3.0 \times 10^8\ \mathrm{m/s}}{1100\ \mathrm{ft/s}} \cdot \frac{1}{(10^{-2}\ \mathrm{m/cm})(2.54\ \mathrm{cm/in})(12\ \mathrm{in/ft})} \approx 8.9 \times 10^5$$

—rounding to two significant digits, in accordance with the given data.

(b) Let v_l and v_s be the speeds of light and sound, respectively. If at time $t = 0$ a lightning strike occurs a distance d away from an observer, then the lightning is seen at $t = d/v_l$ and heard later at $t = d/v_s$; hence, the amount of time that expires between seeing the lightning and hearing its thunder is $d/v_s - d/v_l \approx d/v_s$, since $v_l \gg v_s$, as found in part (a). Therefore, if a lightning strike is seen x seconds before its thunder is heard, then $d/v_s \approx x$ s, and thus

$$d \approx v_s \cdot x\,\text{s} \approx (1100\,\text{ft/s})(x\,\text{s}) \cdot \frac{1}{5280\,\text{ft/mi}} = \frac{x}{4.8}\,\text{mi}.$$

(Note that, as discussed at the end of Subsection 12.2.1, "x" is a mathematical variable, whereas "v_l", "v_s", "t", and "d" are physical variables. However, no unit is required when stating "$t = 0$"—e.g., "$t = 0$ s"—since the numerical value being applied to the unit is then 0.)

Solution 12.19 Although *either* byte or bit can be interpreted as the number 1, it is inconsistent to do so for both, because both measure the same physical feature (computer memory).

Solution 12.20 Using (12.36) with $\Delta a = -0.07$ m and $\Delta b = 0.08$ m, we find that the absolute error in $a+b$ is $\Delta a + \Delta b = 0.01$ m, whereas that in $a-b$ is $\Delta a - \Delta b = -0.15$ m. Also, we find that the relative error in $a \cdot b$ is about $(\Delta a)/a + (\Delta b)/b \approx -0.04$, and that a/b is about $(\Delta a)/a - (\Delta b)/b \approx -0.07$.

Solution 12.21 **(a)** Since for $n \geq 1$ we have $a^n = a \cdot a \cdot \cdots \cdot a$ with n factors of a (and thus $n-1$ multiplications), it follows from $n-1$ applications of (12.36c) that

$$\frac{\Delta a^n}{a^n} = \frac{\Delta a}{a} + \frac{\Delta a}{a} + \cdots + \frac{\Delta a}{a} \approx n\frac{\Delta a}{a},$$

proving (P12.21.1) for $n \geq 1$.

(b) For $n \geq 1$, it follows from the binomial expansion (8.4) that

$$(a + \Delta a)^n = \sum_{k=0}^{n} \binom{n}{k} (\Delta a)^k a^{n-k} = a^n \sum_{k=0}^{n} \binom{n}{k} \left(\frac{\Delta a}{a}\right)^k.$$

Assuming $|\Delta a|$ is sufficiently small, the terms of the last sum decrease so rapidly in magnitude as k increases that we may neglect all but the first two, thereby obtaining

$$(a + \Delta a)^n \approx a^n \left(1 + n\frac{\Delta a}{a}\right).$$

Hence,

$$\frac{\Delta a^n}{a^n} = \frac{(a + \Delta a)^n - a^n}{a^n} \approx n\frac{\Delta a}{a},$$

again proving (P12.21.1) for $n \geq 1$.

(c) For $n \leq -1$ we have $-n \geq 1$; therefore, applying the result of part (a) along with (12.36d) yields

$$\frac{\Delta a^n}{a^n} = \frac{\Delta(1/a)^{-n}}{(1/a)^{-n}} \approx (-n)\frac{\Delta(1/a)}{(1/a)} \approx (-n)\left(-\frac{\Delta a}{a}\right) = n\frac{\Delta a}{a},$$

proving (P12.21.1) for $n \leq -1$.

(d) Assuming $x = \sqrt[n]{a}$ and $x + \Delta x = \sqrt[n]{a + \Delta a}$ for some $a > 0$ and $n \geq 1$, we have $a = x^n$ and $a + \Delta a = (x + \Delta x)^n$; also, $|\Delta x| \lll |x|$ if $|\Delta a|$ is sufficiently small. Therefore, by (P12.21.1), $(\Delta a)/a \approx n[(\Delta x)/x]$, which gives (P12.21.2).

(e) Letting $c = a^2 + b^2$, it follows from (12.36a) and (P12.21.1) that

$$\begin{aligned}\Delta c = \Delta a^2 + \Delta b^2 &= a^2 \cdot \frac{\Delta a^2}{a^2} + b^2 \cdot \frac{\Delta b^2}{b^2} \approx a^2 \cdot 2\frac{\Delta a}{a} + b^2 \cdot 2\frac{\Delta b}{b} \\ &= 2[a(\Delta a) + b(\Delta b)].\end{aligned}$$

Also, $x = c^{1/2}$; therefore, by (P12.21.2) we have $(\Delta x)/x \approx (1/2)(\Delta c)/c = (1/2)(\Delta c)/x^2$, making $\Delta x \approx (\Delta c)/2x$. Hence, $\Delta x \approx [a(\Delta a) + b(\Delta b)]/x$; or, more elegantly, $x(\Delta x) \approx a(\Delta a) + b(\Delta b)$.

Solution 12.22 Let real numbers a, b, Δa, and Δb be such that $|\Delta a| \ll |a|$ and $|\Delta b| \ll |b|$; thus, a, $b \neq 0$. Also, suppose $a \approx b$, thereby making $a + b \neq 0$ as well. Then, mimicking (12.34), the relative error in $(a + \Delta a) + (b + \Delta b)$ with respect to $a + b$ is

$$\frac{(a + \Delta a) + (b + \Delta b)}{a + b} - 1 = \frac{\Delta a}{a + b} + \frac{\Delta b}{a + b} \approx \frac{\Delta a}{a + a} + \frac{\Delta b}{b + b} = \left(\frac{\Delta a}{a} + \frac{\Delta b}{b}\right) \Big/ 2,$$

which is the average of the relative errors $\Delta a/a$ and $\Delta b/b$.

Solution 12.23 Calculated directly in terms of uncertainty intervals, the man lost at least (191.0 − 0.5) pounds − (188.5 + 0.5) pounds = 1.5 pounds, and at most (191.0 + 0.5) pounds − (188.5 − 0.5) pounds = 3.5 pounds. Equivalently, the man lost (191.0 − 188.5) pounds = 2.5 pounds with an uncertainty of (0.5 + 0.5) pounds = 1.0 pounds; therefore, he lost at least (2.5 − 1.0) pounds = 1.5 pounds, and at most (2.5 + 1.0) pounds = 3.5 pounds.

Solution 12.24 For the sake of analysis, let the lengths a, b, and c be variable. The purported right angle must be opposite to the side of length c, since we are to have $c \geq a, b$. Thus, letting the measure of this angle be C, it follows from the law of cosines (11.32) that $c^2 = a^2 + b^2 - 2ab \cos C$; therefore, with $C \approx 90°$, we have $C = \cos^{-1}[(a^2 + b^2 - c^2)/2ab]$. Now, although it is clear that $(a^2 + b^2 - c^2)/2ab$ decreases if c increases with a and b fixed, which in turn causes C to increase since arccosine is a strictly decreasing function, it is not obvious what the effects are of varying a and b. It appears from the triangle in Figure 11.12 that if a is increased with b and c fixed, then C decreases. Indeed, it follows from (12.36) that with b and c fixed,

$$\begin{aligned}\Delta \frac{a^2 + b^2 - c^2}{2ab} &= \Delta \frac{a}{2b} + \Delta \frac{b^2 - c^2}{2ab} \approx \frac{\Delta a}{2b} - \frac{b^2 - c^2}{2b} \cdot \frac{\Delta a}{a^2} \\ &\approx \frac{\Delta a}{2b} - \frac{b^2 - (a^2 + b^2)}{2b} \cdot \frac{\Delta a}{a^2} = \frac{\Delta a}{b},\end{aligned}$$

which shows that $(a^2 + b^2 - c^2)/2ab$ increases if a increases, thereby causing C to decrease; likewise, C decreases if we increase b with a and c fixed. Accordingly, if a, b, and c are each allowed to vary $\pm p\%$ about their desired values, then the uncertainty interval for C is

$$\begin{aligned}\cos^{-1} \frac{[(1 + p\%)a]^2 + [(1 + p\%)b]^2 - [(1 - p\%)c]^2}{2[(1 + p\%)a][(1 + p\%)b]} &\leq C \\ &\leq \cos^{-1} \frac{[(1 - p\%)a]^2 + [(1 - p\%)b]^2 - [(1 + p\%)c]^2}{2[(1 - p\%)a][(1 - p\%)b]}\end{aligned}$$

—where a, b, and c are now the *desired* values—which upon substituting $c^2 = a^2 + b^2$ simplifies to

$$\cos^{-1}\left[\frac{2p\%}{(1 + p\%)^2}(a/b + b/a)\right] \leq C \leq \cos^{-1}\left[\frac{-2p\%}{(1 - p\%)^2}(a/b + b/a)\right].$$

In particular, for $a : b = 3 : 4$ and $p = 0.1$ we obtain $89.76° \lesssim C \lesssim 90.24°$ (i.e., an uncertainty of about one-fourth degree).

Solution 12.25 We first observe that although the numeral expressing x does not contain a decimal point, it has no trailing zeros, so all of its digits are significant; also, the numeral expressing y does contain a decimal point and has no leading zeros, so all of its digits are significant as well. To express $x + y$ and $x - y$, note that $x = 123 \times 10^6$ Bq $= 1.23 \times 10^8$ Bq and $y = 76.54 \times 10^9$ Bq $= 765.4 \times 10^8$ Bq; therefore,

$$\begin{aligned}x + y &= 1.23 \times 10^8 \text{ Bq} + 765.4 \times 10^8 \text{ Bq} = 766.63 \times 10^8 \text{ Bq} \approx 766.6 \times 10^8 \text{ Bq} \\ &= 7.666 \times 10^{10} \text{ Bq}\end{aligned}$$

and

$$x - y = 1.23 \times 10^8 \text{ Bq} - 765.4 \times 10^8 \text{ Bq} = -764.17 \times 10^8 \text{ Bq}$$
$$\approx -764.2 \times 10^8 \text{ Bq} = -7.642 \times 10^{10} \text{ Bq}.$$

Also,

$$xy = (123 \times 10^6 \text{ Bq})(76.54 \times 10^9 \text{ Bq}) = 9.41442 \times 10^{18} \text{ Bq}^2 \approx 9.41 \times 10^{18} \text{ Bq}^2$$

and

$$x/y = (123 \times 10^6 \text{ Bq})/(76.54 \times 10^9 \text{ Bq}) \approx 1.60700 \times 10^{-3} \approx 1.61 \times 10^{-3}.$$

Solution 12.26 To obtain a perfectly uniform distribution of $n + 1 \geq 2$ points x_i on a logarithmic scale from $x = a$ to $x = b$ $(0 < a < b < \infty)$, we let

$$x_i = a(b/a)^{i/n} \quad (i = 0, 1, \ldots, n);$$

for then $\log_{10} x_i = \log_{10} a + i(\log_{10} b - \log_{10} a)/n$ (all i), analogous to (P12.26.1). In particular, for $n = 5$, $a = 100$, and $b = 1000$ we have $x_i = 100(10^{1/5})^i$ (all i), which yields $x_0 = 100$, $x_1 \approx 158.5$, $x_2 \approx 251.2$, $x_3 \approx 398.1$, $x_4 \approx 631.0$, and $x_5 = 1000$. Thus, rounding to two significant digits of resolution, we obtain the values 100, 160, 250, 400, 630, and 1000. (*Aside*: The 1-3 and 1-2-5 sequences obtain by taking $b/a = 10$ with either $n = 2$ or $n = 3$, and rounding to one significant digit; because, $10^{1/2} \approx 3$, whereas $10^{1/3} \approx 2$ and $10^{2/3} \approx 5$.)

Solution 12.27 If $m \geq d_L$, then the least-squares-fit polynomial must be the same as the Lagrange interpolation polynomial; because, the former polynomial is the unique polynomial of degree no greater than m that minimizes the squared-error sum (12.38), whereas the squared-error sum (12.38) of the latter polynomial is zero. Therefore, if $m \geq d_L$ then $d_m = d_L$; whereas if $m < d_L$ then $d_m < d_L$, since $d_m \leq m$. Hence, $d_m = d_L$ if and only if $m \geq d_L$.

Solution 12.28 Since a first-degree polynomial (line) is being fit to the plot, the coefficients a_0 and a_1 obtain from (8.18) upon replacing each occurrence of "y_i" with "$\log_{10} y_i$"; the result is

$$a_0 \approx -290.40 \text{ decades}, \quad a_1 \approx 0.14901 \text{ decade per year}.$$

(The unit of the slope a_1 is decade per year because a base-10 logarithm was used.) Figure S12.28.1 shows a plot of the data along with the fitted line. Since there is $\log_{10} 2 \approx 0.301$ decade per octave, the inverse slope of the line is about

$$\frac{0.301 \text{ decade per octave}}{0.14901 \text{ decade per year}} \approx 2.0 \text{ years per octave};$$

thus, approximately 2.0 years is required for the number of transistors to double.

Solution 12.29 By (12.39), the sum in (12.38) equals

$$(\mathbf{Xa} - \mathbf{y})^{\mathrm{T}}(\mathbf{Xa} - \mathbf{y}) = (\mathbf{a}^{\mathrm{T}}\mathbf{X}^{\mathrm{T}} - \mathbf{y}^{\mathrm{T}})(\mathbf{Xa} - \mathbf{y})$$
$$= \mathbf{a}^{\mathrm{T}}\mathbf{X}^{\mathrm{T}}\mathbf{Xa} - \mathbf{a}^{\mathrm{T}}\mathbf{X}^{\mathrm{T}}\mathbf{y} - \mathbf{y}^{\mathrm{T}}\mathbf{Xa} + \mathbf{y}^{\mathrm{T}}\mathbf{y},$$

where

$$\mathbf{a}^{\mathrm{T}}\mathbf{X}^{\mathrm{T}}\mathbf{Xa} = \mathbf{a}^{\mathrm{T}}\mathbf{X}^{\mathrm{T}}\mathbf{X}[(\mathbf{X}^{\mathrm{T}}\mathbf{X})^{-1}\mathbf{X}^{\mathrm{T}}\mathbf{y}] = \mathbf{a}^{\mathrm{T}}[\mathbf{X}^{\mathrm{T}}\mathbf{X}(\mathbf{X}^{\mathrm{T}}\mathbf{X})^{-1}]\mathbf{X}^{\mathrm{T}}\mathbf{y} = \mathbf{a}^{\mathrm{T}}\mathbf{I}\mathbf{X}^{\mathrm{T}}\mathbf{y}$$
$$= \mathbf{a}^{\mathrm{T}}\mathbf{X}^{\mathrm{T}}\mathbf{y};$$

therefore, the sum in (12.38) equals $-\mathbf{y}^{\mathrm{T}}\mathbf{Xa} + \mathbf{y}^{\mathrm{T}}\mathbf{y} = \mathbf{y}^{\mathrm{T}}(\mathbf{y} - \mathbf{Xa})$.

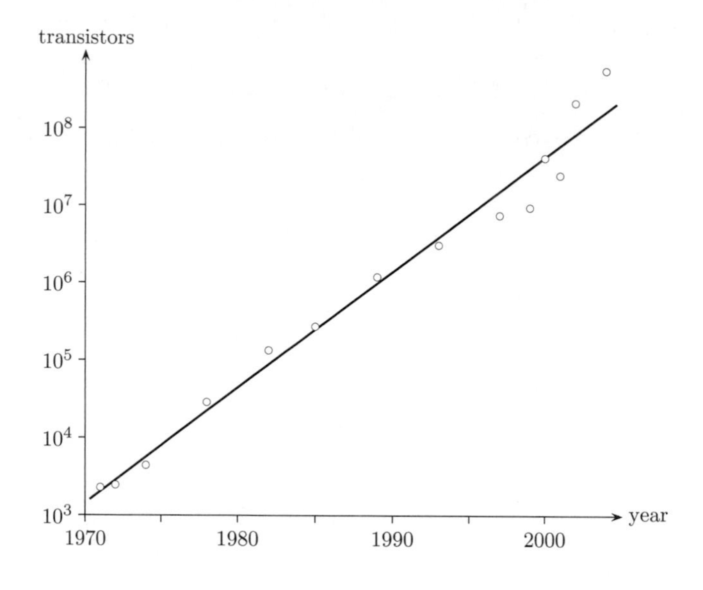

FIGURE S12.28.1 Verification of Moore's Law

Solution 12.30 For $m = n - 1$, the $n \times (m+1)$ matrix $\mathbf{X}$ is $n \times n$. Also, this square matrix must be nonsingular; for otherwise there would exist a nonzero $n \times 1$ matrix $\mathbf{z}$ such that $\mathbf{Xz} = \mathbf{0}$, making

$$\mathbf{0} = \left[(\mathbf{X}^{\mathrm{T}}\mathbf{X})^{-1}\mathbf{X}^{\mathrm{T}}\right]\mathbf{0} = \left[(\mathbf{X}^{\mathrm{T}}\mathbf{X})^{-1}\mathbf{X}^{\mathrm{T}}\right]\mathbf{Xz} = \left[(\mathbf{X}^{\mathrm{T}}\mathbf{X})^{-1}(\mathbf{X}^{\mathrm{T}}\mathbf{X})\right]\mathbf{z} = \mathbf{Iz} = \mathbf{z},$$

which contradicts $\mathbf{z}$ being nonzero. Hence, $\mathbf{X}^{\mathrm{T}}$ is also nonsingular, and we obtain $(\mathbf{X}^{\mathrm{T}}\mathbf{X})^{-1}\mathbf{X}^{\mathrm{T}} = \mathbf{X}^{-1}(\mathbf{X}^{\mathrm{T}})^{-1}\mathbf{X}^{\mathrm{T}} = \mathbf{X}^{-1}$. (Therefore, the polynomial in (12.37) obtained from (12.39) interpolates all of the data points.)

List of Notation

Almost all the notation used in this book is standard, or at least widely used. Each page number indicates the first significant usage (e.g., where the notation is defined, explicitly or implicitly). If a notation has multiple meanings, then each is listed separately. (Notation that is merely mentioned, though—i.e., not subsequently used—is not necessarily listed.)

Notation	Name or Meaning	Page
PROLOGUE		
$\square$	*end of proof*	2
$\therefore$	*therefore*	2
$\because$	*because* or *since*	2
CHAPTER 1		
$\neg A$	*not A*	10
$A \wedge B$	*A and B*	10
$A \vee B$	*A or B*	10
$A \Rightarrow B$	*if A, then B*	10
$A \Leftrightarrow B$	*A if and only if B*	10
T	*true*	10
F	*false*	10
$A \models B$	*A logically implies B*	12
$A \models\!\dashv B$	*A is logically equivalent to B*	12
$A \not\models B$	*A does not logically imply B*	12
$\forall x$	*for all x*	13
$x = y$	*x equals y*	14
$\exists x$	*there exists an x*, or *for some x*	15
$\ni$	*such that*	15n
CHAPTER 2		
$x \in A$	*is an element of*	19
$x \notin A$	*is not an element of*	19
$\{x_1, \ldots, x_n\}$	*set of objects $x_1, \ldots, x_n$*	19
$\emptyset$	*empty set*	19
$x \triangleq y$	*x equals y by definition*	19n
$\lvert A \rvert$	*cardinality of set A*	19
$\{x \mid P(x)\}$	*set of objects possessing property P*	19
$A \neq B$	*set A does not equal set B*	20
$A \subseteq B$	*A is a subset of B*	20
$A \subset B$	*A is a proper subset of B*	20
$A \supseteq B$	*A is a superset of B*	21
$A \supset B$	*A is a proper superset of B*	21
$A \not\subseteq B$	*A is not a subset of B*	21

(Continued)

Notation	Name or Meaning	Page
$A \not\subset B$	*A is not a proper subset of B*	21
$A \not\supseteq B$	*A is not a superset of B*	21
$A \not\supset B$	*A is not a proper superset of B*	21
$A \cap B$	*intersection of sets A and B*	21
$A \cup B$	*union of sets A and B*	21
$A - B$	*difference of sets A and B*	21
A^{c}	*complement of set A*	21
$\bigcap_{i=m}^{n} A_i$	*intersection of sets* $A_m, \ldots, A_n$	22
$\bigcup_{i=m}^{n} A_i$	*union of sets* $A_m, \ldots, A_n$	22
$\bigcap_{A \in \mathcal{S}} A$	*intersection of sets belonging to set* $\mathcal{S}$	22
$\bigcup_{A \in \mathcal{S}} A$	*union of sets belonging to set* $\mathcal{S}$	22
$(x_1, \ldots, x_n)$	*ordered tuple*, or *n-tuple*, *of objects* $x_1, \ldots, x_n$	22
$A \times B$	*Cartesian product of sets A and B*	22
A^n	$A \times \cdots \times A$ *with n occurrences of "A"*	23
$x R y$	$(x, y) \in R$ (for a binary relation R)	24
$x \equiv y$	*x is equivalent to y* (in some sense)	25
CHAPTER 3		
$\mathbb{N}$	*set of natural numbers*	27
$+$	*plus* (as a prefix sign to a numeral)	27
$-$	*minus* (as a prefix sign to a numeral)	27
$\mathbb{Z}$	*set of integers*	27
$\pm$	*plus-or-minus*	27
$\mp$	*minus-or-plus*	27n
$\mathbb{Z}^+$	*set of positive integers*	27
$\mathbb{Z}^-$	*set of negative integers*	27
$\mathbb{Q}$	*set of rational numbers*	28
$\frac{x}{y}$	*fraction with numerator x and denominator y*	28
$\mathbb{R}$	*set of real numbers*	28
$1234\overline{56.789}\ldots$	*repeating digits indicated by a vinculum*	30
123.45_b	*a base-b number*	31
$x < y$	*x is less than y*	31
$x = y$	*x equals y*	31
$x > y$	*x is greater than y*	31
$x \leq y$	*x is less than or equal to y*	32
$x \geq y$	*x is greater than or equal to y*	32
$x \neq y$	*x does not equal y*	32
$x \not< y$	*x is not less than y*	32
$x \not> y$	*x is not greater than y*	32
$x \not\leq y$	*x is not less than or equal to y*	32
$x \not\geq y$	*x is not greater than or equal to y*	32
$x \approx y$	*x approximately equals y*	32

Notation	Name or Meaning	Page
$x \ll y$	*x is much less than y*	32
$x \gg y$	*x is much greater than y*	32
$x \lesssim y$	*x is less than or approximately equal to y*	32
$x \gtrsim y$	*x is greater than or approximately equal to y*	32
∞	*infinity*	32
$\overline{\mathbb{R}}$	*set of extended real numbers*	32
(a, b)	*interval from a (excluded) to b (excluded)*	32
$(a, b]$	*interval from a (excluded) to b (included)*	32
$[a, b)$	*interval from a (included) to b (excluded)*	32
$[a, b]$	*interval from a (included) to b (included)*	32
$\max A$	*maximum of a set A of numbers*	33
$\min A$	*minimum of a set A of numbers*	33
$\sup A$	*supremum of a set A of numbers*	33
$\inf A$	*infimum of a set A of numbers*	33
$x + y$	*x plus y*	33
$x - y$	*x minus y*	33
$x \cdot y$	*x times y*	33
$x \times y$	*x times y*	33
xy	*x times y*	33
x/y	*x divided by y*	33
$x \div y$	*x divided by y*	33
$\frac{x}{y}$	*x divided by y*	33
sgn	*signum function*	35
$\|x\|$	*magnitude*, or *absolute value*, *of real number x*	35
( )	*left and right parentheses* (for grouping)	36
[ ]	*left and right brackets* (for grouping)	36
{ }	*left and right braces* (for grouping)	36
$\mathrm{LCM}(x_1, \ldots, x_n)$	*least common multiple of* $x_1, \ldots, x_n$	38
$\mathrm{GCD}(x_1, \ldots, x_n)$	*greatest common divisor of* $x_1, \ldots, x_n$	38
$x \bmod y$	*x modulo y*	39
$x \oplus y$	*modular addition of x and y* (for some base)	39
$x : y$	*ratio of x to y*	39
$x : y :: x' : y'$	*x is to y as x′ is to y′*	39
%	*percent*	39
i	*the imaginary unit,* $\sqrt{-1}$	40
$\mathbb{C}$	*set of complex numbers*	40
$\mathrm{Re}\, z$	*real part of z*	40
$\mathrm{Im}\, z$	*imaginary part of z*	40
$\|z\|$	*magnitude*, or *modulus*, *of complex number z*	41
$\arg z$	*phase*, or *an argument*, *of complex number z*	41
$\mathrm{Arg}\, z$	*principal value of* arg *z*	42n
$\|z\| \angle \arg z$	*polar form of complex number z*	42
z^*	*conjugate of complex number z*	43
CHAPTER 4		
$\{x_i\}_{i=m}^{n}$	*sequence of objects* x_i *from* $i = m$ *to* $i = n$	46
$\{x_i\}$	*sequence of objects* x_i	46
$\{x_{i,j}\}$	*double sequence of objects* $x_{i,j}$	46

(Continued)

Notation	Name or Meaning	Page
$\max_i x_i$	*maximum of* $\{x_i\}$	47
$\min_i x_i$	*minimum of* $\{x_i\}$	47
$\sup_i x_i$	*supremum of* $\{x_i\}$	47
$\inf_i x_i$	*infimum of* $\{x_i\}$	47
$x_i \to x$	x_i *approaches* x	48
$x_i \xrightarrow[i\to\infty]{} x$	$x_i \to x$ *as* $i \to \infty$	48
$\lim_{i\to\infty} x_i$	*limit of* $\{x_i\}$ *as* $i \to \infty$	48
$x_{i,j} \xrightarrow[i,j\to\infty]{} x$	$x_{i,j} \to x$ *as* $i \to \infty$ *and* $j \to \infty$ *together*	51
$\lim_{i,j\to\infty} x_{i,j}$	*limit of* $\{x_{i,j}\}$ *as* $i \to \infty$ *and* $j \to \infty$ *together*	51
$\sum_{i=m}^{n} x_i$	*sum of values* x_i *for* $i = m$ *to* $i = n$	53
$\prod_{i=m}^{n} x_i$	*product of values* x_i *for* $i = m$ *to* $i = n$	53
$\sum_{(\ldots i \ldots)} x_i$	*sum of values* x_i *versus i, as constrained*	53
$\prod_{(\ldots i \ldots)} x_i$	*product of values* x_i *versus i, as constrained*	53
$\mathbf{1}(\cdot)$	*indicator function (of some condition)*	59
CHAPTER 5		
$f(\cdot)$	*function having one argument*	70n
$x \overset{f}{\longmapsto} y$	*x maps to y under* $f(\cdot)$	70
$x \mapsto y$	*x maps to y*	70
$f : D \to S$	$f(\cdot)$ *maps D into S*	70
$f^{-1}(B)$	*inverse image*, or *pre-image*, *of set B, under* $f(\cdot)$	72
$f^{-1}(\cdot)$	*inverse of* $f(\cdot)$	72
$f(\cdot, \ldots, \cdot)$	*function having several arguments*	72
$\max(\cdot, \ldots, \cdot)$	*maximum function*	72
$\min(\cdot, \ldots, \cdot)$	*minimum function*	72
$\max_{x\in S} f(x)$	*maximum of* $f(\cdot)$ *over S*	73
$\min_{x\in S} f(x)$	*minimum of* $f(\cdot)$ *over S*	73
$\sup_{x\in S} f(x)$	*supremum of* $f(\cdot)$ *over S*	73
$\inf_{x\in S} f(x)$	*infimum of* $f(\cdot)$ *over S*	73
$\max_x f(x)$	*maximum of* $f(\cdot)$ *over its domain*	73
$\min_x f(x)$	*minimum of* $f(\cdot)$ *over its domain*	73
$\sup_x f(x)$	*supremum of* $f(\cdot)$ *over its domain*	73
$\inf_x f(x)$	*infimum of* $f(\cdot)$ *over its domain*	73

Notation	Name or Meaning	Page
Δx	*variation in x*	77n
$y \propto x$	*y is proportional to x*	78
$f(x) \xrightarrow[x \to X]{} Y$	$f(x) \to Y$ *as* $x \to X$	82
$\lim_{x \to X} f(x)$	*limit of* $f(x)$ *as* $x \to X$	82
$f(X+)$	*limit from the right of* $f(\cdot)$	83
$\lim_{x \to X+} f(x)$	*limit from the right of* $f(\cdot)$	83
$\lim_{x \downarrow X} f(x)$	*limit from the right of* $f(\cdot)$	83
$f(X-)$	*limit from the left of* $f(\cdot)$	83
$\lim_{x \to X-} f(x)$	*limit from the left of* $f(\cdot)$	83
$\lim_{x \uparrow X} f(x)$	*limit from the left of* $f(\cdot)$	83
$\sim$	*behaves asymptotically like*	83
$\nsim$	*does not behaves asymptotically like*	83
$O(\cdot), o(\cdot)$	*asymptotic order symbols*	84
CHAPTER 6		
x^n	$x \cdot \cdots \cdot x$ *with n occurrences of "x"*	88
$\sqrt[n]{x}$	*nth root of x*	90
$\sqrt{x}$	*square root of x*	90
x^p	*pth power of x*	91
$\deg f(\cdot)$	*degree of polynomial* $f(\cdot)$	93
CHAPTER 7		
e	*Napier's number*	100
$\exp$	*exponential function*	100
$\log_b$	*base-b logarithm function*	102
$\ln$	*natural logarithm function*	103
CHAPTER 8		
$n!$	*n factorial*	104
$\binom{n}{k}$	*binomial coefficient*, or *n choose k*	104
$\Pr(\ldots)$	*probability of having condition ...*	105
$\overline{x}$	*arithmetic mean of random variable x*	108
σ_x	*standard deviation of random variable x*	109
CHAPTER 9		
$\mathbf{A}$	*a matrix*	115
$[a_{i,j}]_{i,j=1}^{m,n}$	*matrix of elements* $a_{i,j}$ *for* $i = 1$ *to* $i = m$ *and* $j = 1$ *to* $j = n$	115
$[a_{i,j}]$	*matrix of elements* $a_{i,j}$	115
$[a_{i,j}]_{i,j=1}^{n}$	*square matrix* $[a_{i,j}]_{i,j=1}^{n,n}$	115
$\mathbf{A}^{\mathrm{T}}$	*transpose of* $\mathbf{A}$	115
$\mathbf{A}^*$	*adjoint of* $\mathbf{A}$	116
$\overline{\mathbf{A}}$	*complex conjugate of* $\mathbf{A}$	116
$\mathbf{A} + \mathbf{B}$	*sum of matrices* $\mathbf{A}$ *and* $\mathbf{B}$	117
$\mathbf{A} - \mathbf{B}$	*difference of matrices* $\mathbf{A}$ *and* $\mathbf{B}$	117
$\mathbf{AB}$	*product of matrices* $\mathbf{A}$ *and* $\mathbf{B}$	117
$\alpha\mathbf{A}$ or $\mathbf{A}\alpha$	*multiplication of* $\mathbf{A}$ *by scalar* α	117
$-\mathbf{A}$	*negative of* $\mathbf{A}$	118

(Continued)

Notation	Name or Meaning	Page
$\mathbf{0}$	*zero matrix*	118
$\mathbf{I}$	*identity matrix*	119
$\det \mathbf{A}$	*determinant of* $\mathbf{A}$	120
$\mathbf{A}^{-1}$	*inverse of* $\mathbf{A}$	121
$\operatorname{tr} \mathbf{A}$	*trace of* $\mathbf{A}$	126
CHAPTER 10		
$\overleftrightarrow{AB}$	*line through points A and B*	130
$\overrightarrow{AB}$	*ray through point B having endpoint A*	130
$\overleftarrow{AB}$	*ray through point A having endpoint B*	130
$\overline{AB}$	*line segment having endpoints A and B*	130
$f \parallel g$	f *is parallel to g*	131
$\angle ABC$	*angle formed by rays* $\overleftarrow{AB}$ *and* $\overrightarrow{BC}$	132
$\angle B$	*angle having vertex B*	132
$f \sim g$	f *is similar to g*	133
$f \cong g$	f *is congruent to g*	133
$\urcorner$	*right-angle indicator* (on geometric pictures)	133
$f \perp g$	f *is perpendicular,* or *orthogonal, to g*	134
$d(A, B)$	*distance between points A and B*	135
$\lvert\overline{AB}\rvert$	*length of* $\overline{AB}$	135
$\measuredangle ABC$	*measure of* $\angle ABC$	136
$\measuredangle B$	*measure of* $\angle B$	136
12.3°	*angle measure in degrees*	136
$12^\circ\,4.5'$	*angle measure in degrees and minutes*	136n
$12^\circ\,4'\,6.7''$	*angle measure in degrees, minutes, and seconds*	136n
polygon $A_1 \ldots A_n$	*polygon having vertices* $A_1, \ldots, A_n$	138
$\triangle A_1A_2A_3$	*triangle having vertices* A_1, A_2, A_3	140
$\square A_1A_2A_3A_4$	*quadrilateral having vertices* A_1, A_2, A_3, A_4	140
$\widehat{AB}$	*arc having endpoints A and B*	147
$\widehat{ACB}$	*arc having endpoints A and B, and containing point C*	147
π	*pi* (circumference of a unit-diameter circle)	149
$\alpha(f)$	*area of plane figure f*	160
$\nu(f)$	*volume of figure f*	162
$\sigma(f)$	*surface area of solid figure f*	162
CHAPTER 11		
$\mathbb{E}^n$	*n-dimensional Euclidean space*	167
O	*origin* (of a geometric space)	169
sin	*sine function*	174
cos	*cosine function*	174
tan	*tangent function*	174
cot	*cotangent function*	174
sec	*secant function*	174
csc	*cosecant function*	174
$\sin^{-1}$	*arcsine function*	177
$\cos^{-1}$	*arccosine function*	177
$\tan^{-1}$	*arctangent function*	177
sinh	*hyperbolic sine function*	183
cosh	*hyperbolic cosine function*	183
tanh	*hyperbolic tangent function*	183

Notation	Name or Meaning	Page
$\sinh^{-1}$	*inverse hyperbolic sine function*	184
$\cosh^{-1}$	*inverse hyperbolic cosine function*	184
$\tanh^{-1}$	*inverse hyperbolic tangent function*	184
$\vec{x}$	*a vector*	185
$\vec{x}+\vec{y}$	*sum of* $\vec{x}$ *and* $\vec{y}$	185
$a\vec{x}$	*multiplication of* $\vec{x}$ *by scalar a*	186
$\vec{0}$	*zero vector*	186
$-\vec{x}$	*additive inverse of* $\vec{x}$	186
$\vec{x}-\vec{y}$	*difference of* $\vec{x}$ *and* $\vec{y}$	186
$\|\vec{x}\|$	*norm of* $\vec{x}$	187
$[\vec{x},\vec{y}]$	*inner product*, or *scalar product, of* $\vec{x}$ *and* $\vec{y}$	187
(O,Q)	*position vector corresponding to point Q*	192
$\mathbb{V}^n$	*n-dimensional Euclidean vector space*	193
$\mathbf{x}\bullet\mathbf{y}$	*dot product of* **x** *and* **y**	195
$\angle(\mathbf{x},\mathbf{y})$	*angle measure between* **x** *and* **y**	196
$\mathbf{e}_1,\ldots,\mathbf{e}_n$	*standard basis vectors for* $\mathbb{V}^n$	197
$\hat{\mathbf{i}},\hat{\mathbf{j}}$	*standard basis vectors for* $\mathbb{V}^2$	197
$\hat{\mathbf{i}},\hat{\mathbf{j}},\hat{\mathbf{k}}$	*standard basis vectors for* $\mathbb{V}^3$	197
$\mathbf{x}\times\mathbf{y}$	*cross product*, or *vector product, of* **x** *and* **y**	198
CHAPTER 12		
x_{exp}	*experimental value of x*	228
x_{phy}	*true physical value of x*	237
x_{err}	*absolute error in* x_{exp}	237
$x_{\text{rel-err}}$	*relative error in* x_{exp}	237
$x\|_{\text{dB}}$	*x in decibels*	238
x_{ref}	*reference value for x*	238
x_{lb}	*lower bound on* x_{phy}	239
x_{ub}	*upper bound on* x_{phy}	239
x_{unc}	*absolute uncertainty in* x_{exp}	239
$x_{\text{exp}}\pm x_{\text{unc}}$	*uncertainty interval for* x_{exp}	239
$x_{\text{rel-unc}}$	*relative uncertainty in* x_{exp}	239
x_{out}	*equipment output from measuring x*	240
$x_{\text{q-err}}$	*quantization error in* x_{exp}	241
x_{res}	*resolution in recording of* x_{exp}	241
$x_{\text{rel-res}}$	*relative resolution in recording of* x_{exp}	241
$\dim Q$	*dimension of physical quantity Q*	247
EPILOGUE		
dx	*differential of x*	263
$\dfrac{df(x)}{dx}$	*derivative of f (x) with respect to x*	263
$f'(\cdot)$	*derivative of f (·)*	264
$\dfrac{d}{dx}$	*differentiation operator with respect to x*	265
$\int_{x=a}^{b} f(x)\,dx$	*integral of f (x) versus x from* $x=a$ *to* $x=b$	268
$f^{(n)}(\cdot)$	*nth derivative of f(·)*	271
$P(A)$	*probability of random event A*	278
$p_x(\cdot)$	*probability density of random variable x*	278
$E\{x\}$	*mean*, or *expected value, of random variable x*	279
$\aleph_0,\aleph_1,\ldots$	*first transfinite cardinal numbers*	290
ω	*first transfinite ordinal number*	290

(Continued)

Notation	Name or Meaning	Page
Appendix		
$A \veebar B$	*A exclusive-or B*	294
$\exists! x$	*there exists a unique x*	296
$A \triangle B$	*symmetric difference of sets A and B*	297
$\mathbf{1}_A(\cdot)$	*indicator function of a set A*	297
$x \ominus y$	*modular difference of numbers x and y*	300
$\operatorname{argmax}_{x \in S} f(x)$	*inverse maximum of f (·) over S*	306
$\operatorname{argmin}_{x \in S} f(x)$	*inverse minimum of f (·) over S*	306
$\operatorname{argmax}_{x} f(x)$	*inverse maximum of f (·) over its domain*	306
$\operatorname{argmin}_{x} f(x)$	*inverse minimum of f (·) over its domain*	306
$\lfloor x \rfloor$	*floor function, applied to x*	307
$\lceil x \rceil$	*ceiling function, applied to x*	307

The Greek Alphabet

The table below lists all of the letters of the Greek alphabet, a popular source of symbols in mathematical discourse (e.g., to denote constants, variables, and functions). Omitted, though, are a few symbols that receive little use. In particular, the variation "ϵ" of epsilon does not appear because it looks very much like the symbol "$\in$" used for the set-membership relation, and therefore is avoided.

Name	Lowercase	Uppercase	Name	Lowercase	Uppercase
alpha	α	A	nu	ν	N
beta	β	B	xi	ξ	Ξ
gamma	γ	Γ	omicron	o	O
delta	δ	Δ	pi	π	Π
epsilon	ε	E	rho	ρ	P
zeta	ζ	Z	sigma	σ	Σ
eta	η	H	tau	τ	T
theta	θ	Θ	upsilon	υ	Υ
iota	ι	I	phi	ϕ, φ	Φ
kappa	κ	K	chi	χ	X
lambda	λ	Λ	psi	ψ	Ψ
mu	μ	M	omega	ω	Ω

Bibliography

[1] Carl B. Allendoerfer and Cletus O. Oakley. *Principles of Mathematics*. McGraw-Hill, New York, 3rd edition, 1969.

[2] American Mathematical Society (AMS). www.ams.org.

[3] Tom M. Apostol. *Mathematical Analysis*. Addison-Wesley, Reading, Mass., 2nd edition, 1974.

[4] Tom M. Apostol, Gulbank D. Chakerian, Geraldine C. Darden, and John D. Neff, editors. *Selected Papers on Precalculus*. Mathematical Association of America, Washington, D.C., 1977.

[5] William R. Ballard. *Geometry*. W. B. Saunders Company, Philadelphia, 1970.

[6] George David Birkhoff and Ralph Beatley. *Basic Geometry*. Chelsea, New York, 3rd edition, 1959.

[7] Max Black. *The Nature of Mathematics*. Littlefield, Adams & Co., Totowa, N.J., 1965.

[8] George S. Boolos and Richard C. Jeffrey. *Computability and Logic*. Cambridge University Press, Cambridge, 2nd edition, 1980.

[9] Julio F. Caballero and Delphia F. Harris. There seems to be uncertainty about the use of significant figures in reporting uncertainties of results. *Journal of Chemical Education*, 75(8):996, Aug. 1998.

[10] Allan Calder. Constructive mathematics. *Scientific American*, pages 146–71, Oct. 1979.

[11] George S. Carr. *Formulas and Theorems in Pure Mathematics*. Chelsea, New York, 2nd edition, 1970.

[12] Ruel V. Churchill, James W. Brown, and Roger F. Verhey. *Complex Variables and Applications*. McGraw-Hill, New York, 3rd edition, 1974.

[13] Comprehensive TeX Archive Network (CTAN). http://www.ctan.org/.

[14] E. T. Copson. *Asymptotic Expansions*. Cambridge University Press, Cambridge, 1965.

[15] Joseph W. Dauben. Georg Cantor and the origins of transfinite set theory. *Scientific American*, pages 122–31, June 1983.

[16] Philip J. Davis and Reuben Hersh. *The Mathematical Experience*. Houghton Mifflin, Boston, 1981.

[17] Howard DeLong. Unsolved problems in arithmetic. *Scientific American*, pages 50–60, March 1971.

[18] Albert Einstein. *Ideas and Opinions*. Dell, New York, 1954.

[19] H. B. Enderton. *A Mathematical Introduction to Logic*. Academic Press, New York, 1972.

[20] Joel N. Franklin. *Matrix Theory*. Prentice-Hall, Englewood Cliffs, N.J., 1968.

[21] Carl-Erik Fröberg. *Introduction to Numerical Analysis*. Addison-Wesley, Reading, Mass., 2nd edition, 1969.

[22] Jacob F. Golightly. *Precalculus Mathematics: Algebra and Trigonometry*. W. B. Saunders Company, Philadelphia, 1968.

[23] Marvin J. Greenburg. *Euclidean and Non-Euclidean Geometries*. W. H. Freeman and Company, New York, 2nd edition, 1980.

[24] Paul R. Halmos. *Finite-Dimensional Vector Spaces*. Springer-Verlag, New York, 1974.

[25] Paul R. Halmos. *Naive Set Theory*. Springer-Verlag, New York, 1974.

[26] John E. Hopcroft. Turing machines. *Scientific American*, pages 86–98, May 1984.

[27] Intel Corporation. Moore's law: Raising the bar, 2005. ftp://download.intel.com/museum/Moores_Law/Printed_Materials/Moores_Law_Backgrounder.pdf.

[28] International Organization for Standardization (ISO). *Guide to the Expression of Uncertainty in Measurement*. ISO, Geneva, 1995.

[29] D. C. Ipsen. *Units, Dimensions, and Dimensionless Numbers*. McGraw-Hill, New York, 1960.

[30] Shôkichi Iyanaga and Yukiyosi Kawada, editors. *Encyclopedic Dictionary of Mathematics*. MIT Press, Cambridge, Mass., 1980.

[31] Mervin L. Keedy and Charles W. Nelson. *Geometry: A Modern Introduction*. Addison-Wesley, Reading, Mass., 1973.

[32] K. N. King. The case for Java as a first language. *Proceedings of the 35th Annual ACM Southeast Conference*, pages 124–31, April 1997.

[33] Morris Kline. *Mathematics, the Loss of Certainty*. Oxford University Press, Oxford, 1980.

[34] Donald E. Knuth. *The TeXbook*. Addison-Wesley, Reading, Mass., 1992.

[35] Stephan Körner. *The Philosophy of Mathematics: An Introductory Essay*. Dover, New York, 1968.

[36] Leslie Lamport. *LaTeX: A Document Preparation System*. Addison-Wesley, Reading, Mass., 2nd edition, 1985.

[37] Jeff Miller. Earliest known uses of some of the words of mathematics. http://members.aol.com/jeff570/mathword.html.

[38] R. P. Mody. C in education and software engineering. *SIGCSE Bulletin*, 23:45–56, Sept. 1991.

[39] Edwin H. Moise. *Elementary Geometry from an Advanced Standpoint*. Addison-Wesley, Reading, Mass., 2nd edition, 1974.

[40] Christopher Mulliss and Wei Lee. On the standard rounding rule for multiplication and division. *Chinese Journal of Physics*, 36(3):479–87, 1998. Available at http://www.angelfire.com/oh/cmulliss/.

[41] Ernest Nagel and James R. Newman. *Gödel's Proof*. New York University Press, New York, 1958.

[42] Heinz R. Pagels. *The Cosmic Code*. Bantam, New York, 1983.

[43] R. C. Pankhurst. *Dimensional Analysis and Scale Factors*. Reinhold, New York, 1964.

[44] Steven K. Park and Keith W. Miller. Random number generators: Good ones are hard to find. *Communications of the ACM*, 31:1192–201, Oct. 1988.

[45] G. Polya. *Mathematics and Plausible Reading*, Vol. I: *Induction and Analogy in Mathematics*. Princeton University Press, Princeton, N.J., 1954.

[46] G. Polya. *How to Solve It*. Doubleday, Garden City, N.Y., 2nd edition, 1957.

[47] G. Polya. *Mathematics and Plausible Reasoning*, Vol. II: *Patterns of Plausible Inference*. Princeton University Press, Princeton, N.J., 2nd edition, 1968.

[48] W. V. Quine. *Philosophy of Logic*. Prentice-Hall, Englewood Cliffs, N.J., 1970.

[49] W. V. Quine. *Mathematical Logic*. Harvard University Press, Cambridge, Mass., revised edition, 1981.

[50] Rudy Rucker. *Infinity and the Mind*. Bantam, New York, 1982.

[51] Walter Rudin. *Principles of Mathematical Analysis*. McGraw-Hill, New York, 3rd edition, 1976.

[52] Bertrand Russell. *Mysticism and Logic*. Barnes & Noble, Totowa, N.J., 1981. First published 1917 by G. Allen & Unwin.

[53] Bertrand Russell. *Introduction to Mathematical Philosophy*. Simon & Schuster, New York, 1920.

[54] Patrick J. Ryan. *Euclidean and Non-Euclidean Geometry*. Cambridge University Press, Cambridge, 1986.

[55] Abraham Schwartz. *Calculus and Analytic Geometry*. Holt, Reinhart & Winston, New York, 2nd edition, 1967.

[56] Scilab Consortium. Scilab. www.scilab.org.

[57] James R. Smart. *Informal Geometry: An Informal Approach*. Brooks/Cole, Belmont, Calif., 1967.

[58] Barry Spain. *Analytical Conics*. Pergamon Press, New York, 1957.

[59] Murray R. Spiegel. *Theory and Problems of Statistics*. McGraw-Hill, New York, 1961.

[60] Patrick Suppes. *Axiomatic Set Theory*. Dover, New York, 1972.

[61] Alfred Tarski. Truth and proof. *Scientific American*, pages 63–77, June 1969.

[62] Barry N. Taylor, editor. *The International System of Units (SI)*. U.S. Government Printing Office, Washington, D.C., 1991. Available at http://physics.nist.gov/Document/sp330.pdf.

[63] Barry N. Taylor. *Guide for the Use of the International System of Units (SI)*. U.S. Government Printing Office, Washington, D.C., 1995. Available at http://physics.nist.gov/Document/sp811.pdf.

[64] John R. Taylor. *An Introduction to Error Analysis: The Study of Uncertainties in Physical Measurements*. University Science Books, Sausalito, Calif., 2nd edition, 1997.

[65] Eric Weisstein. MathWorld. http://mathworld.wolfram.com.

[66] Wikimedia Foundation. Wikipedia. http://www.wikipedia.org.

Index